LINEAR ALGEBRA

THEORY AND APPLICATIONS

SECOND EDITION

The Jones & Bartlett Learning Series in Mathematics

Geometry

Geometry with an Introduction to Cosmic Topology
Hitchman (978-0-7637-5457-0) © 2009

Euclidean and Transformational Geometry: A Deductive Inquiry
Libeskind (978-0-7637-4366-6) © 2008

A Gateway to Modern Geometry: The Poincaré Half-Plane, Second Edition
Stahl (978-0-7637-5381-8) © 2008

Understanding Modern Mathematics
Stahl (978-0-7637-3401-5) © 2007

Lebesgue Integration on Euclidean Space, Revised Edition
Jones (978-0-7637-1708-7) © 2001

Precalculus

Essentials of Precalculus with Calculus Previews, Fifth Edition
Zill/Dewar (978-1-4496-1497-3) © 2012

Algebra and Trigonometry, Third Edition
Zill/Dewar (978-0-7637-5461-7) © 2012

College Algebra, Third Edition
Zill/Dewar (978-1-4496-0602-2) © 2012

Trigonometry, Third Edition
Zill/Dewar (978-1-4496-0604-6) © 2012

Precalculus: A Functional Approach to Graphing and Problem Solving, Sixth Edition
Smith (978-0-7637-5177-7) © 2012

Precalculus with Calculus Previews (Expanded Volume), Fourth Edition
Zill/Dewar (978-0-7637-6631-3) © 2010

Calculus

Single Variable Calculus: Early Transcendentals, Fourth Edition
Zill/Wright (978-0-7637-4965-1) © 2011

Multivariable Calculus, Fourth Edition
Zill/Wright (978-0-7637-4966-8) © 2011

Calculus: Early Transcendentals, Fourth Edition
Zill/Wright (978-0-7637-5995-7) © 2011

Multivariable Calculus
Damiano/Freije (978-0-7637-8247-4) © 2011

Calculus: The Language of Change
Cohen/Henle (978-0-7637-2947-9) © 2005

Applied Calculus for Scientists and Engineers
Blume (978-0-7637-2877-9) © 2005

Calculus: Labs for Mathematica
O'Connor (978-0-7637-3425-1) © 2005

Calculus: Labs for MATLAB®
O'Connor (978-0-7637-3426-8) © 2005

Linear Algebra

Linear Algebra: Theory and Applications, Second Edition
Cheney/Kincaid (978-1-4496-1352-5) © 2012

Linear Algebra with Applications, Seventh Edition
Williams (978-0-7637-8248-1) © 2011

Linear Algebra with Applications, Alternate Seventh Edition
Williams (978-0-7637-8249-8) © 2011

Advanced Engineering Mathematics

A Journey into Partial Differential Equations
Bray (978-0-7637-7256-7) © 2012

Advanced Engineering Mathematics, Fourth Edition
Zill/Wright (978-0-7637-7966-5) © 2011

An Elementary Course in Partial Differential Equations, Second Edition
Amaranath (978-0-7637-6244-5) © 2009

Complex Analysis

Complex Analysis for Mathematics and Engineering, Sixth Edition
Mathews/Howell (978-1-4496-0445-5) © 2012

A First Course in Complex Analysis with Applications, Second Edition
Zill/Shanahan (978-0-7637-5772-4) © 2009

Classical Complex Analysis
Hahn (978-0-8672-0494-0) © 1996

Real Analysis

Elements of Real Analysis
Denlinger (978-0-7637-7947-4) © 2011

An Introduction to Analysis, Second Edition
Bilodeau/Thie/Keough (978-0-7637-7492-9) © 2010

Basic Real Analysis
Howland (978-0-7637-7318-2) © 2010

Closer and Closer: Introducing Real Analysis
Schumacher (978-0-7637-3593-7) © 2008

The Way of Analysis, Revised Edition
Strichartz (978-0-7637-1497-0) © 2000

Topology

Foundations of Topology, Second Edition
Patty (978-0-7637-4234-8) © 2009

Discrete Mathematics and Logic

Essentials of Discrete Mathematics, Second Edition
Hunter (978-1-4496-0442-4) © 2012

Discrete Structures, Logic, and Computability, Third Edition
Hein (978-0-7637-7206-2) © 2010

Logic, Sets, and Recursion, Second Edition
Causey (978-0-7637-3784-9) © 2006

Numerical Methods

Numerical Mathematics
Grasselli/Pelinovsky (978-0-7637-3767-2) © 2008

Exploring Numerical Methods: An Introduction to Scientific Computing Using MATLAB®
Linz (978-0-7637-1499-4) © 2003

Advanced Mathematics

Mathematical Modeling with Excel®
Albright (978-0-7637-6566-8) © 2010

Clinical Statistics: Introducing Clinical Trials, Survival Analysis, and Longitudinal Data Analysis
Korosteleva (978-0-7637-5850-9) © 2009

Harmonic Analysis: A Gentle Introduction
DeVito (978-0-7637-3893-8) © 2007

Beginning Number Theory, Second Edition
Robbins (978-0-7637-3768-9) © 2006

A Gateway to Higher Mathematics
Goodfriend (978-0-7637-2733-8) © 2006

For more information on this series and its titles, please visit us online at http://www.jblearning.com. Qualified instructors, contact your Publisher's Representative at 1-800-832-0034 or info@jblearning.com to request review copies for course consideration.

The Jones & Bartlett Learning International Series in Mathematics

Linear Algebra: Theory and Applications, Second Edition, International Version
Cheney/Kincaid (978-1-4496-2731-7) © 2012

Multivariable Calculus
Damiano/Freije (978-0-7637-8247-4) © 2012

Complex Analysis for Mathematics and Engineering, Sixth Edition, International Version
Mathews/Howell (978-1-4496-2870-3) © 2012

A Journey into Partial Differential Equations
Bray (978-0-7637-7256-7) © 2012

Functions of Mathematics in the Liberal Arts
Johnson (978-0-7637-8116-3) © 2012

Advanced Engineering Mathematics, Fourth Edition, International Version
Zill/Wright (978-0-7637-7994-8) © 2011

Calculus: Early Transcendentals, Fourth Edition, International Version
Zill/Wright (978-0-7637-8652-6) © 2011

Real Analysis
Denlinger (979-0-7637-7947-4) © 2011

Mathematical Modeling for the Scientific Method
Pravica/Spurr (978-0-7637-7946-7) © 2011

Mathematical Modeling with Excel®
Albright (978-0-7637-6566-8) © 2010

An Introduction to Analysis, Second Edition
Bilodeau/Thie/Keough (978-0-7637-7492-9) © 2010

Basic Real Analysis
Howland (978-0-7637-7318-2) © 2010

For more information on this series and its titles, please visit us online at http://www.jblearning.com. Qualified instructors, contact your Publisher's Representative at 1-800-832-0034 or info@jblearning.com to request review copies for course consideration.

SECOND EDITION

LINEAR ALGEBRA

THEORY AND APPLICATIONS

WARD CHENEY | DAVID KINCAID
University of Texas at Austin

JONES & BARTLETT
LEARNING

World Headquarters
Jones & Bartlett Learning
40 Tall Pine Drive
Sudbury, MA 01776
978-443-5000
info@jblearning.com
www.jblearning.com

Jones & Bartlett Learning Canada
6339 Ormindale Way
Mississauga, Ontario L5V 1J2
Canada

Jones & Bartlett Learning International
Barb House, Barb Mews
London W6 7PA
United Kingdom

Jones & Bartlett Learning books and products are available through most bookstores and online booksellers. To contact Jones & Bartlett Learning directly, call 800-832-0034, fax 978-443-8000, or visit our website, www.jblearning.com.

Production Credits
Publisher: Cathleen Sether
Senior Acquisitions Editor: Timothy Anderson
Associate Editor: Melissa Potter
Production Director: Amy Rose
Production Assistant: Sara Fowles
Senior Marketing Manager: Andrea DeFronzo
V.P., Manufacturing and Inventory Control: Therese Connell
Composition: Laserwords Private Limited, Chennai, India
Cover Design: Kristin E. Parker
Cover Image: © ErickN/ShutterStock, Inc.
Printing and Binding: Malloy, Inc.
Cover Printing: Malloy, Inc.

Library of Congress Cataloging-in-Publication Data
Cheney, E. W. (Elliott Ward), 1929–
    Linear algebra : theory and applications / Ward Cheney, David R. Kincaid.—2nd ed.
        p. cm.
    Includes bibliographical references and index.
    ISBN 978-1-4496-1352-5 (casebound)
    1. Algebras, Linear. I. Kincaid, David (David Ronald) II. Title.
    QA184.2.C54 2011
    512′.5—dc22
                                        2010037146

6048
Printed in the United States of America
14 13 12 11 10   10 9 8 7 6 5 4 3 2 1

Contents

APPENDIX A   DEDUCTIVE REASONING AND PROOFS   563

APPENDIX B   COMPLEX ARITHMETIC   581

ANSWERS/HINTS FOR GENERAL EXERCISES   585

REFERENCES   607

INDEX   611

Preface

In writing *Linear Algebra: Theory and Applications*, we were motivated by our desire to create a source that is flexible: serving diverse groups of students as well as instructors and professionals with various and differing interests and requirements. In particular, we wanted a book that appeals to students in science, engineering, and business (who are primarily interested in applications) as well as to mathematics students (who wish to acquire a mastery of theoretical linear algebra). Some students may be studying linear algebra to satisfy a requirement for an undergraduate degree, whereas others may have different needs and goals to fulfill.

We have tried to keep in mind the basic objective of acquainting students of mathematics, science, and engineering with the potential of using linear algebra as a tool for solving mathematical problems in their professions. For students majoring in natural sciences, engineering, social sciences, and business, the pedagogical emphasis is on understanding the concepts and learning to use vector spaces, matrices, linear transformations, and other tools of applied linear algebra. For these students, class time may not be available for dwelling upon the proofs of theorems. Other students may be more theory-oriented, and they often need practice in the art of devising proofs and in dealing with abstractions. These students can participate in building up their knowledge of the vast imposing logical structure that is linear algebra. This branch of mathematics is well suited for learning the rudiments of logic and proofs. Naturally, students can sharpen their problem-solving skills in a course that emphasizes theoretical linear algebra.

We have tried to achieve an elementary style of presentation for the benefit of students who are not necessarily advanced in the study of mathematics. We have found that many students learn by studying examples. Therefore, we have provided numerous examples to illustrate theorems and applications. In many cases, a well-chosen example may suggest a theorem and proof. That creates an ideal setting for enunciating a formal theorem, with or without a proof. Some simple or straightforward proofs of theorems are left to the reader as exercises.

Our text is designed to allow instructors to make their own choices of emphasis. After covering the introductory material, instructors may skip around and shift emphasis to teach either a theoretical or an application-oriented linear algebra course. In the book and online, there are approximately three thousand exercises of great variety, and the homework assignments can be made to reflect these choices. The instructor may choose to skip material and exercises involving proofs in an applied linear algebra course, while doing just the opposite in a theoretical class.

To enhance the versatility of this book, exercises are supplied in abundance and they are divided into two categories: General Exercises and Computer Exercises. In the first category, students may use a pencil and paper, as well as either a calculator or mathematical software for computations. In the second category, one finds a mixture of computer programming and project-oriented assignments. The text encourages students to learn one of the powerful mathematical software tools such as MATLAB®, Maple™, or Mathematica® to assist in solving the computer exercises.

The *Student Resource Manual* provides worked solutions to selected exercises. It also contains True-False Exercises and Multiple-Choice Exercises, which are available online as well. These exercises are suitable for examinations or for self study. They can test a student's knowledge of numerous topics yet involve only a small amount of pencil-and-paper analysis. On the other hand, the student may be asked to give a full justification for the answers selected.

WebAssign™ is the premier independent online teaching and learning environment, guiding several million students through their academic careers since 1997. With WebAssign, you can: create and distribute algorithmic assignments using questions specific to this textbook; grade, record, and analyze student responses and performance instantly; offer more practice exercises, quizzes, and homework. For more detailed information, please visit:

www.webassign.net.

Some sample syllabi for courses using this book include:

- A one-term course on matrices and matrix applications could emphasize suitable parts of the following: Chapter 1, Systems of Linear Equations; Chapter 3, Matrix Operations; Chapter 4, Determinants; Chapter 6, Eigensystems; and Chapter 8, Additional Topics, as time permits.
- A one-term course emphasizing theoretical linear algebra could be based on the following: Chapter 1, Systems of Linear Equations; Chapter 2,

Vector Spaces; Chapter 3, Matrix Operations; Chapter 5, Vector Subspaces; and Chapter 7, Inner-Product Vector Spaces.

- A two-term course would carefully cover almost all of the chapters.

Second Edition

Instructors and students using the book have been kind enough to notify us of typos and errata. All items on the online errata list have been fixed. Throughout, figures have been improved or redrawn. Some material has been added and some rewritten. For example, the section on Gerschgorin's Theorem has been revised and expanded to clarify the presentation with new examples and figures.

Additional exercises have been added so that all sections have at least fifty General Exercises and five Computer Exercises. There are approximately 1350 General Exercises and 150 Computer Exercises in the text. In an effort to utilize the Internet more effectively, all 1250 True-False Exercises and 650 Multiple-Choice Exercises have been moved online and to the *Student Resource Manual*. Many of these exercises have multiple parts.

This second edition may also be purchased in electronic (eBook) format. Please visit www.jblearning.com for details.

Supplements

For instructors, ancillary material is available online (electronically), including an Instructor Solutions Manual, and Presentation (PDF) Slides for lectures and illustrations. Also, available on the text's catalog page are computer programs, based on material in the textbook, written in MATLAB®, Maple™, and Mathematica®.

Additional supporting material is available at the textbook's website:

www.ma.utexas.edu/CNA/LA2

To view sample computer codes go to:

www.ma.utexas.edu/CNA/LA/la-matlab.html   (**MATLAB**)
www.ma.utexas.edu/CNA/LA/la-maple.html   (**Maple**)
www.ma.utexas.edu/CNA/LA/la-mathematica.html (**Mathematica**)

Links to the sample computer codes are available on the publisher's Web catalog page at:

http://www.jblearning.com/catalog/9781449613525

These sample programs are primarily intended as learning and teaching aids for use with this textbook. We believe that these computer routines are coded in a clear and easy-to-understand style, but we have intentionally excluded some comment statements so that students may read the code and study the algorithms—adding comments as they decipher them. These programs can be used on computer systems with appropriate software, from small personal computers to large scientific computing machines.

Acknowledgments

In both the first and second editions, we have been assisted by many individuals. We wish to express our thanks to those who were generous with their help. In particular, reviewers of the manuscript were Weiming Cao, University of Texas at San Antonio; Douglas B. Meade, Department of Mathematics, University of South Carolina; Antonio Palacios, San Diego State University; Stanley Payne, University of Denver and Health Sciences Center; Matthew Saltzman, Clemson University. Moreover, we thank Margaret Combs and Maorong Zou for their technical assistance. Also, we thank Steve Coomer, Soon Yung-Chien, Daniel Cornforth, Claus Deoring, Greg Lavender, Durene Ngo, as well as our teaching assistants and students who provided useful feedback.

We also offer our heartfelt gratitude to Victor Cheney, Joyce Pfluger, and Martha Wells.

We wish to thank the Department of Mathematics, the Department of Computer Sciences, and the Institute for Computational Engineering and Sciences at The University of Texas at Austin for their excellent computing facilities and support. Some work on the development of the material for this book was done while one the authors was visiting KAUST: King Abdullah University of Science and Technology, CERFACS: European Centre for Research and Advanced Training in Scientific Computation, and LANL: Los Alamos National Laboratory. We wish to express our appreciation to them for their hospitality.

In the production of this book, the staff of Jones & Bartlett Learning and associated individuals have been most understanding and patient. We would especially like to thank Tim Anderson and Amy Rose.

In spite of our best effort to eliminate all errors, there may exist flaws that have escaped us. We would appreciate comments, questions, criticisms, or corrections that readers may wish to send us.

Ward Cheney
University of Texas at Austin
Department of Mathematics
(cheney@ma.utexas.edu)

David Kincaid
University of Texas at Austin
Institute for Computational Engineering and Sciences
(kincaid@cs.utexas.edu)

Authors

Ward Cheney has taught in the Mathematics Department of The University of Texas at Austin for most of his career, as well as at the University of Kansas, Iowa State University, University of California at Los Angeles (UCLA), Michigan State University, and Lund University (Sweden). His research interests are in approximation theory, applied mathematics, and numerical analysis. He has published over a hundred research papers and is the author of eleven mathematical textbooks. In particular, his classical book *Introduction to Approximation Theory* has been in continuous print since it first appeared in 1966. Ward exercises daily at the gym and plays the clarinet in a trio that performs classical music weekly for their own enjoyment and entertainment.

David Kincaid has taught in the Computer Sciences Department of The University of Texas at Austin for most of his career, as well as at Purdue University and at King Abdullah University of Science and Technology in Saudi Arabia. His interests are in computational linear algebra and scientific computing with a specialty in iterative methods for solving large sparse linear systems using supercomputers. He has published over a hundred research papers while doing research in the Center for Numerical Analysis, the Institute for Computational Engineering and Science, and the Computation Center. He was one of the original co-authors of the *BLAS: Basic Linear Algebra Subprograms.* David likes to play tennis several times a week and enjoys travel.

David and Ward have co-authored three books, including *Numerical Mathematics and Computing, 6th Edition* and *Numerical Analysis: Mathematics of Scientific Computing, 3rd Edition.* Their books have been published in multiple editions and have been translated into various languages.

CHAPTER ONE

Systems of Linear Equations

1.1 SOLVING SYSTEMS OF LINEAR EQUATIONS

> *Do not worry about your difficulties in mathematics. I can assure you that mine are still greater.*
> —ALBERT EINSTEIN (1879–1955)

> *Begin at the beginning and go until you come to the end: then stop.*
> —LEWIS CARROLL (1832–1889),
> *ALICE'S ADVENTURES IN WONDERLAND*

The term *linear algebra* applies to a branch of mathematics that studies vectors, matrices, vector spaces, and systems of linear equations. It is hoped that the reader has acquaintance with some of these terms. For example, a vector with three components looks like $(2.5, 3.7, -5.1)$, while a 2×3 matrix looks like this:

$$\begin{bmatrix} 4.1 & -3.2 & 5.4 \\ 1.3 & 2.0 & -5.1 \end{bmatrix}$$

These two building blocks, vectors and matrices, can produce systems of linear equations, and these in turn can often model an applied problem from the real world.

The subject of linear algebra was already being studied sporadically in ancient times, as is known from surviving manuscripts. But the subject blossomed in the early 1800s and thus is approximately 200 years old. It is much younger than calculus, which was already thriving at the time of Newton and Leibniz, in the 1700s.

In this book you will find many examples illustrating how some computational problem originating in engineering, physics, finance, or economic planning (and in other disciplines) becomes a fully understood type of problem in linear algebra, and therefore yields easily to standard techniques already available. In many cases, such problems can be solved by means of well-tested and documented computer software.

Linear Equations

This topic is basic to much of what comes later. We begin our study by discussing a single linear equation containing two variables.

EXAMPLE 1 Consider the equation $-3x + 7y = 21$, which represents a line in the xy-plane. Where does this line cross the two axes?

SOLUTION The question is: What points of the form $(x, 0)$ and $(0, y)$ are on this line? For the x-intercept point, we let $y = 0$, and the resulting equation is $-3x = 21$. Thus, we have $x = -7$ and the point sought is $(-7, 0)$. For the y-intercept point, we let $x = 0$ and the equation now reads $7y = 21$. Hence, we have $y = 3$ and the point wanted is $(0, 3)$. These points are the **intercepts** on the line. See Figure 1.1. ∎

The **point-slope form** of a line is

$$y = mx + b$$

where m is the **slope** and b is the **intercept** on the y-axis. From Example 1, the line $-3x + 7y = 21$ can be written in point-slope form as $y = \frac{3}{7}x + 3$.

EXAMPLE 2 Use the line described in Example 1.
Are the points $(-14, -3)$, $(1, 3)$, and $(7, 6)$ on the line?

SOLUTION In each case, one can substitute the coordinates in the equation $-3x + 7y = 21$ to see whether the equation is satisfied. For the first point, we calculate $(-3)(-14) + (7)(-3) = 21$, for the second point $(-3)(1) + (7)(3) = 18$, and for the third point $(-3)(7) + (7)(6) = 21$. Hence, the first and third points are on the line, but the second is not. See Figure 1.1. ∎

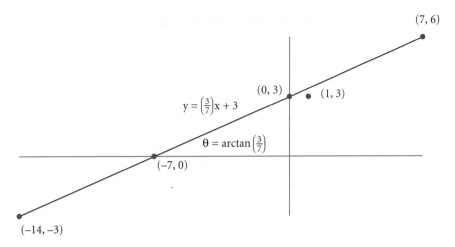

FIGURE 1.1 A point–slope form of a line in $\mathbb{R}^2$.

For two points (x_0, y_0) and (x_1, y_1), the **two-point form** of the line through these points is

$$y - y_0 = m(x - x_0) \quad \text{where} \quad m = \frac{y_1 - y_0}{x_1 - x_0}$$

For example, using the two intercept points $(-7, 0)$ and $(0, 3)$ from Example 2, we obtain the two-point form as $(y - 3) = \frac{3}{7}(x - 0)$.

The equation discussed above is called a **linear equation** precisely because its graph is a line, and the word **linear** derives from the word *line*.

We extend the meaning of a **linear equation** to encompass one of the form

$$a_1 x_1 + a_2 x_2 + a_3 x_3 + \cdots + a_n x_n = b \qquad \text{or} \qquad \sum_{j=1}^{n} a_j x_j = b$$

involving n variables. Here we have named the variables x_1, x_2, and so on because we need the flexibility of handling any number of variables, even hundreds of thousands! The variables x_j may (in some contexts) also be called **unknowns**. The second form of writing the equation employs standard summation notation: the variable j runs through the integers 1 to n, and we are to take the sum of all the resulting terms, $a_j x_j$. When we use summation notation, the variable index (j in the preceding equation) can be almost any convenient variable, such as i, j, k, μ, or ν. However, one must be careful to avoid any conflict with letters already used in a different context. For example, $\sum_{i=1}^{n} x_i = y_i$ is obviously wrong!

Systems of Linear Equations

We are rarely interested in only one such equation in isolation. We usually encounter *systems* of linear equations. We must stretch our notation a little bit further and settle upon the following standard formulation. A completely general system of m linear equations with n unknowns (or **variables**) has equations of this form

$$a_{i1}x_1 + a_{i2}x_2 + a_{i3}x_3 + \cdots + a_{i,n-1}x_{n-1} + a_{in}x_n = b_i \qquad (1 \le i \le m)$$

which is the exact form of the ith generic equation in the system. Sometimes a comma is needed to separate the two subscripts on the letter a if it is not clear otherwise. Each equation in the system involves the same set of variables, $x_1, x_2, x_3, \ldots, x_n$. The entire system can be written succinctly with summation notation:

$$\sum_{j=1}^{n} a_{ij}x_j = b_i \qquad (1 \le i \le m)$$

The symbolism on the left of this equation means a sum of terms, each of the form $a_{ij}x_j$. The index j runs over the integer values from 1 to n. Off to the side we see in parentheses an indication that the index i also runs through a set of integers, in this case $i = 1, 2, 3$, up to and including m. Here and elsewhere, we often expect i and j to be restricted to nonnegative integer values.

An important concept that we will return to in Section 1.2 is the consistency of systems of linear equations.

> **DEFINITION**
>
> *A system of equations is* **consistent** *if it has at least one solution, and* **inconsistent** *if it has no solution.*

Next, we consider a system of two linear equations in two unknowns:

$$L_1 : -x + y = 1, \qquad L_2 : 2x + y = 4$$

These two equations correspond to two lines L_1 and L_2 in $\mathbb{R}^2$. Adding 2 times the first equation to the second equation produces $3y = 6$ or $y = 2$. Substituting this value into the first equation reveals that $x = 1$. Plots of these two lines are shown in Figure 1.2(a), and we see that they intersect at the point $(1, 2)$. In this case, there is exactly *one* solution.

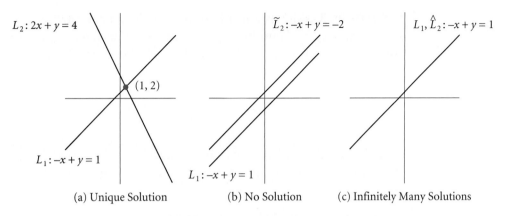

$L_2: 2x + y = 4$

$(1, 2)$

$L_1: -x + y = 1$

$\tilde{L}_2: -x + y = -2$

$L_1: -x + y = 1$

$L_1, \hat{L}_2: -x + y = 1$

(a) Unique Solution   (b) No Solution   (c) Infinitely Many Solutions

FIGURE 1.2 Different cases of two lines in $\mathbb{R}^2$.

Modifying the second equation, we consider the system

$$L_1 : -x + y = 1, \qquad \tilde{L}_2 : 2x - 2y = 4$$

Dividing the second equation by -2, we find that this pair of equations is *inconsistent*, requiring that $-x + y = 1$ and $-x + y = -2$. This means that there is *no* solution. As shown in Figure 1.2(b), these two lines are parallel and therefore do not intersect at all.

Again, modifying the second equation slightly leads us to the third case:

$$L_1 : -x + y = 1, \qquad \hat{L}_2 : 2x - 2y = -2$$

Dividing the second equation by -2, we find that these equations are now duplicates of each other. Letting $y = 1$, we find $x = 0$. Letting $y = 3$, we have $x = 2$, and so on. Giving y *any* value, we have $x = -1 + y$. Consequently, there are *infinitely many* solutions in this case. As shown in Figure 1.2(c), there is now *only* one line and we can think of the two equations as having graphs that lie on top of each other.

EXAMPLE 3 A system of four equations in three unknowns is exemplified by

$$\begin{cases} 3x_1 - 2x_2 + 5x_3 = 7 \\ x_1 + 4x_2 - 3x_3 = 7 \\ 6x_1 - 4x_2 + 2x_3 = -2 \\ x_1 + 2x_2 + x_3 = 9 \end{cases}$$

Is the point $(1, 3, 2)$ a solution of this system?

SOLUTION It is simply a matter of putting the numerical values $x_1 = 1$, $x_2 = 3$, and $x_3 = 2$ into each of the four equations and seeing that the equations are indeed satisfied by the given numbers. We have not revealed how $(1, 3, 2)$ was obtained. Indeed, some effort would have to be invested to discover this solution. However, verifying a purported solution is trivial in contrast. It requires only a substitution and a bit of arithmetic.   ■

In Example 3, the system of equations is a textbook problem. It is not *typical* of one that would arise in an application, where the data given (i.e., the numbers a_{ij} and b_i) are rarely integers (whole numbers). In this text most of the examples and problems employ integers, for simplicity.

General Systems of Linear Equations

Let us return to the general system of m linear equations in n unknowns. The data for this system of equations are all the numbers a_{ij} and b_i. The number a_{ij} is the coefficient of x_j in the ith equation. In a typical problem, all these coefficients would be given to us numerically. The numbers b_i on the righthand side would also be given.

$$\begin{cases} a_{11}x_1 + a_{12}x_2 + \cdots + a_{1n}x_n = b_1 \\ a_{21}x_1 + a_{22}x_2 + \cdots + a_{2n}x_n = b_2 \\ \qquad\qquad\vdots\qquad\qquad\vdots \\ a_{m1}x_1 + a_{m2}x_2 + \cdots + a_{mn}x_n = b_m \end{cases}$$

Then the challenge is to find the values of the *unknowns* $x_1, x_2, \ldots, x_n$ that make the equations true. The coefficient data and the list of unknowns can be exhibited in a number of ways. Consider these four arrays:

$$\mathbf{A} = \begin{bmatrix} a_{11} & a_{12} & \cdots & a_{1n} \\ a_{21} & a_{22} & \cdots & a_{2n} \\ \vdots & \vdots & \ddots & \vdots \\ a_{m1} & a_{m2} & \cdots & a_{mn} \end{bmatrix}, \quad \mathbf{x} = \begin{bmatrix} x_1 \\ x_2 \\ \vdots \\ x_n \end{bmatrix}, \quad \mathbf{b} = \begin{bmatrix} b_1 \\ b_2 \\ \vdots \\ b_m \end{bmatrix}$$

$$[\mathbf{A} \mid \mathbf{b}] = \begin{bmatrix} a_{11} & a_{12} & \cdots & a_{1n} & \bigm| & b_1 \\ a_{21} & a_{22} & \cdots & a_{2n} & \bigm| & b_2 \\ \vdots & \vdots & \ddots & \vdots & \vdots & \vdots \\ a_{m1} & a_{m2} & \cdots & a_{mn} & \bigm| & b_m \end{bmatrix}$$

All of these arrays are examples of **matrices**. (The plural form of the word *matrix* is *matrices*.) The middle two are also examples of **column vectors**.

In this context of a system of equations, the four matrices are called, respectively, the **coefficient matrix**, the **vector of unknowns**, the **righthand-side vector**, and the **augmented matrix**. If we call the coefficient matrix **A**, the righthand-side vector **b**, and the unknown vector **x**, the system of equations can be expressed as

$$\mathbf{Ax} = \mathbf{b}$$

This formalism will be correct if we define the product **Ax** appropriately. This comes later.

The matrix **A** displayed in detail earlier is called an $m \times n$ matrix because it has m rows and n columns. The rows are the horizontal arrays inside **A** and the columns are the vertical arrays inside **A**. We always give the number of rows first and the number of columns second in describing the dimensions of a matrix. Hence, we do *not* call **A** an $n \times m$ matrix because that would be a matrix of a different shape, if $n \neq m$. The indices on the letter a, such as a_{ij}, tell us that the number being referenced is in row i and column j. One can call the first index i the *row index*, and the second j the *column index*. Thus, for example, a_{pq} is the element in row p and column q. These traditions must be followed so that we can understand each other when speaking of matrices!

Gaussian Elimination

It is possible to use a process called **Gaussian elimination** to solve any system of linear equations that has a solution.[1]

EXAMPLE 4 Solve this system of linear equations:

$$\begin{cases} 3x_1 + 2x_2 - 5x_3 = -1 \\ \quad\quad 4x_2 + \ x_3 = 14 \\ \quad\quad\quad\quad - 2x_3 = -4 \end{cases}$$

---

[1] This name honors Johann Karl Friedrich Gauss (1777–1855), one of the greatest mathematicians. In elementary school he demonstrated his mathematical potential and amazed his teachers by inventing a simple method for summing an arithmetic series. Namely, one multiplies the number of terms by the average of the smallest and largest terms. In the subject of linear algebra, when he was 18, Gauss invented the method of least squares—a topic taken up in Sections 2.2 and 7.2 of this book. Also, he was the first to prove the **Fundamental Theorem of Algebra**: Every nonconstant polynomial assumes the value 0 at some point in the complex plane.

SOLUTION We observe that this problem has a certain structure that can be exploited in arriving at the solution: namely, the last equation can be solved at once to obtain $x_3 = 2$. Then, with x_3 known, we can solve the second equation for x_2. To do so, write it as $4x_2 + 2 = 14$, and solve for $x_2 = 3$. Finally, with x_2 and x_3 known, we can compute x_1 from the first equation: $3x_1 + (2)(3) + (-5)(2) = -1$, and $x_1 = 1$. We write the solution

as $\mathbf{x} = \begin{bmatrix} 1 \\ 3 \\ 2 \end{bmatrix}$ or $\mathbf{x} = (1, 3, 2)$; it is a point in three-space (denoted by $\mathbb{R}^3$). ∎

It may appear that the problem in Example 4 is artificial, since in practice we do not expect systems of linear equations to have the structure of which we took advantage. However, as we shall see, there is an algorithm for turning a system of equations into the so-called **triangular** form present in Example 4.

EXAMPLE 5 For a concrete example of modest size to illustrate the techniques for solving systems of linear equations, we use this special case:

$$\begin{cases} 3x_1 + 2x_2 = 4 \\ 9x_1 + 7x_2 = 17 \end{cases}$$

SOLUTION The basic operation that can be used over and over again in solving linear systems is the *addition* of a multiple of one equation to another equation. For reasons that will become clear later, we call this a **replacement operation**. In the example, let us add -3 times the first equation to the second. The result is

$$\begin{cases} 3x_1 + 2x_2 = 4 \\ 0x_1 + 1x_2 = 5 \end{cases}$$

In this process, the first equation itself was not changed, although it played a role in this first step. At this stage in the solution process, we have a choice. First, we can see that the new second equation can be solved immediately to get $x_2 = 5$. Then, as in Example 4, we can solve the first equation for x_1, since x_2 is 5. This yields $x_1 = -2$. The alternative way of proceeding is to carry out another replacement operation to produce a zero coefficient of x_2 in the first equation. In fact, we should add to the first equation -2 times the second equation, getting

$$\begin{cases} 3x_1 + 0x_2 = -6 \\ 0x_1 + 1x_2 = 5 \end{cases}$$

The solution is now clear: $x_1 = -2$ and $x_2 = 5$. (Of course, we could divide the first equation by 3 so that the solution values appear on the righthand side. This is a **scale operation**.) There is an independent manner of verifying the work: Simply substitute the purported solution into the original system of equations to see whether it is actually a solution. The work of doing so yields $(3)(-2) + (2)(5) = 4$ and $(9)(-2) + (7)(5) = 17$. It is a good habit to find independent ways of verifying a solution—that is, methods different from simply checking the steps that led to a solution in the first place.   ■

The process we have just illustrated is often called **Gaussian elimination**. The process is also called the **row-reduction algorithm**. The process whereby the system of equations produces explicit values of each variable is called **Gauss-Jordan elimination** in honor of Gauss and Wilhelm Jordan.[2]

Elementary Replacement and Scale Operations

In the examples, we have been using two elementary row operations: a **replacement** operation of adding a multiple of one equation to another equation, and a **scale** operation of multiplying an equation by a nonzero scalar. How can we be sure that, when we transform a system in the way that we did, we do not introduce spurious solutions or lose genuine solutions? It is simply that one can *add equal quantities to equal quantities* to obtain further equalities, and the process can be reversed. For example, if we write

---

[2] Wilhelm Jordan (1842–1899) is remembered for making improvements in the stability of the Gaussian elimination algorithm when it is applied to least squares problems. In the *Gauss-Jordan elimination* procedure for solving systems of linear equations, *Jordan* is the geodesist Wilhelm Jordan. [Some people have made the mistake of crediting Camille Jordan (1838–1922) in this context. In the *Jordan normal form of a matrix*, it is indeed Camille Jordan who is to be credited.] In the simple Gauss procedure (Gaussian elimination), row operations are used to produce an upper triangular coefficient matrix, whereas in the Gauss-Jordan computation, the row operations are designed to lead to the identity matrix on the left, and the solution vector on the right. Wilhelm Jordan had a brilliant career as a master surveyor and was involved in surveying large areas of Germany. His textbook *Handbook of Geodesy*, in German, went through five editions and was translated into French, Italian, and Russian. See Althoen and McLaughlin [1987].

the original pair of equations as

$$\begin{cases} E_1(\mathbf{x}) = 0 \\ E_2(\mathbf{x}) = 0 \end{cases}$$

then we can proceed to

$$\begin{cases} E_1(\mathbf{x}) = 0 \\ \alpha E_1(\mathbf{x}) + E_2(\mathbf{x}) = 0 \end{cases}$$

where α is any real number—that is, a scalar. Thus, when the first pair of equations is true, the second pair must also be true. In other words, an $\mathbf{x}$ that satisfies the first pair will also satisfy the second pair. Furthermore, we can get back to the first pair by applying a similar process to the second pair: to the second equation in the second pair, we add $-\alpha E_1(\mathbf{x})$. One can say that the row operation we are talking about here is **reversible** by another row operation of the same type. This shows that any solution of the second pair of equations must satisfy the first pair. These two parallel assertions establish that solutions are neither created nor destroyed in the process we are using. To emphasize:

1. Every solution of the first pair of equations satisfies the second pair.
2. Every solution of the second pair of equations satisfies the first pair.

If two systems of m linear equations in n unknowns have precisely the same set of solutions, we can get the solutions to the first system by solving the second, or vice versa. This simple idea is at the heart of our procedure for solving systems of linear equations. The remarks in the preceding paragraph establish that if one system of equations is obtained from another by a sequence of the permitted row operations, then the resulting two systems of equations have precisely the same set of solutions. In other words, the steps that we use do not change the set of solutions. With this in mind, we aim for a simple set of equations derived from the one with which we started. The solutions of the simple system are exactly the solutions of the original. This is the grand strategy, which has a number of variations. As matters stand, two people could work on a system of equations and produce two different sequences of simplified systems. The solutions of the simplified systems, in each of the two sequences, should be the same, however!

Row-Equivalent Pairs of Matrices

It should be pointed out that the work needed to solve the previous system of equations requires only that we keep track of the numerical data. There is no need to write the names of the variables in each step, or the equals sign. Thus, we can set up the data in successive arrays like this:

$$\begin{bmatrix} 3 & 2 & | & 4 \\ 9 & 7 & | & 17 \end{bmatrix} \sim \begin{bmatrix} 3 & 2 & | & 4 \\ 0 & 1 & | & 5 \end{bmatrix} \sim \begin{bmatrix} 3 & 0 & | & -6 \\ 0 & 1 & | & 5 \end{bmatrix} \sim \begin{bmatrix} 1 & 0 & | & -2 \\ 0 & 1 & | & 5 \end{bmatrix}$$

In this display, the symbol $\sim$ means that the matrices on either side of the symbol are connected by allowable row operations. In English, we say that the two matrices are **row equivalent** to each other.

Some readers may wish to describe each of the steps with the following notation:

$$\begin{matrix} \rightarrow \\ -3 \end{matrix} \begin{bmatrix} 3 & 2 & | & 4 \\ 9 & 7 & | & 17 \end{bmatrix} \sim \begin{matrix} -2 \\ \rightarrow \end{matrix} \begin{bmatrix} 3 & 2 & | & 4 \\ 0 & 1 & | & 5 \end{bmatrix} \sim^{\frac{1}{2}} \begin{bmatrix} 3 & 0 & | & -6 \\ 0 & 1 & | & 5 \end{bmatrix}$$

$$\sim \begin{bmatrix} 1 & 0 & | & -2 \\ 0 & 1 & | & 5 \end{bmatrix}$$

Here we have written to the left of these matrices an arrow $\rightarrow$ indicating the so-called **pivot row** and a number indicating the **multiplier** next to the *target* row. A multiple of the pivot row is added to the target row. This is helpful in recalling what was done. We rarely include these symbols in the text, but the reader may wish to add them.

The rectangular arrays of data are again *matrices*. A **matrix** can be any rectangular array of real numbers. In the preceding example, we are using 2×3 matrices, meaning that there are two horizontal **rows** and three vertical **columns**. The symbol $\sim$ indicates that we have proceeded from one matrix to another by one or more **row operations** of the permitted type: addition of a multiple of one row to another row. Later we shall add further row operations to our arsenal, and the notation $\mathbf{A} \sim \mathbf{B}$ will mean that each of the matrices $\mathbf{A}$ and $\mathbf{B}$ can be obtained from the other by applying one or more allowable row operations. We say in this case that the two matrices are **row equivalent** to each other.

This type of relation occurs often in mathematics, especially in linear algebra. A formal definition follows. In this definition we have deliberately chosen a symbol, $\bigstar$, that is not likely to conflict with other notation.

DEFINITION

An **equivalence relation** *on a set of entities is a relation that we denote here by the symbol* $\bigstar$. *It must satisfy these three conditions:*

- $p \bigstar p$.   (**reflexive**)
- *If* $p \bigstar q$, *then* $q \bigstar p$.   (**symmetric**)
- *If* $p \bigstar q$ *and* $q \bigstar r$, *then* $p \bigstar r$.   (**transitive**)

An example of an equivalence relation is $\mathbf{A} \sim \mathbf{B}$. To see why, think of the row operations used. Clearly, $\mathbf{A} \sim \mathbf{A}$ and the reflexive property holds. Why is the symmetric property true for a row-equivalence relation? The row operations that take us from $\mathbf{A}$ to $\mathbf{B}$ are reversible, and the reverse operations lead from $\mathbf{B}$ back to $\mathbf{A}$. For the transitive property, if $\mathbf{A} \sim \mathbf{B}$ and $\mathbf{B} \sim \mathbf{C}$, then $\mathbf{A} \sim \mathbf{C}$ because the respective row operations from $\mathbf{A}$ and $\mathbf{B}$ can be done one after another on $\mathbf{A}$.

When solving textbook problems (which often have integer data), one can avoid divisions and fractions in some row reductions. For example, consider this row reduction where we begin with row two as the first pivot row:

$$
\begin{array}{c} -1 \\ \rightarrow \\ \\ \end{array}
\left[\begin{array}{rr|r} 3 & -1 & 2 \\ 2 & 5 & -4 \\ 7 & -25 & 1 \end{array}\right]
\begin{array}{c} \rightarrow \\ \sim -2 \\ -7 \end{array}
\left[\begin{array}{rr|r} 1 & -6 & 6 \\ 2 & 5 & -4 \\ 7 & 25 & 1 \end{array}\right]
\begin{array}{c} \\ \sim \rightarrow \\ -1 \end{array}
\left[\begin{array}{rr|r} 1 & -6 & 6 \\ 0 & 17 & -16 \\ 0 & 17 & -41 \end{array}\right]
$$

$$
\sim
\left[\begin{array}{rr|r} 1 & -6 & 6 \\ 0 & 17 & -16 \\ 0 & 0 & 15 \end{array}\right]
$$

Here there is *no* solution. Indeed, the third equation states that $0x + 0y = 15$. Recall that a system of linear equations is inconsistent when it has no solution, and consistent when it has one or more solutions. These concepts are studied further in Section 1.2. In the preceding example, each pair of equations has a unique solution, but there exists no point satisfying all three equations simultaneously.

We have inserted a vertical line in a matrix to separate the coefficient matrix from the righthand side, if the matrix is the augmented matrix of a system of equations. On the right side of this line, we have the righthand side of the original system of equations. We will encounter situations where there are multiple columns to the right of this line, and therefore it is a good practice to place this vertical line in any augmented matrix.

If we have several systems with the same lefthand side but different righthand sides, we can use an augmented matrix to solve them simultaneously. For example, we can solve these two systems

$$
\begin{cases} 3x_1 + 2x_2 = 1 \\ 9x_1 + 7x_2 = 0 \end{cases} \quad \text{and} \quad \begin{cases} 3x_1 + 2x_2 = 0 \\ 9x_1 + 7x_2 = 1 \end{cases}
$$

by row-reducing the following augmented matrix, in which there are two righthand sides:

$$
\begin{array}{c} \rightarrow \\ -3 \end{array}
\left[\begin{array}{rr|rr} 3 & 2 & 1 & 0 \\ 9 & 7 & 0 & 1 \end{array}\right]
\begin{array}{c} -2 \\ \sim \rightarrow \end{array}
\left[\begin{array}{rr|rr} 3 & 2 & 1 & 0 \\ 0 & 1 & -3 & 1 \end{array}\right]
\sim
\left[\begin{array}{rr|rr} 3 & 0 & 7 & -2 \\ 0 & 1 & -3 & 1 \end{array}\right]
$$

The solutions are $x_1 = \frac{7}{3}$, $x_2 = -3$ for the first system and $x_1 = -\frac{2}{3}$, $x_2 = 1$ for the second system. This procedure can be useful later in computing inverses of matrices. (See Section 3.2.)

EXAMPLE 6 Another example, requiring a little more thought, calls for solving the following system:

$$\begin{cases} 42x_1 + 30x_2 = 83 \\ 222x_1 + 42x_2 = 361 \end{cases}$$

SOLUTION Our method is the same, and we add a multiple of the first equation to the second equation to create a 0 where the number 222 stands. Denote the unknown multiplier by α. Then we want $222 + \alpha 42 = 0$, and the multiplier should be $\alpha = -222/42$. The resulting system is

$$\begin{cases} 42x_1 + 30x_2 = 83 \\ 0x_1 - 816x_2 = -544 \end{cases}$$

In the same way as before, we now decide that $30/816$ times the second equation should be added to the first. The result of doing so is

$$\begin{cases} 42x_1 + 0x_2 = 21 \\ 0x_1 - 816x_2 = -544 \end{cases}$$

The solution is therefore $x_1 = \frac{3}{2}$ and $x_2 = \frac{2}{3}$. We summarize the work in matrices like this:

$$\begin{bmatrix} 42 & 30 & | & 83 \\ 222 & 42 & | & 361 \end{bmatrix} \sim \begin{bmatrix} 42 & 30 & | & 83 \\ 0 & -816 & | & -544 \end{bmatrix}$$

$$\sim \begin{bmatrix} 42 & 0 & | & 21 \\ 0 & -816 & | & -544 \end{bmatrix} \sim \begin{bmatrix} 1 & 0 & | & \frac{3}{2} \\ 0 & 1 & | & \frac{2}{3} \end{bmatrix} \qquad \blacksquare$$

EXAMPLE 7 Given the following equivalent matrices:

$$\begin{bmatrix} 5 & 3 & 0 & | & 21 \\ 1 & 3 & -1 & | & 15 \\ -2 & 0 & -1 & | & 1 \end{bmatrix} \sim \begin{bmatrix} 0 & -12 & 5 & | & -54 \\ 1 & 3 & -1 & | & 15 \\ 0 & 6 & -3 & | & 31 \end{bmatrix} \sim \begin{bmatrix} 0 & 0 & -1 & | & 8 \\ 1 & 3 & -1 & | & 15 \\ 0 & 6 & -3 & | & 31 \end{bmatrix}$$

What is the solution?

SOLUTION In this example, further reduction can be carried out, but observe that the system of equations can be easily solved by using the work already done. To determine x_3, use the first equation (in the final array), which asserts that $x_3 = -8$. With x_3 in hand, we now use the third equation, which indicates that $6x_2 - 3x_3 = 31$. This becomes $6x_2 + 24 = 31$; therefore, $x_2 = \frac{7}{6}$. Lastly, use equation two, which asserts that $x_1 + 3x_2 - x_3 = 15$. From this, we have $x_1 + 3(\frac{7}{6}) - (-8) = 15$ and $x_1 = \frac{7}{2}$. The procedure we have just illustrated is the one usually used in mathematical software. In brief, we reduce the matrix so that there is one equation containing only one unknown and another equation containing that unknown and one other, and so on. The system is readily solved by starting with the equation having only one unknown, and proceeding through the whole system one equation at a time.

In the preceding example, further row reduction leads to

$$\left[\begin{array}{ccc|c} 0 & 0 & -1 & 8 \\ 2 & 6 & -2 & 30 \\ 0 & 6 & -3 & 31 \end{array}\right] \sim \left[\begin{array}{ccc|c} 0 & 0 & -1 & 8 \\ 2 & 0 & 1 & -1 \\ 0 & 6 & -3 & 31 \end{array}\right]$$

$$\sim \left[\begin{array}{ccc|c} 0 & 0 & -1 & 8 \\ 2 & 0 & 0 & 7 \\ 0 & 6 & 0 & 7 \end{array}\right] \sim \left[\begin{array}{ccc|c} 0 & 0 & 1 & -8 \\ 1 & 0 & 0 & \frac{7}{2} \\ 0 & 1 & 0 & \frac{7}{6} \end{array}\right]$$

Again, we obtain $x_1 = \frac{7}{2}$, $x_2 = \frac{7}{6}$, and $x_3 = -8$.   ∎

The row operations that we have been exploiting are:

1. Add to a row a multiple of another row.
2. Multiply a row by a nonzero constant.

These two operations are sufficiently powerful to solve any system of linear equations by repeated application. (We are not stopping to prove this fact.) Example 7 illustrates how this is done. Here is another example on which to test yourself.

EXAMPLE 8 Solve this system of linear equations by repeated use of the row operation illustrated in the two preceding examples:

$$\begin{cases} 35x_1 + 6x_2 - 7x_3 = 15 \\ -90x_1 - 15x_2 + 21x_3 = -40 \\ 25x_1 + 3x_2 - 7x_3 = 11 \end{cases}$$

SOLUTION We systematically create zeros in strategic positions. Here is one sequence of reductive steps. Start by subtracting row three from row one. Then add 3 times row three to row two. Next add 2 times row one onto row two. Then add -3 times row two to row three. Finally, add -2 times row two to row one and -3 times row two to row three.

$$\begin{bmatrix} 35 & 6 & -7 & | & 15 \\ -90 & -15 & 21 & | & -40 \\ 25 & 3 & -7 & | & 11 \end{bmatrix} \sim \begin{bmatrix} 10 & 3 & 0 & | & 4 \\ -90 & -15 & 21 & | & -40 \\ 25 & 3 & -7 & | & 11 \end{bmatrix}$$

$$\sim \begin{bmatrix} 10 & 3 & 0 & | & 4 \\ -15 & -6 & 0 & | & -7 \\ 25 & 3 & -7 & | & 11 \end{bmatrix} \sim \begin{bmatrix} 10 & 3 & 0 & | & 4 \\ 5 & 0 & 0 & | & 1 \\ 15 & 0 & -7 & | & 7 \end{bmatrix}$$

$$\sim \begin{bmatrix} 0 & 3 & 0 & | & 2 \\ 5 & 0 & 0 & | & 1 \\ 0 & 0 & -7 & | & 4 \end{bmatrix}$$

Therefore, the solution is $x_1 = \frac{1}{5}$, $x_2 = \frac{2}{3}$, and $x_3 = -\frac{4}{7}$. ∎

In the preceding example, we have illustrated the judicious use of row operations to avoid dealing with fractions. Conversely, some may prefer to follow the systematic approach of using the first row as the pivot row and creating zeros in the first column below the first row. Next, using the second row as the pivot row, one can create zeros in the second column below the second row, and so on.

Elementary Row Operations

Next on our agenda is the introduction of another of the **elementary row operations**:

3. Multiplication of a row by a nonzero scalar.

This operation is a **scaling** of an equation. With this operation, we can further reduce some of the preceding matrices as follows:

$$\begin{bmatrix} 7 & 0 & | & 21/2 \\ 0 & -136/7 & | & -272/21 \end{bmatrix} \sim \begin{bmatrix} 1 & 0 & | & \frac{3}{2} \\ 0 & 1 & | & \frac{2}{3} \end{bmatrix}$$

$$\begin{bmatrix} 0 & 0 & -1 & | & 8 \\ 1 & 0 & 0 & | & \frac{7}{2} \\ 0 & 6 & 0 & | & 7 \end{bmatrix} \sim \begin{bmatrix} 0 & 0 & 1 & | & -8 \\ 1 & 0 & 0 & | & \frac{7}{2} \\ 0 & 1 & 0 & | & \frac{7}{6} \end{bmatrix}$$

$$\begin{bmatrix} 0 & 3 & 0 & | & 2 \\ 5 & 0 & 0 & | & 1 \\ 0 & 0 & -7 & | & 4 \end{bmatrix} \sim \begin{bmatrix} 0 & 1 & 0 & | & \frac{2}{3} \\ 1 & 0 & 0 & | & \frac{1}{5} \\ 0 & 0 & 1 & | & -\frac{4}{7} \end{bmatrix}$$

As with the first type of row operation (adding a multiple of one row onto another row), we should be sure that we do not alter the solutions when we use this new row operation. But this fact is obvious: If we look at only one equation, $E(\mathbf{x}) = 0$, then this is equivalent to $\alpha E(\mathbf{x}) = 0$, provided that $\alpha \neq 0$. This last condition is necessary to get back to $E(\mathbf{x}) = 0$ from $\alpha E(\mathbf{x}) = 0$; of course, we must multiply by $1/\alpha$.

Without introducing any really new row operations, we can now perform **interchanges** or **swaps** among the rows in a system of equations. Let $\mathbf{r}_1$ and $\mathbf{r}_2$ be two rows. With a succession of row operations of the two types already described, we can execute a *swap*, that is an *interchange* of two rows:

$$\begin{bmatrix} \mathbf{r}_1 \\ \mathbf{r}_2 \end{bmatrix} \sim \begin{bmatrix} \mathbf{r}_1 \\ \mathbf{r}_1 + \mathbf{r}_2 \end{bmatrix} \sim \begin{bmatrix} -\mathbf{r}_2 \\ \mathbf{r}_1 + \mathbf{r}_2 \end{bmatrix} \sim \begin{bmatrix} \mathbf{r}_2 \\ \mathbf{r}_1 + \mathbf{r}_2 \end{bmatrix} \sim \begin{bmatrix} \mathbf{r}_2 \\ \mathbf{r}_1 \end{bmatrix}$$

Despite the fact that only two types of row operations are needed, it is conventional to define three types of **elementary row operations**:

1. **(replacement)** Add a multiple of one row to another.
2. **(scale)** Multiply a row by a nonzero factor.
3. **(swap)** Interchange a pair of rows.

The rows change positions only in a swap operation. Remember that in any proof involving row operations, the swapping process can be performed with four replacement and scaling operations. Thus, only replacement and scaling operations are essential in proofs.

If it is desired to describe the row operations being used on a matrix, the following notation is suggested:

replacement $\mathbf{r}_i \leftarrow \mathbf{r}_i + a\,\mathbf{r}_j$ $(i \neq j,\ a$ is a scalar$)$

scale $\mathbf{r}_i \leftarrow c\,\mathbf{r}_i$ $($scalar $c \neq 0)$

swap $\mathbf{r}_i \leftrightarrow \mathbf{r}_j$

In this description, the rows are $\mathbf{r}_1, \mathbf{r}_2$, and so forth. The replacement operation is the adding of a multiple of one row onto another row. The scaling operation is simply multiplying a row by a nonzero constant. The swap operation is the interchanging of two rows. (A similar notation is common in computer science and computer programming.) To illustrate, we use Example 5. The operations needed to reduce the given matrix are $(\mathbf{r}_2 \leftarrow \mathbf{r}_2 - 3\mathbf{r}_1)$, $(\mathbf{r}_1 \leftarrow \mathbf{r}_1 - 2\mathbf{r}_2)$, and $(\mathbf{r}_1 \leftarrow \frac{1}{3}\mathbf{r}_1)$. Note that the order of performing these steps must be observed.

In computer programs for solving large systems, swapping of rows can be avoided to reduce so-called *data motion.* An index array may be used to keep track of the order of pivot rows. In addition, it is usually not necessary to make the pivot elements equal to unity. High-quality software usually proceeds as we did in Example 7, resulting in one equation having only the term x_3, another equation involving x_2 and x_3, and a third equation involving all three variables.

Reduced Row Echelon Form

For a deeper understanding of linear equations, we often want the simplest form of a system of equations arrived at by use of all three types of row operations. At this stage, we can summarize our work: With suitable row operations, any matrix can be transformed into a special standard form called *reduced row echelon form.* This form is characterized as follows.

DEFINITION

A matrix is in **reduced row echelon form** *if*

- *All zero rows have been moved to the bottom of the matrix.*
- *Each nonzero row has 1 as its leading nonzero entry, using left-to-right ordering. Each such leading 1 is called a* **pivot**.
- *In each column containing a pivot, there are no other nonzero elements.*
- *The pivot in any row is farther to the right than the pivots in rows above.*

An important theorem, which we prove in Section 1.3 (Theorem 6), asserts the following:

THEOREM 1

Every matrix has one and only one reduced row echelon form.

With the help of this theorem, we see that the pivots are uniquely determined by the given matrix.

DEFINITION

A **pivot position** *in a matrix is a location where a leading 1 (a* **pivot***) appears in the reduced row echelon form of that matrix.*

In general, we do not know the pivot positions until we have found the reduced row echelon form of the matrix, or *any* row echelon form (to be defined later in this section).

Here are four examples of matrices in reduced row echelon form:

$$\begin{bmatrix} 0 & 1 \\ 0 & 0 \end{bmatrix}, \quad \begin{bmatrix} 1 & 0 \\ 0 & 1 \end{bmatrix}, \quad \begin{bmatrix} 0 & 1 & 3 & 0 \\ 0 & 0 & 0 & 1 \\ 0 & 0 & 0 & 0 \end{bmatrix}, \quad \begin{bmatrix} 1 & 5 & 0 & -7 & 0 \\ 0 & 0 & 1 & 3 & 0 \\ 0 & 0 & 0 & 0 & 1 \end{bmatrix}$$

All the 1's in these four matrices happen to be pivots. (See the previous definition to verify this assertion.) Here are four examples of matrices *not* in reduced row echelon form:

$$\begin{bmatrix} 0 & 0 \\ 1 & 0 \end{bmatrix}, \quad \begin{bmatrix} 0 & 1 \\ 1 & 0 \end{bmatrix}, \quad \begin{bmatrix} 0 & 1 & 3 & 2 \\ 0 & 0 & 0 & 1 \\ 0 & 0 & 0 & 0 \end{bmatrix}, \quad \begin{bmatrix} 0 & 1 & 3 & 0 \\ 0 & 0 & 0 & 4 \\ 0 & 0 & 0 & 0 \end{bmatrix}$$

An example of the general structure of a matrix in reduced row echelon form is as follows:

$$\begin{bmatrix} 0 & \boxtimes & \times & 0 & 0 & \times \\ 0 & 0 & 0 & \boxtimes & 0 & \times \\ 0 & 0 & 0 & 0 & \boxtimes & \times \\ 0 & 0 & 0 & 0 & 0 & 0 \\ 0 & 0 & 0 & 0 & 0 & 0 \end{bmatrix}$$

Here the boxed entries are the pivot positions. The symbol $\times$ designates either a zero or nonzero entry. Notice the staircase pattern of the pivots and zeros, and the fact that above and below each pivot the entries are 0. For an $n \times n$ matrix, the reduced row echelon form may be an **identity matrix**. This matrix is square and is denoted by $\mathbf{I}$ or $\mathbf{I}_n$. Its entries are δ_{ij}, where $\delta_{ii} = 1$ and $\delta_{ij} = 0$ when $i \neq j$. This notation, δ_{ij}, is called the **Kronecker delta**.[3] For example, the 5×5 identity matrix $\mathbf{I}_5$ is shown here:

$$\begin{bmatrix} 1 & 0 & 0 & 0 & 0 \\ 0 & 1 & 0 & 0 & 0 \\ 0 & 0 & 1 & 0 & 0 \\ 0 & 0 & 0 & 1 & 0 \\ 0 & 0 & 0 & 0 & 1 \end{bmatrix}$$

---

[3] Leopold Kronecker (1823–1891) concentrated his research on the theory of algebraic numbers, and contributed to several new branches of mathematics, viz. group theory and field theory. (See footnote on p. 193 for more details.)

EXAMPLE 9 Show how to avoid fractions until the last step in finding

the reduced row echelon form for this matrix $\begin{bmatrix} 3 & 2 & -3 & -3 \\ 5 & 1 & 0 & 7 \\ 7 & -4 & 32 & 6 \end{bmatrix}$.

SOLUTION The tactic is to scale two rows so that their leading nonzero entries are the same and then simply subtract one of these rows from the other, creating a zero leading entry in one row.

$$\begin{bmatrix} 3 & 2 & -3 & -3 \\ 5 & 1 & 0 & 7 \\ 7 & -4 & 32 & 6 \end{bmatrix} \sim \begin{bmatrix} 15 & 10 & -15 & -15 \\ 15 & 3 & 0 & 21 \\ 7 & -4 & 32 & 6 \end{bmatrix}$$

$$\sim \begin{bmatrix} 15 & 10 & -15 & -15 \\ 0 & -7 & 15 & 36 \\ 7 & -4 & 32 & 6 \end{bmatrix} \sim \begin{bmatrix} 3 & 2 & -3 & -3 \\ 0 & -7 & 15 & 36 \\ 7 & -4 & 32 & 6 \end{bmatrix}$$

$$\sim \begin{bmatrix} 21 & 14 & -21 & -21 \\ 0 & -7 & 15 & 36 \\ 21 & -12 & 96 & 18 \end{bmatrix} \sim \begin{bmatrix} 21 & 14 & -21 & -21 \\ 0 & -7 & 15 & 36 \\ 0 & -26 & 117 & 39 \end{bmatrix}$$

$$\sim \begin{bmatrix} 3 & 2 & -3 & -3 \\ 0 & -7 & 15 & 36 \\ 0 & -2 & 9 & 3 \end{bmatrix} \sim \cdots \sim \begin{bmatrix} 11 & 0 & 0 & 34 \\ 0 & -11 & 0 & 93 \\ 0 & 0 & 11 & -17 \end{bmatrix}$$

$$\sim \begin{bmatrix} 1 & 0 & 0 & 34/11 \\ 0 & 1 & 0 & -93/11 \\ 0 & 0 & 1 & -17/11 \end{bmatrix}$$

We leave the intermediate steps as an exercise. ∎

These results can be confirmed by using the MATLAB® command `format rat` and `rref(A)`. One can also use `rrefmovie(A)` to see the algorithm working step-by-step. (See the mathematical software examples at the end of Section 1.2.)

Fortunately for us, advanced mathematical software systems have built-in commands for computing the reduced row echelon form of matrices. For example, MATLAB uses `rref(A)`, Maple™ has `ReducedRowEchelon Form(A)`, and Mathematica® has `RowReduce[A]`. Further examples using these commands are found in the subsections on Mathematical Software at the ends of sections.

Row Echelon Form

A partially reduced form of a matrix is often used; it has three weaker properties.

DEFINITION

A matrix is in **row echelon form** *if*

- *All zero rows have been moved to the bottom.*
- *The leading nonzero element in any row is farther to the right than the leading nonzero element in the row just above it.*
- *In each column containing a leading nonzero element, the entries below that leading nonzero element are 0.*

For example, the matrix below is in row echelon form:

$$\begin{bmatrix} 0 & 3 & 5 & 9 & 6 & 4 & 1 \\ 0 & 0 & 0 & 7 & 6 & 3 & 2 \\ 0 & 0 & 0 & 0 & 0 & 4 & 3 \\ 0 & 0 & 0 & 0 & 0 & 0 & 0 \end{bmatrix}$$

An example of the general structure of a matrix in row echelon form is as follows:

$$\begin{bmatrix} 0 & \boxtimes & \times & \times & \times & \times \\ 0 & 0 & 0 & \boxtimes & \times & \times \\ 0 & 0 & 0 & 0 & \boxtimes & \times \\ 0 & 0 & 0 & 0 & 0 & 0 \\ 0 & 0 & 0 & 0 & 0 & 0 \end{bmatrix}$$

Here the boxed entries $\boxtimes$ are the leading nonzero entries in the rows. It is obvious that a row echelon form is obtained with less work than is required for the reduced row echelon form. Notice the staircase pattern of the pivot positions. For some questions about a matrix (such as its rank), the row echelon form gives the answers more quickly. The reduced row echelon form of a matrix is unique, whereas a matrix may have many row echelon forms. Thus, one may ask for *the* reduced row echelon form of a matrix or *a* row echelon form. A proof of the uniqueness of the reduced row echelon form is presented in Section 1.3 (Theorem 6). The concepts of reduced row echelon form and row echelon form are independent of whether the matrix is an augmented matrix. We may ignore the vertical line in an augmented matrix when deciding whether it is in row echelon or reduced row echelon form.

Intuitive Interpretation

There is an intuitive interpretation of a system of linear equations that helps one cope with some of the mysteries in linear algebra.

Let us start with a vector $\mathbf{x} = (x_1, x_2)$ that is free to roam all over $\mathbb{R}^2$. The two variables x_1 and x_2 are not connected. It is often said in such a situation that the point has *two degrees of freedom.*

What happens when we impose a single linear condition on this point? Suppose, for example, that we require $x_1 + 3x_2 = 7$. The point now has only *one degree of freedom.* If we start with a point that satisfies this equation and then decide to allow the point to move, while still obeying that equation, it can move only in such a way that the equation $x_1 + 3x_2 = 7$ remains true. As we know, this means that the points must lie on a line. We can start with $(4, 1)$, a point that already satisfies the equation. If we change the point while maintaining the condition $x_1 + 3x_2 = 7$, we can go to $(1, 2)$ but not to $(2, 2)$, for example.

Suppose now that another condition is imposed, in addition to the first one. Say that we require $2x_1 - 5x_2 = -8$. This in effect restricts our point to a single location, namely $(1, 2)$. The point has lost its degrees of freedom. Each added condition further restricts the point, and we can even make the point *disappear*—meaning that it cannot satisfy all the conditions laid down. In the present example, this occurs if we require further that $x_1 - x_2 = 0$.

Unfortunately, our intuition can be faulty, and that is where the science of linear algebra enters the picture. For example, suppose that we require $x_1 + 3x_2 = 7$, as before, and add the condition $2x_1 + 6x_2 = 14$. This does *not* further restrict the point, because this *new* condition is the same as the first (since it is a simple multiple of the first equation). That is the first *unusual* case. Another unusual case is illustrated by making our second condition $2x_1 + 6x_2 = 8$. Now *no* point satisfies the two conditions. Each equation by itself represents a line, but these two lines are parallel and non-intersecting. Hence, there is no point that satisfies both equations simultaneously.

Next, one can contemplate more complicated situations, such as linear equations with three variables. Again, we can start with an unrestricted point $\mathbf{x} = (x_1, x_2, x_3)$; it has three degrees of freedom. If we impose a single equation, such as $3x_1 - 4x_2 + 5x_3 = 9$, we find that our variable point is now confined to a plane and has only two degrees of freedom. If we add another linear condition, such as $x_1 - 5x_2 - x_3 = 7$, we find that the points satisfying both equations lie on a line. But there are other cases besides this *normal* or *expected* case. Two equations might describe an impossible case (a pair of parallel planes), or they might not actually represent two different conditions. Going on to three equations, one might expect that these three equations

$$\begin{cases} 3x_1 - 4x_2 + 5x_3 = 9 \\ x_1 - 5x_2 - x_3 = 7 \\ 5x_1 - 14x_2 + 3x_3 = 23 \end{cases}$$

would define a single point in $\mathbb{R}^3$, but in fact they describe a plane. Can you see why?

With more variables and more equations the situation becomes more complicated, but we shall develop methods for understanding these problems, no matter how many equations and variables are present. In fact, that is our first major goal: to understand the set of all solutions to any given system of linear equations.

Application: Feeding Bacteria

From time to time in this book, we interrupt the mathematical proceedings to give applications of the theory. Here is such an example.

EXAMPLE 10 A bacteriologist has placed three types of bacteria, labeled B_1, B_2, and B_3, in a culture dish, along with certain quantities of three nutrients, labeled N_1, N_2, and N_3. She knows the amounts of each nutrient that can be consumed by each bacterium in a 24-hour period. These data are collected in a table:

	B_1	B_2	B_3
N_1	4	2	6
N_2	3	1	2
N_3	7	5	2

This table tells us, for example, that each bacterium B_1 in one day can consume 4 units of N_1, 3 units of N_2, and 7 units of N_3. How many bacteria of each type can be supported daily by 4200 units of N_1, 1900 units of N_2, and 4700 units of N_3?

SOLUTION Denote by x_1, x_2, and x_3 the number of bacteria of each type represented in the culture. Considering just the first nutrient and noting how it can be consumed by the three types of bacteria, we have the equation

$4x_1 + 2x_2 + 6x_3 = 4200$. The other two equations governing the nutrients N_2 and N_3 are $3x_1 + x_2 + 2x_3 = 1900$ and $7x_1 + 5x_2 + 2x_3 = 4700$. The augmented matrix for this system is

$$\left[\begin{array}{ccc|c} 4 & 2 & 6 & 4200 \\ 3 & 1 & 2 & 1900 \\ 7 & 5 & 2 & 4700 \end{array} \right]$$

The steps in the row reductions are these:

$$\left[\begin{array}{ccc|c} 1 & 1 & 4 & 2300 \\ 3 & 1 & 2 & 1900 \\ 7 & 5 & 2 & 4700 \end{array} \right] \sim \left[\begin{array}{ccc|c} 1 & 1 & 4 & 2300 \\ 0 & -2 & -10 & -5000 \\ 0 & -2 & -26 & -11400 \end{array} \right]$$

$$\sim \left[\begin{array}{ccc|c} 1 & 1 & 4 & 2300 \\ 0 & 1 & 5 & 2500 \\ 0 & 1 & 13 & 5700 \end{array} \right]$$

$$\sim \left[\begin{array}{ccc|c} 1 & 1 & 4 & 2300 \\ 0 & 1 & 5 & 2500 \\ 0 & 0 & 8 & 3200 \end{array} \right]$$

The calculation of the solution goes like this:

$$\begin{cases} x_3 = \frac{1}{8}(3200) = 400 \\ x_2 = 2500 - 5x_3 = 2500 - 2000 = 500 \\ x_1 = 2300 - 4x_3 - x_2 = 2300 - 1600 - 500 = 200 \end{cases}$$

Hence, the solution is $x_1 = 200$, $x_2 = 500$, and $x_3 = 400$ and we obtain $\mathbf{x} = (200, 500, 400)$. ■

Mathematical Software

Throughout the book, we give examples using MATLAB, Maple, and Mathematica, which are currently the most popular mathematical software packages. The websites for these software systems are MATLAB (www.mathworks.com), Maple (www.maplesoft.com), and Mathematica (www.wolfram.com). There are many other mathematical software systems available. For example, there is a noncommercial system called Octave that is freely redistributable under the terms of the GNU General Public License of the Free Software Foundation (www.gnu.org/software/octave). Unfortunately, each of these systems has its own syntax!

A matrix may undergo Gaussian elimination or a variant of it when put into reduced row echelon form. For example, MATLAB produces the reduced row echelon form using Gauss-Jordan elimination with *partial pivoting* in which a tolerance parameter is used to test for negligible column elements. Partial pivoting is a process by which pivot elements are selected to reduce roundoff error, and may involve interchanging rows.

EXAMPLE 11 Here is a seven decimal-place example of two linear equations with two variables:

$$\begin{cases} 3.215793x_1 + 82.13459x_2 = 5.332873 \\ 9.300567x_1 - 1.776321x_2 = -12.99334 \end{cases}$$

How can we use the computer to find the solution of this system?

SOLUTION In a computer on which MATLAB has been installed, we input the data for our problem by typing this information followed by the commands to solve the system.

```
MATLAB

format long
A = [3.215793,82.13459;9.300567,-1.776321]
b = [5.332873;-12.99334]
x = A\b
r = A*x - b
```

In MATLAB, spaces can be used in place of the commas between array entries. MATLAB does all its computations in full precision but offers various formats for displaying the results. The command `format long` requests answers in full floating point form, with 15 decimal places. The command `format rat` requests answers as ratios of small integers. The command `format short` produces answers with five-digit floating point values; this format is the default. In this example, MATLAB responds by printing all the input data (*echoing*) and the solution of the problem $\mathbf{x} \approx (-1.37437016149456, 0.11873880352654)$. We then ask for an independent verification of the *answer* by substituting the numerical values of x_1 and x_2 into the two equations and comparing to the prescribed values of b_1 and b_2. In fact, there is a very small discrepancy, which is due to roundoff errors. The difference is $(-0.089, -0.18) \times 10^{-14}$. (In many types of problems, one can verify the proffered solution by substitution or some other independent check.)

The Maple system handles the same problem as follows:

```
Maple
with(LinearAlgebra):
A := Matrix([[3.215793,82.13459],[9.300567,-1.776321]]);
b := Vector([5.332873,-12.99334]);
x := LinearSolve(A,b);
r := A.x - b;
```

Here, we illustrate with a user-friendly package called `LinearAlgebra`, designed explicitly for carrying out linear algebraic computations.

In the Mathematica software package, there are similar commands, as follows:

```
Mathematica
A = {{3.215793,82.13459},{9.300567,-1.776321}}
b = {5.332873,-12.99334}
x = LinearSolve(A,b)
r = A.x - b
```

We have not explained all the notational devices (vectors and matrices) that are being used in this example because at this stage we only want to emphasize that high-quality software is available to solve problems such as this. The user has only to type input values for such programs, and then request the solution. ∎

Maple has commands that perform elementary row operations symbolically:

Replacement	$r_i \leftarrow r_i + a\, r_j$	`RowOperation(A,[i,j],a);`
Scale	$r_i \leftarrow c\, r_i$	`RowOperation(A,i,c);`
Swap	$r_i \leftrightarrow r_j$	`RowOperation(A,[i,j]);`

As an illustration, we solve the linear system $3x_1 + 2x_2 = 4$ and $9x_1 + 7x_2 = 17$ from Example 5.

```
Maple
with(LinearAlgebra):
A  := Matrix([[3,2,4],[9,7,17]]);
A1 := RowOperation(A,[2,1],-3);
A2 := RowOperation(A1,[1,2],-2);
A3 := RowOperation(A2,1,1/3);
```

The Basic Linear Algebra Subprograms (BLAS) is a collection or "library" of computer routines. It provides standard building blocks for performing basic vector and matrix operations. For the elementary vector operations, the replacement is _axpy to suggest the operation $\mathbf{y} \leftarrow a\mathbf{x} + \mathbf{y}$ in which a vector $\mathbf{y}$ is replaced by a scalar a times a vector $\mathbf{x}$ plus the original vector $\mathbf{y}$. Also, there are routines called _swap for interchanging two vectors ($\mathbf{x} \leftrightarrow \mathbf{y}$) and _scale for scaling a vector $\mathbf{x}$ ($\mathbf{x} \leftarrow c\mathbf{x}$). The BLAS routines are particularly useful on high-performance computers and have been extended and improved for each new generation of supercomputer and for handling either dense or sparse data.

Algorithm for the Reduced Row Echelon Form

One algorithm for finding the reduced row echelon form of a matrix is presented here. (The steps described are easily translated into a computer program.)

1. Interchange rows if necessary to place all zero rows on the bottom.
2. Identify the leftmost nonzero column. Say it is pivot column j. Interchange rows to bring a nonzero element to the top row and jth column, which is the pivot position. (Computer programs often choose the largest entry in absolute value in an attempt to minimize round-off errors.) Use the row replacement operation to create zeros in all positions in the pivot column below the pivot position.
3. Repeat Steps 1 and 2 on the remaining submatrix until there are no nonzero rows left. (We have found a row echelon form, but it is not unique.)
4. Beginning with the rightmost pivot, working upward and to the left, use row replacement operations to create zeros in all positions in the pivot column above the pivot position. Scale the entry in the pivot row to create a leading 1.
5. Repeat Step 4, ending with the unique reduced row echelon form of the given matrix.

Steps **1–3** are called the *Gaussian* or *forward* portion of this algorithm, and Steps **4–5** are the *backward* portion. An alternative algorithm called *Gauss-Jordan elimination* combines the forward and backward portions of the algorithm by doing the elimination steps above and below the pivot positions. Although students may find the Gauss-Jordan elimination useful for pencil and paper calculations, it is more computationally intensive for computer programs.

SUMMARY 1.1

- Point–slope form of a line: $y = mx + b$ where m is the slope and b is the intercept

- Two-point form of a line through points (x_0, y_0) and (x_1, y_1):
$y - y_0 = m(x - x_0)$ where $m = (y_1 - y_0)/(x_1 - x_0)$

- Lines L_1 and L_2 in $\mathbb{R}^2$: there can be a unique solution (intersect once), no solution (parallel), or infinitely many solutions (co-linear)

- Matrix form of a system of linear equations: $\mathbf{Ax} = \mathbf{b}$

- Augmented matrix: $[\mathbf{A} \,|\, \mathbf{b}]$

- Gaussian elimination: $\mathbf{Ax} = \mathbf{b}$ becomes (after row reduction) $\mathbf{Ux} = \mathbf{c}$, where $\mathbf{U}$ is an upper triangular matrix

- Row-equivalent augmented matrices: $[\mathbf{A} \,|\, \mathbf{b}] \sim [\mathbf{U} \,|\, \mathbf{c}]$

- Elementary row operations:

  - (**replacement**) Add to row i a multiple of row j, where $i \neq j$ ($\mathbf{r}_i \leftarrow \mathbf{r}_i + a\,\mathbf{r}_j$)

- (**scale**) Multiply row i by c, a nonzero scalar ($\mathbf{r}_i \leftarrow c\,\mathbf{r}_i$)

- (**swap**) Interchange the two rows i and j ($\mathbf{r}_i \leftrightarrow \mathbf{r}_j$)

- Consistent system has at least one solution; inconsistent system has no solution

- Pivots: leading 1's in reduced row echelon form; pivot positions: locations where the pivots will be at the end of the reduction process

- The reduced row echelon form (unique): zero rows at bottom; pivot rows form a staircase pattern (with possibly different widths of steps); pivot columns contain only the pivot entry 1

- One row echelon form (non-unique): zero rows are moved to the bottom; the rows containing leading nonzero elements form a staircase pattern; columns containing leading nonzero elements contain only 0's elsewhere

KEY CONCEPTS 1.1

Linear equations, systems of linear equations, lines and planes, coefficient matrix, augmented matrix, vector of unknowns, vector of right-hand sides, Gaussian elimination, triangular systems, elementary row operations (replacement, scale, swap), row equivalence of matrices, consistent and inconsistent systems, row-reduction process, reduced row echelon form, pivots and pivot positions, row echelon form, bacteria-nutrition application, using mathematical software, algorithm for reduced row echelon form

GENERAL EXERCISES 1.1

1. Solve this system of equations and verify your answer:

$$\begin{cases} 2x_2 - 3x_3 = -11 \\ 4x_1 + x_2 + 3x_3 = 34 \\ 5x_3 = 35 \end{cases}$$

2. Solve this system of linear equations and verify your answer:

$$\begin{cases} 3x_1 = 9 \\ 2x_1 - 5x_2 + 6x_3 = -28 \\ -4x_1 + 5x_3 = -32 \end{cases}$$

3. Solve this system by Gaussian elimination and verify your answer:

$$\begin{cases} 2x_1 + 3x_2 = -3 \\ 6x_1 + 4x_2 = -5 \end{cases}$$

4. Solve the system whose augmented matrix is given here, and verify your answer:

$$\begin{bmatrix} -3 & 4 & | & 37 \\ 2 & -5 & | & -41 \end{bmatrix}$$

5. (Continuation.) What are a_{21} and a_{12} in the matrix displayed in the preceding problem?

6. Let $\begin{cases} 3x_2 + 7x_3 + \quad x_1 = 42 \\ \quad x_2 + \ 2x_1 = \ 6 \\ \quad 3x_3 + 11x_1 = 76 \end{cases}$

Solve this system of equations by carrying out the row-reduction process to reduced row echelon form.

7. Carry out Gaussian elimination on the system of equations in Example 3. Does that system have any solutions other than $(1, 3, 2)$?

8. Show that

$$\begin{bmatrix} 0 & 1 & 3 & 0 & 7 & 0 \\ 0 & 0 & 0 & 5 & 2 & 0 \\ 0 & 0 & 0 & 0 & 0 & 1 \end{bmatrix}$$

$$\sim \begin{bmatrix} 0 & 1 & 3 & 15 & 13 & 4 \\ 0 & 2 & 6 & 35 & 28 & 3 \\ 0 & -2 & -6 & 20 & -6 & -3 \end{bmatrix}$$

9. Solve this system in such a way that fractions enter only in the last step:

$$\begin{cases} 3x + \ y = -5 \\ 2x + 4y = \ 7 \end{cases}$$

10. Solve the system of equations whose augmented matrix is $\begin{bmatrix} 1 & 2 & 1 & | & 0 \\ 3 & 1 & 0 & | & 3 \\ 5 & 4 & 3 & | & 10 \end{bmatrix}$

11. Solve the system of equations whose augmented matrix is $\begin{bmatrix} 3 & 3 & 4 & | & 11 \\ 2 & 1 & 1 & | & 6 \\ 1 & 2 & 3 & | & 4 \end{bmatrix}$

12. Solve this system: $\begin{cases} 3x_1 + 6x_2 + 6x_3 = 21 \\ 2x_1 + 4x_2 + 5x_3 = 16 \\ 2x_1 + 5x_2 + 4x_3 = 17 \end{cases}$

13. Use the theory of linear equations to determine whether the lines described by these three equations have a point in common: $x + 2y = 1$, $2x - 3y = 9$, $-3x - 2y = -7$.

14. Solve the system of equations having this augmented matrix, without performing any further row operations:

$$\begin{bmatrix} 0 & 0 & 3 & | & 12 \\ 1 & 2 & 1 & | & 12 \\ 0 & 2 & 2 & | & 6 \end{bmatrix}$$

15. Show how to solve these three systems all at the same time:

a. $\begin{cases} x + 3y + 2z = 1 \\ 2x + \ y + \ z = 0 \\ 4x - \ y + 3z = 0 \end{cases}$

b. $\begin{cases} x + 3y + 2z = 0 \\ 2x + \ y + \ z = 1 \\ 4x - \ y + 3z = 0 \end{cases}$

c. $\begin{cases} x + 3y + 2z = 0 \\ 2x + \ y + \ z = 0 \\ 4x - \ y + 3z = 1 \end{cases}$

16. Draw graphs of the two lines having equations $x + y = 13$ and $-2x + y = 4$. From the graphs, estimate where the lines intersect. Confirm the estimate by solving the system of two equations. Convert the system of equations to reduced row echelon form.

17. Two lines are given in $\mathbb{R}^2$, namely, $-x + y = 1$ and $2x + y = 4$. Find the point of intersection of these two lines. Then investigate what happens if we change the second equation to $2x - 2y = 4$. Finally, find out what happens if we change the second equation to $2x - 2y = -2$. What do you learn from making small changes to the system?

18. It is claimed that the reduced row echelon form of the matrix

$$\begin{bmatrix} 13 & 17 & -31 & 1097 \\ 11 & -19 & 7 & -413 \\ 5 & 3 & 29 & -359 \end{bmatrix}$$

$$= \begin{bmatrix} 1 & 0 & 0 & 13 \\ 0 & 1 & 0 & 23 \\ 0 & 0 & 1 & -17 \end{bmatrix}$$

How can we verify or disprove the claim without going through the complete row-reduction process? (Think about systems of equations represented by the two matrices.)

19. Solve these two systems of linear equations and check your work with an independent verification. In each case the augmented matrix is shown:

$$\begin{bmatrix} 2 & -1 & a \\ 3 & 4 & b \end{bmatrix}, \qquad \begin{bmatrix} 2 & -1 & 11 \\ 3 & 4 & -11 \end{bmatrix}$$

20. Find the reduced row echelon forms of these matrices:

a. $\begin{bmatrix} 0 & 3 & 0 & 5 \\ 4 & 0 & 0 & -3 \\ 0 & 0 & 1 & 7 \\ 8 & 3 & 1 & 6 \end{bmatrix}$ **b.** $\begin{bmatrix} -12 & 0 & -1 & 2 \\ 16 & 3 & 1 & 0 \\ 20 & 3 & 2 & 4 \\ 12 & 3 & 1 & 3 \end{bmatrix}$

c. $\begin{bmatrix} 3 & 1 \\ 4 & 2 \\ -6 & 1 \end{bmatrix}$ **d.** $\begin{bmatrix} 3 & 4 & -6 \\ 1 & 2 & 1 \end{bmatrix}$

21. In an $n \times n$ matrix whose elements are $a_{ij} = (-1)^{i+j}$, how many positive terms are there? (A formula is sought.)

22. Consider $\begin{bmatrix} 1 & 3 & 2 & 1 \\ 0 & -4 & 5 & -23 \\ 2 & 2 & 9 & t \end{bmatrix}$

This is a system of equations in which one element t can change. Find values of the parameter t for which we can obtain solutions to this augmented matrix. Explain the implications of this example in the theory of linear equations.

23. Suppose that we have an equation of the form $f(x) = g(x)$, where f and g are real-valued functions of the real variable x. We want to determine x from this equation. Certainly, we can proceed to the equation $[f(x)]^2 = [g(x)]^2$. Give an example to show that solutions of the second equation are not necessarily solutions of the first. Why is the situation here different from the one discussed in this section?

24. By obtaining the reduced row echelon form, find the solution to this pair of equations: $3x_2 + x_1 = 17$ and $2x_1 + 7x_2 = 39$

25. Use the row-reduction techniques to solve this system, in which you may assume $c \neq 0$ and $3c + 5a \neq 0$:

$$\begin{cases} ax + 3y = 7 \\ cx - 5y = -4 \end{cases}$$

26. Explain: If a system of linear equations has exclusively rational numbers for the data a_{ij} and b_i, and if the system has a solution, then it will have a rational solution. (A real number is said to be *rational* if it can be expressed as the quotient of two integers.)

27. Complete all the steps in Example 9 without involving fractions until the last step.

28. Find all solutions of this nonlinear system:
$$\begin{cases} 3x^2 - 5y^3 = -123 \\ 7x^2 + 4y^3 = 136 \end{cases}$$

29. Let $\begin{cases} 3x^3 - \ln y = 77 \\ 2x^3 + \ln y^3 = 66 \end{cases}$

Find all solutions of this nonlinear system. (Here $\ln = \log_e$.)

30. a. Using the notation $\mathbf{r}_i \leftarrow \alpha\, \mathbf{r}_j + \mathbf{r}_i$, describe the steps used in the row reduction of Example 8.
    b. Use the systematic approach described in the text.

31. Solve the bacteria-nutrition problem when the given data are as in the following table and the nutrients are supplied daily in amounts of 1800 units of N_1, 1500 units of N_2, and 2500 units of N_3. The answer states the number of each type of bacterium to be inserted in the culture.

	B_1	B_2	B_3
N_1	2	1	5
N_2	1	3	1
N_3	3	1	7

32. Consider $\begin{bmatrix} 1 & 3 & 2 & | & 12 \\ 5 & -7 & 1 & | & 7 \\ -11 & 33 & t & | & 23 \end{bmatrix}$

Investigate the system of linear equations having this augmented matrix. Here t is a parameter allowed to run over $\mathbb{R}$. What happens if t approaches 5? The moral here is that the solution of a system of linear equations may be a discontinuous function of the data!

33. Explain why this is really a linear equation:
$$3x(2y + 5) + \log(x^2) - 2y(6 + 3x)$$
$$= 13 + 2\log x$$

34. Explain why this nonlinear system of equations is inconsistent if we allow only real numbers as solutions:
$$\begin{cases} 3x^2 + 4y^3 = 7 \\ -2x^2 + 3y^3 = 18 \end{cases}$$

35. Solve this system of three linear equations:
$$\begin{cases} -5 = x_1 + 2x_3 + 3x_2 \\ 4x_2 - 4x_3 + 2x_1 = 14 \\ x_1 + 2 + x_3 + 3x_2 = 0 \end{cases}$$

36. Consider $\begin{cases} 7\ln(x^3) + 2\ln(x^2) + y^2 = 77 \\ 2\ln x + 5y^2 = 16 \end{cases}$

Solve this pair of equations for x and y using logarithms to base e.

37. You have seen in General Exercises 28, 29, 33, 34, and 36 examples of nonlinear equations that yield to the techniques of changing variables. Linear changes of variables in linear equations can also be used.

Let $\begin{cases} 3x - 2y = 4 \\ 7x + 5y = 21 \end{cases}$

What system of equations results from this system when we change variables like this: $x = u - v$ and $y = u + v$?

38. Let $\begin{cases} 3x_1 - 5x_2 = 17 \\ -x_1 + 2x_2 = 23 \end{cases}$

In this system of equations make the change of variables $x_1 = 2u_1 + 5u_2$ and $x_2 = u_1 + 3u_2$. The new system of equations, involving u_1 and u_2, can be much simpler than the original system of equations. You should find that $x_1 = 149$ and $x_2 = 86$.

39. There exists a matrix in reduced row echelon form such that one column can be removed, leaving a matrix that is also in reduced row echelon form. Find one or more examples of this phenomenon.

40. Consider $\begin{bmatrix} 3 & 6 & 5 \\ 1 & -1 & 0 \end{bmatrix}, \begin{bmatrix} 4 & 5 & 5 \\ 2 & 7 & 5 \end{bmatrix}$.

Are these two matrices row equivalent to each other? Why or why not?

41. Let $\begin{cases} 2x - 3y = -1 \\ 4x + 5y = 53 \end{cases}$

Solve this system of linear equations by converting the augmented matrix to reduced row echelon form.

42. Consider three planes in $\mathbb{R}^3$ whose equations are $3x + y = 6$, $5x + y + z = 6$, and $3x + z = 1$. Do they have a point in common? If so, find that point (or points).

43. Let $\begin{cases} 14x - 21y = -117 \\ 28x + 35y = 371 \end{cases}$

Solve this system of linear equations by converting the augmented matrix to reduced row echelon form.

44. Criticize this solution of a system of three linear equations: $-x + y = 1$, $2x + y = 4$, and $3x - y = 2$. To solve this, we can ignore the third equation because the first two equations by themselves give us $x = 1$ and $y = 2$. Hence, the solution is $(1, 2)$.

45. Let $\begin{cases} 2x_1 + 3x_2 - x_3 = -5 \\ 4x_1 - x_2 + 2x_3 = 24 \\ 3x_1 + x_2 - 3x_3 = -8 \end{cases}$

Solve this system using only integers.

46. Establish that if a matrix has all integer entries, then it is row equivalent to a matrix in row echelon form having only integer entries. Can we make the same assertion for the reduced row echelon form?

47. If possible, give an example of a 1×5 matrix that is not in row echelon form. Then give an example of a 1×5 matrix that is in row echelon form but not in reduced row echelon form.

48. Which of these matrices is in reduced row echelon form: $[0 \quad 2 \quad 2]$, $[2 \quad 1 \quad 3]$, $[4 \quad 2 \quad 2]$, $[0 \quad 0 \quad 1]$?

49. Consider the matrices $\begin{bmatrix} 3 & 4 & 2 \\ 2 & -3 & 1 \\ 5 & 7 & 1 \end{bmatrix}$ and $\begin{bmatrix} 1 & 7 & 1 \\ 0 & 17 & 1 \\ 0 & 7 & 1 \end{bmatrix}$. Are these two matrices row equivalent to each other? Why or why not?

50. Let $\begin{cases} 3x - y = 2 \\ 2x + 5y = -4 \\ 7x - 25y = 1 \end{cases}$

Show that each pair of these equations has a solution but the entire system does not. Interpret geometrically.

51. Consider $\begin{bmatrix} 0.780 & 0.563 \\ 0.913 & 0.659 \end{bmatrix} \begin{bmatrix} x_1 \\ x_2 \end{bmatrix} = \begin{bmatrix} 0.217 \\ 0.254 \end{bmatrix}$

Which solution is better?
$\hat{x} = (0.341, -0.087)$ or
$\tilde{x} = (0.999, -1.001)$. Explain.

COMPUTER EXERCISES 1.1

1. Use mathematical software such as MAT-LAB, Maple, or Mathematica to solve one or more of the General Exercises: 8, 12, 15, 20, 22, 31, 45.

2. Use mathematical software to find the reduced row echelon form of a 20×20 matrix containing random integers in $[-20, 20]$.

3. Write a computer program for computing the reduced row echelon form of a given $m \times n$ matrix **A**. In the first version of the program, do not attempt any *pivoting strategy*. In other words, just use the natural ordering of the rows in the matrix as they are given and assume that no zero pivot entries are encountered. (Don't worry about the code not working if division by zero arises.) By the use of comments in the code, indicate the portion of the code that does the *forward elimination phase*, *backward elimination phase*, and *scaling phase*. Test the code on several matrices.

4. (Continuation.) In the second version of the program, generalize the code to handle matrices with zero pivots and zero rows and columns. Test the code on matrices such as this one:

$$\begin{bmatrix} 4 & 12 & 2 & 0 & 16 & 1 & 7 & 26 \\ 6 & 18 & -6 & 0 & 42 & -9 & -57 & -36 \\ 0 & 0 & 0 & 0 & 0 & 0 & 0 & 0 \\ -2 & -6 & 2 & 0 & -14 & 3 & 19 & 12 \\ 1 & 3 & 3 & 0 & -1 & 2 & 17 & 25 \end{bmatrix}$$

5. (Continuation.) In the third version of the program, modify the general code to use the Basic Linear Algebra Subprograms (BLAS) for carrying out the replacement (_axpy), swap (_swap), and scale (_scale) operations. (As originally proposed by Lawson, Hanson, Kincaid, and Krogh [1979], the BLAS are routines that provide standard building blocks for performing basic vector and matrix operations. They are efficient, portable, and widely available in computer systems. They find common use in the development of high-performance linear algebra software. They can be downloaded at `www.netlib.org/blas`.)

6. (Continuation.) In the final version of the program, use the Gauss-Jordan method.

In the Gauss-Jordan algorithm (without pivoting) at the kth major step, the pivot entry in row k is scaled to 1 and multiples of row k are subtracted from *all* the other rows so that all elements above and below the pivot element are 0. In other words, the scaling phase is done first and the forward and backward elimination phases are done together.

7. After seeing page after page of simple numerical examples, usually involving small integers, let's explore a more realistic example. Find the reduced row echelon form of this augmented matrix:

$$\begin{bmatrix} 1325.9627 & -23.874191 & \bigm| & 4513.1622 \\ -0.31224877 & 531.26915 & \bigm| & -25492.204 \end{bmatrix}$$

1.2 VECTORS AND MATRICES

> Mathematics is the queen of the sciences.
>
> —KARL FRIEDRICH GAUSS (1777–1855),
> *ONE OF THE GREATEST MATHEMATICIANS OF ALL TIMES*

The word mathematics comes from the Greek μάθημα (*máthēma*), meaning *science, knowledge, or learning* and μαθηματικός (mathēmatikós) meaning *fond of learning.*

> —EN.WIKIPEDIA.ORG

We continue the discussion of vectors and matrices begun in Section 1.1. These concepts play a central role in linear algebra, especially the part of the subject that concerns systems of linear equations.

Vectors

A *vector* is conventionally represented as a vertical column of numbers, such as

$$
\mathbf{x} = \begin{bmatrix} x_1 \\ x_2 \\ \vdots \\ x_n \end{bmatrix}, \qquad
\mathbf{v} = \begin{bmatrix} 3 \\ 72 \\ -24 \\ 5221 \end{bmatrix}, \qquad
\mathbf{w} = \begin{bmatrix} 2 \\ 5 \\ 0 \end{bmatrix}
$$

Vectors can also be written as horizontal arrays. In some contexts, either form can be used. However, when sums and products of vectors and matrices occur, we must observe certain conventions. For typographical reasons, such as writing vectors in-line and saving space, we often write a vector as $\mathbf{x} = (x_1, x_2, \ldots, x_n)$ or as $\mathbf{x} = [x_1, x_2, \ldots, x_n]^T$. Here the superscript T means *transpose*. It serves to turn a vertical array into a horizontal one and a horizontal array into a vertical one. The entries in the vector are its *components*.

For a fixed value of n, the set of all vectors having n components is denoted by $\mathbb{R}^n$. Thus, $\mathbb{R}^1$ is just the real numbers $\mathbb{R}$; $\mathbb{R}^2$ is the familiar two-dimensional plane; $\mathbb{R}^3$ is the three-dimensional space of our universe. The remaining cases, $n = 4, 5, \ldots$ do not have familiar geometric interpretations.

The special vector $[0, 0, \ldots, 0]^T$ in $\mathbb{R}^n$ is the zero vector or *origin*. It is also written as $(0, 0, \ldots, 0)$ and is denoted by $\mathbf{0}$. The **addition** of two members of $\mathbb{R}^n$ is effected by the following rule, which is termed **vector addition:**

$$
\mathbf{x} + \mathbf{y} = \begin{bmatrix} x_1 \\ x_2 \\ \vdots \\ x_n \end{bmatrix} + \begin{bmatrix} y_1 \\ y_2 \\ \vdots \\ y_n \end{bmatrix} = \begin{bmatrix} x_1 + y_1 \\ x_2 + y_2 \\ \vdots \\ x_n + y_n \end{bmatrix}
$$

For example, we can write $(1, 3, 7) + (5, -6, 3) = (6, -3, 10)$ or $[1, 3, 7]^T + [5, -6, 3]^T = [6, -3, 10]^T$ or

$$
\begin{bmatrix} 1 \\ 3 \\ 7 \end{bmatrix} + \begin{bmatrix} 5 \\ -6 \\ 3 \end{bmatrix} = \begin{bmatrix} 6 \\ -3 \\ 10 \end{bmatrix}
$$

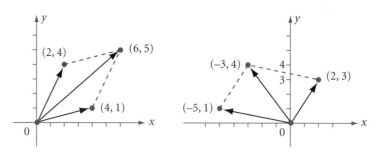

FIGURE 1.3 Addition of pairs of vectors in $\mathbb{R}^2$.

We say that the addition of two (or more) vectors is done *component-wise*—that is, *component-by-component*. Notice that the zero vector in $\mathbb{R}^n$ has this property: $\mathbf{x} + \mathbf{0} = \mathbf{x}$, for all $\mathbf{x}$ in $\mathbb{R}^n$.

The geometry for the addition of two vectors in $\mathbb{R}^2$ is shown in Figure 1.3. We form the parallelogram with the given vectors as two sides; their sum is the diagonal vector in the parallelogram. The diagram on the left shows that $(4, 1) + (2, 4) = (6, 5)$, while the one on the right shows that $(-5, 1) + (2, 3) = (-3, 4)$.

The multiplication of a vector by a **scalar** (i.e., a real number c) is also done component-by-component:

$$c\,\mathbf{x} = c \begin{bmatrix} x_1 \\ x_2 \\ \vdots \\ x_n \end{bmatrix} = \begin{bmatrix} cx_1 \\ cx_2 \\ \vdots \\ cx_n \end{bmatrix}$$

For example, we have

$$4 \begin{bmatrix} 3 \\ 7 \\ -5 \end{bmatrix} = \begin{bmatrix} 12 \\ 28 \\ -20 \end{bmatrix}$$

Figure 1.4 shows scalar multiples of vectors in $\mathbb{R}^2$.

Linear Combinations of Vectors

With these two new definitions, one can form **linear combinations** of vectors, such as

$$3 \begin{bmatrix} 1 \\ 7 \\ 2 \end{bmatrix} - 5 \begin{bmatrix} 2 \\ -3 \\ 4 \end{bmatrix} - \begin{bmatrix} 1 \\ 1 \\ 1 \end{bmatrix} = \begin{bmatrix} -8 \\ 35 \\ -15 \end{bmatrix}$$

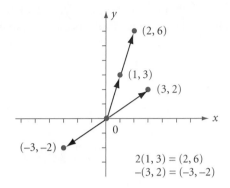

FIGURE 1.4 Scalar multiples of vectors in $\mathbb{R}^2$.

These concepts can be put to immediate use in the subject of linear equations. Observe that the system of equations

$$\begin{cases} 3x_1 - 5x_2 + x_3 = 11 \\ 2x_1 + 4x_2 - 3x_3 = -13 \\ 4x_1 - x_2 + 5x_3 = 4 \end{cases}$$

is the same as

$$x_1 \begin{bmatrix} 3 \\ 2 \\ 4 \end{bmatrix} + x_2 \begin{bmatrix} -5 \\ 4 \\ -1 \end{bmatrix} + x_3 \begin{bmatrix} 1 \\ -3 \\ 5 \end{bmatrix} = \begin{bmatrix} 11 \\ -13 \\ 4 \end{bmatrix}$$

For another example, consider the vectors $\mathbf{u} = (5, 7)$ and $\mathbf{v} = (-1, 3)$. One can easily calculate a particular linear combination of them such as

$$3\mathbf{u} + 4\mathbf{v} = 3 \begin{bmatrix} 5 \\ 7 \end{bmatrix} + 4 \begin{bmatrix} -1 \\ 3 \end{bmatrix} = \begin{bmatrix} 11 \\ 33 \end{bmatrix}$$

It is more complicated to reverse the direction; that is, if the vector $\mathbf{w} = (11, 33)$ is given, find the linear combination of $\mathbf{u}$ and $\mathbf{v}$ that equals $\mathbf{w}$. We then want to solve the following equation for a and b:

$$a\mathbf{u} + b\mathbf{v} = \mathbf{w}$$

Undertaking this task, we have

$$a \begin{bmatrix} 5 \\ 7 \end{bmatrix} + b \begin{bmatrix} -1 \\ 3 \end{bmatrix} = \begin{bmatrix} 11 \\ 33 \end{bmatrix} \qquad \text{or}$$

$$\begin{cases} 5a - b = 11 \\ 7a + 3b = 33 \end{cases} \qquad \text{or} \qquad \begin{bmatrix} 5 & -1 & | & 11 \\ 7 & 3 & | & 33 \end{bmatrix}$$

The augmented matrix can be transformed to reduced row echelon form as follows (where we took special care to avoid fractions):

$$
\left[\begin{array}{cc|c} 5 & -1 & 11 \\ 7 & 3 & 33 \end{array}\right] \sim \left[\begin{array}{cc|c} 5 & -1 & 11 \\ 2 & 4 & 22 \end{array}\right] \sim \left[\begin{array}{cc|c} 1 & 2 & 11 \\ 5 & -1 & 11 \end{array}\right]
$$

$$
\sim \left[\begin{array}{cc|c} 1 & 2 & 11 \\ 0 & -11 & -4 \end{array}\right] \sim \left[\begin{array}{cc|c} 1 & 2 & 11 \\ 0 & 1 & 4 \end{array}\right]
$$

$$
\sim \left[\begin{array}{cc|c} 1 & 0 & 3 \\ 0 & 1 & 4 \end{array}\right]
$$

EXAMPLE 1 Let $e_1 = (1, 0)$ and $e_2 = (0, 1)$. These are two vectors in $\mathbb{R}^2$. Is every point in $\mathbb{R}^2$ a linear combination of e_1 and e_2?

SOLUTION *Yes*, because for any $\mathbf{x} \in \mathbb{R}^2$,

$$
\mathbf{x} = \left[\begin{array}{c} x_1 \\ x_2 \end{array}\right] = x_1 \left[\begin{array}{c} 1 \\ 0 \end{array}\right] + x_2 \left[\begin{array}{c} 0 \\ 1 \end{array}\right] = x_1 \mathbf{e}_1 + x_2 \mathbf{e}_2
$$

$\blacksquare$

Here is a harder problem.

EXAMPLE 2 Is every point of $\mathbb{R}^2$ a linear combination of the vectors $\mathbf{u} = (5, 2)$ and $\mathbf{v} = (7, 3)$?

SOLUTION We try to solve for scalars α and β in the equation

$$
\alpha \mathbf{u} + \beta \mathbf{v} = \alpha \left[\begin{array}{c} 5 \\ 2 \end{array}\right] + \beta \left[\begin{array}{c} 7 \\ 3 \end{array}\right] = \left[\begin{array}{c} b_1 \\ b_2 \end{array}\right] = \mathbf{b}
$$

where $\mathbf{b} = (b_1, b_2)$ is an arbitrary vector in $\mathbb{R}^2$. The augmented matrix and its reduced row echelon form are

$$
\left[\begin{array}{cc|c} 5 & 7 & b_1 \\ 2 & 3 & b_2 \end{array}\right] \sim \left[\begin{array}{cc|c} 5 & 7 & b_1 \\ 4 & 6 & 2b_2 \end{array}\right] \sim \cdots \sim \left[\begin{array}{cc|c} 1 & 0 & 3b_1 - 7b_2 \\ 0 & 1 & -2b_1 + 5b_2 \end{array}\right]
$$

The reader can fill in the omitted steps. The answer is *yes*, and we find these values: $\alpha = 3b_1 - 7b_2$ and $\beta = -2b_1 + 5b_2$ for any vector $\mathbf{b} \in \mathbb{R}^2$. ∎

EXAMPLE 3 Is the vector $\mathbf{w} = (-1, 3, 7)$ a linear combination of the vectors $\mathbf{u} = (4, 2, 7)$ and $\mathbf{v} = (3, 1, 4)$?

SOLUTION We want to solve the vector equation

$$x \begin{bmatrix} 4 \\ 2 \\ 7 \end{bmatrix} + y \begin{bmatrix} 3 \\ 1 \\ 4 \end{bmatrix} = \begin{bmatrix} -1 \\ 3 \\ 7 \end{bmatrix} \quad \text{or} \quad \begin{cases} 4x + 3y = -1 \\ 2x + \ y = \ \ 3 \\ 7x + 4y = \ \ 7 \end{cases}$$

The augmented matrix and its reduced row echelon form are

$$\begin{bmatrix} 4 & 3 & | & -1 \\ 2 & 1 & | & 3 \\ 7 & 4 & | & 7 \end{bmatrix} \sim \begin{bmatrix} 1 & 0 & | & 5 \\ 0 & 1 & | & -7 \\ 0 & 0 & | & 0 \end{bmatrix}$$

Thus, we obtain $x = 5$ and $y = -7$. We can verify the results: $5(4, 2, 7) - 7(3, 1, 4) = (-1, 3, 7)$. Most vectors in $\mathbb{R}^3$ are *not* linear combinations of $\mathbf{u} = (4, 2, 7)$ and $\mathbf{v} = (3, 1, 4)$. The vectors that are linear combinations of $\mathbf{u}$ and $\mathbf{v}$ lie on a plane in $\mathbb{R}^3$. Why? ∎

Matrix–Vector Products

Now let $\mathbf{A}$ be an $m \times n$ matrix, thought of as a collection of column vectors. We can write

$$\mathbf{A} = \begin{bmatrix} a_{11} & a_{12} & \cdots & a_{1n} \\ a_{21} & a_{22} & \cdots & a_{2n} \\ \vdots & \vdots & \ddots & \vdots \\ a_{m1} & a_{m2} & \cdots & a_{mn} \end{bmatrix} = [\mathbf{a}_1 \quad \mathbf{a}_2 \quad \cdots \quad \mathbf{a}_n]$$

Here $\mathbf{a}_j$ denotes the jth column vector in $\mathbf{A}$: $\mathbf{a}_j = \begin{bmatrix} a_{1j} \\ a_{2j} \\ \vdots \\ a_{mj} \end{bmatrix}$.

> ## DEFINITION
>
> *The **matrix–vector product** $\mathbf{Ax}$ of an $m \times n$ matrix $\mathbf{A}$ and a column vector $\mathbf{x} = (x_1, x_2, \ldots, x_n)$ is defined to be*
>
> $$\mathbf{Ax} = x_1\,\mathbf{a}_1 + x_2\,\mathbf{a}_2 + \cdots + x_n\,\mathbf{a}_n$$
>
> *Here the scalars x_j are the components of the column vector $\mathbf{x}$, and the column vectors $\mathbf{a}_j$ are the columns of $\mathbf{A}$.*

We describe $\mathbf{Ax}$ as a linear combination of the columns in $\mathbf{A}$ with coefficients taken to be the components of the vector $\mathbf{x}$. Suppose that $\mathbf{A}$ has dimensions $m \times n$, and let the generic entries in $\mathbf{A}$ be written as a_{ij}, where $1 \leq i \leq m$ and $1 \leq j \leq n$. Using the columns $\mathbf{a}_j$ as above, we write

$$\mathbf{A} = \begin{bmatrix} \mathbf{a}_1 & \mathbf{a}_2 & \cdots & \mathbf{a}_n \end{bmatrix}$$

By the definition above, we have

$$\mathbf{Ax} = \begin{bmatrix} \mathbf{a}_1 & \mathbf{a}_2 & \cdots & \mathbf{a}_n \end{bmatrix} \begin{bmatrix} x_1 \\ x_2 \\ \vdots \\ x_n \end{bmatrix} = \sum_{j=1}^{n} x_j \mathbf{a}_j$$

To recover one component of $\mathbf{Ax}$, we write

$$(\mathbf{Ax})_i = \sum_{j=1}^{n} x_j [\mathbf{a}_j]_i = \sum_{j=1}^{n} x_j a_{ij} = \sum_{j=1}^{n} a_{ij} x_j = \mathbf{r}_i \mathbf{x}$$

This last expression is a genuine matrix product of a $1 \times n$ vector $\mathbf{r}_i = [a_{i1}, a_{i2}, \ldots, a_{in}]$ and an $n \times 1$ vector $\mathbf{x} = [x_1, x_2, \ldots, x_n]^T$. In this situation, we are denoting the ith row of $\mathbf{A}$ by $\mathbf{r}_i$. Thus, to get the ith component of $\mathbf{Ax}$, we compute the vector product

$$\mathbf{r}_i \mathbf{x} = \begin{bmatrix} a_{i1} & a_{i2} & \cdots & a_{in} \end{bmatrix} \begin{bmatrix} x_1 \\ x_2 \\ \vdots \\ x_n \end{bmatrix} = \sum_{j=1}^{n} a_{ij} x_j$$

EXAMPLE 4   Express as a single vector the product $\begin{bmatrix} 1 & 5 \\ 3 & 1 \\ 2 & 4 \end{bmatrix} \begin{bmatrix} 2 \\ 7 \end{bmatrix}$.

SOLUTION We must carry out a calculation to do this:

$$2 \begin{bmatrix} 1 \\ 3 \\ 2 \end{bmatrix} + 7 \begin{bmatrix} 5 \\ 1 \\ 4 \end{bmatrix} = \begin{bmatrix} 2 \\ 6 \\ 4 \end{bmatrix} + \begin{bmatrix} 35 \\ 7 \\ 28 \end{bmatrix} = \begin{bmatrix} 37 \\ 13 \\ 32 \end{bmatrix}$$

∎

EXAMPLE 5 Express as a single vector the product

$$\begin{bmatrix} 1 & 2 & 3 \\ 4 & 5 & 6 \\ 7 & 8 & 9 \\ 10 & 11 & 12 \end{bmatrix} \begin{bmatrix} x_1 \\ x_2 \\ x_3 \end{bmatrix}$$

SOLUTION Again, we must carry out the multiplication of a matrix times a vector, as follows:

$$x_1 \begin{bmatrix} 1 \\ 4 \\ 7 \\ 10 \end{bmatrix} + x_2 \begin{bmatrix} 2 \\ 5 \\ 8 \\ 11 \end{bmatrix} + x_3 \begin{bmatrix} 3 \\ 6 \\ 9 \\ 12 \end{bmatrix} = \begin{bmatrix} x_1 + 2x_2 + 3x_3 \\ 4x_1 + 5x_2 + 6x_3 \\ 7x_1 + 8x_2 + 9x_3 \\ 10x_1 + 11x_2 + 12x_3 \end{bmatrix}$$

∎

The Span of a Set of Vectors

DEFINITION

The set of all linear combinations of a set of vectors is called the **span** *of that set of vectors.*

Our work in the preceding text shows that the span of the set of columns of a matrix **A** is the set of all vectors **Ax**, where **x** runs over all vectors having the right number of entries. Explicitly, if **A** is $m \times n$, then the span of the set of its columns is

$$\mathrm{Col}(\mathbf{A}) = \mathrm{Span}\{\mathbf{a}_1, \mathbf{a}_2, \ldots, \mathbf{a}_n\} = \{\mathbf{Ax} \ : \ \mathbf{x} \in \mathbb{R}^n\}$$

The notation $\mathbb{R}^n$ signifies the set of all vectors having n components. We explore this topic more fully in Section 5.1.

The span of the set of columns in a matrix $\mathbf{A}$ is called the **column space** of $\mathbf{A}$ and is written $\text{Col}(\mathbf{A})$. Here are some relevant examples. From Example 1, the span of the set consisting of the two vectors $\mathbf{e}_1 = (1, 0)$ and $\mathbf{e}_2 = (0, 1)$ is $\mathbb{R}^2$; that is,

$$\text{Span}\{\mathbf{e}_1, \mathbf{e}_2\} = \text{Col}(\mathbf{I}_2) = \mathbb{R}^2$$

From Example 2, the span of the pair $(5, 2)$ and $(7, 3)$ is $\mathbb{R}^2$ as well.

Obviously, the span of the set of three vectors $\mathbf{e}_1 = (1, 0, 0), \mathbf{e}_2 = (0, 1, 0)$, and $\mathbf{e}_3 = (0, 0, 1)$ is $\mathbb{R}^3$; that is,

$$\text{Span}\{\mathbf{e}_1, \mathbf{e}_2, \mathbf{e}_3\} = \text{Col}(\mathbf{I}_3) = \mathbb{R}^3$$

By referring to the calculations already done in Example 3, the span of the pair $(4, 2, 7)$ and $(3, 1, 4)$ contains $(-1, 3, 7)$ but not $(1, 0, 0)$. For the last part of this example, we let

$$x \begin{bmatrix} 4 \\ 2 \\ 7 \end{bmatrix} + y \begin{bmatrix} 3 \\ 1 \\ 4 \end{bmatrix} = \begin{bmatrix} 1 \\ 0 \\ 0 \end{bmatrix}$$

We form the augmented matrix and undertake the row reduction:

$$\begin{bmatrix} 4 & 3 & | & 1 \\ 2 & 1 & | & 0 \\ 7 & 4 & | & 0 \end{bmatrix} \sim \begin{bmatrix} 1 & 0 & | & 0 \\ 0 & 1 & | & 0 \\ 0 & 0 & | & 1 \end{bmatrix}$$

The system has no solution and is therefore characterized as *inconsistent*.

The next example is more complicated.

EXAMPLE 6 Give a simple description for the span of $\{\mathbf{u}, \mathbf{v}\}$, where $\mathbf{u} = (4, 2, 7)$ and $\mathbf{v} = (3, 1, 4)$. (Ideally, there will be a simple test that can be applied to a vector to determine whether it is or is not in the span of a given set.)

SOLUTION Following Example 2, we form an augmented matrix and carry out the row reduction:

$$\begin{bmatrix} \mathbf{u} & \mathbf{v} & | & \mathbf{b} \end{bmatrix} = \begin{bmatrix} 4 & 3 & | & b_1 \\ 2 & 1 & | & b_2 \\ 7 & 4 & | & b_3 \end{bmatrix} \sim \cdots \sim \begin{bmatrix} 1 & 0 & | & 4b_2 - b_3 \\ 0 & 1 & | & b_1 - 2b_2 \\ 0 & 0 & | & b_1 + 5b_2 - 2b_3 \end{bmatrix}$$

For the consistency of this system, we require $b_1 + 5b_2 - 2b_3 = 0$. This condition on vector $\mathbf{b}$ is necessary and sufficient for $\mathbf{b}$ to be in the span of $\mathbf{u}$ and $\mathbf{v}$. The *simple* description asked for could be that the span of $\{\mathbf{u}, \mathbf{v}\}$ consists of all vectors $\mathbf{b}$ whose components satisfy the equation $b_1 + 5b_2 - 2b_3 = 0$. We write

$$\text{Span}\{\mathbf{u}, \mathbf{v}\} = \{\mathbf{b} \in \mathbb{R}^3 : b_1 + 5b_2 - 2b_3 = 0\}$$

Thus, most vectors in $\mathbb{R}^3$ are *not* linear combinations of $\mathbf{u} = (4, 2, 7)$ and $\mathbf{v} = (3, 1, 4)$. The vectors that are linear combinations of $\mathbf{u}$ and $\mathbf{v}$ lie on the plane $b_1 + 5b_2 - 2b_3 = 0$ in $\mathbb{R}^3$. ∎

EXAMPLE 7 Is the vector $(42, 6, 76)$ in the span of this set of three vectors: $(1, 2, 11), (3, 1, 4), (7, -4, 3)$?

SOLUTION It will be advantageous to think of all these vectors as column vectors. The question is whether a solution exists for this system of linear equations:

$$x_1 \begin{bmatrix} 1 \\ 2 \\ 11 \end{bmatrix} + x_2 \begin{bmatrix} 3 \\ 1 \\ 4 \end{bmatrix} + x_3 \begin{bmatrix} 7 \\ -4 \\ 3 \end{bmatrix} = \begin{bmatrix} 42 \\ 6 \\ 76 \end{bmatrix}$$

This equation can be written in equivalent forms as explained previously:

$$\begin{cases} x_1 + 3x_2 + 7x_3 = 42 \\ 2x_1 + x_2 - 4x_3 = 6 \\ 11x_1 + 4x_2 + 3x_3 = 76 \end{cases} \quad \text{or} \quad \begin{bmatrix} 1 & 3 & 7 & | & 42 \\ 2 & 1 & -4 & | & 6 \\ 11 & 4 & 3 & | & 76 \end{bmatrix}$$

Here we see one numerical vector as a presumed linear combination of three other numerical vectors. Solving the system of equations answers the question of whether the numerical vector on the right is in the span of the three numerical vectors on the left. Only by solving the system can one answer that question. Turning this over to mathematical software, such as MATLAB, we find that the answer is *yes*, and that the needed coefficients are $x_1 = 7/2 = 3.5$, $x_2 = 147/19 \approx 7.7368$, and $x_3 = 83/38 \approx 2.1842$. Check: $133(1, 2, 11) + 294(3, 1, 4) + 83(7, -4, 3) = 38(42, 6, 76)$. ∎

Interpreting Linear Systems

In general, we can write a system of linear equations as

$$\mathbf{Ax} = \mathbf{b}$$

Usually, we are given the $m \times n$ **coefficient matrix A** and the m-component **righthand-side vector b** and wish to solve for the n-component **unknown vector x**.

One can interpret **b** as a vector or as a matrix having only one column. The important new idea is that a matrix times a vector will be interpreted as a linear combination of the columns of the matrix, and the coefficients in this linear combination are precisely the components of the vector. The conventions of matrix algebra require that we write the vector **x** in the expression **Ax** as a column vector, that is, an $n \times 1$ matrix. (If we wrote **x** as a row vector, the product **Ax** would not be defined.)

Note that for the product **Ax** to be defined, there must be a match between the number of columns in **A** and the number of components in **x**. If the vectors **x** and **b** are interpreted as matrices (of sizes $n \times 1$ and $m \times 1$, respectively), then the matrix equation **Ax** = **b** can be written in several equivalent forms, as shown on the next page.

Now we have seen various interpretations of a system of linear equations. These interpretations will occur over and over again in our subsequent work. We can use these new concepts to understand a system of equations **Ax** = **b**. If **A** and **b** are given, such a system challenges us to determine whether **b** is in the span of the columns of **A** and, if so, to find the coefficients needed to express **b** as a linear combination of the columns of **A**.

One now has a loftier viewpoint for the problem of solving a system of linear equations. Think of such a problem in the form **Ax** = **b** and apply the general theory appropriate to such problems. They are conceptually much simpler in this form, and the field is now open to applying the vast armamentarium of matrix theory to such a problem! At this moment, we have discussed only the Gaussian elimination method for solving these systems, but later other methods will be introduced. (Some of these methods are based upon a completely different approach, which allows one to obtain solutions of low precision quickly and solutions of high precision with increasing effort and time.) Many investigators are working on ever-more-efficient ways of solving extremely large systems of equations, often taking advantage of special properties of systems that occur in applications. (If you find this subject interesting, you can devote your life to it and be paid for doing so!)

Here are different ways to think of a system of linear equations.

Equivalent Forms of $Ax = b$

1. The matrix form:
$$Ax = b$$

2. As a compact summation:
$$\sum_{j=1}^{n} a_{ij}x_j = b_i \qquad (1 \le i \le m)$$

3. As linear equations in complete detail:
$$\begin{cases} a_{11}x_1 + a_{12}x_2 + \cdots + a_{1n}x_n = b_1 \\ a_{21}x_1 + a_{22}x_2 + \cdots + a_{2n}x_n = b_2 \\ \qquad\qquad\qquad \vdots \qquad\qquad\quad \vdots \\ a_{m1}x_1 + a_{m2}x_2 + \cdots + a_{mn}x_n = b_m \end{cases}$$

4. As a matrix with vectors (arrays) in great detail:
$$\begin{bmatrix} a_{11} & a_{12} & \cdots & a_{1n} \\ a_{21} & a_{22} & \cdots & a_{2n} \\ \vdots & \vdots & \ddots & \vdots \\ a_{m1} & a_{m2} & \cdots & a_{mn} \end{bmatrix} \begin{bmatrix} x_1 \\ x_2 \\ \vdots \\ x_n \end{bmatrix} = \begin{bmatrix} b_1 \\ b_2 \\ \vdots \\ b_m \end{bmatrix}$$

5. As an augmented matrix:
$$\left[\begin{array}{cccc|c} a_{11} & a_{12} & \cdots & a_{1n} & b_1 \\ a_{21} & a_{22} & \cdots & a_{2n} & b_2 \\ \vdots & \vdots & \ddots & \vdots & \vdots \\ a_{m1} & a_{m2} & \cdots & a_{mn} & b_m \end{array} \right]$$

6. As a linear combination of the columns of A:
$$x_1 \begin{bmatrix} a_{11} \\ a_{21} \\ \vdots \\ a_{m1} \end{bmatrix} + x_2 \begin{bmatrix} a_{12} \\ a_{22} \\ \vdots \\ a_{m2} \end{bmatrix} + \cdots + x_n \begin{bmatrix} a_{1n} \\ a_{2n} \\ \vdots \\ a_{mn} \end{bmatrix} = \begin{bmatrix} b_1 \\ b_2 \\ \vdots \\ b_m \end{bmatrix}$$

7. As a linear combination of the column vectors of A, denoted by a_j:
$$x_1 a_1 + x_2 a_2 + \cdots + x_n a_n = b$$

Row-Equivalent Systems

The procedure developed in Section 1.1 now operates as follows. Given the matrix $\mathbf{A}$ and the righthand-side $\mathbf{b}$, form the augmented matrix $[\mathbf{A}\,|\,\mathbf{b}]$ and carry out the row-reduction process on this matrix. The solutions of the original system and the new system (obtained by applying the row-reduction process) are the same. Here is the formal statement.

THEOREM 1

Let two linear systems of equations be represented by their augmented matrices. If these two augmented matrices are row equivalent to each other, then the solutions of the two systems are identical.

If $[\mathbf{A}\,|\,\mathbf{b}] \sim [\mathbf{B}\,|\,\mathbf{c}]$, then $\{\mathbf{x} : \mathbf{A}\mathbf{x} = \mathbf{b}\} = \{\mathbf{x} : \mathbf{B}\mathbf{x} = \mathbf{c}\}$, which is Theorem 1 in symbols.

In Section 1.1, the term **row equivalent** was briefly mentioned. We repeat its definition here:

DEFINITION

*Two matrices are **row equivalent** to each other if each can be obtained from the other by applying a sequence of permitted row operations.*

Recall that the permitted row operations are of the following types: *replacement* ($\mathbf{r}_i \leftarrow c\,\mathbf{r}_j + \mathbf{r}_i$), *scale* ($\mathbf{r}_i \leftarrow c\,\mathbf{r}_i$ with $c \neq 0$), and *swap* ($\mathbf{r}_i \leftrightarrow \mathbf{r}_j$). Remember that in the replacement operation $i \neq j$.

Important facts are these: Two matrices are row equivalent to each other if and only if they have the same reduced row echelon form. Every matrix has one and only one reduced row echelon form. (This fact is proved in Section 1.3, Theorem 6.)

We can show that these two matrices are row equivalent to each other:

$$\mathbf{A} = \begin{bmatrix} 2 & 3 & 1 \\ 5 & 2 & 2 \end{bmatrix} \qquad \mathbf{B} = \begin{bmatrix} 8 & 1 & 3 \\ 10 & 4 & 4 \end{bmatrix}$$

because they are row equivalent to the same matrix in reduced row echelon form:

$$\mathbf{A} = \begin{bmatrix} 2 & 3 & 1 \\ 5 & 2 & 2 \end{bmatrix} \sim \begin{bmatrix} 1 & 0 & 4/11 \\ 0 & 1 & 1/11 \end{bmatrix}$$

$$\mathbf{B} = \begin{bmatrix} 8 & 1 & 3 \\ 10 & 4 & 4 \end{bmatrix} \sim \begin{bmatrix} 1 & 0 & 4/11 \\ 0 & 1 & 1/11 \end{bmatrix}$$

> **EXAMPLE 8** Find all the solutions to the equation $\mathbf{Ax} = \mathbf{b}$, where $\mathbf{A}$ is the matrix in Example 5 and $\mathbf{b}$ is the column vector with entries $[20, 47, 74, 101]^T$.

SOLUTION We form the augmented matrix and undertake the row reduction:

$$
\left[
\begin{array}{ccc|c}
1 & 2 & 3 & 20 \\
4 & 5 & 6 & 47 \\
7 & 8 & 9 & 74 \\
10 & 11 & 12 & 101
\end{array}
\right]
\sim
\left[
\begin{array}{ccc|c}
1 & 2 & 3 & 20 \\
3 & 3 & 3 & 27 \\
6 & 6 & 6 & 54 \\
9 & 9 & 9 & 81
\end{array}
\right]
$$

$$
\sim \cdots \sim
\left[
\begin{array}{ccc|c}
1 & 0 & -1 & -2 \\
0 & 1 & 2 & 11 \\
0 & 0 & 0 & 0 \\
0 & 0 & 0 & 0
\end{array}
\right]
$$

Fortunately to enhance our understanding of linear equations, we have here a new phenomenon: It is not clear exactly what the solution of the problem is. The two nonzero rows in this last matrix stand for these two equations:

$$
\begin{cases}
x_1 - x_3 = -2 \\
x_2 + 2x_3 = 11
\end{cases}
$$

Here there are two conditions placed on a vector having three components, and we expect some arbitrariness in the solution. The easiest way to express the set of all solutions is to write the equations in the more suggestive form

$$
\begin{cases}
x_1 = -2 + x_3 \\
x_2 = 11 - 2x_3
\end{cases}
$$

We are at liberty to assign *any* value we please to x_3 and thereby obtain the two other components of $\mathbf{x}$. (The variable x_3 is therefore called a **free variable**.) For example, if $x_3 = 0$, then $x_1 = -2$ and $x_2 = 11$, giving the solution vector $[-2, 11, 0]^T$. Or we can let $x_3 = 7$, from which $\mathbf{x} = [5, -4, 7]^T$. It is more illuminating to write the solution as follows:

$$
\begin{bmatrix} x_1 \\ x_2 \\ x_3 \end{bmatrix}
=
\begin{bmatrix} -2 \\ 11 \\ 0 \end{bmatrix}
+ s
\begin{bmatrix} 1 \\ -2 \\ 1 \end{bmatrix}
$$

Here s is a **free parameter** standing for the free variable x_3. The reader should study this equation carefully in order to understand where the vectors on the right came from. It is a bit of *sleight of hand* to get from the preceding pair of equations to the single vector equation. (General Exercises 6–9, 16, 23, 32, 38, 40, and 47–49, at the end of this section, give some practice in this art.) This last form of the solution indicates that all the solutions, taken together in $\mathbb{R}^3$, form a line passing through the point $(-2, 11, 0)$ and having the direction vector $(1, -2, 1)$. The variable x_3 now becomes a free parameter s to which we can assign arbitrary values. Each value chosen leads to a solution of the system. The preceding equation gives us the **general solution** of the system of equations. We recommend that solutions to problems such as this be displayed as shown. Usually there will be some constant vectors plus arbitrary multiples of one or more other vectors.

It is customary to treat the variables in their natural order. But in the example just given, we could go counter to this custom and treat x_1 as the free variable. (See General Exercise 19.) Now the general solution would be written

$$
\begin{cases} x_3 = 2 + x_1 \\ x_2 = 7 - 2x_1 \end{cases} \quad \text{or} \quad \begin{bmatrix} x_1 \\ x_2 \\ x_3 \end{bmatrix} = \begin{bmatrix} 0 \\ 7 \\ 2 \end{bmatrix} + t \begin{bmatrix} 1 \\ -2 \\ 1 \end{bmatrix}
$$

In this equation, the free parameter t is used in place of x_1. Notice that the same set of solutions can be expressed as the vector $[0, 7, 2]^T$ plus any scalar multiple of $[1, -2, 1]^T$. It is easy to verify that this gives a solution for every choice of the free parameter. The two forms given above describe the same set of solutions. This general solution is the same line in $\mathbb{R}^3$ as given above because it has the same direction vector $(1, -2, 1)$ and goes through the point $(-2, 11, 0)$ as we see by letting $t = -2$. ■

Let us return to the algebraic construct involved in the product $\mathbf{Ax}$, where $\mathbf{A}$ is an $m \times n$ matrix. This expression has been defined previously. From that definition, we obtain immediately

$$
\mathbf{A}(\mathbf{x} + \mathbf{y}) = \mathbf{Ax} + \mathbf{Ay} \quad \text{and} \quad \mathbf{A}(\alpha\mathbf{x}) = \alpha\mathbf{Ax}
$$

if α is a scalar. Because of these two properties, the mapping $\mathbf{x} \mapsto \mathbf{Ax}$ is said to be **linear**. (This mapping notation is useful when we want to show the effect of a mapping but do not wish to assign a name to it.) Showing more detail in this calculation, we let $\mathbf{a}_i$ denote column i in $\mathbf{A}$ and then write

$$\mathbf{A}(\mathbf{x}+\mathbf{y}) = \sum_{i=1}^{n}(\mathbf{x}+\mathbf{y})_i\mathbf{a}_i = \sum_{i=1}^{n}(x_i+y_i)\mathbf{a}_i = \sum_{i=1}^{n}x_i\mathbf{a}_i + \sum_{i=1}^{n}y_i\mathbf{a}_i = \mathbf{A}\mathbf{x}+\mathbf{A}\mathbf{y}$$

$$\mathbf{A}(\alpha\mathbf{x}) = \sum_{i=1}^{n}(\alpha\mathbf{x})_i\mathbf{a}_i = \sum_{i=1}^{n}\alpha x_i\mathbf{a}_i = \alpha\sum_{i=1}^{n}x_i\mathbf{a}_i = \alpha\mathbf{A}\mathbf{x}$$

With an induction argument, we obtain

$$\mathbf{A}\left(\sum_{i=1}^{k}\alpha_i\mathbf{u}_i\right) = \sum_{i=1}^{k}\alpha_i\mathbf{A}\mathbf{u}_i$$

In this equation, each $\mathbf{u}_i$ is a column vector and the α_i are scalars.

Consistent and Inconsistent Systems

Naturally, when presented with a system, we might ask first whether it is consistent. If it is not, then some more advanced techniques in a later chapter can be invoked to produce an approximate solution to the problem. We recall that when a system of equations has at least one solution, we say that the system is *consistent*; otherwise it is *inconsistent*. Observe that a system of the form $\mathbf{A}\mathbf{x} = \mathbf{0}$ is always consistent because $\mathbf{0}$ is a solution. (This is called the *trivial solution* of a homogeneous system, which we take up in Section 1.3.) No similar remark can be made about the general case $\mathbf{A}\mathbf{x} = \mathbf{b}$, when $\mathbf{b} \neq \mathbf{0}$.

EXAMPLE 9 Here is an example of an inconsistent system of equations. We show the original system and its reduced row echelon form.

$$\begin{bmatrix} 3 & -4 & -8 & 40 \\ 6 & -10 & -26 & 95 \\ 9 & -12 & -24 & 125 \end{bmatrix} \sim \begin{bmatrix} 1 & 0 & 4 & 0 \\ 0 & 1 & 5 & 0 \\ 0 & 0 & 0 & 1 \end{bmatrix}$$

SOLUTION The second of these augmented matrices certainly exhibits the inconsistency, because the third equation reads $0x_1 + 0x_2 + 0x_3 = 1$, and this cannot be true. Because the second augmented matrix is the reduced row echelon form of the first, we conclude that the first system is inconsistent, although that fact is *not obvious* from the original matrix. But the row reduction process reveals the inconsistency concealed in the system. ∎

The example just given illustrates the following theorem.

THEOREM 2

A system of linear equations, $\mathbf{Ax} = \mathbf{b}$, is consistent if and only if the vector $\mathbf{b}$ is in the span of the set of columns of $\mathbf{A}$.

PROOF Suppose that $\mathbf{x}$ is a vector such that $\mathbf{Ax} = \mathbf{b}$. When this is written in detail as a linear combination of columns of $\mathbf{A}$ equaling the vector $\mathbf{b}$, we see that $\mathbf{b}$ is indeed a linear combination of the columns of $\mathbf{A}$. The converse is true: if $\mathbf{b} = \sum_{i=1}^{n} x_i \mathbf{a}_i$ where the $\mathbf{a}_i$ are the columns of $\mathbf{A}$ and the x_i are scalars, then the vector $\mathbf{x}$ having components x_i is a solution of the system $\mathbf{Ax} = \mathbf{b}$. ∎

THEOREM 3

Let $\mathbf{A}$ be an $m \times n$ matrix. The system of equations $\mathbf{Ax} = \mathbf{b}$ is consistent for all $\mathbf{b}$ in $\mathbb{R}^m$ if and only if the columns of $\mathbf{A}$ span $\mathbb{R}^m$. In other words, $\mathrm{Col}(\mathbf{A}) = \mathbb{R}^m$.

PROOF By Theorem 2, consistency of the system for all $\mathbf{b}$ means that every $\mathbf{b}$ in $\mathbb{R}^m$ is a linear combination of columns of $\mathbf{A}$. In other words, the columns of $\mathbf{A}$ span $\mathbb{R}^m$. ∎

THEOREM 4

Let $\mathbf{A}$ be an $m \times n$ matrix. The system of equations $\mathbf{Ax} = \mathbf{b}$ is consistent for all $\mathbf{b}$ in $\mathbb{R}^m$ if and only if each row of the coefficient matrix $\mathbf{A}$ has a pivot position.

PROOF Asserting that each row of the coefficient matrix $\mathbf{A}$ has a pivot position is equivalent to asserting that the reduced row echelon form of $\mathbf{A}$ has a pivot in each row. That, in turn, is equivalent to asserting that the columns of $\mathbf{A}$ span $\mathbb{R}^m$. By Theorem 3, this is equivalent to the system of equations $\mathbf{Ax} = \mathbf{b}$ being consistent for all $\mathbf{b}$ in $\mathbb{R}^m$. ∎

THEOREM 5

A system of linear equations is inconsistent if and only if its augmented matrix has a pivot position in the last column.

PROOF Recall that the last column in the augmented matrix is the right-hand-side vector. If the reduced row echelon form of the augmented matrix has a pivot in the last column, the equation represented by the row containing that pivot is inconsistent, because the coefficients of all the variables are 0, whereas the righthand side has a 1. (It has the form $0x_1 + 0x_2 + \cdots + 0x_n = 1$, and no solution is possible.) Conversely, if there is no pivot element in the last column, values can be assigned to all the variables, creating a solution. For example, assign arbitrary values to all the free variables (if there are any) and use the reduced row echelon form to find the values of all the other variables. ∎

As mentioned previously, every matrix has a unique reduced row echelon form. It is proven in Theorem 6 in Section 1.3. (It's not easy!) This special relationship of **row equivalence** is denoted in this book by the symbol $\sim$. For example, we know that the two following matrices, **A** and **B**, are row equivalent to each other because each is row equivalent to $\mathbf{I}_3$:

$$\mathbf{A} = \begin{bmatrix} 1 & 2 & 3 \\ 4 & 5 & 6 \\ 10 & 8 & 7 \end{bmatrix} \sim \begin{bmatrix} 1 & 0 & 0 \\ 0 & 1 & 0 \\ 0 & 0 & 1 \end{bmatrix} = \mathbf{I}_3$$

$$\mathbf{B} = \begin{bmatrix} 17 & 31 & -11 \\ 2 & 5 & 47 \\ -19 & 3 & 13 \end{bmatrix} \sim \begin{bmatrix} 1 & 0 & 0 \\ 0 & 1 & 0 \\ 0 & 0 & 1 \end{bmatrix} = \mathbf{I}_3$$

This example shows that it may be easier to prove $\mathbf{A} \sim \mathbf{I}$ and $\mathbf{B} \sim \mathbf{I}$ than to find a chain of row operations that go directly from **A** to **B**.

Caution

Theorem 5 involves the reduced row echelon form of an augmented matrix $[\mathbf{A} \,|\, \mathbf{b}]$, whereas Theorem 4 involves only the coefficient matrix **A**. The system of equations $\mathbf{Ax} = \mathbf{b}$ may or may not be consistent for all **b** in $\mathbb{R}^m$. In Example 2, the system was found to be consistent for all righthand sides **b**:

$$\begin{bmatrix} 5 & 7 & | & b_1 \\ 2 & 3 & | & b_2 \end{bmatrix} \sim \begin{bmatrix} 1 & 0 & | & 3b_1 - 7b_2 \\ 0 & 1 & | & -2b_1 + 5b_2 \end{bmatrix}$$

The coefficient matrix has a pivot position in each row.

The system in Example 6 can be consistent or inconsistent, depending on the numbers b_1, b_2, b_3. The row reduction yields

$$\begin{bmatrix} 4 & 3 & | & b_1 \\ 2 & 1 & | & b_2 \\ 7 & 4 & | & b_3 \end{bmatrix} \sim \begin{bmatrix} 1 & 0 & | & 4b_2 - b_3 \\ 0 & 1 & | & b_1 - 2b_2 \\ 0 & 0 & | & b_1 + 5b_2 - 2b_3 \end{bmatrix}$$

This system is consistent only when $b_1 + 5b_2 - 2b_3 = 0$. When that condition is met, there will *not* be a pivot position in the last column of the augmented matrix.

THEOREM 6

A system of linear equations is consistent if and only if the reduced row echelon form of its augmented matrix does not have a pivot position in the last column.

PROOF This is really just a restatement of Theorem 5. ∎

Application: Linear Ordinary Differential Equations

A system of linear ordinary differential equations can be expressed using matrices and vectors. For example, consider the two equations

$$dx/dt = y, \qquad dy/dt = -x$$

with initial values $x(0) = 0$ and $y(0) = 1$. Letting

$$\mathbf{z} = \begin{bmatrix} x \\ y \end{bmatrix}$$

we obtain

$$d\mathbf{z}/dt = \mathbf{A}\mathbf{z}$$

where

$$\mathbf{A} = \begin{bmatrix} 0 & 1 \\ -1 & 0 \end{bmatrix}$$

with initial conditions

$$\mathbf{z}(0) = \begin{bmatrix} 0 \\ 1 \end{bmatrix}$$

The solution is $x(t) = \sin(t)$ and $y(t) = \cos(t)$. When we plot the solutions we obtain two overlapping sine and cosine curves. We can generalize this concept to handle n linear ordinary differential equations by using a vector of size n and a matrix of size $n \times n$.

Application: Bending of a Beam

Studying the elasticity of building materials can bring in problems of linear algebra via Hooke's law, which has a linear nature. Consider a flexible steel beam supported by posts at each end. If a downward force is applied to the beam somewhere between the supports there will be a deflection in the beam. If there are other such forces (called *stresses*), then we would like to compute the expected deflections due to all the stresses. The deflections are called *strains*.

For example, we consider four forces (stresses) w_1, w_2, w_3, w_4 applied to the beam at four different locations. The beam then suffers deflections (strains) x_1, x_2, x_3, x_4 as shown in Figure 1.5. Define the vector $\mathbf{w} = (w_1, w_2, w_3, w_4)$ and the vector $\mathbf{x} = (x_1, x_2, x_3, x_4)$. In 1676, Robert Hooke noted that in an elastic material strain is proportional to stress.[4] By **Hooke's Law**, there is a linear relationship between the forces and the deflections, given by the equation

$$\mathbf{x} = \mathbf{Fw}$$

where the matrix $\mathbf{F}$ is the **flexibility matrix**. The entries in the matrix $\mathbf{F}$ have to be determined by careful experiments.

In this example, $\mathbf{F}$ is a 4×4 matrix. The equation $\mathbf{x} = \mathbf{Fw}$ can be written in full detail as follows:

$$\mathbf{x} = \begin{bmatrix} x_1 \\ x_2 \\ x_3 \\ x_4 \end{bmatrix} = \begin{bmatrix} f_{11} & f_{12} & f_{13} & f_{14} \\ f_{21} & f_{22} & f_{23} & f_{24} \\ f_{31} & f_{32} & f_{33} & f_{34} \\ f_{41} & f_{42} & f_{43} & f_{44} \end{bmatrix} \begin{bmatrix} w_1 \\ w_2 \\ w_3 \\ w_4 \end{bmatrix}$$

$$= w_1 \begin{bmatrix} f_{11} \\ f_{21} \\ f_{31} \\ f_{41} \end{bmatrix} + w_2 \begin{bmatrix} f_{12} \\ f_{22} \\ f_{32} \\ f_{42} \end{bmatrix} + w_3 \begin{bmatrix} f_{13} \\ f_{23} \\ f_{33} \\ f_{43} \end{bmatrix} + w_4 \begin{bmatrix} f_{14} \\ f_{24} \\ f_{34} \\ f_{44} \end{bmatrix}$$

$$= w_1 \mathbf{f}_1 + w_2 \mathbf{f}_2 + w_3 \mathbf{f}_3 + w_4 \mathbf{f}_4$$

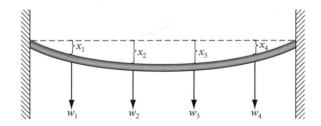

FIGURE 1.5 Bending beam.

[4] Robert Hooke (1635−1703) was at his best when his mind was jumping from one thing to another and not doing just one thing at a time. Throughout his life, Hooke had bitter disputes with fellow scientists, such as claiming that Newton stole some of Hooke's own ideas about the theory of light. Consequently, Newton removed all references to Hooke from *The Principia*.

The first column $\mathbf{f}_1$ in the matrix $\mathbf{F}$ is obtained by measuring the strains that arise from applying a unit stress on the beam at the first point in the diagram. In the same way, the three other columns of $\mathbf{F}$ are found by placing a unit weight at the three other locations on the beam and measuring the strains. Another way to explain this is to imagine vector $\mathbf{w}$ to be $\mathbf{e}_1 = (1, 0, 0, 0)$. Physically it means that a unit weight has been placed at the first point on the elastic beam. The deflections are measured at all four points of the beam and entered as column 1 in the matrix $\mathbf{F}$. The first column of $\mathbf{F}$ gives the deflections due solely to a unit force applied at the first point. Similar interpretations are valid for the other columns of $\mathbf{F}$. For example, if the flexibility matrix is

$$\mathbf{F} = \begin{bmatrix} 0.2 & 0.1 & 0.3 & 0.4 \\ 0.7 & 0.5 & 0.6 & 0.8 \\ 0.9 & 0.2 & 0.2 & 0.5 \\ 0.6 & 0.3 & 0.4 & 0.9 \end{bmatrix}$$

and the forces are given by $\mathbf{w} = [25, 45, 35, 55]^T$ in millimeters per Newton, then the deflection vector would be $\mathbf{x} = [42, 105, 66, 92]^T$ in millimeters measured from the original four points on the unbent beam.

Mathematical Software

One can use sophisticated mathematical software to carry out tedious calculations. To illustrate, consider the augmented matrix in Example 8. Here are the MATLAB commands needed to enter the matrix and to invoke the algorithm for reduced row echelon form:

```
MATLAB
A = [1,2,3,20;4,5,6,47;7,8,9,74;10,11,12,101]
rref(A)
```

In MATLAB, if you wish to see the successive steps in the row-reduction process, replace the command `rref` by `rrefmovie`. Similarly, the commands in Maple are

```
Maple
with(LinearAlgebra):
A := Matrix([[1,2,3,20],[4,5,6,47],[7,8,9,74],[10,11,12,101]]);
ReducedRowEchelonForm(A);
```

Finally, the Mathematica commands are

```
Mathematica
A = {{1,2,3,20},{4,5,6,47},{7,8,9,74},{10,11,12,101}
RowReduced[A]
```

The Maple software package supports symbolic calculations and can carry out the calculations for the general solution.

```
Maple
with(LinearAlgebra):
A := Matrix([[1,2,3,20],[4,5,6,47],[7,8,9,74],[10,11,12,101]]);
B := ReducedRowEchelonForm(A);
C := BackwardSubstitute(B,free='x');
Sol := LinearSolve(A,free='x');
```

One can also use Mathematica to solve systems with symbols in them such as this one with an arbitrary righthand side:

$$\begin{cases} x + 2y = a \\ 3x + 7y = b \end{cases}$$

Maple	Mathematica
with(LinearAlgebra): solve({x +2*y=a,3*y+7*y=b},{x,y});	Solve[{x + 2y == a, 3x + 7y == b},{x,y}]

We find $x = 7a - 2b$ and $y = -3a + b$. Some of the Maple symbolic manipulations can also be carried out in MATLAB, using its *Symbolic Math Toolbox*.

SUMMARY 1.2

- Vectors in $\mathbb{R}^n$: $\mathbf{x} = (x_1, x_2, \ldots, x_n) = [x_1, x_2, \ldots, x_n]^T$; zero vector: $\mathbf{0} = (0, 0, \ldots, 0)$

- Vector addition: $\mathbf{x} + \mathbf{y} = (x_1 + y_1, x_2 + y_2, \ldots, x_n + y_n)$; scalar multiplication: $c\mathbf{x} = (cx_1, cx_2, \ldots, cx_n)$

- Elementary unit vectors: $\mathbb{R}^2$: $\mathbf{e}_1 = (1, 0)$, $\mathbf{e}_2 = (0, 1)$; $\mathbb{R}^3$: $\mathbf{e}_1 = (1, 0, 0), \mathbf{e}_2 = (0, 1, 0), \mathbf{e}_3 = (0, 0, 1)$

- Matrix $\mathbf{A} = [\mathbf{a}_1, \mathbf{a}_2, \ldots, \mathbf{a}_n]_{m \times n}$ where $\mathbf{a}_j = [a_{1j}, a_{2j}, \ldots, a_{nj}]^T$ and is the jth column vector in $\mathbf{A}$

- Matrix–vector product: $\mathbf{Ax} = x_1\mathbf{a}_1 + x_2\mathbf{a}_2 + \cdots + x_n\mathbf{a}_n$, a linear combination of columns in $\mathbf{A}$; $(\mathbf{Ax})_i = \mathbf{r}_i\mathbf{x}$, where $\mathbf{r}_i$ is the ith row of $\mathbf{A}$

- Span$\{\mathbf{a}_1, \mathbf{a}_2, \ldots, \mathbf{a}_n\}$ is the set of all linear combinations of $\{\mathbf{a}_1, \mathbf{a}_2, \ldots, \mathbf{a}_n\}$; Col$(\mathbf{A}) =$ Span$\{\mathbf{a}_1, \mathbf{a}_2, \ldots, \mathbf{a}_n\}$

- Equivalent forms of linear systems:
  - Matrix form: $\mathbf{Ax} = \mathbf{b}$
  - Compact summation: $\sum_{j=1}^{n} a_{ij}x_j = b_i$ for $1 \le i \le m$
  - In complete detail:

$$\begin{cases} a_{11}x_1 + a_{12}x_2 + \cdots + a_{1n}x_n = b_1 \\ a_{21}x_1 + a_{22}x_2 + \cdots + a_{2n}x_n = b_2 \\ \quad\quad\quad\quad\quad \vdots \\ a_{m1}x_1 + a_{m2}x_2 + \cdots + a_{mn}x_n = b_m \end{cases}$$

  - As a matrix with vectors:

$$\begin{bmatrix} a_{11} & a_{12} & \cdots & a_{1n} \\ a_{21} & a_{22} & \cdots & a_{2n} \\ \vdots & \vdots & \ddots & \vdots \\ a_{m1} & a_{m2} & \cdots & a_{mn} \end{bmatrix} \begin{bmatrix} x_1 \\ x_2 \\ \vdots \\ x_n \end{bmatrix} = \begin{bmatrix} b_1 \\ b_2 \\ \vdots \\ b_m \end{bmatrix}$$

  - As an augmented matrix:

$$\left[\begin{array}{cccc|c} a_{11} & a_{12} & \cdots & a_{1n} & b_1 \\ a_{21} & a_{22} & \cdots & a_{2n} & b_2 \\ \vdots & \vdots & \ddots & \vdots & \vdots \\ a_{m1} & a_{m2} & \cdots & a_{mn} & b_m \end{array}\right]$$

- As a linear combination of the columns of $\mathbf{A}$:

$$x_1 \begin{bmatrix} a_{11} \\ a_{21} \\ \vdots \\ a_{m1} \end{bmatrix} + x_2 \begin{bmatrix} a_{12} \\ a_{22} \\ \vdots \\ a_{m2} \end{bmatrix} + \cdots +$$

$$x_n \begin{bmatrix} a_{1n} \\ a_{2n} \\ \vdots \\ a_{mn} \end{bmatrix} = \begin{bmatrix} b_1 \\ b_2 \\ \vdots \\ b_m \end{bmatrix}$$

- $\mathbf{Ax} = \mathbf{b}$ is the same as $x_1\mathbf{a}_1 + x_2\mathbf{a}_2 + \cdots + x_n\mathbf{a}_n = \mathbf{b}$, which is a linear combination of the column vectors $\mathbf{a}_j$ in $\mathbf{A}$

- The mapping $\mathbf{x} \to \mathbf{Ax}$ is linear: $\mathbf{A}(\mathbf{x} + \mathbf{y}) = \mathbf{Ax} + \mathbf{Ay}$ and $\mathbf{A}(c\mathbf{x}) = c\mathbf{Ax}$

- The equation $\mathbf{Ax} = \mathbf{b}$ is consistent if there exists at least one solution; otherwise it is inconsistent

- Theorems:

  - The system $\mathbf{Ax} = \mathbf{b}$ is consistent if and only if $\mathbf{b} \in$ Col$(\mathbf{A})$.
  - Let $\mathbf{A}$ be an $m \times n$ matrix. The system $\mathbf{Ax} = \mathbf{b}$ is consistent for all $\mathbf{b} \in \mathbb{R}^n$ if and only if Col$(\mathbf{A}) = \mathbb{R}^m$.
  - Let $\mathbf{A}$ be an $m \times n$ matrix. The system $\mathbf{Ax} = \mathbf{b}$ is consistent for all $\mathbf{b} \in \mathbb{R}^n$ if and only if there is a pivot position in each row of $\mathbf{A}$. Equivalently, $\mathbf{A}$ has m pivot positions.
  - The system $\mathbf{Ax} = \mathbf{b}$ is inconsistent if and only if there is a pivot in the last column of the row-reduced augmented matrix $[\mathbf{A} \mid \mathbf{b}]$.

KEY CONCEPTS 1.2

Vectors, components, transpose of a vector, linear combinations of vectors, matrix–vector product $\mathbf{Ax}$, span of a set of vectors, various forms of linear systems, row-equivalent matrices, free parameters and variables, general solution of a system of linear equations, consistent and inconsistent systems of equations, bending beam application, using mathematical software

GENERAL EXERCISES 1.2

1. Solve these two systems:

$$\begin{bmatrix} 3 & 2 \\ 1 & 4 \\ 2 & 1 \end{bmatrix} \begin{bmatrix} x_1 \\ x_2 \end{bmatrix} = \begin{bmatrix} 4 \\ 18 \\ 1 \end{bmatrix}$$

$$\begin{bmatrix} 3 & 2 \\ 1 & 4 \\ 2 & 1.1 \end{bmatrix} \begin{bmatrix} x_1 \\ x_2 \end{bmatrix} = \begin{bmatrix} 4 \\ 18 \\ 1 \end{bmatrix}$$

What conclusion can be drawn?

2. Let $\mathbf{A} = \begin{bmatrix} 3 & 7 & -4 \\ 5 & -2 & 6 \\ 2 & 1 & -1 \\ 4 & 1 & 2 \end{bmatrix}$

Let $\mathbf{b}$ be a vector in $\mathbb{R}^4$ such that the system $\mathbf{Ax} = \mathbf{b}$ has a solution. Explain why it has only one.

3. (Continuation.) Let $\mathbf{A}$ be as in General Exercise 2, and let $\mathbf{b} = [68, -32, 15, 4]^T$ and $\mathbf{x} = [2, 6, -5]^T$. The superscript T indicates that these vectors are to be considered as *column vectors*. Determine whether $\mathbf{x}$ is a solution of the system $\mathbf{Ax} = \mathbf{b}$.

4. Find a vector that solves the first of these two systems but not the second. Then find a vector that satisfies the second but not the first.

$$\left[\begin{array}{ccc|c} 1 & 0 & 2 & 5 \\ 0 & 1 & 3 & 7 \end{array} \right], \qquad \left[\begin{array}{ccc|c} 1 & 0 & 3 & 5 \\ 0 & 1 & 2 & 7 \end{array} \right]$$

5. Without doing any calculations, explain why these two matrices are row equivalent to each other:

$$\begin{bmatrix} 7 & 3 & 5 & -8 \\ 0 & 2 & 6 & 11 \\ 0 & 0 & 5 & 3 \\ 0 & 0 & 0 & 11 \end{bmatrix} \sim \begin{bmatrix} 1 & 0 & 0 & 0 \\ 0 & 1 & 0 & 0 \\ 0 & 0 & 1 & 0 \\ 0 & 0 & 0 & 1 \end{bmatrix}$$

6. Let $\begin{cases} x - y + z = 4 \\ 2x + y - 3z = 5 \\ -y + 7x - 3z = 22 \end{cases}$

Find all the solutions of the system.

7. Consider $\begin{bmatrix} 1 & 3 & 0 & 5 & 6 \\ 2 & 6 & 1 & 8 & 14 \\ 3 & 9 & 0 & 15 & 18 \end{bmatrix}$

Find the reduced row echelon form of this matrix.

8. Solve the system whose augmented matrix is the following, by finding the reduced row echelon form of the matrix:

$$\left[\begin{array}{ccc|c} 1 & 3 & 1 & -8 \\ 2 & -3 & 1 & 27 \\ 5 & 2 & 3 & 23 \end{array} \right]$$

9. Find the general solution of this system of equations:

$$\begin{cases} 4x_1 + 12x_2 + 2x_3 + 16x_4 + x_5 + 7x_6 = 26 \\ -2x_1 - 6x_2 + 2x_3 - 14x_4 + 3x_5 + 19x_6 = 12 \\ x_1 + 3x_2 + 3x_3 - 1x_4 + 2x_5 + 17x_6 = 25 \end{cases}$$

10. Show that this system is inconsistent:

$$\begin{cases} x_1 + 3x_2 + 15x_3 = 28 \\ 2x_1 + 4x_2 + 20x_3 = 40 \\ x_1 + 2x_2 + 10x_3 = 27 \end{cases}$$

11. Solve the system of equations whose augmented matrix is

$$\left[\begin{array}{cccc|c} 2 & 0 & 2 & 8 & 11 \\ 7 & 1 & 2 & 8 & 35 \\ 2 & 0 & 1 & 4 & 9 \\ 1 & 0 & 2 & 8 & 7 \end{array} \right]$$

12. Accept the hypothesis that

$$\begin{bmatrix} 2 & 3 & 4 & 1 \\ 4 & 11 & 13 & -1 \\ 2 & 3 & 7 & -8 \end{bmatrix} \sim \begin{bmatrix} 2 & 3 & 4 & 1 \\ 0 & 5 & 2 & 6 \\ 0 & 0 & 1 & -3 \end{bmatrix}$$

Find the solution of the following system:

$$\begin{cases} 2x_1 + 3x_2 + 4x_3 = 1 \\ 4x_1 + 11x_2 + 13x_3 = -1 \\ 2x_1 + 3x_2 + 7x_3 = -8 \end{cases}$$

13. What are the inverses of these four row operations: $\mathbf{r}_p \leftarrow \mathbf{r}_p + \alpha\mathbf{r}_q$, $\mathbf{r}_q \leftarrow \mathbf{r}_p + \alpha\mathbf{r}_q$, $\mathbf{r}_p \leftrightarrow \mathbf{r}_q$, and $\mathbf{r}_q \leftarrow \beta\mathbf{r}_q$, where $\alpha \neq 0$ and $\beta \neq 0$. In which cases must we assume $q \neq p$ or another hypothesis?

14. Describe the span of the set of columns in the matrix

$$\begin{bmatrix} 3 & 1 & 4 & 2 \\ 0 & 2 & 3 & 1 \\ 0 & 0 & 0 & 3 \end{bmatrix}$$

15. Without any calculations, explain why these two matrices are not row equivalent to each other:

$$\begin{bmatrix} 3.2 & 1.3 & 5.5 & 16.1 & 22.8 & 19.2 & 11.5 \\ 1.3 & 3.7 & 2.0 & 19.2 & 11.7 & 16.9 & 12.3 \\ 4.7 & 5.9 & 9.3 & 12.4 & 13.2 & 15.8 & 18.7 \end{bmatrix}$$

$$\begin{bmatrix} 18.5 & 5.8 & 7.7 & 3. & 2.9 & 4. & 11.5 & 8.1 \\ 11.7 & 8.3 & 2.4 & 6. & 1.4 & 2. & 21.3 & 9.8 \\ 37.2 & 9.1 & 5.6 & 3. & 8.2 & 5. & 23.3 & 1.8 \end{bmatrix}$$

16. Sometimes, in applying the row-reduction process, there are zeros already present in the matrix, and one is tempted to take advantage of that fact. However, this may not be possible, and, if so, those zeros will be *sacrificed* in the reduction process. Here is such an example:

$$\left[\begin{array}{ccc|c} 1 & 1 & 3 & 0 \\ -3 & -2 & 0 & -4 \\ 5 & 0 & 4 & 3 \end{array}\right]$$

Taking the given matrix to be the augmented matrix for a system of equations, find the reduced row echelon form and the solution vector.

17. What test can you devise to ascertain that two equations $ax + by = c$ and $rx + sy = t$ define the same line? (Assume that the coefficients a, b, r, s are all nonzero.)

18. Without performing any row operations, solve this augmented matrix

$$\left[\begin{array}{cccc|c} 3 & 2 & 1 & 0 & 6 \\ 5 & -4 & 0 & 1 & 7 \end{array}\right]$$

19. Redo Example 8, using x_1 as the free variable.

20. Let $\mathbf{u} = (1, 0, 1)$ and $\mathbf{v} = (1, 1, 0)$. Test the following four vectors to see which ones are in the span of $\{\mathbf{u}, \mathbf{v}\}$: $\mathbf{w} = (1, -1, 2)$, $\mathbf{x} = (4, 3, 1)$, $\mathbf{y} = (1, 1, 1)$, and $\mathbf{z} = (1, 2, -1)$. Can you devise a simple test for this task, keeping the vectors $\mathbf{u}$ and $\mathbf{v}$ as they are? (The test should be easy to apply to any vector $\mathbf{b}$ in $\mathbb{R}^3$.)

21. Let $\begin{cases} \ln x^{25} + y^2 = 77 \\ \ln x^2 + 5y^2 = 16 \end{cases}$

Solve this system for x and y. The logarithms are based on $e = 2.71828\ldots$.

22. In this problem, we describe matrices by listing their columns, which are vectors in $\mathbb{R}^m$. Explain why if $[\mathbf{a}_1 \ \mathbf{a}_2 \ \cdots \ \mathbf{a}_n] \sim [\mathbf{b}_1 \ \mathbf{b}_2 \cdots \mathbf{b}_n]$ and $k < n$, then $[\mathbf{a}_1 \ \mathbf{a}_2 \ \cdots \ \mathbf{a}_k] \sim [\mathbf{b}_1 \ \mathbf{b}_2 \ \cdots \ \mathbf{b}_k]$. If this turns out to be false, provide a suitable example.

23. Consider

$$\begin{cases} 0x_1 - 4x_2 + 9x_3 + 29x_4 + 14x_5 = 9 \\ 0x_1 - 2x_2 + 3x_3 + 12x_4 + 6x_5 = 2 \\ 0x_1 + 6x_2 - 12x_3 - 41x_4 - 20x_5 = -11 \end{cases}$$

Solve the system. Be sure to identify the free variables (*parameters*) in the general solution. Express the general solution in the manner recommended in the text. (See Example 8.)

24. Explain why the span of the set of columns in an $m \times n$ matrix $\mathbf{A}$ is identical to the set $\{\mathbf{Ax} : \mathbf{x} \in \mathbb{R}^n\}$.

25. Suppose that a system of equations $\mathbf{Ax} = \mathbf{b}$ is consistent and that a set of coefficients c_i has the property $\sum_{i=1}^{m} c_i a_{ij} = 0$ for $j = 1, 2, \ldots, n$. Explain why $\sum_{i=1}^{m} c_i b_i = 0$. Here we have supposed that $\mathbf{A}$ is an $m \times n$ matrix.

26. Determine whether the vector is in the column space of the matrix:

$$\begin{bmatrix} -9 \\ 10 \\ 8 \end{bmatrix}, \begin{bmatrix} 1 & 2 & -4 \\ 2 & 0 & 4 \\ 3 & 1 & 3 \end{bmatrix}$$

27. Give an argument why if one matrix can be obtained from another via allowable row operations, then the two matrices have the same reduced row echelon form.

28. If a system of equations $\mathbf{Ax} = \mathbf{b}$ is inconsistent, can we always restore consistency by changing one entry in vector $\mathbf{b}$?

29. Explain why or provide counterexamples: For a pair of vectors $\mathbf{x}, \mathbf{y}$ interpreted as $n \times 1$ matrices:

a. $\mathbf{x}^T \mathbf{y} = \mathbf{y}^T \mathbf{x}$ **b.** $\mathbf{xy}^T = \mathbf{yx}^T$

30. Let $\begin{cases} 3x_1 + x_2 + x_3 = a \\ -3x_1 + 9x_2 - 5x_3 = b \\ 6x_1 - 3x_2 + 4x_3 = c \end{cases}$

Find the exact condition on (a, b, c) so that this system is consistent.

31. Describe the solutions of the system whose augmented matrix is $\begin{bmatrix} 0 & 1 & 2 & 0 & 3 & | & 4 \\ 0 & 0 & 0 & 1 & 7 & | & 5 \end{bmatrix}$

Indicate which variables are independent (free) and which are dependent.

32. Consider $\begin{bmatrix} 1 & 2 & 2 & 0 & 6 & | & 8 \\ 0 & 2 & 0 & 3 & 9 & | & 7 \end{bmatrix}$

Describe the set of all solutions of the system having this augmented matrix. Indicate which variables are independent (free) and which are dependent.

33. Establish that if the matrix $\mathbf{A}$ having entries a_{ij} is in row echelon form, then $a_{ij} = 0$ when $j < i$.

34. By using row operations, determine whether these two matrices are row equivalent to each other:

$$\begin{bmatrix} 1 & 0 & 0 \\ 0 & 1 & 0 \\ 0 & 0 & 1 \end{bmatrix}, \begin{bmatrix} 2 & 9 & 2 \\ 2 & 6 & 1 \\ 2 & -3 & -1 \end{bmatrix}$$

35. Determine whether these two matrices are row equivalent to each other:

$$\begin{bmatrix} 2 & 6 & 4 \\ 1 & 3 & 3 \end{bmatrix}, \begin{bmatrix} 2 & 4 & 10 \\ 1 & 2 & 10 \end{bmatrix}$$

36. Suppose that the reduced row echelon form of $[\mathbf{A} \mid \mathbf{b}]$ has a pivot in the last column. Explain why the system of equations $\mathbf{Ax} = \mathbf{b}$ is inconsistent, that is, has *no* solution. Is this true for any row echelon form of the augmented matrix?

37. In an $m \times n$ matrix whose elements are $a_{ij} = (-1)^{i+j}$, how many positive terms are there?

38. Find two or more solutions to the system of equations whose augmented matrix is given here, and verify your answer:

$$\begin{bmatrix} 2 & 3 & -5 & | & 20 \\ 3 & -4 & 6 & | & -15 \end{bmatrix}$$

39. (Continuation.) Find the solution of the system in the preceding exercise that minimizes the expression $x_1^2 + x_2^2 + x_3^2$.

40. Find all solutions of this augmented matrix:

$$\begin{bmatrix} 2 & 3 & 5 & | & 0 \\ 6 & -4 & 1 & | & 0 \\ 10 & 2 & 11 & | & 0 \end{bmatrix}$$

41. Let $\mathbf{A}$ and $\mathbf{B}$ be $m \times n$ matrices. Explain why $\mathbf{A} = \mathbf{B}$ if and only if $\mathbf{Ax} = \mathbf{Bx}$ for all $\mathbf{x}$ in $\mathbb{R}^n$. Half of this (the *only if*) part is rather obvious. It is the *if* part that requires an idea!

42. For linear systems $\mathbf{Ax} = \mathbf{b}$ and $\mathbf{By} = \mathbf{c}$, does $[\mathbf{A} \,|\, \mathbf{b}] \sim [\mathbf{B} \,|\, \mathbf{c}]$ imply $\mathbf{x} = \mathbf{y}$? Explain why or why not.

43. Fill in the missing steps in Example 2.

44. Express each of these as a single vector:

a. $\begin{bmatrix} 4 & 2 \\ 1 & 3 \end{bmatrix} \begin{bmatrix} 1 \\ -1 \end{bmatrix}$

b. $\begin{bmatrix} 2 & 0 & 4 \\ 1 & 3 & 0 \\ 0 & 1 & 5 \end{bmatrix} \begin{bmatrix} 1 \\ -2 \\ 1 \end{bmatrix}$

c. $\begin{bmatrix} 1 \\ 3 \\ 7 \end{bmatrix} - 2 \begin{bmatrix} 2 \\ 1 \\ 0 \end{bmatrix} + 2 \begin{bmatrix} 3 \\ 4 \\ 3 \end{bmatrix}$

45. Let

$$\mathbf{A} = \begin{bmatrix} 4 & 0 & 4 & 0 \\ 2 & -1 & 0 & 1 \\ 1 & 0 & 2 & 3 \end{bmatrix}$$

$$\mathbf{x} = \begin{bmatrix} 0 \\ 3 \\ 1 \\ 2 \end{bmatrix}$$

a. Without calculating all the entries, find the first component of $\mathbf{Ax}$.

b. With a minimum amount of work, find the third component of $\mathbf{x}^T \mathbf{A}^T$.

46. Let

$$\mathbf{A} = \begin{bmatrix} 3 & -1 & 4 \\ -1 & 0 & 2 \\ 4 & 2 & 1 \end{bmatrix}$$

$$\mathbf{x} = \begin{bmatrix} 1 \\ -1 \\ 1 \end{bmatrix}$$

$$\mathbf{y} = \begin{bmatrix} -1 \\ 0 \\ 1 \end{bmatrix}$$

a. What is $\mathbf{y}^T \mathbf{Ax}$?

b. With little work, what is $\mathbf{x}^T \mathbf{A}^T \mathbf{y}$?

47. For the linear systems corresponding to these augmented matrices, find their general solutions:

a. $\begin{bmatrix} 1 & -2 & 0 & 0 & 7 & | & -3 \\ 0 & 1 & 0 & 0 & -3 & | & 1 \\ 0 & 0 & 0 & 1 & 5 & | & -4 \end{bmatrix}$

b. $\begin{bmatrix} 0 & 4 & 3 & 4 & 6 & 11 & | & 5 \\ 0 & 0 & 0 & 7 & 2 & 8 & | & 7 \\ 0 & 0 & 0 & 0 & 0 & 3 & | & 2 \end{bmatrix}$

48. Describe all solutions of this linear system

$$\begin{cases} 3x_1 + 6x_2 + x_3 + 3x_4 = -9 \\ 2x_1 + 4x_2 + x_3 + 3x_4 = 7 \end{cases}$$

49. What are the general solutions of the linear systems corresponding to these augmented matrices?

a. $\begin{bmatrix} 2 & 4 & 6 & | & 2 \\ 1 & 2 & 3 & | & 1 \\ 1 & 0 & 1 & | & -3 \\ 2 & 4 & 0 & | & 8 \end{bmatrix}$

b. $\begin{bmatrix} 5 & 4 & 3 & 2 & 1 & | & 0 \\ 0 & 0 & 3 & 4 & 5 & | & 6 \\ 0 & 0 & 0 & 0 & 6 & | & 2 \end{bmatrix}$

50. Under what conditions do these augmented matrices correspond to consistent linear systems?

a. $\begin{bmatrix} 1 & 0 & 2 & | & a \\ 2 & 1 & 5 & | & b \\ 1 & -1 & 1 & | & c \end{bmatrix}$

b. $\begin{bmatrix} 1 & 2 & -1 & | & \alpha \\ 3 & 0 & -3 & | & \beta \\ 2 & 3 & -1 & | & \gamma \end{bmatrix}$

51. For what values of the constants r and s is the first vector a linear combination of the second two vectors?

a. $\mathbf{a} = (r, 20, -6)$
$\mathbf{x} = (3, 6, 9), \qquad \mathbf{y} = (4, -2, 3)$

b. $\mathbf{b} = (-2, 14, s)$
$\mathbf{u} = (2, 6, -1), \qquad \mathbf{v} = (4, 8, 3)$

52. Write as many equivalent forms for the 2×2 linear system as you can:

$$\mathbf{Ax} = \begin{bmatrix} a_{11} & a_{12} \\ a_{21} & a_{22} \end{bmatrix} \begin{bmatrix} x_1 \\ x_2 \end{bmatrix}$$

$$= \begin{bmatrix} b_1 \\ b_2 \end{bmatrix} = \mathbf{b}$$

COMPUTER EXERCISES 1.2

1. (Roundoff Error.) Solve the following system of equations by the Gaussian elimination method, using MATLAB or some other similar system:

$$\begin{cases} 3.277x_1 + 5.113x_2 = 2.237 \\ 1.482x_1 + 2.321x_2 = 4.209 \end{cases}$$

Then do the same when all real numbers in the input data are rounded to three significant figures. Draw conclusions.

2. In MATLAB, one can ask for calculated quantities to be expressed as quotients of integers by the command `format rat`. Using this command, find the solution of this system expressed as quotients of integers.

$$\begin{bmatrix} 5 & 1 & -2 & | & 2 \\ 3 & -1 & 4 & | & 1 \\ 2 & 7 & 6 & | & 3 \end{bmatrix}$$

3. Use mathematical software to find the general solutions to General Exercises 6–9, 16, 23, 32, 38, 40, and 47–49.

4. Find the reduced row echelon forms for the two matrices in General Exercise 15.

5. Use mathematical software to solve General Exercise 21 in its original form without using a change of variable.

6. Compute the general solutions of the linear systems corresponding to these augmented matrices.

a. $\begin{bmatrix} 1 & 1 & 2 & 2 & 1 & | & 1 \\ 2 & 2 & 4 & 4 & 3 & | & 1 \\ 2 & 2 & 4 & 4 & 2 & | & 2 \\ 3 & 5 & 8 & 6 & 5 & | & 3 \end{bmatrix}$

b. $\begin{bmatrix} 1 & 1 & 1 & 1 & 1 & | & 1 \\ -1 & -1 & 0 & 0 & 1 & | & -1 \\ -2 & -2 & 0 & 0 & 1 & | & 1 \\ 0 & 0 & 1 & 1 & 3 & | & -1 \\ 1 & 1 & 2 & 2 & 4 & | & 1 \end{bmatrix}$

c. $\begin{bmatrix} 0 & 3 & 6 & 4 & | & 0 \\ 1 & 3 & 0 & 4 & | & 0 \\ 2 & 3 & 5 & 3 & | & 0 \\ 1 & 4 & 5 & 9 & | & 0 \end{bmatrix}$

d. $\begin{bmatrix} -2 & -1 & 0 & 4 & | & 5 \\ -3 & 2 & 7 & 8 & | & 3 \\ 1 & 0 & 0 & 2 & | & 4 \\ 3 & 0 & 1 & 3 & | & 6 \end{bmatrix}$

> *Each problem that I solved became a rule which served afterwards to solve other problems.*
> —RENÉ DESCARTES (1596–1650)

> *There are no solved problems. There are only problems that are more or less solved.*
> JULES HENRÍ POINCARÉ (1854–1912)

At this point in the book, it is assumed that the reader knows all about the reduced row echelon form of a matrix (in particular how to compute it) and what *pivot* elements are.

Kernel or Null Space of a Matrix

A system of linear equations of the form $\mathbf{Ax} = \mathbf{0}$ is said to be **homogeneous**. (In the study of differential equations the same concept arises.) This is a special case of the general system that we usually write as $\mathbf{Ax} = \mathbf{b}$. A first observation about a homogeneous system is that we can take $\mathbf{x} = \mathbf{0}$ as a solution. This is called the **trivial solution**. The issue, then, is whether there are any other solutions. If so, they are called **nontrivial solutions**. Our goal is to find, for any specific matrix $\mathbf{A}$, a complete description of the set of all solutions of its homogeneous equation. That set of vectors is

$$\text{Ker}(\mathbf{A}) = \{\, \mathbf{x} \,:\, \mathbf{Ax} = \mathbf{0}\,\} = \text{Null}\,(\mathbf{A})$$

The abbreviation **Ker** is for the word **kernel**. The notation Null($\mathbf{A}$) is also used, where **Null** is an abbreviation for **null space**. Thus, the kernel or null space of a matrix is the set of all vectors that are mapped into $\mathbf{0}$ by the mapping $\mathbf{x} \mapsto \mathbf{Ax}$. If $\mathbf{A}$ is an $m \times n$ matrix, the kernel of $\mathbf{A}$ is a subset of $\mathbb{R}^n$. In symbols, Ker($\mathbf{A}$) $\subseteq \mathbb{R}^n$ or Null($\mathbf{A}$) $\subseteq \mathbb{R}^n$. This set is never empty, is it?

EXAMPLE 1 Let $\mathbf{A} = \begin{bmatrix} 1 & 3 & 5 \\ 2 & 1 & 4 \end{bmatrix}$

Is the vector $\mathbf{v} = [7, 6, -5]^T$ in the kernel of this matrix?

SOLUTION Yes, one has only to verify that $\mathbf{Av} = \mathbf{0}$:

$$\begin{bmatrix} 1 & 3 & 5 \\ 2 & 1 & 4 \end{bmatrix} \begin{bmatrix} 7 \\ 6 \\ -5 \end{bmatrix} = 7 \begin{bmatrix} 1 \\ 2 \end{bmatrix} + 6 \begin{bmatrix} 3 \\ 1 \end{bmatrix} - 5 \begin{bmatrix} 5 \\ 4 \end{bmatrix}$$

$$= \begin{bmatrix} 7 \\ 14 \end{bmatrix} + \begin{bmatrix} 18 \\ 6 \end{bmatrix} + \begin{bmatrix} -25 \\ -20 \end{bmatrix} = \begin{bmatrix} 0 \\ 0 \end{bmatrix} \qquad \blacksquare$$

If $\mathbf{A}$ is an $m \times n$ matrix and $\mathbf{x}$ is a nonzero vector such that $\mathbf{Ax} = \mathbf{0}$, then we have a nontrivial equation of the form

$$x_1\mathbf{a}_1 + x_2\mathbf{a}_2 + \cdots + x_n\mathbf{a}_n = \mathbf{0}$$

Here the $\mathbf{a}_i$ terms are the columns of the matrix $\mathbf{A}$. If $\mathbf{A}$ is $m \times n$, then each column is a vector in $\mathbb{R}^m$. The preceding equation exhibits a linear relation among the columns of $\mathbf{A}$. In this case, it is a nontrivial equation, because $\mathbf{x} \neq \mathbf{0}$. (In other words, the vector $\mathbf{x}$ has at least one nonzero component.) When this occurs, we say that the set of columns of $\mathbf{A}$ is **linearly dependent**. Otherwise, we say that the set of columns of $\mathbf{A}$ is **linearly independent**. This important terminology is explained in detail later in this section.

THEOREM 1

If $\mathbf{x}$ and $\mathbf{y}$ are vectors in $\mathrm{Ker}(\mathbf{A})$, and if α is a scalar, then $\mathbf{x} + \mathbf{y}$ and $\alpha\mathbf{x}$ are also in $\mathrm{Ker}(\mathbf{A})$.

PROOF Here is the proof for addition: Assume the hypotheses. Then

$$\mathbf{A}(\mathbf{x} + \mathbf{y}) = \mathbf{Ax} + \mathbf{Ay} = \mathbf{0} + \mathbf{0} = \mathbf{0} \qquad \blacksquare$$

THEOREM 2

If $\mathbf{Ax} = \mathbf{b}$ and $\mathbf{y} \in \mathrm{Ker}(\mathbf{A})$, then $\mathbf{A}(\mathbf{x} + \mathbf{y}) = \mathbf{b}$.

THEOREM 3

If $\mathbf{Ax} = \mathbf{Ay}$, then $\mathbf{x} - \mathbf{y} \in \mathrm{Ker}(\mathbf{A})$.

THEOREM 4

If $\mathbf{u}$ is a vector such that $\mathbf{Au} = \mathbf{b}$, then every solution of the equation $\mathbf{Ax} = \mathbf{b}$ is of the form $\mathbf{x} = \mathbf{u} + \mathbf{z}$, for some vector $\mathbf{z}$ in $\mathrm{Ker}(\mathbf{A})$.

The proofs of Theorems 2, 3, and 4 are straightforward.

THEOREM 5

If two matrices are row equivalent to each other, then their kernels are the same.

PROOF If **A** and **B** are two matrices that are row equivalent to each other, then the solutions of **Ax** = **0** and the solutions of **Bx** = **0** are the same. This is because row operations performed on a system of equations do not alter the solutions. Hence, the kernels of **A** and **B** are the same. ∎

Homogeneous Equations

Theorems 1 to 5 hint at the utility of knowing all solutions to a homogeneous equation **Ax** = **0**. A typical example follows.

EXAMPLE 2 We seek a description of the set of all solutions to the equation **Ax** = **0**, where

$$
\mathbf{A} = \begin{bmatrix} 1 & 3 & -2 & 0 \\ 3 & 10 & -7 & 1 \\ -5 & -5 & 3 & 7 \end{bmatrix}
$$

SOLUTION The row reduction of the augmented matrix leads to

$$
\left[\begin{array}{cccc|c} 1 & 3 & -2 & 0 & 0 \\ 3 & 10 & -7 & 1 & 0 \\ -5 & -5 & 3 & 7 & 0 \end{array}\right] \sim \left[\begin{array}{cccc|c} 1 & 0 & 0 & -2 & 0 \\ 0 & 1 & 0 & 0 & 0 \\ 0 & 0 & 1 & -1 & 0 \end{array}\right]
$$

Notice that x_4 is a free variable. Furthermore, the last column of 0's in the augmented matrix could be omitted in such a calculation because it remains a zero column throughout the row-reduction process. The corresponding system of equations can be written as

$$
\begin{cases} x_1 = 2x_4 \\ x_2 = 0 \\ x_3 = x_4 \end{cases}
$$

We write this in the form

$$
\begin{bmatrix} x_1 \\ x_2 \\ x_3 \\ x_4 \end{bmatrix} = x_4 \begin{bmatrix} 2 \\ 0 \\ 1 \\ 1 \end{bmatrix} \quad \text{or} \quad \begin{bmatrix} x_1 \\ x_2 \\ x_3 \\ x_4 \end{bmatrix} = t \begin{bmatrix} 2 \\ 0 \\ 1 \\ 1 \end{bmatrix}
$$

where t is a free parameter. In words: The kernel of **A** consists precisely of all scalar multiples of the vector $[2, 0, 1, 1]^T$. This is a line in four-space ($\mathbb{R}^4$) passing through **0**. ∎

Uniqueness of the Reduced Row Echelon Form

The logical underpinning of row reduction, as a technique applied to matrices in general, depends on the following theorem.

THEOREM 6

Every matrix has one and only one reduced row echelon form.

PROOF Let $\mathbf{A}$ be $m \times n$. The proof uses induction on n. For $n = 1$, the theorem is obvious. Suppose, now, that the theorem has been established for $m \times (n - 1)$ matrices. Let $\mathbf{A}$ be an $m \times n$ matrix. Denote by $\mathbf{A}_{n-1}$ the matrix obtained from $\mathbf{A}$ by removing its nth column. Any sequence of row operations that brings $\mathbf{A}$ to reduced row echelon form also puts $\mathbf{A}_{n-1}$ in reduced row echelon form. By the induction hypothesis, $\mathbf{A}_{n-1}$ has one and only one reduced row echelon form. If $\mathbf{B}$ and $\mathbf{C}$ are reduced row echelon forms of $\mathbf{A}$, they can differ only in the nth column. Assume $\mathbf{B} \neq \mathbf{C}$. Select i so that row i in $\mathbf{B}$ differs from row i in $\mathbf{C}$. Thus, we have $b_{in} \neq c_{in}$. Because $\mathbf{A} \sim \mathbf{B} \sim \mathbf{C}$, the homogeneous systems $\mathbf{Ax} = \mathbf{0}, \mathbf{Bx} = \mathbf{0}$, and $\mathbf{Cx} = \mathbf{0}$ have the same solutions. Let $\mathbf{Bu} = \mathbf{0}$. Then $\mathbf{Cu} = \mathbf{0}$ and $(\mathbf{B} - \mathbf{C})\mathbf{u} = \mathbf{0}$. The first $n - 1$ columns of $\mathbf{B} - \mathbf{C}$ are all zeros. Hence, we have $[(\mathbf{B}-\mathbf{C})\mathbf{u}]_i = (b_{in}-c_{in})u_n = 0$. Since $b_{in} \neq c_{in}$, we must have $u_n = 0$. Thus, any solution of $\mathbf{Bx} = \mathbf{0}$ or $\mathbf{Cx} = \mathbf{0}$ must have $x_n = 0$. It follows that the nth column of $\mathbf{B}$ and $\mathbf{C}$ must have pivots, for otherwise those columns would be associated with free variables, and we could choose x_n to be nonzero. Because the first $n - 1$ columns of $\mathbf{B}$ and $\mathbf{C}$ are identical, the row in which this pivotal 1 appears must be the same for $\mathbf{B}$ and $\mathbf{C}$; namely, it is the row that is the first zero row of the reduced row echelon form of $\mathbf{A}_{n-1}$. Because the remaining entries in the nth columns of $\mathbf{B}$ and $\mathbf{C}$ must be zero, we have $\mathbf{B} = \mathbf{C}$, which is a contradiction. This proof follows the one given by Yuster [1984]. ∎

Rank of a Matrix

An important concept in linear algebra is the **rank** of a matrix. It is directly related to the number of pivot positions in the matrix, and, thus, not surprisingly, it rests upon the preceding theorem.

DEFINITION

The **rank** of a matrix is the number of nonzero rows in its reduced row echelon form. We use the notation **Rank**($\mathbf{A}$) for this number.

This definition depends logically on our acknowledging that a given matrix has a unique reduced row echelon form. Remember that in mathematics the word *unique* does not mean *unusual* or *noteworthy*; it means one of a kind. (In non-scientific and non-mathematical contexts, casual use of the word *unique* has degraded it to the point where it has almost *no* meaning.)

The computation of the rank of a matrix can be done by obtaining its reduced row echelon form and counting the number of nonzero rows in the result. In fact, it is not necessary to get the *reduced* row echelon form of the given matrix; its rank will be evident from any row echelon form of the matrix. This is true because a row echelon form of a matrix reveals how many nonzero rows there will be in the reduced row echelon form.

EXAMPLE 3 What is the rank of this matrix?

$$\mathbf{A} = \begin{bmatrix} 1 & 3 & 3 & -1 & 2 & 17 \\ 2 & 6 & -2 & 14 & -3 & -19 \\ 4 & 12 & 2 & 16 & 1 & 7 \end{bmatrix}$$

SOLUTION In the row-reduction process, we show a row echelon form and the reduced row echelon form:

$$\mathbf{A} \sim \begin{bmatrix} \boxed{1} & 3 & 3 & -1 & 2 & 17 \\ 0 & 0 & \boxed{2} & -4 & 0 & 8 \\ 0 & 0 & 0 & 0 & \boxed{7} & 21 \end{bmatrix} \sim \begin{bmatrix} \boxed{1} & 3 & 0 & 5 & 0 & -1 \\ 0 & 0 & \boxed{1} & -2 & 0 & 4 \\ 0 & 0 & 0 & 0 & \boxed{1} & 3 \end{bmatrix}$$

That the rank is 3 can be concluded from either of these two row echelon forms. Here the pivot positions are the boxed entries. In particular, one need not carry out the reduction to the *reduced* row echelon form. ∎

Recall that the locations of the pivots are called **pivot positions** for the given matrix.

EXAMPLE 4 What are the pivot positions in the matrix of Example 3?

SOLUTION The pivot positions are $(1, 1)$, $(2, 3)$, and $(3, 5)$. Notice that the definition of pivot positions refers to the reduced row echelon form of the matrix, but the location of those pivot positions is already clear from any row echelon form. ∎

In the reduced row echelon form, each pivot must be to the right of and below the position of the previous pivot. In other words, each new pivot must account for a nonzero row and a nonzero column because of the required pattern.

COROLLARY 1

The rank of a matrix is the number of pivots in its reduced row echelon form, which is the same as the number of pivot positions in the matrix.

EXAMPLE 5 What are the pivot positions in the matrix $\begin{bmatrix} 0 & 2 & 3 & 1 \\ 1 & 4 & 6 & 3 \\ 3 & 3 & 7 & 5 \end{bmatrix}$?

SOLUTION By carrying out a few row operations, we obtain

$$\begin{bmatrix} 0 & 2 & 3 & 1 \\ 1 & 4 & 6 & 3 \\ 3 & 3 & 7 & 5 \end{bmatrix} \sim \begin{bmatrix} 1 & 4 & 6 & 3 \\ 0 & 2 & 3 & 1 \\ 3 & 3 & 7 & 5 \end{bmatrix} \sim \begin{bmatrix} 1 & 4 & 6 & 3 \\ 0 & 2 & 3 & 1 \\ 0 & -9 & -11 & -4 \end{bmatrix}$$

$$\sim \begin{bmatrix} 1 & 4 & 6 & 3 \\ 0 & 2 & 3 & 1 \\ 0 & -1 & 1 & 0 \end{bmatrix} \sim \begin{bmatrix} 1 & 4 & 6 & 3 \\ 0 & 1 & -1 & 0 \\ 0 & 2 & 3 & 1 \end{bmatrix} \sim \begin{bmatrix} \boxed{1} & 4 & 6 & 3 \\ 0 & \boxed{1} & -1 & 0 \\ 0 & 0 & \boxed{5} & 1 \end{bmatrix}$$

$$\sim \begin{bmatrix} 1 & 4 & 0 & \frac{9}{5} \\ 0 & 1 & 0 & \frac{1}{5} \\ 0 & 0 & 1 & \frac{1}{5} \end{bmatrix} \sim \begin{bmatrix} \boxed{1} & 0 & 0 & 1 \\ 0 & \boxed{1} & 0 & \frac{1}{5} \\ 0 & 0 & \boxed{1} & \frac{1}{5} \end{bmatrix}$$

We see that the pivot positions are $(1, 1)$, $(2, 2)$, and $(3, 3)$, which are shown as boxed entries. The matrix has rank 3. Note especially that the pivot positions cannot be predicted solely from the numbers in the original matrix. For example, the entry 0 in the original matrix turns out to occupy a pivot position. ∎

THEOREM 7

The rank of an m × n matrix cannot be greater than n or m. In symbols, we have **Rank(A)** ≤ min(*m, n*).

PROOF Think about the reduced row echelon form of an $m \times n$ matrix. Each nonzero row must have a pivot element, and these must occur in the staircase pattern. In counting the pivots, we note that each new pivot must account for a nonzero row and a nonzero column. The count of all pivots cannot therefore exceed n or m. Because the rank is the number of pivots, the rank can be at most n and at most m. ∎

The next theorem elaborates on this theme. It concerns matrices of size $m \times n$ that have rank less than n. The theorem states that if an $m \times n$ matrix has any one of the properties labeled $\mathbf{a}, \mathbf{b}, \ldots, \mathbf{f}$, then it must have all those properties.

THEOREM 8

These properties of an $\mathbf{m} \times \mathbf{n}$ matrix $\mathbf{A}$ are equivalent to each other.

a. *The rank of $\mathbf{A}$ is less than $\mathbf{n}$.*
b. *The reduced row echelon form of $\mathbf{A}$ has fewer than $\mathbf{n}$ nonzero rows.*
c. *The matrix $\mathbf{A}$ has fewer than $\mathbf{n}$ pivot positions.*
d. *At least one column in $\mathbf{A}$ has no pivot position.*
e. *There is at least one free variable in the system of equations $\mathbf{Ax} = \mathbf{0}$.*
f. *The system $\mathbf{Ax} = \mathbf{0}$ has some nontrivial solutions.*

A useful corollary of Theorem 8 involves only the dimensions of a matrix:

COROLLARY 2

A homogeneous system of linear equations in which there are more variables than equations must have some nontrivial solutions.

PROOF Let the system have the form $\mathbf{Ax} = \mathbf{0}$, where the matrix $\mathbf{A}$ is $m \times n$. The homogeneous system of equations has m equations and n unknowns. Therefore, by hypothesis, $n > m$. It follows from Theorem 7 that

$$\text{Rank}(\mathbf{A}) \leq \min(n, m) = m < n$$

Apply Theorem 8 to conclude that the homogeneous system has nontrivial solutions. ∎

General Solution of a System

The next few examples illustrate techniques for finding the general solution of a system of equations.

EXAMPLE 6 What do the Theorems 7 and 8 and Corollary 2 tell us about this system?

$$\begin{cases} x_1 + 3x_2 + 9x_3 = 0 \\ 2x_1 + 7x_2 + 3x_3 = 0 \end{cases}$$

SOLUTION In this system, there are more variables than equations. Therefore, by Corollary 2, the system has nontrivial solutions. Using Theorem 7, we find that Rank($\mathbf{A}$) $\le$ min$\{2, 3\}$ = 2. The row-reduction process applied to the augmented matrix shows that

$$\left[\begin{array}{ccc|c} 1 & 3 & 9 & 0 \\ 2 & 7 & 3 & 0 \end{array}\right] \sim \left[\begin{array}{ccc|c} \boxed{1} & 0 & 54 & 0 \\ 0 & \boxed{1} & -15 & 0 \end{array}\right]$$

There are two pivots, and the rank of the matrix is 2. There is one column without a pivot, and this indicates the presence of a free variable. (Here, it is x_3.) With x_3 assigned any value, we have $x_1 = -54x_3$ and $x_2 = 15x_3$. This is the *general* solution of the homogeneous system. The recommended form of the *general solution* is

$$\begin{bmatrix} x_1 \\ x_2 \\ x_3 \end{bmatrix} = x_3 \begin{bmatrix} -54 \\ 15 \\ 1 \end{bmatrix} \quad \text{or} \quad \begin{bmatrix} x_1 \\ x_2 \\ x_3 \end{bmatrix} = t \begin{bmatrix} -54 \\ 15 \\ 1 \end{bmatrix}$$

where t is a free parameter. Thus, all solutions to the homogeneous equation are scalar multiples of the vector $[-54, 15, 1]^T$. All these points lie on a line through the origin in three-space ($\mathbb{R}^3$). ∎

EXAMPLE 7 Find all the solutions of this system of linear equations:

$$\begin{cases} 2x_1 - 4x_2 = 3 \\ 4x_1 - x_2 = 2 \\ x_1 - x_2 = 1 \end{cases}$$

SOLUTION The row-reduction algorithm leads to this conclusion:

$$
\begin{bmatrix} 2 & -4 & 3 \\ 4 & -1 & 2 \\ 1 & -1 & 1 \end{bmatrix} \sim \begin{bmatrix} \boxed{1} & 0 & 0 \\ 0 & \boxed{1} & 0 \\ 0 & 0 & \boxed{1} \end{bmatrix}
$$

Of course, we are working with the augmented matrix. The final row of the row-reduced matrix states that $0x_1 + 0x_2 = 1$. This precludes the existence of a solution. ∎

The system of equations in Example 7 is *inconsistent*. Notice that there is a pivot position in the last column of this augmented matrix. A moment's thought will convince us that this is always the sign of an inconsistent system. Indeed, if there is a pivot in the last column, there must be zeros elsewhere in that row. Hence, the equation corresponding to that row is of the form $0x_1 + 0x_2 + \cdots + 0x_n = 1$, which is not possible!

Matrix–Matrix Product

In Section 1.2, the matrix–vector product

$$
\begin{bmatrix} a_{11} & a_{12} & \cdots & a_{1n} \\ a_{21} & a_{22} & \cdots & a_{2n} \\ a_{31} & a_{32} & \cdots & a_{3n} \\ \vdots & \vdots & \ddots & \vdots \\ a_{m1} & a_{m2} & \cdots & a_{mn} \end{bmatrix} \begin{bmatrix} x_1 \\ x_2 \\ x_3 \\ \vdots \\ x_n \end{bmatrix}
$$

was defined. Now we give meaning to the matrix–matrix product **AB** whenever **A** is an $m \times n$ matrix and **B** is an $n \times k$ matrix. Notice the requirement that the number of columns in **A** matches the number of rows in **B**.

> **DEFINITION**
>
> *If* **A** *is an* $m \times n$ *matrix and* **B** *is an* $n \times k$ *matrix, then the* **matrix–matrix product AB** *is defined to be the* $m \times k$ *matrix whose columns are* $\mathbf{Ab}_1, \mathbf{Ab}_2,$ *...,* $\mathbf{Ab}_k$. *Here the vectors* $\mathbf{b}_i$ *are the columns of matrix* **B**.

One can write the defining equation in this form:

$$
\mathbf{AB} = \mathbf{A} \begin{bmatrix} \mathbf{b}_1 & \mathbf{b}_2 & \cdots & \mathbf{b}_k \end{bmatrix} = \begin{bmatrix} \mathbf{Ab}_1 & \mathbf{Ab}_2 & \cdots & \mathbf{Ab}_k \end{bmatrix}
$$

where each column of **AB** can be computed as a linear combination of the columns of **A**.

$$
\mathbf{Ab}_i = b_{i1}\mathbf{a}_1 + b_{i2}\mathbf{a}_2 + \cdots + b_{in}\mathbf{a}_n
$$

Special cases of the definition arise when $m = 1$ or $k = 1$. In these cases, we get formulas for $\mathbf{y}^T\mathbf{B}$ and $\mathbf{Ax}$ where $\mathbf{y}^T$ is a $1 \times n$ row vector and $\mathbf{x}$ is an $n \times 1$ column vector. When $m = k = 1$, we get the dot product $\mathbf{y}^T\mathbf{x}$, which is a 1×1 scalar.

EXAMPLE 8 As an illustration, we carry out the important process of matrix–matrix multiplication of a 2×2 matrix $\mathbf{A}$ and a 2×3 matrix $\mathbf{B}$ in full detail:

$$\mathbf{AB} = \begin{bmatrix} 3 & 2 \\ 1 & 1 \end{bmatrix} \begin{bmatrix} 1 & 4 & -1 \\ 2 & 2 & 3 \end{bmatrix}$$

$$= \begin{bmatrix} 3 & 2 \\ 1 & 1 \end{bmatrix} \begin{bmatrix} \begin{bmatrix} 1 \\ 2 \end{bmatrix}, & \begin{bmatrix} 4 \\ 2 \end{bmatrix}, & \begin{bmatrix} -1 \\ 3 \end{bmatrix} \end{bmatrix}$$

SOLUTION

$$= \begin{bmatrix} \begin{bmatrix} 3 & 2 \\ 1 & 1 \end{bmatrix} \begin{bmatrix} 1 \\ 2 \end{bmatrix}, & \begin{bmatrix} 3 & 2 \\ 1 & 1 \end{bmatrix} \begin{bmatrix} 4 \\ 2 \end{bmatrix}, & \begin{bmatrix} 3 & 2 \\ 1 & 1 \end{bmatrix} \begin{bmatrix} -1 \\ 3 \end{bmatrix} \end{bmatrix}$$

$$= \begin{bmatrix} 1\begin{bmatrix} 3 \\ 1 \end{bmatrix} + 2\begin{bmatrix} 2 \\ 1 \end{bmatrix}, & 4\begin{bmatrix} 3 \\ 1 \end{bmatrix} + 2\begin{bmatrix} 2 \\ 1 \end{bmatrix}, & -1\begin{bmatrix} 3 \\ 1 \end{bmatrix} + 3\begin{bmatrix} 2 \\ 1 \end{bmatrix} \end{bmatrix}$$

$$= \begin{bmatrix} \begin{bmatrix} 7 \\ 3 \end{bmatrix}, & \begin{bmatrix} 16 \\ 6 \end{bmatrix}, & \begin{bmatrix} 3 \\ 2 \end{bmatrix} \end{bmatrix} = \begin{bmatrix} 7 & 16 & 3 \\ 3 & 6 & 2 \end{bmatrix}$$

Here we have inserted some commas for clarity. ∎

EXAMPLE 9 What is the numerical value of the following product?

$$\mathbf{AB} = \begin{bmatrix} 1 & 3 & 2 & -5 \\ 2 & 2 & -3 & 4 \\ 5 & 1 & 1 & 6 \end{bmatrix} \begin{bmatrix} 2 & 3 \\ -5 & 0 \\ 4 & 1 \end{bmatrix}$$

SOLUTION Such a product is *not* defined because the matrices $\mathbf{A}$ and $\mathbf{B}$ are incompatible! The product $\mathbf{AB}$ of two matrices will exist if and only if the number of columns in $\mathbf{A}$ equals the number of rows in $\mathbf{B}$. Here we have a 3×4 matrix $\mathbf{A}$ and a 2×2 matrix $\mathbf{B}$. Remember, if $\mathbf{A}$ is $m \times n$, then $\mathbf{B}$ must be $n \times k$, for some value of k. The values of m and k are unrestricted. ∎

In Section 3.1, the topic of matrix–matrix multiplication is explored in more detail.

EXAMPLE 10 How can we efficiently solve two systems, $\mathbf{Ax} = \mathbf{b}$ and $\mathbf{Ay} = \mathbf{c}$, when the coefficient matrix $\mathbf{A}$ is the same in the two systems?

SOLUTION This example shows how to solve several systems of equations that differ only in their righthand sides. To solve $\mathbf{Ax} = \mathbf{b}$ and $\mathbf{Ay} = \mathbf{c}$, we can set up two augmented matrices $[\mathbf{A} \mid \mathbf{b}]$ and $[\mathbf{A} \mid \mathbf{c}]$. Then we carry out the row reduction of both augmented matrices. However, because $\mathbf{A}$ is the same in both, it is more efficient to create this augmented matrix $[\mathbf{A} \mid \mathbf{b} \mid \mathbf{c}]$ and carry out the row reduction on it. ∎

EXAMPLE 11 For a concrete case of this technique, solve the systems $\mathbf{Ax} = \mathbf{b}$ and $\mathbf{Ay} = \mathbf{c}$, when

$$\mathbf{A} = \begin{bmatrix} 1 & 3 & 7 & -11 \\ 2 & -4 & 1 & 1 \\ 1 & 2 & -5 & 2 \end{bmatrix}, \quad \mathbf{b} = \begin{bmatrix} 6 \\ 9 \\ -5 \end{bmatrix}, \quad \mathbf{c} = \begin{bmatrix} 1 \\ 9 \\ -9 \end{bmatrix}$$

SOLUTION The augmented matrix and its reduced row echelon form are

$$\begin{bmatrix} \mathbf{A} & \mid & \mathbf{b} & \mid & \mathbf{c} \end{bmatrix} = \begin{bmatrix} 1 & 3 & 7 & -11 & 6 & 1 \\ 2 & -4 & 1 & -1 & 9 & 9 \\ 1 & 2 & -5 & 2 & -5 & -9 \end{bmatrix}$$

$$\sim \begin{bmatrix} \boxed{1} & 0 & 0 & -1 & 2 & 0 \\ 0 & \boxed{1} & 0 & -1 & -1 & -2 \\ 0 & 0 & \boxed{1} & -1 & 1 & 1 \end{bmatrix}$$

We find the general solution for $\mathbf{Ax} = \mathbf{b}$ to be

$$\mathbf{x} = \mathbf{u} + s\mathbf{z} = \begin{bmatrix} 2 \\ -1 \\ 1 \\ 0 \end{bmatrix} + s \begin{bmatrix} 1 \\ 1 \\ 1 \\ 1 \end{bmatrix}$$

and the general solution for $\mathbf{Ay} = \mathbf{c}$ to be

$$\mathbf{y} = \mathbf{v} + t\mathbf{z} = \begin{bmatrix} 0 \\ -2 \\ 1 \\ 0 \end{bmatrix} + t \begin{bmatrix} 1 \\ 1 \\ 1 \\ 1 \end{bmatrix}$$

Here s and t are free parameters. The work can be verified by these three equations:

$$\begin{bmatrix} 1 & 3 & 7 & -11 \\ 2 & -4 & 1 & 1 \\ 1 & 2 & -5 & 2 \end{bmatrix} \begin{bmatrix} 2 \\ -1 \\ 1 \\ 0 \end{bmatrix} = \begin{bmatrix} 6 \\ 9 \\ -5 \end{bmatrix}$$

$$\begin{bmatrix} 1 & 3 & 7 & -11 \\ 2 & -4 & 1 & 1 \\ 1 & 2 & -5 & 2 \end{bmatrix} \begin{bmatrix} 0 \\ -2 \\ 1 \\ 0 \end{bmatrix} = \begin{bmatrix} 1 \\ 9 \\ -9 \end{bmatrix}$$

$$\begin{bmatrix} 1 & 3 & 7 & -11 \\ 2 & -4 & 1 & 1 \\ 1 & 2 & -5 & 2 \end{bmatrix} \begin{bmatrix} 1 \\ 1 \\ 1 \\ 1 \end{bmatrix} = \begin{bmatrix} 0 \\ 0 \\ 0 \end{bmatrix}$$

■

EXAMPLE 12 Solve the matrix equation $\mathbf{AX} = \mathbf{B}$, where

$$\mathbf{A} = \begin{bmatrix} 1 & 3 & 1 \\ 2 & 1 & 4 \\ -3 & 2 & 5 \end{bmatrix}, \qquad \mathbf{B} = \begin{bmatrix} 6 & 13 & 15 & -5 \\ 9 & 15 & 14 & 2 \\ 1 & 10 & 15 & 1 \end{bmatrix}$$

Also, verify the solution in an independent manner.

SOLUTION This again illustrates the important technique of solving systems of equations with a single coefficient matrix but multiple righthand sides. In this example, we face a problem of solving four linear systems $\mathbf{Ax}^{(i)} = \mathbf{b}^{(i)}$, for $i = 1, 2, 3, 4$, each with the same coefficient matrix $\mathbf{A}$:

$$\mathbf{AX} = \mathbf{A}[\mathbf{x}^{(1)},\ \mathbf{x}^{(2)},\ \mathbf{x}^{(3)},\ \mathbf{x}^{(4)}] = [\mathbf{b}^{(1)},\ \mathbf{b}^{(2)},\ \mathbf{b}^{(3)},\ \mathbf{b}^{(4)}] = \mathbf{B}$$

Because the matrix $\mathbf{A}$ is 3×3 and the matrix $\mathbf{B}$ is 3×4, the matrix $\mathbf{X}$ must be 3×4. The augmented matrix for the problem is

$$[\mathbf{A} \mid \mathbf{B}] = \left[\begin{array}{ccc|cccc} 1 & 3 & 1 & 6 & 13 & 15 & -5 \\ 2 & 1 & 4 & 9 & 15 & 14 & 2 \\ -3 & 2 & 5 & 1 & 10 & 15 & 1 \end{array} \right]$$

The row-reduction process leads to these matrices:

$$[\mathbf{A} \mid \mathbf{B}] \sim \left[\begin{array}{ccc|cccc} 1 & 0 & 0 & 2 & 2 & 1 & 0 \\ 0 & 1 & 0 & 1 & 3 & 4 & -2 \\ 0 & 0 & 1 & 1 & 2 & 2 & 1 \end{array} \right] = [\mathbf{I} \mid \mathbf{X}]$$

where

$$X = \begin{bmatrix} 2 & 2 & 1 & 0 \\ 1 & 3 & 4 & -2 \\ 1 & 2 & 2 & 1 \end{bmatrix}$$

An independent verification is possible as follows:

$$AX = \begin{bmatrix} 1 & 3 & 1 \\ 2 & 1 & 4 \\ -3 & 2 & 5 \end{bmatrix} \begin{bmatrix} 2 & 2 & 1 & 0 \\ 1 & 3 & 4 & -2 \\ 1 & 2 & 2 & 1 \end{bmatrix} = \begin{bmatrix} 6 & 13 & 15 & -5 \\ 9 & 15 & 14 & 2 \\ 1 & 10 & 15 & 1 \end{bmatrix} = B$$

∎

In Section 3.2, the techniques used in this problem will be called upon again, for computing right and left inverses of matrices.

Indexed Sets of Vectors: Linear Dependence and Independence

In describing sets of vectors, we usually think of the vectors as having **indices** attached to them. For example, we might write

$$\left\{ u_1 = (7,3), \quad u_2 = (6,-4), \quad u_3 = (4,11) \right\}$$

Here, the first vector has the index number 1 associated with it, and the other two vectors have index numbers 2 and 3 associated with them. Trouble arises, however, if we have repetitions in the definition of a set, such as this:

$$\left\{ (7,3), \quad (6,-4), \quad (7,3) \right\}$$

Is this a set of two vectors or three vectors? As an ordinary set it has only two elements because the third one mentioned is the same as the first. The *set* is the same as

$$\left\{ (7,3), \quad (6,-4) \right\}$$

The fact that one vector is mentioned twice does not mean that we have three vectors, because one is equal to another. However, in this example, the difficulty is avoided, since these three indexed vectors are deemed to be different (because their indices differ) in the *indexed set:*

$$\left\{ u_1 = (7,3), \quad u_2 = (6,-4), \quad u_3 = (7,3) \right\}$$

As a consequence, the columns of the matrix

$$\begin{bmatrix} u_1 & u_2 & u_3 \end{bmatrix} = \begin{bmatrix} 7 & 6 & 7 \\ 3 & -4 & 3 \end{bmatrix}$$

are regarded as being different, since they have (invisible) indices 1, 2, 3. In a few moments, we will be saying that the columns of this matrix form a linearly dependent set, and the validity of this assertion depends on our thinking of the columns forming an indexed set.

DEFINITION

*An indexed set of vectors, $\{\mathbf{u}_1, \mathbf{u}_2, \ldots, \mathbf{u}_n\}$, is **linearly dependent** if there exists a nontrivial equation of the form $c_1\mathbf{u}_1 + c_2\mathbf{u}_2 + \cdots + c_n\mathbf{u}_n = \mathbf{0}$. In the contrary case, the indexed set is **linearly independent**. Nontrivial in this context means $\sum_{i=1}^{n} |c_i| > 0$.*

EXAMPLE 13 Let $\mathbf{u}_1 = (3, 7, 4)$, $\mathbf{u}_2 = (-4, 2, 2)$, $\mathbf{u}_3 = (0, 17, 11)$. Is this set of three vectors linearly dependent?

SOLUTION We are asking whether there is a nontrivial equation of the form

$$c_1\mathbf{u}_1 + c_2\mathbf{u}_2 + c_3\mathbf{u}_3 = c_1 \begin{bmatrix} 3 \\ 7 \\ 4 \end{bmatrix} + c_2 \begin{bmatrix} -4 \\ 2 \\ 2 \end{bmatrix} + c_3 \begin{bmatrix} 0 \\ 17 \\ 11 \end{bmatrix} = \begin{bmatrix} 0 \\ 0 \\ 0 \end{bmatrix}$$

This equation is the same as

$$\begin{bmatrix} \mathbf{u}_1 & \mathbf{u}_2 & \mathbf{u}_3 \end{bmatrix} \mathbf{c} = \begin{bmatrix} 3 & -4 & 0 \\ 7 & 2 & 17 \\ 4 & 2 & 11 \end{bmatrix} \begin{bmatrix} c_1 \\ c_2 \\ c_3 \end{bmatrix} = \begin{bmatrix} 0 \\ 0 \\ 0 \end{bmatrix}$$

A row reduction on the coefficient matrix leads to a row echelon form:

$$\begin{bmatrix} 3 & -4 & 0 \\ 7 & 2 & 17 \\ 4 & 2 & 11 \end{bmatrix} \sim \begin{bmatrix} 3 & -4 & 0 \\ 1 & 10 & 17 \\ 1 & 6 & 11 \end{bmatrix} \sim \begin{bmatrix} 0 & -34 & -51 \\ 1 & 10 & 17 \\ 0 & -4 & -6 \end{bmatrix}$$

$$\sim \begin{bmatrix} 1 & 10 & 17 \\ 0 & 34 & 51 \\ 0 & 4 & 6 \end{bmatrix} \sim \begin{bmatrix} 1 & 10 & 17 \\ 0 & 2 & 3 \\ 0 & 2 & 3 \end{bmatrix}$$

$$\sim \begin{bmatrix} 1 & 6 & 11 \\ 0 & 2 & 3 \\ 0 & 0 & 0 \end{bmatrix} \sim \begin{bmatrix} 1 & 0 & 2 \\ 0 & 2 & 3 \\ 0 & 0 & 0 \end{bmatrix}$$

Solving the homogeneous system with the resulting matrix, we have $c_1 = -2c_3$ and $2c_2 = -3c_3$. This reveals that there exist nontrivial solutions to the homogeneous problem. If we set the free variable c_3 equal to 2, for example, we get a nontrivial solution $c_1 = -4$, $c_2 = -3$, and $c_3 = 2$. Thus, the original set of three vectors is linearly dependent: $-4\mathbf{u}_1 - 3\mathbf{u}_2 + 2\mathbf{u}_3 = -4(3, 7, 4) - 3(-4, 2, 2) + 2(0, 17, 11) = (0, 0, 0) = \mathbf{0}$. ∎

EXAMPLE 14 By direct use of the definition of linear dependence, determine whether these (indexed) sets are linearly independent or linearly dependent. In the case of linear dependence, give the coefficients that establish that fact.

a. $\{\mathbf{u}_1 = (1, 3, 6),\ \mathbf{u}_2 = (2, 7, 5),\ \mathbf{u}_3 = (0, 0, 0)\}$

b. $\{\mathbf{z}_1 = (7, 6, 3),\ \mathbf{z}_2 = (5, 2, 1),\ \mathbf{z}_3 = (7, 6, 3)\}$

c. $\{\mathbf{x}_1 = (7, 6),\ \mathbf{x}_2 = (5, 4),\ \mathbf{x}_3 = (14, 12)\}$

d. $\{\mathbf{v}_1 = (1, 3),\ \mathbf{v}_2 = (2, 7),\ \mathbf{v}_3 = (4, 13)\}$

e. $\{\mathbf{w}_1 = (7, 2, 3),\ \mathbf{w}_2 = (-1, 1, 0),\ \mathbf{w}_3 = (1, 3, -1)\}$

SOLUTION The first set, **a**, is linearly dependent because it contains the zero vector and $0\mathbf{u}_1 + 0\mathbf{u}_2 + 1\mathbf{u}_3 = \mathbf{0}$. The second set, **b**, illustrates a repeated entry in an indexed set. The set is linearly dependent because $\mathbf{z}_1 - \mathbf{z}_3 = \mathbf{0}$. The third set, **c**, is linearly dependent by inspection because $2\mathbf{x}_1 + 0\mathbf{x}_2 - \mathbf{x}_3 = \mathbf{0}$. The fourth set, **d**, is also linearly dependent because $2\mathbf{v}_1 + \mathbf{v}_2 - \mathbf{v}_3 = \mathbf{0}$. If you did not immediately notice this relationship, the reduced row echelon form could be used to reveal it. However, the linear dependence (without the coefficients) can be predicted most efficiently by Corollary 2. It asserts that an indexed set of $n + 1$ vectors in $\mathbb{R}^n$ is necessarily linearly dependent. The fifth set, **e**, is linearly independent because the reduction process yields

$$\begin{bmatrix} \mathbf{w}_2 & \mathbf{w}_3 & \mathbf{w}_1 \end{bmatrix} = \begin{bmatrix} -1 & 1 & 7 \\ 1 & 3 & 2 \\ 0 & -1 & 3 \end{bmatrix} \sim \begin{bmatrix} 1 & -1 & -7 \\ 0 & 4 & 9 \\ 0 & -1 & 3 \end{bmatrix} \sim \begin{bmatrix} 1 & -1 & -7 \\ 0 & 1 & -3 \\ 0 & 0 & 1 \end{bmatrix}$$

There is a pivot position in each column of the original matrix and the only solution of the homogeneous system is the trivial one. ∎

Here is an algorithm for determining whether a set of vectors in $\mathbb{R}^m$ is linearly independent.

ALGORITHM

Given an (indexed) set of vectors $\mathbf{v}_1, \mathbf{v}_2, \ldots, \mathbf{v}_n$ in $\mathbb{R}^m$, form a matrix $\mathbf{A}$ using these n vectors as columns. If the equation $\mathbf{Ax} = \mathbf{0}$ has a nonzero solution, then the given set of vectors is linearly dependent. If the equation has only the $\mathbf{0}$ solution, then the set is linearly independent.

Another example to illustrate the testing for linear independence follows. It also shows that the property of linear dependence can be sensitive to small changes in the data. The set of three vectors $\{(1, 2, 1), (3, 1, 1), (5, 5, 3)\}$ is linearly dependent because

$$\begin{bmatrix} 1 & 3 & 5 \\ 2 & 1 & 5 \\ 1 & 1 & 3 \end{bmatrix} \sim \begin{bmatrix} 1 & 3 & 5 \\ 0 & -5 & -5 \\ 0 & -2 & -2 \end{bmatrix} \sim \begin{bmatrix} 1 & 3 & 5 \\ 0 & 1 & 1 \\ 0 & 0 & 0 \end{bmatrix}$$

Now we make a small change in one vector. The new set of vectors is $\{(1, 2, 1), (3, 1, 1), (5, 5, 3.01)\}$ and it is linearly independent because

$$\begin{bmatrix} 1 & 3 & 5 \\ 2 & 1 & 5 \\ 1 & 1 & 3.01 \end{bmatrix} \sim \begin{bmatrix} 1 & 3 & 5 \\ 0 & -5 & -5 \\ 0 & -2 & -1.99 \end{bmatrix} \sim \begin{bmatrix} 1 & 3 & 5 \\ 0 & 1 & 1 \\ 0 & 0 & 1 \end{bmatrix}$$

THEOREM 9

If an indexed set of two or more vectors is linearly dependent, then some vector in the set is a linear combination of the others.

PROOF Consider the equation $\sum_{i=1}^{k} c_i \mathbf{v}_i = \mathbf{0}$. If the equation has a nontrivial solution, then some coefficient is nonzero. For simplicity, assume $c_1 \neq 0$. Then $\mathbf{v}_1 = -(1/c_1) \sum_{i=2}^{k} c_i \mathbf{v}_i$. ∎

THEOREM 10

If an indexed set of two or more vectors in $\mathbb{R}^m$ is linearly dependent, then some vector in the list is a nontrivial linear combination of vectors preceding it in the list.

PROOF If $\{\mathbf{v}_1, \mathbf{v}_2, \ldots, \mathbf{v}_p\}$ is linearly dependent, we have $\sum_{i=1}^{p} c_i \mathbf{v}_i = \mathbf{0}$ in a nontrivial equation. Let c_q be the last nonzero coefficient. Then $\sum_{i=1}^{q} c_i \mathbf{v}_i = \mathbf{0}$ and $c_q \neq 0$. Solving for $\mathbf{v}_q$, we have $\mathbf{v}_q = (-1/c_q) \sum_{i=1}^{q-1} c_i \mathbf{v}_i$. ∎

EXAMPLE 15 Consider the matrix $\begin{bmatrix} 1 & -2 & 2 \\ -5 & 10 & -9 \\ -3 & 6 & h \end{bmatrix}$

Can the parameter h be chosen so that the three columns form a linearly independent set?

SOLUTION In this matrix, let the columns be $\mathbf{v}_1, \mathbf{v}_2$, and $\mathbf{v}_3$. Obviously, we have $2\mathbf{v}_1 + \mathbf{v}_2 = \mathbf{0}$, and the set of columns is linearly dependent. No value of h makes $\{\mathbf{v}_1, \mathbf{v}_2, \mathbf{v}_3\}$ linearly independent. ∎

THEOREM 11

The rows of a matrix form a linearly dependent set if and only if there is a zero row in any row echelon form of that matrix.

PROOF Let $\mathbf{A}$ be the matrix under consideration, and let its rows be denoted by $\mathbf{r}_1, \mathbf{r}_2, \ldots, \mathbf{r}_m$. In the following list of assertions, each implies the one following:

  a. The set $\{\mathbf{r}_1, \mathbf{r}_2, \ldots, \mathbf{r}_m\}$ is linearly dependent.
  b. A nontrivial equation $\sum_{i=1}^{m} c_i \mathbf{r}_i = \mathbf{0}$ is true.
  c. For some index k, we have $c_k \neq 0$ and $\mathbf{r}_k + \sum_{i \neq k} (c_i/c_k) \mathbf{r}_i = \mathbf{0}$.
  d. Row $\mathbf{r}_k$ becomes $\mathbf{0}$ if we add to it suitable multiples of the other rows.
  e. Any row echelon form of $\mathbf{A}$ has a zero row.

For the converse, assume that $\mathbf{A}$ is row equivalent to a matrix $\mathbf{B}$ that has a zero row, say row $\mathbf{r}_k$. That zero row arises from adding multiples of rows in $\mathbf{A}$ onto row $\mathbf{r}_k$ and from moving the rows so that the zero rows are at the bottom. This process is the same as the one described in the first half of the proof of this theorem, and we can conclude that the rows of $\mathbf{A}$ form a linearly dependent set. ∎

Using the Row-Reduction Process

By applying the row-reduction process to a matrix, we can easily decide whether the set of rows is linearly dependent or linearly independent. If a row echelon form has a zero row, then the rows form a linearly dependent set, and vice versa. If we want the coefficients that make the equation

$\sum_{i=1}^{n} c_i \mathbf{r}_i = \mathbf{0}$ true, we can get them from the row-reduction process. Here is an example of this technique.

EXAMPLE 16 Determine whether the rows of this matrix form a linearly dependent set, and if so, find the coefficients in a nontrivial combination that is $\mathbf{0}$.

$$\mathbf{A} = \begin{bmatrix} \mathbf{r}_1 \\ \mathbf{r}_2 \\ \mathbf{r}_3 \end{bmatrix} = \begin{bmatrix} 1 & 3 & 2 & 2 \\ -1 & 1 & 3 & -3 \\ -2 & -2 & 1 & -5 \end{bmatrix}$$

SOLUTION We immediately consider $\mathbf{A}^T$, whose columns are the rows of $\mathbf{A}$. Its reduced row echelon form is shown here:

$$\mathbf{A}^T = \begin{bmatrix} 1 & -1 & -2 \\ 3 & 1 & -2 \\ 2 & 3 & 1 \\ 2 & -3 & -5 \end{bmatrix} \sim \begin{bmatrix} 1 & 0 & -1 \\ 0 & 1 & 1 \\ 0 & 0 & 0 \\ 0 & 0 & 0 \end{bmatrix}$$

We have obtained the linear combination $c_1\mathbf{r}_1 + c_2\mathbf{r}_2 + c_3\mathbf{r}_3 = \mathbf{0}$, where $c_1 = c_3$ and $c_2 = -c_3$. Consequently, one nontrivial solution is $c_1 = 1, c_2 = -1$, $c_3 = 1$. Check: $(1, 3, 2, 2) - (-1, 1, 3, -3) + (-2, -2, 1, -5) = \mathbf{0}$. ∎

The proofs of the following two theorems are left as exercises.

THEOREM 12

The column vectors of a matrix form a linearly dependent set if and only if there is a column having no pivot.

THEOREM 13

The column vectors of a matrix form a linearly independent set if and only if there is a pivot position in each column of the matrix.

EXAMPLE 17 Use Theorems 11 and 12 to determine whether this set of vectors is linearly dependent or linearly independent:

$$\{(1, 3, 7), (2, 5, -4), (-5, -11, 37)\}$$

SOLUTION First, form a matrix with the three indicated vectors as its rows. Then carry out a row-reduction process to see whether the reduced row echelon form has a zero row. (Any row echelon form of the matrix will serve to answer this question.) We have

$$
\begin{bmatrix} 1 & 3 & 7 \\ 2 & 5 & -4 \\ -5 & -11 & 37 \end{bmatrix}
\sim
\begin{bmatrix} 1 & 3 & 7 \\ 0 & -1 & -18 \\ 0 & 4 & 72 \end{bmatrix}
\sim
\begin{bmatrix} 1 & 3 & 7 \\ 0 & 1 & 18 \\ 0 & 1 & 18 \end{bmatrix}
$$

$$
\sim
\begin{bmatrix} 1 & 3 & 7 \\ 0 & 1 & 18 \\ 0 & 0 & 0 \end{bmatrix}
$$

We conclude from Theorem 11 that the given set of three vectors is linearly dependent.

Next, we carry out the row-reduction process of a matrix with the three indicated vectors as its columns:

$$
\begin{bmatrix} 1 & 2 & -5 \\ 3 & 5 & -11 \\ 7 & -4 & 37 \end{bmatrix}
\sim
\begin{bmatrix} 1 & 2 & -5 \\ 0 & -1 & 4 \\ 0 & -18 & 72 \end{bmatrix}
\sim
\begin{bmatrix} 1 & 2 & -5 \\ 0 & -1 & 4 \\ 0 & -1 & 4 \end{bmatrix}
$$

$$
\sim
\begin{bmatrix} 1 & 0 & 3 \\ 0 & 1 & -4 \\ 0 & 0 & 0 \end{bmatrix}
$$

Because there is a column without a pivot, these column vectors form a linearly dependent set, by Theorem 12. In particular, we find $3(1, 3, 7) - 4(2, 5, -4) = (-5, -11, 37)$. We can find these coefficients for this linear combination only from the second system—*not* the first one! Why? Notice that

$$3(1, 3, 7) - 4(2, 5, -4) = (-5, -11, 37)$$

and

$$
3\begin{bmatrix} 1 \\ 3 \\ 7 \end{bmatrix} - 4\begin{bmatrix} 2 \\ 5 \\ -4 \end{bmatrix} = \begin{bmatrix} -5 \\ -11 \\ 37 \end{bmatrix}
$$

In determining whether a set of vectors is linearly dependent, the vectors can be taken to be either the rows or the columns of a matrix. ∎

Determining Linear Dependence or Independence

When solving small problems by hand, one can determine whether a set of n vectors $\{v_1, v_2, \ldots, v_n\}$ in $\mathbb{R}^m$ is linearly independent or linearly dependent as follows.

1. By inspection, determine whether the set contains **0**. If so, one concludes immediately that the set is linearly dependent.
2. Does the set contain two vectors, of which one is a multiple of the other? If so, the set is linearly dependent.
3. If it is evident that some vector in the set is a linear combination of other vectors in the set, then the set is linearly dependent.
4. Use Theorems 8, 12, and 13 or Corollary 2. Typically, this will involve putting the vectors as columns in an $m \times n$ matrix. If $n > m$, then the set of columns is linearly dependent, by Corollary 2. In this case, no calculation is needed, only counting!
5. Look at the general case if none of the preceding is true. Put the vectors $v_1, v_2, \ldots, v_n$ as columns into a matrix **A**. Carry out the row-reduction process on **A** to obtain a row echelon form. Either the system $Ax = 0$ has a nonzero solution or it does not. In the first case, the set of columns is linearly dependent. Otherwise, the set of columns in **A** is linearly independent.

In other words, first try to find special cases that immediately solve the problem before launching into the general case, which may be long and tedious.

Application: Chemistry

A typical *unbalanced* equation describing a chemical reaction is

$$B_2S_3 + H_3N \rightarrow B_3N_2 + S_3H_4$$

(This equation may not describe a possible reaction: it is used only to illustrate the principles.) The capital letters refer to various chemical elements: B = Boron, N = Nitrogen, S = Sulfur, and H = Hydrogen. The equation can be put into words as follows: If the compound B_2S_3 reacts with the compound H_3N, the result will be two compounds, B_3N_2 and S_3H_4. What is missing in this assertion is information about the relative quantities of each compound involved in the reaction. To balance the equation, numerical factors must be associated with each compound, and when that has been done we will know the relative amounts of each compound participating in the chemical reaction. Here we are balancing the number of atoms in each element. Once these numbers have been determined, the relative masses of the elements in the reaction can be computed. Thus, after the balancing process, the number of atoms of each element should be the same on the two sides of the equation.

To proceed, we associate a factor with each compound. These factors are unknown at the beginning of our analysis, and are denoted by x, y, z, w. Now we write

$$x\mathbf{B}_2\mathbf{S}_3 + y\mathbf{H}_3\mathbf{N} = z\mathbf{B}_3\mathbf{N}_2 + w\mathbf{S}_3\mathbf{H}_4$$

Counting the number of atoms for each element, $\mathbf{B}, \mathbf{S}, \mathbf{H}, \mathbf{N}$, leads to four equations as follows:

$$\mathbf{B}: \ 2x = 3z \qquad \mathbf{S}: \ 3x = 3w \qquad \mathbf{H}: \ 3y = 4w \qquad \mathbf{N}: \ y = 2z$$

If all terms are placed on the lefthand side, we obtain four homogeneous equations with four unknowns. The coefficient matrix and its partially reduced form are

$$\begin{bmatrix} 2 & 0 & -3 & 0 \\ 3 & 0 & 0 & -3 \\ 0 & 3 & 0 & -4 \\ 0 & 1 & -2 & 0 \end{bmatrix} \sim \begin{bmatrix} 1 & 0 & 0 & -1 \\ 0 & 1 & -2 & 0 \\ 0 & 0 & 3 & -2 \\ 0 & 0 & 0 & 0 \end{bmatrix}$$

A convenient solution for the homogeneous system is $x = 3$, $y = 4$, $z = 2$, and $w = 3$. A balanced equation is

$$3\mathbf{B}_2\mathbf{S}_3 + 4\mathbf{H}_3\mathbf{N} = 2\mathbf{B}_3\mathbf{N}_2 + 3\mathbf{S}_3\mathbf{H}_4$$

This work reveals the ratios of the compounds in the reaction. For example, the number of molecules of $\mathbf{H}_3\mathbf{N}$ should be four-thirds the number of molecules of $\mathbf{B}_2\mathbf{S}_3$, if the chemical reaction is to use all the material provided. After the reaction has taken place, the number of molecules of $\mathbf{S}_3\mathbf{H}_4$ should be three-halves the number of molecules of $\mathbf{B}_3\mathbf{N}_2$, and so on.

SUMMARY 1.3

- Homogeneous systems: $\mathbf{Ax} = \mathbf{0}$
- Trivial solution $\mathbf{x} = \mathbf{0}$; nontrivial solution $\mathbf{x} \neq \mathbf{0}$
- Kernel or Null Space: $\text{Ker}(\mathbf{A}) = \text{Null}(\mathbf{A}) = \{\mathbf{x} : \mathbf{Ax} = \mathbf{0}\}$
- $\text{Ker}(\mathbf{A}) = \text{Null}(\mathbf{A}) \subseteq \mathbb{R}^n$ if $\mathbf{A}$ is $m \times n$

- The columns of $\mathbf{A}$ form a linearly independent set if and only if the equation $\mathbf{Ax} = \mathbf{0}$ has only the trivial solution

- The columns of $\mathbf{A}$ form a linearly dependent set if and only if the equation $\mathbf{Ax} = \mathbf{0}$ has a nontrivial solution

- A linear combination of columns of **A** can be written as $x_1\mathbf{a}_1 + x_2\mathbf{a}_2 + \cdots + x_n\mathbf{a}_n$ or as **Ax**. The vectors $\mathbf{a}_i$ are the columns of **A**.

- Theorems on kernel (or null space):
  - If **x** and **y** are in Null(**A**), then so are **x** + **y** and $\alpha\mathbf{x}$
  - If **Ax** = **b** and $\mathbf{y} \in \text{Ker}(\mathbf{A})$, then **A**(**x**+**y**) = **b**
  - If **Ax** = **b** and **Ay** = **b**, then **x** − **y** is in Ker(**A**)
  - If $\mathbf{A} \sim \mathbf{B}$, then Ker(**A**) = Ker(**B**)

- A given matrix has one and only one reduced row echelon form

- The rank of a matrix is the number of rows that have pivot positions

- If **A** is an $m \times n$ matrix, then
  $$\text{Rank}(\mathbf{A}) \leq \min\{m, n\}$$

- One can solve the equations **Ax** = **b** and **Ay** = **c** by using the augmented matrix $[\mathbf{A} \mid \mathbf{b}\,\mathbf{c}]$

- If **A** is an $m \times n$ matrix, then the number of nonzero rows in its reduced row echelon form is at most n

- If **A** is an $m \times n$ matrix, then **A** has at most n pivot positions

- If **A** is $m \times n$ and **B** is $n \times q$, then
  $$\mathbf{AB} = [\mathbf{Ab}_1\ \mathbf{Ab}_2\ \cdots, \mathbf{Ab}_q]$$

KEY CONCEPTS 1.3

Homogeneous systems, trivial solutions and nontrivial solutions, kernel of a matrix, null space of a matrix, row-equivalent matrices, rank, pivot position, unicity of the reduced row echelon form, pivot positions in a matrix, upper bounds on the rank of a matrix, some equivalent properties of matrices, matrix–matrix multiplication, an inconsistent system and its reduced row echelon form, indexed set, linear independence, linear dependence, chemical application

GENERAL EXERCISES 1.3

1. Solve this system of equations by carrying out the reduction of the augmented matrix to reduced row echelon form:
$$\begin{cases} 5x_1 - 2x_2 = 9 \\ 3x_1 - x_2 = 8 \\ 11x_1 - 3x_2 = 33 \end{cases}$$

2. (Continuation.) Compute the rank of the coefficient matrix of this system.

3. Find the general solution of this system of equations and express it in the manner recommended in the text:

$$\begin{cases} x_1 + 3x_2 + 9x_3 = 6 \\ 2x_1 + 7x_2 + 3x_3 = -5 \\ x_1 + 4x_2 - 6x_3 = -11 \end{cases}$$

4. For the matrix shown here, compute its rank and find a set of vectors whose span is its kernel:

$$\begin{bmatrix} 1 & 4 & -5 & 10 \\ 3 & 1 & 7 & -3 \\ 2 & 2 & 2 & 2 \\ 1 & 3 & -3 & 7 \end{bmatrix}$$

5. Let $A = \begin{bmatrix} 7 & 2 & 0 & 49 & -37 \\ 3 & 1 & 0 & 22 & 16 \\ 4 & 2 & 2 & 31 & 20 \end{bmatrix}$

Find a pair of vectors whose span is $\text{Ker}(A)$.

6. Find a matrix whose kernel is spanned by the two vectors $\mathbf{u} = (1, 3, 2)$ and $\mathbf{v} = (-2, 0, 4)$.

7. Let $A = \begin{bmatrix} 1 & 3 & 7 \\ 2 & -4 & 1 \\ 1 & 2 & 5 \end{bmatrix}$

$B = \begin{bmatrix} 11 & 6 & 2 & -12 & 27 \\ -1 & 9 & -20 & -5 & -19 \\ -2 & -5 & 21 & 22 & 9 \end{bmatrix}$

Solve the system $AX = B$. Then do the same for the following augmented matrix:

$\begin{bmatrix} 1 & 3 & 7 & 1 & 0 & 0 \\ 2 & -4 & 1 & 0 & 1 & 0 \\ 1 & 2 & 5 & 0 & 0 & 1 \end{bmatrix}$

8. This problem is solved readily with the technique explained in Example 10. We ask: What are the vectors $\mathbf{u}$ and $\mathbf{v}$ if $(-12, 10, 20) = 3\mathbf{u} + 5\mathbf{v}$ and $(17, -8, 21) = 5\mathbf{u} - 4\mathbf{v}$?

9. Explain why the system $A\mathbf{x} = \mathbf{0}$ is always consistent.

10. Without any calculations, provide nontrivial solutions to the equation $A\mathbf{x} = \mathbf{0}$, when A is in turn each of the following matrices:

a. $\begin{bmatrix} 1 & 1 \\ 3 & 3 \\ 7 & 7 \end{bmatrix}$ **b.** $\begin{bmatrix} 1 & -2 \\ 3 & -6 \\ 7 & -14 \end{bmatrix}$

c. $\begin{bmatrix} 1 & 4 & 6 \\ 2 & 5 & 9 \\ 2 & 9 & 13 \end{bmatrix}$

11. (Continuation.) For the three matrices, find nontrivial vectors in their kernels. No calculations are necessary.

12. Explain why the following two matrices are *not* row equivalent to each other by showing that the corresponding systems of homogeneous equations have different solutions (assume $a \neq b$):

$A = \begin{bmatrix} 1 & a & 0 \\ 0 & 0 & 1 \end{bmatrix}$, $B = \begin{bmatrix} 1 & b & 0 \\ 0 & 0 & 1 \end{bmatrix}$

13. Establish that if the set $\{\mathbf{v}_1, \mathbf{v}_2, \mathbf{v}_3\}$ is linearly independent, then so is $\{\mathbf{v}_1 + \mathbf{v}_2, \mathbf{v}_2 + \mathbf{v}_3, \mathbf{v}_3 + \mathbf{v}_1\}$.

14. Find two vectors whose span is the kernel of the matrix $\begin{bmatrix} 7 & 3 & 5 & 37 \\ 2 & 1 & 1 & 11 \end{bmatrix}$

15. Efficiently solve these two systems:

$\begin{bmatrix} 1 & 3 \\ 2 & 1 \end{bmatrix} \begin{bmatrix} x_1 \\ x_2 \end{bmatrix} = \begin{bmatrix} 14 \\ 8 \end{bmatrix}$

$\begin{bmatrix} 1 & 3 \\ 2 & 1 \end{bmatrix} \begin{bmatrix} y_1 \\ y_2 \end{bmatrix} = \begin{bmatrix} -4 \\ 7 \end{bmatrix}$

16. Let $A = \begin{bmatrix} x & 1 & 0 \\ -9 & y & 7 \\ -1 & 4 & z \end{bmatrix}$

If the kernel of A contains the vector $\begin{bmatrix} 1 \\ -2 \\ 3 \end{bmatrix}$,

what are x, y, and z?

17. Determine whether this is true:

$\begin{bmatrix} 1 & 3 & 0 & 5 & 0 & 4 \\ 2 & 6 & 1 & 8 & 0 & 5 \\ 2 & 6 & 2 & 6 & 1 & 9 \end{bmatrix} \sim \begin{bmatrix} 1 & 3 & 0 & 5 & 0 & 4 \\ 0 & 0 & 1 & -2 & 0 & -3 \\ 0 & 0 & 0 & 0 & 1 & 7 \end{bmatrix}$

18. Determine whether this set of vectors is linearly dependent:
$\{(3, 2, 7), (4, 1, -3), (6, -1, -23)\}$

19. Explain why a system of equations $A\mathbf{x} = \mathbf{b}$ has either no solution, exactly one solution, or infinitely many solutions. Explain how these three outcomes are easily distinguished after the row reduction of the augmented matrix has been carried out.

20. Justify, without appealing to the reduced row echelon form, the assertion that if a system of equations $\mathbf{Ax} = \mathbf{b}$ has two solutions then it has infinitely many solutions.

21. (Continuation.) Consider a system of equations $\mathbf{Ax} = \mathbf{b}$, and assume that it has two solutions, say $\mathbf{u}$ and $\mathbf{v}$. Explain why, for all real values of t, $t\mathbf{v} + (1 - t)\mathbf{u}$ is also a solution. Establish then that the solution set of the system contains a line.

22. Establish the validity of Theorems 1–4.

23. Explain why two matrices that are row equivalent to each other must have the same rank.

24. Explain why a set of n vectors in $\mathbb{R}^m$ is linearly independent if and only if the matrix having these vectors as its columns has rank n.

25. Establish directly that if $\mathbf{Ax} = \mathbf{0}$ for some nonzero vector $\mathbf{x}$, then the rank of $\mathbf{A}$ is less than n. (Here $\mathbf{A}$ is $m \times n$.)

26. Let the matrix $\mathbf{A}$ be in reduced row echelon form. Explain why each nonzero row contains a pivot element. Is the same assertion true for the columns of $\mathbf{A}$?

27. Explain why the rank of $\mathbf{A}$ is the number of pivot positions in $\mathbf{A}$.

28. Establish that the rank of $\mathbf{A}$ is the number of columns that contain pivot positions.

29. Consider a consistent system of equations $\mathbf{Ax} = \mathbf{b}$, in which $\mathbf{A}$ is $m \times n$ and $m < n$.

Explain why the system must have many solutions.

30. Compute the ranks of these matrices:

a.
$$\begin{bmatrix} 5 & 2 & 0 & 18 \\ 2 & 1 & 0 & 8 \\ 3 & 3 & -1 & 15 \\ 1 & 0 & 0 & 2 \end{bmatrix}$$

b.
$$\begin{bmatrix} 4 & 3 & 7 & 5 & 4 \\ 0 & 2 & 2 & -1 & 6 \\ 0 & 0 & 5 & 2 & 3 \\ 0 & 0 & 0 & 1 & 5 \\ 0 & 0 & 0 & 0 & 3 \end{bmatrix}$$

c.
$$\begin{bmatrix} 0 & 3 & 7 & 5 & 0 \\ 2 & 2 & -1 & 6 & 0 \\ 5 & 2 & 3 & 0 & 0 \\ 1 & 5 & 0 & 0 & 0 \\ 3 & 0 & 0 & 0 & 0 \end{bmatrix}$$

31. Let $\mathbf{A}$ be an $m \times n$ matrix whose kernel is $\mathbf{0}$; that is, the only solution of the equation $\mathbf{Ax} = \mathbf{0}$ is $\mathbf{x} = \mathbf{0}$. What is the rank of $\mathbf{A}$?

32. a. The linear system $x - y - z = 0$, $x + y - z = 0$ has infinitely many solutions. These are the points on the line of intersection of the two given planes in $\mathbb{R}^3$. Find the equation for this line.
b. Find a simple description of the set of points satisfying these three equations: $y + 2z = 0$, $2x - y + 8z = 0$, $x - y + 3z = 0$.

33. Let $\mathbf{A} = \begin{bmatrix} 0 & 0 & 0 \\ 0 & 4 & 0 \\ 0 & 2 & 3 \end{bmatrix}$

What are the pivot positions in this matrix?

34. If $\mathbf{A}$ is an $m \times n$ matrix and $m \geq n$, what is the maximum number of pivot positions in $\mathbf{A}$? Explain. What is the maximum number of pivotal rows that $\mathbf{A}$ can have?

35. Let $\mathbf{A}$ be an $m \times n$ matrix, where $m < n$. What is the maximum number of pivot positions in $\mathbf{A}$? What is the least number of nonpivotal columns in $\mathbf{A}$? What is the least number of free variables in solving the equation $\mathbf{Ax} = \mathbf{0}$?

36. (Continuation.) Adopt the hypotheses on **A** as in the preceding question. Explain why the equation $\mathbf{Ax} = \mathbf{0}$ has a nontrivial solution. Explain why the columns of **A** form a linearly dependent set of vectors.

37. Let $\mathbf{C} = \begin{bmatrix} 1 & 0 & 0 & 3 \\ 0 & 1 & 0 & 2 \\ 0 & 0 & 1 & 7 \end{bmatrix}$

If each of two matrices **A** and **B** is row equivalent to **C**, does it follow that $\mathbf{A} = \mathbf{B}$?

38. If two matrices **A** and **B** are *not* row equivalent to each other, can they have the same reduced row echelon form? Explain.

39. Let $\mathbf{u} = (1, 3, 2)$, $\mathbf{v} = (2, -1, 4)$, and $\mathbf{w} = (-3, 26, -6)$. Determine whether the set $\{\mathbf{u}, \mathbf{v}, \mathbf{w}\}$ is linearly independent.

40. Consider this infinite sequence of matrices:

$$\mathbf{A}_1 = [1], \quad \mathbf{A}_2 = \begin{bmatrix} 1 & 2 \\ 3 & 4 \end{bmatrix}$$

$$\mathbf{A}_3 = \begin{bmatrix} 1 & 2 & 3 \\ 4 & 5 & 6 \\ 7 & 8 & 9 \end{bmatrix}$$

$$\mathbf{A}_4 = \begin{bmatrix} 1 & 2 & 3 & 4 \\ 5 & 6 & 7 & 8 \\ 9 & 10 & 11 & 12 \\ 13 & 14 & 15 & 16 \end{bmatrix} \text{ and so on}$$

(We have shown only the first four of them.) Find the ranks of all of them.

41. If the rank of an augmented matrix $[\mathbf{A} \mid \mathbf{b}]$ is greater than the rank of **A**, what conclusion can be drawn? Is there an implication in both directions?

42. What are the ranks of the matrices in this infinite sequence?

$$\mathbf{A}_1 = [1], \ \mathbf{A}_2 = \begin{bmatrix} 1 & 2 \\ 4 & 3 \end{bmatrix}, \ \mathbf{A}_3 = \begin{bmatrix} 1 & 2 & 3 \\ 6 & 5 & 4 \\ 7 & 8 & 9 \end{bmatrix}$$

$$\mathbf{A}_4 = \begin{bmatrix} 1 & 2 & 3 & 4 \\ 8 & 7 & 6 & 5 \\ 9 & 10 & 11 & 12 \\ 16 & 15 & 14 & 13 \end{bmatrix} \text{ and so on.}$$

43. Describe all the 3×4 matrices of rank 1.

44. Explain why, for any system of linear equations, the rank of the augmented matrix is at least as great as the rank of the coefficient matrix.

45. Let $n \geq 3$, and create an $n \times n$ matrix **A** by defining $\mathbf{A}_{ij} = \alpha i + \beta j + \gamma$, where α, β, and γ are three arbitrary positive numbers. What is the rank of **A**?

46. Define a family of functions f_n by the equation $f_n(x) = 1$ when $x \geq n$ and $f_n(x) = 0$ if $x < n$. Is this family linearly independent? (Here $n = 0, 1, 2$ and so on.)

47. If $\{\mathbf{v}_1, \mathbf{v}_2\}$ is a linearly independent pair of vectors, is the same true for $\{\mathbf{v}_1, \mathbf{v}_2 + \lambda \mathbf{v}_1\}$ when λ is an arbitrary constant?

48. If $\{\mathbf{v}_1, \mathbf{v}_2 + \lambda \mathbf{v}_1\}$ is linearly independent for some nonzero scalar λ, does it follow that $\{\mathbf{v}_1, \mathbf{v}_2\}$ is linearly independent?

49. If $\{\mathbf{v}_1, \mathbf{v}_2 + \lambda \mathbf{v}_1\}$ is linearly dependent for some nonzero scalar λ, does it follow that $\{\mathbf{v}_1, \mathbf{v}_2\}$ is linearly dependent?

50. Balance this chemical reaction:

$$AgNO_3 + NaCl \rightarrow AgCl + NaNO_3$$

51. Balance this chemical reaction:

$$H_2 + NO_2 \rightarrow NH_3 + H_2O$$

52. Balance this hypothetical chemical equation:

$$NHCO_3 + HC \rightarrow NC + H_2O + CO_2$$

53. Consider $\begin{bmatrix} 1 & 3 & 4 \\ 2 & 5 & 1 \\ 3 & 6 & h \end{bmatrix}$

For what value of h does this matrix have a nontrivial kernel?

54. Find the general solution of this system and display it in the recommended form:

$$\begin{cases} x_1 + 2x_2 + x_3 = 0 \\ 3x_1 + 4x_2 - 3x_3 = 0 \end{cases}$$

55. Consider the vectors $\mathbf{u}_1 = (1, 3, 2)$, $\mathbf{u}_2 = (-2, 1, 4)$, and $\mathbf{u}_3 = (8, 3, -8)$. Taken alone, each of these vectors is linearly independent (which means in this case that each vector is nonzero). Hence, the English language allows us to say that they are linearly independent. Reconcile this conclusion with the easily verified fact that $2\mathbf{u}_1 - 3\mathbf{u}_2 - \mathbf{u}_3 = \mathbf{0}$. What is the remedy for this apparent inconsistency?

56. Consider three vectors in $\mathbb{R}^2$: $\mathbf{u} = (u_1, u_2)$, $\mathbf{v} = (v_1, v_2)$, and $\mathbf{w} = (w_1, w_2)$. Show that the following six conditions are incompatible: $u_1 v_1 + u_2 v_2 = 0$, $v_1 w_1 + v_2 w_2 = 0$, $w_1 u_1 + w_2 v_2 = 0$, $u_1^2 + u_2^2 > 0$, $v_1^2 + v_2^2 > 0$, $w_1^2 + w_2^2 > 0$.

57. Consider $\begin{bmatrix} 1 & 3 & 2 \\ 0 & 1 & 4 \\ 1 & 4 & 6 \end{bmatrix}$

Find the rank of this matrix. What can be said of the equation $\mathbf{Ax} = \mathbf{0}$?

58. What conditions must be placed on $\mathbf{A}$ in order that the system $\mathbf{Ax} = \mathbf{0}$ be consistent?

59. Consider the system

$$\begin{cases} x_1 + 3x_2 + x_3 = 6 \\ 2x_1 + 6x_2 + 3x_3 = 16 \\ 3x_1 + 9x_2 + 4x_3 = 22 \end{cases}$$

Show the original augmented matrix. Obtain its reduced row echelon form.

Give the rank of the coefficient matrix. Describe the solution of the system. Identify the independent (free) variables. Find all solutions of $\mathbf{Ax} = \mathbf{0}$ when $\mathbf{A}$ is the coefficient matrix.

60. Let $\mathbf{v}_1 = (1, 5, -2, 4)$, $\mathbf{v}_2 = (-1, 2, 2, -4)$, $\mathbf{v}_3 = (2, 12, -1, 12)$, $\mathbf{v}_4 = (0, 1, 1, 2)$, and $\mathbf{v}_5 = (3, -1, 4, -2)$. Is this set of vectors linearly independent? Explain fully.

61. Explain that if the equation $\mathbf{AX} = \mathbf{B}$ has more than one solution, then the equation $\mathbf{AX} = \mathbf{0}$ has a nontrivial solution. Establish that this equation has infinitely many solutions. Here $\mathbf{A}$ is an $m \times n$ matrix, $\mathbf{X}$ is an $n \times q$ matrix, and $\mathbf{B}$ is an $m \times q$ matrix.

62. If possible, express $(7, -1, 0)$ as a linear combination of $(1, 3, 2)$ and $(4, 1, 1)$. Explain how you solve this.

63. Suppose that the equation $\mathbf{Ax} = \mathbf{b}$ has more than one solution. Explain why the equation $\mathbf{Ax} = \mathbf{0}$ has infinitely many solutions.

64. For each n there is a matrix $\mathbf{A}_n$ following this pattern:

$$\mathbf{A}_1 = [1], \mathbf{A}_2 = \begin{bmatrix} 1 & 3 \\ 2 & 4 \end{bmatrix}, \mathbf{A}_3 = \begin{bmatrix} 1 & 4 & 7 \\ 2 & 5 & 8 \\ 3 & 6 & 9 \end{bmatrix}$$

$$\mathbf{A}_4 = \begin{bmatrix} 1 & 5 & 9 & 13 \\ 2 & 6 & 10 & 14 \\ 3 & 7 & 11 & 15 \\ 4 & 8 & 12 & 16 \end{bmatrix} \text{ and so on.}$$

What are the ranks of these matrices?

65. Consider the system

$$\begin{cases} 4x_1 + 12x_2 + 6x_3 = 32 \\ 3x_1 + 9x_2 + 4x_3 = 22 \\ 4x_1 + 12x_2 + 4x_3 = 24 \end{cases}$$

Show the accompanying augmented matrix. Obtain the reduced row echelon form. Give the rank of the coefficient matrix. Describe the solutions of the system. Identify the independent (free) variables, if there are any.

66. (Continuation.) Find all solutions of $\mathbf{Ax} = \mathbf{0}$ when $\mathbf{A}$ is the coefficient matrix of the preceding problem.

67. Give a simple example where a system has a free variable and yet *no* solutions.

68. If $\mathrm{Rank}(\mathbf{A}) = k$, what is $\mathrm{Rank}([\mathbf{A} \mid \mathbf{b}])$?

69. Can our theory of linear equations be built up with just this one row operation: $\mathbf{r}_i \leftarrow \alpha \mathbf{r}_i + \beta \mathbf{r}_j$ for $i \neq j$ and for nonzero scalars α and β?

70. Establish that every matrix of rank r is a sum of r matrices of rank 1.

71. Establish the validity of
   a. Theorem 8.
   b. Theorem 12.
   c. Theorem 13.

72. Consider $\{p_0, p_1, p_2, p_3\}$. Determine whether this set of polynomials is linearly independent or linearly dependent. The definitions are $p_0(t) = 1$, $p_1(t) = t$, $p_2(t) = 4 - t$, $p_3(t) = t^3$.

73. Let $f(t) = \sin t$ and $g(t) = \cos t$. Determine whether the pair $\{f, g\}$ is linearly dependent or linearly independent.

74. Let $f(t) = 1$, $g(t) = \cos 2t$, and $h(t) = \sin^2 t$. Determine whether the set

$\{f, g, h\}$ is linearly dependent or independent. The domain of the functions is taken to be $\mathbb{R}$.

75. Test each of these three sets of functions $\{u_1, u_2, u_3\}$ for linear dependence or linear independence:
   a. $u_1(t) = 1$, $u_2(t) = \sin t$, $u_3(t) = \cos t$
   b. $u_1(t) = 1$, $u_2(t) = \sin^2 t$, $u_3(t) = \cos^2 t$
   c. $u_1(t) = \cos 2t$, $u_2(t) = \sin^2 t$, $u_3(t) = \cos^2 t$

76. Verify that there exists no 2×2, noninvertible, nonsymmetric matrix $\mathbf{A}$ such that $\mathrm{Ker}(\mathbf{A}) = \mathrm{Ker}(\mathbf{A}^T)$.

77. If the rank of an $m \times n$ matrix is less than m, can we conclude that the rows form a linearly dependent set? (The term *rank* is defined on p. 63.)

78. Establish that if a set of at least two vectors is linearly dependent, then one element of the set is a linear combination of the others.

79. Argue that a pair of vectors is linearly dependent if and only if one of the vectors is a multiple of the other.

80. Explain why a pair of vectors is linearly independent if the two vectors are not colinear with the $\mathbf{0}$-vector.

81. Explain why the rank of $\mathbf{A}$ and the rank of $[\mathbf{A} \mid \mathbf{b}]$ can differ by at most 1. Here $\mathbf{b}$ is a column vector.

82. A set is linearly independent if and only if its indexed set is linearly independent. Explain why or find a counterexample.

COMPUTER EXERCISES 1.3

1. Consider the following scenario. An unidentified object is observed in the night sky during a period of several days, and accurate coordinates of this object have been made available by astronomers. The orbit of such an object should be a *conic section*: circle, ellipse, hyperbola, parabola, or a straight line. This means that if the locations are plotted on a plane, one of these types of conic sections should be evident. In particular, if the orbit is an ellipse, the object will return after a certain number of years, whereas if the orbit is a parabola or hyperbola it will not return. An arbitrary conic section in (x, y) coordinates should have the form $ax^2 + bxy + cy^2 + dx + ey = 1$.

a. Find the values of a, b, c, d, and e from these five points on the orbit: $(1.8, 3.1)$, $(1.4, 1.9)$, $(2.5, 1.2)$, $(4.0, 1.6)$, $(4.8, 2.5)$.

b. Determine whether the orbit is elliptical, parabolic, or hyperbolic.

c. Find a formula by which the y-values can be computed from the x-values in this orbit. Remember, there may be two y-values for a given x, since there is a quadratic equation to be solved.

d. If you have suitable facilities at your disposal, obtain a plot of the orbit.

2. Using floating-point arithmetic (without scaling), compute the reduced row echelon form for the linear systems represented by this augmented matrix:

$$\left[\begin{array}{cc|c} 0.835 & 0.667 & 0.168 \\ 0.333 & 0.266 & 0.067 \end{array} \right]$$

a. Use single-precision arithmetic.
b. Use double-precision arithmetic.

Explain the results.

3. Repeat Computer Exercise 2 using this perturbed augmented matrix:

$$\left[\begin{array}{cc|c} 0.835 & 0.667 & 0.1669995 \\ 0.333 & 0.266 & 0.0666010 \end{array} \right]$$

This illustrates that the solution of an **ill-conditioned** linear system may undergo a radical change with even a slight pertubation of the data. See Meyer [2000].

4. Repeat Computer Exercise 2 using this augmented matrix:

$$\left[\begin{array}{ccc|c} 0.8350 & 0.6670 & 0.5000 & 0.1680 \\ 0.3330 & 0.2660 & 0.1994 & 0.0670 \\ 1.6700 & 1.3340 & 1.1000 & 0.4360 \end{array} \right]$$

5. Repeat Computer Exercise 3 using this perturbed augmented matrix:

$$\left[\begin{array}{ccc|c} 0.8350 & 0.6670 & 0.5000 & 0.1670 \\ 0.3330 & 0.2660 & 0.1994 & 0.0666 \\ 1.6700 & 1.3340 & 1.1000 & 0.4360 \end{array} \right]$$

CHAPTER | **TWO**

Vector Spaces

2.1 EUCLIDEAN VECTOR SPACES

> *I hear and I forget. I see and I remember. I do and I understand.*
>
> CHINESE PROVERB
>
> *Mathematics is not a spectator sport.*
>
> —ANONYMOUS

At a later point in this book, we shall consider and see many examples of the abstract concept of a vector space. Here, however, we introduce an important example of a vector space—one that can serve as a model for all the others! In addition, we discuss applications involving electrical networks subject to Ohm's law and Kirchhoff's laws, or other types of networks such as traffic grids.

n-Tuples and Vectors

First, let us review some terminology from previous discussions, especially Section 1.2. We fix a positive integer n. An n-**tuple** of real numbers is a list of n real numbers put into an array like this $(x_1, x_2, \ldots, x_n)$. This is not just a simple set because the order of the entries is important. Also, in an n-tuple, there may be repeated entries such as in $(7, 4, 5, 7)$. In the special case $n = 2$, a typical array of this sort would be $(1.732, -5.002)$. As everyone knows, a pair of real numbers such as this can be used to represent a *point* in two-dimensional space. In the same way, an array $(3.44, 2.89, -3.41)$ can be used to specify a point in three-space. In general, an n-tuple can represent a **point** in n-dimensional space, and this space (consisting of all n-tuples) will be denoted by $\mathbb{R}^n$. Thus, $\mathbb{R}^2$ is the two-dimensional plane, and $\mathbb{R}^3$ is the familiar three-space. We are using the word *dimension* in an intuitive sense; later, it will receive a careful definition, in harmony with our intuition. These n-tuples will usually be called **vectors**. The individual numbers x_i that make up a vector are called the **components** of that vector. We often use boldface type to signify a vector, and use the same letter with subscripts to denote the components of that vector. For example, $\mathbf{x} = (x_1, x_2, \ldots, x_n)$. Alternative notations are $\mathbf{x} = [x_1, x_2, \ldots, x_n]$ and $\mathbf{x} = [x_1, x_2, \ldots, x_n]^T$. When the vector is being interpreted as a matrix, either $1 \times n$ or $n \times 1$, these notational distinctions become important.

For the moment, $\mathbb{R}^n$ is nothing but a set. As such, the equality of two members means that they are one and the same entity. That is, if $\mathbf{x} = (x_1, x_2, \ldots, x_n)$ and $\mathbf{y} = (y_1, y_2, \ldots, y_n)$, we will mean by $\mathbf{x} = \mathbf{y}$ that $x_i = y_i$ for $i = 1, 2, \ldots, n$. For example, letting $n = 3$, we have $(\frac{1}{2}, \frac{5}{8}, -\frac{7}{16}) = (0.5, 0.625, -0.4375)$, whereas for $n = 2$, we offer these examples: $(61325, 48598) \neq (61325, 48589)$, $(0.67, 0.75) \neq (\frac{2}{3}, \frac{3}{4})$, and $(\pi, 0) \neq (3.14159, 0)$.

Vector Addition and Multiplication by Scalars

We progress beyond the simple concept of $\mathbb{R}^n$ being just a set by introducing two algebraic operations in $\mathbb{R}^n$ as follows. Using $\mathbf{x}$ and $\mathbf{y}$ as displayed previously, we define the sum of $\mathbf{x}$ and $\mathbf{y}$ by the equation

$$\mathbf{x} + \mathbf{y} = \left(x_1 + y_1, \, x_2 + y_2, \ldots, x_n + y_n\right)$$

This operation in $\mathbb{R}^n$ is called **vector addition**. We say that the addition of vectors is carried out **componentwise** or *component-by-component*. Notice that we have no definition for $\mathbf{x} + \mathbf{y}$ if $\mathbf{x}$ and $\mathbf{y}$ have different numbers of components. Thus, if $\mathbf{x} \in \mathbb{R}^n$, $\mathbf{y} \in \mathbb{R}^m$, and $n \neq m$, then $\mathbf{x} + \mathbf{y}$ is *not* defined.

If $\mathbf{x}$ is as before and α is a real number, $\alpha\mathbf{x}$ is defined by the equation

$$\alpha\mathbf{x} = \left(\alpha x_1, \, \alpha x_2, \ldots, \alpha x_n\right)$$

This operation is called **multiplication by scalars**. Again, we say that the multiplication of a vector by a scalar is done *component-by-component*. As numerical examples, we offer:

$$\left(2, \, 3, \, -4, \, 7\right) + \left(3, \, -2, \, 6, \, 11\right) = \left(5, \, 1, \, 2, \, 18\right)$$
$$5\left(2, \, 3, \, -4, \, 7\right) = \left(10, \, 15, \, -20, \, 35\right)$$

The term **scalar** means just a real number. Later there will be situations where we allow vectors to have complex numbers as components, and we define multiplication of a vector by a complex number. In that case, *scalar* will mean any complex number, that is, a number of the form $\alpha + i\beta$, where α and β are real numbers and $i = \sqrt{-1}$.

Properties of $\mathbb{R}^n$ as a Vector Space

Vector spaces are characterized by two algebraic operations: vector addition and multiplication of a vector by a scalar (usually a real number).

DEFINITION

*The space $\mathbb{R}^n$ consists of all n-tuples of real numbers. In $\mathbb{R}^n$, there are two algebraic operations: vector addition and multiplication of vectors by scalars, as described previously. This space is called the n-**dimensional Euclidean vector space** or **real n-space**.*

Here we shall record a list of properties that the space $\mathbb{R}^n$ has when the preceding algebraic operations are adopted. These are the properties that will eventually justify our calling $\mathbb{R}^n$ a **vector space**. Our list is nearly a minimal one and all other useful properties of $\mathbb{R}^n$ follow logically from these.

BASIC PROPERTIES OF THE VECTOR SPACES $\mathbb{R}^n$

1. The sum of any two elements of $\mathbb{R}^n$ is well defined and is a member of $\mathbb{R}^n$.
2. If **x** and **y** belong to $\mathbb{R}^n$, then $\mathbf{x} + \mathbf{y} = \mathbf{y} + \mathbf{x}$.
3. If **x**, **y**, and **z** belong to $\mathbb{R}^n$, then $(\mathbf{x} + \mathbf{y}) + \mathbf{z} = \mathbf{x} + (\mathbf{y} + \mathbf{z})$.
4. There is an element **0** in $\mathbb{R}^n$ such that $\mathbf{x} + \mathbf{0} = \mathbf{x}$ for all **x** in $\mathbb{R}^n$.
5. Corresponding to any **x** in $\mathbb{R}^n$, there is an element $-\mathbf{x}$ having the property $-\mathbf{x} + \mathbf{x} = \mathbf{0}$.
6. If **x** is in $\mathbb{R}^n$ and α is a real number, then $\alpha\mathbf{x}$ (or $\alpha \cdot \mathbf{x}$) is well defined and is an element of $\mathbb{R}^n$.
7. For real numbers α and β and for any **x** in $\mathbb{R}^n$, $\alpha(\beta\mathbf{x}) = (\alpha\beta)\mathbf{x}$.
8. For any real number α and for any two elements **x** and **y** in $\mathbb{R}^n$, $\alpha(\mathbf{x} + \mathbf{y}) = \alpha\mathbf{x} + \alpha\mathbf{y}$.
9. For real numbers α and β and any **x** in $\mathbb{R}^n$, $(\alpha + \beta)\mathbf{x} = \alpha\mathbf{x} + \beta\mathbf{x}$.
10. For any **x** in $\mathbb{R}^n$, $1 \cdot \mathbf{x} = \mathbf{x}$.

These properties of $\mathbb{R}^n$ are all quickly verified. Later, when we consider abstract vector spaces, we will interpret the Properties 1–10 as *axioms* (or *postulates*) of a vector space. To justify calling some system a vector space, one will then have to verify that the axioms are true for the elements and operations in that system. It is important to have an economical set of axioms to minimize the effort of testing whether a specific example is a genuine vector space.[1]

Let us elaborate on the properties enumerated previously. Property 1 tells us that in combining two members of $\mathbb{R}^n$ by addition, we do not leave the set $\mathbb{R}^n$. This is called **closure under addition**. Likewise, there is Property 6, **closure under multiplication by real numbers**. In this context, the real numbers are called **scalars** and the elements of $\mathbb{R}^n$ are called **vectors**. Property 2 is called **commutativity of addition**. It asserts that the order of summands in a sum is immaterial. Property 3 is called **associativity of addition**. The vector **0** mentioned in Property 4 is the **zero vector** $\mathbf{0} = (0, 0, \ldots, 0)$. Note that this equation has two different zeros. One is the vector **0** in $\mathbb{R}^n$ and the others are the real number 0. Sometimes no notational distinction is made between these two zeros, thereby violating the principle that the same symbol should *not* have two *different* meanings! But we try always to use **0** for the vector and 0 for the scalar. Properties 8 and 9 are called **distributive laws**. By repeated application of Properties 2 and 3, we arrive at properties such as this:

$$\mathbf{x} + \mathbf{y} + \mathbf{z} \ = \ \mathbf{x} + \mathbf{z} + \mathbf{y} \ = \ \mathbf{z} + \mathbf{x} + \mathbf{y} \ = \ \mathbf{z} + \mathbf{y} + \mathbf{x} \ = \ \mathbf{y} + \mathbf{z} + \mathbf{x} \ = \ \mathbf{y} + \mathbf{x} + \mathbf{z}$$

We will not stop to prove general theorems of this sort. It suffices to say that in $\mathbb{R}^n$ the sum of any finite collection of vectors is the same no matter what order is chosen for carrying out the calculation.

In some situations, it will be better to write a vector in $\mathbb{R}^n$ as a vertical array. For example, if we wish to compute a linear combination of vectors, it is visually simpler to use **column vectors**:

$$3 \begin{bmatrix} 1 \\ 7 \\ 5 \end{bmatrix} - 2 \begin{bmatrix} -3 \\ 4 \\ 6 \end{bmatrix} = \begin{bmatrix} 9 \\ 13 \\ 3 \end{bmatrix}$$

More important is the convention established in Section 1.2: In multiplying a matrix and a vector, we usually want $\mathbf{Ax}$, where **x** is a column vector. If **A** is $m \times n$, then **x** should be an $n \times 1$ column vector.

---

[1] Some mathematicians would argue that the term *Euclidean vector space* should include the dot product of two vectors. Then all concepts that depend on the dot product would automatically be included. The dot product of vectors **x** and **y** in $\mathbb{R}^n$ is defined by

$$\mathbf{x} \bullet \mathbf{y} = x_1 y_1 + x_2 y_2 + \ldots + x_n y_n$$

The **length** of a vector **x** would have meaning because it is $\sqrt{\mathbf{x} \bullet \mathbf{x}}$. The notion of **orthogonality** of a pair of vectors **x** and **y** is defined by the condition $\mathbf{x} \bullet \mathbf{y} = 0$. We postpone these important matters to Section 7.1.

Linear Combinations

The term **linear combination** was defined in Section 1.2; it refers to any sum of vectors multiplied by constants, such as

$$\alpha_1 \mathbf{u}_1 + \alpha_2 \mathbf{u}_2 + \cdots + \alpha_m \mathbf{u}_m = \sum_{j=1}^{m} \alpha_j \mathbf{u}_j$$

In this equation, all the terms $\mathbf{u}_j$ are vectors in the same space, for example, in $\mathbb{R}^n$. Usually the context will reveal whether an entity $\mathbf{u}_j$ is a vector or a component u_j of a vector named $\mathbf{u}$. One could also adopt the convention that a set of vectors would be indexed with superscripts, such as $\mathbf{u}^{(1)}, \mathbf{u}^{(2)}, \ldots, \mathbf{u}^{(m)}$. Here the parentheses are required to prevent the superscripts from being interpreted as exponents! In some textbooks, vectors are distinguished by an arrow over the letter (without boldface), like this: $\vec{x}$. If a set of vectors is given in $\mathbb{R}^n$, say $\{\mathbf{x}_1, \mathbf{x}_2, \ldots, \mathbf{x}_m\}$, then the jth components of these vectors can be labeled $x_{1j}, x_{2j}, \ldots x_{mj}$. If we use this notation, then $\mathbf{x}_1 = [x_{11}, x_{12}, \ldots, x_{1n}]^T$ and $\mathbf{x}_2 = [x_{21}, x_{22}, \ldots, x_{2n}]^T$ and so on. Thus, the jth component of the vector $\mathbf{x}_i$ is x_{ij}. In this book, we use a boldface font to indicate vectors and matrices. In handwritten work, one may find it convenient to use arrows over any letter that designates a vector. Often the context will reveal which entities are scalars and which are vectors.

Span of a Set of Vectors

The term *span* was first introduced in Section 1.2, and we now explore this important concept further.

DEFINITION

The **span** *of a set of vectors is the collection of all linear combinations of vectors in the given set. If the set is S, its span is denoted by* Span(*S*).

EXAMPLE 1 Describe the span of the set consisting of these two points in $\mathbb{R}^2$: $\mathbf{e}_1 = (1, 0), \mathbf{e}_2 = (0, 1)$.

SOLUTION If we take an arbitrary linear combination of these two points (vectors), we have

$$\alpha \mathbf{e}_1 + \beta \mathbf{e}_2 = \alpha \begin{bmatrix} 1 \\ 0 \end{bmatrix} + \beta \begin{bmatrix} 0 \\ 1 \end{bmatrix} = \begin{bmatrix} \alpha \\ \beta \end{bmatrix}$$

Obviously, we get all vectors in $\mathbb{R}^2$. The span of the given set is all of $\mathbb{R}^2$. The vectors $\mathbf{e}_1$ and $\mathbf{e}_2$ are often called the standard unit vectors in $\mathbb{R}^2$. ∎

EXAMPLE 2 Is the vector $(4, 5, 1)$ in the span of the set consisting of $(3, 5, -4)$, $(2, 1, -5)$, and $(-2, 1, 3)$?

SOLUTION This question asks whether there are three numbers x_1, x_2, x_3 such that

$$x_1 \begin{bmatrix} 3 \\ 5 \\ -4 \end{bmatrix} + x_2 \begin{bmatrix} 2 \\ 1 \\ -5 \end{bmatrix} + x_3 \begin{bmatrix} -2 \\ 1 \\ 3 \end{bmatrix} = \begin{bmatrix} 4 \\ 5 \\ 1 \end{bmatrix}$$

This can be written as a set of three linear equations in three unknowns or as a matrix–vector system:

$$\begin{cases} 3x_1 + 2x_2 - 2x_3 = 4 \\ 5x_1 + 1x_2 + 1x_3 = 5 \\ -4x_1 - 5x_2 + 3x_3 = 1 \end{cases} \quad \text{or} \quad \begin{bmatrix} 3 & 2 & -2 \\ 5 & 1 & 1 \\ -4 & -5 & 3 \end{bmatrix} \begin{bmatrix} x_1 \\ x_2 \\ x_3 \end{bmatrix} = \begin{bmatrix} 4 \\ 5 \\ 1 \end{bmatrix}$$

This system has the solution $\mathbf{x} = (2, -3, -2)$. Therefore, the answer to the posed question is *yes*! ∎

EXAMPLE 3 Is the vector $(3, 7, 23, 41)$ in the span of this set of four vectors: $\mathbf{u}_1 = (13, -41, 12, 3)$, $\mathbf{u}_2 = (2, -4, 0, 0)$, $\mathbf{u}_3 = (9, 11, 17, 0)$, $\mathbf{u}_4 = (8, 0, 0, 0)$?

SOLUTION We are asking whether this system of equations has a solution:

$$x_1 \begin{bmatrix} 13 \\ -41 \\ 12 \\ 3 \end{bmatrix} + x_2 \begin{bmatrix} 2 \\ -4 \\ 0 \\ 0 \end{bmatrix} + x_3 \begin{bmatrix} 9 \\ 11 \\ 17 \\ 0 \end{bmatrix} + x_4 \begin{bmatrix} 8 \\ 0 \\ 0 \\ 0 \end{bmatrix} = \begin{bmatrix} 3 \\ 7 \\ 23 \\ 41 \end{bmatrix}$$

The augmented matrix for the system is

$$\begin{bmatrix} 13 & 2 & 9 & 8 & | & 3 \\ -41 & -4 & 11 & 0 & | & 7 \\ 12 & 0 & 17 & 0 & | & 23 \\ 3 & 0 & 0 & 0 & | & 41 \end{bmatrix}$$

From the fourth equation, we can get x_1 immediately. With x_1 now known, the third equation gives us x_3. With x_1 and x_3 now known, the second equation gives us x_2. Finally, the first equation will give us x_4. The answer to the question is *yes*! Here we need not do any complicated arithmetic to arrive at this conclusion. ∎

EXAMPLE 4 What is the span of the set of vectors $\{\mathbf{u}_1, \mathbf{u}_2, \mathbf{u}_3, \mathbf{u}_4\}$ as described in Example 3?

SOLUTION We can replace the vector $(3, 7, 23, 41)$ by any vector in $\mathbb{R}^4$. The argument given in Example 3 can be repeated and applied to any such vector. Hence, no matter what vector in $\mathbb{R}^4$ is chosen, the answer is that *it* is in the span of the given set. Hence, the span of this set of four vectors is the entire Euclidean space $\mathbb{R}^4$. ∎

Geometric Interpretation of Vectors

The geometric interpretation of vectors in $\mathbb{R}^2$ and $\mathbb{R}^3$ will aid one in visualizing vector manipulations in higher dimensional spaces. The simplest case of this is multiplication of a vector by a scalar. Taking a typical point, say $(2, 3)$ in $\mathbb{R}^2$, plot it as a point in the plane. The multiples $3(2, 3) = (6, 9)$, $\frac{1}{2}(2, 3) = (1, 1.5)$, $-1(2, 3) = (-2, -3)$ are easily plotted and are seen to lie on a line through $\mathbf{0}$. This is true in $\mathbb{R}^n$ as well. We observe that the span of a set consisting of just one nonzero vector is the line through $\mathbf{0}$ and the point in question. (Two points determine a line.)

Vector addition, too, has a simple geometric interpretation. When we add two vectors, they and their sum will lie in a plane passing through $\mathbf{0}$. Thus, it is a two-dimensional construction. See Figure 2.1, which shows the addition of $(7, 3)$ and $(2, 5)$ yielding $(9, 8)$. With the help of some congruent triangles, one can see that the sum is the fourth vertex of a parallelogram having vertices $\mathbf{0}$, $(7, 3)$, and $(2, 5)$. This description of the sum will be valid if the two prescribed vectors are in a higher dimensional space. In $\mathbb{R}^3$, for

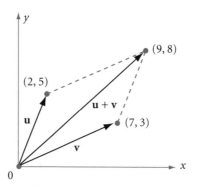

FIGURE 2.1   Vector addition: $(7, 3) + (2, 5) = (9, 8)$.

example, the sum of two vectors (neither a multiple of the other) lies in having the plane that goes through zero and the two given vectors. Similarly, we conclude that the span of a set of two vectors (neither a multiple of the other) is a plane containing the origin and the two given vectors. (Three points determine a plane.)

If we select two vectors **u** and **v** in $\mathbb{R}^n$, and if $\mathbf{v} \neq \mathbf{0}$, then we can describe a line through the point **u** having the direction **v**. Start at the point **u** and add arbitrary scalar multiples of **v**. We define the **line** L as this set of vectors:

$$L = \{\mathbf{u} + t\mathbf{v} : t \in \mathbb{R}\}$$

Of course, we get the point **u** by using the scalar $t = 0$, so the geometrical object we are constructing contains the point **u**. The added multiples of **v** produce a line through **u** parallel to **v**. Figure 2.2 shows this construction in $\mathbb{R}^2$, where we have taken $\mathbf{u} = (8, 4)$ and $\mathbf{v} = (5, -2)$. The subject of lines is taken up again in Section 2.2.

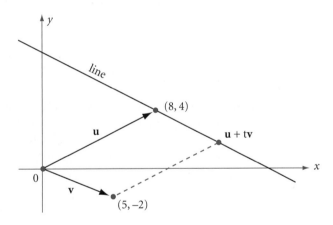

FIGURE 2.2 Line through (8, 4) parallel to (5, −2).

Application: Elementary Mechanics

Here, again, we interrupt the mathematical theory to see an application of linear algebra. This has to do with centers of mass of a system of particles or weights.

We are all familiar with some principles of balancing. For example, Figure 2.3 shows two weights of 5 kilos each in a state of balance or **equilibrium** about a fulcrum. As long as the two weights are placed equidistant from the fulcrum, the beam will be in balance.

5 kilos   5 kilos

FIGURE 2.3 Balanced weights.

The next stage in complexity is shown in Figure 2.4, where again we have equilibrium. In this configuration, the 3-kilo weight should be twice as far from the fulcrum as the 6-kilo weight. The product of the 3 kilos and the length of the *arm* from weight to fulcrum is the **moment** of the weight in that position. The moment of the other weight is the product of the 6 kilos with the length of the arm (distance from weight to fulcrum). It is the equality of these two numbers that allows us to conclude that the system is **stable**, or in balance. These matters (especially levers) were already understood by Archimedes![2]

3 kilos   6 kilos

FIGURE 2.4 More balanced weights.

Another way to interpret the *equilibrium* in Figure 2.4 is to recognize that the moments of the two weights are equal. Better yet, we can attach signs (+ or −) to the distance from the fulcrum. Think of the fulcrum as being placed at position zero on the horizontal axis. Then the 3-kilo weight will have a moment of $3x$, where x is negative, because that weight is located at a point to the left of the origin. The 6-kilo weight will have a moment of $6y$, where y is positive. The equation for a balanced situation is $3x + 6y = 0$.

---

[2] The Greek mathematician and inventor Archimedes (287–212 B.C.) was the son of an astronomer, Phidias, in the city of Syracuse. He died at the hands of the Romans during their sacking of Syracuse. One of his early works was *The Equilibriums of Planes* in two volumes. In this work, he computed the center of gravity of many plane figures and developed the theory of the lever. Archimedes declared "*Give me a place to stand, and I will move the world!*" A modern riposte to which was "*Give me a place to sit and I'll watch*" in [*Maybe He's Dead*, ed. by Mary Ann Madden [1981], response submitted by Mariel Bossert.]

For similar problems in higher dimensions, say $\mathbb{R}^n$, we use vectors to specify the locations of a set of weights. The weights themselves can be denoted by $w_1, w_2, \ldots, w_m$. (These are scalars, and the units can be kilograms, for example.) Let $\mathbf{x}_i$ be the location vector of the weight denoted by w_i. The **center of mass** for this system is defined to be the point (vector)

$$\mathbf{c} = \frac{1}{W}(w_1\mathbf{x}_1 + w_2\mathbf{x}_2 \cdots + w_m\mathbf{x}_m)$$

where W is the sum of all the weights, $W = \sum_{i=1}^{m} w_i$. If a fulcrum or support is located at the point $\mathbf{c}$, the system will be in balance. The displayed equation is, of course, a vector equation.

It is not hard to see that when the origin is moved to the point $\mathbf{c}$, the sums of the moments of the weights with respect to $\mathbf{c}$ are all 0. To carry out this calculation, we think of all the position vectors being changed from $\mathbf{x}_i$ to $\mathbf{x}_i - \mathbf{c}$. These are position vectors relative to a new *origin*, $\mathbf{c}$. The sum of the moments now is given by the vector equation

$$\sum_{i=1}^{m} w_i(\mathbf{x}_i - \mathbf{c}) = \sum_{i=1}^{m} w_i\mathbf{x}_i - \left(\sum_{i=1}^{m} w_i\right)\mathbf{c} = \mathbf{c}W - \mathbf{c}W = \mathbf{0}$$

EXAMPLE 5 Find the center of mass for three weights located at the points $(1, 3)$, $(2, -2)$, and $(3, 2)$, the weights being 5, 6, and 2 kilos, respectively. (See Figure 2.5.)

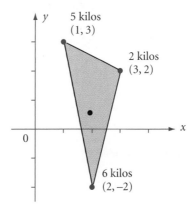

FIGURE 2.5 Three weights.

SOLUTION The sum of the weights is $W = 5 + 6 + 2 = 13$. The center of mass is therefore

$$
(1/13)\big[5(1,3) + 6(2,-2) + 2(3,2)\big] = (1/13)\big[(5,15) + (12,-12) + (6,4)\big]
$$
$$
= (23/13, 7/13) = (1.8, 0.54) \qquad \blacksquare
$$

EXAMPLE 6 Let points be prescribed: $(6,6)$, $(11,-4)$, and $(-5,-2)$. Weights of 3 kilos are placed at the first two points. What weight should be placed on the third point if the center of mass is to be at the point $(4,0)$? (See Figure 2.6.)

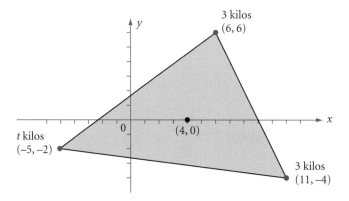

FIGURE 2.6 Center of mass.

SOLUTION Let t be the unknown weight to be placed at the point $(-5,-2)$. Then the total mass will be $W = 6 + t$. The requirement about the center of mass is

$$
\frac{1}{W}\big[3(6,6) + 3(11,-4) + t(-5,-2)\big] = (4,0)
$$

Equivalently, we have

$$
(18,18) + (33,-12) + t(-5,-2) = [6+t](4,0)
$$

After simplification, this equation becomes $(27,6) = t(9,2)$, which gives $t = 3$. So 3 kilos of weight placed at the point $(-5,-2)$ will cause the center of mass to locate at $(4,0)$. $\qquad \blacksquare$

Application: Network Problems, Traffic Flow

Another application of linear algebra arises in the study of networks with nodes and branches as shown in Figure 2.7.

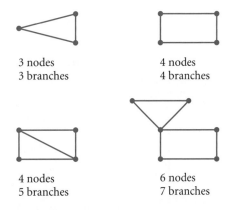

3 nodes
3 branches

4 nodes
4 branches

4 nodes
5 branches

6 nodes
7 branches

FIGURE 2.7 Nodes and branches.

FLOW AXIOM

The flow into a node equals the flow away from that node. In the network as a whole, the total flow in equals the total flow out.

In Figure 2.8, labeled as Traffic Problem I, we see a grid of one-way streets in a mythical city. The intersections (**nodes**) are labeled with numbers in small squares, and the hourly flow of traffic is indicated at various points by numbers such as 200, 400, etc. The traffic densities on the streets inside the grid are not known. They are designated by $x_1, x_2, \ldots, x_7$. Specifically, x_1 is the flow from node 1 to node 2, x_2 is the flow from node 1 to node 4, x_3 is for node 3 to 2, x_4 is for 3 to 4, x_5 is for 4 to 5, x_6 is for 5 to 6, and x_7 is for node 6 to 3. Traffic is constrained to move in the directions indicated by the arrows. The guiding principle for the analysis is that the hourly traffic into an intersection (*node*) equals the traffic out of the node. Balancing the inward and outward traffic at each node gives us six linear equations as follows:

$$\begin{cases} x_1 + x_2 &= 300 \\ x_1 + x_3 &= 300 \\ x_7 &= x_3 + x_4 \\ x_2 + x_4 &= 500 + x_5 \\ x_5 + 400 &= x_6 + 200 \\ x_6 + 300 &= x_7 \end{cases}$$

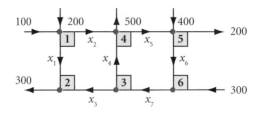

FIGURE 2.8 Traffic problem I.

The augmented matrix for the system of equations is

$$\begin{bmatrix} 1 & 1 & 0 & 0 & 0 & 0 & 0 & 300 \\ 1 & 0 & 1 & 0 & 0 & 0 & 0 & 300 \\ 0 & 0 & 1 & 1 & 0 & 0 & -1 & 0 \\ 0 & 1 & 0 & 1 & -1 & 0 & 0 & 500 \\ 0 & 0 & 0 & 0 & 1 & -1 & 0 & -200 \\ 0 & 0 & 0 & 0 & 0 & 1 & -1 & -300 \end{bmatrix}$$

The reduced row echelon form of this matrix is

$$\begin{bmatrix} 1 & 0 & 0 & -1 & 0 & 0 & 1 & 300 \\ 0 & 1 & 0 & 1 & 0 & 0 & -1 & 0 \\ 0 & 0 & 1 & 1 & 0 & 0 & -1 & 0 \\ 0 & 0 & 0 & 0 & 1 & 0 & -1 & -500 \\ 0 & 0 & 0 & 0 & 0 & 1 & -1 & -300 \\ 0 & 0 & 0 & 0 & 0 & 0 & 0 & 0 \end{bmatrix}$$

We note that the free variables are x_4 and x_7. A simple solution is obtained by setting the free variables equal to zero and solving for the remaining variables. This produces the solution

$$\mathbf{x} = [300, 0, 0, 0, -500, -300, 0]^T$$

This is not satisfactory, however, because the negative values indicate traffic flowing in the direction opposite to that permitted by the one-way streets. All the variables must be nonnegative. In order to have $x_5 \geq 0$, we must have $x_7 \geq 500$, as we see from the fourth equation. Setting $x_7 = 500$, we can let $x_4 = 200$; thus, we get the solution

$$\mathbf{x} = [0, 300, 300, 200, 0, 200, 500]^T$$

Perhaps the city planning department should be advised that the street connecting node 6 to node 3 in the sketch will have a heavy burden of traffic!

Application: Electrical Circuits

The flow of electrical current in a simple circuit is similar to the flow of traffic in a grid of one-way streets. However, different rules apply. Typically, we are interested in the currents, measured in amperes, through each branch of the circuit. The theory rests on three principles (Ohm's law[3] and Kirchhoff's laws[4]), as follows.

OHM'S LAW

The voltage drop across a resistor is the product of the resistance and the current.

KIRCHHOFF'S FIRST LAW (CURRENT LAW)

The sum of currents into any node equals the flow of currents out of the node.

KIRCHHOFF'S SECOND LAW (VOLTAGE LAW)

The total voltage produced by sources in a closed loop is the sum of the voltage drops in that loop.

---

[3] George Simon Ohm (1789–1854): His father, a locksmith and self-taught man, was able to give his son a fine mathematical and scientific education through his own teaching—a remarkable achievement. In fact, Ohm's father became angry when his son seemed to be preoccupied with student life at the university (dancing, ice skating, and playing billiards) more than his studies. He demanded that his son leave after only three semesters and take a post as a mathematics teacher. Throughout his life, Ohm had some difficulty convincing others that his mathematical approach to physics was the right one. (They were still using a nonmathematical approach.)

[4] Gustav Robert Kirchhoff (1824–1887) first announced his laws in 1845 and revised them a few years later, correcting an error. Kirchhoff's laws extended the work of Ohm and reduced the calculation of the current in each loop of an electrical network to the solution of algebraic equations. Kirchhoff's fundamental work on black body radiation was important in the development of quantum theory.

From Ohm's law, when a current I, measured in amperes, flows through a resistor having resistance R, measured in ohms, the voltage V, measured in volts, will drop by an amount IR. The equation is $V = IR$ and the slang version is *volts equals ohms times amperes.* One thinks of the resistor as having absorbed some of the energy, and the voltage lost is $V = IR$. This is intuitively correct: the greater the resistance, the greater the loss of voltage. The greater the current, the greater the loss of voltage when the current passes through the resistor. For example, in a simple circuit having a battery that produces 12 volts and a resistor of 4 ohms, we have $R = 4$ and $V = 12$. Because $V = IR$, I must be 3 amperes.

When the current flows around a loop in a designated direction, there will be a sum of voltage drops equal to the sum of the voltage sources. Across a resistor, the voltage is *positive* if the flow is in the designated direction and *negative* if the flow is opposite to the indicated direction. Across a battery, the source is *positive* if the direction of the flow is from the short side to the long side (negative to positive terminals) and otherwise it is *negative,* as shown in Figure 2.9.

FIGURE 2.9 Signs for flow across a battery.

We can easily set up and solve a linear system for the circuit with one loop shown in Figure 2.10. From Kirchhoff's second law, we obtain $(4 + 2)I = 12$. So $I = 2$ amps.

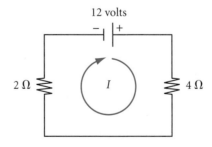

FIGURE 2.10 Circuit with one loop.

A circuit with two loops is a bit more complicated, and the analysis leads to a linear system to be solved. We set up and solve a linear system for the circuit in Figure 2.11. Kirchhoff's second law gives us these three equations:

$$\text{Voltage Drop} = \text{Voltage Source}$$
$$\text{Right Loop} : (10 + 20)I_1 + \quad (-10)I_2 = 60$$
$$\text{Left Loop} : \qquad (-10)I_1 + (5 + 10)I_2 = 50$$

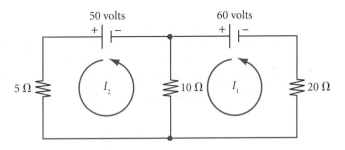

FIGURE 2.11 Circuit with two loops.

We form the augmented matrix and transform it to reduced row echelon form:

$$\begin{bmatrix} 30 & -10 & 60 \\ -10 & 15 & 50 \end{bmatrix} \sim \begin{bmatrix} 1 & 0 & 4 \\ 0 & 1 & 6 \end{bmatrix}$$

Therefore, the solution is $I_1 = 4$ and $I_2 = 6$.

In Figure 2.12 with the circuit labeled Circuit Problem I, let the current induced by the 16-volt battery be designated as I_1. Let the current in the line containing the 3-ohm resistor be named I_3. Finally, let I_2 designate

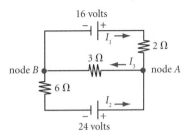

FIGURE 2.12 Circuit problem I.

the current in the part of the circuit containing the 24-volt battery and the 6-ohm resistance. Notice that we must allow for the currents to divide wherever three wires come together. The greater the current, the greater the loss of voltage when the current passes through the resistor. Recall that currents created by a battery are assumed to flow out of the side marked with a negative sign into the one with the plus sign in the battery symbol. One needs also *Kirchhoff's first law*, which states that the sum of voltage drops in a loop equals the sum of the voltage sources in that loop. Here the quantities in question have plus or minus values, and the sign depends on which direction is chosen for calculating these quantities. Apply Kirchhoff's first law to the points where the currents I_1, I_2, and I_3 meet. The result is $I_3 = I_1 + I_2$. (Current I_3 runs between Node A and Node B.) Next, apply Kirchhoff's second law to the top loop in which I_1 goes clockwise. We get $16 = (2 + 3)I_1 + 3I_2$. Doing the same for the bottom loop in which I_2 goes counterclockwise, we get $24 = (3 + 6)I_2 + 3I_1$. Putting all of these equations together, we are led to a system of three linear equations:

$$\begin{cases} I_1 + I_2 - I_3 = 0 \\ 5I_1 + 3I_2 = 16 \\ 3I_1 + 9I_2 = 24 \end{cases}$$

The augmented matrix (with its reduced row echelon form) for the system of three equations with three unknowns is

$$\left[\begin{array}{ccc|c} 1 & 1 & -1 & 0 \\ 5 & 3 & 0 & 16 \\ 3 & 9 & 0 & 24 \end{array}\right] \sim \left[\begin{array}{ccc|c} 1 & 0 & 0 & 2 \\ 0 & 1 & 0 & 2 \\ 0 & 0 & 1 & 4 \end{array}\right]$$

The solution is $I_1 = 2$, $I_2 = 2$, and $I_3 = 4$.

A simpler approach in solving the circuit problem of Figure 2.12 would be to ignore the I_3 segment and focus on only the two circuit loops involving I_1 and I_2. That leads to the linear system

$$\begin{cases} (2 + 3)I_1 + (3)I_2 = 16 \\ (3)I_1 + (3 + 6)I_2 = 24 \end{cases} \quad \text{or} \quad \begin{cases} 5I_1 + 3I_2 = 16 \\ 3I_1 + 9I_2 = 24 \end{cases}$$

The solution is $I_1 = 2$ and $I_2 = 2$, which is the same as found previously.

If negative solutions arise in solving a circuit problem, it indicates that the current in the loop is actually going in the direction opposite to the one assumed. Notice that the computation does not require the equation arising from a loop going around the outside of the entire circuit.

SUMMARY 2.1

- Vector equality: The equation $(x_1, x_2, \ldots, x_n) = (y_1, y_2, \ldots, y_n)$ means that $x_i = y_i$ for all indices i.

- Vector addition:
  $\mathbf{x} + \mathbf{y} = (x_1 + y_1, x_2 + y_2, \ldots, x_n + y_n)$

- Scalar multiplication:
  $\alpha\mathbf{x} = (\alpha x_1, \alpha x_2, \ldots, \alpha x_n)$

- Linear combination:
  $\alpha_1\mathbf{u}_1 + \alpha_2\mathbf{u}_2 + \cdots + \alpha_m\mathbf{u}_m = \sum_{j=1}^{m} \alpha_j\mathbf{u}_j$

- $\mathbb{R}^n$ is the space of all n-tuples of real numbers with two operations: vector addition and scalar multiplication. (This is real n-space or n-dimensional coordinate space.)

- Span(S): The span of a set S of vectors is the collection of all linear combinations of vectors in the set S.

- Basic properties of the vector spaces $\mathbb{R}^n$:

  - If $\mathbf{x}, \mathbf{y} \in \mathbb{R}^n$, then $\mathbf{x} + \mathbf{y} \in \mathbb{R}^n$.
    (Closure under vector addition)

  - If $\mathbf{x}, \mathbf{y} \in \mathbb{R}^n$, then $\mathbf{x} + \mathbf{y} = \mathbf{y} + \mathbf{x}$.
    (Commutativity of vector addition)

  - If $\mathbf{x}, \mathbf{y}, \mathbf{z} \in \mathbb{R}^n$, then $(\mathbf{x} + \mathbf{y}) + \mathbf{z} = \mathbf{x} + (\mathbf{y} + \mathbf{z})$.
    (Associativity of vector addition)

  - The zero vector, $\mathbf{0} \in \mathbb{R}^n$, obeys this equation: $\mathbf{x} + \mathbf{0} = \mathbf{x}$ for all $\mathbf{x}$ in $\mathbb{R}^n$.

  - For each $\mathbf{x} \in \mathbb{R}^n$, there is a vector $-\mathbf{x}$ such that $-\mathbf{x} + \mathbf{x} = \mathbf{0}$. (Negative of $\mathbf{x}$)

  - If $\mathbf{x} \in \mathbb{R}^n$ and $\alpha \in \mathbb{R}$, then $\alpha\mathbf{x} \in \mathbb{R}^n$.
    (Closure under scalar multiplication)

- For $\alpha, \beta \in \mathbb{R}$ and $\mathbf{x} \in \mathbb{R}^n$, $\alpha(\beta\mathbf{x}) = (\alpha\beta)\mathbf{x}$. (Associative law for scalars multiplying a vector)

- For $\alpha \in \mathbb{R}$ and $\mathbf{x}, \mathbf{y} \in \mathbb{R}^n$, $\alpha(\mathbf{x}+\mathbf{y}) = \alpha\mathbf{x}+\alpha\mathbf{y}$ (Distributive law for a scalar multiplying vectors)

- For $\alpha, \beta \in \mathbb{R}$ and $\mathbf{x} \in \mathbb{R}^n$, $(\alpha + \beta)\mathbf{x} = \alpha\mathbf{x} + \beta\mathbf{x}$. (Distributive law of scalars times a vector)

- For $\mathbf{x} \in \mathbb{R}^n$, $1 \cdot \mathbf{x} = \mathbf{x}$. (Multiplication by unit scalar)

- **Flow Axiom** (for networks): For each node, the flow in equals the flow out. For the network as a whole, the total flow in equals the total flow out.

- **Ohm's Law**: $V = RI$, where V is the voltage drop across a resistor, R is the resistance, and I is the current.

- **Kirchhoff's First Law**: For any node, the sum of currents in equals the sum of currents out.

- **Kirchhoff's Second Law**: In a closed loop the total voltage produced by sources equals the sum of the voltage drops.

- The line through the point $\mathbf{u}$ and parallel to a line containing $\mathbf{v}$ is $\{\mathbf{u} + t\mathbf{v} : t \in \mathbb{R}\}$.

- Center of mass for a set of weights:
  $\mathbf{c} = (w_1\mathbf{x}_1 + w_2\mathbf{x}_2 + \cdots + w_m\mathbf{x}_m)/W$, where $W = \sum_{i=1}^{m} w_i$.

KEY CONCEPTS 2.1

Vector addition, vector multiplication by scalars; real coordinate space, properties of vector space $\mathbb{R}^n$ (closure, commutativity, associativity, distributivity); linear combinations; lines and planes in real n-space; span of set; applications: balancing weights, flow of traffic, electrical circuits, Ohm's law, Kirchhoff's laws

GENERAL EXERCISES 2.1

1. Devise a test to determine whether three given points in $\mathbb{R}^2$ are **co-linear**. Apply your test to the three points $(-530, 699)$, $(510, -661)$, and $(380, -491)$.

2. Let $\mathbf{u} = (1, 0, 0)$, $\mathbf{v} = (1, 1, 0)$, and $\mathbf{w} = (1, 1, 1)$. What is the span of the set $\{\mathbf{u}, \mathbf{v}, \mathbf{w}\}$? Explain.

3. What is the missing coefficient c in this equation?

$$3 \begin{bmatrix} 4 \\ 5 \\ -2 \end{bmatrix} + c \begin{bmatrix} 6 \\ 2 \\ 1 \end{bmatrix} - 4 \begin{bmatrix} -2 \\ 7 \\ -3 \end{bmatrix}$$

$$= \begin{bmatrix} -10 \\ -23 \\ 1 \end{bmatrix}$$

4. Is the vector $(10, 27, 4, 2, -3)$ in the span of the pair $(1, 3, 7, 1, 0)$ and $(4, 11, 6, 0, 1)$?

5. If possible, express $(7, -1, 0)$ as a linear combination of $(1, 3, 2)$ and $(4, 1, 1)$. Explain how you solve this.

6. Find the center of mass of the system $\mathbf{x}_1 = (1, 3, 2)$, $\mathbf{x}_2 = (-2, 1, 0)$, $\mathbf{x}_3 = (-3, 2, 2)$, if the weights are $3, 7, 5$, respectively.

7. Let weights of 11 kilos, 2 kilos, and 7 kilos be situated at these three points: $(2, 1, 3)$, $(1, -1, 0)$, and $(3, -2, 1)$. Where should a weight of 5 kilos be situated so that the four weights would be in equilibrium?

8. For n points $\mathbf{x}_1, \mathbf{x}_2, \ldots, \mathbf{x}_n$ in $\mathbb{R}^m$, show that if all their associated masses are equal, then the center of mass equals the **centroid**, which is given by $\left(\sum_{i=1}^{n} \mathbf{x}_i \right) / n$. This formula gives the average of any finite set

of vectors. It therefore extends the notion of averaging a set of *numbers* to averaging a set of *vectors*.

9. (Continuation.) Find the centroids for
  a. $(2, -4), (1, 2), (-1, 7)$
  b. $(1, 1, -1), (0, 0, 0), (1, 1, 1), (-2, 1, 0)$

10. Trace the current around the outside loop in the circuit of Figure 2.12. What can you conclude?

11. Define four polynomials by these formulas: $p_1(x) = x^3 - 2x^2 + x + 1$, $p_2(x) = x^2 - x + 2$, $p_3(x) = 2x^3 + 3x + 4$, $p_4(x) = 3x^2 + 2x + 1$. Does the set $\{p_1, p_2, p_3, p_4\}$ span $\mathbb{P}_3$ (the space of all polynomials of degree at most 3)? Explain fully.

12. Let S and T be subsets of a vector space V and assume that $S \subseteq \text{Span}(T)$. Establish that $\text{Span}(S) \subseteq \text{Span}(T)$.

13. Devise a test to determine whether two lines in $\mathbb{R}^n$ are the same. Let the lines be described as $L_1 = \{\mathbf{p} + t\mathbf{q} \: : \: t \in \mathbb{R}\}$ and $L_2 = \{\mathbf{v} + t\mathbf{w} : t \in \mathbb{R}\}$. Use your test on this special case: $\mathbf{p} = (4, 2, 1)$, $\mathbf{q} = (-1, 3, 2)$, $\mathbf{v} = (1, 11, 7)$, and $\mathbf{w} = (3, -9, -6)$.

14. Let $\mathbf{x}$ be a nonzero vector in $\mathbb{R}^n$, and let α and β be real numbers such that $\alpha \mathbf{x} = \beta \mathbf{x}$. Explain why it follows that $\alpha = \beta$.

15. Use your knowledge of analytic geometry to see that a line in $\mathbb{R}^2$ can be described in set notation as $\{\mathbf{u} + t\mathbf{v} \: : \: t \in \mathbb{R}\}$. Here, $\mathbf{u}$ and $\mathbf{v}$ are two prescribed vectors in $\mathbb{R}^2$, and $\mathbf{v} \neq \mathbf{0}$. The parameter t is running through all the real numbers. Answer the following questions:

a. What is the **slope** of that line?
b. What is the **vertical intercept** of that line? (In the terminology of analytic geometry, that is the point where the line intersects the y-axis.)
c. Give the **slope–intercept** form of the line. That means $y = ax + b$.
d. Give the **intercepts** form of the line. That means $x/a + y/b = 1$.
e. Give the **point–slope** form of the line. That means $y - y_0 = m(x - x_0)$.

16. In $\mathbb{R}^n$, define vectors $\mathbf{e}_1 = [1, 0, 0, \ldots, 0]^T$, $\mathbf{e}_2 = [0, 1, 0, \ldots, 0]^T$, $\mathbf{e}_3 = [0, 0, 1, \ldots, 0]^T$, and so on. There are n of these special vectors. Show that every vector in $\mathbb{R}^n$ is a linear combination of these vectors, $\mathbf{e}_1, \mathbf{e}_2, \mathbf{e}_3, \ldots, \mathbf{e}_n$.

17. (Continuation.) In the preceding exercise, prove that the coefficients needed to express a given vector in terms of the special vectors $\mathbf{e}_j$ are unique. Thus, in the equation $\mathbf{u} = \sum_{j=1}^{n} \alpha_j \mathbf{e}_j$, the n-tuple $(\alpha_1, \alpha_2, \ldots, \alpha_n)$ is uniquely determined by $\mathbf{u}$.

18. Show that if $\mathbf{u}$, $\mathbf{v}$, and $\mathbf{w}$ are three points in $\mathbb{R}^2$, then for suitable real numbers α, β, and γ, not all zero, we have $\alpha\mathbf{u} + \beta\mathbf{v} + \gamma\mathbf{w} = \mathbf{0}$.

19. Let $\mathbf{x} \in \mathbb{R}^n, c \in \mathbb{R}$, and $c\mathbf{x} = \mathbf{0}$. Can we divide by $\mathbf{x}$ to conclude that $c = 0$?

20. Let $\mathbf{x}$ be the vector $(x_1, x_2, \ldots, x_n)$. Assume that $\sum_{i=1}^{n} c_i x_i = \mathbf{0}$ for every vector $\mathbf{c} = (c_1, c_2, \ldots, c_n)$. Explain why $\mathbf{x}$ must be $\mathbf{0}$. *Note*: We call the expression $\sum_{i=1}^{n} c_i x_i$ the **dot product** or **inner product** of the two vectors $\mathbf{c} = (c_1, c_2, \ldots, c_n)$ and $\mathbf{x} = (x_1, x_2, \ldots, x_n)$.

21. If we changed the definition of scalar multiplication in $\mathbb{R}^2$ by defining $\alpha(x_1, x_2) = (\alpha x_1, 0)$, would the Properties 1–10 enumerated for $\mathbb{R}^2$ still be true?

22. How does an n-tuple of real numbers $(x_1, x_2, \ldots, x_n)$ differ from a real-valued function defined on the set $\{1, 2, \ldots, n\}$?

23. Consider two vectors, $\mathbf{x} = [x_1, x_2, \ldots, x_n]^T$ and $\mathbf{y} = [y_1, y_2, \ldots, y_n]^T$. We say that $\mathbf{x} = \mathbf{y}$ if $x_i = y_i$ for all i. Explain why $\mathbf{x} \neq \mathbf{y}$ does *not* mean $x_i \neq y_i$ for all i.

24. Establish that for any set U in $\mathbb{R}^n$ (or in any vector space) we have $\mathrm{Span}\big(\mathrm{Span}(U)\big) = \mathrm{Span}(U)$.

25. Answer and explain these questions:
a. Can the span of a set be the empty set?
b. Can the span of a set contain one and only one vector?
c. If $\mathrm{Span}(S) = \mathrm{Span}(T)$, does it follow that $S = T$?
d. If $\mathrm{Span}(S) \subseteq \mathrm{Span}(T)$, does it follow that $S \subseteq T$?
e. If $S \subseteq T$, does it follow that $\mathrm{Span}(S) \subseteq \mathrm{Span}(T)$?

26. Let $\mathbf{x}$, $\mathbf{y}$, and $\mathbf{z}$ be the vertices of a triangle in $\mathbb{R}^2$. Find the formula for the center of mass when equal weights are situated at each vertex.

27. Study the motion of the center of mass when the location vectors remain fixed and all masses remain fixed except for one. In particular, what happens to the center of mass when the free weight becomes infinite?

28. Consider Traffic Problem I. Let $\mathbf{u} = (0, 300, 300, 200, 0, 200, 500)$. Is there any vector $\mathbf{x}$ that solves the set of equations and satisfies the vector inequalities $\mathbf{0} \le \mathbf{x} \le \mathbf{u}$, $\mathbf{x} \neq \mathbf{u}$?

29. Briefly, discuss the basic properties of the vector space of all $m \times n$ matrices.

30. Find the values of α so that $(3, -1, \alpha)$ is a linear combination of this set of vectors $\{(1, 3, -1), (-5, 5, 2)\}$.

31. Find the center of mass of three equal weights situated at the vertices of an equilateral triangle.

32. Set up and solve this circuit problem:

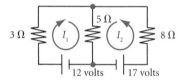

33. Find the dot product of the vectors $(-1, 2, 3, -1)$ and $(1, -3, -2, 1)$.

34. Briefly, discuss the basic properties of the vector space $\mathbb{V}$ of all directed line segments (portrayed as **arrows**).

35. Find the values of β so that $(\beta, 3, -5)$ is a linear combination of this set of vectors $\{(1, 3, -1), (-5, -5, 2)\}$.

36. Briefly, discuss the basic properties of the vector space of all continuous functions that are defined on the closed interval $[0, 1]$.

37. Set up and solve this traffic flow problem:

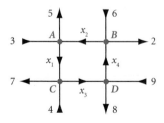

38. Find the dot product of the vectors $(0, 1, -2, 3, 4, -5, 6, 7, -8, 9)$ and $(9, -8, 7, -6, 5, 4, 3, 2, -1, 0)$.

39. Briefly, discuss the basic properties of the vector space $\mathbb{S}$ of all forward and backward infinite sequences of the form $(\ldots, x_{-2}, x_{-1}, x_0, x_1, x_2, \ldots)$. It arises in **discrete signal processing**.

40. Set up and solve this traffic flow problem:

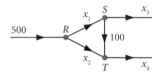

41. Find the center of mass of the weights 10, 15, 20, and 25 kilos located at the points $(1, 1)$ $(1, -1)$, $(-1, -1)$, and $(-1, 1)$, respectively.

42. Find the values of γ so that the vector $(2, \gamma, -3)$ is in the span of this set of vectors $\{(2, 4, -1), (-1, 0, 1), (4, 8, 2)\}$.

43. Set up and solve this circuit problem:

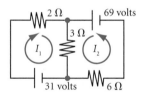

44. Briefly, discuss the basic properties of the vector space $\mathbb{P}_n$ of all polynomials of degree at most n (an integer) of the form

$$\mathbf{p}_n(x) = a_0 + a_1 x + a_2 x^2 + \cdots + a_n x^n$$

where the variable x and the coefficients $a_0, a_1, a_2, \ldots, a_n$ are real numbers. It arises in the **statistical trend analysis** of data.

45. Determine the values of a, b, and c so that the vector (a, b, c) is not in the span of the set of three vectors $\{(3, 5, 1), (2, 1, 2), (-1, -2, 1)\}$

46. Set up and solve this circuit problem:

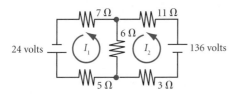

47. Is the vector $(0, 4, -3, -2, -1)$ in the span of this set of vectors?
$$\{(0, -1, 0, 1, 0), (-1, -4, -1, 4, 1)$$
$$(-2, -3, 1, 4, 1), (-1, 1, -1, 1, 3)\}$$

48. Set up and solve this circuit problem:

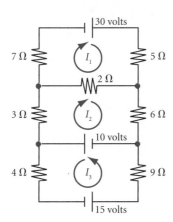

49. Suppose that the points $(5, -4, 4)$, $(4, 3, -2)$, $(-4, -3, -1)$, and $(-7, 4, 9)$ have weights of x, 5, y, and 1, respectively. Find the unknown weights so that the center of mass is $(\frac{3}{2}, \frac{1}{2}, \frac{1}{2})$.

50. Set up and solve this circuit problem:

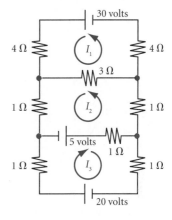

51. Briefly, discuss the basic properties of the **infinite-dimensional vector space** $\mathbb{R}^\infty$.

COMPUTER EXERCISES 2.1

Instruction for Computer Exercises: For each, find the linear system of equations that represents the situation described and compute its solution.

1. a. Set up and solve the linear system for this traffic problem:

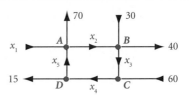

b. Find a solution so that the maximum number of cars involved at any intersection is a minimum.

c. Analyze the situation that arises if the datum 70 is replaced by the variable w.

2. Set up and solve the traffic problem shown:

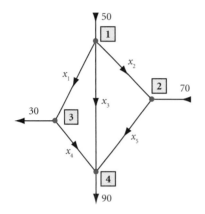

3. Set up and solve the traffic problem shown:

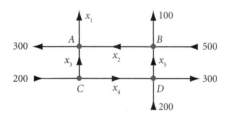

4. Set up and solve the linear system for this grid problem:

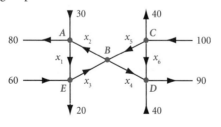

5. Set up and solve the circuit problem shown:

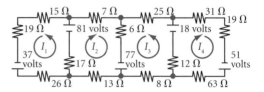

6. Set up and solve the circuit problem shown:

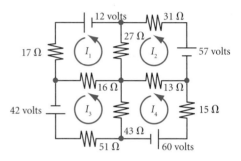

7. Set up and solve the circuit problem shown:

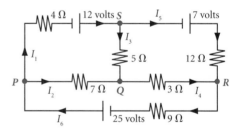

8. Set up and solve the grid problem shown:

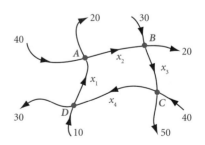

9. Set up and solve the grid problem shown:

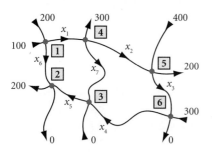

10. Set up and solve for the general solution of this traffic-flow problem of one-way streets:

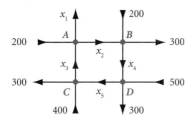

11. Assuming no returns, set up and solve this flow problem of products from manufacturers M_i to wholesalers W_i to retailers R_i to consumers C_i:

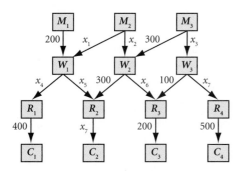

12. Construct and solve the system of linear equations that describes this traffic-flow problem. What are the values when $x_6 = 150$, $x_7 = 50$, and $x_8 = 100$?

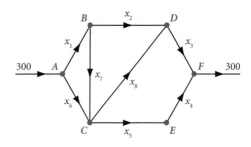

13. Set up and solve the linear system for the flow around this traffic circle. What are the values when $x_4 = 200$?

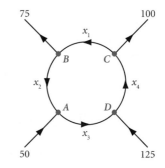

14. Find and solve the linear system for this electrical circuit:

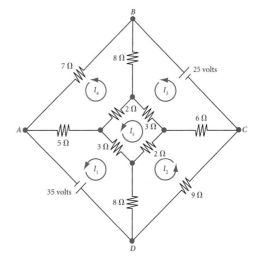

15. Describe the linear system for this traffic-flow problem and solve it. What are the values when $x_4 = 300$ and $x_5 = 100$?

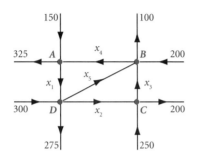

16. Set up and solve the linear system for this electrical circuit:

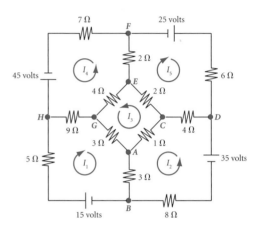

2.2 LINES, PLANES, AND HYPERPLANES

> *Whenever there is a number, there is beauty.*
>
> —DIADOCHUS PROCLUS (411–485)

> *Mathematics is the science that uses easy words for hard ideas.*
>
> —EDWARD KASNER (1878–1955) AND —JAMES R. NEWMAN

> *Everything should be made as simple as possible, but not simpler.*
>
> —ALBERT EINSTEIN (1879–1955)

In the low-dimensional spaces $\mathbb{R}$, $\mathbb{R}^2$, and $\mathbb{R}^3$, we can visualize many of the objects that we shall want to study in the higher-dimensional spaces $\mathbb{R}^n$. Very often what we see happening in $\mathbb{R}^2$, for example, will lead us to insights in $\mathbb{R}^n$.

Line Passing Through Origin

Let us begin with the simple concept of a line in $\mathbb{R}^2$ passing through the origin. Select a point **v** on the line, other than **0**. These two points, **0** and **v**, determine a line. The other points on the line are simply scalar multiples of the point **v**, as shown in Figure 2.13, where we took **v** to be the point

$(3, 2)$. Other points indicated are $(-3, -2)$, $(0, 0)$, $(\frac{3}{2}, 1)$, and $(6, 4)$. Thus, the **line** is a set, L, in $\mathbb{R}^2$ defined as follows:

$$L = \{c\mathbf{v} : c \in \mathbb{R}\} \qquad (\mathbf{v} \text{ is a prescribed point in } \mathbb{R}^2)$$

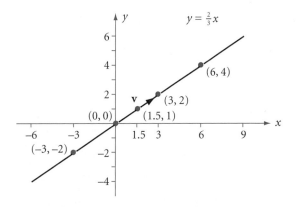

FIGURE 2.13 Line passing through origin.

This is read as "*The set of all points of the form $c\mathbf{v}$ as c runs over $\mathbb{R}$.*" This set obviously contains $\mathbf{0}$ and $\mathbf{v}$, because these arise from taking $c = 0$ and $c = 1$. Some would prefer to describe the set in this way:

$$L = \{\mathbf{x} : \text{ for some } c \text{ in } \mathbb{R}, \ \mathbf{x} = c\,\mathbf{v}\}$$

If there is still a possibility of misinterpretation, we can write

$$L = \{\mathbf{x} : \mathbf{x} \in \mathbb{R}^2 \text{ and, for some constant } c \text{ in } \mathbb{R}, \ \mathbf{x} = c\,\mathbf{v}\}$$

As concrete examples of lines through the origin in $\mathbb{R}^2$, one can let

$$L_1 = \{c\,(5, -2) : c \in \mathbb{R}\}, \qquad L_2 = \{r\,(2.718, 3.1416) : r \in \mathbb{R}\}$$

Lines in $\mathbb{R}^2$

Next, consider a line in $\mathbb{R}^2$ that does not necessarily pass through the origin. Select two points $\mathbf{u}$ and $\mathbf{v}$ on the line. One way to describe the **line** is like this:

$$L = \{t\mathbf{u} + (1 - t)\mathbf{v} : t \in \mathbb{R}\}$$

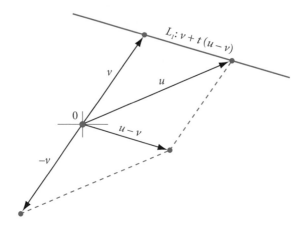

FIGURE 2.14   Parametric line.

This is a linear expression in the variable t, and the vectors $\mathbf{u}$ and $\mathbf{v}$ are fixed. Taking $t = 0$, we see that $\mathbf{v} \in L$. Taking $t = 1$, we see that $\mathbf{u} \in L$. The midpoint of the line segment joining $\mathbf{u}$ and $\mathbf{v}$ is obtained by letting $t = \frac{1}{2}$. By taking all values of t between 0 and 1, we get all the points on the line segment joining $\mathbf{u}$ and $\mathbf{v}$. Points on the line that are not on this line segment are obtained by taking either $t > 1$ or $t < 0$. Simple algebra shows that this same line can be written in the form

$$L = \{\mathbf{v} + t(\mathbf{u} - \mathbf{v}) \,:\, t \in \mathbb{R}\}$$

This suggests that the line is a **translation** (by the vector $\mathbf{v}$) of a line through the origin. Thus, the line can be described as

$$L = \mathbf{v} + \{t\,(\mathbf{u} - \mathbf{v}) \,:\, t \in \mathbb{R}\}$$

See Figure 2.14. Here a vector $\mathbf{v}$ plus a set S is denoted by $\mathbf{v} + S$. It contains vectors of the form $\mathbf{v} + \mathbf{s}$ where $\mathbf{s} \in S$. We conclude that every line in $\mathbb{R}^2$ can be described as follows, where $\mathbf{z} \neq \mathbf{0}$:

$$\{\mathbf{w} + t\,\mathbf{z} \,:\, t \in \mathbb{R}\} \qquad \text{or} \qquad \mathbf{w} + \{t\,\mathbf{z} \,:\, t \in \mathbb{R}\}$$

EXAMPLE 1 The expression $(12, 28) + t\,(9, 21)$ describes a line in $\mathbb{R}^2$ when t runs over all of $\mathbb{R}$. Does this line pass through $\mathbf{0}$?

SOLUTION The question is whether $(0, 0) = (12, 28) + t(9, 21)$ for some value of t. Looking at the first components, we see that $0 = 12 + 9t$, whence $t = -12/9 = -\frac{4}{3}$. From the second components, we have that $0 = 28 + 21t$, whence $t = -28/21 = -\frac{4}{3}$. Because these are the same, the answer to the question is *yes!* The value of t that proves this is $t = -\frac{4}{3}$. ∎

EXAMPLE 2 Is the point $(3, 7)$ on the line described parametrically by the expression $(4, -3) + t(11, 2)$?

SOLUTION We try to solve the equation $(3, 7) = (4, -3) + t(11, 2)$ for t. Proceeding as in Example 1, we make the first components correct by solving $3 = 4 + 2t$, getting $t = \frac{1}{2}$. For the second components, we solve $7 = -3 + 21t$, getting $t = 10/21$. Therefore, the answer is *no!* ∎

In describing a line L **parametrically** with an expression

$$\mathbf{x} = \mathbf{w} + t\mathbf{z}$$

we treat t as a parameter that is understood to run over all of $\mathbb{R}$. The roles of $\mathbf{w}$ and $\mathbf{z}$ are different. First, we note that $\mathbf{w}$ is a point on the line, because $t = 0$ gives us $\mathbf{w}$. The vector $\mathbf{z}$, on the other hand, is used to specify the direction of the line. That being the case, we cannot use $\mathbf{z} = \mathbf{0}$ in the description. However, we should be able to replace $\mathbf{z}$ by any nonzero multiple of $\mathbf{z}$ without changing the line. Therefore, we ask, "*Do* $\mathbf{w} + t\mathbf{z}$ *and* $\mathbf{w} + s(\alpha\mathbf{z})$ *describe the same line?*" *Yes!* Simply use $t = \alpha s$ or $s = t/\alpha$ to get the point expressed in either way.

Let L be the line described by the expression

$$\mathbf{w} + t\mathbf{z}$$

where $\mathbf{w}$ and $\mathbf{z}$ belong to $\mathbb{R}^2$, while t is a parameter. To get a line parallel to L but not equal to L, select any point $\mathbf{p}$ not on L and use

$$\mathbf{p} + t\mathbf{z}$$

to describe a new line, $\widetilde{L}$. One can verify the correctness of this by proving that the two lines do not meet. Indeed, if they do meet, we will have $\mathbf{p} + t\mathbf{z} = \mathbf{w} + s\mathbf{z}$ for some t and for some s. But this leads to $\mathbf{p} = \mathbf{w} + (s - t)\mathbf{z}$, and this means that $\mathbf{p}$ is on the line L, contrary to our choice of $\mathbf{p}$. See Figure 2.15.

Lines in $\mathbb{R}^3$

Now, we turn our attention to lines in $\mathbb{R}^3$. The only change in our previous description is that the two vectors $\mathbf{w}$ and $\mathbf{z}$ now will have three components.

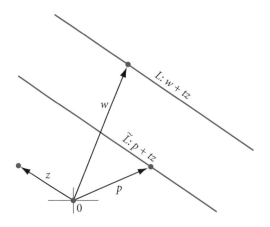

FIGURE 2.15 Parallel lines L and $\tilde{L}$.

Thus, a typical line in $\mathbb{R}^3$ can be described by

$$\mathbf{x} = \mathbf{w} + t\mathbf{z}$$

where $\mathbf{w}$ and $\mathbf{z}$ are in $\mathbb{R}^3$, and $\mathbf{z} \neq \mathbf{0}$. For a specific example, we take $(3, 2, -5) + t\,(1, -3, 2)$. Does the point $(-2, 17, -15)$ lie on this line? *Yes!* We solve $(-2, 17, -15) = (3, 2, -5) + t\,(1, -3, 2)$ and find $t = 5$.

EXAMPLE 3 Do these two lines in $\mathbb{R}^3$ intersect: $(3, 2, -5) + t\,(1, -3, 2)$ and $(2, -23, 5) + t\,(2, 1, 1)$.

SOLUTION Without thinking, we write down

$$(3, 2, -5) + t\,(1, -3, 2) = (2, -23, 5) + t\,(2, 1, 1)$$

What is wrong with this? For these two lines to intersect it is only necessary that some point on the first line should equal some point on the second. There is no justification for assuming that the values of t will be the same for the two points. Hence, we should attempt to solve instead

$$(3, 2, -5) + t\,(1, -3, 2) = (2, -23, 5) + s\,(2, 1, 1)$$

This leads to $t\,(-1, 3 - 2) + s\,(2, 1, 1) = (1, 25, -10)$ and in column form

$$t \begin{bmatrix} -1 \\ 3 \\ -2 \end{bmatrix} + s \begin{bmatrix} 2 \\ 1 \\ 1 \end{bmatrix} = \begin{bmatrix} 1 \\ 25 \\ -10 \end{bmatrix}$$

The augmented matrix for this system of three equations in two unknowns is

$$\begin{bmatrix} -1 & 2 & | & 1 \\ 3 & 1 & | & 25 \\ -2 & 1 & | & -10 \end{bmatrix} \sim \begin{bmatrix} 1 & 0 & | & 7 \\ 0 & 1 & | & 4 \\ 0 & 0 & | & 0 \end{bmatrix}$$

Hence, we obtain $t = 7$ and $s = 4$, and the answer to the original question is again *yes*. An independent verification can be made to see that the common point is $(10, -19, 9)$. ∎

EXAMPLE 4 Are these two lines in $\mathbb{R}^3$, described parametrically, the same?

$$(4, 7, 2) + t\,(3, 11, 5) \quad \text{and} \quad (223, 810, 367) + s\,(153, 561, 255)$$

SOLUTION We notice that the direction vectors for the two lines are multiples of each other: $(153, 561, 255) = 51\,(3, 11, 5)$. This tells us that, at least, the two lines are parallel. If they share a point, they must be identical. Taking $s = 0$ in the description of the second line, we get the point $(223, 810, 367)$. Is this point on the first line also? We set $(223, 810, 367) = (4, 7, 2) + t\,(3, 11, 5)$ and find that $t = 73$. The answer is *yes*. ∎

THEOREM 1

Let L_1 and L_2 be two lines in $\mathbb{R}^n$ described as

$$L_1 = \{\mathbf{u} + t\mathbf{v} : t \in \mathbb{R}\}, \qquad L_2 = \{\mathbf{w} + s\mathbf{z} : s \in \mathbb{R}\}$$

These lines are the same if and only if $\mathbf{u} - \mathbf{w}$ and $\mathbf{v}$ are multiples of $\mathbf{z}$.

PROOF (**Part 1.**) Assume that $L_1 = L_2$. We wish to prove that $\mathbf{u} - \mathbf{w}$ is a multiple of $\mathbf{z}$. Since $\mathbf{u} \in L_1$ and $L_1 = L_2$, we have $\mathbf{u} \in L_2$. Therefore, we obtain $\mathbf{u} = \mathbf{w} + s\mathbf{z}$ for some s. Then $\mathbf{u} - \mathbf{w} = s\mathbf{z}$ and $\mathbf{u} - \mathbf{w}$ is a multiple of $\mathbf{z}$.

(**Part 2.**) Assume $L_1 = L_2$. We wish to prove that $\mathbf{v}$ is a multiple of $\mathbf{z}$. We have $\mathbf{u} + \mathbf{v} \in L_1$ by the definition of L_1. We have $\mathbf{u} + \mathbf{v} \in L_2$ because $L_1 = L_2$. Hence, we find $\mathbf{u} + \mathbf{v} = \mathbf{w} + t\mathbf{z}$, for some t. Because $\mathbf{u} \in L_1$ and $L_1 = L_2$, we have $\mathbf{u} \in L_2$ and $\mathbf{u} = \mathbf{w} + p\mathbf{z}$. Since $\mathbf{u} + \mathbf{v} = \mathbf{w} + t\mathbf{z}$ and $\mathbf{u} = \mathbf{w} + p\mathbf{z}$, we get, by subtraction, $\mathbf{v} = t\mathbf{z} - p\mathbf{z} = (t - p)\mathbf{z}$.

(**Part 3.**) We want to prove that if $\mathbf{u} - \mathbf{w}$ and $\mathbf{v}$ are multiples of $\mathbf{z}$, then $L_1 = L_2$. The line L_1 is defined by the two points on it: $\mathbf{u}$ and $\mathbf{u} + \mathbf{v}$. We shall prove that the two points $\mathbf{u}$ and $\mathbf{u} + \mathbf{v}$ are also on L_2. This will prove that $L_1 = L_2$. By hypothesis, $\mathbf{u} - \mathbf{w} = \alpha\mathbf{z}$. The equation $\mathbf{u} = \mathbf{w} + \alpha\mathbf{z}$ shows that $\mathbf{u} \in L_2$. Finally, we show that $\mathbf{u} + \mathbf{v} \in L_2$. From our hypotheses, $\mathbf{u} + \mathbf{v} = \mathbf{w} + \alpha\mathbf{z} + \beta\mathbf{z} = \mathbf{w} + \gamma\mathbf{z} \in L_2$. ∎

Theorem 1 provides a simple test to determine whether two lines in $\mathbb{R}^n$ described parametrically are the same:

- Two lines given parametrically by $\mathbf{u} + t\mathbf{v}$ and $\mathbf{w} + s\mathbf{z}$ are the same if and only if $\mathbf{v} = \alpha\mathbf{z}$ and $\mathbf{u} - \mathbf{w} = \beta\mathbf{z}$ for suitable constants α and β.

For the two lines to be parallel and distinct, the first condition would hold but not the second. We obtain another simple test for distinct parallel lines.

- The two lines described as $\mathbf{u} + t\mathbf{v}$ and $\mathbf{w} + s\mathbf{z}$ are parallel and distinct if $\mathbf{v} = \alpha\mathbf{z}$ and $\mathbf{u} - \mathbf{w} \neq \beta\mathbf{z}$ for suitable constants α and β.
- The two lines $\mathbf{u} + t\mathbf{v}$ and $\mathbf{w} + s\mathbf{z}$ intersect if $\mathbf{u} - \mathbf{w}$ is in the span of the set $\{\mathbf{z}, \mathbf{v}\}$.

EXAMPLE 5 Show that these two lines, described parametrically in $\mathbb{R}^3$, are parallel and distinct: $(3, 2, -5) + t(1, -3, 2)$ and $(-1, -8, 2) + s(-4, 12, -8)$.

SOLUTION Let $\mathbf{u} = (3, 2, -5)$, $\mathbf{v} = (1, -3, 2)$, $\mathbf{w} = (-1, -8, 2)$, and $\mathbf{z} = (-4, 12, -8)$. We obtain $\mathbf{u} = (1, -3, 2) = -\frac{1}{4}(-4, 12, -8) = -\frac{1}{4}\mathbf{w}$ and $\mathbf{u} - \mathbf{w} = (3, 2, -5) - (-1, -8, 2) = (4, 10, -7) \neq \beta(-4, 12, -8) = \beta\mathbf{z}$, for any β. ∎

Planes in $\mathbb{R}^3$

Another linear object in $\mathbb{R}^3$ is the plane. One way to describe a plane through the origin is to say that it is the span of a pair of vectors, neither of which is a multiple of the other. For example, the span of the pair $(1, 6, -3)$ and $(2, 11, 0)$ is a plane passing through the origin and the two named points. Because three points determine a plane, this description is complete.

EXAMPLE 6 Let a plane in $\mathbb{R}^3$ be defined as the span of the set consisting of these two vectors $\mathbf{u} = (0, 2, 6)$ and $\mathbf{v} = (3, 9, 0)$. Are the points $\mathbf{y} = (1, 4, 3)$ and $\mathbf{z} = (2, -1, 5)$ in this plane?

SOLUTION We are asking whether the vectors $\mathbf{y}$ and $\mathbf{z}$ are in the span of the pair of vectors $\mathbf{u}$ and $\mathbf{v}$. This means that we want to know whether

$\alpha\mathbf{u}+\beta\mathbf{v} = \mathbf{y}$ and $\gamma\mathbf{u}+\delta\mathbf{v} = \mathbf{z}$ for some constants α, β, γ, and δ. It is necessary to solve the system

$$\left[\begin{array}{cc|cc}\mathbf{u} & \mathbf{v} & \mathbf{y} & \mathbf{z}\end{array}\right] = \left[\begin{array}{cc|cc} 0 & 3 & 1 & 2 \\ 2 & 9 & 4 & -1 \\ 6 & 0 & 3 & 5 \end{array}\right] \sim \left[\begin{array}{cc|cc} 2 & 9 & 4 & -1 \\ 0 & 3 & 1 & 2 \\ 0 & 0 & 0 & 26 \end{array}\right]$$

Using the usual row-reduction process, we are led to $\alpha = \frac{1}{2}$ and $\beta = \frac{1}{3}$ for the righthand-side $\mathbf{y}$. When $\mathbf{z}$ is the righthand side, no solution exists for γ and δ because the system is inconsistent. Thus, the answer to the original question is *yes* for the vector $\mathbf{y}$ but *no* for the vector $\mathbf{z}$. In fact, we know that

$$\mathbf{y} = (1, 4, 3) = \tfrac{1}{2}(0, 2, 6) + \tfrac{1}{3}(3, 9, 0) = \tfrac{1}{2}\mathbf{u} + \tfrac{1}{3}\mathbf{v}$$

See Figure 2.16. Later, we will say that $\frac{1}{2}$ and $\frac{1}{3}$ are the *coordinates* of the given vector $(1, 4, 3)$ in terms of the two vectors that span the plane. ∎

We have just seen that a plane, P, containing the origin can be described parametrically as follows:

$$P = \{s\mathbf{u} + t\mathbf{v} : s \in \mathbb{R}, t \in \mathbb{R}\}$$

Here $\mathbf{u}$ and $\mathbf{v}$ are fixed vectors in $\mathbb{R}^3$, neither a multiple of the other. It is obvious that $\mathbf{0} \in P$, $\mathbf{u} \in P$, and $\mathbf{v} \in P$; simply take appropriate values of s and t. For a plane in general (not necessarily containing $\mathbf{0}$), we use the expression

$$P = \{\mathbf{w} + s\mathbf{u} + t\mathbf{v} : s, t \in \mathbb{R}\}$$

A plane through the origin can also be described as follows:

$$P = \{(x_1, x_2, x_3) : a_1 x_1 + a_2 x_2 + a_3 x_3 = 0\}$$

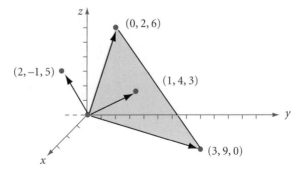

FIGURE 2.16 Plane through the origin in $\mathbb{R}^3$.

Here we must require the vector $\mathbf{a} = (a_1, a_2, a_3)$ to be nonzero. The equation that the vector $\mathbf{x}$ must satisfy is an example of a *system* of homogeneous equations, there being in actuality only one equation. But the theory of row reduction and obtaining the general solution still applies. Suppose that $a_1 \neq 0$. Then the row-reduction process leads in one step to the reduced echelon form

$$\left[\begin{array}{ccc|c} 1 & a_2/a_1 & a_3/a_1 & 0 \end{array}\right]$$

There is one pivot element and there are two free variables. Thus, as usual, we write $x_1 = (-a_2/a_1)x_2 + (-a_3/a_1)x_3$ and

$$\begin{bmatrix} x_1 \\ x_2 \\ x_3 \end{bmatrix} = x_2 \begin{bmatrix} -a_2/a_1 \\ 1 \\ 0 \end{bmatrix} + x_3 \begin{bmatrix} -a_3/a_1 \\ 0 \\ 1 \end{bmatrix}$$

$$= (x_2/a_1) \begin{bmatrix} -a_2 \\ a_1 \\ 0 \end{bmatrix} + (x_3/a_1) \begin{bmatrix} -a_3 \\ 0 \\ a_1 \end{bmatrix} = s \begin{bmatrix} -a_2 \\ a_1 \\ 0 \end{bmatrix} + t \begin{bmatrix} -a_3 \\ 0 \\ a_1 \end{bmatrix}$$

This is the **parametric form** of the plane. It shows that the plane is the span of a pair of vectors, which we are free to modify by scalar multiples, getting $(-a_2, a_1, 0)$ and $(-a_3, 0, a_1)$. See Figure 2.17. One verifies easily that these two vectors lie on the plane as originally described.

EXAMPLE 7 Find a parametric form for the plane whose generic point $\mathbf{x}$ satisfies the equation $7x_1 - 11x_2 + 13x_3 = 5$.

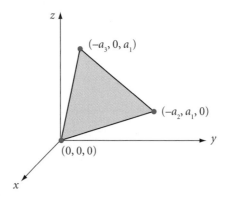

FIGURE 2.17 Plane through the origin in $\mathbb{R}^3$.

SOLUTION The given equation has two free variables, say, x_2 and x_3. Proceeding as in the previous paragraph, we find that the general solution of this equation is described thusly:

$$\begin{bmatrix} x_1 \\ x_2 \\ x_3 \end{bmatrix} = \begin{bmatrix} \frac{5}{7} \\ 0 \\ 0 \end{bmatrix} + x_2 \begin{bmatrix} \frac{11}{7} \\ 1 \\ 0 \end{bmatrix} + x_3 \begin{bmatrix} -\frac{13}{7} \\ 0 \\ 1 \end{bmatrix}$$

$$= \begin{bmatrix} \frac{5}{7} \\ 0 \\ 0 \end{bmatrix} + s \begin{bmatrix} 11 \\ 7 \\ 0 \end{bmatrix} + t \begin{bmatrix} -13 \\ 0 \\ 7 \end{bmatrix}$$ ∎

The *standard* equation of a plane in $\mathbb{R}^3$ not passing through **0** has the form

$$a_1 x_1 + a_2 x_2 + a_3 x_3 = 1$$

Here, a_1, a_2, and a_3 would be known coefficients and the variable is the vector **x**. Notice that the same plane is described if we multiply all terms in the equation by one nonzero factor. A plane through **0** will of necessity have **0** on the righthand side. One *normal* to the plane is the vector (a_1, a_2, a_3); it is perpendicular to the plane, but its length is unspecified. (The concept of perpendicularity or *orthogonality*, or *normality*, is the subject of Section 7.2.) The equation of the plane through three given points in $\mathbb{R}^3$ can be found by solving a system of three equations, as illustrated in the next example.

EXAMPLE 8 A plane in $\mathbb{R}^3$ contains these three points:

$$\mathbf{u} = (15, 5, 2), \qquad \mathbf{v} = (6, 2, 1), \qquad \mathbf{w} = (10, 3, 2)$$

What is the standard or normal form of this plane?

SOLUTION We wish to describe the plane in question as the set of all solutions to a single linear equation

$$P = \big\{ (x_1, x_2, x_3) \ : \ a_1 x_1 + a_2 x_2 + a_3 x_3 = 1 \big\}$$

This equation will be the normal form for the plane. The unknowns are the three coefficients a_1, a_2, and a_3. In the preceding equation, we substitute each point known to be on the plane. In this way, we arrive at three equations in three unknowns, as shown:

$$\begin{cases} 15a_1 + 5a_2 + 2a_3 = 1 \\ 6a_1 + 2a_2 + a_3 = 1 \\ 10a_1 + 3a_2 + 2a_3 = 1 \end{cases}$$

The augmented matrix and its reduced row echelon form are

$$\begin{bmatrix} 15 & 5 & 2 & 1 \\ 6 & 2 & 1 & 1 \\ 10 & 3 & 2 & 1 \end{bmatrix} \sim \begin{bmatrix} 1 & 0 & 0 & -2 \\ 0 & 1 & 0 & 5 \\ 0 & 0 & 1 & 3 \end{bmatrix}$$

The answer to the question posed is therefore $-2x_1 + 5x_2 + 3x_3 = 1$. ∎

Lines and Planes in $\mathbb{R}^n$

The description of lines and planes in higher-dimensional spaces proceeds very much in the same way as previously outlined. A line in $\mathbb{R}^n$ can be given parametrically as

$$\mathbf{x} = \mathbf{w} + t\mathbf{z}$$

where $\mathbf{w}$ and $\mathbf{z}$ are vectors in $\mathbb{R}^n$. We remind the reader that this is shorthand for the more formal definition: $\{\mathbf{w} + t\mathbf{z} \ : \ t \in \mathbb{R}\}$. The point $\mathbf{w}$ can be any point on the line, and $\mathbf{z}$ is a nonzero vector specifying the direction of the line.

EXAMPLE 9 In $\mathbb{R}^4$, we are given two lines, having the parametric representations

$$\mathbf{x} = (15, -3, 5, 1) + t(-1, 3, 5, 0), \qquad \mathbf{x} = (2, 6, 4, -17) + s(3, 1, 7, 6)$$

Do these lines meet?

SOLUTION Take the generic points on the two lines and determine whether they can be the same point when s and t are appropriately chosen. This leads to the equations

$$(15, -3, 5, 1) + t(-1, 3, 5, 0) = (2, 6, 4, -17) + s(3, 1, 7, 6)$$

$$t(1, -3, -5, 0) + s(3, 1, 7, 6) = (13, -9, 1, 18)$$

$$\begin{bmatrix} 1 & 3 & \bigm| & 13 \\ -3 & 1 & \bigm| & -9 \\ -5 & 7 & \bigm| & 1 \\ 0 & 6 & \bigm| & 18 \end{bmatrix} \sim \begin{bmatrix} 1 & 0 & \bigm| & 4 \\ 0 & 1 & \bigm| & 3 \\ 0 & 0 & \bigm| & 0 \\ 0 & 0 & \bigm| & 0 \end{bmatrix}$$

The system of four equations in the two unknowns t and s has a solution, $t = 4$ and $s = 3$, and the pair of lines has the point $(11, 9, 25, 1)$ in common.

∎

As for planes in $\mathbb{R}^n$, we first agree to define **plane** as a two-dimensional object, reserving the term **hyperplane** for similar but higher-dimensional objects. Thus, a plane in $\mathbb{R}^n$ is given parametrically by the expression

$$\mathbf{x} = \mathbf{w} + s\mathbf{u} + t\mathbf{v}$$

We assume that the pair $\{\mathbf{u}, \mathbf{v}\}$ is linearly independent. In this simple case, it means that neither $\mathbf{u}$ nor $\mathbf{v}$ is a scalar multiple of the other.

General Solution of a System of Equations

We turn now to a standard format for expressing the general solution of a system of equations. The idea is that the solution of any system of equations can be written in vector form as

$$\mathbf{x} = \mathbf{w} + c_1\mathbf{u}_1 + c_2\mathbf{u}_2 + \cdots + c_k\mathbf{u}_k$$

Sometimes there is no solution to a given system. That is the *inconsistent* case, and there is nothing further to say about it now. The next interesting case occurs when there is one and only one solution. Here, too, there is no further discussion. The final case occurs when there are infinitely many solutions. Here, we want to establish a standard way of expressing all the solutions. Instead of a verbal description, an example will be used, as follows.

EXAMPLE 10 Consider these three linear equations involving points in $\mathbb{R}^5$:

$$\begin{cases} x_1 + x_2 + 2x_3 + 3x_4 + 2x_5 = 1 \\ 2x_1 + 2x_2 + 6x_3 + 10x_4 + 10x_5 = 0 \\ 2x_1 + 2x_2 + 4x_3 + 8x_4 + 9x_5 = 0 \end{cases}$$

Find a simpler description of the set of points that satisfy all three equations.

SOLUTION We write down the augmented matrix, use row reductions to get an echelon form, and continue to the reduced echelon form:

$$\left[\, \mathbf{A} \,\middle|\, \mathbf{b} \,\right] = \begin{bmatrix} 1 & 1 & 2 & 3 & 2 & \big| & 1 \\ 2 & 2 & 6 & 10 & 10 & \big| & 0 \\ 2 & 2 & 4 & 8 & 9 & \big| & 0 \end{bmatrix}$$

$$\sim \begin{bmatrix} 1 & 1 & 2 & 3 & 2 & \big| & 1 \\ 0 & 0 & 2 & 4 & 6 & \big| & -2 \\ 0 & 0 & 0 & 2 & 5 & \big| & -2 \end{bmatrix}$$

$$\sim \begin{bmatrix} 1 & 1 & 0 & 0 & -\frac{3}{2} & \big| & 2 \\ 0 & 0 & 1 & 0 & -2 & \big| & 1 \\ 0 & 0 & 0 & 1 & \frac{5}{2} & \big| & -1 \end{bmatrix}$$

Notice that two columns (the second and fifth) have no pivot elements. The corresponding variables (x_2 and x_5) are *free* variables. (Free variables were first encountered in Section 1.2.) The remaining variables (x_1, x_3, and x_4) are then *basic* variables. The system of equations represented by the reduced echelon form is

$$\begin{cases} x_1 + x_2 - \frac{3}{2}x_5 = \;\; 2 \\ \quad\quad x_3 - 2x_5 = \;\; 1 \\ \quad\quad x_4 + \frac{5}{2}x_5 = -1 \end{cases} \quad \text{or} \quad \begin{cases} x_1 = \;\; 2 - x_2 + \frac{3}{2}x_5 \\ x_3 = \;\; 1 \quad\quad + 2x_5 \\ x_4 = -1 \quad\quad - \frac{5}{2}x_5 \end{cases}$$

The final step is designed to express all solutions as the sum of a fixed numerical vector and arbitrary scalar multiples of two other vectors, as follows:

$$\begin{bmatrix} x_1 \\ x_2 \\ x_3 \\ x_4 \\ x_5 \end{bmatrix} = \begin{bmatrix} 2 \\ 0 \\ 1 \\ -1 \\ 0 \end{bmatrix} + x_2 \begin{bmatrix} -1 \\ 1 \\ 0 \\ 0 \\ 0 \end{bmatrix} + x_5 \begin{bmatrix} \frac{3}{2} \\ 0 \\ 2 \\ -\frac{5}{2} \\ 1 \end{bmatrix}$$

$$= \begin{bmatrix} 2 \\ 0 \\ 1 \\ -1 \\ 0 \end{bmatrix} + s \begin{bmatrix} -1 \\ 1 \\ 0 \\ 0 \\ 0 \end{bmatrix} + t \begin{bmatrix} 3 \\ 0 \\ 4 \\ -5 \\ 2 \end{bmatrix}$$

In words, we have expressed all solutions in the form of a single equation. This equation is

$$\mathbf{x} = \mathbf{c} + s\mathbf{u} + t\mathbf{v}$$

and consists of one fixed numerical vector $\mathbf{c}$ plus arbitrary scalar multiples of two other vectors $\mathbf{u}$ and $\mathbf{v}$. All of this is taking place in $\mathbb{R}^5$. If you are skeptical (as you should be), take any values of x_2 and x_5, say $x_2 = 3$ and $x_5 = 2$. The vector on the right will then be $\mathbf{u} = [2, 3, 5, -6, 2]^T$ and $\mathbf{Au} = \mathbf{b}$, as it should be. The set of solutions is a two-dimensional plane in the five-dimensional Euclidean vector space $\mathbb{R}^5$. ∎

Application: The Predator–Prey Simulation

In this topic, we attempt to analyze the interaction of two animal species as they compete in the wild. Only the simple action of multiplying a vector by a matrix is needed to simulate this contest. A more advanced discussion of the same problem is postponed to Section 6.1, where eigenvalues can play a role.

Suppose that the two competing species under study are mice and coyotes. The predator here would be the coyotes and the prey would be the mice. If the coyotes are very efficient in feeding off the mice, the mouse population will decline, and the coyotes eventually will cease to flourish simply because their food supply has been diminished. When the population of coyotes decreases, the mice, in turn, will thrive, as their main adversary is reduced in number.

We will use a vector having two components, the first being the number of coyotes in the area considered. The second component of this vector will be the number of mice in the same area. We think of this vector as changing, say once per year, to reflect the changes in population of these two species. At the beginning of the process, this vector will be denoted by $\mathbf{x}^{(0)} = (c_0, m_0)$. Its first component is the number of coyotes present and the second component is the number of mice present. This vector represents the state of affairs at the beginning of the time period being considered. Suppose that ecological studies have revealed that after one year the two populations will differ according to this formula:

$$\mathbf{x}^{(1)} = \mathbf{Ax}^{(0)} \qquad \text{or} \qquad \begin{bmatrix} x_1^{(1)} \\ x_2^{(1)} \end{bmatrix} = \begin{bmatrix} 0.6 & 0.4 \\ -0.1 & 1.2 \end{bmatrix} \begin{bmatrix} x_1^{(0)} \\ x_2^{(0)} \end{bmatrix}$$

Some weak conclusions can be drawn immediately from these data. For example, if the mouse population dropped to zero, then the number of coyotes would quickly decline because of the factor 0.6 in the matrix. That reflects the fact that the mice are an important part of the coyotes' diet. Conversely, if there were no coyotes, the mouse population would increase by a factor of 1.2 in each time interval, simply because our model contains no other threat to the mice.

For a concrete illustration, suppose that at the beginning of the year the initial $\mathbf{x}^{(0)}$-vector is $\begin{bmatrix} 50 \\ 150 \end{bmatrix}$. At the end of the year, the two populations will be $\begin{bmatrix} 90 \\ 175 \end{bmatrix}$, because

$$\begin{bmatrix} 0.6 & 0.4 \\ -0.1 & 1.2 \end{bmatrix} \begin{bmatrix} 50 \\ 150 \end{bmatrix} = \begin{bmatrix} 90 \\ 175 \end{bmatrix}$$

To repeat the calculation, we write $\mathbf{x}^{(k)} = \mathbf{A}\mathbf{x}^{(k-1)}$, and let the computer do the work with suitable mathematical software. In successive years, we find these values for the coyotes and mouse populations:

$$\mathbf{x}^{(0)} = \begin{bmatrix} 50 \\ 150 \end{bmatrix}, \quad \mathbf{x}^{(1)} = \begin{bmatrix} 90 \\ 175 \end{bmatrix}, \quad \mathbf{x}^{(2)} = \begin{bmatrix} 124 \\ 201 \end{bmatrix}, \quad \mathbf{x}^{(3)} = \begin{bmatrix} 155 \\ 229 \end{bmatrix},$$

$$\mathbf{x}^{(4)} = \begin{bmatrix} 184 \\ 259 \end{bmatrix}, \quad \mathbf{x}^{(8)} = \begin{bmatrix} 316 \\ 417 \end{bmatrix}, \quad \cdots, \quad \mathbf{x}^{(32)} = \begin{bmatrix} 5233 \\ 6850 \end{bmatrix}$$

The ratio of coyotes to mice is settling down to approximately $5233/6850 \approx 0.76$, that is, approximately three coyotes to every four mice.

Application: Partial-Fraction Decomposition

In various applications, we need to be able to break a complicated fraction into a sum of simpler fractions. For example, we might want to find A and B so that

$$\frac{10x - 5}{x^2 + x - 6} = \frac{A}{x + 3} + \frac{B}{x - 2}$$

We can rewrite this as

$$\frac{10x - 5}{(x + 3)(x - 2)} = \frac{A(x - 2) + B(x + 3)}{(x + 3)(x - 2)}$$

For the numerators to be equal, we require that $10x - 5 = (A + B)x + (-2A + 3B)$. This leads to a linear system:

$$\begin{cases} A + B = 10 \\ -2A + 3B = -5 \end{cases}$$

The augmented matrix and its row echelon form are

$$\begin{bmatrix} 1 & 1 & | & 10 \\ -2 & 3 & | & -5 \end{bmatrix} \sim \begin{bmatrix} 1 & 0 & | & 7 \\ 0 & 1 & | & 3 \end{bmatrix}$$

The solution is $A = 7$ and $B = 3$. Consequently, we have found that

$$\frac{10x - 5}{x^2 + x - 6} = \frac{7}{x + 3} + \frac{3}{x - 2}$$

For example, this can be used as follows in an integration problem:

$$\int \frac{10x - 5}{x^2 + x - 6}\,dx = \int \frac{7}{x + 3}\,dx + \int \frac{3}{x - 2}\,dx$$
$$= 7\log(x + 3) + 3\log(x - 2) + C$$

Application: Method of Least Squares

We consider the problem of finding a straight line that represents, approximately, a table of n points (x_i, y_i), in the xy-plane. (In general, the line does not necessarily go through the data points.) These are given numerically. The formula sought will be

$$y = ax + b$$

and the two parameters a and b are to be chosen so that the resulting line fits the data as well as possible. There are many different criteria that can be used on this problem, but the most important of these is to minimize the sum of the squares of the errors. What this means can be explained as follows. For any particular choice of the parameters a and b, the line $y = ax + b$ differs from the tabular data. The difference is called the *residual* at that point. For example, at the point x_i the residual is $r_i = ax_i + b - y_i$. We want the sum of the squares of the residuals to be as small as possible. In other words, the coefficients a and b are to be chosen so that $\sum_{i=1}^{n} r_i^2$ is a minimum. This procedure is known as the **method of least squares.**

Define $E = \sum_{i=1}^{n}(ax_i + b_i - y_i)^2$. To find the minimum point of E, we appeal to the calculus, where we learned to set the partial derivatives equal to 0 and solve for the points thus obtained. This leads to

$$\partial E/\partial a = \sum_{i=1}^{n} 2(ax_i + b - y_i)x_i = 0$$

$$\partial E/\partial b = \sum_{i=1}^{n} 2(ax_i + b - y_i) = 0$$

Rearranging and simplifying, we obtain the following system of equations:

$$\begin{bmatrix} a_{11} & a_{12} \\ a_{12} & n \end{bmatrix} \begin{bmatrix} a \\ b \end{bmatrix} = \begin{bmatrix} b_1 \\ b_2 \end{bmatrix}$$

where

$$a_{11} = \sum_{i=1}^{n} x_i^2, \qquad b_1 = \sum_{i=1}^{n} x_i y_i$$

$$a_{12} = \sum_{i=1}^{n} x_i, \qquad b_2 = \sum_{i=1}^{n} y_i$$

Using Gaussian elimination, we solve this system and obtain the best coefficients for the line $y = ax + b$ where

$$a = \frac{nb_1 - a_{12}b_2}{na_{11} - a_{12}^2}, \qquad b = \frac{a_{11}b_2 - a_{12}b_1}{na_{11} - a_{12}^2}$$

EXAMPLE 11   Here is a concrete example (drawn from physics) that illustrates the method of least squares. We assume as known that the surface tension y in a liquid is a linear function of temperature x. Measurements have been made of the surface tension at certain temperatures for a particular liquid. These data are as given in the following table:

x	1	2	3	4	5	6	7	8	9
y	6.96	6.61	6.74	6.76	6.21	6.48	6.15	6.28	6.02

SOLUTION   To use the formulas presented earlier, we must compute $u = 285$, $v = 45$, $w = 284.84$, $z = 58.21$, $n = 9$, $a = -0.1035$, and $b = 6.9853$. The plot of these data points and this line are shown in Figure 2.18 on the next page. More general least squares problems are considered in Sections 7.2 and 8.2. Techniques developed there provide a way of finding best approximate solutions to any inconsistent system of linear equations. ∎

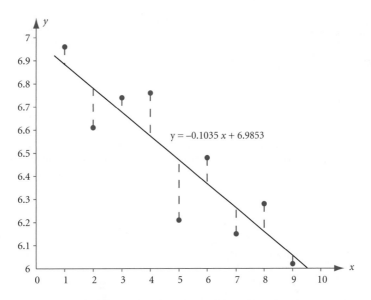

FIGURE 2.18 Linear least squares example.

SUMMARY 2.2

- A line in $\mathbb{R}^n$ through the zero point $\mathbf{0}$ is a set of the form $\{c\mathbf{v} \; : \; c \in \mathbb{R}\}$.

- A line in $\mathbb{R}^n$ is a set of the form $\{\mathbf{w} + t\mathbf{z} \; : \; t \in \mathbb{R}\}$, where $\mathbf{z} \neq \mathbf{0}$.

- If two points $\mathbf{u}$ and $\mathbf{v}$ on a line are known, then the line is of the form $\{t\mathbf{u} + (1 - t)\mathbf{v} \; : \; t \in \mathbb{R}\}$.

- Let L_1 and L_2 be two lines in $\mathbb{R}^n$, given by the formulas $L_1 = \{\mathbf{u} + t\mathbf{v} \; : \; t \in \mathbb{R}\}$ and $L_2 = \{\mathbf{w} + s\mathbf{z} \; : \; s \in \mathbb{R}\}$. Then L_1 and L_2 are the same if and only if $\mathbf{v}$ and $\mathbf{u} - \mathbf{w}$ are multiples of $\mathbf{z}$. Moreover, L_1 and L_2 are parallel and distinct if $\mathbf{v}$ is a multiple of $\mathbf{z}$ but $\mathbf{u} - \mathbf{w}$ is not.

- A plane in $\mathbb{R}^3$ is a set of the form $\{\mathbf{w} + s\mathbf{u} + t\mathbf{v} : s, t \in \mathbb{R}\}$. (Here $\mathbf{w}, \mathbf{u}, \mathbf{v} \in \mathbb{R}^3$.)

- A plane containing $\mathbf{0}$ in $\mathbb{R}^3$ is given by $\{(x_1, x_2, x_3) \; : \; a_1 x_1 + a_2 x_2 + a_3 x_3 = 0\}$.

- A two-dimensional plane in $\mathbb{R}^n$ is a set of the form $\mathbf{x} = \mathbf{w} + s\mathbf{u} + t\mathbf{v}$, where $\mathbf{w}, \mathbf{u}, \mathbf{v} \in \mathbb{R}^n$.

- Iteration of a linear mapping uses the equation $\mathbf{x}^{(k)} = \mathbf{A}\mathbf{x}^{(k-1)}$, for $k = 1, 2, \ldots$, and a given initial vector $\mathbf{x}^{(0)}$.

- Linear least squares solution: $y = ax + b$ where $a = (nb_1 - a_{12}b_2)/(na_{11} - a_{12}^2)$ and $b = (a_{11}b_2 - a_{12}b_1)/(na_{11} - a_{12}^2)$ with $a_{11} = \sum_{i=1}^{n} x_i^2$, $a_{12} = \sum_{i=1}^{n} x_i$, $b_1 = \sum_{i=1}^{n} x_i y_i$, and $b_2 = \sum_{i=1}^{n} y_i$.

KEY CONCEPTS 2.2

Lines passing through the origin; lines and planes in $\mathbb{R}^n$; parametric representation of lines and planes; other ways of describing lines and planes; parallel lines; translation operation; expressing the general solution of a system of equations; applications: predator–prey simulation model, partial-fraction decomposition, least squares problems

GENERAL EXERCISES 2.2

1. Find a parametric form for the line in $\mathbb{R}^2$ that passes through the points $(7,3)$ and $(-5,6)$. Is the answer unique?

2. In $\mathbb{R}^5$ does the line described parametrically by $(3,4,-5,6,2)+t\,(2,-2,1,3,6)$ intersect the line represented by $(17,-10,2,27,44)+t\,(-3,2,-5,1,4)$?

3. For what value of the variable h will the line $(3,5,2)+t\,(1,-2,h)$ intersect the line $(4,2,1)+t\,(-1,0,-2)$? Here, the lines are described parametrically in the usual manner.

4. Is the point $(3,2,5,-4)$ on the plane described parametrically by $(1,2,4,-3)+s\,(2,1,-2,0)+t\,(4,3,-7,1)$?

5. Let $\mathbf{u}=(3,7)$, $\mathbf{v}=(5,-4)$, $\mathbf{w}=(18,-5)$, and $\mathbf{z}=(-20,16)$. Verify the equality of these two sets:
$$\{\mathbf{u}+t\mathbf{v}:t\in\mathbb{R}\}=\{\mathbf{w}+t\mathbf{z}:t\in\mathbb{R}\}$$

6. A plane in $\mathbb{R}^3$ is represented parametrically by the equation $\mathbf{x}=\mathbf{w}+s\mathbf{u}+t\mathbf{v}$, where $\mathbf{w}=(\frac{5}{3},0,0)$, $\mathbf{u}=(-\frac{7}{3},1,0)$, and $\mathbf{v}=(3,0,1)$. Represent this plane in the form $a_1x_1+a_2x_2+a_3x_3=b$, where $\mathbf{x}=(x_1,x_2,x_3)$.

7. Consider a plane in $\mathbb{R}^3$ described by the equation $3x_2-5x_3=7$. Describe this plane in the parametric form $\mathbf{u}+t\mathbf{v}+s\mathbf{w}$.

8. Let P be the set of all vectors $\mathbf{x}=(x_1,x_2,x_3,x_4)$ such that
$$\begin{cases}3x_1+2x_2-\;\;x_3+\;\;x_4=2\\ x_1-\;\;x_2+3x_3-4x_4=4\end{cases}$$

Is P a plane in $\mathbb{R}^4$? In other words, is P given parametrically by an expression of the form $\mathbf{w}+s\mathbf{u}+t\mathbf{v}$? If so, determine the vectors $\mathbf{w}$, $\mathbf{u}$, and $\mathbf{v}$.

9. Without doing any computation, determine whether these two lines, given in parametric form, are identical:
$$L_1:(3,7,-2)+t(3,5,-4)$$
$$L_2:(4,2,1)+s(-9,-15,11)$$

10. Is there a plane in $\mathbb{R}^3$ that contains the two lines described parametrically by $(1,-2,3)+t(1,0,0)$ and $(-2,5,-7)+s(4,-7,10)$?

11. Two lines are given parametrically as $(4,7,2)+t(3,1,0)$ and $(13,10,2)+t(4,2,0)$. Do they have a point in common?

12. Find the general solution of the system whose augmented matrix is
$$\begin{bmatrix}1 & 3 & 2 & 2 & | & -1\\ -2 & -5 & 1 & 2 & | & 1\end{bmatrix}$$
Verify your result by substituting your solution in the original system of equations.

13. (Continuation.) Solve the system in General Exercise 12 when the righthand side of the system is $\begin{bmatrix}a\\b\end{bmatrix}$ in place of $\begin{bmatrix}-1\\1\end{bmatrix}$.

14. Determine whether these two lines (described parametrically) are the same: $(3,7)+t(12,-10)$ and $(-3,12)+t(6,-5)$.

15. The equation $2x_1-3x_2+7x_3=4$ represents a plane in $\mathbb{R}^3$. Give a parametric form for this plane.

16. Use the techniques of *partial fractions* to find these indefinite integrals:

a. $\displaystyle\int \frac{1}{x^2 - 1}\,dx$

b. $\displaystyle\int \frac{x + 2}{x^3 - x^2}\,dx$ **c.** $\displaystyle\int \frac{1 + x}{x - x^3}\,dx$

17. Use your knowledge of analytic geometry to compute the minimum distance from the point $(3, 7)$ to the line given by $2x - 11y = 5$.

18. Find the standard form for the equation of the plane passing through these three points: $\mathbf{u} = (3, 0, 1)$, $\mathbf{v} = (0, 2, 3)$, and $\mathbf{w} = (3, 4, 0)$.

19. Let lines be described as $L_1 = \{\mathbf{p} + t\mathbf{q} : t \in \mathbb{R}\}$ and $L_2 = \{\mathbf{v} + t\mathbf{w} : t \in \mathbb{R}\}$, where $\mathbf{p} = (4, 2, 1)$, $\mathbf{q} = (-1, 3, 2)$, $\mathbf{v} = (1, 11, 7)$, and $\mathbf{w} = (3, -9, -6)$. Determine whether these two lines in $\mathbb{R}^n$ are the same.

20. Are the two lines $(11, 11, 40) + t(12, 8, 28)$ and $(-4, 1, 5) + s(3, 2, 7)$ (given parametrically) the same?

21. Are these lines the same? Are they parallel?
$L_1 = \{\mathbf{u} + t\mathbf{v} : t \in \mathbb{R}\}$
$L_2 = \{\mathbf{w} + t\mathbf{z} : t \in \mathbb{R}\}$
where $\mathbf{u} = (7, 11, 5)$, $\mathbf{v} = (9, 15, -6)$, $\mathbf{w} = (-8, -14, 15)$, $\mathbf{z} = (3, 5, -2)$

22. Use Theorem 1 to determine whether these two lines are identical:
$L_1 = \{(2, 1) + t(-1, -1) : t \in \mathbb{R}\}$
$L_2 = \{(7, 6) + s(3, 3) : s \in \mathbb{R}\}$

23. Recall from the study of analytic geometry that the distance between two points $\mathbf{u} = (u_1, u_2)$ and $\mathbf{v} = (v_1, v_2)$ in $\mathbb{R}^2$ is given by $\sqrt{(u_1 - v_1)^2 + (u_2 - v_2)^2}$. Use this fact to find the parametric representation of the **perpendicular bisector** of a line segment in $\mathbb{R}^2$ whose endpoints are $\mathbf{a} = (a_1, a_2)$ and

$\mathbf{b} = (b_1, b_2)$. The midpoint of this line segment is the point $\frac{1}{2}(\mathbf{a} + \mathbf{b})$.

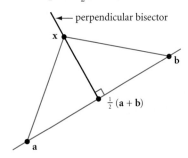

24. Establish this assertion or find a counterexample: For two lines in $\mathbb{R}^n$ given parametrically by $\mathbf{v} + t\mathbf{w}$ and $\mathbf{x} + s\mathbf{y}$ to intersect, it is necessary and sufficient that $\mathbf{x} - \mathbf{v}$ be in the span of $\{\mathbf{w}, \mathbf{y}\}$.

25. Establish this assertion or find a counterexample: A necessary and sufficient condition for the line given parametrically by $t\mathbf{u} + (1 - t)\mathbf{v}$ to contain the point $\mathbf{0}$ is that $\mathbf{v}$ be a scalar multiple of $\mathbf{u} - \mathbf{v}$.

26. In $\mathbb{R}^2$, two random lines will likely have a point in common. Is the same to be expected in $\mathbb{R}^3$? Answer the related question about a line and a plane in $\mathbb{R}^3$.

27. Is the origin on these lines?
  a. $\mathbf{r} = (4, -12) + s(2, -6)$
  b. $\mathbf{s} = (-8, -6, -18) + s(4, 3, 6)$

28. Is the given point on the line shown?
  a. $(0, -1)$
    $\mathbf{u} = (4, 5) + a(2, 3)$
  b. $(-2, 1, 4)$
    $\mathbf{v} = (3, -4, -1) + b(5, 5, 5)$
  c. $(-11, 11)$
    $\mathbf{p} = (4, -5) + c(5, 4)$
  d. $(7, 5)$
    $\mathbf{q} = (3, 3) + d(2, 1)$

29. Do these lines intersect?

a. $\begin{cases} \mathbf{p} = (0,7) + s(14,-7) \\ \mathbf{q} = (0,3) + t(3,6) \end{cases}$

b. $\begin{cases} \mathbf{x} = (3,2,-5) + s(1,-3,2) \\ \mathbf{y} = (0,-3,-5) + t(2,1,1) \end{cases}$

30. Are these lines the same?

a. $\begin{cases} \mathbf{u} = (2,1) + s(3,3) \\ \mathbf{v} = (8,4) + t(2,2) \end{cases}$

b. $\begin{cases} \mathbf{p} = (3,7,2) + s(3,5,4) \\ \mathbf{q} = (-3,-3,-6) + t(9,15,12) \end{cases}$

31. Are these lines parallel and distinct?

a. $\begin{cases} \mathbf{x} = (3,7,2) + s(3,5,-4) \\ \mathbf{y} = (4,2,1) + t(-9,-15,12) \end{cases}$

b. $\begin{cases} \mathbf{u} = (2,4,1) + s(-1,3,3) \\ \mathbf{v} = (10,20,7) + t(2,6,6) \end{cases}$

32. Consider the three lines $x - 2y = -1$, $3x + 5y = 8$, and $4x + 3y = 7$.
a. Do they intersect?
b. Write each in parametric form.

33. Are the points $(2,-3,0)$, $(0,10,9)$ in the plane defined by the span of this set of vectors $\{(1,3,5),(2,-4,1)\}$?

34. Find the equation of a plane that contains the point $(2,4,-1)$ and that is normal to the vector $\mathbf{v} = (3,5,-2)$. *Hint:* Vectors $\mathbf{v} = (v_1,v_2,v_3)$ and $\mathbf{u} = (u_1,u_2,u_3)$ are **normal** if $\mathbf{v}^T\mathbf{u} = 0$.

35. What is the normal form of the plane in $\mathbb{R}^3$ containing the points $\mathbf{u} = (1,1,1)$, $\mathbf{v} = (1,2,0)$, and $\mathbf{w} = (-1,2,1)$.

36. Find the intersection of the two lines, if there is one:
$$\begin{cases} \mathbf{x} = (4,3,9) + s(2,-3,7) \\ \mathbf{y} = (3,2,0) + t(-1,4,2) \end{cases}$$

37. Find an alternative description of the set of points that satisfy this system of linear equations.
$$\begin{cases} 2x_1 + 2x_2 + 3x_3 + 4x_4 = -1 \\ 4x_1 + 6x_2 - 3x_3 - 4x_4 = -2 \end{cases}$$

38. Use partial fraction decompositions to evaluate these integrals.

a. $\displaystyle\int \frac{x-1}{x^2+x}\,dx$

b. $\displaystyle\int \frac{8x-42}{x^2+3x-18}\,dx$

c. $\displaystyle\int \frac{x^2-1}{x^2-16}\,dx$

39. Find a plane in $\mathbb{R}^3$ containing both of these lines
$$\begin{cases} \mathbf{u} = (1,-2,3) + s(1,0,0) \\ \mathbf{v} = (-2,5,-7) + t(4,-7,10) \end{cases}$$

40. What are the characteristics of these lines?
$$\begin{cases} \mathbf{u} = (3,7,2) + s(3,5,4) \\ \mathbf{v} = (-3,-3,-6) + t(-9,-15,-12) \end{cases}$$

41. Is it possible for two lines to be parallel and intersect? Consider these lines
$$\begin{cases} \mathbf{u} = (2,4,1) + s(1,-3,2) \\ \mathbf{v} = (0,10,7) + t(-2,6,4) \end{cases}$$

42. Explain why points $(0,0,0)$ and $(1,1,1)$ lie in the plane $3x_1 + 2x_2 - 5x_3 = 0$. In fact, the line (c,c,c) is in this plane.

43. Find values of the unknowns so that the given points are on the line shown.
a. $(a,-1),(-4,b)$
$\mathbf{p} = (6,5) + t(10,-2)$
b. $(-4,7,c),(d,-1,2)$
$\mathbf{q} = (6,3,2) + t(-5,2,2)$

44. Do these lines $2x_1 - 4x_2 = -6$, $4x_1 - 3x_2 = 2$, and $-2x_1 + 5x_2 = 6$ have a point in common?

45. Does the line $(x, y) = (17, 7) + t(5, 2)$ go through the points $(7, 3)$ and $(2, 1)$?

46. Is the point $(21, 10, 0, 27)$ on the plane $\mathbf{p} = (9, -5, 7, 4) + s(3, -4, 5, -1) + t(6, 2, 1, 7)$?

47. Does the plane $4x_1 - x_2 + 5x_3 = 0$ have this alternative parametric form $(x_1, x_2, x_3) = t(3, 4, 0) + s(-5, 0, 4)$?

48. From the parametric representation of the line through the point given and parallel to the vector, find the linear equation of form $ax_1 + bx_2 = c$.
  a. $(1, 5)$, $\mathbf{v} = (-3, 2)$
  b. $(7, 1)$, $\mathbf{u} = (2, 5)$

49. Find three points in $\mathbb{R}^3$ that are on the plane $\mathbf{v} = (1, 4, 3) + s(0, 2, 1) + t(8, 4, -2)$. What is the equation for this plane?

50. In the linear least squares method, if the line is $y = a + bx$ and the data points (x_i, y_i) are $(1, c_1), (2, c_2), \ldots, (n, c_n)$, then the **overdetermined linear system** is

$$\mathbf{A}\mathbf{x} = \mathbf{b}$$

$$\begin{bmatrix} 1 & 1 \\ 1 & 2 \\ 1 & 3 \\ \vdots & \vdots \\ 1 & n \end{bmatrix} \begin{bmatrix} a \\ b \end{bmatrix} = \begin{bmatrix} c_1 \\ c_2 \\ c_3 \\ \vdots \\ c_n \end{bmatrix}$$

Multiply this system on both sides by $\mathbf{A}^T$ obtaining

$$\mathbf{A}^T \mathbf{A}\mathbf{x} = \mathbf{A}^T \mathbf{b}$$

Show that this is the 2×2 system of **normal equations** that is used in the linear least squares method.

51. (Continuation.) Show that when the least squares method is modified so that the data $(1, d_1), (2, d_2), \ldots, (n, d_n)$ fits a parabola $y = a + bx + cx^2$, the **overdetermined linear system** is

$$\mathbf{A}\mathbf{x} = \mathbf{b}$$

$$\begin{bmatrix} 1 & 1 & 1 \\ 1 & 2 & 4 \\ 1 & 3 & 9 \\ \vdots & \vdots & \vdots \\ 1 & n & n^2 \end{bmatrix} \begin{bmatrix} a \\ b \\ c \end{bmatrix} = \begin{bmatrix} d_1 \\ d_2 \\ d_3 \\ \vdots \\ d_n \end{bmatrix}$$

Find the associated 3×3 normal equations.

COMPUTER EXERCISES 2.2

1. Here is a situation where three biological species compete against each other. We use C for cod, H for herring, and P for plankton. In each case, suitable units have been introduced to represent the population of each species in a certain area of the North Sea. Studies of the local fishery have revealed that in one unit of time, the original population vector $[C, H, P]^T$ has become

$$\begin{bmatrix} 1.0 & -0.2 & 0.4 \\ -0.1 & 1.0 & 0.3 \\ -0.1 & -0.3 & 1.0 \end{bmatrix} \begin{bmatrix} C \\ H \\ P \end{bmatrix}$$

Let the initial vector be $[1, 1, 1]^T$. Using mathematical software such as MATLAB, compute eight generations of this vector, allowing the negative numbers to remain in the calculation. Then recompute the values when any negative value is replaced by 0. Interpret the results. Do the given data predict the herring winning out over the cod?

2. In the predator–prey model described in the text, change the matrix slightly as follows:

$$A = \begin{bmatrix} 0.5 & 0.4 \\ -0.2 & 1.1 \end{bmatrix}$$

Use the same initial vector. What differences are seen and how do you account for them?

3. An amateur runner has kept an account of his best times for running the mile during each year from 1941 to 1974. These times are, in order, 5.94, 6.44, 5.94, 5.94, 5.94, 5.94, 5.96, 6.42, 6.25, 7.00, 6.18, 6.29, 6.83, 6.82, 6.23, 6.20, 6.70, 6.84, 7.09, 7.11, 7.31, 7.30, 7.39, 7.94, 7.92, 7.96, 8.15, 8.25, 8.90, 8.96, 8.90, 9.15, 9.46, 10.07. (The times given are in minutes and are in decimal form.) Find the least squares line for these data, assuming that $T = cY + b$, where T is the time and Y is the year. The parameters c and b are to be determined by the least squares criterion.

Does your work support the conclusion that for this runner T increases by about 6 seconds each year?

4. Consider

x	1	2	3	4	5	6	7	8	9	10
y	441	458	478	493	506	516	523	531	543	551

    a. Using the method of linear least squares, find the linear equation $y = ax + b$ that best fits the data above in the least squares sense. Plot the graph of the line as well as the original data points. *Hint:* See General Exercise 50.

    b. Use a quadratic equation of the form $y = a + bx + cx^2$ to best fit the data in the least squares sense and plot the results. *Hint:* See General Exercise 51.

5. For each set of data, plot the points and compare the linear least squares fit and the quadratic least squares fit.

    a. $(-7, 110), (-3, 70), (1, 21), (5, -35)$
    b. $(-2, -13), (-1, -9), (1, -1), (4, 11)$
    c. $(-1, 22), (1, 28), (3, 30), (5, 28)$

    Hint: See General Exercises 50–51.

2.3   LINEAR TRANSFORMATIONS

> *Cogito, Ergo Sum. I think, therefore I am.*
> —RENÉ DESCARTES (1596–1650)

> *Mathematics seems to endow one with something like a new sense.*
> —CHARLES DARWIN (1809–1882)

Functions, Mappings, and Transformations

The terms *function, mapping, map,* and *transformation* are synonymous. Let us describe what is meant by a **mapping** f from a set X to a set Y. The symbolism $f : X \rightarrow Y$ signifies that f is a mapping from X to Y.

This means that with each **x** in X, a single, particular, element $f(\mathbf{x})$ in Y is associated. For example, one can define a mapping f from $\mathbb{R}$ to $\mathbb{R}$ by writing $f(x) = e^x$. It is essential that the definition of $f(x)$ be unambiguous. For example, one should *not* say: "*Let $f(a,b,c)$ be a root of the quadratic equation $ax^2 + bx + c = 0$.*" Nor should we say: "*Let $f(x)$ be the real number whose sine is x.*" These two descriptions are ambiguous. In the first of these faulty definitions, it is not clear which of the two roots of a quadratic expression is intended. In the second faulty definition, there can be many angles whose sine is x or there may be no angles whose sine is x (if $|x| > 1$). Examples of properly defined functions are given in the following.

These concepts are (or should be) familiar from the calculus, where one is constantly working with mappings $f : \mathbb{R} \to \mathbb{R}$. These are *real-valued functions of a real variable.* In the calculus, one also works with functions of two variables. These are mappings $f : \mathbb{R}^2 \to \mathbb{R}$. In principle, we can have mappings from any set X to any set Y. It is only necessary that for each **x** in X, a unique image $f(\mathbf{x})$ be somehow specified in Y. Sometimes, we do not wish to assign a name to a function but only to describe its effect. The notation $x \mapsto x^3$ illustrates this alternative. One should assiduously avoid confusing a function with one of its values. We do *not* refer to the function x^3. We could refer to a function f defined by $f(x) = x^3$, or to the function $x \mapsto x^3$. In other words, in this situation, one should distinguish the function f from the real number $f(x)$. The point $f(\mathbf{x})$ is called the **image** of **x** created by the function f or, simply, the *image* of **x** by the mapping f.

Domain, Co-domain, and Range

If $f : X \to Y$, we call X the **domain** of f, and write Domain(f) to signify the domain. The set Y is the **co-domain** of f. The **range** of f is the set

$$\text{Range}(f) = \{ f(\mathbf{x}) : \mathbf{x} \in X \}$$

In words, the range of f is the set of points in Y that actually occur as images (under f) of points in X. Obviously, the range of a mapping is a subset of its co-domain. Sometimes the range of a function is the same as the co-domain. In that case, we say that the function maps X **onto** Y. A formal definition is given later in this section for the term "*onto.*"

The essential ingredients of a function $f : X \to Y$ are

- Its domain: X
- Its co-domain: Y
- The description of the image of each point in the domain of f

Various Examples

Here are some examples of functions (mappings, transformations).

Definition	Domain	Co-domain
$f(x) = e^{x^2}$	$\mathbb{R}$	$\mathbb{R}$
$g(x) = \begin{bmatrix} \sin x \\ \cos x \end{bmatrix}$	$\mathbb{R}$	$\mathbb{R}^2$
$h(x) = (\sin x,\ \cos x)$	$\mathbb{R}$	$\mathbb{R}^2$
$R(x, y) = x^2 - 2xy + 7$	$\mathbb{R}^2$	$\mathbb{R}$
$S(x, y) = (3xy,\ 2x^3,\ \tan y)$	$\mathbb{R}^2$	$\mathbb{R}^3$
$T(\mathbf{x}) = \begin{bmatrix} 3x_1 - 2x_2 \\ x_1 + 3x_2 \end{bmatrix}$	$\mathbb{R}^2$	$\mathbb{R}^2$
$\mathbf{x} \mapsto \mathbf{Ax},\ \mathbf{A} \in \mathbb{M}^{m \times n}$	$\mathbb{R}^n$	$\mathbb{R}^m$

EXAMPLE 1 For x in the interval from 0 to π, define $f(x) = \sin(x)$. What are the domain, co-domain, and range of f?

SOLUTION The domain is specified to be the closed interval $[0, \pi]$. The co-domain can be taken to be $\mathbb{R}$. The range is the interval $[0, 1]$. There are many other choices for the co-domain in this example, such as $[0, \infty)$ or $[-1, 5]$ or $[0, 1]$. This last choice coincides with the range of f. Of course, the range of a function is always a subset of its co-domain. ∎

EXAMPLE 2 For $\mathbf{x}$ in $\mathbb{R}^2$, define $f(\mathbf{x}) = f(x_1, x_2) = x_1 x_2$. What are the domain, the co-domain, and the range of f?

SOLUTION The domain is specified as $\mathbb{R}^2$. The co-domain is $\mathbb{R}$, and the range is $\mathbb{R}$. Here we note that any real number can be obtained as the product $x_1 x_2$. ∎

EXAMPLE 3 Let $\mathbb{Z}^+$ denote the set of all positive integers (*Zahlen* is German for *numbers*). Define f by saying that for any n in $\mathbb{Z}^+$, $f(n)$ is the largest proper factor of n. (A **proper factor** of a positive integer m is any positive divisor of m that is smaller than m.) Thus, we have $f(7) = 1$, $f(8) = 4, f(18) = 9$, and so on. What are the domain, the co-domain, and the range of f?

SOLUTION The domain of f is given as $\mathbb{Z}^+$, and we can take the co-domain to be the set of all integers. The range is the set of all positive integers. ∎

EXAMPLE 4 Let $T(\mathbf{x}) = \mathbf{Ax}$ where $\mathbf{A} = \begin{bmatrix} 3 & 1 \\ 2 & 5 \\ 7 & -2 \end{bmatrix}$, $\mathbf{x} \in \mathbb{R}^2$, and $T(\mathbf{x}) \in \mathbb{R}^3$. Let $\mathbf{u} = \begin{bmatrix} 1 \\ 3 \end{bmatrix}$ and $\mathbf{v} = \begin{bmatrix} 3 \\ 1 \end{bmatrix}$.

What are $T(\mathbf{u})$, $T(\mathbf{v})$, $T(\mathbf{u}+\mathbf{v})$, and $T(\mathbf{u}) + T(\mathbf{v})$?

SOLUTION We have

$$T(\mathbf{u}) = \mathbf{Au} = \begin{bmatrix} 3 & 1 \\ 2 & 5 \\ 7 & -2 \end{bmatrix} \begin{bmatrix} 1 \\ 3 \end{bmatrix} = \begin{bmatrix} 6 \\ 17 \\ 1 \end{bmatrix}$$

$$T(\mathbf{v}) = \mathbf{Av} = \begin{bmatrix} 3 & 1 \\ 2 & 5 \\ 7 & -2 \end{bmatrix} \begin{bmatrix} 3 \\ 1 \end{bmatrix} = \begin{bmatrix} 10 \\ 11 \\ 19 \end{bmatrix}$$

Observe that

$$T(\mathbf{u}) + T(\mathbf{v}) = \begin{bmatrix} 6 \\ 17 \\ 1 \end{bmatrix} + \begin{bmatrix} 10 \\ 11 \\ 19 \end{bmatrix} = \begin{bmatrix} 16 \\ 28 \\ 20 \end{bmatrix}$$

Also, we obtain $\mathbf{u} + \mathbf{v} = \begin{bmatrix} 1 \\ 3 \end{bmatrix} + \begin{bmatrix} 3 \\ 1 \end{bmatrix} = \begin{bmatrix} 4 \\ 4 \end{bmatrix}$. Hence, we have

$$T(\mathbf{u}+\mathbf{v}) = T\left(\begin{bmatrix} 4 \\ 4 \end{bmatrix}\right) = \begin{bmatrix} 3 & 1 \\ 2 & 5 \\ 7 & -2 \end{bmatrix} \begin{bmatrix} 4 \\ 4 \end{bmatrix} = \begin{bmatrix} 16 \\ 28 \\ 20 \end{bmatrix}$$

This function T has the property $T(\mathbf{x} + \mathbf{y}) = T(\mathbf{x}) + T(\mathbf{y})$. ∎

Injective and Surjective Mappings

Here the concepts of a mapping being injective or surjective enter. (Synonyms for these technical terms are *one-to-one* and *onto*, respectively.)

DEFINITION

A mapping is **injective** *(or* **one-to-one***) if it maps distinct entities in the domain to distinct points in the range.*

Equivalently, a mapping $f : X \to Y$ is injective (one-to-one) if the equation $f(\mathbf{u}) = f(\mathbf{v})$ implies $\mathbf{u} = \mathbf{v}$. A third way of saying it is that each element of the co-domain can be the image of at most one element of the domain.

None of the functions in the preceding Examples 1, 2, and 3 is injective. For the function in Example 1, $f(0) = f(\pi)$. For the function in Example 2, $f(1, 2) = f(2, 1)$. For the function in Example 3, $f(10) = f(15)$. The function $x \mapsto x^2$ is injective on the domain $[0, \infty)$ but not on the domain $\mathbb{R}$. (Why?) The function $x \mapsto x^3$ is injective on $\mathbb{R}$. (Why?)

> **DEFINITION**
>
> *A mapping is* **surjective** *(or* **onto***) if its range and co-domain are the same.*

In other words, a mapping $f : X \to Y$ is surjective (onto) if every element of the co-domain is the image of some point in the domain: for each $\mathbf{y} \in Y$ there is an $\mathbf{x} \in X$ such that $f(\mathbf{x}) = \mathbf{y}$.

> **DEFINITION**
>
> *A mapping is* **bijective** *if it is both injective and surjective (one-to-one and onto).*

An injective mapping need not be surjective. Conversely, a surjective mapping need not be injective. In Figure 2.19, diagrams illustrate the four possible combinations of injective and surjective.

The mapping $x \mapsto \sin(x)$ is surjective if we define its domain to be $\mathbb{R}$ and its co-domain to be the interval $[-1, 1]$, but not if we take its co-domain

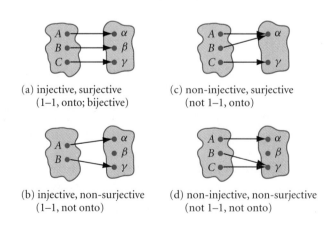

(a) injective, surjective
(1–1, onto; bijective)

(c) non-injective, surjective
(not 1–1, onto)

(b) injective, non-surjective
(1–1, not onto)

(d) non-injective, non-surjective
(not 1–1, not onto)

FIGURE 2.19 Injective and surjective mappings.

to be $\mathbb{R}$. The mapping $x \mapsto x^3$ is surjective if we define its domain and range to be $\mathbb{R}$. The mapping $x \mapsto x^2$ is surjective if we define its domain to be $\mathbb{R}$ and its co-domain is $[0, \infty)$. Common terminology would be that the function $x \mapsto x^2$ maps $\mathbb{R}$ onto $[0, \infty)$. The preposition *onto* emphasizes that every point in $[0, \infty)$ is an image.

The preceding example shows that the definition of a function is *not* complete until we have specified its domain, its co-domain, and the *rule* by which the images are computed. (Being traditionalists, we prefer to keep the word *onto* grammatically as a preposition, and seldom use it as an adjective.)

We will discover that a transformation T defined by $T(\mathbf{x}) = \mathbf{Ax}$ maps $\mathbb{R}^n$ onto $\mathbb{R}^m$ if and only if the columns of $\mathbf{A}$ span $\mathbb{R}^m$. If every point in $\mathbb{R}^m$ is an image, then we say that T maps $\mathbb{R}^n$ **onto** $\mathbb{R}^m$. For example, the transformation T defined by

$$T(\mathbf{x}) = \begin{bmatrix} 1 & 3 \\ 3 & 4 \end{bmatrix} \begin{bmatrix} x_1 \\ x_2 \end{bmatrix}$$

maps $\mathbb{R}^2$ onto $\mathbb{R}^2$. Every point $\mathbf{b}$ in $\mathbb{R}^2$ is an image because

$$\left[\begin{array}{cc|c} 1 & 3 & b_1 \\ 3 & 4 & b_2 \end{array} \right] \sim \left[\begin{array}{cc|c} 1 & 0 & \tilde{b}_1 \\ 0 & 1 & \tilde{b}_2 \end{array} \right]$$

The pair of columns $\begin{bmatrix} 1 \\ 3 \end{bmatrix}, \begin{bmatrix} 3 \\ 4 \end{bmatrix}$ spans $\mathbb{R}^2$.

Linear Transformations

The concept of linearity is pervasive in linear algebra. When a mapping has the property of linearity, matters that are otherwise difficult can become simple. For this reason, *linearity* has become a broad area of study in applied mathematics, containing many theories, facts, and formulas—probably far beyond what any one person can comprehend!

> **DEFINITION**
>
> *A mapping f from $\mathbb{R}^n$ to $\mathbb{R}^m$ is* **linear** *if $f(a\mathbf{x} + b\mathbf{y}) = af(\mathbf{x}) + bf(\mathbf{y})$ for all vectors $\mathbf{x}$ and $\mathbf{y}$ in $\mathbb{R}^n$ and for all scalars a and b.*

The same definition will serve when we study general vector spaces in Section 2.4. The domain and co-domain of a linear mapping must be vector spaces in order that the equation of linearity shall make sense. Notice that if we let $a = b = 0$ in the definition of linearity, then the result is $f(\mathbf{0}) = \mathbf{0}$.

By induction, we can prove that a linear mapping f satisfies the following equation, for all $k \geq 1$:

$$f\left(\sum_{i=1}^{k} \alpha_i \mathbf{x}_i\right) = \sum_{i=1}^{k} \alpha_i f(\mathbf{x}_i)$$

THEOREM 1

Let $\mathbf{A}$ be an $m \times n$ matrix. The mapping $\mathbf{x} \mapsto \mathbf{Ax}$ is linear from $\mathbb{R}^n$ to $\mathbb{R}^m$.

PROOF This is a property pointed out in Section 1.2. Recall that $\mathbf{Ax}$ is defined to be $\sum_{j=1}^{n} x_j \mathbf{a}_j$, where $\mathbf{a}_j$ denotes the jth column in $\mathbf{A}$. Thus, we obtain

$$\mathbf{A}(\alpha\mathbf{x} + \beta\mathbf{y}) = \sum_{j=1}^{n} (\alpha\mathbf{x} + \beta\mathbf{y})_j \mathbf{a}_j = \sum_{j=1}^{n} [(\alpha\mathbf{x})_j + (\beta\mathbf{y})_j]\mathbf{a}_j = \sum_{j=1}^{n} (\alpha x_j + \beta y_j)\mathbf{a}_j$$

$$= \sum_{j=1}^{n} \alpha x_j \mathbf{a}_j + \sum_{j=1}^{n} \beta y_j \mathbf{a}_j = \alpha \sum_{j=1}^{n} x_j \mathbf{a}_j + \beta \sum_{j=1}^{n} y_j \mathbf{a}_j = \alpha\mathbf{Ax} + \beta\mathbf{Ay} \qquad \blacksquare$$

Here are some additional examples in which we ask the question: Are these transformations linear? First, consider the mapping $T : \mathbb{R}^3 \to \mathbb{R}^3$ defined by

$$T(x_1, x_2, x_3) = T\left(\begin{bmatrix} x_1 \\ x_2 \\ x_3 \end{bmatrix}\right) = \begin{bmatrix} 2x_1 + x_3 \\ 3x_1 + x_2 + 4 \\ 5x_1 + x_2 - x_3 \end{bmatrix}$$

This mapping is not linear because $T(\mathbf{0}) = (0, 4, 0)$. Next, look at the mapping $T : \mathbb{R}^2 \to \mathbb{R}^2$ given by

$$T(x_1, x_2) = T\left(\begin{bmatrix} x_1 \\ x_2 \end{bmatrix}\right) = \begin{bmatrix} x_1^2 + x_2^2 \\ x_2 \end{bmatrix}$$

It is not linear, since

$$T(2\mathbf{x}) = ((2x_1)^2 + (2x_2)^2, 2x_2) \neq (2x_1^2 + 2x_2^2, 2x_2) = 2T(\mathbf{x})$$

Finally, another mapping $T : \mathbb{R}^2 \to \mathbb{R}^2$ is specified by

$$T(x_1, x_2) = T\left(\begin{bmatrix} x_1 \\ x_2 \end{bmatrix}\right) = \begin{bmatrix} x_2 \\ x_1 \end{bmatrix}$$

Yes, it is a linear transformation because $T\left(\begin{bmatrix} x_1 \\ x_2 \end{bmatrix}\right) = \begin{bmatrix} 0 & 1 \\ 1 & 0 \end{bmatrix}\begin{bmatrix} x_1 \\ x_2 \end{bmatrix}$

and therefore $T(\mathbf{x}) = \mathbf{Ax}$, where $\mathbf{A} = \begin{bmatrix} 0 & 1 \\ 1 & 0 \end{bmatrix}$. (Theorem 1 applies.)

EXAMPLE 5 Define a mapping f from $\mathbb{R}^2$ to $\mathbb{R}^3$ by the formula

$$f(s,t) = f\left(\begin{bmatrix} s \\ t \end{bmatrix}\right) = \begin{bmatrix} 2s + 3t \\ -s + 5t \\ 4s - 3t \end{bmatrix}$$

Is it a linear map?

SOLUTION One way to find out is to set $\mathbf{x} = (x_1, x_2)$ and $\mathbf{y} = (y_1, y_2)$. Then we have

$$\begin{aligned}
f(\alpha\mathbf{x} + \beta\mathbf{y}) &= f\big(\alpha(x_1,\ x_2) + \beta(y_1,\ y_2)\big) \\
&= f\big((\alpha x_1,\ \alpha x_2) + (\beta y_1,\ \beta y_2)\big) \\
&= f\big((\alpha x_1 + \beta y_1,\ \alpha x_2 + \beta y_2)\big) \\
&= \big(2(\alpha x_1 + \beta y_1) + 3(\alpha x_2 + \beta y_2), \\
&\quad\ - (\alpha x_1 + \beta y_1) + 5(\alpha x_2 + \beta y_2), \\
&\quad\ 4(\alpha x_1 + \beta y_1) - 3(\alpha x_2 + \beta y_2)\big) \\
&= \alpha(2x_1 + 3x_2,\ -x_1 + 5x_2,\ 4x_1 - 3x_2) \\
&\quad\ + \beta(2y_1 + 3y_2,\ -y_1 + 5y_2,\ 4y_1 - 3y_2) \\
&= \alpha f(\mathbf{x}) + \beta f(\mathbf{y})
\end{aligned}$$

Surely there is a better way to do this! Could it be a special case of Theorem 1? *Yes!* Here is how to see this connection. Simply write the defining equation like this:

$$f(s,t) = f\left(\begin{bmatrix} s \\ t \end{bmatrix}\right) = \begin{bmatrix} 2s + 3t \\ -s + 5t \\ 4s - 3t \end{bmatrix} = s\begin{bmatrix} 2 \\ -1 \\ 4 \end{bmatrix} + t\begin{bmatrix} 3 \\ 5 \\ -3 \end{bmatrix} = \begin{bmatrix} 2 & 3 \\ -1 & 5 \\ 4 & -3 \end{bmatrix}\begin{bmatrix} s \\ t \end{bmatrix}$$

This equation has the form $f(\mathbf{x}) = \mathbf{Ax}$, and Theorem 1 applies. ∎

Using Matrices to Define Linear Maps

The preceding example suggests that every linear transformation from $\mathbb{R}^n$ to $\mathbb{R}^m$ is given by a matrix. This formal result is proved next.

THEOREM 2

Let T be a linear transformation from $\mathbb{R}^n$ to $\mathbb{R}^m$. Then there is an $m \times n$ matrix $\mathbf{A}$ such that $T(\mathbf{x}) = \mathbf{A}\mathbf{x}$ for all $\mathbf{x}$ in $\mathbb{R}^n$.

PROOF Define the **standard unit vectors** in $\mathbb{R}^n$ by

$$\mathbf{e}_1 = \begin{bmatrix} 1 \\ 0 \\ 0 \\ \vdots \\ 0 \end{bmatrix}, \quad \mathbf{e}_2 = \begin{bmatrix} 0 \\ 1 \\ 0 \\ \vdots \\ 0 \end{bmatrix}, \quad \cdots, \quad \mathbf{e}_n = \begin{bmatrix} 0 \\ 0 \\ 0 \\ \vdots \\ 1 \end{bmatrix}$$

Then any $\mathbf{x}$ in $\mathbb{R}^n$ can be written as

$$\mathbf{x} = \sum_{j=1}^{n} x_j \mathbf{e}_j$$

By the linearity of T, we have

$$T(\mathbf{x}) = T\left(\sum_{j=1}^{n} x_j \mathbf{e}_j\right) = \sum_{j=1}^{n} x_j T(\mathbf{e}_j)$$

Let $\mathbf{A}$ be the $m \times n$ matrix whose columns are $T(\mathbf{e}_j)$, for $j = 1, 2, \ldots, n$. Then our previous equation shows that $T(\mathbf{x}) = \mathbf{A}\mathbf{x}$, where $\mathbf{x}$ is written as a column vector. ∎

EXAMPLE 6 Let $\mathbf{u} = (1, 1, 3)$, $\mathbf{v} = (3, 2, -2)$, $L(\mathbf{u}) = (4, 1, 1, 1)$, and $L(\mathbf{v}) = (-5, 1, -3, 3)$. Assume further that L is a linear transformation from $\mathbb{R}^3$ to $\mathbb{R}^4$. If $\mathbf{w} = (5, 4, 4)$ and $\mathbf{y} = (2, 1, 7)$, what are $L(\mathbf{w})$ and $L(\mathbf{y})$?

SOLUTION We must express **w** and **y** as linear combinations of **u** and **v**. The work of doing so is as follows:

$$
[\ \mathbf{u} \quad \mathbf{v} \ | \ \mathbf{w} \quad \mathbf{y}\ ] =
\left[\begin{array}{rr|rr}
1 & 3 & 5 & 2 \\
1 & 2 & 4 & 1 \\
3 & -2 & 4 & 7
\end{array}\right]
\sim
\left[\begin{array}{rr|rr}
1 & 0 & 2 & -1 \\
0 & 1 & 1 & 1 \\
0 & 0 & 0 & 1
\end{array}\right]
$$

This shows that $\mathbf{w} = 2\mathbf{u} + \mathbf{v}$; whence

$$
L(\mathbf{w}) = L(2\mathbf{u} + \mathbf{v}) = 2
\begin{bmatrix} 4 \\ 1 \\ 1 \\ 1 \end{bmatrix}
+
\begin{bmatrix} -5 \\ 1 \\ -3 \\ 3 \end{bmatrix}
=
\begin{bmatrix} 3 \\ 4 \\ -1 \\ 5 \end{bmatrix}
$$

Our work also shows that **y** is not a linear combination of **u** and **v**. Thus, we cannot compute $L(\mathbf{y})$ from the information given. (The vector **y** is not in Span{**u**, **v**}.) ∎

THEOREM 3

A linear transformation from $\mathbb{R}^n$ to $\mathbb{R}^m$ is completely determined by the images of the standard unit vectors in $\mathbb{R}^n$, which we have written as $\mathbf{e}_1, \mathbf{e}_2, \ldots, \mathbf{e}_n$.

PROOF The proof of Theorem 2 contains all the ingredients for the proof of Theorem 3. An extension of this theorem occurs as Theorem 15 in Section 5.2. ∎

Notice that the proof of Theorem 2 gives a prescription for constructing the matrix that accompanies a linear transformation from $\mathbb{R}^n$ to $\mathbb{R}^m$. Namely, we compute the vectors $T(\mathbf{e}_j)$, for $1 \le j \le n$, and put them into the matrix **A** as columns, preserving the correct order.

If **A** is the matrix of a linear transformation T, so that $T(\mathbf{x}) = \mathbf{Ax}$, then the columns of **A** should be $T(\mathbf{e}_1)$, $T(\mathbf{e}_2)$, and so on. Consider this example:

$$
T(\mathbf{x}) =
\begin{bmatrix} x_1 - x_2 \\ -2x_1 + x_2 \\ x_1 \end{bmatrix}
= x_1
\begin{bmatrix} 1 \\ -2 \\ 1 \end{bmatrix}
+ x_2
\begin{bmatrix} -1 \\ 1 \\ 0 \end{bmatrix}
$$

Can we find the matrix **A** of this transformation? *Yes*, because

$$T(\mathbf{e}_1) = T\left(\begin{bmatrix} 1 \\ 0 \end{bmatrix}\right) = \begin{bmatrix} 1 \\ -2 \\ 1 \end{bmatrix}$$

$$T(\mathbf{e}_2) = T\left(\begin{bmatrix} 0 \\ 1 \end{bmatrix}\right) = \begin{bmatrix} -1 \\ 1 \\ 0 \end{bmatrix}$$

Hence, we obtain $\mathbf{A} = \begin{bmatrix} 1 & -1 \\ -2 & 1 \\ 1 & 0 \end{bmatrix}$

> **EXAMPLE 7** Find the matrix that accompanies the linear transformation of Example 5.

SOLUTION We must compute $f(1,0) = (2, -1, 4)$ and $f(0, 1) = (3, 5, -3)$. Then the matrix having these vectors as columns is the matrix sought: $\mathbf{A} = \begin{bmatrix} 2 & 3 \\ -1 & 5 \\ 4 & -3 \end{bmatrix}$. It occurs also in the second solution in Example 5.

∎

Injective and Surjective Linear Transformations

In this subsection, the properties of injectivity and surjectivity for linear transformations are explored.

> **THEOREM 4**
>
> *Let **A** be an $m \times n$ matrix. For the linear map $\mathbf{x} \mapsto \mathbf{Ax}$ to be surjective (onto), it is necessary and sufficient that the columns of **A** span $\mathbb{R}^m$.*

PROOF It is obvious that the map $\mathbf{x} \mapsto \mathbf{Ax}$ maps $\mathbb{R}^n$ into $\mathbb{R}^m$. Each image, **Ax**, is a linear combination of the columns of **A**. For the range to be all of $\mathbb{R}^m$, it is necessary and sufficient that every point in $\mathbb{R}^m$ be an image—that is, a linear combination of columns of **A**. ∎

> **DEFINITION**
>
> *The **kernel** of a linear map T is $\{\mathbf{x} : T(\mathbf{x}) = \mathbf{0}\}$ and is denoted by $\mathrm{Ker}(T)$.*

This definition is in harmony with the definition of the kernel of a matrix in Section 1.3. We simply identify any matrix $\mathbf{A}$ with the linear map $\mathbf{x} \mapsto \mathbf{Ax}$. Later, after discussing vector spaces and subspaces we will see that if T is not injective, then there exist two different vectors that have the same image under T. Say, $T(\mathbf{x}) = T(\mathbf{y})$. By linearity, $T(\mathbf{x} - \mathbf{y}) = \mathbf{0}$. Hence, the kernel of T contains the nonzero vector $\mathbf{x} - \mathbf{y}$. Conversely, if the kernel of T is not $\mathbf{0}$, then there exists a nonzero element that is mapped to $\mathbf{0}$. Because $\mathbf{0}$ is certainly mapped to $\mathbf{0}$, we have here a violation of the injective property.

THEOREM 5

In order that the linear map $\mathbf{x} \mapsto \mathbf{Ax}$ be injective (one-to-one), it is necessary and sufficient that the kernel of $\mathbf{A}$ contain only the $\mathbf{0}$-vector.

PROOF We use the elementary idea that $\mathbf{Ax} = \mathbf{Ay}$ if and only if $\mathbf{A}(\mathbf{x} - \mathbf{y}) = \mathbf{0}$. If the map $\mathbf{x} \mapsto \mathbf{Ax}$ is not injective, we find a pair $\mathbf{u} \neq \mathbf{v}$ such that $\mathbf{Au} = \mathbf{Av}$. Then $\mathbf{u} - \mathbf{v}$ is a nonzero vector in the kernel of $\mathbf{A}$. Conversely, if the map $\mathbf{x} \mapsto \mathbf{Ax}$ is injective, then the equation $\mathbf{Au} = \mathbf{Av}$ implies $\mathbf{u} = \mathbf{v}$. In particular, by taking $\mathbf{v} = \mathbf{0}$, we see that the equation $\mathbf{Au} = \mathbf{0}$ implies that $\mathbf{u} = \mathbf{0}$. ∎

Effects of Linear Transformations

In the space $\mathbb{R}^2$, it is possible to draw sketches showing how a linear transformation behaves. It is at first surprising that for a linear transformation $T : \mathbb{R}^2 \to \mathbb{R}^2$, the effect of T on any point of $\mathbb{R}^2$ can be predicted from a knowledge of $T(\mathbf{e}_1)$ and $T(\mathbf{e}_2)$, where $\mathbf{e}_1 = (1, 0)$ and $\mathbf{e}_2 = (0, 1)$. This important fact is the preceding Theorem 3. It has its analog in all vector spaces. (See Theorem 15 in Section 5.2.) In $\mathbb{R}^2$, an arbitrary vector $\mathbf{x}$ can be written

$$\mathbf{x} = (x_1, x_2) = x_1(1, 0) + x_2(0, 1) = x_1 \mathbf{e}_1 + x_2 \mathbf{e}_2$$

Then, if T is linear, we have

$$T(\mathbf{x}) = T(x_1 \mathbf{e}_1 + x_2 \mathbf{e}_2) = x_1 T(\mathbf{e}_1) + x_2 T(\mathbf{e}_2)$$

The matrix whose columns are $T(\mathbf{e}_1)$ and $T(\mathbf{e}_2)$ will be referred to here as the **transformation matrix** corresponding to T. Refer to the preceding Theorem 2 and its proof. It is worth noting that $T(\mathbf{e}_1)$ and $T(\mathbf{e}_2)$ can be points in $\mathbb{R}^2$ assigned quite arbitrarily, and then T will still be linear.

The next few examples illustrate the effects of various linear transformations on $\mathbb{R}^2$. In higher dimensions, similar mappings are available, but they are more difficult to illustrate. The accompanying figures go with these examples. In each case, we show the effects of the transformation on the two standard unit elementary vectors, $\mathbf{e}_1 = (1, 0)$ and $\mathbf{e}_2 = (0, 1)$. The effects on the unit square having vertices $(0, 0), (1, 0), (0, 1)$, and $(1, 1)$ are also shown.

EXAMPLE 8 Explain the identity map from $\mathbb{R}^2$ to $\mathbb{R}^2$.

SOLUTION It leaves every point of $\mathbb{R}^2$ unchanged. We won't waste ink drawing a picture of that! It can be defined by the expression $(x_1, x_2) \mapsto (x_1, x_2)$, and it is certainly linear. Its matrix is, of course, the so-called *identity matrix*,

$$\mathbf{I}_2 = \begin{bmatrix} 1 & 0 \\ 0 & 1 \end{bmatrix}$$

∎

EXAMPLE 9 Describe the linear mapping that has the matrix $\begin{bmatrix} 0 & -1 \\ 1 & 0 \end{bmatrix}$

SOLUTION It is a **rotation** transformation that rotates every vector through the angle $90°$ counterclockwise. Notice that the first column in the matrix is the image of $\mathbf{e}_1$ and the second column is the image of $\mathbf{e}_2$. Another description of this mapping is $(x_1, x_2) \mapsto (-x_2, x_1)$. This transformation is one of a family, since the angle through which the vectors are rotated can be any angle, positive or negative. See Figure 2.20 and the future Theorem 7 (p. 150). ∎

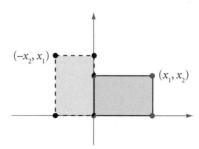

FIGURE 2.20 Rotation of 90° counterclockwise.

EXAMPLE 10 Discuss the transformation whose matrix is $\begin{bmatrix} 1 & 0 \\ 0 & 0 \end{bmatrix}$

SOLUTION It projects every point onto the horizontal axis. This is an example of an **orthogonal projection**, studied in greater detail in Section 7.1. Notice that in applying this transformation we lose information: $(x_1, x_2) \mapsto (x_1, 0)$. From a knowledge of $T(\mathbf{x})$, one cannot recover $\mathbf{x}$, because the value of the second coordinate is lost in the application of T to $\mathbf{x}$. See Figure 2.21. The effect of this transformation is similar to the shadow of the geometric object at high noon. ∎

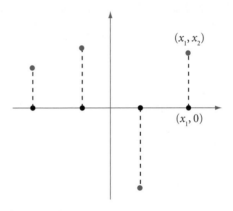

FIGURE 2.21 Projection onto horizontal axis (orthogonal projection).

EXAMPLE 11 Find the transformation for the matrix $\begin{bmatrix} 2 & 0 \\ 0 & 2 \end{bmatrix}$

SOLUTION This linear map moves all points farther from the origin by a factor of 2. It is called a **dilation** because the transformation produces an image that is the same shape as the original but is a different size. (In particular, it is called **enlargement** when the image is larger and a **reduction** when it is smaller.) It is described by $(x_1, x_2) \mapsto (2x_1, 2x_2)$ or $\mathbf{x} \mapsto 2\mathbf{x}$. See Figure 2.22. ∎

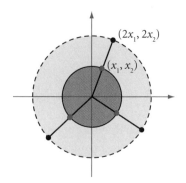

FIGURE 2.22 Dilation by a factor of 2.

EXAMPLE 12 Explain the linear map whose matrix is $\begin{bmatrix} 1 & 0 \\ 1 & 1 \end{bmatrix}$

SOLUTION It is a **shear** along the vertical axis. This is a technical term from elasticity theory. Another description is $(x_1, x_2) \mapsto (x_1, x_1 + x_2)$.

See Figure 2.23. The effect of a shear transformation looks as if the geometric object is being pushed over in a stiff wind. ■

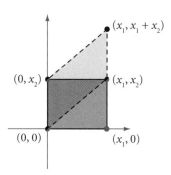

FIGURE 2.23 Shear along vertical axis.

EXAMPLE 13 Discuss the mapping whose matrix is $\begin{bmatrix} 1 & 0 \\ 0 & -1 \end{bmatrix}$

SOLUTION It is a **reflection** about the horizontal axis. The first coordinate of a point is unchanged, but the second component changes sign: $(x_1, x_2) \mapsto (x_1, -x_2)$. See Figure 2.24. ■

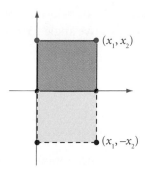

FIGURE 2.24 Reflection about horizontal axis.

EXAMPLE 14 Characterize the mapping whose matrix is $\begin{bmatrix} 0 & 1 \\ 1 & 0 \end{bmatrix}$

SOLUTION It is another **reflection**, this time a reflection across the diagonal line $x_1 = x_2$. A formula is $(x_1, x_2) \mapsto (x_2, x_1)$. See Figure 2.25. ∎

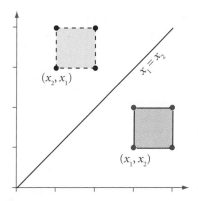

FIGURE 2.25 Reflection across diagonal line.

Figures 2.20–2.25 illustrate the above transformations, and each linear transformation is completely determined by the images of some key points such as $(1, 0)$ and $(0, 1)$. (Why?)

EXAMPLE 15 Describe the mapping whose effect is to rotate points counterclockwise through an angle such as $\varphi = 45°$.

SOLUTION The transformation that rotates all vectors counterclockwise through the positive angle 45° has the following matrix:

$$\begin{bmatrix} 1/\sqrt{2} & -1/\sqrt{2} \\ 1/\sqrt{2} & 1/\sqrt{2} \end{bmatrix}$$

A formula is $(x_1, x_2) \mapsto \left((x_1 - x_2)/\sqrt{2}, (x_1 + x_2)/\sqrt{2}\right)$. For details, see Theorem 7 below. ∎

Effects of Transformations on Geometrical Figures

In trying to understand how a linear transformation acts on a geometrical figure, it is helpful to use this easy result about linear transformations: They always *preserve* line segments.

THEOREM 6

A linear transformation maps one line segment into another.

PROOF Let L be a linear transformation from $\mathbb{R}^n$ into $\mathbb{R}^m$. Let S be a line segment in $\mathbb{R}^n$. Is $L(S)$ a line segment, too? Suppose that $S = \{t\mathbf{x} + (1-t)\mathbf{y} : 0 \le t \le 1\}$. (This is the best way to describe a line segment.) As t traverses the interval $[0, 1]$, the point $t\mathbf{x} + (1-t)\mathbf{y}$ traverses the line segment joining $\mathbf{x}$ and $\mathbf{y}$. For example, when $t = \frac{1}{2}$, we get the midpoint of that line segment, and when $t = 0$ or $t = 1$, we get $\mathbf{x}$ or $\mathbf{y}$. The image of S by the linear transformation L consists of the points $L(t\mathbf{x} + (1-t)\mathbf{y})$, where $0 \le t \le 1$. Since this is the same as $tL(\mathbf{x}) + (1-t)L(\mathbf{y})$, by the linearity of L, we get the line segment joining $L(\mathbf{x})$ to $L(\mathbf{y})$. It may reduce to a single point. ∎

THEOREM 7

In $\mathbb{R}^2$, a counterclockwise rotation of every point by an angle φ is a linear transformation whose matrix is

$$\begin{bmatrix} \cos\varphi & -\sin\varphi \\ \sin\varphi & \cos\varphi \end{bmatrix}$$

PROOF Recall the polar coordinate system, whereby a point in $\mathbb{R}^2$ can be described by giving its distance, r, from the origin and an angular parameter, θ:

$$\begin{bmatrix} x_1 \\ x_2 \end{bmatrix} = \begin{bmatrix} r\cos\theta \\ r\sin\theta \end{bmatrix}$$

If the point is rotated by an angle φ, the new point will be $\begin{bmatrix} r\cos(\theta + \varphi) \\ r\sin(\theta + \varphi) \end{bmatrix}$.
At this stage, we require these ancient formulas from trigonometry:

$$\begin{cases} \cos(\theta + \varphi) = \cos\theta\cos\varphi - \sin\theta\sin\varphi \\ \sin(\theta + \varphi) = \cos\theta\sin\varphi + \sin\theta\cos\varphi \end{cases}$$

By using these formulas, we obtain

$$T(\mathbf{x}) = r\begin{bmatrix} \cos\theta\cos\varphi - \sin\theta\sin\varphi \\ \cos\theta\sin\varphi + \sin\theta\cos\varphi \end{bmatrix} = r\begin{bmatrix} \cos\varphi & -\sin\varphi \\ \sin\varphi & \cos\varphi \end{bmatrix}\begin{bmatrix} \cos\theta \\ \sin\theta \end{bmatrix}$$

$$= \begin{bmatrix} \cos\varphi & -\sin\varphi \\ \sin\varphi & \cos\varphi \end{bmatrix}\begin{bmatrix} r\cos\theta \\ r\sin\theta \end{bmatrix} = \begin{bmatrix} \cos\varphi & -\sin\varphi \\ \sin\varphi & \cos\varphi \end{bmatrix}\begin{bmatrix} x_1 \\ x_2 \end{bmatrix}$$

∎

Notice that when φ is a positive angle the rotation is counterclockwise and when φ is a negative angle the rotation is clockwise. For example, when $\varphi = 90°$, the rotation matrix from Theorem 7 is the matrix of Example 9. See Figure 2.26.

Composition of Two Linear Mappings

For functions f and g whose domains and ranges are properly related, we define $f \circ g$ by the equation

$$(f \circ g)(\mathbf{x}) = f(g(\mathbf{x}))$$

The construction $f \circ g$ is called the **composition** of f and g. It is meaningful if the range of g is contained in the domain of f. In the calculus, you have seen many examples of this construction. In particular, the **chain**

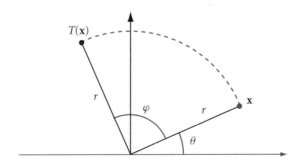

FIGURE 2.26 Rotate counterclockwise through angle $\varphi = 90°$.

rule tells us how to differentiate the composition of two functions. Notice that $f \circ g$ is usually not the same as $g \circ f$. For example, $\sin(x^2)$ is not the same as $\sin^2(x)$. The relation $f \circ (g \circ h) = (f \circ g) \circ h$ is easily established:

$$[f \circ (g \circ h)](x) = f[(g \circ h)(x)] = f[g(h(x))]$$
$$= (f \circ g)(h(x)) = [(f \circ g) \circ h](x)$$

If S and T are linear transformations such that the range of T is contained in the domain of S, then the composition of these two maps is defined by

$$(S \circ T)(x) = S(T(x))$$

THEOREM 8

If S and T are linear maps obeying the preceding equation, then $S \circ T$ is linear.

PROOF We have $(S \circ T)(a\mathbf{x} + b\mathbf{y}) = S(T(a\mathbf{x} + b\mathbf{y})) = S(aT(\mathbf{x}) + bT(\mathbf{y})) = aS(T(\mathbf{x})) + bS(T(\mathbf{y})) = a(S \circ T)(\mathbf{x}) + b(S \circ T)(\mathbf{y})$. ∎

The following theorem shows the relationship between the composition of two linear transformations and matrix–matrix multiplication, a subject that will be taken up again in Section 3.1.

THEOREM 9

Let $\mathbf{A}$ be an $m \times n$ matrix, and let $\mathbf{B}$ be an $n \times k$ matrix. Define $S(\mathbf{x}) = \mathbf{A}\mathbf{x}$ and $T(\mathbf{y}) = \mathbf{B}\mathbf{y}$. Then $(S \circ T)(\mathbf{x}) = (\mathbf{AB})\mathbf{x}$.

PROOF We have $(S \circ T)(\mathbf{x}) = S(T(\mathbf{x})) = \mathbf{A}(\mathbf{B}\mathbf{x}) = (\mathbf{AB})(\mathbf{x})$. ∎

This elegant state of affairs was understood by Arthur Cayley in 1850. The definition of matrix multiplication was framed precisely to make this equation valid. See Section 3.1 for details.

Application: Data Smoothing

Data collected from measured quantities usually include noise, which may manifest itself in the form of erratic points or sudden disturbances in the data. To minimize the impact of noise, we can smooth out these local fluctuations in the raw data using **averaging**. For example, suppose that we have an infinite sequence of data points $x_0, x_1, x_2, x_3, x_4, \dots$. We can transform it into a new sequence by successive averaging: $y_i = \frac{1}{2}(x_i + x_{i+1})$ for $i = 0, 1, 2, \ldots$. This can be viewed as a matrix–vector multiplication, $\mathbf{y} = \mathbf{A}\mathbf{x}$ with vectors $\mathbf{x} = (x_0, x_1, x_2, \ldots)$, $\mathbf{y} = (y_0, y_1, y_2, \ldots)$, and the matrix $\mathbf{A}$ has infinitely many rows and columns, obeying the formula $a_{ii} = a_{i,i+1} = \frac{1}{2}$. All other entries in $\mathbf{A}$ are zero. Such a matrix is **upper bi-diagonal**. If the data sequence is finite and has exactly n terms, we can use the replacement given previously, except that we define $y_n = x_n$.

To smooth the data further, we can average twice. This double smoothing can be written in terms of the original data as $z_i = \frac{1}{2}(y_i + y_{i-1}) = \frac{1}{4}(x_i + 2x_{i+1} + x_{i+2})$ for $i = 0, 1, 2, \ldots$, which is equivalent to a matrix–vector multiplication $\mathbf{z} = \mathbf{B}\mathbf{x}$ with the vector $\mathbf{x} = (z_0, z_1, z_2, \ldots)$ and the matrix $\mathbf{B} = $ upper-tridiagonal$(\frac{1}{4}, \frac{1}{2}, \frac{1}{4})$. A moving average can be taken over n points given by $y_{k-n} = (1/n)\sum_{i=k-n}^{k-1} x_i$ for $k = n, n+1, n+2, \ldots$. Averaging works well if there is no trend or cyclic pattern in the data, but many other schemes are used in practice. For more details on this and other examples of matrix transformations, see P. Davis [1965].

SUMMARY 2.3

- Let f be a function that maps X into Y. Then the domain of f is X and the co-domain of f is Y. The range of f is $\{f(\mathbf{x}) : \mathbf{x} \in X\}$.

- A mapping f is injective if the equation $f(\mathbf{x}) = f(\mathbf{y})$ occurs only when $\mathbf{x} = \mathbf{y}$.

- A mapping is surjective if its range and co-domain are the same.

- If f is injective and surjective (one-to-one and onto), then f is said to be bijective.

- A mapping f from $\mathbb{R}^n$ to $\mathbb{R}^m$ is linear if $f(a\mathbf{x} + b\mathbf{y}) = af(\mathbf{x}) + bf(\mathbf{y})$ for all $\mathbf{x}, \mathbf{y} \in \mathbb{R}^n$ and for all $a, b \in \mathbb{R}$.

- If $\mathbf{A}$ is an $m \times n$ matrix, then the mapping $\mathbf{x} \mapsto \mathbf{A}\mathbf{x}$ is linear from $\mathbb{R}^n$ to $\mathbb{R}^m$.

- If T is a linear transformation from $\mathbb{R}^n$ to $\mathbb{R}^m$, then there is an $m \times n$ matrix $\mathbf{A}$ such that $T(\mathbf{x}) = \mathbf{A}\mathbf{x}$ for all $\mathbf{x} \in \mathbb{R}^n$.

- A linear transformation T from $\mathbb{R}^n$ to $\mathbb{R}^m$ is completely determined by the images of the n elementary vectors $\mathbf{e}_1, \mathbf{e}_2, \ldots, \mathbf{e}_n$ in $\mathbb{R}^n$.

- Let $\mathbf{A}$ be an $m \times n$ matrix, and let $T(\mathbf{x}) = \mathbf{A}\mathbf{x}$. The mapping T will be surjective if and only if the columns of $\mathbf{A}$ span $\mathbb{R}^m$.

- The kernel of a linear transformation, T, is $Ker(T) = \{\mathbf{x} : T(\mathbf{x}) = \mathbf{0}\}$.

- A linear transformation T maps line segments into line segments.

- In $\mathbb{R}^2$, a rotation of every point by an angle φ is a linear transformation of the matrix

$$\begin{bmatrix} \cos\varphi & -\sin\varphi \\ \sin\varphi & \cos\varphi \end{bmatrix}$$

- If $S(\mathbf{x}) = \mathbf{A}\mathbf{x}$ and $T(\mathbf{y}) = \mathbf{B}\mathbf{y}$, then the matrix for $S \circ T$ is $\mathbf{A}\mathbf{B}$. Here $\mathbf{A}$ is an $m \times n$ matrix, $\mathbf{B}$ is an $n \times k$ matrix, and $\mathbf{A}\mathbf{B}$ is an $m \times k$ matrix.

- If S and T are linear maps, then $S \circ T$ is linear.

- A linear transformation T is injective if and only if its kernel is $\mathbf{0}$.

- In $\mathbb{R}^2$ some basic linear transformations are these: Identity: $\mathbf{I}_2 = \begin{bmatrix} 1 & 0 \\ 0 & 1 \end{bmatrix}$; Rotation:

$\begin{bmatrix} 0 & -1 \\ 1 & 0 \end{bmatrix}$; Orthogonal projection: $\begin{bmatrix} 1 & 0 \\ 0 & 0 \end{bmatrix}$;

Dilation (double): $\begin{bmatrix} 2 & 0 \\ 0 & 2 \end{bmatrix}$; Shear: $\begin{bmatrix} 1 & 0 \\ 1 & 1 \end{bmatrix}$;

Reflection about horizontal axis: $\begin{bmatrix} 1 & 0 \\ 0 & -1 \end{bmatrix}$;

Reflection about $x_1 = x_2$ line: $\begin{bmatrix} 0 & 1 \\ 1 & 0 \end{bmatrix}$

KEY CONCEPTS 2.3

Function, map, linear mapping, image, domain, co-domain, range, injective (one-to-one), surjective (onto), bijective (one-to-one, onto) kernel of a linear map, effects of transformations on geometrical figures (rotation, orthogonal projection, dilation, shear, reflection), data smoothing

GENERAL EXERCISES 2.3

1. Determine whether there is a linear map T such that $T(1, 3, 2) = (5, 2, 2)$, $T(2, -1, 3) = (11, -6, 3)$, and $T(4, -9, 5) = (23, -22, 6)$.

2. Find the matrix for the transformation $T : \mathbb{R}^3 \to \mathbb{R}^3$ defined by this formula:

$$T(\mathbf{x}) = (3x_2 - 2x_3) \begin{bmatrix} 3 \\ -4 \\ 2 \end{bmatrix}$$

$$+ (2x_1 + 4x_3) \begin{bmatrix} -2 \\ 5 \\ -3 \end{bmatrix}$$

$$+ (x_1 - 3x_2) \begin{bmatrix} 1 \\ -3 \\ 2 \end{bmatrix}$$

3. The transformations $\mathbf{A} = \begin{bmatrix} 1 & 0 \\ 0 & 1 \end{bmatrix}$ and

$\mathbf{B} = \begin{bmatrix} 0 & 1 \\ 1 & 0 \end{bmatrix}$ seem to have the same effect on the unit square shown in Examples 8 and 14. Explain this phenomenon.

4. Is there a linear map that sends $(1, 3, 2)$ to $(5, 0, 1)$, $(2, -1, 3)$ to $(4, 2, 1)$, and $(6, 1, 1)$ to $(7, 3, 9)$?

5. Is there a linear map that sends $(1, 0)$ and $(0, 1)$ to $(1, 1)$?

6. Give as much information as you can for the linear transformation T defined by

$$T(\mathbf{x}) = \mathbf{Ax}, \text{ where } \mathbf{A} = \begin{bmatrix} 1 & 3 & 2 \\ 2 & 1 & 4 \\ -1 & 0 & 2 \\ 4 & 1 & 1 \end{bmatrix}$$

For example, identify its domain; identify its co-domain; describe its kernel; describe its range; determine whether it is injective; determine whether it is surjective.

7. (Continuation.) Repeat the preceding exercise with the matrix $\mathbf{A}^T$.

8. Let $A = \begin{bmatrix} 1 & -3 & 2 \\ 0 & 1 & -4 \\ 3 & -5 & -9 \end{bmatrix}$

Is there some vector $\mathbf{x}$ such that $T(\mathbf{x}) = (6, -7, -9)$ for the linear transformation $T(\mathbf{x}) = A\mathbf{x}$?

9. Let $A = \begin{bmatrix} 1 & 3 & 9 & 2 \\ 1 & 0 & 3 & -4 \\ 0 & 1 & 2 & 3 \\ -2 & 3 & 0 & 5 \end{bmatrix}$

Is there some vector $\mathbf{x}$ such that $T(\mathbf{x}) = (-1, 3, -1, 4)$ for the linear transformation $T(\mathbf{x}) = A\mathbf{x}$?

10. Let $A = \begin{bmatrix} 1 & 2 \\ 3 & 4 \\ 0 & 0 \end{bmatrix}$

Establish that the mapping T defined by $T(\mathbf{x}) = A\mathbf{x}$ is one-to-one but not surjective, whereas the mapping $\mathbf{x} \mapsto A^T\mathbf{x}$ is surjective but not one-to-one.

11. Define $f : \mathbb{R}^3 \rightarrow \mathbb{R}^3$ by the formula $f(\mathbf{x}) = \left(x_1 + 2(x_2 + 7), 3x_1 - x_2 + x_3, 5x_1 - x_3\right)$ where $\mathbf{x} = (x_1, x_2, x_3)$. Is f linear? Explain.

12. Explain what is meant by saying that a mapping is *one-to-one*. Let $A = \begin{bmatrix} 1 & 2 \\ 3 & 4 \\ 5 & 6 \end{bmatrix}$.

Define a linear transformation from $\mathbb{R}^2$ to $\mathbb{R}^3$ by the equation $T(\mathbf{x}) = A\mathbf{x}$. Determine whether T is one-to-one. Explain your approach to this problem. Cite any theorems you use.

13. Define $f : \mathbb{R}^3 \rightarrow \mathbb{R}^3$ by $f(\mathbf{x}) = \left(9x_1 + 5(x_2 + 7x_3), 5x_1 - 12x_2 + 27x_3, 55x_1 - 42x_3\right)$ where $\mathbf{x} = (x_1, x_2, x_3)$. Is f linear? Explain.

14. Let f be continuous and real-valued on $[0, 1]$ and have the property $f(0) = f(1) = 0$. Assume that $\max_{0 \le t \le 1} f(t) > 0$.

Do there exist points t_1 and t_2 such that $f(t_1) = f(t_2)$ and $0 < t_1 < t_2 < 1$?

15. Determine whether there are linear transformations L and M that have these properties:
a. $L(1, 2) = (2, 3)$; $L(3, 4) = (-5, 1)$; $L(5, 6) = (-12, -1)$
b. $M(1, 3, 7) = (4, 2, 1, 5)$; $M(2, -1, 3) = (5, 1, 0, 1)$; $M(-7, 14, 6) = (-13, 1, 3, 11)$

16. Find, if possible, an injective linear map from $\mathbb{R}^5$ to $\mathbb{R}^4$.

17. Let $A = \begin{bmatrix} a & b & c \\ d & e & f \end{bmatrix}$

Find values of a, b, c, d, e, f so that the mapping $\mathbf{x} \mapsto A\mathbf{x}$ is one-to-one.

18. Let $A = \begin{bmatrix} a & b \\ c & d \\ e & f \end{bmatrix}$

Find values of a, b, c, d, e, f so that the transformation $\mathbf{x} \mapsto A\mathbf{x}$ maps $\mathbb{R}^2$ onto $\mathbb{R}^3$.

19. Identify these mappings as being one-to-one or not. Explain why or why not.
a. $f : \mathbb{R} \rightarrow \mathbb{R}, f(x) = x^3$
b. $f : \mathbb{R} \rightarrow \mathbb{R}, f(x) = x^2$
c. $f : [0, \infty) \rightarrow \mathbb{R}, f(x) = x^2$
d. $L : \mathbb{R}^2 \rightarrow \mathbb{R}^3, L(\mathbf{x}) = \begin{bmatrix} 1 & 2 \\ 2 & 1 \\ 0 & 0 \end{bmatrix} \begin{bmatrix} x_1 \\ x_2 \end{bmatrix}$

20. Identify these mappings as being surjective or not. Explain why or why not.
a. $f : \mathbb{R} \rightarrow \mathbb{R}, f(x) = \sin x$
b. $f : \mathbb{R} \rightarrow [-1, 1], f(x) = \sin x$
c. $L : \mathbb{R}^2 \rightarrow \mathbb{R}^3, L(\mathbf{x}) = \begin{bmatrix} 1 & 2 \\ 2 & 1 \\ 0 & 0 \end{bmatrix} \begin{bmatrix} x_1 \\ x_2 \end{bmatrix}$
d. $f : \mathbb{R} \rightarrow \mathbb{R}, f(x) = x^2$

21. The following 2×2 matrices correspond to linear transformations that are **reflections** in $\mathbb{R}^2$. Describe them geometrically.

a. $\begin{bmatrix} 1 & 0 \\ 0 & -1 \end{bmatrix}$ **b.** $\begin{bmatrix} -1 & 0 \\ 0 & 1 \end{bmatrix}$

c. $\begin{bmatrix} 0 & 1 \\ 1 & 0 \end{bmatrix}$ **d.** $\begin{bmatrix} 0 & -1 \\ 1 & 0 \end{bmatrix}$

e. $\begin{bmatrix} -1 & 0 \\ 0 & -1 \end{bmatrix}$

22. The following 2×2 matrices correspond to linear transformations that are **contractions** or **expansions** in $\mathbb{R}^2$. Describe them geometrically.

a. $\begin{bmatrix} c & 0 \\ 0 & 1 \end{bmatrix}, 0 < c < 1$ **b.** $\begin{bmatrix} c & 0 \\ 0 & 1 \end{bmatrix}, c > 1$

c. $\begin{bmatrix} 1 & 0 \\ 0 & c \end{bmatrix}, 0 < c < 1$ **d.** $\begin{bmatrix} 1 & 0 \\ 0 & c \end{bmatrix}, c > 1$

23. The following 2×2 matrices correspond to linear transformations that are **shears** in $\mathbb{R}^2$. Describe them geometrically.

a. $\begin{bmatrix} 1 & c \\ 0 & 1 \end{bmatrix}, c < 0$ **b.** $\begin{bmatrix} 1 & 0 \\ c & 1 \end{bmatrix}, c < 0$

c. $\begin{bmatrix} 1 & c \\ 0 & 1 \end{bmatrix}, c > 0$ **d.** $\begin{bmatrix} 1 & 0 \\ c & 1 \end{bmatrix}, c > 0$

24. The following 2×2 matrices correspond to linear transformations that are **projections** in $\mathbb{R}^2$. Describe them geometrically.

a. $\begin{bmatrix} 1 & 0 \\ 0 & 0 \end{bmatrix}$ **b.** $\begin{bmatrix} 0 & 0 \\ 0 & 1 \end{bmatrix}$ **c.** $\begin{bmatrix} \frac{1}{2} & \frac{1}{2} \\ \frac{1}{2} & \frac{1}{2} \end{bmatrix}$

25. The following 2×2 matrices correspond to linear transformations that are **contractions** or **dilations** in $\mathbb{R}^2$. Describe them geometrically.

a. $\begin{bmatrix} c & 0 \\ 0 & c \end{bmatrix}, 0 < c < 1$ **b.** $\begin{bmatrix} c & 0 \\ 0 & c \end{bmatrix}, c > 1$

26. The following 2×2 matrices correspond to linear transformations that are **rotations** in $\mathbb{R}^2$. Describe them geometrically.

a. $\begin{bmatrix} 0 & -1 \\ 1 & 0 \end{bmatrix}$ **b.** $\begin{bmatrix} 0 & 1 \\ -1 & 0 \end{bmatrix}$

c. $\begin{bmatrix} \sqrt{2}/2 & -\sqrt{2}/2 \\ \sqrt{2}/2 & \sqrt{2}/2 \end{bmatrix}$ **d.** $\begin{bmatrix} \sqrt{2}/2 & \sqrt{2}/2 \\ -\sqrt{2}/2 & \sqrt{2}/2 \end{bmatrix}$

27. Let $T(x_1, x_2, x_3) = (x_1 + x_2 + 3, 3x_2 - 4x_1, 3x_3)$. Determine whether T is a linear transformation.

28. Let two functions (or *mappings*) be given, say T and S. Their **composition** is denoted by $S \circ T$ and is defined by the equation $(S \circ T)(x) = S(T(x))$. This is meaningful if the range of T is contained in the domain of S. Give examples to illustrate that $S \circ T$ is usually not the same as $T \circ S$. Explain why, if S and T are linear, then so is $S \circ T$.

29. Is there a matrix $\mathbf{A}$ such that the corresponding linear transformation T has the property that $Ker(T)$ and $Range(T)$ have a nonzero vector in common? (The linear map T is defined by $T(\mathbf{x}) = \mathbf{Ax}$.)

30. Find a linear map T from $\mathbb{R}^2$ to $\mathbb{R}^2$ such that T is neither $\mathbf{0}$ nor the identity map, but $T^2 = T$. (The notation T^2 denotes $T \circ T$, as in Exercise 28.)

31. Find a linear map T from $\mathbb{R}^2$ to $\mathbb{R}^2$ such that $T \neq \mathbf{0}$ and $T^2 = \mathbf{0}$.

32. Establish that a linear transformation will map a linearly dependent set into a linearly dependent set. Will it map a linearly independent set into a linearly independent set?

33. Let $\mathbf{r}, \mathbf{s}, \mathbf{u}, \mathbf{v}, \mathbf{w}$, and $\mathbf{z}$ be six prescribed points in $\mathbb{R}^3$. How can we find a linear transformation T such that $T(\mathbf{r}) = \mathbf{s}$, $T(\mathbf{u}) = \mathbf{v}$, and $T(\mathbf{w}) = \mathbf{z}$? When is this problem insoluble? Test your method on a specific case. One could start with the case in which the set $\{\mathbf{r}, \mathbf{u}, \mathbf{w}\}$ is linearly independent.

34. If a mapping T from $\mathbb{R}^n$ to $\mathbb{R}^m$ satisfies the equation $T(\alpha\mathbf{x} - \mathbf{y}) = \alpha T(\mathbf{x}) - T(\mathbf{y})$ for all scalars α and all vectors $\mathbf{x}$ and $\mathbf{y}$, does it follow that T is linear?

35. Consider the map T from $\mathbb{R}^2$ to $\mathbb{R}^2$ defined as follows: For each $\mathbf{x}$, $T(\mathbf{x})$ is its reflection in the line given by the formula $x_1 = x_2$. Is this map linear? What is the formula for T?

36. Let T project every point of $\mathbb{R}^2$ perpendicularly onto the horizontal axis. Is T linear? Find a formula for T. (The projection is along a line perpendicular to the horizontal axis.)

37. Let T project every point of $\mathbb{R}^2$ onto the horizontal axis, but the line of projection meets the horizontal axis at an angle of $45°$. Find a formula for T.

38. Find the matrix for the transformation that projects every point in $\mathbb{R}^3$ perpendicularly onto the x_2x_3-plane.

39. Find the matrix for the transformation that projects each point of $\mathbb{R}^3$ perpendicularly onto the vertical axis.

40. Find the matrix for the transformation that projects each point in $\mathbb{R}^3$ perpendicularly onto the x_1x_2-plane and then rotates it in that plane through an angle of $+90°$.

41. Consider a linear transformation T and a pair of two different points, $\mathbf{u}$ and $\mathbf{v}$. Assume that $T(\mathbf{u}) = T(\mathbf{v})$ although $\mathbf{u} \neq \mathbf{v}$. Then there are infinitely many such pairs of points. Establish this. Is this assertion true for continuous real-valued functions defined on the real line?

42. Give an argument why (for any mapping f) if $f(x) = f(y)$ for some pair of distinct points, x and y, then f cannot have an inverse.

43. Define $T : \mathbb{R}^3 \to \mathbb{R}^4$ by the equation $T(x_1, x_2, x_3) = (x_1 - 3x_3, 2x_2 + x_1, x_3 + x_1 + x_2, x_3 - x_1)$. Is T one-to-one? Is T linear? Is T surjective? Justify your answers in such a way that you demonstrate your understanding of these terms. Find the matrix $\mathbf{A}$ such that $T(\mathbf{x}) = \mathbf{A}\mathbf{x}$ for all $\mathbf{x} \in \mathbb{R}^3$.

44. Define a mapping f from $\mathbb{R}^3$ to $\mathbb{R}^4$ by the equation $f(\mathbf{x}) = (x_1 + x_2 - 3x_3, 2x_1 - x_2 + x_3, 3x_1 + 2x_2, x_1 + 3x_2 + x_3)$. What are the domain and the co-domain of f? Is f surjective? Is f injective? What is $Ker(f)$? Express the kernel of f as the span of a small set of vectors. Do the same for the range of f. If S is the line segment joining $(1, 3, -2, 1)$ to $(0, -1, 3, 2)$, what is the image $f[S]$?

45. Consider the shear in Example 12. Verify that the image of the unit square is the parallelogram shown in Figure 2.23 and that its area is 1.

46. Return to General Exercise 31 and find all the matrices such that $\mathbf{A}^2 = \mathbf{0}$.

47. Consider the operation on a vector that rearranges the components in increasing order. For example, $T(3, 2, 6) = (2, 3, 6)$. Is this a linear mapping?

48. For each of the following exercises, plot the original sequence. Average once and then twice, plotting the new sequence each time. Produce a graph by connecting the dots with straight lines. Compare the resulting plots. Do they yield smoother graphs each time?

a. $2, 1, 3, 3, 4, 5, 3, 4, 3, 2, 1, 2$

b. $2, 3, 7, 2, 3, 9, 1, 10$

c. $25, 15, 45, 15, 20, 30, 20, 50$

49. (Continuation.) Repeat, using matrix–vector multiplication to do the data smoothing.

50. (Continuation.) Determine the recurrence relations and the matrices for
   a. Averaging three times.
   b. Moving average with $n = 3$ done once and then twice.

51. If F and G are two maps, their composition (in that order) is denoted by the symbol $F \circ G$ and is defined by the equation $(F \circ G)(\mathbf{x}) = F(G(\mathbf{x}))$. Is the composition of two injective maps also injective?

52. Establish that if L is a one-to-one linear transformation from $\mathbb{R}^k$ to $\mathbb{R}^p$, then $p \geq k$.

53. Show the reasoning behind this assertion: If L is a linear transformation from $\mathbb{R}^n$ onto $\mathbb{R}^m$, then $m \leq n$.

54. Let L be a linear transformation from one vector space into another, say, $L : V \to W$. Suppose that there exists a map f, not assumed to be linear, such that $f(L(\mathbf{v})) = \mathbf{v}$ for all $\mathbf{v}$ in V and $L(f(\mathbf{w})) = \mathbf{w}$ for all $\mathbf{w}$ in W. Explain why f is linear. (The map f has domain W and co-domain V.) Also, explain why L is surjective and injective.

55. Give an example of a function whose co-domain is different from its range. Give an example of a mapping that is not one-to-one but becomes one-to-one if the domain is changed. Is this always possible?

56. Establish that if $X = \text{Span}\{\mathbf{u}_1, \mathbf{u}_2, \dots, \mathbf{u}_k\}$ and $Y = \text{Span}\{\mathbf{u}_{k+1}, \mathbf{u}_{k+2}, \dots, \mathbf{u}_m\}$, then $X + Y = \text{Span}\{\mathbf{u}_1, \mathbf{u}_2, \dots, \mathbf{u}_m\}$. The notation $X + Y$ denotes the set $\{\mathbf{x} + \mathbf{y} : \mathbf{x} \in X, \mathbf{y} \in Y\}$.

57. Let $\mathbf{A}$ be an $m \times n$ matrix and $\mathbf{B}$ an $n \times k$ matrix. These give rise to linear transformations T and S defined by $T(\mathbf{x}) = \mathbf{Ax}$ and $S(\mathbf{y}) = \mathbf{By}$. Establish that $(T \circ S)(\mathbf{y}) = (\mathbf{AB})\mathbf{y}$.

58. Let $L : \mathbb{P}_3 \mapsto \mathbb{P}_2$ be as given in the following, where there are two cases. In each case find the kernel of L.
   a. $(Lp)(t) = 2p'(t) + 3tp''(t)$
   b. $Lp = 2p' + 3p''$

59. If $\{\mathbf{u}_1, \mathbf{u}_2, \dots, \mathbf{u}_n\}$ is a linearly dependent set of vectors and if L is a linear transformation, can it happen that the set $\{L(\mathbf{u}_1), L(\mathbf{u}_2), \dots, L(\mathbf{u}_n)\}$ is a linearly independent set?

60. Discuss the function $f : \mathbb{R} \to \mathbb{R}$ defined by $f(x) = \cos(x)$. What reasonable restrictions of the domain and co-domain make f injective (*one-to-one*) and surjective (*onto*)?

61. For any transformation, T, linear or not, we call a point $\mathbf{w}$ a fixed point of T if $T(\mathbf{w}) = \mathbf{w}$. Find all the fixed points of the linear transformations in Examples 8–15.

62. Find vectors $\mathbf{u}, \mathbf{v}$, and $\mathbf{w}$ in $\mathbb{R}^2$ and a matrix $\mathbf{A}$ such that $\{\mathbf{Au}, \mathbf{Av}, \mathbf{Aw}\}$ is linearly independent.

63. A function F is said to be **one-to-one** or **injective** if the condition $F(x) = F(y)$ implies that $x = y$. Show that in solving an equation $h(x) = g(x)$ it is permissible to replace it by $F(h(x)) = F(g(x))$ if F is one-to-one.

64. Give some examples where $f \circ g = g \circ f$.

65. Show that if we wish to solve $f(x) = g(x)$, where f and g are given functions, then we may get *extraneous* solutions by squaring both sides: $[f(x)]^2 = [g(x)]^2$.

66. For $a, b \in \mathbb{R}$, the complex number $a + bi$ corresponds to the point (a, b) in $\mathbb{C}$. Show that we can write a 2×2 matrix $\mathbf{A}$ as follows:

$$\mathbf{A} = \begin{bmatrix} a & -b \\ b & a \end{bmatrix} = \begin{bmatrix} r & 0 \\ 0 & r \end{bmatrix} \begin{bmatrix} \cos\theta & -\sin\theta \\ \sin\theta & \cos\theta \end{bmatrix}$$

where $r = \sqrt{a^2 + b^2}$. Give a geometric explanation of the effect of the transformation involving $\mathbf{A}$. (See Appendix B on complex arithmetic.)

67. Explain why the function $x \mapsto x^2$ is injective on the domain $[0, \infty)$, but not on the domain $\mathbb{R}$. Also, explain why the function $x \mapsto x^3$ is injective on $\mathbb{R}$.

68. For Figures 2.20–2.25, explain why the associated linear transformations are completely determined by the images of some key points. Identify these vectors for each figure.

69. Explain why the chain rule is a composition mapping and give some examples of it. Are $f \circ g$ and $g \circ f$ the same for this rule? Explain why or give a counterexample.

70. Establish and display the matrices $\mathbf{A}$ and $\mathbf{B}$ in the subsection on data smoothing.

COMPUTER EXERCISES 2.3

Consider the following linear transformations $T(\mathbf{x}) = \mathbf{A}\mathbf{x}$ from $\mathbb{R}^2$ to $\mathbb{R}^2$. What are the corresponding linear transformations from $\mathbb{R}^3$ to $\mathbb{R}^3$? Explore mathematical software packages and routines that carry out these transformations. Plot the effects of applying them to some ridged bodies such as squares, circles, and triangles, for example.

1. Zero/Identity: $\begin{bmatrix} 0 & 0 \\ 0 & 0 \end{bmatrix}$, $\begin{bmatrix} 1 & 0 \\ 0 & 1 \end{bmatrix}$

2. Reflection

    a. about x-axis: $\begin{bmatrix} 1 & 0 \\ 0 & -1 \end{bmatrix}$

    b. about y-axis: $\begin{bmatrix} -1 & 0 \\ 0 & 1 \end{bmatrix}$

    c. about line $x = y$: $\begin{bmatrix} 0 & 1 \\ 1 & 0 \end{bmatrix}$

    d. about origin: $\begin{bmatrix} -1 & 0 \\ 0 & -1 \end{bmatrix}$

3. Rotation

    a. by $90°$ counterclockwise:

$$\begin{bmatrix} 0 & -1 \\ 1 & 0 \end{bmatrix}$$

    b. by $\theta°$ counterclockwise:

$$\begin{bmatrix} \cos\theta & -\sin\theta \\ \sin\theta & \cos\theta \end{bmatrix}$$

4. Orthogonal Projection

    a. on x-axis: $\begin{bmatrix} 1 & 0 \\ 0 & 0 \end{bmatrix}$

    b. on y-axis: $\begin{bmatrix} 0 & 0 \\ 0 & 1 \end{bmatrix}$

5. a. Contraction/Dilation by α:

$$\begin{bmatrix} \alpha & 0 \\ 0 & \alpha \end{bmatrix}$$

b. Scaling by α and β:

$$\begin{bmatrix} \alpha & 0 \\ 0 & \beta \end{bmatrix}$$

c. Squeezing by $\gamma \neq 0$:

$$\begin{bmatrix} \gamma & 0 \\ 0 & 1/\gamma \end{bmatrix}$$

d. Shear by ω parallel to x-axis:

$$\begin{bmatrix} 1 & \omega \\ 0 & 1 \end{bmatrix}$$

e. Shear by ω parallel to y-axis:

$$\begin{bmatrix} 1 & 0 \\ \omega & 1 \end{bmatrix}$$

2.4 GENERAL VECTOR SPACES

> *Mathematics is a game played according to certain simple rules and meaningless marks on paper.*
>
> —DAVID HILBERT (1862–1943)

> *When you try to prove a theorem, you don't just list the hypotheses, and start to reason. What you do is trial and error, experimentation, guesswork.*
>
> —PAUL R. HALMOS (1916–2006)

Vector Spaces

There are many vector spaces other than the spaces $\mathbb{R}^n$ that we have been using heretofore. What type of mathematical constructs will be called "*vector spaces*"? The structure they must have is set out in the following formal definition.

DEFINITION

*A **vector space** is a set V of elements (called **vectors**), together with two algebraic operations, called **vector addition** and **scalar multiplication**. The following axioms must be fulfilled:*

1. *If **u** and **v** are vectors, then **u** + **v** is defined and is an element of V. (Closure axiom for addition)*

(continued)

2. *For all* **u** *and* **v** *in V,* **u** + **v** = **v** + **u**.
   (Commutativity of vector addition)
3. *For any three vectors* **u**, **v**, *and* **w**, (**u** + **v**) + **w** = **u** + (**v** + **w**).
   (Associativity of vector addition)
4. *There is an element* **0** *in V such that for all* **u** *in V,* **u** + **0** = **u**.
   (Existence of a zero vector)
5. *For each vector* **u** *there is at least one element* $\tilde{\mathbf{u}}$ *in V such that* $\tilde{\mathbf{u}}$ + **u** = **0**.
   (Existence of additive inverses)
6. *If* **u** *is a vector and* α *is a scalar, then the product* α**u** *is defined and is an element of V.*
   (Closure axiom for scalar–vector product)
7. *For any scalar* α *and vectors* **u** *and* **v**, α(**u** + **v**) = α**u** + α**v**.
   (Distributive law: scalar times vectors)
8. *If* α *and* β *are scalars and* **u** *is a vector, then* $(\alpha + \beta)$**u** = α**u** + β**u**.
   (Distributive law: scalar sum times a vector)
9. *If* α *and* β *are scalars and* **u** *is a vector,* $\alpha(\beta$**u**$)$ = $(\alpha\beta)$**u**.
   (Associativity of scalar–vector product)
10. *For each* **u** *in V,* $1 \cdot$ **u** = **u**.
    (Multiplication of unit scalar times vector)

The reader will notice that these requirements are exactly the salient properties of the spaces $\mathbb{R}^n$ that we called attention to in Section 2.1.

We have already noticed that, with the definitions adopted in Section 2.1, $\mathbb{R}^n$ is a vector space. Thus, we have at once an infinite number of different vector spaces, because n can be any natural number.

The word *scalar* occurs frequently in the context of vector spaces. It almost always will mean a real number, but in some situations there is good reason for allowing complex numbers as scalars. (A complex number is of the form $\alpha + i\beta$, where α and β are real numbers and $i^2 = -1$. See Appendix B.)

It turns out that the element $\tilde{\mathbf{u}}$ in Axiom 5 is unique for each vector **u**. It is usually denoted by $-$**u**. The axiom then states that $(-\mathbf{u}) + \mathbf{u} = \mathbf{0}$ for all **u**. We further drop the parentheses and use Axiom 2 in this last equation to arrive at **u** $-$ **u** = 0.

Theorems on Vector Spaces

To illustrate the deductions that can be drawn from the preceding axioms, we consider five theorems valid *in all vector spaces*. We can use only the axioms to prove Theorem 1. But then we can use the axioms *and* Theorem 1 to prove Theorem 2, and so on.

THEOREM 1

In any vector space, if c is a scalar, then $c\,\mathbf{0} = \mathbf{0}$.

PROOF Let $\mathbf{x} = c\,\mathbf{0}$. We want to prove that $\mathbf{x} = \mathbf{0}$.

$$
\begin{aligned}
\mathbf{x} &= \mathbf{x} + \mathbf{0} & \text{(Axiom 4)} \\
&= \mathbf{x} + (\widetilde{\mathbf{x}} + \mathbf{x}) & \text{(Axiom 5)} \\
&= \mathbf{x} + (\mathbf{x} + \widetilde{\mathbf{x}}) & \text{(Axiom 2)} \\
&= (\mathbf{x} + \mathbf{x}) + \widetilde{\mathbf{x}} & \text{(Axiom 3)} \\
&= (c\,\mathbf{0} + c\,\mathbf{0}) + \widetilde{\mathbf{x}} & \text{(Definition of } \mathbf{x}) \\
&= c\,(\mathbf{0} + \mathbf{0}) + \widetilde{\mathbf{x}} & \text{(Axiom 7)} \\
&= c\,\mathbf{0} + \widetilde{\mathbf{x}} & \text{(Axiom 4)} \\
&= \mathbf{x} + \widetilde{\mathbf{x}} & \text{(Definition of } \mathbf{x}) \\
&= \widetilde{\mathbf{x}} + \mathbf{x} & \text{(Axiom 2)} \\
&= \mathbf{0} & \text{(Axiom 5)}
\end{aligned}
$$

See General Exercise 19 for another proof of Theorem 1. ∎

THEOREM 2

In a vector space, if $\mathbf{x}$ is a vector and c is a scalar such that $c\mathbf{x} = \mathbf{0}$, then either $c = 0$ or $\mathbf{x} = \mathbf{0}$.

PROOF Assume the hypotheses, and suppose that $c \neq 0$. Then c^{-1} exists, and we can multiply the equation $c\mathbf{x} = \mathbf{0}$ by c^{-1}, arriving at $1 \cdot \mathbf{x} = c^{-1}\mathbf{0}$. By Theorem 1 and Axiom 10, this becomes $\mathbf{x} = \mathbf{0}$. ∎

THEOREM 3

In a vector space, for each $\mathbf{x}$, the point $\widetilde{\mathbf{x}}$ is uniquely determined.

PROOF The crucial property of $\widetilde{\mathbf{x}}$ is that $\widetilde{\mathbf{x}} + \mathbf{x} = \mathbf{0}$. Suppose, then, that for some $\mathbf{x}$ we have both $\mathbf{u} + \mathbf{x} = \mathbf{0}$ and $\mathbf{v} + \mathbf{x} = \mathbf{0}$. Can we conclude that $\mathbf{u}$ must equal $\mathbf{v}$? It is probably *not* obvious, but one proof goes like this:

$$
\begin{aligned}
\mathbf{u} &= \mathbf{u} + \mathbf{0} && \text{(Axiom 4)} \\
&= \mathbf{u} + (\mathbf{v} + \mathbf{x}) && \text{(Hypothesis)} \\
&= \mathbf{u} + (\mathbf{x} + \mathbf{v}) && \text{(Axiom 2)} \\
&= (\mathbf{u} + \mathbf{x}) + \mathbf{v} && \text{(Axiom 3)} \\
&= \mathbf{0} + \mathbf{v} && \text{(Hypothesis)} \\
&= \mathbf{v} + \mathbf{0} && \text{(Axiom 2)} \\
&= \mathbf{v} && \text{(Axiom 4)}
\end{aligned}
$$

Notice that in this proof each equality requires specific justification and the axioms are used in the order indicated. ∎

THEOREM 4

In a vector space, every vector $\mathbf{x}$ satisfies the equation $0\mathbf{x} = \mathbf{0}$.

PROOF One proof proceeds as follows, and the axioms used are indicated in each step.

$$
\begin{aligned}
0\mathbf{x} &= 0\mathbf{x} + \mathbf{0} && \text{(Axiom 4)} \\
&= 0\mathbf{x} + (\mathbf{x} + \widetilde{\mathbf{x}}) && \text{(Axioms 5, 2)} \\
&= (0\mathbf{x} + 1 \cdot \mathbf{x}) + \widetilde{\mathbf{x}} && \text{(Axioms 3, 10)} \\
&= \mathbf{x} + \widetilde{\mathbf{x}} && \text{(Axioms 8, 10)} \\
&= \mathbf{0} && \text{(Axioms 2, 5)}
\end{aligned}
$$

Verify the use of the axioms in the order indicated. ∎

THEOREM 5

In any vector space, we have $(-1)\mathbf{x} = \widetilde{\mathbf{x}}$.

PROOF Following the proof above, we have

$$
\begin{aligned}
[(-1)\mathbf{x}] + \mathbf{x} &= [(-1)\mathbf{x}] + 1\mathbf{x} && \text{(Axiom 10)} \\
&= [(-1) + 1]\mathbf{x} && \text{(Axiom 8)} \\
&= 0\mathbf{x} && \text{(Ordinary arithmetic)} \\
&= \mathbf{0} && \text{(Theorem 4)}
\end{aligned}
$$

Consequently $\widetilde{\mathbf{x}} = (-1)\mathbf{x}$ by Theorem 3. ∎

Before going on, some remarks should be made. First, the element $\tilde{\mathbf{x}}$ is written as $-\mathbf{x}$ or $(-1)\mathbf{x}$. Theorem 5 justifies this. Second, the structure that is described by the ten axioms is technically a *real vector space*. This means that the constants appearing in Axioms 6–10 are real numbers. The real numbers in this context are often called **scalars** to distinguish them from the elements of V, which are then called **vectors**. (Thus, the answer to the question "*What is a vector?*" can be given truthfully as "*It is an element of a vector space.*")

Later in the book, we shall broaden our perspective and discuss vector spaces with complex numbers as the scalars. Certain problems in vector space theory (notably eigenvalue problems) may have no solution if we restrict ourselves to the real numbers as scalars. When one speaks of a *real vector space*, it means only that the scalars are taken to be real numbers. In principle, other *fields* can be used for the scalars. If a particular field, F, is used, one speaks of a *vector space over the field F*. For information about fields, consult a textbook on abstract algebra or search the World Wide Web on the Internet.

Various Examples

We present a number of examples to illustrate these concepts.

EXAMPLE 1 Is the set of all polynomials having degree no greater than 3 a vector space?

SOLUTION Some elementary polynomials are $p_0(t) = 1$, $p_1(t) = t$, $p_2(t) = t^2$, and $p_3(t) = t^3$. Notice that any polynomial of degree at most 3 is a linear combination of the four polynomials p_0, p_1, p_2, and p_3. Thus, we can write

$$a_0 + a_1 t + a_2 t^2 + a_3 t^3 = a_0 p_0(t) + a_1 p_1(t) + a_2 p_2(t) + a_3 p_3(t)$$

The set we have described is a vector space—it being assumed that the usual algebraic operations are used. One can add two polynomials of degree at most 3, and the result is another polynomial of degree at most 3. The zero element in this vector space is the polynomial defined by $p(t) = 0$ for all t. Notice that the notation p_2 is the name of the function whose value at t is given by $p_2(t) = t^2$. One should *not* refer to t^2 as a polynomial. It is nothing but a real number, because t is understood to be a real number. The notation $t \mapsto t^2$ can also be used to specify the function p_2. We do not stop to verify all the vector-space axioms for this vector space. (They are very easy.) ∎

In the same way as in Example 1, for each nonnegative integer n, we have a space consisting of all polynomials of degree not exceeding n. This space is usually written $\mathbb{P}_n$. Now we have two infinite lists of distinct vector spaces, $\mathbb{R}^n$ and $\mathbb{P}_{n-1}$, where $n = 1, 2, 3, \ldots$.

EXAMPLE 2 Consider the set of all continuous real-valued functions defined on the interval $[-1, 1]$. Is this a vector space if we adopt the standard definitions for the algebraic operations?

$$(f + g)(t) = f(t) + g(t), \qquad (\alpha f)(t) = \alpha f(t)$$

SOLUTION Why is the first axiom true in this space? If we add two continuous functions, the result is another continuous function, by a theorem in calculus. A similar remark concerns the multiplication of a continuous function by a scalar. Thus, the two closure axioms are true by some theorems in calculus. The remaining axioms are all easily verified. ∎

The space in Example 2 is customarily given the designation $C[-1, 1]$. More properly, one should write $C([-1, 1])$, because for an arbitrary domain, D, we would write $C(D)$. As in Example 1, one should make a careful distinction between a function f and one of its values, $f(t)$. In this book, we try to adhere to these standards, and the reader is encouraged to do the same.

EXAMPLE 3 Consider the set of all infinite sequences

$$\mathbf{x} = [x_1, x_2, x_3, \ldots] \qquad (x_i \in \mathbb{R})$$

If we use the natural definitions of vector addition and scalar multiplication, is this a vector space?

SOLUTION Yes. The closure axioms are true, just as they are for $\mathbb{R}^n$. In fact, we have here an obvious extension of those familiar spaces. This new space should be named $\mathbb{R}^\infty$. Notice that a sequence $\mathbf{x}$ is really a function. Its domain is the set of natural numbers, $\mathbb{N} = \{1, 2, 3, \ldots\}$. We could use either the notation $\mathbf{x}(k)$ or $\mathbf{x}_k$ or x_k for a generic component of the vector $\mathbf{x}$. In set theory, this space would also be denoted by $\mathbb{R}^\mathbb{N}$. (In general, the set of all functions from one set B into another set A is denoted by A^B. For example, $S^\mathbb{N}$ is the set of all maps from the natural numbers $\mathbb{N}$ to the set S.) The vectors in $\mathbb{R}^\infty$ can be interpreted as signals in electrical engineering. ∎

EXAMPLE 4 Consider the set of all continuous real-valued functions f defined on the entire real line, $(-\infty, +\infty)$, and having the property

$$\int_{-\infty}^{\infty} \left| f(t) \right| dt < \infty$$

Is this a vector space?

SOLUTION It is understood that the natural definitions of the algebraic operations are used. Let us verify the first axiom for this vector space. Suppose that f and g are two elements of this space. Is their sum also in this space? The answer is *yes* by the triangle inequality:

$$\int_{-\infty}^{\infty} \left| f(t) + g(t) \right| dt \le \int_{-\infty}^{\infty} \left[\left| f(t) \right| + \left| g(t) \right| \right] dt$$

$$= \int_{-\infty}^{\infty} \left| f(t) \right| dt + \int_{-\infty}^{\infty} \left| g(t) \right| dt < \infty \qquad \blacksquare$$

It sometimes happens that one vector space is a subset of a larger vector space and shares the definitions of vector addition, scalar multiplication, **0**, and so on. For example, the set, X, of all vectors in $\mathbb{R}^3$ that have the form $(x_1, x_2, 0)$ is a vector space, as is easily verified, and it is obviously a subset of $\mathbb{R}^3$. It has the same vector addition and scalar multiplication that $\mathbb{R}^3$ has. We say that X is a **subspace** of $\mathbb{R}^3$. (Being a subspace is more than being a subset. Do you see why?) The next example illustrates this same phenomenon, and the topic arises again in Section 5.1.

EXAMPLE 5 Let **A** be an $m \times n$ matrix. Is the kernel of **A** (or *null space* of **A**) a vector space?

SOLUTION Recall from Section 1.3 that

$$\text{Ker}(\mathbf{A}) = \{ \mathbf{x} : \mathbf{x} \in \mathbb{R}^n \text{ and } \mathbf{A}\mathbf{x} = \mathbf{0} \}$$

The kernel of **A** is obviously a subset of $\mathbb{R}^n$. But is it a vector subspace? The only axioms that require verification are the two *closure* Axioms 1 and 6, and Axiom 4 concerning the zero element. All of the other axioms are automatically fulfilled because we know that $\mathbb{R}^n$ is a vector space. To establish Axioms 1 and 6 together, simply write

$$\mathbf{A}(\alpha\mathbf{x} + \beta\mathbf{y}) = \alpha\mathbf{A}\mathbf{x} + \beta\mathbf{A}\mathbf{y}$$

Thus, if $\mathbf{x}$ and $\mathbf{y}$ are in the kernel of $\mathbf{A}$, then so is $\alpha\mathbf{x} + \beta\mathbf{y}$. Axiom 4 is obviously true, since $\mathbf{A0} = \mathbf{0}$. We say that the kernel of $\mathbf{A}$ is a *subspace* of $\mathbb{R}^n$. ■

Linearly Dependent Sets

The concepts of linear dependence and linear independence play a crucial role in linear algebra, as noted in Section 1.3. These terms apply to *sets* of vectors, not to single vectors, and that fact by itself makes for difficulties in comprehension.

> **DEFINITION**
>
> *A subset S in a vector space is* **linearly dependent** *if there exists a nontrivial equation of the form $c_1\mathbf{u}_1 + c_2\mathbf{u}_2 + \cdots + c_m\mathbf{u}_m = \mathbf{0}$, where the points $\mathbf{u}_1, \mathbf{u}_2, \ldots, \mathbf{u}_m$ are elements of S and different from each other. The set S is* **linearly independent** *if no equation of the type described exists.*

(The cited equation has only a finite number of terms.) The term *nontrivial* in this context means that at least one of the coefficients c_i is not zero. That condition can be expressed by writing $\sum_{i=1}^{m}|c_i| > 0$.

A number of examples follow, to help in assimilating this concept.

> **EXAMPLE 6** What can be said about the linear independence of a set containing the zero vector $\mathbf{0}$?

SOLUTION It is automatically linearly dependent, because $1 \cdot \mathbf{0} = \mathbf{0}$. This is a nontrivial equation by our definition (at least one coefficient is not zero). ■

EXAMPLE 7 Consider the set of rows in the matrix

$$\begin{bmatrix} \mathbf{r}_1 \\ \mathbf{r}_2 \\ \mathbf{r}_3 \end{bmatrix} = \begin{bmatrix} 1 & 7 & 3 \\ 5 & -2 & 4 \\ 7 & -25 & -1 \end{bmatrix}$$

Is it linearly dependent?

SOLUTION Yes. We observe that $-3\mathbf{r}_1 + 2\mathbf{r}_2 - \mathbf{r}_3 = 0$, where $\mathbf{r}_i$ is the ith row in the matrix. ∎

EXAMPLE 8 Consider the three polynomials defined as follows: $p_1(t) = 7t^5 - 4t^2 + 3$, $p_2(t) = 2t^5 + 5t^2$, $p_3(t) = 8t^5 - 23t^2 + 6$. Is the set $\{p_1, p_2, p_3\}$ linearly independent?

SOLUTION No. We notice that $2p_1 - 3p_2 - p_3 = 0$. ∎

EXAMPLE 9 Is the set of these three matrices linearly independent?

$$A = \begin{bmatrix} 1 & 3 \\ 2 & 7 \end{bmatrix}, \qquad B = \begin{bmatrix} 4 & 2 \\ -1 & 2 \end{bmatrix}, \qquad C = \begin{bmatrix} -1 & 7 \\ 7 & 19 \end{bmatrix}$$

SOLUTION The answer is *no*, because $\mathbf{C} = 3\mathbf{A} - \mathbf{B}$. ∎

EXAMPLE 10 Let $\mathbf{u}_1 = \begin{bmatrix} 1 \\ 1 \\ 0 \end{bmatrix}$, $\mathbf{u}_2 = \begin{bmatrix} 1 \\ 0 \\ 1 \end{bmatrix}$, $\mathbf{u}_3 = \begin{bmatrix} 0 \\ 1 \\ 1 \end{bmatrix}$

Is $\{\mathbf{u}_1, \mathbf{u}_2, \mathbf{u}_3\}$ linearly independent? What about $\{\mathbf{u}_1, \mathbf{u}_2\}$?

SOLUTION For the first set, we consider the equation $c_1\mathbf{u}_1 + c_2\mathbf{u}_2 + c_3\mathbf{u}_3 = \mathbf{0}$. The issue is whether this equation has a nontrivial solution. The coefficient matrix for the problem is

$$\begin{bmatrix} \mathbf{u}_1 & \mathbf{u}_2 & \mathbf{u}_3 \end{bmatrix} = \begin{bmatrix} 1 & 1 & 0 \\ 1 & 0 & 1 \\ 0 & 1 & 1 \end{bmatrix} \sim \begin{bmatrix} 1 & 0 & 0 \\ 0 & 1 & 0 \\ 0 & 0 & 1 \end{bmatrix}$$

Here we do not include the righthand side because the system of equations is homogeneous and the righthand side consists of zeros. Hence, we obtain $c_1 = c_2 = c_3 = 0$, and the set is linearly independent. For the second set, we look at just the first two columns in the preceding matrix. Of course, the conclusion is the same. ∎

EXAMPLE 11 Let $A = \begin{bmatrix} 1 & 2 \\ 1 & 1 \end{bmatrix}$, $B = \begin{bmatrix} 1 & 0 \\ 1 & 0 \end{bmatrix}$, $C = \begin{bmatrix} 0 & 1 \\ 0 & 0 \end{bmatrix}$

Is the set $\{A, B, C\}$ linearly independent?

SOLUTION How can we identify these 2×2 matrices with column vectors so that we can treat them in the same manner as in Example 10? We make them into vectors, which exactly correspond to each matrix:

$$\begin{bmatrix} 1 & 1 & 0 \\ 2 & 0 & 1 \\ 1 & 1 & 0 \\ 1 & 0 & 0 \end{bmatrix} \sim \begin{bmatrix} 1 & 0 & 0 \\ 0 & 1 & 0 \\ 0 & 0 & 1 \\ 0 & 0 & 0 \end{bmatrix}$$

Because the only solution is $[0, 0, 0]^T$, the given set of matrices is linearly independent. ∎

Later, in Section 5.2, we shall justify fully the method used in Example 11. It hinges on the fact that $\mathbb{R}^{2 \times 2}$ is isomorphic to $\mathbb{R}^4$.

DEFINITION

*Consider a finite, indexed set of vectors $\{\mathbf{u}_1, \mathbf{u}_2, \ldots, \mathbf{u}_m\}$ in a vector space. We say that the indexed set is **linearly dependent** if there exist scalars c_i such that*

$$\sum_{i=1}^{m} c_i \mathbf{u}_i = 0, \qquad \sum_{i=1}^{m} |c_i| > 0$$

*If the indexed set is not linearly dependent, we say that it is **linearly independent**. The expression $\sum_{i=1}^{m} |c_i| > 0$ implies that at least one c_i is nonzero.*

EXAMPLE 12 Is the indexed set of rows in the matrix linearly dependent?

$$A = \begin{bmatrix} 2 & 5 & 7 \\ 4 & 1 & -5 \\ 2 & 5 & 7 \end{bmatrix}$$

SOLUTION Label the rows of A as $r_1, r_2,$ and r_3. This is understood to be an indexed set. (Each vector has an **index** attached to it.) In the definition of linear dependence, we can take $c_1 = 1$, $c_2 = 0$, and $c_3 = -1$ because $r_1 - r_3 = 0$. The indexed set of rows contains three rows with the vector $(2, 5, 7)$ appearing twice. If we are interested only in the set, we need not mention this vector twice. Thus, the *set* consists of two vectors, $(2, 5, 7)$ and $(4, 1, -5)$. There is a difference between the set and the indexed set. In this example, the latter is linearly dependent, whereas the former is linearly independent. In an indexed set, we consider the vector $u_1 = (2, 5, 7)$ to be different from $u_3 = (2, 5, 7)$ because their indices (1 and 3) are different. ∎

In an effort to avoid confusion we always consider the rows (or the columns) of a matrix to be *indexed* sets.

EXAMPLE 13 Is the set of columns in the following matrix linearly dependent?

$$A = \begin{bmatrix} 1 & 4 & 2 \\ 3 & 1 & 1 \\ 7 & -5 & 1 \end{bmatrix}$$

SOLUTION It is linearly independent, because when we try to solve the equation $Ax = 0$ for x, we discover that $x = 0$ is the only solution. ∎

EXAMPLE 14 How can we test to determine whether the rows of the matrix in Example 13 form a linearly independent set?

SOLUTION Apply the technique of Example 13 to the transposed matrix

$$\mathbf{A}^T = \begin{bmatrix} 1 & 3 & 7 \\ 4 & 1 & -5 \\ 2 & 1 & 1 \end{bmatrix}$$

The equation $\mathbf{A}^T\mathbf{x} = \mathbf{0}$ has only the trivial solution ($\mathbf{x} = \mathbf{0}$), and the set in question is linearly independent. Later we will prove theorems that make such questions easier to answer. ∎

THEOREM 6

Each vector in the span of a linearly independent set (in a vector space) has a unique representation as a linear combination of elements of that set.

PROOF Let $\mathbf{x}$ be a point in the span of S, where we suppose S to be a linearly independent set. Then $\mathbf{x} = \sum_{i=1}^{n} a_i\mathbf{u}_i$, for appropriate $a_i \in \mathbb{R}$ and $\mathbf{u}_i \in S$. Suppose there is another such representation, $\mathbf{x} = \sum_{i=1}^{m} b_i\mathbf{v}_i$, where $b_i \in \mathbb{R}$ and $\mathbf{v}_i \in S$. Put $U = \{\mathbf{u}_1, \mathbf{u}_2, \ldots, \mathbf{u}_n\}$ and $V = \{\mathbf{v}_1, \mathbf{v}_2, \ldots, \mathbf{v}_m\}$. Consider

$$U \cup V = \{\mathbf{w}_1, \mathbf{w}_2, \ldots, \mathbf{w}_k\}$$

(Here, k is not necessarily $n + m$.) All the points $\mathbf{w}_i$ are different from each other and belong to S. Because S is linearly independent, so is $U \cup V$. By supplying zero coefficients, if necessary, we can write $\mathbf{x} = \sum_{i=1}^{k} c_i\mathbf{w}_i$ and $\mathbf{x} = \sum_{i=1}^{k} d_i\mathbf{w}_i$. By subtraction, we get $\sum_{i=1}^{m}(c_i - d_i)\mathbf{w}_i = 0$. Because $U \cup V$ is linearly independent, all the coefficients in this last equation are zero. Hence, we obtain $c_i = d_i$ for all i. ∎

EXAMPLE 15 In two different ways, express the vector $[5, 3, 5]^T$ in terms of the three columns in the matrix of Example 12.

SOLUTION If we take a linear combination of these columns with coefficients $(-3, 5, -2)$, we get the same vector as when we use coefficients $(13, -14, 7)$. (In both cases the result is the vector $[5, 3, 5]^T$.) By Theorem 6, the columns in question must form a linearly dependent set. ∎

THEOREM 7

If a matrix has more columns than rows, then the (indexed) set of its columns is linearly dependent.

PROOF Let A be $m \times n$, and $n > m$. Think of what happens if we try to solve the equation $Ax = 0$. This is a homogeneous equation with more variables (n) than equations (m). By Corollary 2 in Section 1.3, such a system must have nontrivial solutions. ∎

EXAMPLE 16 Let $v_1 = \begin{bmatrix} 1 \\ 0 \\ 1 \end{bmatrix}$, $v_2 = \begin{bmatrix} 0 \\ 1 \\ 1 \end{bmatrix}$

$v_3 = \begin{bmatrix} 1 \\ 1 \\ 0 \end{bmatrix}$, $v_4 = \begin{bmatrix} 1 \\ 0 \\ 0 \end{bmatrix}$

Is the set $\{v_1, v_2, v_3, v_4\}$ linearly independent?

SOLUTION Put the vectors into a matrix as columns. The resulting matrix will have three rows and four columns:

$$\begin{bmatrix} v_1 & v_2 & v_3 & v_4 \end{bmatrix} = \begin{bmatrix} 1 & 0 & 1 & 1 \\ 0 & 1 & 1 & 0 \\ 1 & 1 & 0 & 0 \end{bmatrix}$$

By Theorem 7, the set of columns is linearly dependent. ∎

THEOREM 8

Any (indexed) set of more than n vectors in $\mathbb{R}^n$ is necessarily linearly dependent.

PROOF Put the vectors as columns into an $n \times m$ matrix, and apply Theorem 7. ∎

THEOREM 9

Let $\{u_1, u_2, \ldots, u_m\}$ be a linearly dependent indexed set of at least two vectors (in a vector space). Then some vector in the indexed list is a linear combination of preceding vectors in that list.

PROOF By hypothesis, there is a relation $\sum_{i=1}^{m} c_i \mathbf{u}_i = \mathbf{0}$, in which the coefficients c_i are not all 0. Let r be the last index for which $c_r \neq 0$. Then $c_i = 0$ for $i > r$ and $\sum_{i=1}^{r} c_i \mathbf{u}_i = \mathbf{0}$. Because $c_r \neq 0$, this last equation can be solved for $\mathbf{u}_r$, showing that $\mathbf{u}_r$ is a linear combination of the preceding vectors in the set. ∎

THEOREM 10

If a vector space is spanned by some set of n vectors, then every set of more than n vectors in that space must be linearly dependent.

PROOF Let $\{\mathbf{v}_1, \mathbf{v}_2, \dots, \mathbf{v}_n\}$ span a vector space V, and let $\{\mathbf{u}_1, \mathbf{u}_2, \dots, \mathbf{u}_m\}$ be any subset of V such that $m > n$. By the definition of *span*, there exist coefficients a_{ij} such that $\mathbf{u}_i = \sum_{j=1}^{n} a_{ij} \mathbf{v}_j$ for $1 \leq i \leq m$. Consider the homogeneous system of equations $\sum_{i=1}^{m} a_{ij} x_i = \mathbf{0}$ $(1 \leq j \leq n)$. This system has fewer equations than unknowns. Therefore, Corollary 2 in Section 1.3 is applicable, and there must exist a nontrivial solution, $\mathbf{x} = (x_1, x_2, \cdots, x_m)$. Now we see that the set of vectors $\mathbf{u}_i$ is linearly dependent, because

$$\sum_{i=1}^{m} x_i \mathbf{u}_i = \sum_{i=1}^{m} x_i \sum_{j=1}^{n} a_{ij} \mathbf{v}_j = \sum_{j=1}^{n} \left(\sum_{i=1}^{m} a_{ij} x_i \right) \mathbf{v}_j = \sum_{j=1}^{n} 0 \mathbf{v}_j = \mathbf{0} \qquad ∎$$

THEOREM 11

Let S be a linearly independent set in a vector space V. If $\mathbf{x} \in V$ and $\mathbf{x} \notin \mathrm{Span}(S)$, then $S \cup \{x\}$ is linearly independent.

PROOF If that conclusion is false, there must exist a nontrivial equation of the form $c_0 \mathbf{x} + \sum_{i=1}^{n} c_i \mathbf{v}_i = \mathbf{0}$, where the points $\mathbf{v}_i$ belong to S and the c_i are scalars, not all zero. If $c_0 = 0$, then that equation will contradict the linear independence of the set S. Hence, we obtain $c_0 \neq 0$ and $\mathbf{x} = -c_0^{-1} \sum_{i=1}^{n} c_i \mathbf{v}_i$, showing that $\mathbf{x}$ is in the span of S. ∎

EXAMPLE 17 Here is an example involving an infinite set of vectors. Let $p_j(t) = t^j$ for $j = 0, 1, 2, 3, \dots$. These functions are called **monomials** and they are the building blocks of all polynomials. They also lie in the vector space of *all* continuous functions $\mathbb{R} \to \mathbb{R}$. Is the set $\{p_j : j = 0, 1, 2, \dots\}$ linearly independent?

SOLUTION The answer is *yes* and deserves a proof. We shall use the method of contradiction to carry out this proof. Suppose therefore that the set of functions p_j is linearly dependent. (Then we hope to arrive at a contradiction.) From the definition of linear dependence, there must exist a nontrivial equation of the form

$$c_0 p_0 + c_1 p_1 + c_2 p_2 + \cdots + c_m p_m = 0$$

(In this context, *nontrivial* means that the coefficients c_j are not all zero.) Think about this displayed equation. Does it not say that a certain nontrivial polynomial of degree at most m is, in fact, equal everywhere to 0? One need only recall from the study of algebra that a nontrivial polynomial of degree k can have at most k zeros (or *roots*). See Appendix B, Theorem B.1. We have reached a contradiction.　　　　　■

Observe that linear dependence always involves a linear combination of vectors, and a linear combination is always restricted to a finite set of terms. We cannot consider infinite series of vectors without bringing convergence questions into play, and that is another story altogether.

Linear Mapping

The concept of a linear mapping was introduced in Section 2.3. It is meaningful for maps between any two vector spaces. The characteristic property is $T(a\mathbf{x} + b\mathbf{y}) = aT(\mathbf{x}) + bT(\mathbf{y})$. Let us illustrate this matter with a linear space of polynomials.

Consider the space $\mathbb{P}_3$, consisting of all polynomials whose degree does not exceed 3. (This is an important example of a vector space.) We can use the following natural basic polynomials for this space: $p_j(t) = t^j$, where $0 \le j \le 3$. Thus, we have $p_0(t) = 1$, $p_1(t) = t$, $p_2(t) = t^2$, and $p_3(t) = t^3$. Every element of $\mathbb{P}_3$ is a unique linear combination of these four basic polynomials. For example, the polynomial f defined by $f(t) = 7 - 3t + 9t^2 - 5t^3$ is expressible as $f = 7p_0 - 3p_1 + 9p_2 - 5p_3$. Consequently, if g is a polynomial of degree at most 3, and if $g = \sum_{j=0}^{3} c_j p_j$, then we can represent g by the vector (c_0, c_1, c_2, c_3). If T is a linear transformation from $\mathbb{P}_3$ into some other vector space, we will have

$$T(g) = T\left(\sum_{j=0}^{3} c_j p_j\right) = \sum_{j=0}^{3} c_j T(p_j)$$

This easy calculation shows that a linear map is completely determined as soon as its values have been prescribed on a set that spans the domain. *Caution:* The values assigned are not necessarily unrestricted.

EXAMPLE 18 For a concrete case of the preceding idea, suppose that T maps $\mathbb{P}_3$ into $\mathbb{R}^5$ according to the following rules:

$$T(p_0) = [7, 2, 5, -3, 0]^T, \quad T(p_1) = [4, 0, -3, 2, 1]^T$$
$$T(p_2) = [3, 11, 8, 2, 5]^T, \quad T(p_3) = [3, 0, 2, 1, 6]^T$$

What is the matrix for this transformation?

SOLUTION For $\mathbf{u} = c_0 p_0 + c_1 p_1 + c_2 p_2 + c_3 p_3$, we have

$$T(\mathbf{u}) = T(c_0 p_0 + c_1 p_1 + c_2 p_2 + c_3 p_3)$$
$$= c_0 T(p_0) + c_1 T(p_1) + c_2 T(p_2) + c_3 T(p_3)$$

$$= c_0 \begin{bmatrix} 7 \\ 2 \\ 5 \\ -3 \\ 0 \end{bmatrix} + c_1 \begin{bmatrix} 4 \\ 0 \\ -3 \\ 2 \\ 1 \end{bmatrix} + c_2 \begin{bmatrix} 3 \\ 11 \\ 8 \\ 2 \\ 5 \end{bmatrix} + c_3 \begin{bmatrix} 3 \\ 0 \\ 2 \\ 1 \\ 6 \end{bmatrix}$$

$$= \begin{bmatrix} 7 & 4 & 3 & 3 \\ 2 & 0 & 11 & 0 \\ 5 & -3 & 8 & 2 \\ -3 & 2 & 2 & 1 \\ 0 & 1 & 5 & 6 \end{bmatrix} \begin{bmatrix} c_0 \\ c_1 \\ c_2 \\ c_3 \end{bmatrix} \qquad \blacksquare$$

EXAMPLE 19 Use the information in Example 18 to find the effect of applying T to the polynomial $\mathbf{u}(t) = 4t^3 - 2t^2 + 3t - 5$.

SOLUTION We note that the representation of this polynomial by a vector in $\mathbb{R}^4$ is $(-5, 3, -2, 4)$. (One must write the terms in the correct order.) The answer to the question is therefore

$$T(\mathbf{u}) = \begin{bmatrix} 7 & 4 & 3 & 3 \\ 2 & 0 & 11 & 0 \\ 5 & -3 & 8 & 2 \\ -3 & 2 & 2 & 1 \\ 0 & 1 & 5 & 6 \end{bmatrix} \begin{bmatrix} -5 \\ 3 \\ -2 \\ 4 \end{bmatrix} = \begin{bmatrix} -17 \\ -32 \\ -42 \\ 21 \\ 17 \end{bmatrix} \qquad \blacksquare$$

A theorem to formalize what we have seen in the preceding examples is as follows.

THEOREM 12

Let $\{\mathbf{u}_1, \mathbf{u}_2, \ldots \mathbf{u}_n\}$ be a linearly independent set in some vector space. Let $\mathbf{v}_1, \mathbf{v}_2, \ldots, \mathbf{v}_n$ be arbitrary vectors in another vector space. Then there is a linear transformation T such that $T(\mathbf{u}_i) = \mathbf{v}_i$ for $1 \leq i \leq n$.

PROOF The linear transformation will be defined on the set $U = \text{Span}\{\mathbf{u}_1, \mathbf{u}_2, \ldots \mathbf{u}_n\}$ by this equation:

$$T(a_1\mathbf{u}_1 + a_2\mathbf{u}_2 + \cdots + a_n\mathbf{u}_n) = a_1 T(\mathbf{u}_1) + a_2 T(\mathbf{u}_2) + \cdots + a_n T(\mathbf{u}_n)$$

$$= a_1\mathbf{v}_1 + a_2\mathbf{v}_2 + \cdots + a_n\mathbf{v}_n$$

We must recall that any vector in U has one and only one representation as a linear combination of the vectors $\mathbf{u}_i$. ∎

EXAMPLE 20   Is there a linear transformation that maps $(1,0)$ to $(5,3,4)$ and maps $(3,0)$ to $(1,3,2)$?

SOLUTION Let T be such a map. Being linear, T must obey the equation $T(c\mathbf{x}) = cT(\mathbf{x})$. Hence, we have

$$T(3,0) = 3T(1,0) = 3(5,3,4) = (15,9,12) \neq (1,3,2)$$

The answer to the question posed is *no*. ∎

Theorem 12 does not apply in Example 20 because the pair of vectors $(1,0)$ and $(3,0)$ is not linearly independent. A slightly different point of view is that $(3,0) - 3(1,0) = \mathbf{0}$ and therefore by linearity we must have $T(3,0) - 3T(1,0) = \mathbf{0}$. This leads to the following theorem.

THEOREM 13

If T is a linear transformation, then any linear dependence of the type $\sum_{i=1}^{n} a_i\mathbf{u}_i = \mathbf{0}$ must imply $\sum_{i=1}^{n} a_i T(\mathbf{u}_i) = \mathbf{0}$.

Consequently, if $\{\mathbf{u}_i\}$ is linearly dependent, then $\{T(\mathbf{u}_i)\}$ is linearly dependent, and if $\{T(\mathbf{u}_i)\}$ is linearly independent, then $\{\mathbf{u}_i\}$ is linearly independent.

Application: Models in Economic Theory

Linear algebra has a large role to play in economic theory. Two simple, yet basic, economic models are discussed here.

Think of a nation's economy as being divided into *n* *sectors*, or *industries*. For example, we could have the steel industry, the shoe industry, the medical industry, and so on. (In a realistic model, there could be 1000 or more sectors.) An $n \times n$ matrix $\mathbf{A} = (a_{ij})$ is given; it contains data concerning the productivity of the *n* sectors. Specifically, a_{ij} is the fraction of the output from industry *j* that goes to industry *i* as input in one year. We assume in a *closed model* that all the output of the various industries remains in the system and is purchased by the sectors in the system. Because of this assumption (that the system is *closed*), the column sums in the matrix $\mathbf{A}$ are 1:

$$\sum_{i=1}^{n} a_{ij} = 1 \qquad (1 \le j \le n)$$

Let x_j be the value placed on the output from the *j*th sector in one year. The **price vector** $\mathbf{x} = (x_1, x_2, \ldots, x_n)$ is unknown but will be determined after certain requirements have been imposed on it.

An equilibrium state occurs if the price vector $\mathbf{x}$ is a nonzero vector $\mathbf{x} = (x_1, x_2, \ldots, x_n)$ such that $\mathbf{Ax} = \mathbf{x}$. Notice that if $\mathbf{x}$ has this property, then any nonzero scalar multiple of $\mathbf{x}$ will also have that property.

THEOREM 14

If $\mathbf{A}$ *is a square matrix whose column sums are all equal to 1, then the equation* $\mathbf{Ax} = \mathbf{x}$ *has nontrivial solutions.*

PROOF The equation in question is $(\mathbf{I} - \mathbf{A})\mathbf{x} = \mathbf{0}$. The coefficient matrix here is singular (noninvertible) because its rows add up to $\mathbf{0}$. Consequently, the kernel of $\mathbf{I} - \mathbf{A}$ contains nonzero vectors. ∎

This model is called the **Leontief Closed Model** in honor of Wassily Leontief. [5] The model described is said to be *closed* because the sectors in

---

[5] Professor Wassily Leontief (1906–1999) was awarded the 1973 Nobel Prize in Economics for his work on mathematical models in economic theory. Leontief was born in St. Petersburg, Russia. He was arrested several times for criticizing the control of intellectual and personal freedom under communism. In 1925, he was allowed to leave the country because the authorities believed that he had a cancerous growth on his neck that would soon kill him. The growth turned out to be benign, and Leontief earned a Ph.D. in Berlin prior to his immigration to the United States in 1931. He served as a professor at Harvard University and later at New York University. He began compiling the data for the input–output model of the United States economy in 1932, and in 1941 his paper *Structure of the American Economy, 1919–1929* was published. Leontief was an early user of computers, beginning in 1935 with large-scale mechanical computing machines and, in 1943, moving on to the Mark I, the first large-scale electronic computer.

the model satisfy the needs of all sectors, but no goods leave or enter the system.

All of the elements of the matrix $\mathbf{A}$ are nonnegative, and the elements in each column sum to 1. The matrix $\mathbf{A}$ occurring here is called an **input–output matrix** or **consumption matrix**. If $\mathbf{x}$ is a solution of the equation, then scalar multiples of $\mathbf{x}$ will also be solutions. The equation $\mathbf{A}\mathbf{x} = \mathbf{x}$ indicates that $\mathbf{x}$ is a fixed point of the mapping $\mathbf{x} \mapsto \mathbf{A}\mathbf{x}$. Solving the equation $(\mathbf{I} - \mathbf{A})\mathbf{x} = \mathbf{0}$ is easily done: we are looking for a nontrivial solution to a system of homogeneous equations.

To illustrate the Leontief Closed Model, we consider an example involving three sectors S_1, S_2, and S_3. The pertinent information is given in this table:

Buy \ Sell	S_1	S_2	S_3
S_1	$\frac{1}{4}$	$\frac{1}{3}$	$\frac{1}{2}$
S_2	$\frac{1}{4}$	$\frac{1}{3}$	$\frac{1}{4}$
S_3	$\frac{1}{2}$	$\frac{1}{3}$	$\frac{1}{4}$

The entries in the table form the matrix $\mathbf{A}$. Notice that the column sums are all 1. The consumption matrix, $\mathbf{I} - \mathbf{A}$, is shown here along with its reduced echelon form:

$$\begin{bmatrix} \mathbf{I} - \mathbf{A} & | & \mathbf{0} \end{bmatrix} \sim \left[\begin{array}{ccc|c} \frac{3}{4} & -\frac{1}{3} & -\frac{1}{12} & 0 \\ -\frac{1}{4} & \frac{2}{3} & -\frac{1}{4} & 0 \\ -\frac{1}{2} & -\frac{1}{3} & \frac{3}{2} & 0 \end{array}\right]$$

$$\sim \left[\begin{array}{ccc|c} 1 & 0 & -1 & 0 \\ 0 & 1 & -\frac{3}{4} & 0 \\ 0 & 0 & 0 & 0 \end{array}\right]$$

The general solution of this homogeneous system is $x_1 = x_3$ and $x_2 = \frac{3}{4}x_3$, where x_3 is a free variable. One simple solution is $\mathbf{x} = (4, 3, 4)$. (All solutions are multiples of this one.) The rows in the input–output table above show how the three commodities go from source to user (seller to buyer) in the economy. The flow of commodities between the different sectors can be shown in a graph as in Figure 2.27. Row 1 of the table indicates that for sector S_1 to produce one unit of its product it must consume one-fourth of its own product, one-third of sector S_2's product, and one-half of sector S_3's product. There are similar flows in the other sectors.

Usually, an economy has to satisfy some outside demands from nonproducing sectors such as government agencies. In this case, there is also

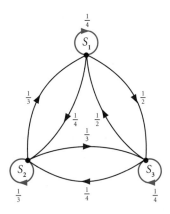

FIGURE 2.27 Flow among three sectors.

a **final demand vector** $\mathbf{b} = (b_i)$, which lists the demands placed upon the various industries for output unconnected with the industries mentioned previously. With x_i and c_{ij} as previously defined, we are led to the equations: $a_{i1}x_1 + a_{i2}x_2 + \cdots + a_{in}x_n + b_i = x_i$. The linear system for the **Leontief Open Model** is

$$\mathbf{Ax} + \mathbf{b} = \mathbf{x} \qquad \text{or} \qquad (\mathbf{I} - \mathbf{A})\mathbf{x} = \mathbf{b}$$

The term $\mathbf{Ax}$ is called the **intermediate demand vector** and $\mathbf{b}$ is the **final demand vector**. In this case, the elements of the matrix $\mathbf{A}$ are nonnegative and the column sums are less than 1.

To illustrate the Leontief Open Model, we consider a simple example involving two industrial sectors S_1 and S_2 as given by this table:

Buy \ Sell	S_a	S_b
S_a	0.15	0.25
S_b	0.20	0.05

Suppose that the initial demand vector is $\mathbf{b} = (350, 1700)$. Then we solve the system

$$(\mathbf{I} - \mathbf{A})\mathbf{x} = \mathbf{b} \qquad \text{or} \qquad \begin{bmatrix} 0.85 & -0.25 \\ -0.20 & 0.95 \end{bmatrix} \begin{bmatrix} x_1 \\ x_2 \end{bmatrix} = \begin{bmatrix} 350 \\ 1700 \end{bmatrix}$$

and obtain the solution $\mathbf{x} = (1000, 2000)$. The flow between the different sectors in this input–output table can be illustrated by the graph shown in Figure 2.28.

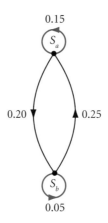

FIGURE 2.28 Flow example between two sectors.

A common question in input–output analysis is: Given a forecasted demand, how much output from each of the sectors would be needed to satisfy the final demand? If we increase the final demand to $400 for sector S_1 and decrease it to $1600 for sector S_2, we have the new demand vector $\mathbf{b} = (400, 1600)$. Here the change in final demand is $\Delta\mathbf{b} = (50, -100)$. Solving the system

$$(\mathbf{I} - \mathbf{A})(\Delta\mathbf{x}) = (\Delta\mathbf{b})$$

we find $\Delta\mathbf{x} = (29.70, -99.01)$, which is the change in the output vector. This change in demand requires an increase in production by sector S_1 and a decrease in S_2. Consequently, the new output vector is $\mathbf{x} \approx (1030, 1900)$. This process can be repeated for different possible situations.

In Section 8.3, the Leontief models are studied in greater detail.

SUMMARY 2.4

- A vector space is a set V of vectors, together with vector addition and scalar multiplication that satisfy the following axioms:

- If $\mathbf{u}, \mathbf{v} \in V$, then $\mathbf{u} + \mathbf{v} \in V$.
  (Closure axiom for addition)

- For all $\mathbf{u}, \mathbf{v} \in V, \mathbf{u} + \mathbf{v} = \mathbf{v} + \mathbf{u}$.
  (Commutativity of vector addition)

- For any $\mathbf{u}, \mathbf{v}, \mathbf{w} \in V, (\mathbf{u}+\mathbf{v})+\mathbf{w} = \mathbf{u}+(\mathbf{v}+\mathbf{w})$.
  (Associativity of vector addition)

- There is a zero element $\mathbf{0} \in V$ such that $\mathbf{u} + \mathbf{0} = \mathbf{u}$, for all $\mathbf{u} \in V$.
  (Existence of a zero vector)

- For each vector $\mathbf{u} \in V$ there is at least one element $\widetilde{\mathbf{u}} \in V$ such that $\widetilde{\mathbf{u}} + \mathbf{u} = 0$.
  (Existence of additive inverses)

- If $\mathbf{u} \in V$ and $\alpha \in \mathbb{R}$, then $\alpha\mathbf{u} \in V$.
  (Closure axiom for scalar–vector product)

- For any $\alpha \in \mathbb{R}$ and $\mathbf{u}, \mathbf{v} \in V$, $\alpha(\mathbf{u} + \mathbf{v}) = \alpha\mathbf{u} + \alpha\mathbf{v}$. (Distributive law: scalar times vectors)
- If $\alpha, \beta \in \mathbb{R}$ and $\mathbf{u} \in V$, then $(\alpha + \beta)\mathbf{u} = \alpha\mathbf{u} + \beta\mathbf{u}$. (Distributive law: scalar sum times a vector)
- If $\alpha, \beta \in \mathbb{R}$ and $\mathbf{u} \in V$, then $\alpha(\beta\mathbf{u}) = (\alpha\beta)\mathbf{u}$. (Associativity of scalar–vector product)
- For each $\mathbf{u} \in V$, $1 \cdot \mathbf{u} = \mathbf{u}$. (Multiplication of unit scalar times vector)
- In any vector space, if $c \in \mathbb{R}$, then $c\mathbf{0} = \mathbf{0}$.
- In a vector space V, if $\mathbf{x} \in V$, $c \in \mathbb{R}$, and $c\mathbf{x} = \mathbf{0}$, then either $c = 0$ or $\mathbf{x} = \mathbf{0}$.
- In a vector space, for each $\mathbf{x} \in V$, the point $\widetilde{\mathbf{x}}$ is uniquely determined. (Recall that $\mathbf{x} + \widetilde{\mathbf{x}} = \mathbf{0}$.)
- In a vector space, every vector $\mathbf{x} \in V$ satisfies the equation $0\mathbf{x} = \mathbf{0}$.
- In any vector space, for each $\mathbf{x} \in V$, we have $(-1)\mathbf{x} = \widetilde{\mathbf{x}}$.
- A subset S in a vector space is linearly dependent if there exists a nontrivial equation of the form $c_1\mathbf{u}_1 + c_2\mathbf{u}_2 + \cdots + c_m\mathbf{u}_m = \mathbf{0}$, where the distinct points $\mathbf{u}_1, \mathbf{u}_2, \ldots, \mathbf{u}_m$ come from S.
- Each vector in the span of a linearly independent set (in a vector space) has a unique representation as a linear combination of elements of that set.
- If a matrix has more columns than rows, then the indexed set of its columns is linearly dependent.

- Any indexed set of more than n vectors in $\mathbb{R}^n$ is necessarily linearly dependent.
- Let $\{\mathbf{u}_1, \mathbf{u}_2, \ldots, \mathbf{u}_m\}$ be a linearly dependent indexed set of at least two vectors (in a vector space). Then some vector in the indexed list is a linear combination of preceding vectors in that list.
- If a matrix has more columns than rows, then the indexed set of its columns is linearly dependent.
- If a vector space is spanned by some set of n vectors, then every set of more than n vectors in that space must be linearly dependent.
- Let S be a linearly independent set in a vector space V. If $\mathbf{x} \in V$ and $\mathbf{x} \notin \text{Span}(S)$, then $S \cup \{x\}$ is linearly independent.
- Let $\{\mathbf{u}_1, \mathbf{u}_2, \ldots \mathbf{u}_n\}$ be a linearly independent set in some vector space. Let $\mathbf{v}_1, \mathbf{v}_2, \ldots, \mathbf{v}_n$ be arbitrary vectors in another vector space. Then there is a linear transformation T such that $T(\mathbf{u}_i) = \mathbf{v}_i$ for $1 \leq i \leq n$.
- If T is a linear transformation, then any linear dependence of the type $\sum_{i=1}^{n} a_i\mathbf{u}_i = \mathbf{0}$ must imply $\sum_{i=1}^{n} a_i T(\mathbf{u}_i) = \mathbf{0}$.
- If $\mathbf{A}$ is a square matrix whose column sums are all 1, then the equation $\mathbf{Ax} = \mathbf{x}$ has nontrivial solutions.
- Leontief Closed Model: $(\mathbf{I} - \mathbf{A})\mathbf{x} = \mathbf{0}$; Leontief Open Model: $(\mathbf{I} - \mathbf{A})\mathbf{x} = \mathbf{b}$.

KEY CONCEPTS 2.4

Vector spaces, axioms for vector spaces, spaces of sequences, spaces of functions, spaces of polynomials, linear independence, linear dependence, spaces of matrices, vector subspaces, indexed set, Leontief Closed Model

GENERAL EXERCISES 2.4

1. Determine whether this set of four vectors is linearly dependent or linearly independent: $\{(1, 3, 2, -1), (4, -1, 5, 1), (-2, 1, 1, 7), (12, 2, 6, -22)\}$

2. Consider the set $\{p_1, p_2, p_3, p_4\}$. Determine whether this set of polynomials is linearly independent or linearly dependent. The definitions are $p_1(t) = 1$, $p_2(t) = t$, $p_3(t) = 4 - t$, $p_4(t) = t^3$.

3. Let $f(t) = \sin t$ and $g(t) = \cos t$. Determine whether the pair $\{f, g\}$ is linearly dependent or linearly independent.

4. Let $f(t) = 1$, $g(t) = \cos 2t$, and $h(t) = \sin^2 t$. Determine whether the set $\{f, g, h\}$ is linearly dependent or independent.

5. Test each of these sets of functions for linear dependence or linear independence:
  a. $u_1(t) = 1$, $u_2(t) = \sin t$, $u_3(t) = \cos t$
  b. $w_1(t) = 1$, $w_2(t) = \sin^2 t$, $w_3(t) = \cos^2 t$
  c. $v_1(t) = \cos 2t$, $v_2(t) = \sin^2 t$, $v_3(t) = \cos^2 t$

6. Define polynomials $p_0(t) = t + t^2 + t^3$, $p_1(t) = t^3 + t^4 + t^5$, $p_2(t) = t^5 + t^6 + t^7$. Verify the assertion or find a counterexample: $\{p_0, p_1, p_2\}$ is linearly dependent.

7. Consider the matrix $\mathbf{A} = \begin{bmatrix} 1 & 3 & 7 \\ 4 & -2 & -1 \\ 7 & 7 & 20 \end{bmatrix}$

Do its rows form a linearly dependent set?

8. Consider $\begin{bmatrix} 1 & 3 \\ 2 & 7 \end{bmatrix}, \begin{bmatrix} 4 & 2 \\ -1 & 2 \end{bmatrix}, \begin{bmatrix} -1 & 7 \\ 7 & 19 \end{bmatrix}$

Is this set of three matrices linearly independent?

9. Let $\mathbf{u} = (7, 1, 2)$, $\mathbf{v} = (3, -2, 4)$, and $\mathbf{w} = (6, 13, -14)$. Find a simple description of $\text{Span}\{\mathbf{u}, \mathbf{v}\} \cap \text{Span}\{\mathbf{w}\}$ and $\text{Span}\{\mathbf{u}, \mathbf{v}\} + \text{Span}\{\mathbf{w}\}$.

10. Let $\{\mathbf{x}, \mathbf{y}, \mathbf{z}\}$ be a linearly independent set of three vectors in some vector space. Justify the assertion or find a counterexample: $\{2\mathbf{x} + 3\mathbf{y} - \mathbf{z}, 4\mathbf{y} + 2\mathbf{z}, 3\mathbf{z}\}$ is linearly independent.

11. Is the following example a vector space? $X = \mathbb{R}, x \oplus y = x + y + 1, \alpha \otimes x = \alpha x + \alpha$

12. Consider any set of four vectors in $\mathbb{R}^3$. It must be linearly dependent. Why?

13. Let $L : \mathbb{P}_3 \mapsto \mathbb{P}_2$ be as given. In each case, find the kernel of L.
  a. $(Lp)(t) = 2p'(t) + 3tp''(t)$
  b. $Lp = 2p' + 3p''$

14. Define three polynomials as follows: $p_1(t) = t^3 + t$, $p_2(t) = t^2 + 1$, and $p_3(t) = 3t^3 - 2t^2 + 3t - 2$. Is the set $\{p_1, p_2, p_3\}$ linearly independent?

15. Define these polynomials: $p_1(t) = 1 + t^2$, $p_2(t) = 2t - 4$, $p_3(t) = 1 + t + t^2$. Does $\{p_1, p_2, p_3\}$ span $\mathbb{P}_2$?

16. Consider $p_0(t) = 1$, $p_1(t) = 1 + t$, $p_2(t) = 1 + t + t^2$. Give an argument why $\text{Span}\{p_0, p_1, p_2\} = \mathbb{P}_2$, the set of all polynomials of degree less than or equal to 2.

17. Does the set of all vectors in $\mathbb{R}^4$ that have exactly two zero entries span $\mathbb{R}^4$?

18. Let $\mathbf{v}_1 = [1, 0, 0, 0]^T$, $\mathbf{v}_2 = [0, 1, 0, 0]^T$, and $\mathbf{v}_3 = [0, 0, 1, 0]^T$. Is the set of these three vectors linearly independent? Why?

19. Justify each step in this alternative proof of Theorem 1: $0 = 0 + 0$, $c\,0 = c\,0 + c\,0$, $c\,0 - c\,0 = c\,0, 0 = c\,0$.
(Use only the axioms of a vector space.)

20. Define the standard monomial functions: $p_0(t) = 1$, $p_1(t) = t$, $p_2(t) = t^2$, and so on. Is this set linearly independent if we take the domain of these functions to be the set of all positive integers?

21. Discuss the question of whether the set of all discontinuous functions defined on $[-1, 1]$ forms a vector space. (Refer to Example 2.)

22. Verify that if $\{v_1, v_2, v_3\}$ is linearly independent, then so is the set $\{v_3 + v_2, \ v_1 + v_3, \ v_2 + v_1\}$.

23. Establish that a set of two vectors is linearly dependent if and only if one of the vectors is a multiple of the other.

24. Consider a sequence of polynomials p_k, where $k = 0, 1, 2, \ldots$. Assume that for all k, the polynomial p_k has degree k. Thus, the highest power present in p_k is t^k, and it occurs with a nonzero coefficient. Explain why the set $\{p_k \ : \ k = 0, 1, 2, \ldots\}$ is linearly independent. *Suggestion:* Begin with a simple case: $\{p_0, p_1, p_2\}$.

25. Establish that if a set of two or more nonzero vectors is linearly dependent, then one element of the set is a linear combination of the others.

26. Let $\{u_1, u_2, \ldots, u_n\}$ be a linearly independent set in a vector space. Explain that if $\sum_{i=1}^{n} \alpha_i u_i = \sum_{i=1}^{n} \beta_i u_i$, then $(\alpha_1, \alpha_2, \ldots, \alpha_n) = (\beta_1, \beta_2, \ldots, \beta_n)$.

27. Let G and H be subsets of a vector space, and suppose that $G \subseteq H$. (Thus, G is a subset of H: each element of G is in H.) Establish the following:
a. If H is linearly independent, then so is G.
b. If G spans the vector space, then so does H.
c. In some cases, G can be linearly independent while H is not.
d. In some cases, H can span the vector space while G does not.

28. In the axioms for a vector space, can Axioms 2 and 3 be replaced by a single axiom that states $(u + v) + w = v + (u + w)$?

29. Establish that if $\{u_1, u_2, \cdots, u_m\}$ is a linearly independent set in $\text{Span}\{v_1, v_2, \ldots, v_n\}$, then $m \leq n$.

30. In the axioms for a vector space, can we dispense with the fifth one, and still prove that $u + (-1)u = 0$?

31. Let V be the set of all positive real numbers. In V, define $u \oplus v = uv$ and $\alpha \otimes v = v^\alpha$ for $\alpha \in \mathbb{R}, u \in V, v \in V$. Establish that V, with the operations just defined, is a vector space. (This example is found in many texts, such as that of Andrilli and Hecker, [1998] and Ikramov [1975]. We call it an **exotic vector space**, since it is so different from the familiar examples. See also the next exercise.)

32. This is another **exotic vector space**, similar to one in the textbook of Andrilli and Hecker [1998]. Verify all the axioms for a vector space. Let V be the set $\mathbb{R}^2$ with new definitions of vector addition and multiplication of scalars: $(x_1, x_2) \oplus (y_1, y_2) = (x_1 + y_1 + a, \ x_2 + y_2 + b)$ and $\lambda \odot (x_1, x_2) = (\lambda x_1 + \lambda a - a, \ \lambda x_2 + \lambda b - b)$. Here, a and b are arbitrary constants, fixed at the outside.

33. Establish that if V is a vector space, then $V^{\mathbb{R}}$ is a vector space with the standard definitions of the two algebraic operations. The notation Y^X denotes the set of all functions from X to Y.

34. Is $\text{Span}(S) + \text{Span}(T) = \text{Span}(S \cup T)$ true for subsets S and T in a vector space? A proof or counterexample is required. See Section 1.2, for the definition of the span of a set. The sum of two sets, U and V, is the set $\{u + v : u \in U, v \in V\}$. This is not the same as their union.

35. The symbol $\mathbb{C}$ denotes the set of all complex numbers, $x + iy$, where $i^2 = -1$, and x and y are real. The notation $\mathbb{C}^n$ denotes a space like $\mathbb{R}^n$ except that we allow the vectors to have complex components. Verify that $\mathbb{C}^n$ together with $\mathbb{R}$ as the field of scalars is a vector space. (See Appendix B.)

36. Explain this: If S is a linearly dependent set, then there is an element $\mathbf{v}$ in S such that the span of S is the same as the span of $S \backslash \{\mathbf{v}\}$. If there are two such elements in S, can we remove both without changing the span? (Here $S \backslash T = \{\mathbf{x} : \mathbf{x} \in S \text{ and } \mathbf{x} \notin T\}$.)

37. Establish that in a vector space the zero element is unique.

38. Let S be a linearly independent set in a vector space V. Let $\mathbf{x}$ be a point in V that is not in $\text{Span}(S)$. Explain why the set $S \cup \{\mathbf{x}\}$ is linearly independent.

39. Let $\{\mathbf{x}, \mathbf{y}, \mathbf{z}\}$ be a linearly independent set of three vectors in some vector space. Explain the assertion or find a counterexample: $\{\mathbf{x}+\mathbf{y}+\mathbf{z}, \ \mathbf{x}+\mathbf{y}, \ \mathbf{x}\}$ is linearly independent.

40. Consider a finite set of polynomials, not containing the 0-polynomial, and having the property that any two polynomials in the set will have different degrees. Establish that the set is linearly independent. (In case of difficulty, look at small examples.) Perhaps you can improve on this theorem.

41. Let X and Y be two sets of vectors in some linear space. Assume that each x in X is in the span of Y, and each y in Y is in the span of X. Explain why X and Y have the same span.

42. Define what is meant by saying that a set in a vector space is linearly dependent.

43. Let S and T be subsets of a vector space, and assume that $S \subseteq T$. (This means that S is a subset of T: each member of S is a member of T.) Establish that $\text{Span}(S) \subseteq \text{Span}(T)$.

44. A function $f : \mathbb{R} \to \mathbb{R}$ is continuous at a point x_0 if for any $\varepsilon > 0$ there corresponds a $\delta > 0$ such that $|x - x_0| < \delta$ implies $|f(x) - f(x_0)| < \varepsilon$. Explain why the hypothesis f and g are continuous at x_0 leads to $f + g$ being continuous at x_0.

45. Verify that the set of all 2×2 matrices under the usual operations is a vector space.

46. Explain these: if $\mathbf{A}$ is $m \times n$ and $m > n$ (more rows than columns), then the rows of $\mathbf{A}$ form a linearly dependent set. Equivalently, if $\mathbf{B}$ is $m \times n$ and $n > m$ (more columns than rows), then the columns of $\mathbf{B}$ form a linearly dependent set.

47. If $n < m$, is $\mathbb{R}^n$ a subspace of $\mathbb{R}^m$? Under what conditions is $\mathbb{P}_n$ a subspace of $\mathbb{P}_m$? ($\mathbb{P}_n$ is the vector space of all polynomials whose degrees are at most n.)

48. Let $\{\mathbf{u}_1, \mathbf{u}_2, \dots, \mathbf{u}_n\}$ be a linearly independent set in a vector space. Establish that if $\sum_{i=1}^{n} \alpha_i \mathbf{u}_i = \sum_{i=1}^{n} \beta_i \mathbf{u}_i$, then $(\alpha_1, \alpha_2, \dots, \alpha_n) = (\beta_1, \beta_2, \dots, \beta_n)$.

49. If S is a set (without any algebraic structure) and if V is a vector space, is V^S a vector space? Explain. (The notation V^S denotes the **set of all mappings** from S to V.)

50. (Continuation.) With the hypotheses of the preceding problem, is S^V a vector space? Explain.

51. Are these vector spaces?
  a. The vectors are in $\mathbb{R}^n$ and the scalars can be in $\mathbb{C}$.
  b. The vectors are members of $\mathbb{C}^n$, but the scalars are confined to be in $\mathbb{R}$. (See Appendix B on complex arithmetic.)

52. Explain and write out in detail the following.
  a. Example 1. (Verify all of the vector-space axioms.)
  b. Why is a subspace more than a subset?
  c. Example 13.
  d. Example 14.

COMPUTER EXERCISES 2.4

1. Use mathematical software to solve this Leontief system involving

$$A = \begin{bmatrix} 0.1 & 0.3 & 0.2 \\ 0.3 & 0.4 & 0.5 \\ 0.1 & 0.2 & 0.1 \end{bmatrix}, \qquad b = \begin{bmatrix} 50 \\ 75 \\ 125 \end{bmatrix}$$

2. Suppose that an economy consists of four sectors, labeled S_a, S_b, S_c, and S_d. The interactions between the sectors are governed by a table as follows:

	S_a	S_b	S_c	S_d
S_a	0.5	0.2	0.1	0.2
S_b	0.2	0.2	0.4	0.1
S_c	0.1	0.5	0.1	0.3
S_d	0.2	0.1	0.4	0.3

Solve the corresponding system when the final demand vector is $\mathbf{b} = (25, 10, 30, 45)$.

3. Let $A = \begin{bmatrix} 0.2 & 0.2 & 0.4 \\ 0.6 & 0.6 & 0.0 \\ 0.0 & 0.0 & 0.2 \end{bmatrix}$

Using a mathematical software system, solve the Leontief Open Model with the following data for the different final demand vectors: $(6, 9, 8), (9, 12, 16), (12, 18, 32)$.

4. By generating a large number of two component random integer vectors, determine experimentally the percentage of times they are linearly independent and linearly dependent.

5. (Continuation.) Repeat with three component random integer vectors.

CHAPTER THREE

Matrix Operations

3.1 MATRICES

If an assertion about matrices is false, there is usually a 2×2 matrix that reveals this.

—OLGA TAUSKY-TODD (1906–1995)

I am sorry to say that the subject that I most disliked was mathematics. ... I think the reason was that mathematics leave no room for argument. If you made a mistake, that was all there was to it.

—MALCOLM X (1925–1965)

Matrix Addition and Scalar Multiplication

Matrices play a central role in linear algebra. We can add matrices to each other, multiply a matrix by a scalar, and multiply matrices. Some matrices have inverses; square matrices have eigenvalues and eigenvectors. Many of these matters lie ahead of us in the unfolding of linear algebra.

The word *matrix* now has many commercial and linguistic spinoffs for movies, computer games, computer memory, and medical search engines. There is even a *matrix market*, which is a repository for matrix test data that can be used in comparative studies of linear algebra algorithms.

It is customary to write $\mathbf{A} = (a_{ij})$ to signify that the generic elements in $\mathbf{A}$ are named a_{ij}. Sometimes it is more convenient to write the generic elements of $\mathbf{A}$ as $\mathbf{A}_{ij}$ or $(\mathbf{A})_{ij}$. If $\mathbf{A}$ is $m \times n$, then $1 \leq i \leq m$ and $1 \leq j \leq n$ in this discussion. The notation $\mathbb{R}^{m \times n}$ denotes the set of all $m \times n$ matrices. For a fixed pair (m, n), the set $\mathbb{R}^{m \times n}$ becomes a vector space if we agree to use the ordinary definitions of adding matrices and multiplying matrices by scalars. Thus, if $\mathbf{A}$ and $\mathbf{B}$ are in $\mathbb{R}^{m \times n}$, then the definitions of matrix addition and multiplication by a scalar are

$$(\mathbf{A} + \mathbf{B})_{ij} = \mathbf{A}_{ij} + \mathbf{B}_{ij}, \qquad (c\mathbf{A})_{ij} = c\mathbf{A}_{ij}, \quad \text{for scalar } c$$

The zero element in this space is designated $\mathbf{0}$, and defined by $\mathbf{0}_{ij} = 0$. Notice that $\mathbb{R}^{m \times n}$ is a doubly indexed infinite sequence of distinct vector spaces. In fact, the parameters n and m range independently over the set $\mathbb{N} = \mathbb{Z}^{+}$ of all positive integers.

EXAMPLE 1 To review the addition of two matrices, use a pair of matrices from $\mathbb{R}^{2 \times 3}$.

SOLUTION

$$\begin{bmatrix} 1 & 3 & 2 \\ 12 & -1 & 1 \end{bmatrix} + \begin{bmatrix} 2 & 1 & 5 \\ -8 & 2 & 6 \end{bmatrix} = \begin{bmatrix} 3 & 4 & 7 \\ 4 & 1 & 7 \end{bmatrix} \qquad \blacksquare$$

EXAMPLE 2 Using addition and multiplication by scalars in the space $\mathbb{R}^{2 \times 3}$, form an arbitrary linear combination in the same space.

SOLUTION

$$5 \begin{bmatrix} 1 & 3 & 2 \\ 12 & -1 & 1 \end{bmatrix} - 3 \begin{bmatrix} 2 & 1 & 5 \\ -8 & 2 & 6 \end{bmatrix} + 2 \begin{bmatrix} 1 & 7 & 2 \\ -3 & 4 & 1 \end{bmatrix}$$

$$= \begin{bmatrix} 1 & 26 & -1 \\ 78 & -3 & -11 \end{bmatrix} \qquad \blacksquare$$

In summary, the following are the salient properties of the vector space $\mathbb{R}^{m \times n}$:

- $\mathbf{A} = \mathbf{B}$ if and only if $\mathbf{A}_{ij} = \mathbf{B}_{ij}$ for all i and j. (**Equality** of two matrices.)
- $(\mathbf{A} + \mathbf{B})_{ij} = \mathbf{A}_{ij} + \mathbf{B}_{ij}$ (**Addition** of two matrices.)
- $(c\mathbf{A})_{ij} = c\mathbf{A}_{ij}$ for any scalar c (**Scalar multiplication** of a matrix.)

Here we take all values such that $1 \leq i \leq m$ and $1 \leq j \leq n$. Once these definitions have been made, it is an easy matter to prove all the vector–space properties in Section 2.4. For example, $\mathbf{A} + \mathbf{0} = \mathbf{A}$, and so forth. Thus, $\mathbb{R}^{m \times n}$ is a vector space.

We have already gone far beyond this stage in matrix algebra. Recall that the products $\mathbf{Ax}$ and $\mathbf{AB}$ have been defined in Sections 1.2 and 1.3. Specifically, if $\mathbf{A}$ is $m \times n$ and $\mathbf{x}$ is $n \times 1$, then

$$
\begin{bmatrix} a_{11} & a_{12} & \cdots & a_{1n} \\ a_{21} & a_{22} & \cdots & a_{2n} \\ \vdots & \vdots & \ddots & \vdots \\ a_{m1} & a_{m1} & \cdots & a_{mn} \end{bmatrix} \begin{bmatrix} x_1 \\ x_2 \\ \vdots \\ x_n \end{bmatrix}
$$

$$
= x_1 \begin{bmatrix} a_{11} \\ a_{21} \\ \vdots \\ a_{m1} \end{bmatrix} + x_2 \begin{bmatrix} a_{12} \\ a_{22} \\ \vdots \\ a_{m2} \end{bmatrix} + \cdots + x_n \begin{bmatrix} a_{1n} \\ a_{2n} \\ \vdots \\ a_{mn} \end{bmatrix}
$$

Notice that the product in this case is an $m \times 1$ matrix, or in other words a column vector containing m components.

EXAMPLE 3 What is the product $\begin{bmatrix} 1 & 3 & 2 \\ 4 & -1 & 1 \end{bmatrix} \begin{bmatrix} 6 \\ 2 \\ 5 \end{bmatrix}$

SOLUTION This is an example of a matrix–vector product, obeying the preceding rule. The calculation produces

$$
6 \begin{bmatrix} 1 \\ 4 \end{bmatrix} + 2 \begin{bmatrix} 3 \\ -1 \end{bmatrix} + 5 \begin{bmatrix} 2 \\ 1 \end{bmatrix} = \begin{bmatrix} 6 \\ 24 \end{bmatrix} + \begin{bmatrix} 6 \\ -2 \end{bmatrix} + \begin{bmatrix} 10 \\ 5 \end{bmatrix} = \begin{bmatrix} 22 \\ 27 \end{bmatrix}
$$ ∎

Matrix–Matrix Multiplication

The matrix–matrix product was defined in Section 1.3, and goes like this: The general matrix–matrix product $\mathbf{AB}$ exists whenever the number of columns in $\mathbf{A}$ matches the number of rows in $\mathbf{B}$. Let the columns in $\mathbf{B}$ be denoted by $\mathbf{b}_1, \mathbf{b}_2, \ldots, \mathbf{b}_k$. Then $\mathbf{AB}$ is the matrix whose columns are $\mathbf{Ab}_1, \mathbf{Ab}_2, \ldots, \mathbf{Ab}_k$. Notice that the matrix–vector product is a special case of the matrix–matrix product, since a column vector is an $n \times 1$ matrix.

If $\mathbf{A}$ is $m \times n$ and if $\mathbf{B}$ is $n \times k$, then $\mathbf{AB}$ will have k columns, each being a vector in $\mathbb{R}^m$. In short, $\mathbf{AB}$ is $m \times k$. The pattern of dimensions here should be remembered like this:

$$
(m \times n) \cdot (n \times k) = (m \times k)
$$

The common index n is necessitated by the fact that we must multiply each column $\mathbf{b}_j$ in $\mathbf{B}$ by the matrix $\mathbf{A}$, and $\mathbf{A}\mathbf{b}_j$ makes sense only when the number of columns in $\mathbf{A}$ matches the number of components in each column vector $\mathbf{b}_j$.

EXAMPLE 4 What is the product

$$\mathbf{AB} = \begin{bmatrix} 1 & 2 & 3 \\ 4 & 5 & 6 \end{bmatrix} \begin{bmatrix} 1 & 2 & -3 \\ -2 & -1 & 2 \\ 1 & 3 & 1 \end{bmatrix}$$

SOLUTION We denote the columns of $\mathbf{B}$ by the notation $\mathbf{b}_1, \mathbf{b}_2, \mathbf{b}_3$ and obtain

$$\mathbf{AB} = \begin{bmatrix} \mathbf{A}\mathbf{b}_1, & \mathbf{A}\mathbf{b}_2, & \mathbf{A}\mathbf{b}_3 \end{bmatrix}$$

$$= \begin{bmatrix} \begin{bmatrix} 1 & 2 & 3 \\ 4 & 5 & 6 \end{bmatrix} \begin{bmatrix} 1 \\ -2 \\ 1 \end{bmatrix}, \begin{bmatrix} 1 & 2 & 3 \\ 4 & 5 & 6 \end{bmatrix} \begin{bmatrix} 2 \\ -1 \\ 3 \end{bmatrix}, \\ \begin{bmatrix} 1 & 2 & 3 \\ 4 & 5 & 6 \end{bmatrix} \begin{bmatrix} -3 \\ 2 \\ 1 \end{bmatrix} \end{bmatrix}$$

$$= \begin{bmatrix} 0 & 9 & 4 \\ 0 & 21 & 4 \end{bmatrix}$$ ∎

There are several ways to interpret matrix multiplication.[1] We have just seen the first way, which is taken as the definition. If it is desired to know

---

[1] James Joseph Sylvester (1814–1897) and Arthur Cayley (1821–1895) are considered joint founders of matrix theory because of their many years of contributing to this branch of mathematics. Cayley was the first to recognize how matrices should be multiplied so that the correspondence between linear transformations and matrices given by $S \leftrightarrow \mathbf{A}$ and $T \leftrightarrow \mathbf{B}$ resulted in $(S \circ T)\mathbf{x} = (\mathbf{AB})\mathbf{x}$. We will mention Cayley again in Section 8.1.

The term *matrix* was first used by Sylvester in 1850 to mean a rectangular array of numbers. (By this time, the study of determinants had been going on for more than 150 years. See Sections 4.1 and 4.2.) Sylvester was the *Second Wrangler* in the Tripos exams at Cambridge University in the year 1837. He was, however, barred from taking a degree or teaching at Cambridge because he was Jewish! (Being a religious entity, Cambridge University required all degree candidates and professors to belong to the Church of England.) He immigrated to the United States

the (i, j)-element in the product $\mathbf{AB}$, we must first concentrate on the jth column of the product, which is $\mathbf{Ab}_j$. Here (as previously) we use $\mathbf{b}_j$ to denote the jth column in $\mathbf{B}$. Now, what is this vector $\mathbf{Ab}_j$? It is a linear combination of the columns of $\mathbf{A}$ with coefficients taken from the vector $\mathbf{b}_j$. (This was the matrix-product definition given in Section 1.3.) Suppose $\mathbf{A}$ is an $m \times n$ matrix and $\mathbf{B}$ is an $n \times k$ matrix. Note the critical entry of n in this calculation. Thus, using $\mathbf{a}_r$ for column r in $\mathbf{A}$, we have

$$\mathbf{Ab}_j = b_{1j}\mathbf{a}_1 + b_{2j}\mathbf{a}_2 + \cdots + b_{nj}\mathbf{a}_n = \sum_{r=1}^{n} b_{rj}\mathbf{a}_r$$

Here, the entries $\mathbf{a}_1, \mathbf{a}_2, \ldots, \mathbf{a}_n$ are the columns of $\mathbf{A}$. Because we asked for the ith element in this vector, we have

$$(\mathbf{Ab}_j)_i = b_{1j}a_{i1} + b_{2j}a_{i2} + \cdots + b_{nj}a_{in} = \sum_{r=1}^{n} b_{rj}a_{ir}$$

We prefer to write this as

$$(\mathbf{AB})_{ij} = \sum_{r=1}^{n} a_{ir}b_{rj} \qquad \text{or} \qquad (\mathbf{AB})_{ij} = \sum_{r=1}^{n} \mathbf{A}_{ir}\mathbf{B}_{rj}$$

Here, $1 \leq i \leq m$ and $1 \leq j \leq k$. This formula provides a second way of interpreting the product $\mathbf{AB}$.

Each summation in computing the product $\mathbf{AB}$ contains n multiplications, and there are mk of them because we need all values of (i, j). Consequently, matrix–matrix multiplication can be a computationally *expensive* operation, involving nmk scalar multiplications. If the matrices are square, say $n \times n$, then the number of multiplications is n^3. (This does not include the additions involved.) Of course, these considerations come into play only when the matrices are enormous, as occurs

<hr/>

in 1841 and taught at the University of Virginia from 1841 to 1845. Sylvester was a fiery and passionate person who once had an altercation with a badly behaved student during a lecture! In 1845, Sylvester returned to England and for ten years was an actuary and lawyer in London. Between 1854 and 1871 he taught at the Royal Military Academy at Woolwich. Then he became a professor at Johns Hopkins University in Baltimore, where he wrote many papers, especially on quadratic forms (which we treat in a later chapter). In 1878 (at the age of 64), he founded the *American Journal of Mathematics*, the first mathematical journal published in the United States. From 1884 to the end of his life, he was a professor at Oxford University.

in many practical problems. Think of $n = 10^8$, for which we obtain $n^3 = 10^{24}$.

Pre-multiplication and Post-multiplication

Another example is given here to illustrate in complete detail that we can carry out a matrix–matrix product as either a post-multiplication by columns or a pre-multiplication by rows. First, let us use **post-multiplication** of the **columns b_i** of the matrix **B** into the matrix **A**. In this example, **A** is 3×2 and **B** is 2×2.

$$\mathbf{AB} = \begin{bmatrix} 3 & 2 \\ 1 & 4 \\ 0 & -1 \end{bmatrix} \begin{bmatrix} 1 & 3 \\ 2 & 1 \end{bmatrix} = \begin{bmatrix} \begin{bmatrix} 3 & 2 \\ 1 & 4 \\ 0 & -1 \end{bmatrix} \begin{bmatrix} 1 \\ 2 \end{bmatrix}, & \begin{bmatrix} 3 & 2 \\ 1 & 4 \\ 0 & -1 \end{bmatrix} \begin{bmatrix} 3 \\ 1 \end{bmatrix} \end{bmatrix}$$

$$= \begin{bmatrix} \mathbf{Ab}_1, & \mathbf{Ab}_2 \end{bmatrix}$$

$$= \begin{bmatrix} 1 \begin{bmatrix} 3 \\ 1 \\ 0 \end{bmatrix} + 2 \begin{bmatrix} 2 \\ 4 \\ -1 \end{bmatrix}, & 3 \begin{bmatrix} 3 \\ 1 \\ 0 \end{bmatrix} + 1 \begin{bmatrix} 2 \\ 4 \\ -1 \end{bmatrix} \end{bmatrix}$$

$$= \begin{bmatrix} \begin{bmatrix} 7 \\ 9 \\ -2 \end{bmatrix}, & \begin{bmatrix} 11 \\ 7 \\ -1 \end{bmatrix} \end{bmatrix} = \begin{bmatrix} 7 & 11 \\ 9 & 7 \\ -2 & -1 \end{bmatrix}$$

Next, we illustrate **pre-multiplication** by the **rows r_i** of matrix **A** with the same matrices **B**.

$$\mathbf{AB} = \begin{bmatrix} 3 & 2 \\ 1 & 4 \\ 0 & -1 \end{bmatrix} \begin{bmatrix} 1 & 3 \\ 2 & 1 \end{bmatrix} = \begin{bmatrix} \begin{bmatrix} 3 & 2 \end{bmatrix} \begin{bmatrix} 1 & 3 \\ 2 & 1 \end{bmatrix} \\ \begin{bmatrix} 1 & 4 \end{bmatrix} \begin{bmatrix} 1 & 3 \\ 2 & 1 \end{bmatrix} \\ \begin{bmatrix} 0 & -1 \end{bmatrix} \begin{bmatrix} 1 & 3 \\ 2 & 1 \end{bmatrix} \end{bmatrix} = \begin{bmatrix} \mathbf{r}_1\mathbf{B} \\ \mathbf{r}_2\mathbf{B} \\ \mathbf{r}_3\mathbf{B} \end{bmatrix}$$

$$= \begin{bmatrix} 3 \begin{bmatrix} 1 & 3 \end{bmatrix} + 2 \begin{bmatrix} 2 & 1 \end{bmatrix} \\ 1 \begin{bmatrix} 1 & 3 \end{bmatrix} + 4 \begin{bmatrix} 2 & 1 \end{bmatrix} \\ 0 \begin{bmatrix} 1 & 3 \end{bmatrix} - 1 \begin{bmatrix} 2 & 1 \end{bmatrix} \end{bmatrix} = \begin{bmatrix} \begin{bmatrix} 7 & 11 \end{bmatrix} \\ \begin{bmatrix} 9 & 7 \end{bmatrix} \\ \begin{bmatrix} -2 & -1 \end{bmatrix} \end{bmatrix} = \begin{bmatrix} 7 & 11 \\ 9 & 7 \\ -2 & -1 \end{bmatrix}$$

Of course, the two procedures lead to identical results!

Dot Product

The **dot product** between two vectors $\mathbf{u} = (u_1, u_2, u_3, \ldots, u_n)$ and $\mathbf{v} = (v_1, v_2, v_3, \ldots, v_n)$ is

$$\mathbf{u} \cdot \mathbf{v} = \sum_{j=1}^{n} u_j v_j$$

which is the standard inner product in the Euclidean vector space $\mathbb{R}^n$. In matrix notation, the dot product can also be written as

$$\mathbf{u} \cdot \mathbf{v} = \mathbf{u}^T \mathbf{v} = [u_1, u_2, \ldots, u_n] \begin{bmatrix} v_1 \\ v_2 \\ \vdots \\ v_n \end{bmatrix}$$

If we use this notation, then we can say that $(\mathbf{AB})_{ij}$ is the dot product of row i in $\mathbf{A}$ with column j in $\mathbf{B}$. This is a third way of interpreting $\mathbf{AB}$. Let us use $\mathbf{r}_i$ to designate row i in $\mathbf{A}$. Then we have the formula

$$(\mathbf{AB})_{ij} = \mathbf{r}_i \mathbf{b}_j$$

(Here we continue to denote column j in $\mathbf{B}$ by $\mathbf{b}_j$.) This formula enables us to compute any element in the product $\mathbf{AB}$ with one simple dot product.

THEOREM 1

The (i, j)-element in the product of two matrices $\mathbf{A}$ and $\mathbf{B}$ can be written in summation notation or as the dot product of row i in $\mathbf{A}$ with column j in $\mathbf{B}$:

$$(\mathbf{AB})_{ij} = \sum_{r=1}^{n} a_{ir} b_{rj} = \mathbf{r}_i \mathbf{b}_j$$

EXAMPLE 5 What is the $(2, 4)$-element in this product? That is, what is $(\mathbf{AB})_{2,4}$?

$$\mathbf{AB} = \begin{bmatrix} 1 & 4 & 3 & 2 \\ 2 & 0 & -3 & 5 \\ -3 & 5 & 7 & 1 \end{bmatrix} \begin{bmatrix} 3 & 5 & 4 & 3 & 2 \\ -4 & 1 & 11 & 7 & 13 \\ 2 & -5 & 1 & 4 & 6 \\ 3 & -7 & 6 & 2 & 1 \end{bmatrix}$$

SOLUTION We need not compute the entire product **AB** to answer this question. We simply compute the dot product of row 2 in **A** with column 4 in **B**, which is

$$(\mathbf{AB})_{24} = \mathbf{r}_2\mathbf{b}_4 = (2)(3) + (0)(7) + (-3)(4) + (5)(2) = 4 \qquad \blacksquare$$

Special Matrices

A number of special categories of matrices will be enumerated here. A matrix is **square** if it has the same number of rows as columns. The **diagonal** of a square matrix **A** is the sequence of elements $\mathbf{A}_{ii}$. The other elements are called **off-diagonal** elements. If all off-diagonal elements are 0, we say that **A** is a **diagonal matrix**. Among the diagonal matrices, the **identity** matrix is especially important: it has 1's on the diagonal, and 0's elsewhere. Here are examples of a square matrix, a diagonal matrix, and an identity matrix:

$$\begin{bmatrix} 3 & 2 & -4 \\ 5 & -7 & 2 \\ 0 & 6 & -3 \end{bmatrix}, \quad \begin{bmatrix} 3 & 0 & 0 \\ 0 & -5 & 0 \\ 0 & 0 & 8 \end{bmatrix}, \quad \begin{bmatrix} 1 & 0 & 0 \\ 0 & 1 & 0 \\ 0 & 0 & 1 \end{bmatrix}$$

The $n \times n$ identity matrix is denoted by $\mathbf{I}_n$ or by $\mathbf{I}$, if its size is clear from the context. We often denote the elements of an identity matrix by δ_{ij}. This is called the **Kronecker delta**.[2] The formal definition is

$$\delta_{ij} = \begin{cases} 1 & \text{if } i = j \\ 0 & \text{if } i \neq j \end{cases}$$

EXAMPLE 6 If we compute the product of a 3×3 matrix and a 3×3 identity matrix, what do we expect to get?

---

[2] Leopold Kronecker (1823–1891) was born in Liegniz, near Breslau (in modern-day Poland). The family was independently wealthy by its ownership and management of a large estate. Kronecker himself participated in this activity but reserved time for his mathematical research. This saved him from having to hold a regular academic position. When he did teach, his lectures were demanding and stimulating for the best students but not popular with the average students. He is famous for the dictum concerning mathematics: ''*God made the integers; all else is the work of man.*''

SOLUTION Here is a case in point:

$$
\begin{bmatrix} 2 & 4 & 3 \\ 2 & -1 & 5 \\ 1 & 2 & -3 \end{bmatrix}
\begin{bmatrix} 1 & 0 & 0 \\ 0 & 1 & 0 \\ 0 & 0 & 1 \end{bmatrix}
=
\begin{bmatrix} 2 & 4 & 3 \\ 2 & -1 & 5 \\ 1 & 2 & -3 \end{bmatrix}
$$

$$
\begin{bmatrix} 1 & 0 & 0 \\ 0 & 1 & 0 \\ 0 & 0 & 1 \end{bmatrix}
\begin{bmatrix} 2 & 4 & 3 \\ 2 & -1 & 5 \\ 1 & 2 & -3 \end{bmatrix}
=
\begin{bmatrix} 2 & 4 & 3 \\ 2 & -1 & 5 \\ 1 & 2 & -3 \end{bmatrix}
$$
∎

THEOREM 2

For any $n \times n$ matrix $\mathbf{A}$, we have $\mathbf{A}\mathbf{I}_n = \mathbf{I}_n\mathbf{A} = \mathbf{A}$.

PROOF We compute the (i, j)-element in the product $\mathbf{A}\mathbf{I}_n$ using methods discussed previously:

$$
(\mathbf{AI})_{ij} = \sum_{r=1}^{n} a_{ir}\delta_{rj} = a_{ij}
$$

Note that in the summation, the term δ_{rj} will be zero except when $r = j$. Then its value is 1. Hence, only one term in the sum survives, namely the one where $r = j$. The other equation is proved in the same manner. ∎

Matrices with special structure occur in many applications. A square matrix $\mathbf{A}$ is said to be **upper triangular** if $a_{ij} = 0$ when $i > j$. Similarly, we have a concept of **lower triangular**: $a_{ij} = 0$ when $i < j$. (These definitions are also meaningful for nonsquare matrices.) See Figure 3.1 for illustrations of some matrices with special structure.

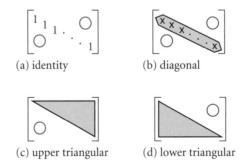

(a) identity   (b) diagonal

(c) upper triangular   (d) lower triangular

FIGURE 3.1 Matrices with special structure.

EXAMPLE 7 If we add two upper triangular matrices together, what do we expect to find?

SOLUTION Here is a numerical case:

$$
\begin{bmatrix} 3 & 2 & 5 \\ 0 & 4 & -3 \\ 0 & 0 & 7 \end{bmatrix} + \begin{bmatrix} -2 & 4 & 5 \\ 0 & 6 & -2 \\ 0 & 0 & 3 \end{bmatrix} = \begin{bmatrix} 1 & 6 & 10 \\ 0 & 10 & -5 \\ 0 & 0 & 10 \end{bmatrix} \quad \blacksquare
$$

THEOREM 3

The set of upper triangular matrices in $\mathbb{R}^{n \times n}$ is a subspace of $\mathbb{R}^{n \times n}$. That is, it is a vector space in its own right.

Subspaces are taken up systematically in Section 5.1.

EXAMPLE 8 If we multiply two upper triangular matrices of the same size, what do we expect to obtain?

SOLUTION Here is a simple case to guide us:

$$
\begin{bmatrix} 3 & 2 & 5 \\ 0 & 4 & -3 \\ 0 & 0 & 7 \end{bmatrix} \begin{bmatrix} -2 & 4 & 5 \\ 0 & 6 & -2 \\ 0 & 0 & 3 \end{bmatrix} = \begin{bmatrix} -6 & 24 & 26 \\ 0 & 24 & -17 \\ 0 & 0 & 21 \end{bmatrix} \quad \blacksquare
$$

THEOREM 4

The product of two upper triangular matrices (of the same size) is also upper triangular.

PROOF Let $\mathbf{A}$ and $\mathbf{B}$ be $n \times n$ upper triangular matrices. Let $1 \leq j < i \leq n$. It is to be proved that $(\mathbf{AB})_{ij} = 0$. We have

$$
(\mathbf{AB})_{ij} = \sum_{r=1}^{n} a_{ir} b_{rj} = \sum_{r=i}^{j} a_{ir} b_{rj} = 0
$$

In this calculation, use first the fact that $a_{ir} = 0$ if $r < i$. Then only terms having $r \geq i$ enter the sum. Next, use the fact that $b_{rj} = 0$ when $r > j$. The restriction on r is then $i \leq r \leq j$, which is impossible, because we assumed $j < i$. ∎

A similar result is true for lower triangular matrices, and is left as General Exercise 62.

Matrix Transpose

When we exchange the rows and columns of a matrix, we obtain the transpose of the matrix, which is of particular importance.

> **DEFINITION**
>
> *The **transpose** of a matrix $\mathbf{A}$ is the matrix $\mathbf{A}^T$ whose rows are the columns of $\mathbf{A}$, in the same order. In symbols: $(\mathbf{A}^T)_{ij} = \mathbf{A}_{ji}$.*

> **EXAMPLE 9** Give a 3×4 matrix and its 4×3 transpose.

SOLUTION

$$\mathbf{A} = \begin{bmatrix} 1 & 3 & 7 & 2 \\ 6 & 1 & 9 & 5 \\ 11 & -8 & 3 & 1 \end{bmatrix}, \quad \mathbf{A}^T = \begin{bmatrix} 1 & 6 & 11 \\ 3 & 1 & -8 \\ 7 & 9 & 3 \\ 2 & 5 & 1 \end{bmatrix} \quad \blacksquare$$

> **THEOREM 5**
>
> *If the product $\mathbf{AB}$ of two matrices $\mathbf{A}$ and $\mathbf{B}$ exists, then $(\mathbf{AB})^T = \mathbf{B}^T \mathbf{A}^T$.*

PROOF We use the subscript notation for elements of matrices and (not unexpectedly) the definition of a matrix product:

$$[(\mathbf{AB})^T]_{ij} = (\mathbf{AB})_{ji} = \sum_{r=1}^{n} \mathbf{A}_{jr} \mathbf{B}_{ri} = \sum_{r=1}^{n} (\mathbf{B}^T)_{ir} (\mathbf{A}^T)_{rj} = (\mathbf{B}^T \mathbf{A}^T)_{ij} \quad \blacksquare$$

There is another way to describe what is happening in the preceding proof: To determine the (i, j)-element in $\mathbf{B}^T \mathbf{A}^T$, we should take the dot product of row i in $\mathbf{B}^T$ with column j in $\mathbf{A}^T$. But this is the same as the dot product of column i in $\mathbf{B}$ with row j in $\mathbf{A}$. That is the same as the dot product of row j in $\mathbf{A}$ with column i in $\mathbf{B}$. Finally, this is the (j, i)-element in $\mathbf{AB}$ or the (i, j)-element in $(\mathbf{AB})^T$.

Here are four computational rules involving the matrix transpose:

$$(\mathbf{A}^T)^T = \mathbf{A}, \qquad (\mathbf{A} + \mathbf{B})^T = \mathbf{A}^T + \mathbf{B}^T$$
$$(c\mathbf{A})^T = c\mathbf{A}^T, \qquad (\mathbf{AB})^T = \mathbf{B}^T \mathbf{A}^T$$

Symmetric Matrices

A matrix $\mathbf{A}$ such that $\mathbf{A} = \mathbf{A}^T$ is said to be **symmetric**. Here is a symmetric matrix whose elements are random numbers from the interval $[0, 1]$, generated by mathematical software:

$$\mathbf{A} = \begin{bmatrix} 0.8992206 & 0.7876846 & 0.0086799 \\ 0.7876846 & 0.2811037 & 0.0327878 \\ 0.0086799 & 0.0327878 & 0.6004940 \end{bmatrix}$$

The symmetric property here depends upon three equations being true: $a_{21} = a_{12}, a_{31} = a_{13},$ and $a_{32} = a_{23}$.

Skew–Symmetric Matrices

While a **symmetric** matrix $\mathbf{A}$ has the property $\mathbf{A}^T = \mathbf{A}$, a **skew–symmetric** matrix has the property $\mathbf{A}^T = -\mathbf{A}$, and this contains a minus sign! If a matrix is simultaneously symmetric and skew–symmetric, then it must be the zero matrix. (If $\mathbf{A}^T = \mathbf{A} = -\mathbf{A}^T$, then $\mathbf{A} = \mathbf{0}$.) Moreover, there is a representation of a square matrix as the sum of a symmetric matrix and a skew–symmetric matrix. The equation that proves this is

$$\mathbf{A} = \tfrac{1}{2}(\mathbf{A} + \mathbf{A}^T) + \tfrac{1}{2}(\mathbf{A} - \mathbf{A}^T)$$

The first matrix on the right is symmetric and the second matrix is skew–symmetric.

The unicity of the splitting of $\mathbf{A}$ into symmetric and skew–symmetric parts can be easily established. Suppose that

$$\mathbf{A} = \mathbf{B} + \mathbf{C}$$

where $\mathbf{B}$ is symmetric and $\mathbf{C}$ is skew–symmetric. We shall prove first that $\mathbf{B} = \tfrac{1}{2}(\mathbf{A} + \mathbf{A}^T)$. By the first hypothesis, we obtain

$$\tfrac{1}{2}(\mathbf{A} + \mathbf{A}^T) = \tfrac{1}{2}(\mathbf{B} + \mathbf{C} + \mathbf{B}^T + \mathbf{C}^T) = \tfrac{1}{2}(\mathbf{B} + \mathbf{B}^T) + \tfrac{1}{2}(\mathbf{C} + \mathbf{C}^T)$$

Because **C** is skew–symmetric by hypotheses, $\frac{1}{2}(\mathbf{C} + \mathbf{C}^T) = 0$. Using the symmetry of **B**, we see that

$$\tfrac{1}{2}(\mathbf{A} + \mathbf{A}^T) = \tfrac{1}{2}(\mathbf{B} + \mathbf{B}^T) = \mathbf{B}$$

Then **C** is determined by the equation

$$\mathbf{C} = \mathbf{A} - \mathbf{B} = \mathbf{A} - \tfrac{1}{2}(\mathbf{A} + \mathbf{A}^T) = \tfrac{1}{2}(\mathbf{A} - \mathbf{A}^T)$$

EXAMPLE 10 Express this matrix as the sum of a symmetric matrix and a skew–symmetric matrix:

$$\mathbf{A} = \begin{bmatrix} 3 & 2 & 1 \\ 4 & 7 & 5 \\ 6 & 8 & 9 \end{bmatrix}$$

SOLUTION We let the symmetric part be

$$\tfrac{1}{2}(\mathbf{A} + \mathbf{A}^T) = \tfrac{1}{2} \begin{bmatrix} 6 & 6 & 7 \\ 6 & 14 & 13 \\ 7 & 13 & 18 \end{bmatrix}$$

and let the skew–symmetric part be

$$\tfrac{1}{2}(\mathbf{A} - \mathbf{A}^T) = \tfrac{1}{2} \begin{bmatrix} 0 & -2 & -5 \\ 2 & 0 & -3 \\ 5 & 3 & 0 \end{bmatrix}$$

Verifying, we find that $\mathbf{A} = \tfrac{1}{2}(\mathbf{A} + \mathbf{A}^T) + \tfrac{1}{2}(\mathbf{A} - \mathbf{A}^T)$. ∎

Non-commutativity of Matrix Multiplication

Now we come to the question of commutativity of matrix multiplication. Let us choose at random several square matrices **A** and **B** and see whether **AB** = **BA**.

$$\begin{bmatrix} 1 & 2 \\ 5 & -3 \end{bmatrix} \begin{bmatrix} 11 & 4 \\ 10 & 3 \end{bmatrix} = \begin{bmatrix} 31 & 10 \\ 25 & 11 \end{bmatrix} = \begin{bmatrix} 11 & 4 \\ 10 & 3 \end{bmatrix} \begin{bmatrix} 1 & 2 \\ 5 & -3 \end{bmatrix}$$

$$\begin{bmatrix} 7 & 3 \\ 5 & -8 \end{bmatrix} \begin{bmatrix} 18 & 4 \\ 5 & 3 \end{bmatrix} = \begin{bmatrix} 141 & 30 \\ 50 & -9 \end{bmatrix} = \begin{bmatrix} 18 & 4 \\ 5 & 3 \end{bmatrix} \begin{bmatrix} 7 & 3 \\ 5 & -8 \end{bmatrix}$$

Because these matrices seem typical, can we not assume that, in general, **AB** = **BA**? Most emphatically, the answer is *no*! In the real world, *Baconian* *inductive reasoning* is often used. It does proceed by looking at many examples, from which some tentative conclusion can be drawn. But this can be used in mathematics only to suggest something that *may* be true. In the present case, the suspected theorem is *false*, and generally $\mathbf{AB} \neq \mathbf{BA}$. For example, we have

$$\begin{bmatrix} 0 & 1 \\ 0 & 0 \end{bmatrix}\begin{bmatrix} 1 & 0 \\ 0 & 0 \end{bmatrix} = \begin{bmatrix} 0 & 0 \\ 0 & 0 \end{bmatrix} \neq \begin{bmatrix} 0 & 1 \\ 0 & 0 \end{bmatrix} = \begin{bmatrix} 1 & 0 \\ 0 & 0 \end{bmatrix}\begin{bmatrix} 0 & 1 \\ 0 & 0 \end{bmatrix}$$

Of course, one example is sufficient to demolish the conjecture that matrix multiplication is *commutative*. The apparently *random* integer matrices displayed at the beginning of this section show that *sometimes* a pair of matrices will commute with each other. But those matrices are not actually random; they were carefully constructed to illustrate the existence of commuting pairs. Maybe you can find others.

CAUTION

Matrix multiplication is not commutative; that is, in general, one must expect that

$$\mathbf{AB} \neq \mathbf{BA}$$

Associativity Law for Matrix Multiplication

What can be said about associativity of matrix multiplication? Here the result is simple (and very useful).

THEOREM 6

Matrix multiplication is associative. Thus, if the products exist, we have

$$(\mathbf{AB})\mathbf{C} = \mathbf{A}(\mathbf{BC})$$

PROOF We calculate (and get practice in handling summations):

$$[(\mathbf{AB})\mathbf{C}]_{ij} = \sum_{r=1}^{n}(\mathbf{AB})_{ir}(\mathbf{C})_{rj} = \sum_{r=1}^{n}\left[\sum_{s=1}^{m}\mathbf{A}_{is}\mathbf{B}_{sr}\right]\mathbf{C}_{rj}$$

$$= \sum_{s=1}^{m}\mathbf{A}_{is}\sum_{r=1}^{n}\mathbf{B}_{sr}\mathbf{C}_{rj} = \sum_{s=1}^{m}(\mathbf{A})_{is}(\mathbf{BC})_{sj} = [\mathbf{A}(\mathbf{BC})]_{ij} \qquad \blacksquare$$

An example of the associative law for matrix multiplication is given here:

$$(\mathbf{AB})\mathbf{C} = \left(\begin{bmatrix} 1 & 2 & 3 \\ 2 & 1 & 2 \end{bmatrix} \begin{bmatrix} 2 & -1 & 0 \\ 1 & 3 & 1 \\ 4 & 1 & -1 \end{bmatrix} \right) \begin{bmatrix} 1 & 2 & 3 & 3 \\ 0 & 1 & 2 & 1 \\ 1 & 0 & 1 & -2 \end{bmatrix}$$

$$= \begin{bmatrix} 16 & 8 & -1 \\ 13 & 3 & -1 \end{bmatrix} \begin{bmatrix} 1 & 2 & 3 & 3 \\ 0 & 1 & 2 & 1 \\ 1 & 0 & 1 & -2 \end{bmatrix} = \begin{bmatrix} 15 & 40 & 63 & 58 \\ 12 & 29 & 44 & 44 \end{bmatrix}$$

Calculating the product in a different order, we have

$$\mathbf{A}(\mathbf{BC}) = \begin{bmatrix} 1 & 2 & 3 \\ 2 & 1 & 2 \end{bmatrix} \left(\begin{bmatrix} 2 & -1 & 0 \\ 1 & 3 & 1 \\ 4 & 1 & -1 \end{bmatrix} \begin{bmatrix} 1 & 2 & 3 & 3 \\ 0 & 1 & 2 & 1 \\ 1 & 0 & 1 & -2 \end{bmatrix} \right)$$

$$= \begin{bmatrix} 1 & 2 & 3 \\ 2 & 1 & 2 \end{bmatrix} \begin{bmatrix} 2 & 3 & 4 & 5 \\ 2 & 5 & 10 & 4 \\ 3 & 9 & 13 & 15 \end{bmatrix} = \begin{bmatrix} 15 & 40 & 63 & 58 \\ 12 & 29 & 44 & 44 \end{bmatrix}$$

Linear Transformations

The next theorem shows why matrix multiplication is defined in the manner explained in Section 1.3.

THEOREM 7

Let S and T be two linear transformations defined by $S(\mathbf{y}) = \mathbf{Ay}$ and $T(\mathbf{x}) = \mathbf{Bx}$. ($\mathbf{A}$ and $\mathbf{B}$ are matrices.) If the composition $S \circ T$ exists, then its matrix is $\mathbf{AB}$.

PROOF We use the associativity of matrix multiplication in this quick calculation:

$$(S \circ T)(\mathbf{x}) = S(T(\mathbf{x})) = S(\mathbf{Bx}) = \mathbf{A}(\mathbf{Bx}) = (\mathbf{AB})\mathbf{x} \qquad \blacksquare$$

Another interpretation of this proof is as follows. Column i of matrix $\mathbf{A}$, say $\mathbf{a}_i$, is the image of the ith standard unit vector $\mathbf{e}_i = (0, \ldots, 1, 0, \ldots, 0)$ under the linear transformation S; namely, $S(\mathbf{e}_i) = \mathbf{a}_i$. We obtain similar results for matrix $\mathbf{B}$ and linear transformation T; that is, $T(\mathbf{e}_i) = \mathbf{b}_i$, where $\mathbf{b}_i$ is the ith column of matrix $\mathbf{B}$. Consequently, we obtain

$$\mathbf{Ab}_i = S(\mathbf{b}_i) = S(T(\mathbf{e}_i)) = (S \circ T)(\mathbf{e}_i)$$

which is the ith column of matrix $\mathbf{AB}$. Now associativity of matrix multiplication is a corollary of associativity of functional composition.

Elementary Matrices

When a single row operation is applied to the identity matrix, the result is called an **elementary matrix**. Thus, if $\mathcal{R}$ denotes any row operation, we can set $\mathbf{E} = \mathcal{R}(\mathbf{I})$ and get an elementary matrix, $\mathbf{E}$. It is easy to recognize an elementary matrix because it differs very little from an identity matrix. Here, we interpret a row operation as a function that applies to a matrix and produces another matrix. Examples of elementary matrices are

$$\mathbf{E}_1 = \begin{bmatrix} 1 & 0 & 0 \\ 0 & 1 & 0 \\ 0 & 3 & 1 \end{bmatrix} \quad \textbf{(replacement)}$$

$$\mathbf{E}_2 = \begin{bmatrix} 1 & 0 & 0 \\ 0 & 7 & 0 \\ 0 & 0 & 1 \end{bmatrix} \quad \textbf{(scale)}$$

$$\mathbf{E}_3 = \begin{bmatrix} 0 & 1 & 0 \\ 1 & 0 & 0 \\ 0 & 0 & 1 \end{bmatrix} \quad \textbf{(swap)}$$

Why are elementary matrices useful? They enable us to interpret row operations on a matrix as multiplication on the left by one or more elementary matrices. For example, using $\mathbf{E}_1$, $\mathbf{E}_2$, and $\mathbf{E}_3$ as in the preceding display, we have

$$\mathbf{E}_1\mathbf{A} = \begin{bmatrix} 1 & 0 & 0 \\ 0 & 1 & 0 \\ 0 & 3 & 1 \end{bmatrix} \begin{bmatrix} 1 & 2 & 3 \\ 4 & 5 & 6 \\ 7 & 8 & 9 \end{bmatrix} = \begin{bmatrix} 1 & 2 & 3 \\ 4 & 5 & 6 \\ 19 & 23 & 27 \end{bmatrix}$$

$$\mathbf{E}_2\mathbf{A} = \begin{bmatrix} 1 & 0 & 0 \\ 0 & 7 & 0 \\ 0 & 0 & 1 \end{bmatrix} \begin{bmatrix} 1 & 2 & 3 \\ 4 & 5 & 6 \\ 7 & 8 & 9 \end{bmatrix} = \begin{bmatrix} 1 & 2 & 3 \\ 28 & 35 & 42 \\ 7 & 8 & 9 \end{bmatrix}$$

$$\mathbf{E}_3\mathbf{A} = \begin{bmatrix} 0 & 1 & 0 \\ 1 & 0 & 0 \\ 0 & 0 & 1 \end{bmatrix} \begin{bmatrix} 1 & 2 & 3 \\ 4 & 5 & 6 \\ 7 & 8 & 9 \end{bmatrix} = \begin{bmatrix} 4 & 5 & 6 \\ 1 & 2 & 3 \\ 7 & 8 & 9 \end{bmatrix}$$

Now we have a more sophisticated way of thinking about row equivalence. If $\mathbf{A} \sim \mathbf{B}$, then a sequence of row operations can be applied to $\mathbf{A}$, producing $\mathbf{B}$. Hence, there will exist elementary matrices $\mathbf{E}_i$ such that

$$\mathbf{E}_k\mathbf{E}_{k-1} \cdots \mathbf{E}_2\mathbf{E}_1 \, \mathbf{A} = \mathbf{B}$$

> **THEOREM 8**
>
> *If the matrix $\widetilde{\mathbf{A}}$ is the result of applying a row operation $\mathcal{R}$ to the matrix $\mathbf{A}$, and if $\mathbf{E}$ is the matrix that results from applying $\mathcal{R}$ to $\mathbf{I}$, then $\widetilde{\mathbf{A}} = \mathbf{EA}$.*

PROOF We give the proof for one type of row operation—namely, the addition of a scalar multiple of one row onto another row. (This is the most tiresome of the three cases.) Suppose specifically that $\mathcal{R}$ is the operation of adding c times row s in $\mathbf{A}$ onto row r. (Of course, $s \neq r$.) The action of $\mathcal{R}$ on the $n \times n$ identity matrix produces the elementary matrix $\mathbf{E}$ with these entries:

$$\mathbf{E}_{ij} = \delta_{ij} + c\,\delta_{ir}\,\delta_{sj}$$

(To verify this, begin by noticing that $\mathbf{E}$ and $\mathbf{I}$ differ only in row r because of the factor δ_{ir}.) The matrix $\widetilde{\mathbf{A}}$ that results from applying $\mathcal{R}$ to $\mathbf{A}$ has this definition:

$$\widetilde{\mathbf{A}}_{ij} = \mathbf{A}_{ij} + c\,\delta_{ir}\,\mathbf{A}_{sj}$$

Next, we compute $\mathbf{EA}$, element-by-element:

$$(\mathbf{EA})_{ij} = \sum_{t=1}^{n} \mathbf{E}_{it}\mathbf{A}_{tj} = \sum_{t=1}^{n} [\delta_{it} + c\delta_{ir}\delta_{st}]\mathbf{A}_{tj} = \mathbf{A}_{ij} + c\delta_{ir}\mathbf{A}_{sj} = \widetilde{\mathbf{A}}_{ij} \qquad \blacksquare$$

More on the Matrix–Matrix Product

We have been using the product $\mathbf{Ax}$ in many situations involving systems of equations. This product exists if $\mathbf{A}$ is, say, $m \times n$ and $\mathbf{x}$ is $n \times 1$. An analogous product $\mathbf{y}^T\mathbf{A}$ exists when $\mathbf{y}^T$ is $1 \times m$. This will certainly arise naturally if we find it necessary to proceed from $\mathbf{Ax} = \mathbf{b}$ to $(\mathbf{Ax})^T = \mathbf{b}^T$, because this is the same as $\mathbf{x}^T\mathbf{A}^T = \mathbf{b}^T$. (Remember that taking the transpose of a product requires the order of the factors to be reversed.) We can prove theorems about this product $\mathbf{y}^T\mathbf{A}$ that are logically parallel to the results about the product $\mathbf{Ax}$. We have

$$\mathbf{Ax} = \mathbf{b} \qquad (\textbf{post-multiplication} \text{ by } \mathbf{x})$$
$$\mathbf{y}^T\mathbf{A} = \mathbf{c}^T \qquad (\textbf{pre-multiplication} \text{ by } \mathbf{y}^T)$$

where the column vector $\mathbf{b}$ is $m \times 1$ and the row vector $\mathbf{c}^T$ is $1 \times n$.

Recall, from Section 1.2, that the product $\mathbf{Ax}$ was defined to be

$$\mathbf{Ax} = x_1\mathbf{a}_1 + x_2\mathbf{a}_2 + \cdots + x_n\mathbf{a}_n$$

where $\mathbf{a}_j$ denotes the jth column of $\mathbf{A}$. The vector $\mathbf{x}$ is a **column vector**, and n is the number of components in $\mathbf{x}$ as well as the number of columns in $\mathbf{A}$. This definition of a matrix multiplying a vector led us to the appropriate definition of $\mathbf{AB}$. Namely, if the matrix $\mathbf{B}$ has columns $\mathbf{b}_j$, then

$$\mathbf{AB} = \mathbf{A}[\mathbf{b}_1 \quad \mathbf{b}_2 \quad \cdots \quad \mathbf{b}_k] = [\mathbf{Ab}_1 \quad \mathbf{Ab}_2 \quad \cdots \quad \mathbf{Ab}_k]$$

> **EXAMPLE 11** Use the preceding definition of matrix multiplication to determine the entry occupying position (i, j) in the matrix $\mathbf{AB}$.

SOLUTION The desired entry is in column j of $\mathbf{AB}$, and this column is $\mathbf{Ab}_j$, by the definition of the product $\mathbf{AB}$ (as quoted previously). The ith component in this vector is the desired entry in $\mathbf{AB}$. Thus, it is

$$(\mathbf{AB})_{ij} = (\mathbf{Ab}_j)_i = \mathbf{r}_i \mathbf{b}_j = \sum_{k=1}^{n} \mathbf{A}_{ik} \mathbf{B}_{kj}$$

Just as we expect, it is the dot product of row i of matrix $\mathbf{A}$ (denoted here by $\mathbf{r}_i$) and column j of matrix $\mathbf{B}$ (written here as $\mathbf{b}_j$). ■

Vector–Matrix Product

Suppose that we have the $m \times n$ matrix $\mathbf{A}$, the $1 \times n$ row vector $\mathbf{c}$, and the $1 \times m$ row vector $\mathbf{y}^T$. Formulas similar to those obtained previously can be given for products

$$\mathbf{y}^T \mathbf{A} = \mathbf{c}^T$$

when $\mathbf{y}^T$ is a row vector. Namely, we write

$$\mathbf{y}^T \mathbf{A} = y_1 \mathbf{r}_1 + y_2 \mathbf{r}_2 + \cdots + y_m \mathbf{r}_m$$

In this equation, the rows of $\mathbf{A}$ are the vectors $\mathbf{r}_i$. The matrix $\mathbf{A}$ is $m \times n$. We can verify the correctness of the displayed formula by computing the jth component of the vectors on the right and on the left. The one on the left is

$$(\mathbf{y}^T \mathbf{A})_j = \sum_{i=1}^{m} y_i \mathbf{A}_{ij} = y_1 \mathbf{A}_{1j} + y_2 \mathbf{A}_{2j} + \cdots + y_m \mathbf{A}_{mj}$$

The one on the right is

$$(y_1 \mathbf{r}_1 + y_2 \mathbf{r}_2 + \cdots + y_m \mathbf{r}_m)_j = y_1 (\mathbf{r}_1)_j + y_2 (\mathbf{r}_2)_j + \cdots + y_m (\mathbf{r}_m)_j$$
$$= y_1 \mathbf{A}_{1j} + y_2 \mathbf{A}_{2j} + \cdots + y_m \mathbf{A}_{mj}$$

Consequently, we have the following two important theorems.

THEOREM 9

If $\mathbf{y}^T$ is a row vector, then $\mathbf{y}^T\mathbf{A}$ is a linear combination of the rows of $\mathbf{A}$, with coefficients taken to be the components of the row vector $\mathbf{y}^T$:

$$\mathbf{y}^T\mathbf{A} = [y_1, y_2, \ldots, y_n]\mathbf{A} = y_1\mathbf{r}_1 + y_2\mathbf{r}_2 + \cdots + y_n\mathbf{r}_n$$

where $\mathbf{r}_i$ are the rows of $\mathbf{A}$.

THEOREM 10

If the rows of $\mathbf{A}$ are denoted by $\mathbf{r}_1, \mathbf{r}_2, \ldots, \mathbf{r}_m$, then

$$\mathbf{AB} = \begin{bmatrix} \mathbf{r}_1 \\ \mathbf{r}_2 \\ \vdots \\ \mathbf{r}_m \end{bmatrix} \mathbf{B} = \begin{bmatrix} \mathbf{r}_1\mathbf{B} \\ \mathbf{r}_2\mathbf{B} \\ \vdots \\ \mathbf{r}_m\mathbf{B} \end{bmatrix}$$

PROOF The (i, j)-element in $\mathbf{AB}$ is $\sum_{k=1}^{n} \mathbf{A}_{ik}\mathbf{B}_{kj}$.

The (i, j)-element in $\begin{bmatrix} \mathbf{r}_1 \\ \mathbf{r}_2 \\ \vdots \\ \mathbf{r}_m \end{bmatrix} \mathbf{B}$ is $(\mathbf{r}_i\mathbf{B})_j = \mathbf{r}_i\mathbf{b}_j$, where $\mathbf{b}_j$ is column j in $\mathbf{B}$.

This is just the dot-product formula for any entry in the product $\mathbf{AB}$. ■

EXAMPLE 12 If $\mathbf{A}$ and $\mathbf{B}$ are $n \times n$ nonzero matrices, can it happen that $\mathbf{AB} = \mathbf{0}$? If so, this would be contrary to what can happen with real numbers. In that more elementary setting the equation $xy = 0$ implies that at least one of the factors x and y is zero.

SOLUTION As an example to exhibit this phenomenon in matrix theory, we offer

$$\mathbf{AB} = \begin{bmatrix} 1 & 2 \\ 1 & 2 \end{bmatrix} \begin{bmatrix} -2 & 6 \\ 1 & -3 \end{bmatrix} = \begin{bmatrix} 0 & 0 \\ 0 & 0 \end{bmatrix} = \mathbf{0}$$

Thus, one must resist the temptation in matrix algebra to go from the equation $\mathbf{AB} = \mathbf{AC}$ to the equation $\mathbf{B} = \mathbf{C}$, solely on the strength of the fact that $\mathbf{A} \neq \mathbf{0}$. See General Exercises 3, 12, 34, and 44. ∎

Application: Diet Problems

Linear algebra can play a role in dietetics, whether for humans, animals, or bacteria, among others. The objective might be to achieve some nutritional goal through a suitably compounded mixture of foods and dietary supplements. These problems often lead to systems of linear inequalities, a topic treated briefly here.

We assume that there are foods labeled $f_1, f_2, \ldots, f_n$, and that the nutritional content of each is known. There are m different nutrients being considered, and they are labeled $v_1, v_2, \ldots, v_m$. (Think of these as vitamins in a typical case.) The nutritional values are displayed in a table.

	f_1	f_2	$\cdots$	f_n
v_1	a_{11}	a_{12}	$\cdots$	a_{1n}
v_2	a_{21}	a_{22}	$\cdots$	a_{2n}
$\vdots$	$\vdots$	$\vdots$	$\ddots$	$\vdots$
v_m	a_{m1}	a_{m2}	$\cdots$	a_{mn}

The entries a_{ij} signify the amounts of nutrients in each gram of each food. Specifically, a_{ij} is the amount of nutrient v_i in food f_j. (Standard units must have been established.) In a given row, the units should be the same, as they pertain to a single nutrient. Units (for example, grams) would be assigned to each food, f_j.

Looking down a column, say column j, we see the amounts of each nutrient contained in one unit of food f_j. Looking across a row, say row i, we see what foods contain nutrient v_i and in what quantities. For example, if f_j is orange juice and v_i is vitamin C, then a_{ij} might indicate how many milligrams of vitamin C are contained in each gram of orange juice.

A *diet* will be a vector $\mathbf{x} = (x_1, x_2, \ldots, x_n)$. If each day a patient ingests x_1 units of food f_1, x_2 units of food f_2, and so on, then he or she will get $\sum_{j=1}^{n} x_j a_{ij}$ units of nutrient v_i. The problem is to formulate a diet that provides on a daily basis certain prescribed amounts of each nutrient. We can define a vector $\mathbf{b}$ containing these *minimum daily requirements* and ask for a *diet*, $\mathbf{x}$, such that

$$\mathbf{Ax} = \mathbf{b}$$

For example, suppose that we are considering a diet that involves three nutrients and four foods. The matrix of nutritional values is known, and we have a target for minimal nutrition. The table of nutrients is the following matrix:

$$A = \begin{bmatrix} 10 & 3 & 1 & 51 \\ 5 & 1 & 1 & 25 \\ 8 & 1 & 2 & 39 \end{bmatrix}$$

A diet is a vector $\mathbf{x}$ in $\mathbb{R}^4$ whose components indicate the number of units of each food to be made available to the patient or animal, as the case may be. The product $\mathbf{Ax}$ is a vector of three components, and we want these components to be at least 84, 43, 68. We can start by solving the system

$$\begin{bmatrix} 10 & 3 & 1 & 51 \\ 5 & 1 & 1 & 25 \\ 8 & 1 & 2 & 39 \end{bmatrix} \begin{bmatrix} x_1 \\ x_2 \\ x_3 \\ x_4 \end{bmatrix} = \begin{bmatrix} 84 \\ 43 \\ 68 \end{bmatrix}$$

In the usual way, we have

$$\begin{bmatrix} 10 & 3 & 1 & 51 & | & 84 \\ 5 & 1 & 1 & 25 & | & 43 \\ 8 & 1 & 2 & 39 & | & 69 \end{bmatrix} \sim \begin{bmatrix} 1 & 0 & 0 & 4 & | & 5 \\ 0 & 1 & 0 & 3 & | & 8 \\ 0 & 0 & 1 & 2 & | & 10 \end{bmatrix}$$

We find the general solution to be

$$\begin{bmatrix} x_1 \\ x_2 \\ x_3 \\ x_4 \end{bmatrix} = \begin{bmatrix} 5 \\ 8 \\ 10 \\ 0 \end{bmatrix} + x_4 \begin{bmatrix} -4 \\ -3 \\ -2 \\ 1 \end{bmatrix}$$

Here we have one free variable, x_4. We do not want any component of the vector $\mathbf{x}$ to be negative, and this restricts x_4 to the interval $0 \le x_4 \le \frac{5}{4}$. The resulting extreme cases are $\mathbf{x} = (5, 8, 10, 0)$ and $(0, 17/4, 15/2, \frac{5}{4})$. Different choices of pivot rows lead to different extreme solutions.

This is probably a good opportunity to raise the issue of systems of linear **inequalities**. We might relax the requirement $\mathbf{Ax} = \mathbf{b}$ to be

$$\mathbf{Ax} \ge \mathbf{b}$$

(For two vectors $\mathbf{u}$ and $\mathbf{v}$ in $\mathbb{R}^n$, we write $\mathbf{u} \ge \mathbf{v}$ when $u_i \ge v_i$ for $1 \le i \le n$.) That would certainly be better if there exists no solution to the first problem, $\mathbf{Ax} = \mathbf{b}$.

In the diet problems, it can often happen that the system of linear equations is inconsistent. (Remember that a system of equations in which the equations outnumber the variables is likely to be inconsistent.) Conversely, it may not be necessary to meet the requirements exactly because overnourishment is usually not serious—it is merely wasteful. What is the right formulation of the problem if it is inconsistent as it originally stands? Obviously we can require simply $\mathbf{Ax} \geq \mathbf{b}$, as explained previously. Is this a numerical problem that we can easily solve with our trustworthy row-reduction algorithm? Here, we tread on dangerous ground. Consider a very simple case

$$\begin{cases} 3x_1 + 5x_2 \geq 7 \\ 6x_1 + 7x_2 \geq 16 \end{cases}$$

Yielding to the temptation of using row operations, we proceed to

$$\begin{cases} 3x_1 + 5x_2 \geq 7 \\ 0x_1 - 3x_2 \geq 2 \end{cases}$$

This is *wrong!* For example, it seems to say that $x_2 < 0$. Where is the error? The manipulation on the inequalities is completely unjustified. A better approach is to turn the inequalities into equations by inserting positive parameters p_1, p_2 like this:

$$\begin{cases} 3x_1 + 5x_2 = 7 + p_1 \\ 6x_1 + 7x_2 = 16 + p_2 \end{cases}$$

One step of the standard row-reduction leads to

$$\begin{cases} 3x_1 + 5x_2 = 7 + p_1 \\ 0x_1 - 3x_2 = 2 - 2p_1 + p_2 \end{cases}$$

Notice that the term $-2p_1 + p_2$ can be either positive or negative. In order that $x_2 \geq 0$, we must have $2 - 2p_1 + p_2 \leq 0$. For example, we could let $p_1 = 1$, $p_2 = 0$, $x_2 = 0$, and $x_1 = \frac{8}{3}$. In problems of this type, there can be many solutions, and usually there is a cost function involved. In that case, the variables should have low values, in general. This leads to a type of problem called **linear programming**, in which we seek to minimize a linear objective function $c_1 x_1 + \cdots + c_n x_n$ subject to constraints of the form $\mathbf{x} \geq 0$ and $\mathbf{Ax} \geq 0$. Special codes are available for this type of problem. Row-reduction techniques by themselves are virtually useless in these optimization problems.

EXAMPLE 13 A *diet* problem leads to this system of linear inequalities:

$$\begin{bmatrix} 7 & 3 \\ 2 & 1 \\ 8 & 2 \end{bmatrix} \begin{bmatrix} x_1 \\ x_2 \end{bmatrix} \geq \begin{bmatrix} 8 \\ 5 \\ 12 \end{bmatrix}$$

Find one or more good solutions.

SOLUTION　We change from a system of inequalities to a system of equations $\mathbf{Ax} \geq \mathbf{b}$ by introducing a vector $\mathbf{p}$ whose coordinates are to be nonnegative: $\mathbf{Ax} = \mathbf{b} + \mathbf{p}$. There are now five unknowns: the two components of $\mathbf{x}$ and the three components of $\mathbf{p}$. Here is the augmented matrix for the problem:

$$\left[\begin{array}{ccccc|c} 7 & 3 & -1 & 0 & 0 & 8 \\ 2 & 1 & 0 & -1 & 0 & 5 \\ 8 & 2 & 0 & 0 & -1 & 12 \end{array} \right]$$

The reduced echelon form of this matrix is

$$\left[\begin{array}{ccccc|c} 1 & 0 & 0 & \frac{1}{2} & -\frac{1}{4} & \frac{1}{2} \\ 0 & 1 & 0 & -2 & \frac{1}{2} & 4 \\ 0 & 0 & 1 & -\frac{5}{2} & -\frac{1}{4} & 7.5 \end{array} \right]$$

We obtain the general solution

$$\begin{cases} x_1 = \frac{1}{2} - \frac{1}{2}p_2 + \frac{1}{4}p_3 \\ x_2 = 4 + 2p_2 - \frac{1}{2}p_3 \\ p_1 = 7.5 + \frac{5}{2}p_2 + \frac{1}{4}p_3 \end{cases}$$

From the third equation we see that the smallest possible value for p_1 is 7.5, obtained by selecting $p_2 = p_3 = 0$. With these values of $\mathbf{p}$, we compute an acceptible solution $(x_1, x_2) = (\frac{1}{2}, 4)$. Hence, we obtain $(15.5, 5, 12) \geq (8, 5, 12)$ on the righthand side of the preceding inequality.　■

Dangerous Pitfalls

We end this section with some words of caution.

1. Usually $\mathbf{AB} \neq \mathbf{BA}$. For example, we see that

$$\mathbf{AB} = \begin{bmatrix} 1 & 2 \\ 3 & 4 \end{bmatrix} \begin{bmatrix} 2 & 1 \\ 0 & 2 \end{bmatrix} = \begin{bmatrix} 2 & 5 \\ 6 & 11 \end{bmatrix}$$

$$\mathbf{BA} = \begin{bmatrix} 2 & 1 \\ 0 & 2 \end{bmatrix} \begin{bmatrix} 1 & 2 \\ 3 & 4 \end{bmatrix} = \begin{bmatrix} 5 & 8 \\ 6 & 8 \end{bmatrix}$$

Thus, the factors in matrix multiplication do not *commute*, in general.

2. If $\mathbf{AB} = \mathbf{AC}$, we cannot infer that $\mathbf{B} = \mathbf{C}$. For example, we have

$$\mathbf{AB} = \begin{bmatrix} 1 & 2 \\ 1 & 2 \end{bmatrix} \begin{bmatrix} 1 & 3 \\ 5 & 7 \end{bmatrix} = \begin{bmatrix} 11 & 17 \\ 11 & 17 \end{bmatrix}$$

$$= \begin{bmatrix} 1 & 2 \\ 1 & 2 \end{bmatrix} \begin{bmatrix} 3 & -3 \\ 4 & 10 \end{bmatrix} = \mathbf{AC}$$

where cancellation of the matrix factor $\mathbf{A}$ is not correct!

3. If $\mathbf{AB} = \mathbf{0}$, we cannot conclude that $\mathbf{A} = \mathbf{0}$ or $\mathbf{B} = \mathbf{0}$. For example,

$$\mathbf{AB} = \begin{bmatrix} 3 & 4 \\ 9 & 12 \end{bmatrix} \begin{bmatrix} 4 & -4 \\ -3 & 3 \end{bmatrix} = \begin{bmatrix} 0 & 0 \\ 0 & 0 \end{bmatrix}$$

but neither factor $\mathbf{A}$ nor $\mathbf{B}$ is the zero matrix!

4. If a practical problem results in a system of linear inequalities, say $\mathbf{Ax} \leq \mathbf{b}$, do not expect the row-reduction techniques to work. See the discussion in the subtopic of diets, in this section, for one way of proceeding.

SUMMARY 3.1

- Matrix properties:
  - The equation $\mathbf{A} = \mathbf{B}$ means that $\mathbf{A}_{ij} = \mathbf{B}_{ij}$ for all appropriate indices. (Equality of matrices)
  - $(\mathbf{A} + \mathbf{B})_{ij} = \mathbf{A}_{ij} + \mathbf{B}_{ij}$ (Matrix addition)
  - $(c\mathbf{A})_{ij} = c\mathbf{A}_{ij}$ when c is a scalar (Scalar–matrix multiplication)
  - $\mathbf{AB} = [\mathbf{Ab}_1, \mathbf{Ab}_2, \ldots, \mathbf{Ab}_n]$, where $\mathbf{b}_j$ are the column vectors of $\mathbf{B}$. Also, $\mathbf{Ab}_j = \sum_{r=1}^{n} b_{rj}\mathbf{a}_r$
  - $(\mathbf{AB})_{ij} = \sum_{r=1}^{n} a_{ir}b_{rj}$ (Matrix–matrix multiplication)
- Special matrices:
  - $\mathbf{I} = \mathbf{I}_n = (\delta_{ij})$ where $\delta_{ij} = 1$ if $i = j$ and $\delta_{ij} = 0$ otherwise. (Identity matrix)
  - For a square matrix $\mathbf{A}$, Diag($\mathbf{A}$) has the diagonal of $\mathbf{A}$, but all other elements of the matrix are zero.

- A lower triangular matrix $\mathbf{L}$ has the property $\ell_{ij} = 0$ when $j > i$.
- An upper triangular matrix $\mathbf{U}$ has the property $u_{ij} = 0$ when $j < i$.
- $(\mathbf{A}^T)_{ij} = \mathbf{A}_{ji}$ (Matrix transpose)
- If $\mathbf{A} = \mathbf{A}^T$, then $\mathbf{A}$ is said to be symmetric.
- If $\mathbf{A}^T = -\mathbf{A}$, then $\mathbf{A}$ is said to be skew–symmetric.
- Every square matrix is the sum of a symmetric matrix and a skew–symmetric matrix: $\mathbf{A} = \frac{1}{2}(\mathbf{A} + \mathbf{A}^T) + \frac{1}{2}(\mathbf{A} - \mathbf{A}^T)$
- Examples of elementary matrices:
  - $\mathbf{E}_1 = \begin{bmatrix} 1 & 0 & 0 \\ 0 & 1 & 0 \\ c & 0 & 1 \end{bmatrix}$ (Replacement $\mathbf{r}_i \leftarrow \mathbf{r}_i + c\mathbf{r}_j; n = 3, i = 3, j = 1$)

- $\mathbf{E}_2 = \begin{bmatrix} 1 & 0 & 0 \\ 0 & c & 0 \\ 0 & 0 & 1 \end{bmatrix}$ (Scale $\mathbf{r}_i \leftarrow c\mathbf{r}_i$; here $n = 3, i = 2$)

- $\mathbf{E}_3 = \begin{bmatrix} 0 & 1 & 0 \\ 1 & 0 & 0 \\ 0 & 0 & 1 \end{bmatrix}$ (Swap $\mathbf{r}_i \leftrightarrow \mathbf{r}_j$; here $n = 3$, $i = 1, j = 2$)

- If $\mathbf{A} \sim \mathbf{B}$, then $\mathbf{E}_k \mathbf{E}_{k-1} \cdots \mathbf{E}_2 \mathbf{E}_1 \, \mathbf{A} = \mathbf{B}$, where all $\mathbf{E}_i$ are elementary matrices.

- The product $\mathbf{Ax}$ can be computed with a post-multiplication of $\mathbf{A}$ by column vector $\mathbf{x}$.

- The product $\mathbf{y}^T\mathbf{A}$ can be computed by a pre-multiplication of $\mathbf{A}$ by the row vector $\mathbf{y}^T$.

- We can sometimes solve a system of linear *inequalities* by introducing a nonnegative vector so as to get a system of *equations*.

- Theorems about matrices:

  - If $\mathbf{A}$ is an $n \times n$ matrix, then $\mathbf{AI}_n = \mathbf{I}_n\mathbf{A} = \mathbf{A}$.

  - The set of all upper triangular matrices is a subspace of $\mathbb{R}^{n \times n}$.

  - The product of two $n \times n$ upper triangular matrices is also upper triangular. The same is true for lower triangular matrices.

  - $(\mathbf{AB})^T = \mathbf{B}^T\mathbf{A}^T$

  - In general, $\mathbf{AB} \neq \mathbf{BA}$. (Matrix multiplication is usually not commutative.)

- $(\mathbf{AB})\mathbf{C} = \mathbf{A}(\mathbf{BC})$ (Matrix multiplication is associative.)

- If $S(\mathbf{y}) = \mathbf{Ay}$ and $T(\mathbf{x}) = \mathbf{Bx}$, then the composition of S and T obeys the rule $(S \circ T)(\mathbf{x}) = \mathbf{ABx}$.

- If $\mathcal{R}(\mathbf{A})$ is the result of applying a row operation $\mathcal{R}$ to $\mathbf{A}$, and $\mathcal{R}(\mathbf{I})$ is the result of applying the same row operation to $\mathbf{I}$, then $\mathcal{R}(\mathbf{A}) = \mathcal{R}(\mathbf{I})\mathbf{A}$.

- $\mathbf{y}^T\mathbf{A} = [y_1, y_2, \dots, y_n]\mathbf{A} = \sum_{i=1}^{n} y_i\mathbf{r}_i$ where $\mathbf{r}_i$ are the rows of $\mathbf{A}$. (Vector–matrix product)

- $\mathbf{AB} = \begin{bmatrix} \mathbf{r}_1 \\ \mathbf{r}_2 \\ \vdots \\ \mathbf{r}_m \end{bmatrix} \quad \mathbf{B} = \begin{bmatrix} \mathbf{r}_1\mathbf{B} \\ \mathbf{r}_2\mathbf{B} \\ \vdots \\ \mathbf{r}_m\mathbf{B} \end{bmatrix}$

  where $\mathbf{r}_i$ are rows of $\mathbf{A}$.

- Cautions:

  - Usually $\mathbf{AB} \neq \mathbf{BA}$. Factors in matrix multiplication do *not* commute, in general.

  - If $\mathbf{AB} = \mathbf{AC}$, we cannot infer that $\mathbf{B} = \mathbf{C}$. Cancellation of the matrix factor $\mathbf{A}$ is usually incorrect!

  - If $\mathbf{AB} = \mathbf{0}$, we cannot conclude that $\mathbf{A} = \mathbf{0}$ or $\mathbf{B} = \mathbf{0}$.

  - Do not expect the row-reduction techniques to work on a system of linear inequalities such as $\mathbf{Ax} \leq \mathbf{b}$ or $\mathbf{Ax} \geq \mathbf{b}$.

KEY CONCEPTS 3.1

The space $\mathbb{R}^{m \times n}$ of all $m \times n$ matrices, addition and multiplication of matrices, dot product of two vectors, multiplication of a vector by a matrix, diagonal matrices, identity matrices, transpose of a matrix, symmetric matrices, skew–symmetric matrices, elementary matrix, triangular matrices, associativity, commutativity, upper triangular, lower triangular, Kronecker delta, outer products, diet problems

GENERAL EXERCISES 3.1

1. Refer to Example 4. Explain why **BA** cannot be computed.

2. If **A** is a square matrix, the product **AA** exists, and is denoted by $\mathbf{A}^2$. Compute the square of the matrix

$$\mathbf{A} = \begin{bmatrix} 1 & 2 & 3 \\ 4 & 5 & 6 \\ 7 & 8 & 9 \end{bmatrix}$$

3. For real numbers r, we know that if $r^2 = 0$, then $r = 0$. Is this true also for square matrices? A theorem or example is required.

4. Is this the correct definition of *upper triangular*, for a matrix **A**:
$\mathbf{A}_{ij} \neq 0$ when $j \geq i$.

5. Find all the 2×2 matrices that commute with the matrix $\mathbf{A} = \begin{bmatrix} 1 & 2 \\ 3 & 4 \end{bmatrix}$

6. Explain why, in defining row operations, we must insist on $i \neq j$ in the operation $\mathbf{r}_i \leftarrow \mathbf{r}_i + \alpha \mathbf{r}_j$.

7. Which of these matrices is not an *elementary* matrix?

a. $\begin{bmatrix} 0 & 1 \\ 1 & 0 \end{bmatrix}$
b. $\begin{bmatrix} 1 & 0 \\ 0 & 3 \end{bmatrix}$

c. $\begin{bmatrix} 1 & 0 & 0 \\ 0 & 2 & 0 \\ 0 & 0 & 1 \end{bmatrix}$
d. $\begin{bmatrix} 2 & 0 & 0 \\ 0 & 0 & 0 \\ 0 & 0 & 1 \end{bmatrix}$

e. $\begin{bmatrix} 1 & 0 & 0 \\ 3 & 1 & 0 \\ 0 & 0 & 1 \end{bmatrix}$
f. $\begin{bmatrix} 1 & 2 & 0 \\ 0 & 1 & 0 \\ 0 & 0 & 1 \end{bmatrix}$

8. Define $\mathbf{A} = \begin{bmatrix} 1 & 1 \\ 2 & 1 \\ 3 & 1 \end{bmatrix}$

$\mathbf{B} = \begin{bmatrix} 1 & -3 & 2 \\ 5 & -7 & 3 \end{bmatrix}$ and $\mathbf{c} = \begin{bmatrix} 2 \\ 3 \\ 1 \end{bmatrix}$

Verify that $\mathbf{BA} = \mathbf{I}$. Then solve the system $\mathbf{Ax} = \mathbf{c}$ by multiplying both sides of this equation by **B**. This produces $\mathbf{BAx} = \mathbf{Bc}$ or $\mathbf{x} = \mathbf{Bc}$. Verify by an independent calculation that $\mathbf{Ax} = \mathbf{c}$.

9. Let $\mathbf{A} = \begin{bmatrix} 1 & 2 & 3 \\ 4 & 5 & 6 \end{bmatrix}$. What is $\mathbf{A}^T$? What is $\mathbf{Ax}$

if $\mathbf{x} = \begin{bmatrix} 1 \\ -1 \\ 1 \end{bmatrix}$?

10. Let $\mathbf{A} = \begin{bmatrix} 1 & 3 & 2 \\ -1 & -2 & 1 \\ 1 & 5 & 7 \end{bmatrix}$, $\mathbf{B} = \begin{bmatrix} 1 & 0 & 4 \\ 2 & 1 & 0 \\ 0 & 1 & 1 \end{bmatrix}$

$\mathbf{C} = \mathbf{AB}$
Compute these: $\mathbf{C}_{23}, \mathbf{A}^T, \mathbf{A} + \mathbf{B}$, and $\mathbf{B}^2$.

11. Find all the matrices that commute with

$\begin{bmatrix} 1 & 3 \\ 2 & -1 \end{bmatrix}$

12. Certainly we can find $n \times n$ matrices that are nonzero, yet satisfy the equation $\mathbf{AB} = \mathbf{0}$. Is this possible with $\mathbf{B} = \mathbf{A}$? Is it possible with $\mathbf{B} = \mathbf{A}^T$?

13. Explain why this diet problem leads to an inconsistent system of linear equations:

$\mathbf{A} = \begin{bmatrix} 1 & 3 \\ 2 & 1 \\ 4 & 1 \end{bmatrix}$, $\mathbf{b} = \begin{bmatrix} 8 \\ 5 \\ 12 \end{bmatrix}$

14. (Continuation.) Find a *good* solution of the inequality $\mathbf{Ax} \geq \mathbf{b}$, using $\mathbf{A}$ and $\mathbf{b}$ from the preceding exercise. Ideally, there will exist a vector $\mathbf{x}$ such that if the vector $\mathbf{y}$ satisfies $\mathbf{b} \leq \mathbf{Ay} \leq \mathbf{Ax}$, then $\mathbf{x} = \mathbf{y}$. Try to find such an $\mathbf{x}$.

15. Solve this diet problem:

$$\mathbf{A} = \begin{bmatrix} 1 & 5 & 3 & 2 \\ 4 & 1 & 3 & 6 \\ 2 & 3 & 5 & 1 \end{bmatrix}, \quad \mathbf{b} = \begin{bmatrix} 49 \\ 82 \\ 54 \end{bmatrix}$$

As usual, there may be some solutions that are better than others. Try to find an *optimal* solution, using the least quantities of foodstuffs to accomplish the nutritional objective.

16. Criticize this solution to a diet problem:

$$\mathbf{Ax} \geq \mathbf{b} \text{ where } \mathbf{A} = \begin{bmatrix} 1 & 2 \\ 1 & 3 \\ 3 & 4 \end{bmatrix} \text{ and } \mathbf{b} = \begin{bmatrix} 8 \\ 9 \\ 17 \end{bmatrix}$$

We find $\begin{bmatrix} 1 & 2 & | & 8 \\ 1 & 3 & | & 9 \\ 3 & 4 & | & 17 \end{bmatrix} \sim \begin{bmatrix} 1 & 2 & | & 8 \\ 0 & 1 & | & 1 \\ 0 & -2 & | & -7 \end{bmatrix}$

$$\sim \begin{bmatrix} 1 & 0 & | & 1 \\ 0 & 1 & | & 1 \\ 0 & 2 & | & 7 \end{bmatrix}.$$

Thus, we have $\mathbf{x}_1 = 1$ and $\mathbf{x}_2$ is either 1 or $\frac{7}{2}$.

17. Show in detail the operations for the matrix–matrix multiplication $\mathbf{AB}^T$ with

$$\mathbf{A} = \begin{bmatrix} 0 & -1 \\ 2 & 1 \\ 1 & 0 \end{bmatrix}, \quad \mathbf{B} = \begin{bmatrix} 2 & 0 \\ -1 & 0 \\ 0 & 1 \end{bmatrix}$$

a. post-multiplication
b. pre-multiplication

18. Split the matrix $\mathbf{A} = \begin{bmatrix} 1 & 0 & -4 \\ 3 & 3 & -1 \\ 4 & -1 & 0 \end{bmatrix}$ into the sum of a symmetric and a skew–symmetric matrix.

19. Find all 2×2 matrices that commute with

$$\begin{bmatrix} 3 & 2 \\ 1 & 5 \end{bmatrix}$$

20. Explain the flaw in this reasoning: We are given $\mathbf{BA} = \mathbf{I}$. To solve $\mathbf{Ax} = \mathbf{b}$, we multiply by $\mathbf{B}$, getting $\mathbf{BAx} = \mathbf{Bb}$ and $\mathbf{x} = \mathbf{Bb}$.

21. Establish that the product of two upper triangular matrices is upper triangular. (The two matrices are square and of the same size.)

22. Each expression in the following list represents a certain element in a product, such as $\mathbf{A} = (a_{ij})$, $(\mathbf{B}^T\mathbf{A})_{ks}$, or $(\mathbf{A}^2)_{st}$. Identify each. (In each example, the index of summation is indicated; it runs from 1 to some upper limit appropriate for the matrices in question.)

a. $\sum_{t=1}^{n} A_{kt} B_{ti}$ **h.** $\sum_{j=1}^{n} B_{ij} A_{js}$

b. $\sum_{r=1}^{n} B_{rj} A_{ri}$ **i.** $\sum_{s=1}^{n} B_{si} A_{ks}$

c. $\sum_{s=1}^{n} B_{js} A_{is}$ **j.** $\sum_{i=1}^{n} A_{ji} A_{ti}$

d. $\sum_{k=1}^{n} B_{ks} B_{kt}$ **k.** $\sum_{j=1}^{n} B_{sj} B_{jt}$

e. $\sum_{s=1}^{n} A_{is} A_{sj}$ **l.** $\sum_{i=1}^{n} A_{is} A_{ij}$

f. $\sum_{t=1}^{n} (A^T)_{jt} B_{tk}$ **m.** $\sum_{j=1}^{n} (B^T)_{sj} A_{ji}$

g. $\sum_{i=1}^{n} A_{ji} (B^T)_{it}$ **n.** $\sum_{k=1}^{n} (A^T)_{rk} B_{ik}$

23. Explain why, for any square matrix $\mathbf{A}$, $\mathbf{A} + \mathbf{A}^T$ and $\mathbf{AA}^T$ are symmetric. Explain why $\mathbf{A}$ must be square in these assertions.

24. (Continuation.) Explain why every square matrix is the sum of a symmetric matrix and a skew–symmetric matrix. General Exercise 3 may be helpful.

25. Explain why every square matrix is the sum of a lower and an upper triangular matrix.

26. Establish that the diagonal matrices, the upper triangular matrices, and the symmetric matrices form three subspaces in $\mathbb{R}^{n \times n}$.

27. Explain why, for any matrix $\mathbf{A}$, this equation holds: $(\mathbf{A}^T)^T = \mathbf{A}$.

28. Establish that if $(m, n) \neq (\alpha, \beta)$, then $\mathbb{R}^{m \times n} \neq \mathbb{R}^{\alpha \times \beta}$.

29. Explain why or find a counterexample: The product of two symmetric $n \times n$ matrices is symmetric.

30. Let $\mathbf{A}$ be a fixed matrix in $\mathbb{R}^{n \times n}$. Consider the set of all matrices that commute with $\mathbf{A}$. Is this set a subspace of $\mathbb{R}^{n \times n}$?

31. Let $\mathbf{A}$ be an $n \times n$ matrix. Establish that $\mathbf{A}$ commutes with every polynomial in $\mathbf{A}$. (That terminology means a matrix of the form $\alpha_0 \mathbf{I} + \alpha_1 \mathbf{A} + \alpha_2 \mathbf{A}^2 + \cdots + \alpha_m \mathbf{A}^m$.)

32. Is it possible for a 2×2 matrix to have the property that its powers (including $\mathbf{A}^0 = \mathbf{I}$) span $\mathbb{R}^{2 \times 2}$?

33. Explain why, for 2×2 matrices, $\mathbf{A}^2$ is a linear combination of $\mathbf{I}$ and $\mathbf{A}$.

34. Find a 2×2 matrix, none of whose entries is zero, such that its square is zero.

35. Is there a 7×7 matrix of rank 7 all of whose entries are either 5 or 8?

36. Establish that $(\mathbf{A} + \mathbf{B})^T = \mathbf{A}^T + \mathbf{B}^T$.

37. Explain why or find a counterexample: The product of two skew–symmetric matrices is skew–symmetric.

38. Give an example of two matrices, $\mathbf{A}$ and $\mathbf{B}$, that are not row equivalent to each other, but their transposes are row equivalent to each other.

39. Give an argument why every 2×2 matrix $\mathbf{A}$ satisfies the equation
$\mathbf{A}^2 - (\mathbf{A}_{11} + \mathbf{A}_{22})\mathbf{A} + (\mathbf{A}_{11}\mathbf{A}_{22} - \mathbf{A}_{12}\mathbf{A}_{21})\mathbf{I} = \mathbf{0}$
(This is one case of the Cayley–Hamilton Theorem. See Section 8.1.)

40. Establish the remaining case of Theorem 8. (Recall that the interchange of two rows can be executed by row operations of the other two types.)

41. If $\mathbf{x}$ and $\mathbf{y}$ are two vectors in $\mathbb{R}^n$, their **outer product** is
$$\mathbf{x}\mathbf{y}^T = \begin{bmatrix} x_1 \\ x_2 \\ \vdots \\ x_n \end{bmatrix} \begin{bmatrix} y_1 & y_2 & \cdots & y_n \end{bmatrix}$$
Here we assume that $\mathbf{x}$ and $\mathbf{y}$ are column vectors. The outer product then is an $n \times n$ matrix $\mathbf{A}$, specified by the equation $\mathbf{A}_{ij} = x_i y_j$. Explain why the rows of this matrix are scalar multiples of each other. What is the rank of the matrix $\mathbf{x}\mathbf{y}^T$? (The rank of a matrix is defined on p. 63.)

42. Is every square matrix the product of a lower triangular matrix and an upper triangular matrix?

43. Explain why the dot product has this property: $\mathbf{x} \cdot \mathbf{y} = \mathbf{y} \cdot \mathbf{x}$.

44. Find all the 2×2 matrices $\mathbf{A}$ such that $\mathbf{A}^2 = \mathbf{0}$.

45. If $\mathbf{A}$ is an $m \times n$ matrix, what can be said of $\mathbf{A}\mathbf{I}_n$, $\mathbf{A}\mathbf{I}_m$, $\mathbf{I}_m\mathbf{A}$, and $\mathbf{I}_n\mathbf{A}$?

46. If $\mathbf{A}$ is a 2×2 upper triangular matrix and $\mathbf{B}$ is a 2×2 lower triangular matrix, what special properties does the product $\mathbf{A}\mathbf{B}$ have? Is it possible to factor any 2×2 matrix into the product of an upper and a lower triangular matrix?

47. Explain this apparent contradiction: For some pair of matrices $\mathbf{A}$ and $\mathbf{B}$ we can have $\mathbf{A}\mathbf{B} \neq \mathbf{B}\mathbf{A}$, yet in this calculation we have reversed the terms $\mathbf{A}_{ik}\mathbf{B}_{kj}$:
$$(\mathbf{A}\mathbf{B})_{ij} = \sum_{k=1}^{n} \mathbf{A}_{ik}\mathbf{B}_{kj} = \sum_{k=1}^{n} \mathbf{B}_{kj}\mathbf{A}_{ik}$$

48. Find two different 2×2 matrices (other than $\pm\mathbf{I}$) such that $\mathbf{A}\mathbf{A}^T = \mathbf{I}$.

49. Define $\mathbf{A} \oplus \mathbf{B} = \mathbf{A}\mathbf{B} + \mathbf{B}\mathbf{A}$. Is this operation commutative? Is it associative? Answer the same questions for the operation $\mathbf{A} \boxplus \mathbf{B} = \mathbf{A}\mathbf{B} - \mathbf{B}\mathbf{A}$.

50. Establish Theorem 9 by using transposes in the equation $\mathbf{A}\mathbf{x} = \sum_k x_k \mathbf{c}_k$.

51. Let $\mathbf{A}$ be a square matrix. Assume that the homogeneous system of equations $\mathbf{A}^2\mathbf{x} = \mathbf{0}$ has a nontrivial solution. Establish that the system $\mathbf{A}\mathbf{y} = \mathbf{0}$ has a nontrivial solution.

52. Let $\mathbf{A}$, $\mathbf{B}$, and $\mathbf{P}$ be $n \times n$ matrices. If $\mathbf{A} = \mathbf{P}\mathbf{B}\mathbf{P}^T$, does it follow that $\mathbf{A}$ is symmetric?

53. Establish that if $\mathbf{A} \sim \mathbf{B}$, then there is a nonsingular matrix $\mathbf{C}$ such that $\mathbf{A} = \mathbf{C}\mathbf{B}$.

54. Suppose that the column sums of a square matrix $\mathbf{A}$ are all equal to 1. Does it follow that the same property is true for $\mathbf{A}^k$ for all integers 2, 3, and so on?

55. (Continuation.) Repeat the argument in the preceding exercise when the column sums are all equal but not necessarily 1.

56. Establish the statements that are true and give counterexamples for the others. In each case we are dealing with $n \times n$ matrices.

    a. $(\mathbf{A} + \mathbf{B})^2 = \mathbf{A}^2 + 2\mathbf{A}\mathbf{B} + \mathbf{B}^2$
    b. If $\mathbf{A}\mathbf{B} = \mathbf{0}$, then either $\mathbf{A} = \mathbf{0}$ or $\mathbf{B} = \mathbf{0}$
    c. $(\mathbf{A} - \mathbf{B})(\mathbf{A} + \mathbf{B}) = \mathbf{A}^2 - \mathbf{B}^2$
    d. $\mathbf{A}^2\mathbf{B}^2 = (\mathbf{A}\mathbf{B})^2$
    e. $\mathbf{C}(\mathbf{A} + \mathbf{B}) = \mathbf{C}\mathbf{B} + \mathbf{C}\mathbf{A}$
    f. If $\mathbf{A}^2 = \mathbf{0}$, then $\mathbf{A} = \mathbf{0}$
    g. $\mathbf{A}\mathbf{B} = \mathbf{B}\mathbf{A}$
    h. $(\mathbf{A} + \mathbf{B})\mathbf{C} = \mathbf{B}\mathbf{C} + \mathbf{A}\mathbf{C}$

57. Suppose that $\mathbf{A}$ and $\mathbf{B}$ are $n \times n$ symmetric matrices. Does it follow that $(\mathbf{A}\mathbf{B})^T = \mathbf{B}\mathbf{A}$? Is the converse true?

58. Establish that for any matrix $\mathbf{A}$ (not necessarily square), $\mathbf{A}\mathbf{A}^T$ is symmetric.

59. Explain why the hypothesis "$\mathbf{A}$ is square" leads to the conclusion "$\mathbf{A} + \mathbf{A}^T$ is symmetric."

60. Refer to the elements in the matrices $\mathbf{A} = (a_{ij})$ and $\mathbf{B} = (b_{ij})$. Compute these quantities: $\sum_{k=1}^{3} a_{2k}$, $\sum_{k=1}^{3} a_{k3}$, $\sum_{j=1}^{3} a_{j3}\delta_{j3}$, $\sum_{j=1}^{3} a_{2j}b_{j3}$. Recall the **Kronecker delta**: δ_{ij} is 1 when $i = j$; otherwise it is 0.

61. Suppose that a matrix $\mathbf{A}$ has the property $\mathbf{A}^T\mathbf{A} = \mathbf{A}\mathbf{A}^T$. Does it follow that $\mathbf{A}$ is square? Is the converse true? That is, if $\mathbf{A}$ is square, does it follow that $\mathbf{A}^T\mathbf{A} = \mathbf{A}\mathbf{A}^T$?

62. Establish that the product of two $n \times n$ lower triangular matrices is lower triangular.

63. For a vector $\mathbf{x}$, we write $\mathbf{x} \geq \mathbf{0}$ if all components of $\mathbf{x}$ are nonnegative. Suppose that $\mathbf{A}$ is an $m \times n$ matrix such that $\mathbf{A}\mathbf{x} \geq \mathbf{0}$ whenever $\mathbf{x} \geq \mathbf{0}$. Does it follow that all entries in $\mathbf{A}$ are nonnegative?

64. Let $\mathbf{A}$ be an $m \times n$ matrix in which each entry is positive (i.e., greater than zero). Establish that, for any $\mathbf{b}$ in $\mathbb{R}^n$, the inequality $\mathbf{A}\mathbf{x} \geq \mathbf{b}$ has a solution. Also, show that if we drop the assumption on $\mathbf{A}$ that each entry be positive, the result is not true.

65. Let

$$\mathbf{A} = \begin{bmatrix} 3 & 2 & 1 \\ 4 & 1 & 2 \\ 3 & 1 & 5 \\ 0 & 2 & 1 \end{bmatrix}, \qquad \mathbf{b} = \begin{bmatrix} 2 \\ 3 \\ 2 \\ 1 \end{bmatrix}$$

What can be done with this diet problem? (We want $\mathbf{A}\mathbf{x} \geq \mathbf{b}$.)

66. If $\mathbf{x}$ and $\mathbf{y}$ are vectors, does it follow from $\mathbf{x} > \mathbf{y}$ that $\mathbf{A}\mathbf{x} > \mathbf{A}\mathbf{y}$? What hypothesis on $\mathbf{A}$ would be useful here? Establish a theorem of your own devising, and give examples.

67. Find the necessary and sufficient conditions on two symmetric 2×2 matrices so that their product is symmetric.

68. (Continuation.) Find the most general pair of symmetric 2×2 matrices whose product is not symmetric.

69. Let Ω be the operator that produces from $\mathbf{A}$ its reduced row echelon form $\Omega(\mathbf{A})$. Explain why $\Omega(\mathbf{A}\mathbf{B}) = \Omega(\mathbf{A})\mathbf{B}$.

70. In $\mathbb{R}^{n \times n}$, suppose addition is defined by the equation $(\mathbf{A} \oplus \mathbf{B})_{ij} = \mathbf{A}_{ij} + \mathbf{B}_{ji}$. Would $\mathbb{R}^{n \times n}$ be a vector space? (A simple *yes* or *no* is not sufficient.)

71. Establish that the operation $\mathbf{A} \rightarrow \mathbf{A}^T$ is linear.

72. In Example 12, we found that if $\mathbf{A}$ and $\mathbf{B}$ have no zero entries, then it can happen that $\mathbf{A}\mathbf{B} = \mathbf{0}$. For 2×2 matrices, find general conditions for this to be true and give another simple example.

COMPUTER EXERCISES 3.1

1. Use mathematical software such as MATLAB, Maple, or Mathematica to generate random matrices of various sizes and types with elements that are general, upper triangular, lower triangular, symmetric, and skew–symmetric. Repeat making the entries random integers.

2. (Continuation.) Multiply two random integer matrices of the same type (upper triangular, lower triangular, symmetric, and skew–symmetric). In each case, is the resulting matrix of the same type?

3. (Continuation.) Illustrate the cautions in Summary 3.1 using random integer matrices.

4. (Continuation.) Use random matrices to illustrate that $(\mathbf{AB})^T = \mathbf{B}^T\mathbf{A}^T$.

5. Consider a general 2×2 matrix $\begin{bmatrix} a & b \\ c & d \end{bmatrix}$

Using mathematical software such as Maple, determine when these expressions are correct:

a. Trace$(\mathbf{AB})$ = Trace$(\mathbf{BA})$

b. Trace$(\mathbf{AB})$ = Trace$(\mathbf{A})$ · Trace$(\mathbf{B})$

c. Trace$(\mathbf{A}^2) > 0$

The **trace** of matrix $\mathbf{A} = (a_{ij})$ is the sum of the diagonal elements: Trace$(\mathbf{A})$ = $a_{11} + a_{22} + a_{33} + \cdots + a_{nn} = \sum_{i=1}^{n} a_{ii}$

3.2 MATRIX INVERSES

> *Life is good for only two things, discovering mathematics and teaching mathematics.*
>
> —SIMÉON POISSON (1781–1840)

> *God created the integers; all else is the work of man.*
>
> —LEOPOLD KRONECKER (1823–1891)

Would it not be advantageous if, to solve a system of equations, $\mathbf{Ax} = \mathbf{b}$, we simply could *divide* by $\mathbf{A}$ and get the solution $\mathbf{x} = \mathbf{A}^{-1}\mathbf{b}$? Something similar to this turns out to be possible in many cases! (In MATLAB, we get the solution with the single command A\b. How this works will be explained later.)

Solving Systems with a Left Inverse

Suppose it is known that a given system, $\mathbf{Ax} = \mathbf{b}$, has a solution. Instead of treating $\mathbf{x}$ as an unknown, we can let $\mathbf{x}$ denote one such concrete solution. Then $\mathbf{Ax}$ really is $\mathbf{b}$. Suppose also that we are in possession of a matrix $\mathbf{B}$ such that $\mathbf{BA} = \mathbf{I}$. Then we can multiply both sides of the equation $\mathbf{Ax} = \mathbf{b}$ by $\mathbf{B}$ and arrive at $\mathbf{BAx} = \mathbf{Bb}$, or $\mathbf{Ix} = \mathbf{Bb}$, or $\mathbf{x} = \mathbf{Bb}$. We have concluded that if $\mathbf{x}$ is a solution of the equation, then $\mathbf{x}$ must equal $\mathbf{Bb}$.

Because this statement is often misinterpreted, we state it again as a theorem.

THEOREM 1

If the system of equations **Ax** = **b** *is consistent, and if a matrix* **B** *exists such that* **BA** = **I**, *then the system of equations has a unique solution, namely* **x** = **Bb**.

EXAMPLE 1 Consider the system **Ax** = **b**, in which

$$A = \begin{bmatrix} -5 & 3 \\ 2 & -1 \\ 0 & 0 \end{bmatrix}, \qquad b = \begin{bmatrix} -4 \\ 3 \\ 0 \end{bmatrix}$$

Fortune has smiled upon us and given us a matrix **B** such that **BA** = **I**, namely $B = \begin{bmatrix} 1 & 3 & -2 \\ 2 & 5 & 1 \end{bmatrix}$. The verification is

$$BA = \begin{bmatrix} 1 & 3 & -2 \\ 2 & 5 & 1 \end{bmatrix} \begin{bmatrix} -5 & 3 \\ 2 & -1 \\ 0 & 0 \end{bmatrix} = \begin{bmatrix} 1 & 0 \\ 0 & 1 \end{bmatrix} = I_2$$

How can we use this information to solve the given system?

SOLUTION If the system of equations has a solution, it must be **Bb**. So we compute $x = Bb = \begin{bmatrix} 5 \\ 7 \end{bmatrix}$ and then test it by *substitution*:

$$Ax = \begin{bmatrix} -5 & 3 \\ 2 & -1 \\ 0 & 0 \end{bmatrix} \begin{bmatrix} 5 \\ 7 \end{bmatrix} = \begin{bmatrix} -4 \\ 3 \\ 0 \end{bmatrix}$$

Notice that we do not talk about *verifying* the solution; we test it to see whether it *is* a solution. (There is a subtle difference.) Our work prior to the testing does not prove that the given **x** is a solution. Remember that we started by assuming that **x** was a solution! That assumption could have been *false*. A false assumption can get you in trouble! ■

EXAMPLE 2 Here we illustrate an incorrect use of Theorem 1. Consider this system of equations:

$$\begin{cases} -y_1 + y_2 = -4 \\ 0y_1 - y_2 = 3 \\ 2y_1 + y_2 = 0 \end{cases}$$

The coefficient matrix is $\mathbf{C} = \begin{bmatrix} -1 & 1 \\ 0 & -1 \\ 2 & 1 \end{bmatrix}$. Let $\mathbf{D} = \begin{bmatrix} 1 & 2 & 1 \\ 4 & 5 & 2 \end{bmatrix}$.

Then $\mathbf{DC} = \mathbf{I}$, as we quickly verify:

$$\mathbf{DC} = \begin{bmatrix} 1 & 2 & 1 \\ 4 & 5 & 2 \end{bmatrix} \begin{bmatrix} -1 & 1 \\ 0 & -1 \\ 2 & 1 \end{bmatrix} = \begin{bmatrix} 1 & 0 \\ 0 & 1 \end{bmatrix} = \mathbf{I}_2$$

How can we use this information to solve the given system?

SOLUTION The original system of equations is $\mathbf{Cy} = \mathbf{g}$, where $\mathbf{g} = \begin{bmatrix} -4 \\ 3 \\ 0 \end{bmatrix}$. Multiply both sides of this equation by $\mathbf{D}$, getting $\mathbf{DCy} = \mathbf{Dg}$.

(There is no question that this is legitimate.) The equation reduces to $\mathbf{y} = \mathbf{Dg}$, so the solution must be

$$\mathbf{y} = \begin{bmatrix} 1 & 2 & 1 \\ 4 & 5 & 2 \end{bmatrix} \begin{bmatrix} -4 \\ 3 \\ 0 \end{bmatrix} = \begin{bmatrix} 2 \\ -1 \end{bmatrix}$$

Just to be sure, we verify the solution by substitution and get a surprise:

$$\mathbf{Cy} = \begin{bmatrix} -1 & 1 \\ 0 & -1 \\ 2 & 1 \end{bmatrix} \begin{bmatrix} 2 \\ -1 \end{bmatrix} = \begin{bmatrix} -3 \\ 1 \\ 3 \end{bmatrix} \neq \mathbf{g}$$

Where is the error?

The procedure used depends for its validity on the consistency of the system of equations. That is, we must know in advance that there *exists* at least one solution. In this example, such knowledge is not present, and in fact the given system is *inconsistent*. This is revealed by the row-reduction process of the augmented matrix:

$$[\; C \; | \; g \;] = \begin{bmatrix} -1 & 1 & | & -4 \\ 0 & -1 & | & 3 \\ 2 & 1 & | & 0 \end{bmatrix} \sim \begin{bmatrix} 1 & 0 & | & 1 \\ 0 & 1 & | & -3 \\ 0 & 0 & | & 1 \end{bmatrix}$$

Similar fallacies can arise outside of linear algebra. (For example, see General Exercise 27.) ■

Solving Systems with a Right Inverse

To pursue these ideas further, suppose that again we want to solve a system of linear equations, $Ax = b$. Assume now that we have another matrix, B, such that $AB = I$. Then we can write $A(Bb) = (AB)b = Ib = b$; whence Bb solves the equation $Ax = b$. This conclusion did *not* require an *a priori* assumption that a solution exist; we have *produced* a solution. The argument does not reveal whether Bb is the only solution. There may be others.

THEOREM 2

Suppose that two matrices A and B have the property that $AB = I$. Then the system of linear equations $Ax = b$ has at least one solution, namely $x = Bb$.

EXAMPLE 3 Consider another system of linear equations:

$$\begin{bmatrix} 1 & 2 & 1 \\ 4 & 5 & 2 \end{bmatrix} \begin{bmatrix} x_1 \\ x_2 \\ x_3 \end{bmatrix} = \begin{bmatrix} -3 \\ -6 \end{bmatrix}$$

How can we solve this system using information from Example 2?

SOLUTION The coefficient matrix here is the matrix D from Example 2. Hence, the problem to be solved is $Dx = f$, where $f = \begin{bmatrix} -3 \\ -6 \end{bmatrix}$. We know that $\mathbf{DC} = \mathbf{I}$, where $\mathbf{C}$ is as in Example 2. Hence, by Theorem 2, one solution is $\mathbf{x} = \mathbf{Cf} = [-3, 6, -12]^T$. To check these results, we find that

$$\begin{bmatrix} 1 & 2 & 1 \\ 4 & 5 & 2 \end{bmatrix} \begin{bmatrix} -3 \\ 6 \\ -12 \end{bmatrix} = \begin{bmatrix} -3 \\ -6 \end{bmatrix}$$

∎

Analysis

There are two phenomena here, illustrated by Examples 1 and 3. In the first case, if we know from some external information that the system $\mathbf{Ax} = \mathbf{b}$ has a solution, then it must be $\mathbf{Bb}$. That is *not* the same as saying flatly that $\mathbf{Bb}$ is a solution, because there may be *no* solution. There is that annoying additional hypothesis that there must exist a solution. In the second case, we are more fortunate, as the calculation shows that $\mathbf{Cb}$ is a solution, without further hypotheses. However, we do not know whether, in fact, it is the only solution.

DEFINITION

If $\mathbf{AB} = \mathbf{I}$, *then* $\mathbf{B}$ *is a* **right inverse** *of* $\mathbf{A}$ *and* $\mathbf{A}$ *is a* **left inverse** *of* $\mathbf{B}$.

EXAMPLE 4 Are these matrices related with regard to left and right inverses?

$$\mathbf{C} = \begin{bmatrix} -1 & 1 \\ 0 & -1 \\ 2 & 1 \end{bmatrix}, \qquad \mathbf{D} = \begin{bmatrix} 1 & 2 & 1 \\ 4 & 5 & 2 \end{bmatrix}$$

SOLUTION In Example 2, we had $\mathbf{DC} = \mathbf{I}$. Therefore, $\mathbf{D}$ is a left inverse of $\mathbf{C}$ and $\mathbf{C}$ is a right inverse of $\mathbf{D}$. ∎

EXAMPLE 5 Find all the solutions of the system $\mathbf{Ax} = \mathbf{d}$, where $\mathbf{A}$ and $\mathbf{d}$ are as in Example 3.

SOLUTION We carry out a row reduction, getting

$$
\begin{bmatrix} 1 & 2 & 1 & | & -3 \\ 4 & 5 & 2 & | & -6 \end{bmatrix}
\sim
\begin{bmatrix} 1 & 0 & -\frac{1}{3} & | & 1 \\ 0 & 1 & \frac{2}{3} & | & -2 \end{bmatrix}
$$

As usual, we write this in a form with the free variable on the right:

$$
\begin{cases} x_1 = 1 + \frac{1}{3}x_3 \\ x_2 = -2 - \frac{2}{3}x_3 \end{cases}
\quad \text{or} \quad
\begin{bmatrix} x_1 \\ x_2 \\ x_3 \end{bmatrix}
=
\begin{bmatrix} 1 \\ -2 \\ 0 \end{bmatrix}
+ x_3
\begin{bmatrix} \frac{1}{3} \\ -\frac{2}{3} \\ 1 \end{bmatrix}
$$

The problem has infinitely many solutions. The solution of the equation obtained in Example 3 arises by taking $x_3 = -4$. ∎

To summarize in slightly different language, suppose that we wish to solve the system of linear equations $Ax = b$, and have a right inverse of A, say $AB = I$. Then Bb is a solution, as is verified by noting $A(Bb) = (AB)b = Ib = b$. Conversely, if we have a left inverse of A, say $CA = I$, then we can only conclude that Cb is the *sole candidate* for a solution; however, it must be checked by substitution to determine whether, in fact, it *is* a solution.

THEOREM 3

If the matrix A has a right inverse, then for each b the equation $Ax = b$ has at least one solution. If A has a left inverse, then that equation has at most one solution.

Square Matrices

The situation for square matrices is much simpler, and we turn to it next with a theorem of great importance. For example, in Section 7.1, we shall use the following theorem to deduce that if the rows of a square matrix form an orthonormal set, then the same is true of its columns.

THEOREM 4

If A and B are square matrices such that $BA = I$, then $AB = I$.

PROOF **Step 1.** Let A and B be $n \times n$ matrices, and assume that $BA = I$. The equation $Ax = 0$ has only the trivial solution because if $Ax = 0$, then $BAx = 0$ and $x = 0$. (Use the equation $BA = I$.)

Step 2. The reduced echelon form of **A** must have n pivots, for if there were fewer there would exist free variables and there would exist nontrivial solutions to the homogeneous equation $\mathbf{Ax} = \mathbf{0}$, contrary to the conclusion in Step 1.

Step 3. Because the matrix is square, each row has a pivot, and the reduced echelon form of **A** is $\mathbf{I}_n$.

Step 4. We can solve any equation $\mathbf{Ax} = \mathbf{b}$ by the row-reduction process, because **A** reduces to **I** in this process. We can also solve the equation $\mathbf{AX} = \mathbf{I}$, where **I** is the identity matrix and **X** is an $n \times n$ matrix. (This calculation can be done using one column at a time from the matrix **I**. It is more efficient to use the technique explained in Example 8 in Section 1.3.)

Step 5. $\mathbf{AB} = (\mathbf{AB})\mathbf{I} = (\mathbf{AB})(\mathbf{AX}) = \mathbf{A}(\mathbf{BA})\mathbf{X} = \mathbf{AIX} = \mathbf{AX} = \mathbf{I}.$ ■

THEOREM 5

If **A**, **B**, *and* **C** *are square matrices such that* $\mathbf{AB} = \mathbf{AC} = \mathbf{I}$, *then* $\mathbf{B} = \mathbf{C}$.

PROOF By Theorem 4, $\mathbf{BA} = \mathbf{I}$. Multiply each side of the equation in the hypothesis on the left by **B**, getting $\mathbf{BAB} = \mathbf{BAC}$. This reduces to $\mathbf{B} = \mathbf{C}$.   ■

Invertible Matrices

Theorem 5 asserts that if a matrix has an inverse, then it has only one. We need this fact to justify the following definition.

DEFINITION

A square matrix **A** *is* **invertible** *if there is a matrix* **B** *such that* $\mathbf{BA} = \mathbf{I}$. *In this case,* **B** *is the* **inverse** *of* **A** *and we write* $\mathbf{B} = \mathbf{A}^{-1}$. *The equation* $\mathbf{AB} = \mathbf{I}$ *is also true.*

An invertible matrix is also said to be **nonsingular**. If a matrix has no inverse, it is said to be **singular** or **noninvertible**.

EXAMPLE 6 Here are two matrices

$$\mathbf{A} = \begin{bmatrix} 3 & 7 & 5 \\ 5 & 11 & 8 \\ 3 & 4 & 3 \end{bmatrix}, \qquad \mathbf{B} = \begin{bmatrix} 1 & -1 & 1 \\ 9 & -6 & 1 \\ -13 & 9 & -2 \end{bmatrix}$$

Is one of these the inverse of the other? If so, which is which?

SOLUTION Do not make the mistake of thinking that it is necessary to compute the inverse of one of these two matrices. It is much easier to compute one of the products $\mathbf{AB}$ or $\mathbf{BA}$. It turns out that $\mathbf{AB} = \mathbf{I}$. Several conclusions can be drawn, namely, $\mathbf{BA} = \mathbf{I}$, $\mathbf{A}^{-1} = \mathbf{B}$, and $\mathbf{B}^{-1} = \mathbf{A}$. Thus, each matrix is the inverse of the other. Theorem 4 applies here. If you compute the products to see whether they equal the identity matrix, the work will be quite different for $\mathbf{AB}$ and $\mathbf{BA}$. But Theorem 4 asserts that both of the products will be $\mathbf{I}_3$. ∎

THEOREM 6

If $\mathbf{A}$ *is invertible, then for any prescribed* $\mathbf{b}$ *the equation* $\mathbf{Ax} = \mathbf{b}$ *has one and only one solution, namely* $\mathbf{x} = \mathbf{A}^{-1}\mathbf{b}$.

THEOREM 7

If $\mathbf{A}$, $\mathbf{B}$, *and* $\mathbf{C}$ *are square matrices such that* $\mathbf{AC} = \mathbf{BA} = \mathbf{I}$, *then* $\mathbf{B} = \mathbf{C}$.

PROOF $\mathbf{B} = \mathbf{BI} = \mathbf{B(AC)} = \mathbf{(BA)C} = \mathbf{IC} = \mathbf{C}$ ∎

EXAMPLE 7 The inverse of $\begin{bmatrix} 3 & 1 \\ 11 & 4 \end{bmatrix}$ is $\begin{bmatrix} 4 & -1 \\ -11 & 3 \end{bmatrix}$

SOLUTION It is verified by multiplying the matrices, in either order, to get $\mathbf{I}_2$. Recall Theorem 4, which asserts that in verifying that a square matrix $\mathbf{B}$ is the inverse of a square matrix $\mathbf{A}$, we need check only one of the two equations $\mathbf{AB} = \mathbf{I}$ or $\mathbf{BA} = \mathbf{I}$. ∎

Do not be misled by this example into thinking that the inverse of an integer matrix is always an integer matrix, because it rarely is! For a 2×2 matrix, the inverse is

$$\begin{bmatrix} a & b \\ c & d \end{bmatrix}^{-1} = (ad - bc)^{-1} \begin{bmatrix} d & -b \\ -c & a \end{bmatrix}$$

where $ad - bc \neq 0$.

EXAMPLE 8 Solve this equation using an inverse matrix:

$$\begin{bmatrix} 3 & 1 \\ 11 & 4 \end{bmatrix} \begin{bmatrix} x_1 \\ x_2 \end{bmatrix} = \begin{bmatrix} 5 \\ 2 \end{bmatrix}$$

SOLUTION This equation has the form $\mathbf{Ax} = \mathbf{b}$, where $\mathbf{A}$ is the matrix in Example 7. To solve it, multiply both sides of the equation by $\mathbf{A}^{-1}$, getting $\mathbf{x} = \mathbf{A}^{-1}\mathbf{b}$. Thus, we find the solution

$$\begin{bmatrix} x_1 \\ x_2 \end{bmatrix} = \begin{bmatrix} 4 & -1 \\ -11 & 3 \end{bmatrix} \begin{bmatrix} 5 \\ 2 \end{bmatrix} = \begin{bmatrix} 18 \\ -49 \end{bmatrix}$$

We can check our result by substitution into the original equation.   ■

THEOREM 8

For invertible $n \times n$ matrices $\mathbf{A}$ and $\mathbf{B}$, we have $(\mathbf{AB})^{-1} = \mathbf{B}^{-1}\mathbf{A}^{-1}$. Thus, the product of two invertible matrices is invertible. The result extends to the product of any number of invertible matrices.

PROOF We wish to know that $\mathbf{B}^{-1}\mathbf{A}^{-1}$ is the inverse of $\mathbf{AB}$. It only requires direct verification, and this gives us a chance to use our new knowledge that matrix multiplication is associative (Theorem 7 in Section 3.1):

$$(\mathbf{AB})(\mathbf{B}^{-1}\mathbf{A}^{-1}) = \mathbf{A}(\mathbf{BB}^{-1})\mathbf{A}^{-1} = \mathbf{AIA}^{-1} = \mathbf{AA}^{-1} = \mathbf{I} \qquad ■$$

Elementary Matrices and *LU* Factorization

Some matrices have inverses that are easy to find. For example, the inverse of the diagonal matrix $\mathbf{D} = \text{Diag}(d_i)$ is clearly $\mathbf{D}^{-1} = \text{Diag}(d_i^{-1})$. Here $\text{Diag}(\alpha_i)$ is notation for a square diagonal matrix with diagonal elements $\alpha_1, \alpha_2, \ldots, \alpha_n$. Also, the inverse matrix of an elementary matrix that corresponds to a replacement operation such as

$$\mathbf{E} = \begin{bmatrix} 1 & 0 & 0 \\ 0 & 1 & 0 \\ c & 0 & 1 \end{bmatrix}$$

is found by simply changing the sign of the multiplier c in the matrix:

$$\mathbf{E}^{-1} = \begin{bmatrix} 1 & 0 & 0 \\ 0 & 1 & 0 \\ -c & 0 & 1 \end{bmatrix}$$

Inverses of elementary matrices arise naturally when we recall that elementary row operations on a matrix can be interpreted as pre-multiplication of **A** by elementary matrices. (Each elementary matrix carries out one row operation, as explained in Section 3.1, Theorem 8). Thus, the row-reduction process ending at **U** can be expressed like this:

$$\mathbf{E}_k\mathbf{E}_{k-1}\cdots\mathbf{E}_2\mathbf{E}_1\mathbf{A} = \mathbf{U}$$

This equation shows immediately that

$$\mathbf{A} = \mathbf{LU}$$

where

$$\mathbf{L} = (\mathbf{E}_k\mathbf{E}_{k-1}\cdots\mathbf{E}_2\mathbf{E}_1)^{-1} = \mathbf{E}_1^{-1}\mathbf{E}_2^{-1}\cdots\mathbf{E}_{k-1}^{-1}\mathbf{E}_k^{-1}$$

By Theorem 8, any product of elementary matrices is invertible.

EXAMPLE 9 Suppose we have a reduction process that uses only replacement operations such as this one:

$$\mathbf{A} = \begin{bmatrix} 1 & 2 & -3 & 1 & 2 \\ 2 & 4 & -4 & 6 & 10 \\ 3 & 6 & -6 & 9 & 13 \end{bmatrix} \sim \begin{bmatrix} 1 & 2 & -3 & 1 & 2 \\ 0 & 0 & 2 & 4 & 6 \\ 0 & 0 & 3 & 6 & 7 \end{bmatrix}$$

$$\sim \begin{bmatrix} 1 & 2 & -3 & 1 & 2 \\ 0 & 0 & 2 & 4 & 6 \\ 0 & 0 & 0 & 0 & -2 \end{bmatrix} = \mathbf{U}$$

Establish the *LU* factorization of **A**.

SOLUTION By writing down the elementary matrices

$$\mathbf{E}_1 = \begin{bmatrix} 1 & 0 & 0 \\ -2 & 1 & 0 \\ 0 & 0 & 1 \end{bmatrix}, \quad \mathbf{E}_2 = \begin{bmatrix} 1 & 0 & 0 \\ 0 & 1 & 0 \\ -3 & 0 & 1 \end{bmatrix}, \quad \mathbf{E}_3 = \begin{bmatrix} 1 & 0 & 0 \\ 0 & 1 & 0 \\ 0 & -\frac{3}{2} & 1 \end{bmatrix}$$

we obtain

$$\mathbf{E}_3\mathbf{E}_2\mathbf{E}_1\mathbf{A} = \mathbf{U}$$

$$\mathbf{A} = \mathbf{E}_1^{-1}\mathbf{E}_2^{-1}\mathbf{E}_3^{-1}\mathbf{U}$$

where

$$\mathbf{L} = \mathbf{E}_1^{-1}\mathbf{E}_2^{-1}\mathbf{E}_3^{-1} = \begin{bmatrix} 1 & 0 & 0 \\ 2 & 1 & 0 \\ 0 & 0 & 1 \end{bmatrix} \begin{bmatrix} 1 & 0 & 0 \\ 0 & 1 & 0 \\ 3 & 0 & 1 \end{bmatrix} \begin{bmatrix} 1 & 0 & 0 \\ 0 & 1 & 0 \\ 0 & \frac{3}{2} & 1 \end{bmatrix}$$

$$= \begin{bmatrix} 1 & 0 & 0 \\ 2 & 1 & 0 \\ 3 & \frac{3}{2} & 1 \end{bmatrix}$$

Notice that the multiplication of these elementary matrices is particularly easy! We multiply the elementary matrices together to form $\mathbf{E} = \mathbf{E}_3\mathbf{E}_2\mathbf{E}_1$ rather than finding the inverse of $\mathbf{E}$, which would be more difficult. Finally, we obtain

$$\mathbf{A} = \mathbf{LU} = \begin{bmatrix} 1 & 0 & 0 \\ 2 & 1 & 0 \\ 3 & \frac{3}{2} & 1 \end{bmatrix} \begin{bmatrix} 1 & 2 & -3 & 1 & 2 \\ 0 & 0 & 2 & 4 & 6 \\ 0 & 0 & 0 & 0 & -2 \end{bmatrix}$$

The row-reduction process can include swapping and scaling of rows to avoid small pivots, which brings in pivoting and a permutation matrix. We explore matrix factorizations further in Section 8.2. ∎

Computing an Inverse

We turn now to an algorithm for finding the inverse of a square matrix, if it has one. From Theorem 2, we know that if $\mathbf{AX} = \mathbf{I}$ then the equation $\mathbf{XA} = \mathbf{I}$ follows, it being assumed that all three matrices involved are square and of the same size. Thus, if the equation $\mathbf{AX} = \mathbf{I}$ has a solution $\mathbf{X}$, then $\mathbf{X}$ is the inverse of $\mathbf{A}$.

In Section 1.3, systems of equations of this sort were discussed. One simply forms an augmented matrix $\begin{bmatrix} \mathbf{A} & | & \mathbf{I} \end{bmatrix}$ and carries out the row-reduction process. If all goes well, the result will be $\begin{bmatrix} \mathbf{I} & | & \mathbf{X} \end{bmatrix}$. In other words, the result will be $\begin{bmatrix} \mathbf{I} & | & \mathbf{A}^{-1} \end{bmatrix}$.

EXAMPLE 10 Compute an inverse of

$$\mathbf{A} = \begin{bmatrix} 1 & 2 & 3 \\ -2 & -1 & 1 \\ 3 & 2 & 0 \end{bmatrix}$$

SOLUTION We consider the augmented matrix

$$[\mathbf{A} \mid \mathbf{I}] \sim \begin{bmatrix} 1 & 0 & 0 & -2 & 6 & 5 \\ 0 & 1 & 0 & 3 & -9 & -7 \\ 0 & 0 & 1 & -1 & 4 & 3 \end{bmatrix}$$

We remind the reader that this calculation is not dispositive, by itself, in concluding that we have the inverse. As established by the calculation, we know that $\mathbf{AX} = \mathbf{I}$ and $\mathbf{X}$ is a left inverse. An appeal to Theorem 4 is needed so that we can also infer $\mathbf{XA} = \mathbf{I}$ and $\mathbf{X}$ is a right inverse. ∎

More on Left and Right Inverses of Non-square Matrices

The method suggested for computing an inverse can be modified for finding right inverses or left inverses, as shown in the next examples.

EXAMPLE 11 Find all the right inverses of the matrix

$$\mathbf{A} = \begin{bmatrix} 1 & 2 & 3 \\ 4 & 5 & 6 \end{bmatrix}$$

SOLUTION To do this, we solve $\mathbf{AX} = \mathbf{I}_2$, where $\mathbf{A}$ is 2×3, $\mathbf{X}$ is 3×2, and $\mathbf{I}_2$ is 2×2. The augmented matrix for this problem and its reduced row echelon form are

$$\begin{bmatrix} 1 & 2 & 3 & 1 & 0 \\ 4 & 5 & 6 & 0 & 1 \end{bmatrix} \sim \begin{bmatrix} 1 & 0 & -1 & -\frac{5}{3} & \frac{2}{3} \\ 0 & 1 & 2 & \frac{4}{3} & -\frac{1}{3} \end{bmatrix}$$

Because of column 3, we have one free variable for each of the two subproblems, and thus there are many solutions. Setting the free variables equal to zero, we get—as one possible right inverse—the 3×2 matrix

$$\begin{bmatrix} -\frac{5}{3} & \frac{2}{3} \\ \frac{4}{3} & -\frac{1}{3} \\ 0 & 0 \end{bmatrix}$$

Because there are two columns and ultimately two free variables, there is a two-parameter family of solutions. Concentrating on the first column of $\mathbf{X}$, we find that

$$\begin{bmatrix} x_1 \\ x_2 \\ x_3 \end{bmatrix} = \begin{bmatrix} -\frac{5}{3} \\ \frac{4}{3} \\ 0 \end{bmatrix} + s \begin{bmatrix} 1 \\ -2 \\ 1 \end{bmatrix}$$

We can treat the second column of **X** in the same way and obtain the general form of the right inverse:

$$\mathbf{X} = \begin{bmatrix} -\frac{5}{3} + s & \frac{2}{3} + t \\ \frac{4}{3} - 2s & -\frac{1}{3} - 2t \\ 0 + s & 0 + t \end{bmatrix}$$

$$= \begin{bmatrix} -\frac{5}{3} & \frac{2}{3} \\ \frac{4}{3} & -\frac{1}{3} \\ 0 & 0 \end{bmatrix} + s \begin{bmatrix} 1 & 0 \\ -2 & 0 \\ 1 & 0 \end{bmatrix} + t \begin{bmatrix} 0 & 1 \\ 0 & -2 \\ 0 & 1 \end{bmatrix}$$

We have left to General Exercise 50 a verification of this result.   ■

EXAMPLE 12 For the matrix in Example 11, find a left inverse, if it has one.

SOLUTION To execute the search, we begin by setting up the equation to be solved: $\mathbf{YA} = \mathbf{I}_3$, where **A** is 2×3 and **Y** must be 3×2, since $\mathbf{I}_3$ is 3×3. To make this into a familiar sort of problem, take transposes, getting $\mathbf{A}^T\mathbf{Y}^T = \mathbf{I}_3$. Here is the result of the row-reduction process:

$$[\mathbf{A}^T \mid \mathbf{Y}^T] = \left[\begin{array}{cc|ccc} 1 & 4 & 1 & 0 & 0 \\ 2 & 5 & 0 & 1 & 0 \\ 3 & 6 & 0 & 0 & 1 \end{array} \right] \sim \left[\begin{array}{cc|ccc} 1 & 0 & 0 & -2 & \frac{5}{3} \\ 0 & 1 & 0 & 1 & -\frac{2}{3} \\ 0 & 0 & 1 & -2 & 1 \end{array} \right]$$

Clearly, this system is inconsistent and **A** has *no* left inverse!   ■

Invertible Matrix Theorem

The set $\mathbb{R}^{n \times n}$ of all $n \times n$ matrices is divided into two classes: the **invertible** matrices and the **noninvertible** matrices. The latter are also called **singular**, and we have noted previously that the word **nonsingular** is synonymous with "*invertible.*" The favorable and useful properties of an invertible matrix are many in number. Let us contemplate a few of them here, with proofs. Later in this section, a more comprehensive list of equivalent conditions is stated (Theorem 10).

> **THEOREM 9**
>
> *The following properties are equivalent for an n × n matrix* **A**. *That is, each property implies all the others.*
>
> 1. **A** *is invertible.*
> 2. *For each* **b** *in* $\mathbb{R}^n$, *the equation* **Ax** = **b** *has a unique solution.*
> 3. *The kernel of* **A** *consists of only the zero vector* **0**.
> 4. *Every column of* **A** *has a pivot position.*
> 5. *Every row of* **A** *has a pivot position.*

PROOF The efficient way to prove such a theorem is to establish the circle of implications $1 \Rightarrow 2 \Rightarrow 3 \Rightarrow 4 \Rightarrow 5 \Rightarrow 1$.

For $1 \Rightarrow 2$, note that invertibility implies that an inverse exists, $\mathbf{A}^{-1}$. Because $\mathbf{AA}^{-1}\mathbf{b} = \mathbf{b}$, the system $\mathbf{Ax} = \mathbf{b}$ has a solution. To see that it is unique, suppose that $\mathbf{Ax} = \mathbf{b}$. Multiply on the left by $\mathbf{A}^{-1}$ to get $\mathbf{x} = \mathbf{A}^{-1}\mathbf{b}$.

For $2 \Rightarrow 3$, observe that **0** is the only solution of the equation $\mathbf{Ax} = \mathbf{0}$. In other words, **0** is the only member of the kernel of **A**.

For $3 \Rightarrow 4$, recall that if any column of **A** has no pivot position, then the homogeneous set of equations will have free variables and many solutions. By **3**, this does not happen, and every column of **A** must have a pivot position.

For $4 \Rightarrow 5$, observe that **4** forces there to be *n* pivot positions in **A**. Hence, each row must have a pivot.

For $5 \Rightarrow 1$, recall that if each row of **A** has a pivot, then the row-reduction process applied to the augmented system $\begin{bmatrix} \mathbf{A} & | & \mathbf{I} \end{bmatrix}$ ends with $\begin{bmatrix} \mathbf{I} & | & \mathbf{A}^{-1} \end{bmatrix}$, thereby establishing the invertibility of **A**. This argument gives us $\mathbf{AX} = \mathbf{I}$, and that suffices because of Theorem 4. ∎

The following important theorem has many parts, each part asserting some property that an invertible matrix must possess. Each property in the theorem is equivalent logically to all the others. You can think of this as a list of important properties of any invertible matrix. But it is more than that, because a matrix that has any one of the listed properties must be invertible, without any further evidence. *Caution*: The theorem applies only to square matrices.

This theorem and other versions of it are collectively known as the *Invertible Matrix Theorem*. Some of the notation and terminology used in this theorem have *not* yet been dealt with in this text, but it is convenient to have all these properties listed together. For example, column spaces and row spaces arise in Section 5.1 and eigenvalues are covered in Section 6.1. To avoid almost doubling the number of equivalent statements in the Invertible Matrix Theorem, most statements involving $\mathbf{A}^T$ have been omitted. To help make the statement of this theorem more digestible for the reader, we intersperse it with some labels.

THEOREM 10   INVERTIBLE MATRIX THEOREM

The following conditions on an $n \times n$ matrix $\mathbf{A}$ are logically equivalent. In other words, if the matrix has any one of the given properties, then it has all of them.

Coefficient Matrix
1. $\mathbf{A}$ *is invertible.*
2. $\mathbf{A}^T$ *is invertible.*

Linear System
3. *The system* $\mathbf{Ax} = \mathbf{b}$ *has at least one solution for every b in* $\mathbb{R}^n$.
4. *For every* $\mathbf{b}$ *in* $\mathbb{R}^n$, *the equation* $\mathbf{Ax} = \mathbf{b}$ *has a unique solution.*
5. *The homogeneous system* $\mathbf{Ax} = \mathbf{0}$ *has only the trivial solution* $\mathbf{x} = \mathbf{0}$.

Pivots
6. $\mathbf{A}$ *has n pivot positions.*
7. *Every column of* $\mathbf{A}$ *has a pivot position.*
8. *Every row of* $\mathbf{A}$ *has a pivot position.*

Rank and Row Equivalence
9. $\mathbf{A}$ *is row equivalent to* $\mathbf{I}_n$.
10. $\text{Rank}(\mathbf{A}) = n$.

Linear Transformations
11. *The transformation* $\mathbf{x} \mapsto \mathbf{Ax}$ *is one-to-one (injective).*
12. *The transformation* $\mathbf{x} \mapsto \mathbf{Ax}$ *maps* $\mathbb{R}^n$ *onto* $\mathbb{R}^n$ *(surjective).*

Left Inverse and Right Inverse
13. *There is an* $n \times n$ *matrix* $\mathbf{B}$ *such that* $\mathbf{BA} = \mathbf{I}_n$.
14. *There is an* $n \times n$ *matrix* $\mathbf{C}$ *such that* $\mathbf{AC} = \mathbf{I}_n$.

Column Vectors
15. *The columns of* $\mathbf{A}$ *form a linearly independent set.*
16. *The set of columns in* $\mathbf{A}$ *spans* $\mathbb{R}^n$.
17. *The columns of* $\mathbf{A}$ *form a basis for* $\mathbb{R}^n$.
18. *The column space of* $\mathbf{A}$ *is* $\mathbb{R}^n$: $\text{Col}(\mathbf{A}) = \mathbb{R}^n$.
19. $\text{Dim}(\text{Col}(\mathbf{A})) = n$.

(continued)

Row Vectors

20. *The rows of A form a linearly independent set.*
21. *The set of rows in A spans $\mathbb{R}^n$.*
22. *The row space of A is $\mathbb{R}^n$: Row(A) = $\mathbb{R}^n$.*
23. The rows of **A** form a basis for $\mathbb{R}^n$.
24. Dim(Row(**A**)) = n.

Eigenvalues, Singular Values, and Determinant

25. 0 is not an eigenvalue of **A**.
26. 0 is not a singular value of **A**.
27. **A** has n nonzero singular values.
28. Det(**A**) $\neq 0$.

Kernel or Null Space

29. The kernel of **A** contains only the zero vector **0**.
30. Null(**A**) = {**0**}.

Application: Interpolation

One application of linear algebra that arises in many scientific enterprises involves the reconstruction of a function from knowledge of some of its values. For example, we might have empirical data that we wish to reproduce by means of a formula. Or, an experiment might lead to a table of data consisting of points in the plane (t_i, y_i) for $i = 0, 1, 2, \ldots, n$. The variable t_i is thought of as the *independent variable* and y_i is the *dependent variable*. These data are discrete, meaning that we have no curve but only individual points. If a continuous function f is sought such that $y_i = f(t_i)$ for all of the points in the given table, then this problem is called **interpolation**. Frequently we use polynomials for this task of interpolation. (Spline functions are usually better, but their theory is beyond the scope of this book.)

EXAMPLE 13 Find a polynomial of least degree that reproduces these data: $(1, 3)$, $(2, 7)$, and $(3, 9)$.

SOLUTION The polynomial will have these values: $p(1) = 3$, $p(2) = 7$, $p(3) = 9$. Thus, three conditions are placed on p. Degree 2 is chosen, because a polynomial of degree 2 has three available coefficients. Let $p(t) = a + bt + ct^2$. Then

$$\begin{cases} a + b + c = 3 \\ a + 2b + 4c = 7 \\ a + 3b + 9c = 9 \end{cases}$$

The augmented matrix and its reduced echelon form are

$$
\begin{bmatrix}
1 & 1 & 1 & 3 \\
1 & 2 & 4 & 7 \\
1 & 3 & 9 & 9
\end{bmatrix}
\sim
\begin{bmatrix}
1 & 0 & 0 & -3 \\
0 & 1 & 0 & 7 \\
0 & 0 & 1 & -1
\end{bmatrix}
$$

The answer to the problem is then the polynomial p defined by $p(t) = -3 + 7t - t^2$. One verifies this result by evaluating the polynomial at the three given points and getting the values 3, 7, and 9. ■

The polynomial interpolation problem, if carefully stated, will always have a unique solution, as established in the next theorem.

THEOREM 11

If $t_0, t_1, \ldots, t_n$ are distinct real numbers and if $y_0, y_1, \ldots, y_n$ are arbitrary real numbers, then there is a unique polynomial p of degree at most n such that $p(t_i) = y_i$ for $0 \le i \le n$.

PROOF We have seen in the preceding example that such a polynomial is governed by a system of equations. There will be $n + 1$ equations and $n + 1$ unknown coefficients in the general case considered here. The matrix that arises will be $(n + 1) \times (n + 1)$. Consider the homogeneous problem, in which all the values y_i are zero. The polynomial p must satisfy the equation $p(t_i) = 0$ for $0 \le i \le n$. However, a nontrivial polynomial of degree n or less can have at most n zeros in the complex plane (see Fundamental Theorem of Algebra in Appendix B). Hence, we conclude that $p = 0$ and that the homogeneous problem has only the zero solution. The matrix mentioned previously is therefore invertible, and the system of equations in the nonhomogeneous problem will have a unique solution. (We are using one part of the Invertible Matrix Theorem: **5 ⇒ 4.**) ■

EXAMPLE 14 Find a linear combination of these three functions

$$
g_1(t) = t, \qquad g_2(t) = t^2, \qquad g_3(t) = t^3
$$

that takes values $4, 3, -4$ at the points $0, 1, 2$, respectively.

SOLUTION This is obviously impossible because each basic function takes the value zero at the point zero. This impossibility does not contradict the preceding theorem because that theorem does not apply in this situation. (It requires all the powers of the variable t, from 0 to n, inclusive.) ■

Interpolation by other functions is also possible, as illustrated in the next example.

> **EXAMPLE 15** Find a linear combination of these three functions
>
> $$h_1(t) = (t-1)^{-1}, \qquad h_2(t) = (t-2)^{-1}, \qquad h_3(t) = (t-3)^{-1}$$
>
> that takes values $1, 7/12, 13/30$ at the points $4, 5, 6$.

SOLUTION Let the unknown function be $f = a_1 h_1 + a_2 h_2 + a_3 h_3$. When we impose the interpolation conditions, we obtain

$$\begin{cases} f(4) = \frac{1}{3}a_1 + \frac{1}{2}a_2 + \ \ a_3 = 1 \\ f(5) = \frac{1}{4}a_1 + \frac{1}{3}a_2 + \frac{1}{2}a_3 = 7/12 \\ f(6) = \frac{1}{5}a_1 + \frac{1}{4}a_2 + \frac{1}{3}a_3 = 13/30 \end{cases}$$

We multiply each equation by a suitable factor to eliminate the fractions, thus arriving at the augmented matrix and its row reduced echelon form:

$$\left[\begin{array}{ccc|c} 2 & 3 & 6 & 6 \\ 3 & 4 & 6 & 7 \\ 12 & 15 & 20 & 26 \end{array} \right] \sim \left[\begin{array}{ccc|c} 1 & 0 & 0 & 3 \\ 0 & 1 & 0 & -2 \\ 0 & 0 & 1 & 1 \end{array} \right]$$

This work reveals that $a_1 = 3, a_2 = -2$, and $a_3 = 1$. ∎

The preceding example has some interesting theoretical background. The relevant theorem is as follows.

> **THEOREM 12**
>
> If $t_1, t_2, \ldots, t_n$ are distinct points in an interval $[a, b]$, and if distinct points $s_1, s_2, \ldots s_n$ are outside that interval, then the matrix having elements $a_{ij} = (t_i - s_j)^{-1}$ is invertible.

(For a proof of this, see Cheney [1982, p. 195].)

Before leaving the subject of interpolation, we should point out that it is a topic in all elementary textbooks on numerical analysis. Interpolation by polynomials is emphasized, and efficient algorithms are developed for finding the polynomial that takes on prescribed values. The subject has a rich history, going back to Newton and some of his predecessors.

Mathematical Software

Mathematical software systems contain procedures useful in linear algebra such as computing the inverse of a given matrix. They are very easy to use in the following three well-known mathematical software packages, as will be shown here. We use a 3×3 matrix randomly generated by a MATLAB command.

$$\mathbf{A} = \begin{bmatrix} 4.50129285147175 & -0.14017531290700 & -0.43532334831659 \\ -2.68861486425712 & 3.91298966148902 & -4.81496356751776 \\ 1.06842583541787 & 2.62096833027395 & 3.21407164295253 \end{bmatrix}$$

$$\mathbf{A}^{-1} = \begin{bmatrix} 0.21386634449984 & -0.00586037256690 & 0.02018733245119 \\ 0.02968206538651 & 0.12674691286975 & 0.19389831126468 \\ -0.09529846100873 & -0.10140977157452 & 0.14630352138346 \end{bmatrix}$$

MATLAB commands are given for these four tasks: (1) displaying numbers to full precision, (2) constructing a random matrix $\mathbf{A}$, (3) invoking a program to compute the inverse of $\mathbf{A}$, and (4) checking the results. This program and similar ones in Maple and Mathematica are shown in subsequent text. The programs may not automatically display numbers with the full precision in which they are computed and stored.

MATLAB	Maple
`format long`	`with(LinearAlgebra):`
`A = -5.0+10*rand(3)`	`A := RandomMatrix(3,3,`
`inv(A)`	`        generator=-5.0..5.0);`
`B*A`	`B := MatrixInverse(A);`
	`B.A;`

Mathematica
`MatrixForm[A = Table[Random[Real,-5,5,15],3,3]]`
`MatrixForm[B = Inverse[A]]`
`B.A`

SUMMARY 3.2

- If $\mathbf{AB} = \mathbf{I}$, then $\mathbf{B}$ is a right inverse of $\mathbf{A}$, and $\mathbf{A}$ is a left inverse of $\mathbf{B}$.

- If $\mathbf{A}$ is square and either $\mathbf{BA} = \mathbf{I}$ or $\mathbf{AB} = \mathbf{I}$, then $\mathbf{B}$ is the inverse of $\mathbf{A}$ and $\mathbf{A}$ is the inverse of $\mathbf{B}$.

- Computing an inverse: use row reduction to solve $[\mathbf{A} \mid \mathbf{I}] \sim [\mathbf{I} \mid \mathbf{A}^{-1}]$.

- If the equation $\mathbf{Ax} = \mathbf{b}$ has a solution and if $\mathbf{BA} = \mathbf{I}$, then $\mathbf{Bb}$ is the unique solution of the system.

- If $\mathbf{AB} = \mathbf{I}$, then the system of equations $\mathbf{Ax} = \mathbf{b}$ has at least one solution, namely $\mathbf{x} = \mathbf{Bb}$.

- If $\mathbf{A}$ has a right inverse, then the equation $\mathbf{Ax} = \mathbf{b}$ has at least one solution.

- If $\mathbf{A}$ has a left inverse, then the equation $\mathbf{Ax} = \mathbf{b}$ has at most one solution.

- For square matrices only: If $\mathbf{BA} = \mathbf{I}$, then $\mathbf{AB} = \mathbf{I}$.

- For square matrices only: If $\mathbf{AB} = \mathbf{AC} = \mathbf{I}$, then $\mathbf{B} = \mathbf{C}$.

- If $\mathbf{A}$ is an $n \times n$ invertible matrix, then the equation $\mathbf{Ax} = \mathbf{b}$ has a unique solution: $\mathbf{x} = \mathbf{A}^{-1}\mathbf{b}$.

- For square matrices only: If $\mathbf{AC} = \mathbf{BA} = \mathbf{I}$, then $\mathbf{B} = \mathbf{C}$.

- $(\mathbf{AB})^{-1} = \mathbf{B}^{-1}\mathbf{A}^{-1}$ (Here $\mathbf{A}$ and $\mathbf{B}$ are $n \times n$ invertible matrices.)

- Let $\mathbf{A}$ be an arbitrary 2×2 matrix, $\begin{bmatrix} a & b \\ c & d \end{bmatrix}$, and let $\Delta = ad - bc$. The invertibility of $\mathbf{A}$ is equivalent to the condition $\Delta \neq 0$.

- For distinct points $t_0, t_1, \ldots, t_n$ and arbitrary points $y_0, y_1, \ldots, y_n$, there is a unique polynomial p of degree at most n such that $p(t_i) = y_i$ for $0 \leq i \leq n$.

- Invertible Matrix Theorem.

KEY CONCEPTS 3.2

Inverse, right inverse, left inverse, invertible, noninvertible, singular, nonsingular, an inverse as the product of elementary matrices, Invertible Matrix Theorem, interpolation

GENERAL EXERCISES 3.2

1. Compute the inverses of

a. $\begin{bmatrix} 1 & 0 & -2 \\ 3 & 1 & -2 \\ -5 & -2 & 3 \end{bmatrix}$

b. $\begin{bmatrix} -2 & -2 & 7 \\ 0 & 1 & -1 \\ 1 & 3 & -5 \end{bmatrix}$

c. $\begin{bmatrix} 1 & 2 & 3 \\ -2 & -1 & 1 \\ 3 & 2 & 0 \end{bmatrix}$

d. $\begin{bmatrix} 3 & 2 & 1 \\ -2 & -1 & 2 \\ 1 & 2 & 3 \end{bmatrix}$

2. Find another right inverse for the matrix $\mathbf{D}$ in Example 2, and verify its correctness.

3. Find a left inverse for each of the following matrices, if they have one:

a. $\mathbf{A} = \begin{bmatrix} 1 & 4 \\ 2 & 5 \\ 3 & 6 \end{bmatrix}$ b. $\mathbf{B} = \begin{bmatrix} 1 & 4 \\ 2 & 5 \\ 4 & 6 \end{bmatrix}$

4. Establish that if the quantity $\Delta = ad - bc$ is not zero, then the inverse of $\begin{bmatrix} a & b \\ c & d \end{bmatrix}$ is

$$\Delta^{-1} \begin{bmatrix} d & -b \\ -c & a \end{bmatrix}$$

5. (Continuation.) Establish the converse of the assertion in General Exercise 4. That is, show that the invertibility of the given matrix implies $\Delta \neq 0$.

6. Without doing any computing, explain why the following equation is not true.

$$\begin{bmatrix} 0.9501 & -0.8913 & 0.8214 \\ 0.2311 & -0.7621 & 0.4447 \\ 0.6068 & -0.4565 & 0.6154 \\ 0.4860 & -0.0185 & 0.7919 \end{bmatrix}$$

$$\times \begin{bmatrix} 0.9218 & 0.4057 & 0.4103 & 0.3529 \\ 0.7382 & 0.9355 & 0.8936 & 0.8132 \\ 0.1763 & 0.9169 & 0.0579 & 0.0099 \end{bmatrix}$$

$$= \begin{bmatrix} 1 & 0 & 0 & 0 \\ 0 & 1 & 0 & 0 \\ 0 & 0 & 1 & 0 \\ 0 & 0 & 0 & 1 \end{bmatrix}$$

7. Explain what problems are solved by row-reducing these three augmented matrices.

a. $\begin{bmatrix} 1 & 2 & 1 & 1 & 0 \\ 3 & 5 & 1 & 0 & 1 \end{bmatrix}$

b. $\begin{bmatrix} 1 & 2 & 1 & 1 & 0 \\ 3 & 5 & 1 & 0 & 1 \end{bmatrix}$

c. $\begin{bmatrix} 1 & 2 & 1 & 1 & 0 \\ 3 & 5 & 1 & 0 & 1 \end{bmatrix}$

8. Find all the right inverses of the matrix

$$\begin{bmatrix} -5 & 2 & 0 \\ 3 & -1 & 0 \end{bmatrix}$$

9. Find the inverses of these matrices. (Two of them consist of integers.)

$$A = \begin{bmatrix} 1 & 3 & -7 \\ -3 & 4 & 1 \\ 2 & -5 & 3 \end{bmatrix}$$

$$B = \begin{bmatrix} 1 & 0 & -2 \\ -3 & 1 & 4 \\ 2 & -3 & 4 \end{bmatrix}$$

$$C = \begin{bmatrix} 1 & 2 & 0 \\ 2 & 1 & 1 \\ 2 & -3 & 2 \end{bmatrix}$$

10. Critique the solution offered for the system

of equations $\begin{cases} x_1 + 2x_2 = 3 \\ 3x_1 + 7x_2 = 5 \\ 2x_1 + 5x_2 = 4 \end{cases}$

First write $\begin{bmatrix} 1 & 2 \\ 3 & 7 \\ 2 & 5 \end{bmatrix} \begin{bmatrix} x_1 \\ x_2 \end{bmatrix} = \begin{bmatrix} 3 \\ 5 \\ 4 \end{bmatrix}$

Next, multiply both sides of the equation on the left by the matrix

$$\begin{bmatrix} 13 & -8 & 6 \\ -7 & 5 & -4 \end{bmatrix},$$

getting the solution $\begin{bmatrix} 23 \\ -12 \end{bmatrix}$.

Verify your solution by substitution.

11. Find the inverse of $\begin{bmatrix} -2 & -1 & 1 \\ 3 & 2 & 0 \\ 1 & 2 & 3 \end{bmatrix}$. Verify your answer in an independent manner.

12. If A and B are $n \times n$ matrices such that $BA = I$, does it follow that $AB = BA$? If so, explain.

13. Is there an invertible 2×2 matrix $A = \begin{bmatrix} a & b \\ c & d \end{bmatrix}$ such that $A^{-1} = \begin{bmatrix} 1/a & 1/b \\ 1/c & 1/d \end{bmatrix}$?

14. Derive the formula for the inverse of a 2×2 matrix by using the equation

$$\begin{bmatrix} a & b \\ c & d \end{bmatrix} \begin{bmatrix} e & f \\ g & h \end{bmatrix} = \begin{bmatrix} 1 & 0 \\ 0 & 1 \end{bmatrix}$$

and solving for the unknowns e, f, g, h.

15. Find all the right inverses of the matrix
$$\begin{bmatrix} 1 & 2 & 3 \\ 4 & 5 & 6 \end{bmatrix}$$

16. Let $\mathbf{A} = \begin{bmatrix} -5 & 3 \\ 2 & -1 \\ 0 & 0 \end{bmatrix}$, $\mathbf{B} = \begin{bmatrix} 1 & 3 & -2 \\ 2 & 5 & 1 \end{bmatrix}$.

Verify that $\mathbf{BA} = \mathbf{I}_2$. Next, consider the equation $\mathbf{Ax} = \mathbf{c}$, where $\mathbf{c} = [-4, 3, 2]^T$. Solve this equation by multiplying both sides by $\mathbf{B}$, on the left, getting $\mathbf{BAx} = \mathbf{Bc}$ or $\mathbf{x} = \mathbf{Bc} = [1, 9]^T$. Verify that the solution is correct.

17. Consider $\begin{bmatrix} 1 & 2 & 2 \\ 2 & 1 & -3 \\ 0 & 1 & 2 \end{bmatrix}$. The inverse of this

matrix consists of integers. Find it.

18. Find the polynomial of least degree that interpolates these data: $(0, -5), (1, -3), (2, 1)$, and $(-1, -11)$.

19. Find the polynomial of least degree whose graph passes through these points: $(-2, 29), (-1, 12), (0, 3)$, and $(1, 4)$.

20. Find all left inverses of $\begin{bmatrix} 1 & 3 \\ 2 & -4 \\ 3 & 4 \end{bmatrix}$.

(These make a two-parameter family of left inverses.)

21. Find all left inverses of $\begin{bmatrix} 1 & 1 \\ 2 & 1 \\ 3 & 1 \end{bmatrix}$

22. Establish the validity of Theorem 1. Explain all details.

23. Explain why every invertible matrix is row equivalent to $\mathbf{I}$.

24. Establish that if $\mathbf{A}$ is a nonsquare matrix, then it cannot have both a right inverse and a left inverse.

25. Find the general right inverse of the matrix in Example 2 in the form $\mathbf{C} + t\mathbf{E} + s\mathbf{F}$, where t and s are arbitrary real numbers, while $\mathbf{C}$, $\mathbf{E}$, and $\mathbf{F}$ are 3×2 matrices.

26. Establish that if a matrix has a right inverse but has no left inverse, then the right inverse is not unique.

27. Explain why this solution of the equation $\sqrt{x+8} = 1 + \sqrt{2x+2}$ is faulty: square both sides: $x + 8 = 1 + 2\sqrt{2x+2} + 2x + 2$; collect similar terms: $5 - x = 2\sqrt{2x+2}$; square again: $25 - 10x + x^2 = 4(2x+2)$; collect similar terms: $x^2 - 18x + 17 = 0$; factor: $(x-1)(x-17) = 0$; get a root from each factor showing that the solutions are $x = 1$ and $x = 17$.

28. Let $\mathbf{A}$ and $\mathbf{B}$ be $n \times n$ matrices. Assume that $\mathbf{B} \neq \mathbf{0}$ and $\mathbf{AB} = \mathbf{0}$. Does it follow that $\mathbf{A}$ is not invertible?

29. Try to generalize Theorem 4 as follows: If $\mathbf{A}$ and $\mathbf{B}$ are square matrices such that $\mathbf{AB}$ is a diagonal matrix with nonzero values on its diagonal, then $\mathbf{AB} = \mathbf{BA}$.

30. Establish that if $\mathbf{A}$ is invertible, then $(\mathbf{A}^{-1})^{-1} = \mathbf{A}$.

31. Explain why the invertibility of $\mathbf{A}$ implies the invertibility of $\mathbf{A}^T$.

32. For an invertible matrix $\mathbf{A}$, the negative integer powers of $\mathbf{A}$ are defined by $\mathbf{A}^{-j} = (\mathbf{A}^{-1})^j$ where j is a positive integer. Explain why, for such a matrix, $(\mathbf{A}^{-1})^j = \mathbf{A}^{-j}$ for all integers j (positive and negative).

33. If $\mathbf{A}^4 = \mathbf{I}$, what is $\mathbf{A}^{-1}$? Generalize.

34. Give an argument to show that if a symmetric matrix is invertible, then its inverse is symmetric.

35. For an invertible matrix $\mathbf{A}$, establish that $(\mathbf{A}^T)^{-1} = (\mathbf{A}^{-1})^T$.

36. Let $\mathbf{A}$ and $\mathbf{B}$ be $n \times n$ invertible matrices. Let the reduced echelon form of the augmented matrix $[\mathbf{A} \,|\, \mathbf{B}]$ be denoted by $[\mathbf{C} \,|\, \mathbf{E}]$. What are $\mathbf{C}$ and $\mathbf{E}$ explicitly?

37. Establish that if a matrix has more rows than columns, then it has no right inverse.

38. (Continuation.) State and establish the form of a left inverse.

39. Explain why the nonzero scalar multiples of an invertible matrix are invertible.

40. Establish that if $\mathbf{A}$ is noninvertible, then there is a nonzero row vector, $\mathbf{x}$, such that $\mathbf{x}\mathbf{A}^T = \mathbf{0}$.

41. Explain why the inverse of an invertible lower triangular matrix is lower triangular.

42. Find some square matrices, $\mathbf{A}$, having the property $\mathbf{A}^{-1} = \mathbf{A}^T$. Later, we shall see such matrices arising naturally. Perhaps you can find all the 2×2 matrices having this property.

43. Is there a valid computational rule that states $(\mathbf{A}^T\mathbf{B}^T)^{-1} = (\mathbf{A}^{-1}\mathbf{B}^{-1})^T$? Explain.

44. Suppose that $\mathbf{A}$ is an $m \times n$ matrix that has a left inverse and a right inverse. These inverses are not assumed to be identical. Establish that, in fact, they are identical and $m = n$.

45. (Research project.) Considering only $n \times n$ matrices $\mathbf{A}$ and $\mathbf{B}$, find out when $\mathbf{AB} = \mathbf{BA}$.

46. Define a linear transformation T by the equation $T(\mathbf{x}) = \mathbf{Ax}$, where $\mathbf{A}$ is an $n \times n$ matrix. Explain why T is injective if and only if it is surjective.

47. Verify the result in Example 6 by computing $\mathbf{AB}$ and $\mathbf{BA}$.

48. Give an example other than the one in the text in which $\mathbf{BA} = \mathbf{I}$ and $\mathbf{ABb} \neq \mathbf{b}$ for some vector $\mathbf{b}$.

49. Establish that if an $m \times n$ matrix has a right inverse, then $m \le n$.

50. Verify that $\mathbf{X}$, as given in Example 11, is indeed a right inverse of $\mathbf{A}$.

51. Let $\mathbf{A}$, $\mathbf{B}$, and $\mathbf{C}$ be matrices such that $\mathbf{AB} = \mathbf{I}_m$ and $\mathbf{CA} = \mathbf{I}_n$. Establish that $n = m$ and that $\mathbf{C} = \mathbf{B}$.

52. Let $\mathbf{A}$, $\mathbf{B}$, and $\mathbf{C}$ be square matrices such that $\mathbf{BA} = \mathbf{CA} = \mathbf{I}$. Does it follow that $\mathbf{B} = \mathbf{C}$?

53. Do the invertible matrices in $\mathbb{R}^{n \times n}$ form a subspace of $\mathbb{R}^{n \times n}$? What about the noninvertible matrices? Look ahead to Section 5.1 for the definition of *subspace*.

54. Is there a matrix $\mathbf{A}$ that has one and only one right inverse, but has no left inverse?

55. A beginner writes $\mathbf{A} = \begin{bmatrix} a & b & c \\ d & e & f \\ g & h & i \end{bmatrix}$ and

concludes that $\mathbf{A}^{-1} = \begin{bmatrix} 1/a & 1/b & 1/c \\ 1/d & 1/e & 1/f \\ 1/g & 1/h & 1/i \end{bmatrix}$

Is this ever true? Find all the cases when it is true, allowing the dimension of the square matrix to vary.

56. Let $\mathbf{A}$, $\mathbf{B}$, and $\mathbf{C}$ be $n \times n$ matrices such that $\mathbf{BA} = \mathbf{CA}$, and $\mathbf{BA}$ is invertible. Does it follow that $\mathbf{B} = \mathbf{C}$?

57. Establish that from the equation $\mathbf{AB} = \mathbf{CA}$, we cannot conclude, in general, that $\mathbf{B} = \mathbf{C}$, even if $\mathbf{A}$ is invertible.

58. Use the four numbers $2, 3, 5, 7$ to create a 2×2 matrix whose inverse has integers for all its entries. Do the same for $4, 5, 6, 11$ and $5, 6, 9, 11$ and $2, 95, 117, 5557$.

59. Let $\mathbf{A}$, $\mathbf{B}$, and $\mathbf{I}$ be $n \times n$ matrices, where $\mathbf{I}$ is the identity matrix. What conclusion can be drawn from the hypothesis $[\mathbf{I} \mid \mathbf{A}] \sim [\mathbf{B} \mid \mathbf{I}]$?

60. Critique the following procedure for finding the inverse of a matrix $\mathbf{A}$: Select a nonzero vector $\mathbf{x}$ and set $\mathbf{b} = \mathbf{Ax}$. Define $\gamma = \mathbf{x}^T\mathbf{b}$ and $\mathbf{C} = \gamma^{-1}\mathbf{xx}^T$. It follows that $1 = \gamma^{-1}\gamma = \gamma^{-1}\mathbf{x}^T\mathbf{b}$. Furthermore, we have $\mathbf{Cb} = \gamma^{-1}\mathbf{xx}^T\mathbf{b} = \mathbf{x}\gamma^{-1}\mathbf{x}^T\mathbf{b} = \mathbf{x}(\gamma^{-1}\mathbf{x}^T\mathbf{b}) = \mathbf{x}$. From the two equations $\mathbf{Cb} = \mathbf{x}$ and $\mathbf{x} = \mathbf{A}^{-1}\mathbf{b}$, we conclude that $\mathbf{Cb} = \mathbf{A}^{-1}\mathbf{b}$ and $\mathbf{C} = \mathbf{A}^{-1}$. (This example is from the book by Barbeau [2000].)

61. If $\mathbf{A}$ and $\mathbf{B}$ are $n \times n$ invertible matrices, can the equation $\mathbf{A} + \mathbf{B} = \mathbf{AB}$ be true?

62. Let $\mathbf{A}$, $\mathbf{B}$, and $\mathbf{C}$ be $n \times n$ matrices such that $\mathbf{B} \neq \mathbf{C}$ and $\mathbf{AB} = \mathbf{AC}$. Find the inverse of $\mathbf{A}$.

63. Let $\mathbf{A}$ be a square matrix whose rows form a linearly independent set. Do the columns of $\mathbf{A}$ necessarily form a linearly independent set?

64. For square matrices, prove that if $\mathbf{AB} = \mathbf{I}$, then $\mathbf{AB} = \mathbf{BA}$. Is this assertion true for matrices other than $\mathbf{I}$?

65. Suppose that we have $n \times n$ matrices such that $\mathbf{AE}_1\mathbf{E}_2 \cdots \mathbf{E}_m = \mathbf{I}$. What is $\mathbf{A}^{-1}$? (It is not being assumed that the matrices $\mathbf{E}_i$ are elementary matrices.)

66. Justify these implications by use of the Invertible Matrix Theorem. (All matrices here are $n \times n$.)
   a. If $\mathbf{A}$ is invertible and $\mathbf{Ax} = \mathbf{Ay}$, then $\mathbf{x} = \mathbf{y}$.
   b. If $\mathbf{BA} = \mathbf{I}$, then the columns of $\mathbf{B}$ span $\mathbb{R}^n$.
   c. If $\mathbf{Ax} \neq \mathbf{0}$ whenever $\mathbf{x} \neq \mathbf{0}$, then the equation $\mathbf{Ay} = \mathbf{b}$ has a solution for any $\mathbf{b} \in \mathbb{R}^n$.
   d. If $\mathbf{A}$ is row equivalent to $\mathbf{I}_n$, then $\mathbf{A}$ is invertible.
   e. If $\mathbf{A}$ has n pivot positions, then the rows of $\mathbf{A}$ span $\mathbb{R}^n$.
   f. If the columns of $\mathbf{A}$ form a linearly independent set, then the rows also form a linearly independent set.
   g. If the equation $\mathbf{Ax} = \mathbf{b}$ has a solution for every $\mathbf{b} \in \mathbb{R}^n$, then $\mathbf{0}$ is the only solution of the equation $\mathbf{Ax} = \mathbf{0}$.
   h. If the mapping $\mathbf{x} \mapsto \mathbf{Ax}$ is one-to-one, then $\mathbf{A}$ is invertible.

67. Find a linear combination of the three functions $t \mapsto 1, \cos t, \sin t$ that interpolates these data: $(0, 3)$, $(\pi/2, -2)$, and $(3\pi/2, 4)$.

68. Present an argument to explain why a matrix that has both a right inverse and a left inverse is invertible.

69. Verify that with the three functions $f_0(t) = 1$, $f_1(t) = t^3$, and $f_2(t) = t^6$, we can interpolate arbitrary data at any three distinct points.

70. Explain why with the three functions $g_0(t) = 1, g_1(t) = t$, and $g_2(t) = t^3$, we cannot necessarily interpolate arbitrary data at three distinct points. Are there conditions that can be placed on the nodes (the points t_i) to ensure that the interpolation problem can be solved for arbitrary values of y_0, y_1, and y_2?

71. If the data in an interpolation problem come from a polynomial p, will the resulting interpolating polynomial be p? Discover the correct theorem in the answer to this question.

72. The values of t that are involved in an interpolation problem are often called **nodes**. Suppose that our nodes are ordered like this: $0 < t_0 < t_1 < t_2, \ldots < t_n$. Can we interpolate arbitrary data at the nodes with the functions $t \mapsto t^{j/2}$, where $0 \le j \le n$?

73. Explain that if interpolation nodes are specified in this order $0 < t_0 < t_1 < \ldots < t_n$, then arbitrary data at these points can be interpolated by a function of the form $f(t) = \sum_{j=0}^{n} a_j (t + t_j)^{-1}$.

74. Find the coefficients a_1, a_2, a_3 so that the function $f(t) = a_1(1 + t)^{-1} + a_2(2 + t)^{-1} + a_3(3 + t)^{-1}$ will interpolate arbitrary values y_1, y_2, y_3 at the points $1, 2, 3$. In this situation, the inverse matrix would be useful so that we could easily compute the coefficients a_j from the data y_i.

75. Let $f(t) = a_1(t + 1)^{-1} + a_2(t - 2)^{-1} + a_3(t + 2)^{-1}$. Find the coefficients a_j so that $f(0) = 15/2, f(1) = 43/6$, and $f(3) = -\frac{9}{4}$. Draw a rough graph of the resulting function f. Is this function a satisfactory interpolating function for the given data?

76. In Theorem 12, it is asserted that the polynomial of degree at most n that takes assigned values at $n + 1$ distinct points on the real line is unique. How then can you account for the fact that $p(x) = 4x^3 - 22x^2 + 35x - 12$ and $q(x) = 3 + 5(x - 3) - 2x(x - 3) + 4x(x - 3)(x - 2)$ take the same values at the points $0, 1, 2, 3$? Do these two polynomials agree in value at any other points?

77. Find a function of two variables, $f(x, y) = ax + by + c$, that takes these values: $f(2, -1) = 0, f(3, 2) = 11$, and $f(-1, 4) = 9$. This is an example of **bivariate interpolation**, there being two independent variables.

78. What is the maximum number of zeros that can be present in an invertible $n \times n$ matrix?

79. Establish that if A and B are $n \times n$ matrices such that AB is invertible, then A and B are invertible.

80. Establish this string of implications, assuming (a) to start. (a) A is invertible. (b) The system of equations $Ax = b$ has a solution for every $b \in \mathbb{R}^n$. (c) The set of columns in A spans $\mathbb{R}^n$. (d) The transformation $x \mapsto Ax$ maps $\mathbb{R}^n$ onto $\mathbb{R}^n$. (e) The columns of A form a linearly independent set.

81. Establish that if A and B are $n \times n$ matrices, and if one of A and B is invertible, then AB is row equivalent to BA.

82. Explain why if all the elements of an invertible matrix are positive, then the same cannot be true of its inverse. Generalize. What can you say about an invertible matrix whose elements are nonnegative?

83. Establish that $\mathbf{A}^k$ is invertible for some value of k greater than zero, then $\mathbf{A}$ is invertible.

84. Suppose that a matrix $\mathbf{A}$ has the property that $\mathbf{AA}^T$ has a left inverse. Explain why the rows of $\mathbf{A}$ must form a linearly independent set.

85. Argue that if $\mathbf{A}$ is an $m \times n$ matrix of rank m, and if $\mathbf{B}$ is an $n \times n$ invertible matrix, then $\mathbf{ABB}^T\mathbf{A}^T$ is invertible. You should try to generalize to $\mathbf{ABA}^T$ when $\mathbf{B}$ is invertible.

86. In Example 9, form $\mathbf{E}$ and then calculate $\mathbf{E}^{-1}$ using the row-reduction process to illustrate that this approach is more difficult than the method shown in the text for finding the inverse.

87. Solve Example 5 using a different free variable.

88. In Example 6, verify that $\mathbf{AB} = \mathbf{I}$, $\mathbf{A}^{-1} = \mathbf{B}$, and $\mathbf{AB} = \mathbf{I}$.

89. Carry out the details in solving Example 10 and verify that $\mathbf{AX} = \mathbf{I}$ and $\mathbf{XA} = \mathbf{I}$.

90. Explain why the system in Example 12 is inconsistent after carrying out the calculations. In general, does every entry on the righthand side of the last row have to be nonzero for the system to be inconsistent?

COMPUTER EXERCISES 3.2

1. Find out how quickly your software package can invert a random matrix. If you decide to use MATLAB, you will find these commands useful:

   a. To generate a random matrix of size 250×250, use the command `A=rand(250,250);` Notice the symbol ";". It countermands the display of the matrix $\mathbf{A}$. You certainly do not want to see all rows and columns of a huge matrix. The MATLAB command `rand` produces random numbers uniformly distributed in the interval from 0 to 1.

   b. To time the execution of the inversion process, use `tic, inv(A); toc`. This will carry out the inversion of the matrix and report the elapsed time for the process. The output of the inversion command is suppressed by the symbol ";" as described previously.

   c. If you have access to different machines or different operating systems, time several of them. You may prefer to use larger matrices, such as 1000×1000. In our experimentation we found a time of 40 seconds on one machine and 12 seconds on another; in both cases the matrices were of order 1000.

2. Devise an algorithm for inverting a unit lower triangular matrix. Program and test the computer code for inverting some matrices. For example, invert

$$\mathbf{A} = \begin{bmatrix} 1 & 0 & 0 & 0 \\ 3 & 1 & 0 & 0 \\ 5 & 2 & 1 & 0 \\ 7 & 4 & -3 & 1 \end{bmatrix}$$

3. Using rational integer arithmetic, compute $\mathbf{A}^{-1}$ and $\mathbf{x} = \mathbf{A}^{-1}\mathbf{b}$:

   a. $\mathbf{A} = \begin{bmatrix} 33 & 16 & 72 \\ -24 & -10 & -57 \\ -8 & -4 & -17 \end{bmatrix}$

$$\mathbf{b} = \begin{bmatrix} -359 \\ 281 \\ 85 \end{bmatrix}$$

b. $\mathbf{A} = \begin{bmatrix} 1 & -2 & 3 & 1 \\ -2 & 1 & -2 & -1 \\ 3 & -2 & 1 & 5 \\ 1 & -1 & 5 & 3 \end{bmatrix}$

$\mathbf{b} = \begin{bmatrix} 3 \\ -4 \\ 7 \\ 8 \end{bmatrix}$

c. $\mathbf{A} = \begin{bmatrix} 5 & 7 & 6 & 5 \\ 7 & 10 & 8 & 7 \\ 6 & 8 & 10 & 9 \\ 5 & 7 & 9 & 10 \end{bmatrix}$

$\mathbf{b} = \begin{bmatrix} 23 \\ 32 \\ 33 \\ 31 \end{bmatrix}$

4. Using rational integer arithmetic, compute $\mathbf{A}^{-1}$:

a. Consider $\mathbf{A} = \begin{bmatrix} 1 & 1+2i & 2+10i \\ 1+i & 3i & -5+14i \\ 1+i & 5i & -8+20i \end{bmatrix}$

Here $i = \sqrt{-1}$ and complex numbers are involved. (See Appendix B.)

b. $\mathbf{A} = \begin{bmatrix} 4 & 3 & 2 & 1 \\ 3 & 4 & 3 & 2 \\ 2 & 3 & 4 & 3 \\ 1 & 2 & 3 & 4 \end{bmatrix}$

c. $\mathbf{A} = \frac{1}{7} \begin{bmatrix} 1 & 1 & 1 & 1 \\ 1 & 2 & 3 & 4 \\ 1 & 3 & 6 & 10 \\ 1 & 4 & 10 & 20 \end{bmatrix}$

d. $\mathbf{A} = \begin{bmatrix} 1 & 0 & 0 & 0 & 0 & 1 \\ 1 & 1 & 0 & 0 & 0 & -1 \\ -1 & 1 & 1 & 0 & 0 & 1 \\ 1 & -1 & 1 & 1 & 0 & 1 \\ -1 & 1 & -1 & 1 & 1 & 1 \\ 1 & -1 & 1 & -1 & 1 & -1 \end{bmatrix}$

e. $\mathbf{A} = \begin{bmatrix} 2 & -1 & 0 & 0 \\ -1 & 2 & -1 & 0 \\ 0 & -1 & 2 & -1 \\ 0 & 0 & -1 & 2 \end{bmatrix}$

f. $\mathbf{A} = \begin{bmatrix} -3 & 1 & 0 & 0 & 0 \\ 1 & -2 & 1 & 0 & 0 \\ 0 & 1 & -2 & 1 & 0 \\ 0 & 0 & 1 & -2 & 1 \\ 0 & 0 & 0 & 1 & -1 \end{bmatrix}$

g. $\mathbf{A} = \begin{bmatrix} 3 & 1 & 1 & 1 & 1 \\ 1 & 3 & 1 & 1 & 1 \\ 1 & 1 & 3 & 1 & 1 \\ 1 & 1 & 1 & 3 & 1 \\ 1 & 1 & 1 & 1 & 3 \end{bmatrix}$

5. Let $\mathbf{A} = \begin{bmatrix} -73 & 78 & 24 \\ 92 & 66 & 25 \\ -80 & 37 & 10 \end{bmatrix}$

Explore how changing only one entry in this matrix by the addition of 0.01 affects the values in $\mathbf{A}^{-1}$. For example, replace either 92 by 92.01 or 78 by 78.01 or 10 by 10.01.

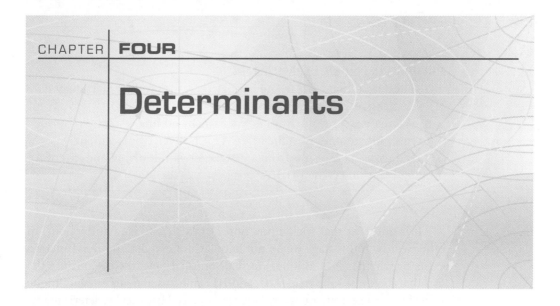

CHAPTER FOUR

Determinants

4.1 DETERMINANTS: INTRODUCTION

> *I was x years old in the year x^2.*
>
> —AUGUSTUS DE MORGAN (1806–1871),
> RESPONDING TO THE QUESTION OF HOW OLD HE WAS

> *To speak algebraically, Mr. M. is execrable, but Mr. C. is x plus 1 - ecrable.*
>
> —EDGAR ALLAN POE (1809–1849)

Properties of Determinants

With every square matrix there is associated a real number called the **determinant** of that matrix. The notation used here is Det(**A**) or Det **A** for the determinant of **A**. There are various ways to present postulates for determinants from which all other properties follow. We have chosen to start with the following three basic properties of determinants because they lead directly to their efficient calculation.

PROPERTIES OF DETERMINANTS

  I. The determinant of a triangular matrix is the product of its diagonal elements. (**triangular matrix**)

  II. If a matrix $\widetilde{\mathbf{A}}$ results from a matrix $\mathbf{A}$ by adding a multiple of one row onto another row, then $\text{Det}(\widetilde{\mathbf{A}}) = \text{Det}(\mathbf{A})$. (**replacement**)

  III. If the matrix $\widetilde{\mathbf{A}}$ results from a matrix $\mathbf{A}$ by multiplying one row of $\mathbf{A}$ by a constant c, then $\text{Det}(\widetilde{\mathbf{A}}) = c\,\text{Det}(\mathbf{A})$. (**scaling**)

It is not clear that such a function *Det* exists and is unambiguous. In other words, at this stage we do not know whether a determinant function exists or whether it is unique. We shall proceed nevertheless and use the fact that the three Properties **I**, **II**, and **III** can be used to arrive at a unique value for the determinant of any (square) matrix. Historically, determinants preceded matrices by more than 150 years, and many facts in matrix theory were originally discovered and developed from determinant theory. See Thomas Muir [1960] for the historical development of determinants from the time of Leibniz to 1920.

EXAMPLE 1 We can already compute the determinants of some matrices with almost no effort. Test your understanding of the Properties **I**, **II**, and **III**, by evaluating the determinants of these four matrices.

$$\begin{bmatrix} 4 & 7 & -11 \\ 0 & 3 & 12 \\ 0 & 0 & 10 \end{bmatrix}, \quad \begin{bmatrix} 3 & 2 & -7 \\ 6 & 11 & -12 \\ 0 & 0 & 5 \end{bmatrix}$$

$$\begin{bmatrix} 3 & 2 & 7 \\ 6 & 11 & -12 \end{bmatrix}, \quad \begin{bmatrix} 3 & 0 & 0 \\ 2 & 4 & 0 \\ 5 & 6 & 2 \end{bmatrix}$$

SOLUTION The first matrix is upper triangular, and, by Property **I**, its determinant is $(4)(3)(10) = 120$. In the second matrix, we can add -2 times row 1 to row 2 without changing the determinant. This brings us

to the matrix $\begin{bmatrix} 3 & 2 & 7 \\ 0 & 7 & 2 \\ 0 & 0 & 5 \end{bmatrix}$. The determinant of this upper triangular matrix is $(3)(7)(5) = 105$. The third matrix, being nonsquare, has no determinant. (The determinant function is defined only for square matrices.) The fourth matrix is lower triangular, and, by Property **I**, its determinant is 24. ∎

The list of Properties **I**, **II**, and **III** is almost a *minimal* set of assumptions: they do allow the whole theory of determinants to be deduced from them. In a complete development of the subject, we would prove that there exists one and only one such function, Det. The assumptions postulated in Properties **I**, **II**, and **III** are stronger than those given in the pioneering textbook of Stoll [1952]. For example, Stoll assumes (instead of Property **I**) that the determinant of every identity matrix is 1. The later revision of this book, Stoll and Wang [1968], is another source for this theory. The books by Strang [1988, 2003] have lucid accounts of this topic based on a different set of postulates: Det(**A**) depends linearly on row 1, Det(**A**) changes sign when two rows are swapped, and Det(**I**) = 1.

First we draw an easy but useful conclusion from the Properties **I**, **II**, and **III**. The property in the following theorem is not an assumption but rather a consequence of the Properties **I**, **II**, and **III**.

THEOREM 1

If the matrix $\widetilde{\mathbf{A}}$ is obtained from the matrix $\mathbf{A}$ by **swapping** *a pair of rows, then*

$$\mathrm{Det}(\widetilde{\mathbf{A}}) = -\,\mathrm{Det}(\mathbf{A})$$

PROOF Recall from Section 1.1, that a swap (or interchange) of two rows is accomplished by this sequence of the two other types of row operations:

$$\begin{bmatrix} \mathbf{u} \\ \mathbf{v} \end{bmatrix} \sim \begin{bmatrix} \mathbf{u} \\ \mathbf{v}+\mathbf{u} \end{bmatrix} \sim \begin{bmatrix} -\mathbf{v} \\ \mathbf{v}+\mathbf{u} \end{bmatrix} \sim \begin{bmatrix} -\mathbf{v} \\ \mathbf{u} \end{bmatrix} \sim \begin{bmatrix} \mathbf{v} \\ \mathbf{u} \end{bmatrix}$$

The first three row operations have no effect on the determinant, by Property **II**. The fourth row operation changes the sign of the determinant, by Property **III**. Hence, the total effect is to change the sign of the determinant when two rows are interchanged. ∎

EXAMPLE 2 Explain why each of these equations is valid:

$$
\text{Det} \begin{bmatrix} 3 & 7 & -5 \\ 6 & 4 & 2 \\ 9 & 1 & 8 \end{bmatrix} = -\text{Det} \begin{bmatrix} 6 & 4 & 2 \\ 3 & 7 & -5 \\ 9 & 1 & 8 \end{bmatrix} = -2\,\text{Det} \begin{bmatrix} 3 & 2 & 1 \\ 3 & 7 & -5 \\ 9 & 1 & 8 \end{bmatrix}
$$

$$
= -2\,\text{Det} \begin{bmatrix} 3 & 2 & 1 \\ 0 & 5 & -6 \\ 9 & 1 & 8 \end{bmatrix} = -2\,\text{Det} \begin{bmatrix} 3 & 2 & 1 \\ 0 & 5 & -6 \\ 0 & -5 & 5 \end{bmatrix}
$$

$$
= -2\,\text{Det} \begin{bmatrix} 3 & 2 & 1 \\ 0 & 5 & -6 \\ 0 & 0 & -1 \end{bmatrix} = (-2)(3)(5)(-1) = 30
$$

SOLUTION The first equation is true by Theorem 1. The next is justified by Property **III**. The next three equalities follow from Property **II**. The last equality uses Property **I**. ∎

We can compute many determinants with only Properties **I** and **II**. The purpose of the Properties **II** and **III**, and Theorem 1 is to keep track of the effect on a determinant when row operations are used to arrive at a triangular form.

Example 2, and the working out of the determinant therein, raises the question hinted at earlier: "*How can we be sure that no matter what sequence of row operations we use to compute the determinant, the result will be the same?*" Eventually, it is necessary to establish the existence and uniqueness of a determinant function that has the properties expressed in the three Properties **I**, **II**, and **III**.

THEOREM 2

There is one and only one determinant function having Properties **I**, **II**, *and* **III**.

Theorem 2 is stated without proof. It is needed to justify the algorithm for computing determinants.

An Algorithm for Computing Determinants

We are now in a position to calculate the determinant of any square matrix, by using Properties **I**, **II**, and **III** and Theorem 1. The idea is very simple: Given a (square) matrix, **A**, use the Properties **I**, **II**, and **III**, as well as Theorem 1 to find a triangular matrix **A** that is row equivalent to **A**. Along the way, keep track of any changes in the determinant caused by the use of Property **III** and Theorem 1. At the end, apply Property **I**. The deeper truth, which we are not going to prove, is that a unique value arises, despite the fact that different sequences of row operations may have been used.

EXAMPLE 3 Compute this determinant, using the suggested algorithm:

$$\mathrm{Det} \begin{bmatrix} 1 & 3 & 4 \\ 2 & 1 & 2 \\ -4 & 2 & 1 \end{bmatrix}$$

SOLUTION Here are the steps, using Property **II** repeatedly and finally Property **I**:

$$\mathrm{Det} \begin{bmatrix} 1 & 3 & 4 \\ 2 & 1 & 2 \\ -4 & 2 & 1 \end{bmatrix} = \mathrm{Det} \begin{bmatrix} 1 & 3 & 4 \\ 0 & -5 & -6 \\ -4 & 2 & 1 \end{bmatrix} = \mathrm{Det} \begin{bmatrix} 1 & 3 & 4 \\ 0 & -5 & -6 \\ 0 & 14 & 17 \end{bmatrix}$$

$$= \mathrm{Det} \begin{bmatrix} 1 & 3 & 4 \\ 0 & -5 & -6 \\ 0 & 0 & \frac{1}{5} \end{bmatrix} = (1)(-5)(\tfrac{1}{5}) = -1 \quad \blacksquare$$

EXAMPLE 4 Using the matrix in Example 3, compute its determinant, but with a different sequence of operations. Start by using Theorem 1.

SOLUTION

$$\mathrm{Det} \begin{bmatrix} 1 & 3 & 4 \\ 2 & 1 & 2 \\ -4 & 2 & 1 \end{bmatrix} = -\mathrm{Det} \begin{bmatrix} 2 & 1 & 2 \\ 1 & 3 & 4 \\ -4 & 2 & 1 \end{bmatrix} = -\mathrm{Det} \begin{bmatrix} 2 & 1 & 2 \\ 0 & \frac{5}{2} & 3 \\ -4 & 2 & 1 \end{bmatrix}$$

$$= -\mathrm{Det} \begin{bmatrix} 2 & 1 & 2 \\ 0 & \frac{5}{2} & 3 \\ 0 & 4 & 5 \end{bmatrix} = -\mathrm{Det} \begin{bmatrix} 2 & 1 & 2 \\ 0 & \frac{5}{2} & 3 \\ 0 & 0 & \frac{1}{5} \end{bmatrix} = -1 \quad \blacksquare$$

In theory, we could establish formulas for determinants of arbitrary square matrices having any size: 1×1, 2×2, 3×3, and so on. The formulas quickly become unmanageable, however. If a matrix is 1×1, then it is triangular, and $\mathrm{Det}[a] = a$ by Property **I**. The next case is the general 2×2 matrix. To get a formula for its determinant, let $\mathbf{A} = \begin{bmatrix} a & b \\ c & d \end{bmatrix}$. If $c = 0$, then the matrix is already upper triangular, and $\text{Det}(\mathbf{A}) = ad = ad - bc$. If $c \neq 0$, then we get the same result by first performing a row swap followed by another row operation:

$$\text{Det}(\mathbf{A}) = \text{Det} \begin{bmatrix} a & b \\ c & d \end{bmatrix} = -\text{Det} \begin{bmatrix} c & d \\ a & b \end{bmatrix} = -\text{Det} \begin{bmatrix} c & d \\ 0 & b - (a/c)d \end{bmatrix}$$

$$= -c[b - (ad)/c] = ad - bc$$

Thus, in both cases, we get $ad - bc$. This formula is worth remembering, since in many calculations involving determinants the 2×2 case is needed.

THEOREM 3

The determinant of a 2×2 matrix follows this rule:

$$\text{Det} \begin{bmatrix} a & b \\ c & d \end{bmatrix} = ad - bc$$

EXAMPLE 5 Calculate (as efficiently as possible) the determinant of this matrix:

$$\begin{bmatrix} 0 & 3 & 5 \\ 4 & 6 & -2 \\ 10 & 2 & 7 \end{bmatrix}$$

SOLUTION One way to do it is shown here:

$$\text{Det} \begin{bmatrix} 0 & 3 & 5 \\ 4 & 6 & -2 \\ 10 & 2 & 7 \end{bmatrix} = -\text{Det} \begin{bmatrix} 4 & 6 & -2 \\ 0 & 3 & 5 \\ 10 & 2 & 7 \end{bmatrix}$$

$$= -2\,\text{Det} \begin{bmatrix} 2 & 3 & -1 \\ 0 & 3 & 5 \\ 10 & 2 & 7 \end{bmatrix} = -2\,\text{Det} \begin{bmatrix} 2 & 3 & -1 \\ 0 & 3 & 5 \\ 0 & -13 & 12 \end{bmatrix}$$

$$= -2\,\text{Det} \begin{bmatrix} 2 & 3 & -1 \\ 0 & 3 & 5 \\ 0 & -1 & 32 \end{bmatrix} = 2\,\text{Det} \begin{bmatrix} 2 & 3 & -1 \\ 0 & -1 & 32 \\ 0 & 0 & 101 \end{bmatrix}$$

$$= (2)(2)(-1)(101) = -404$$

Here we have performed two swaps and some extra steps to avoid fractions.

∎

Algorithm without Scaling

Here is a slightly different algorithm for computing the determinant of any square matrix **A**. Use the row-reduction process without scaling and keep track of the number of row interchanges. To reduce **A** to upper triangular form, $\widetilde{\mathbf{A}}$, we use Properties **I** and **II** and Theorem 1, but *not* Property **III**. Let k be the number of row swaps employed in this process. Then the determinant of the given matrix is $(-1)^k$ times the product of all the diagonal elements in $\widetilde{\mathbf{A}}$.

EXAMPLE 6 Use the matrix in Example 5 to illustrate this more concise algorithm.

SOLUTION

$$
\begin{bmatrix} 0 & 3 & 5 \\ 4 & 6 & -2 \\ 10 & 2 & 7 \end{bmatrix} \sim \begin{bmatrix} 10 & 2 & 7 \\ 0 & 3 & 5 \\ 4 & 6 & -2 \end{bmatrix} \sim \begin{bmatrix} 2 & -10 & 11 \\ 0 & 3 & 5 \\ 4 & 6 & -2 \end{bmatrix}
$$

$$
\sim \begin{bmatrix} 2 & -10 & 11 \\ 0 & 3 & 5 \\ 0 & 26 & -24 \end{bmatrix} \sim \begin{bmatrix} 2 & -10 & 11 \\ 0 & 3 & 5 \\ 0 & 2 & -64 \end{bmatrix} \sim \begin{bmatrix} 2 & -10 & 11 \\ 0 & 1 & 69 \\ 0 & 2 & -64 \end{bmatrix}
$$

$$
\sim \begin{bmatrix} 2 & -10 & 11 \\ 0 & 1 & 69 \\ 0 & 0 & -202 \end{bmatrix}
$$

Because two swaps were employed, the determinant is found to be

$$
\text{Det} \begin{bmatrix} 0 & 3 & 5 \\ 4 & 6 & -2 \\ 10 & 2 & 7 \end{bmatrix} = (-1)^2(2)(1)(-202) = -404 \qquad \blacksquare
$$

EXAMPLE 7 Carry out the row-reduction algorithm on the following matrix and compute its determinant. Is there any connection between these two tasks?

$$
\mathbf{A} = \begin{bmatrix} 2 & 3 & 0 & 2 \\ 2 & 6 & -1 & 5 \\ 4 & 12 & -3 & 7 \\ 2 & 3 & -1 & -1 \end{bmatrix}
$$

SOLUTION A row reduction leads to these equivalent forms:

$$
\begin{bmatrix}
2 & 3 & 0 & 2 \\
2 & 6 & -1 & 5 \\
4 & 12 & -3 & 7 \\
2 & 3 & -1 & -1
\end{bmatrix}
\sim
\begin{bmatrix}
2 & 3 & 0 & 2 \\
0 & 3 & -1 & 3 \\
0 & 6 & -3 & 3 \\
0 & 0 & -1 & -3
\end{bmatrix}
$$

$$
\sim
\begin{bmatrix}
2 & 3 & 0 & 2 \\
0 & 3 & -1 & -3 \\
0 & 0 & -1 & -3 \\
0 & 0 & -1 & -3
\end{bmatrix}
\sim
\begin{bmatrix}
2 & 3 & 0 & 2 \\
0 & 3 & -1 & -3 \\
0 & 0 & -1 & -3 \\
0 & 0 & 0 & 0
\end{bmatrix}
$$

We obtain $\text{Det}(\mathbf{A}) = 0$.

The presence of a zero row shows that the original set of rows is linearly dependent. Here, Theorem 13 in Section 1.3 is helpful: If there is a zero row in the row echelon form of a matrix, then the rows form a dependent set. The zero row also indicates that the determinant of $\mathbf{A}$ is zero. These facts are intimately related, and the theorem asserting this is next. ∎

Zero Determinant

In Section 4.2 (Theorem 3), it is proved that the determinant of a matrix equals the determinant of its transpose. Consequently, if the rows of $\mathbf{A}$ constitute a linearly dependent set, then the same is true of the columns because

$$
\text{Det}(\mathbf{A}) = \text{Det}(\mathbf{A}^T) = 0
$$

THEOREM 4

The determinant of a matrix is 0 if and only if the rows (or columns) of the matrix form a linearly dependent set.

PROOF Assume first that the rows of the $n \times n$ matrix $\mathbf{A}$ form a linearly dependent set. Then a nontrivial equation exists having the form

$$
c_1 \mathbf{r}_1 + c_2 \mathbf{r}_2 + \cdots + c_n \mathbf{r}_n = \mathbf{0}
$$

In this equation $\mathbf{r}_i$ are the row vectors in $\mathbf{A}$. Because this equation is nontrivial, there is at least one nonzero coefficient. For convenience, suppose $c_1 \neq 0$. Write

$$
\mathbf{r}_1 + \left(\frac{c_2}{c_1}\right) \mathbf{r}_2 + \left(\frac{c_3}{c_1}\right) \mathbf{r}_3 + \cdots + \left(\frac{c_n}{c_1}\right) \mathbf{r}_n = \mathbf{0}
$$

In words, this means that we can add multiples of rows $2, 3, 4, \ldots, n$ onto row 1 and get a zero row. Our algorithm for computing $\text{Det}(\mathbf{A})$ produces 0, because there is a zero on the diagonal in the row echelon form of $\mathbf{A}$.

For the other half of the proof, assume that $\text{Det}(\mathbf{A}) = 0$. Then there must be a zero on the diagonal of the upper-triangular form of $\mathbf{A}$. This zero informs us that the number of pivots is at most $n - 1$. Hence, there is a zero row in the row echelon form of $\mathbf{A}$. This zero row is the result of forming linear combinations of the rows of $\mathbf{A}$, and, consequently, there is a nontrivial linear combination of the rows that is $\mathbf{0}$. In other words, the set of rows is linearly dependent. ■

From Theorem 4, the rows (or columns) of $\mathbf{A}$ form a linearly independent set if and only if $\text{Det}(\mathbf{A}) \neq 0$. Theorem 4 holds for rows or columns because $\text{Det}(\mathbf{A}) = \text{Det}(\mathbf{A}^T)$.

The next theorem turns out to be crucial in the theory of eigenvalues of a matrix, a topic taken up in Section 6.1.

THEOREM 5

A square matrix is invertible (nonsingular) if and only if its determinant is nonzero.

PROOF A matrix $\mathbf{A}$ is invertible (nonsingular) if and only if its reduced echelon form is the identity matrix. This, in turn, is equivalent to the assertion that the rows (or columns) of $\mathbf{A}$ constitute a linearly independent set. By Theorem 4, this is equivalent to the assertion that $\text{Det}(\mathbf{A}) \neq 0$. ■

By Theorem 5, $\text{Det}(\mathbf{A}) = 0$ if and only if $\mathbf{A}$ is noninvertible (singular).

EXAMPLE 8 Use the determinant criterion to discover whether the columns of this matrix form a linearly independent set. Also, compute $\text{Det}(\mathbf{A}^T)$.

$$\mathbf{A} = \begin{bmatrix} 1 & -3 & 2 \\ 4 & 7 & 5 \\ 6 & 1 & 10 \end{bmatrix}$$

SOLUTION The determinant of **A** can be computed from the upper-triangular echelon form that results from the row-reduction process. Two full steps reveal the answer:

$$\mathbf{A} = \begin{bmatrix} 1 & -3 & 2 \\ 4 & 7 & 5 \\ 6 & 1 & 10 \end{bmatrix} \sim \begin{bmatrix} 1 & -3 & 2 \\ 0 & 19 & -3 \\ 0 & 19 & -2 \end{bmatrix} \sim \begin{bmatrix} 1 & -3 & 2 \\ 0 & 19 & -3 \\ 0 & 0 & 1 \end{bmatrix}$$

The determinant is 19, and the matrix is invertible by Theorem 5. Because $\text{Det}(\mathbf{A}) \neq 0$, the set of columns in **A** is linearly independent. As an alternative, we can compute the determinant from the upper-triangular echelon form that results from the row-reduction process:

$$\mathbf{A}^T = \begin{bmatrix} 1 & 4 & 6 \\ -3 & 7 & 1 \\ 2 & 5 & 10 \end{bmatrix} \sim \begin{bmatrix} 1 & 4 & 6 \\ 0 & 19 & 19 \\ 0 & -3 & -2 \end{bmatrix} \sim \begin{bmatrix} 1 & 4 & 6 \\ 0 & 19 & 19 \\ 0 & 0 & 1 \end{bmatrix}$$

Again, the determinant is nonzero, and, by Theorem 4, the rows of this matrix form a linearly independent set. ∎

EXAMPLE 9 Are the matrices in Examples 6 and 7 invertible or noninvertible?

SOLUTION The matrix in Example 6 is *invertible* (nonsingular) because its determinant is nonzero. The matrix in Example 7 is *noninvertible* (singular) because its determinant is 0. ∎

EXAMPLE 10 Determine whether this matrix is invertible:

$$\mathbf{A} = \begin{bmatrix} 2 & 9 & 3 & 5 \\ 8 & 16 & 5 & 5 \\ 7 & 9 & 3 & 5 \\ 17 & 34 & 11 & 15 \end{bmatrix}$$

SOLUTION We happen to notice that row 4 is the sum of rows 1, 2, and 3. Thus, if we subtract rows 1, 2, and 3 from row 4, the result is that row 4 becomes a zero row. Hence, the determinant will be zero, and the matrix is noninvertible by Theorem 5 (or by various other theorems). ∎

Calculating Areas and Volumes

An application of determinants occurs in calculating the area of a triangle in $\mathbb{R}^2$. (The theory extends to volumes of three-dimensional figures and similar applications in still higher dimensions.) If the triangle has one of its vertices at **0**, then we label the other two vertices as **u** and **v**. The notation $\Delta(\mathbf{0}, \mathbf{u}, \mathbf{v})$ specifies a triangle having vertices at **0**, **u**, and **v**.

THEOREM 6

In $\mathbb{R}^2$, if a triangle has vertices **0**, **u**, and **v**, then the area of the triangle is one-half of the absolute value of the 2×2 determinant having rows (or columns) **u** and **v**:

$$\text{Area}[\Delta(\mathbf{0}, \mathbf{u}, \mathbf{v})] = \tfrac{1}{2} \left| \text{Det} \begin{bmatrix} u_1 & u_2 \\ v_1 & v_2 \end{bmatrix} \right|$$

PROOF In general, a triangle whose vertices in $\mathbb{R}^2$ are **u**, **v**, and **w** will be denoted by $\Delta(\mathbf{u}, \mathbf{v}, \mathbf{w})$. In the present situation, we note first that $\Delta(\mathbf{0}, \mathbf{u}, \mathbf{v})$ has the same area as $\Delta(\mathbf{0}, \mathbf{u}, \mathbf{v} + \lambda \mathbf{u})$, for any scalar λ. This is true because the two triangles have the same base and the same altitude (height). (See Figure 4.1 to convince yourself of this.) Recall that the area of a triangle is one-half the base, b, times the altitude (height), h, or $\text{Area}[\Delta(\mathbf{u}, \mathbf{v}, \mathbf{w})] = \tfrac{1}{2}bh$. Now apply, to the matrix in the statement of the theorem, the row operations of interchanging rows and adding a multiple of one row to another, ending with a diagonal matrix. For example, if $u_1 \neq 0$ we can write

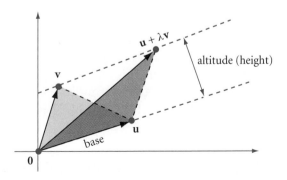

FIGURE 4.1 Triangles having the same base and altitude (height).

$$\text{Det}\begin{bmatrix} u_1 & u_2 \\ v_1 & v_2 \end{bmatrix} = \text{Det}\begin{bmatrix} u_1 & u_2 \\ 0 & v_2' \end{bmatrix} = \text{Det}\begin{bmatrix} u_1 & 0 \\ 0 & v_2' \end{bmatrix}$$

$$= u_1 v_2' = \pm 2\,\text{Area}[\Delta(\mathbf{0}, \mathbf{u}', \mathbf{v}')]$$

$$= \pm 2\,\text{Area}[\Delta(\mathbf{0}, \mathbf{u}, \mathbf{v})]$$

Here we have used Property **II** twice and Property **I** once. Furthermore, we have

$$v_2' = [v_2 - (v_1/u_1)u_2], \qquad \mathbf{v}' = (0, v_2'), \qquad \mathbf{u}' = (u_1, 0)$$

Figures 4.2 to 4.4 show the details of moving the vertices without changing the areas of the triangles. In going from Figure 4.2 to Figure 4.3, the vector **v** moves along a line parallel to **u** into a new position **v′** on the vertical axis. In going from Figure 4.3 to Figure 4.4, **u** moves on a line parallel to **v′** into a new position **u′** on the horizontal axis. These motions do not affect the area of the triangle. (Explain why.) As shown in Figure 4.4, $\Delta(\mathbf{0}, \mathbf{u}', \mathbf{v}')$ is a right triangle because one vertex $\mathbf{u}' = (u_1, 0)$ is on the horizontal axis and another $\mathbf{v}' = (0, v_2')$ is on the vertical axis. So the area of each of these triangles is $\frac{1}{2}u_1 v_2' = \frac{1}{2}u_1[v_2 - (v_1/u_1)u_2]$. ∎

EXAMPLE 11 What is the area of the triangle whose vertices are $(0, 0)$, $(2, 11)$, and $(8, 3)$? Find the answer using determinants and then obtain the answer by drawing a sketch and using right triangles.

SOLUTION The area is one-half the absolute value of the determinant having rows $(2, 11)$ and $(8, 3)$:

$$\frac{1}{2}\left|\text{Det}\begin{bmatrix} 2 & 11 \\ 8 & 3 \end{bmatrix}\right| = \frac{1}{2}|6 - 88| = 41$$

Hence, the area sought is 41. To compute this area in another way, we draw a rectangle (with sides parallel to the coordinate axes) that contains the triangle in question. This rectangle will have area $(8)(11) = 88$. Next we remove three right triangles to leave only the triangle of interest. These triangles have areas 11, 12, and 24. Their total is 47, and $88 - 47 = 41$. See Figure 4.5. ∎

EXAMPLE 12 What is the area of the triangle whose vertices are $(4, 7)$, $(-2, 11)$, and $(12, -6)$?

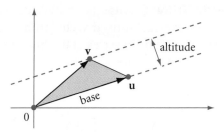

FIGURE 4.2 Step 0: Triangles having the same base and altitude.

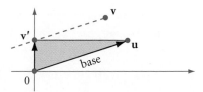

FIGURE 4.3 Step 1: Move **v**, parallel to **u**, into the new position $\mathbf{v}' = (0, v_2)$.

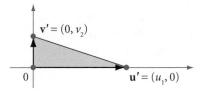

FIGURE 4.4 Step 2: Move **u** vertically into the new position $\mathbf{u}' = (u_1, 0)$.

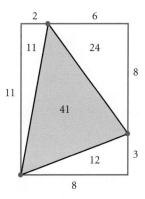

FIGURE 4.5 Triangle in a rectangle.

SOLUTION Because the triangle does not have a vertex at the origin, we simply subtract a fixed vector from each point of the object so that the new triangle has a vertex at **0**. Thus, with a rigid motion, we can move this triangle so that one of its vertices is $(0, 0)$. For example, subtract $(4, 7)$ from each vertex, getting new vertices $(0, 0)$, $(-6, 4)$, and $(8, -13)$. Then use the determinant formula to obtain

$$\frac{1}{2}\left|\text{Det}\begin{bmatrix} -6 & 4 \\ 8 & -13 \end{bmatrix}\right| = \frac{1}{2}(78 - 32) = 23$$ ∎

THEOREM 7

The area of a parallelogram generated by the three points **0**, **u**, *and* **v** *in* $\mathbb{R}^2$ *is the absolute value of the determinant having rows (or columns)* **u** *and* **v**:

$$\text{Area}(\mathbf{0}, \mathbf{u}, \mathbf{v}, \mathbf{u} + \mathbf{v}) = \left|\text{Det}\begin{bmatrix} u_1 & u_2 \\ v_1 & v_2 \end{bmatrix}\right|$$

PROOF This follows from Theorem 6, because the parallelogram can be thought of as consisting of two congruent triangles.   ∎

In Section 4.2 (Theorem 3), we learn that a matrix and its transpose have the same determinant. Consequently, in Theorems 6 and 7, the areas of a triangle or parallelogram are unchanged by using column vectors instead of row vectors.

THEOREM 8

Let S be a triangle or parallelogram in $\mathbb{R}^2$ *having a vertex at* **0**. *Let T be a linear mapping from* $\mathbb{R}^2$ *to* $\mathbb{R}^2$, *given by matrix* **A**, *so that* $T(\mathbf{x}) = \mathbf{Ax}$. *Then*

$$\text{Area}[T(S)] = |\text{Det}(\mathbf{A})| \cdot \text{Area}(S)$$

PROOF If S is a parallelogram having vertices **0**, **u**, **v**, and **u** + **v**, then by Theorem 7, the area of S is $|\text{Det}[\mathbf{u}, \mathbf{v}]|$. The parallelogram $T(S)$ has vertices **0**, $T(\mathbf{u})$, $T(\mathbf{v})$, and $T(\mathbf{u} + \mathbf{v})$, or, in other terms, **0**, **Au**, **Av**, and **Au** + **Av**. Its area is

$$\text{Area}(\mathbf{0}, \mathbf{Au}, \mathbf{Av}, \mathbf{Au} + \mathbf{Av}) = |\text{Det}[\mathbf{Au}, \mathbf{Av}]| = |\text{Det}(\mathbf{A}[\mathbf{u}, \mathbf{v}])|$$

$$= |\text{Det}(\mathbf{A})| \cdot |\text{Det}[\mathbf{u}, \mathbf{v}]| = |\text{Det}(\mathbf{A})| \cdot \text{Area}(S)$$ ∎

EXAMPLE 13 Let $T(\mathbf{x}) = \mathbf{Ax}$ where $\mathbf{A} = \begin{bmatrix} 1 & 3 \\ 2 & 4 \end{bmatrix}$. Let the vertices of a parallelogram S be $(0,0)$, $(2,5)$, $(7,3)$, and $(9,8)$. What is the area of $T(S)$?

SOLUTION By Theorem 8, we have

$$\text{Area}[T(S)] = |\,\text{Det}(\mathbf{A})| \cdot \text{Area}(S) = \left| \text{Det} \begin{bmatrix} 1 & 3 \\ 2 & 4 \end{bmatrix} \right| \cdot \left| \text{Det} \begin{bmatrix} 2 & 7 \\ 5 & 3 \end{bmatrix} \right|$$

$$= |-2| \cdot |-29| = 58$$

Alternatively, we see that the new parallelogram has vertices

$$\mathbf{A} \begin{bmatrix} 0 \\ 0 \end{bmatrix} = \begin{bmatrix} 0 \\ 0 \end{bmatrix}, \quad \mathbf{A} \begin{bmatrix} 2 \\ 5 \end{bmatrix} = \begin{bmatrix} 17 \\ 24 \end{bmatrix}, \quad \mathbf{A} \begin{bmatrix} 7 \\ 3 \end{bmatrix} = \begin{bmatrix} 16 \\ 26 \end{bmatrix}$$

Thus, the area of the new parallelogram is

$$\left| \text{Det} \begin{bmatrix} 17 & 16 \\ 24 & 26 \end{bmatrix} \right| = 58 \qquad \blacksquare$$

Mathematical Software

We can use mathematical software to evaluate determinants. For example, we can compute the determinant of a typical 4×4 matrix by using these commands in MATLAB, Maple, or Mathematica:

```
MATLAB
A = [1,3,7,2;4,-2,1,5;3,1,2,0;1,9,15,-1]
det(A)
```

```
Maple
with(LinearAlgebra):
A : = Matrix([[1,3,7,2],[4,-2,1,5],[3,1,2,0],
        [1,9,15,-1]]);
Determinant(A);
```

```
Mathematica
A = {{1,3,7,2},{4,-2,1,5},{3,1,2,0},{1,9,15,-1}}
Det[A]
```

SUMMARY 4.1

- $\text{Det}(\mathbf{U}) = u_{11}u_{22}\cdots u_{nn}$ if $\mathbf{U}$ is an $n \times n$ triangular matrix.

- $\text{Det}(\mathbf{A}) = \text{Det}(\widetilde{\mathbf{A}})$ if $\widetilde{\mathbf{A}}$ is obtained from $\mathbf{A}$ by adding a multiple of one row onto another row.

- $\text{Det}(\mathbf{A}) = (1/c)\,\text{Det}(\widetilde{\mathbf{A}})$ if $\widetilde{\mathbf{A}}$ is obtained from $\mathbf{A}$ by multiplying a row of $\mathbf{A}$ by a constant c.

- $\text{Det}(\mathbf{A}) = -\,\text{Det}(\widetilde{\mathbf{A}})$ if $\widetilde{\mathbf{A}}$ is obtained by swapping a pair of rows in $\mathbf{A}$.

- There is one and only one determinant function satisfying the rules **I**, **II**, and **III**.

- $\text{Det}\begin{bmatrix} a & b \\ c & d \end{bmatrix} = ad - bc$

- $\text{Det}(\mathbf{A}) = 0$ if and only if $\mathbf{A}$ is singular. (Equivalently, the rows of $\mathbf{A}$ form a linearly dependent set.)

- The matrix $\mathbf{A}$ is noninvertible (singular) if and only if $\text{Det}(\mathbf{A}) = 0$.

- The area of the triangle in $\mathbb{R}^2$ having vertices $\mathbf{0} = (0,0)$, $\mathbf{u} = (u_1, u_2)$, and $\mathbf{v} = (v_1, v_2)$ is

$$\frac{1}{2}\left|\text{Det}\begin{bmatrix} u_1 & u_2 \\ v_1 & v_2 \end{bmatrix}\right| = \frac{1}{2}\left|\text{Det}\begin{bmatrix} u_1 & v_1 \\ u_2 & v_2 \end{bmatrix}\right|$$

- The area of a parallelogram in $\mathbb{R}^2$ generated by points $\mathbf{0} = (0,0)$, $\mathbf{u} = (u_1, u_2)$, and $\mathbf{v} = (v_1, v_2)$ is

$$\left|\text{Det}\begin{bmatrix} u_1 & u_2 \\ v_1 & v_2 \end{bmatrix}\right| \text{ or } \left|\text{Det}\begin{bmatrix} u_1 & v_1 \\ u_2 & v_2 \end{bmatrix}\right|$$

- If S is a triangle or parallelogram in $\mathbb{R}^2$ with a vertex at $\mathbf{0}$ and if T is a linear transformation $T : \mathbb{R}^2 \mapsto \mathbb{R}^2$ given by $T(\mathbf{x}) = \mathbf{A}\mathbf{x}$, then $\text{Area}[T(S)] = |\,\text{Det}(\mathbf{A})| \cdot \text{Area}(S)$.

KEY CONCEPTS 4.1

Determinant of a matrix, computing determinants using standard row operations (replacement, scale, swap), determinant of triangular matrices, formula for the determinant of a 2×2 matrix, a matrix whose columns form a linearly dependent set has determinant 0, every noninvertible (singular) matrix has determinant 0, computing determinants without scaling, using row operations to evaluate determinants, calculating areas of triangles and parallelograms

GENERAL EXERCISES 4.1

1. Calculate the determinants of these matrices:

  a. $\begin{bmatrix} 1 & 3 & 7 \\ 2 & 8 & 19 \\ 3 & 9 & 27 \end{bmatrix}$ **b.** $\begin{bmatrix} 1 & 2 & 3 \\ 3 & 8 & 9 \\ 7 & 19 & 27 \end{bmatrix}$

  c. $\begin{bmatrix} 4 & 2 & 2 \\ 2 & 6 & 1 \\ 1 & 2 & 2 \end{bmatrix}$ **d.** $\begin{bmatrix} 1 & 3 & 0 & 2 \\ 1 & -1 & 2 & -3 \\ 2 & 5 & -7 & -4 \\ 3 & 5 & 2 & 1 \end{bmatrix}$

2. Let $\begin{cases} 3\beta x_1 + 7x_2 = 5 \\ 4x_1 + 2\beta x_2 = -3 \end{cases}$

For what values of the parameter β will the system have a unique solution?

3. Find the area of a triangle whose vertices are $(-1, 3)$, $(-4, -4)$, and $(5, -2)$.

4. Consider $A = \begin{bmatrix} 3 & 7 & 2 \\ -2 & 4 & 1 \\ 6 & -2 & 4 \end{bmatrix}$

Determine whether **A** is invertible or non-invertible by applying the determinant test.

5. Let $A = \begin{bmatrix} 2 & 3 & -3 \\ -4 & 1 & 6 \\ 0 & 7 & 0 \end{bmatrix}$

Determine whether the rows of **A** form a linearly independent set by using the determinant criterion.

6. Verify that these four points are coplanar by showing that the parallelepiped having these vertices has volume 0: $(0,0,0)$, $(3,2,1), (4,-1,2)$, and $(6,-7,4)$.

7. Compute Det $\begin{bmatrix} -2 & 1 & 2 \\ 2 & 0 & 4 \\ 3 & 1 & 0 \end{bmatrix}$

8. Use the determinantal criterion for non-invertibility (singularity) to find all the values of t for which the matrix $\begin{bmatrix} 1 & 3 \\ 2 & 7 \end{bmatrix} + t \begin{bmatrix} 1 & 1 \\ 1 & 1 \end{bmatrix}$ is noninvertible (singular).

9. Find the value of x so that

Det $\begin{bmatrix} x & -1 \\ x+1 & x-2 \end{bmatrix} = 7$

10. Let $A = \begin{bmatrix} 3 & 2 & -1 & -1 \\ 2 & 1 & 1 & 2 \\ -1 & 3 & 2 & -9 \\ 4 & -5 & 2 & 25 \end{bmatrix}$

Determine whether the rows of **A** form a linearly dependent set. If so, find the coefficients involved in a linear combination of rows that is zero. Use the techniques in Section 1.3.

11. Let $A = \begin{bmatrix} 2 & 20 & 4 \\ 1 & 9 & 1 \\ -3 & 2 & 1 \end{bmatrix}$

Using Properties **I, II**, and **III** and Theorem 1, calculate the determinant of **A**.

12. Let $A = \begin{bmatrix} 2 & 5 & 4 & 1 \\ 6 & -2 & -4 & 0 \\ 4 & 7 & 6 & 2 \\ 6 & -7 & -7 & 0 \end{bmatrix}$

Compute the determinant of **A** by using the row replacement operations only (*no* scaling or swapping).

13. Let $A = \begin{bmatrix} 3 & 1 & 1 & 5 \\ 1 & 4 & 2 & 1 \\ 2 & -1 & 3 & 6 \\ 2 & 3 & 5 & 2 \end{bmatrix}$

Compute the determinant of **A** by using row operations.

14. Let $A = \begin{bmatrix} 0 & 2 & 5 \\ 3 & 5 & 4 \\ 2 & 4 & 6 \end{bmatrix}$

Compute the determinant of **A** by row operations leading to an upper triangular matrix.

15. Consider $t \begin{bmatrix} 12 & 12 \\ 12 & 12 \end{bmatrix} + \begin{bmatrix} 1 & 3 \\ 2 & 7 \end{bmatrix}$

Use the determinantal criterion for non-invertibility (singularity) to find all the values of t for which this matrix is non-invertible (singular).

16. Compute the determinant of $\begin{bmatrix} 1 & 2 & -1 \\ 3 & 0 & 1 \\ 4 & 2 & 1 \end{bmatrix}$

17. Find the area of a parallelogram whose vertices are $(0,0), (3,7), (4,-1)$, and $(7,6)$.

18. Compute Det $\begin{bmatrix} 2 & 20 & 3 \\ 1 & 9 & 1 \\ -3 & 2 & 1 \end{bmatrix}$

19. Calculate Det $\begin{bmatrix} 3 & 2 & 4 \\ 3 & 3 & 9 \\ 6 & 7 & 16 \end{bmatrix}$ using two different sets of elementary operations.

20. Using determinants, find the area of the triangle and the parallelogram formed by the origin and these vectors:

  a. $\mathbf{x} = (5, 3), \mathbf{y} = (4, 7)$

  b. $\mathbf{x} = (-4, 3), \mathbf{y} = (-2, 6)$

21. We shall see later that a determinant is a linear function of any one of its rows. Consider a 2×2 matrix $\begin{bmatrix} a & b \\ c & d \end{bmatrix}$ and prove that these are linear functions:

$$f(\mathbf{x}) = \text{Det} \begin{bmatrix} a & b \\ x_1 & x_2 \end{bmatrix} \quad g(x) = \text{Det} \begin{bmatrix} x_1 & x_2 \\ c & d \end{bmatrix}$$

22. We shall establish later that the formula $\text{Det}(\mathbf{AB}) = \text{Det}(\mathbf{A})\,\text{Det}(\mathbf{B})$ is universally true. Explain it for the 2×2 case.

23. If the entries of a matrix are integers, does it follow that the determinant is an integer? A proof or an example is needed. Answer the analogous question for rational numbers.

24. The formula for the area of a parallelogram implies the formula for the area of a triangle, and vice versa. Another way to arrive at the area formula for a triangle is discussed in this problem. First draw a figure for a typical case as guidance. For example, plot these points in the plane:

$\mathbf{o} = (0, 0), \mathbf{a} = (3, 5), \mathbf{b} = (6, 2), \mathbf{a}' = (3, 0)$, and $\mathbf{b}' = (6, 0)$. Let us use the traditional method of describing polygons by listing their vertices, in order. We wish to calculate the area of the triangle **oab** (see Dörrie H. [1940]). We observe that the area of the quadrilateral **oabb**$'$ is the sum of the triangular area **oaa**$'$ and the trapezoid **aa**$'$**b**$'$**b**. Each of these two areas just mentioned is easily calculated because the altitudes are measured on horizontal lines. The area we want is the area of **oabb**$'$ minus the area of the triangle **obb**$'$. Carry out the calculation that leads to

$$\pm \tfrac{1}{2} \text{Det} \begin{bmatrix} a_1 & a_2 \\ b_1 & b_2 \end{bmatrix}$$

Here $\mathbf{a} = (a_1, a_2)$ and $\mathbf{b} = (b_1, b_2)$.

25. Explain why if a parallelogram in the plane has integers for all the coordinates of its vertices, then its area is an integer.

26. Is there an equilateral triangle in the plane all of whose vertices have integer coordinates? If you can, answer the same question for vertices required to have rational coordinates.

27. A parallelogram has vertices $(0, 0), (u_1, u_2)$, (v_1, v_2), and $(u_1 + v_1, u_2 + v_2)$. What is its area?

28. Derive the formula for the area of a triangle by moving $\mathbf{u}$ parallel to $\mathbf{v}$ to a new position on the horizontal axis. Then move $\mathbf{v}$ horizontally to a position on the vertical axis.

29. In $\mathbb{R}^3$, the analog of a parallelogram in $\mathbb{R}^2$ is a **parallelepiped**. A cube is one example of this. A parallelepiped can be described by giving four vertices. A cube, for example, can be described by specifying these vertices: $(0,0,0), (1,0,0), (0,1,0),$ and $(0,0,1)$. Let a parallelepiped be described by giving these vertices: $\mathbf{0}, \mathbf{u}, \mathbf{v},$ and $\mathbf{w}$. We accept without proof the fact that the volume of that figure is $|\operatorname{Det}(\mathbf{u},\mathbf{v},\mathbf{w})|$. (The formula is certainly correct if $\mathbf{u} = (a,0,0)$, $\mathbf{v} = (0,b,0)$, and $\mathbf{w} = (0,0,c)$.) Find the volume of the parallelepiped having vertices $(0,0,0), (2,3,4), (1,-1,2),$ and $(3,0,5)$.

30. (Continuation.) Use the information about parallelepipeds in the preceding problem. Find the volume of the parallelepiped whose vertices are $(1,2,-3), (2,2,1), (2,3,5),$ and $(3,0,2)$.

31. Let $\mathbf{u} = (u_1,u_2)$, $\mathbf{v} = (v_1,v_2)$, $\mathbf{p} = (u_1,v_1)$, and $\mathbf{q} = (u_2,v_2)$. Do the triangles $\Delta(\mathbf{0},\mathbf{u},\mathbf{v})$ and $\Delta(\mathbf{0},\mathbf{p},\mathbf{q})$ have the same area? (Verify or give a counterexample.) Draw the triangles involved here in a concrete case, such as $\mathbf{u} = (5,1)$ and $\mathbf{v} = (4,3)$, and compute the two areas in question.

32. Refer to the proof of Theorem 6. Establish that the description of $\mathbf{v}'$ is correct if we write it as $\mathbf{v} + t\mathbf{u}$ where $t = -v_1/u_1$.

33. (Continuation.) Use the same strategy to get a formula for the area, but begin by moving $\mathbf{u}$ parallel to $\mathbf{v}$ into a position on the first coordinate axis.

34. Explain why a 2×2 matrix $\mathbf{A}$ has this property: $\operatorname{Det}(\alpha\mathbf{A}) = \alpha^2 \operatorname{Det}(\mathbf{A})$ for α, a scalar. Then prove the corresponding result for $n \times n$ matrices.

35. Critique this proof that every nonsquare matrix is invertible: Let $\mathbf{A}$ be a nonsquare matrix. Then $\mathbf{A}$ does not have a determinant. Hence, its determinant is certainly *not* zero. Now apply the theorem that asserts that a matrix is invertible if and only if its determinant is not 0.

36. Explain why $\pm\frac{1}{2}\operatorname{Det}\begin{bmatrix} u_1 & u_2 & 1 \\ v_1 & v_2 & 1 \\ w_1 & w_2 & 1 \end{bmatrix}$

is the area of a triangle whose vertices are three points $\mathbf{u},\mathbf{v},\mathbf{w}$ in $\mathbb{R}^2$. (*Hint:* Move the triangle rigidly so that one vertex becomes the origin, and then apply the formula already established.)

37. Establish that $\operatorname{Det}\begin{bmatrix} x_1 & x_2 & 1 \\ v_1 & v_2 & 1 \\ w_1 & w_2 & 1 \end{bmatrix} = 0$

is the equation of a line in $\mathbb{R}^2$ passing through two distinct points $\mathbf{v}$ and $\mathbf{w}$.

38. What happens in Theorem 6 when vectors $\mathbf{u}$ and $\mathbf{v}$ are multiples of each other? What happens when $u_1 = 0$?

39. Establish that if a 2×2 matrix is invertible (nonsingular), then by changing one entry in the matrix we can make it noninvertible (singular).

40. Explore another approach to determinants that uses these axioms:

I. $\operatorname{Det}(\mathbf{A})$ depends linearly on row 1.

II. $\operatorname{Det}(\mathbf{A})$ changes sign when two rows are swapped.

III. $\operatorname{Det}(\mathbf{I}) = 1$.

See Strang [2003].

41. Derive the formula for the determinant of a 2×2 matrix based on assuming $a = 0$ and $a \neq 0$.

42. Explain why the proof of Theorem 8 is correct when S is a triangle.

43. Consider $\begin{bmatrix} 2\beta & 8 \\ 3 & 4\beta \end{bmatrix}$

Find all values of the parameter β for which the matrix is invertible.

44. Find the area of the triangle whose vertices are given:

a. $(0,0), (-2,2), (4,-1)$

b. $(1,-1), (5,-5), (7,-3)$

45. Find the values of the unknowns when:

a. Det $\begin{bmatrix} 4 & 2 & 0 \\ 1 & 3 & 0 \\ x & 1 & x \end{bmatrix} = 10$

b. Det $\begin{bmatrix} 1 & 1 & 1 \\ 1 & 2 & 6 \\ 1 & t & t^2 \end{bmatrix} = -2$

46. Show that

a. Det $\begin{bmatrix} 1 & a & a^2 \\ 1 & b & b^2 \\ 1 & c & c^2 \end{bmatrix}$

$= (a - b)(a - c)(c - b)$

b. Det $\begin{bmatrix} 1 & x & y \\ 1 & x_0 & y_0 \\ 0 & 1 & m \end{bmatrix} = 0$

is the equation of the line in $\mathbb{R}$ through point (x_0, y_0) with slope m.

c. Det $\begin{bmatrix} 1 & x & y \\ 1 & x_0 & y_0 \\ 1 & x_1 & y_1 \end{bmatrix} = 0$

is the equation of the line in $\mathbb{R}^2$ through distinct points (x_0, y_0) and (x_1, y_1).

47. Find all values of x and y so that this matrix is not invertible.

$\begin{bmatrix} x & x + y \\ x + 2y & x + 3y \end{bmatrix}$

48. Determine values of the integer n so that

Det $\begin{bmatrix} n & n + 1 \\ n + 3 & n + 2 \end{bmatrix} < 0$

49. a. Describe the conic section when

Det $\begin{bmatrix} 2 & x & 2 \\ 1 & -1 & -2 \\ y & 2 & 3 \end{bmatrix} = 0$

b. Under what conditions on the parameters a, b, and c do we have

Det $\begin{bmatrix} 1 & 2 & a \\ 2 & 1 & b \\ 1 & -1 & c \end{bmatrix} \neq 0$

50. What is the locus of points (x, y) in $\mathbb{R}^2$ that satisfy the equation

Det $\begin{bmatrix} 3 & x & 5 \\ 2 & 1 & 0 \\ y & 4 & 7 \end{bmatrix} = 0$

COMPUTER EXERCISES 4.1

1. Show that Det $\begin{bmatrix} -73 & 78 & 24 \\ 92 & 66 & 25 \\ -80 & 37 & 10 \end{bmatrix} = 1.$

Explore how changing only one entry in this matrix by adding 0.01 to it affects the value of the determinant. For example, $92 \leftarrow 92.01$ or $78 \leftarrow 78.01$ or $10 \leftarrow 10.01$. Would you expect the value of the determinant to change very much? Explain what happens.

2. Let $A = (a_{ij})$ be the $n \times n$ matrix defined by $a_{ij} = |i - j|$. Using various values of n, numerically determine if this formula is correct: $\text{Det}(A) = (-1)^{n-1}2^{n-2}(n - 1)$. For example, with $n = 2$, Det $\begin{bmatrix} 0 & 1 \\ 1 & 0 \end{bmatrix} = -1.$

3. Consider the $n \times n$ matrix

$$A = \begin{bmatrix} (r+s) & s & s & \cdots & s \\ s & (r+s) & s & \cdots & s \\ s & s & (r+s) & \cdots & s \\ \vdots & \vdots & \vdots & \ddots & \vdots \\ s & s & s & \cdots & (r+s) \end{bmatrix}$$

Using various values of n, r, and s, numerically determine if $\text{Det}(A) = r^{n-1}(r + ns)$. For example, if $n = 3, r = 2, s = 1$, then

$$\text{Det}(A) = \text{Det} \begin{bmatrix} 3 & 1 & 1 \\ 1 & 3 & 1 \\ 1 & 1 & 3 \end{bmatrix} = 20.$$

4. Let $A = \begin{bmatrix} 1 & 0 & 0 & \cdots & 0 & 1 \\ 0 & 1 & 0 & \cdots & 0 & 2 \\ 0 & 1 & & \cdots & 0 & 3 \\ \vdots & \vdots & \vdots & \ddots & \vdots & \vdots \\ 0 & 0 & 0 & \cdots & 1 & n-1 \\ 1 & 2 & 3 & \cdots & n-1 & n \end{bmatrix}$

Using various values of n, numerically determine whether

$$\text{Det}(A) = -n(n + 1)(2n - 5)/6.$$

For example, if $n = 4$, then

$$\text{Det}(A) = -10.$$

5. Compute the numerical values of each:

a. Det $\begin{bmatrix} 0 & 3 & 6 & 4 & 1 \\ 1 & 3 & 0 & 4 & 0 \\ 2 & 3 & 5 & 3 & 1 \\ 1 & 4 & 5 & 9 & 0 \\ 0 & 1 & 3 & 2 & 1 \end{bmatrix}$

b. Det $\begin{bmatrix} 1 & 1 & 2 & 2 & 1 & 1 \\ 2 & 2 & 4 & 4 & 3 & 1 \\ 2 & 2 & 4 & 4 & 2 & 2 \\ 3 & 5 & 8 & 6 & 5 & 3 \\ 3 & 1 & 7 & 5 & 1 & 9 \\ 5 & 0 & 1 & 3 & 0 & 7 \end{bmatrix}$

c. Det $\begin{bmatrix} -2 & -1 & 0 & 4 & 5 \\ -3 & 2 & 7 & 8 & 3 \\ 1 & 0 & 0 & 2 & 4 \\ 3 & 0 & 1 & 3 & 6 \\ 0 & 3 & 6 & 0 & 9 \end{bmatrix}$

d. Det $\begin{bmatrix} 1 & 1 & 1 & 1 & 1 & 1 \\ -1 & -1 & 0 & 0 & 1 & -1 \\ -2 & -2 & 0 & 0 & 1 & 1 \\ 0 & 0 & 1 & 1 & 3 & -1 \\ 1 & 1 & 2 & 2 & 4 & 1 \\ 0 & 1 & 2 & 6 & -1 & 3 \end{bmatrix}$

4.2 DETERMINANTS: PROPERTIES

No human investigation can be called real science if it cannot be demonstrated mathematically.

—LEONARDO DA VINCI (1452–1519)

Like the crest of a peacock so is mathematics at the head of all knowledge.

—AN OLD SAYING FROM INDIA.

In this unit, the determinant function will be further explored and more of its important properties will be revealed. Recall that in Section 4.1, we started with three important properties of this determinant function. These led to an algorithmic definition. Another definition can be given, of a completely different type. For us, this alternative definition is logically a *theorem*, because we have already defined determinants. This theorem leads to another algorithm for computing determinants, and it has many other applications.

Minors and Cofactors

If $\mathbf{A}$ is an $n \times n$ matrix (where $n \geq 2$), we can form submatrices of $\mathbf{A}$ by deleting one column and one row. The resulting $(n-1) \times (n-1)$ matrix is called a **minor** of the original matrix $\mathbf{A}$. Let us denote by $\mathbf{M}^{ij}$ the minor resulting from the removal of row i and column j from $\mathbf{A}$. If we want to emphasize that the minor comes from the matrix $\mathbf{A}$, we can write $\mathbf{M}^{ij}(\mathbf{A})$, which is an $(n-1) \times (n-1)$ matrix.

EXAMPLE 1 For this 3×3 matrix, what are $\mathbf{M}^{21}(\mathbf{A})$ and $\mathbf{M}^{13}(\mathbf{A})$?

$$\mathbf{A} = \begin{bmatrix} 1 & 7 & 6 \\ -3 & 2 & 9 \\ 11 & -13 & 4 \end{bmatrix}$$

SOLUTION We have

$$\mathbf{M}^{21} = \mathbf{M}^{21}(\mathbf{A}) = \begin{bmatrix} 7 & 6 \\ -13 & 4 \end{bmatrix}, \qquad \mathbf{M}^{13} = \mathbf{M}^{13}(\mathbf{A}) = \begin{bmatrix} -3 & 2 \\ 11 & -13 \end{bmatrix} \quad \blacksquare$$

The next theorem shows how to compute determinants numerically by using a recursive algorithm. This method can be useful for small matrices, but it is inefficient for large matrices. See the following comments.

THEOREM 1

Let $\mathbf{A}$ be an $n \times n$ matrix whose minors are denoted by $\mathbf{M}^{ij}$. Then these formulas are valid:

$$\text{Det}(\mathbf{A}) = \sum_{j=1}^{n} (-1)^{i+j} a_{ij} \, \text{Det}(\mathbf{M}^{ij}) \qquad \textbf{(expansion using row } i)$$

$$\text{Det}(\mathbf{A}) = \sum_{i=1}^{n} (-1)^{i+j} a_{ij} \, \text{Det}(\mathbf{M}^{ij}) \qquad \textbf{(expansion using column } j)$$

We will not prove this theorem but prefer simply to adopt it. The two formulas show how the determinant of an $n \times n$ matrix can be computed as a linear combination of determinants of $(n-1) \times (n-1)$ matrices. For that reason, the preceding formulas can be used in a recursive definition of determinants. A number of textbooks follow this route. See, for example, Andrilli and Hacker [2003] and Lay [2003]. The reader should note that the preceding displayed formulas are $2n$ in number, because the indices i and j can range over the integers from 1 to n.

EXAMPLE 2 In computing the determinant of a 6×6 matrix $\mathbf{A} = (a_{ij})$, show the first step in the expansion using the fourth column.

SOLUTION First, we show the signs in the expansion going down column 4:

$$\begin{bmatrix} + & - & + & - & \cdot & \cdot \\ \cdot & \cdot & \cdot & + & \cdot & \cdot \\ \cdot & \cdot & \cdot & - & \cdot & \cdot \\ \cdot & \cdot & \cdot & + & \cdot & \cdot \\ \cdot & \cdot & \cdot & - & \cdot & \cdot \\ \cdot & \cdot & \cdot & + & \cdot & \cdot \end{bmatrix}_{6 \times 6}$$

To calculate the determinant of this matrix, we begin with

$$\text{Det}(\mathbf{A}) = -a_{14} \, \text{Det}(\mathbf{M}^{14}) + a_{24} \, \text{Det}(\mathbf{M}^{24}) - a_{34} \, \text{Det}(\mathbf{M}^{34})$$
$$+ a_{44} \, \text{Det}(\mathbf{M}^{44}) - a_{54} \, \text{Det}(\mathbf{M}^{54}) + a_{64} \, \text{Det}(\mathbf{M}^{64}) \qquad \blacksquare$$

EXAMPLE 3 Compute the determinant of the matrix $\mathbf{A}$ in Example 1 by using Theorem 1.

SOLUTION Because $n = 3$ in this example, there are six different applicable formulas in Theorem 1. For instance, we can use the first formula with $i = 2$, and apply it to the matrix $\mathbf{A}$. If it is not ambiguous, we write $\text{Det}\,\mathbf{A}$ in place of $\text{Det}(\mathbf{A})$.

$$\text{Det}(\mathbf{A}) = \text{Det}\begin{bmatrix} 1 & 7 & 6 \\ -3 & 2 & 9 \\ 11 & -13 & 4 \end{bmatrix}$$

$$= -a_{21}\,\text{Det}(\mathbf{M}^{21}) + a_{22}\,\text{Det}(\mathbf{M}^{22}) - a_{23}\,\text{Det}(\mathbf{M}^{23})$$

$$= -(-3)\,\text{Det}\begin{bmatrix} 7 & 6 \\ -13 & 4 \end{bmatrix} + 2\,\text{Det}\begin{bmatrix} 1 & 6 \\ 11 & 4 \end{bmatrix} - 9\,\text{Det}\begin{bmatrix} 1 & 7 \\ 11 & -13 \end{bmatrix}$$

$$= (3)(28 + 78) + (2)(4 - 66) - (9)(-13 - 77)$$

$$= (3)(106) + (2)(-62) - (9)(-90) = 1004 \qquad \blacksquare$$

DEFINITION

The first formula in Theorem 1 is called a **cofactor expansion** *of Det($\mathbf{A}$) using row i, and the second formula is a* **cofactor expansion** *of Det($\mathbf{A}$) using column j. The term* **cofactor** *applies to the quantities*

$$C_{ij} = (-1)^{i+j}\,\text{Det}(\mathbf{M}^{ij})$$

Because we are denoting the cofactors by C_{ij}, the formulas in Theorem 1 take this simpler form:

THEOREM 1′

Let $\mathbf{A}$ be an $n \times n$ matrix whose cofactors are denoted by C_{ij}. Then these formulas are valid:

$$\text{Det}(\mathbf{A}) = \sum_{j=1}^{n} a_{ij}C_{ij} \qquad \textbf{(expansion using row } i\textbf{)}$$

$$\text{Det}(\mathbf{A}) = \sum_{i=1}^{n} a_{ij}C_{ij} \qquad \textbf{(expansion using column } j\textbf{)}$$

Notice that the terms $\mathbf{M}^{ij}$ are matrices of size $(n-1) \times (n-1)$, whereas the terms C_{ij} are numbers. The cofactor matrix $\mathbf{C}$ having elements C_{ij} is of some interest, as we shall see later. (It is *almost* the inverse of the matrix $\mathbf{A}$!)

In some examples, the presence of zeros in strategic places can make the calculation of a determinant very easy. Here is one case of this phenomenon:

EXAMPLE 4 Using only Theorem 1′, calculate the value of this 5×5 determinant:

$$
\text{Det}
\begin{bmatrix}
-53 & 0 & 71 & 0 & -3 \\
0 & 0 & 2 & 0 & 0 \\
79 & 4 & -83 & 67 & -51 \\
-95 & 0 & 63 & 7 & -91 \\
3 & 0 & -51 & 0 & 0
\end{bmatrix}
$$

SOLUTION In this calculation, we used alternately a column expansion and a row expansion. We are free to do so, since Theorem 1 asserts the validity of both types of expansion on the successive matrices.

$$
(-4)\,\text{Det}
\begin{bmatrix}
-53 & 71 & 0 & -3 \\
0 & 2 & 0 & 0 \\
-95 & 63 & 7 & -91 \\
3 & -51 & 0 & 0
\end{bmatrix}
$$

$$
= (-4)(2)\,\text{Det}
\begin{bmatrix}
-53 & 0 & -3 \\
-95 & 7 & -91 \\
3 & 0 & 0
\end{bmatrix}
= (-4)(2)(7)\,\text{Det}
\begin{bmatrix}
-53 & -3 \\
3 & 0
\end{bmatrix}
$$

$$
= (-4)(2)(7)(-3)(-3) = -504 \qquad\blacksquare
$$

Work Estimate

The cost of evaluating the determinant of an $n \times n$ matrix $\mathbf{A}$ using cofactor expansions is taken to be the number of long operations (multiplications). It can be denoted by $f(n)$. We can show that

$$
f(n) \approx nf(n-1)
$$

If $\mathbf{A}$ is 1×1, then $f(1) = 0$, since $\text{Det}(\mathbf{A}) = a_{11}$ and no arithmetic is needed. If $\mathbf{A}$ is 2×2, then expanding using the first row, we have $\text{Det}(\mathbf{A}) = a_{11}C_{11} + a_{12}C_{12} = a_{11}a_{22} - a_{12}a_{21}$ and $f(2) = 2$. If $\mathbf{A}$ is 3×3, then again expanding along the first row, we have $\text{Det}(\mathbf{A}) = a_{11}C_{11} + a_{12}C_{12} + a_{13}C_{13}$. Hence, we obtain $f(3) = 3f(2) + 3 = 9$, since there are three matrices of order two and three multiplications. In general, it follows that $f(n) = nf(n-1) + n$ and $f(n) \approx nf(n-1)$.

The amount of work involved in evaluating an $n \times n$ determinant using the recursive algorithm quickly becomes excessive. If we count only the multiplications and ignore the additions, we quickly reach astronomical levels of arithmetic! Let $f(n)$ be the number of multiplications involved in computing the determinant of an $n \times n$ matrix using cofactor expansions. Then $f(2) \geq 2, f(3) \geq 9$, and $f(n) \geq nf(n-1) + n$. Using that inequality repeatedly, one obtains the lower bound

$$f(n) \geq n(n-1)(n-2)\cdots 3 \cdot 2 \cdot 1 = n!$$

This is the nth factorial function.

The burden of work is overwhelming when n is large. For example, when $n = 25$, we obtain $f(25) > 25! \approx 10^{25}$. Suppose that we have a computer that can perform 1 trillion operations per second (1 teraflop per second). How long will this computer take to compute 10^{25} operations? It will take $10^{25} \div 10^{12} = 10^{13}$ seconds. Converting from seconds to years, we find that it would take 300,000 years to finish 10^{25} operations even with such a supercomputer. (Life is too short to wait for the results!)

Of the various ways to compute the determinant of a matrix, the method using row operations (Gaussian elimination) involves much less work than other methods. It requires approximately $\frac{1}{3}n^3$ long operations. For example, with $n = 25$, this is about 5212, *not* 10^{25}.

Direct Methods for Computing Determinants

Direct application of the formulas in Theorem 1 gives us the special formula for a 2×2 matrix:

$$\mathrm{Det} \begin{bmatrix} a_{11} & a_{12} \\ a_{21} & a_{22} \end{bmatrix} = a_{11} \, \mathrm{Det}(\mathbf{M}^{11}) - a_{12} \, \mathrm{Det}(\mathbf{M}^{12}) = a_{11}a_{22} - a_{12}a_{21}$$

Here, we have used the first of the two formulas in the theorem, and have taken $i = 1$. But there are three other choices we could have made, each leading to the same formula. Figure 4.6 illustrates the direct evaluation of a 2×2 determinant.

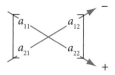

FIGURE 4.6 Direct evaluation of a 2×2 determinant.

Similarly, we can get the determinant of a 3×3 matrix:

$$\text{Det} \begin{bmatrix} a_{11} & a_{12} & a_{13} \\ a_{21} & a_{22} & a_{23} \\ a_{31} & a_{32} & a_{33} \end{bmatrix}$$

$$= a_{11} \text{Det} \begin{bmatrix} a_{22} & a_{23} \\ a_{32} & a_{33} \end{bmatrix} - a_{12} \text{Det} \begin{bmatrix} a_{21} & a_{23} \\ a_{31} & a_{33} \end{bmatrix} + a_{13} \text{Det} \begin{bmatrix} a_{21} & a_{22} \\ a_{31} & a_{32} \end{bmatrix}$$

$$= a_{11}(a_{22}a_{33} - a_{23}a_{32}) - a_{12}(a_{21}a_{33} - a_{23}a_{31}) + a_{13}(a_{21}a_{32} - a_{22}a_{31})$$

$$= a_{11}a_{22}a_{33} - a_{11}a_{23}a_{32} - a_{12}a_{21}a_{33} + a_{12}a_{23}a_{31} + a_{13}a_{21}a_{32} - a_{13}a_{22}a_{31}$$

There are six different row and column expansions of a 3×3 matrix. It is a miraculous fact that all six lead eventually to the same formula. (That is, each formula should be expressed in simple form, as previously, without any 2×2 determinants and without parentheses.) The determinant of a 3×3 matrix can be calculated quickly via direct methods. Figure 4.7(a) illustrates the direct evaluation of a 3×3 determinant in a *round-robin* fashion: follow the arrows, multiply the elements together, and then add or subtract them according to the sign. Figure 4.7(b) is an alternative version of this method. There are no such simple algorithms for determinants of order greater than 3.

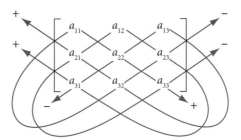

FIGURE 4.7(a) First version for the direct evaluation of a 3×3 determinant.

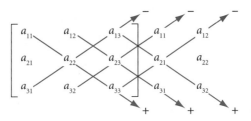

FIGURE 4.7(b) Second version for the direct evaluation of a 3×3 determinant.

Properties of Determinants

We explore next the many ways in which determinants interact with the other topics in linear algebra. This theme will arise again in the study of eigenvalues in Section 6.1.

LEMMA 1

If $\mathbf{E}$ is an $n \times n$ elementary matrix (as in Section 3.1) and if $\mathbf{A}$ is any $n \times n$ matrix, then

$$\text{Det}(\mathbf{EA}) = \text{Det}(\mathbf{E})\,\text{Det}(\mathbf{A})$$

PROOF Remember from Section 1.1 that in verifying such a statement involving elementary row operations, only two types of row operations need be utilized. Consider the row operation of multiplying row k in $\mathbf{A}$ by a nonzero constant c. This is effected by multiplying $\mathbf{A}$ on the left by an elementary matrix, $\mathbf{E}$. The matrices involved are

$$\mathbf{E} = \begin{bmatrix} 1 & 0 & 0 & \cdots & 0 & 0 & 0 & \cdots & 0 \\ 0 & 1 & 0 & \cdots & 0 & 0 & 0 & \cdots & 0 \\ 0 & 0 & 1 & \cdots & 0 & 0 & 0 & \cdots & 0 \\ \vdots & \vdots & \vdots & \ddots & \vdots & \vdots & \vdots & \ddots & \vdots \\ 0 & 0 & 0 & \cdots & 1 & 0 & 0 & \cdots & 0 \\ 0 & 0 & 0 & \cdots & 0 & c & 0 & \cdots & 0 \\ 0 & 0 & 0 & \cdots & 0 & 0 & 1 & \cdots & 0 \\ \vdots & \vdots & \vdots & \ddots & \vdots & \vdots & \vdots & \ddots & \vdots \\ 0 & 0 & 0 & \cdots & 0 & 0 & 0 & \cdots & 1 \end{bmatrix}$$

$$\mathbf{EA} = \begin{bmatrix} a_{11} & a_{12} & \cdots & a_{1n} \\ \vdots & \vdots & \ddots & \vdots \\ a_{k-1,1} & a_{k-1,2} & \cdots & a_{k-1,n} \\ ca_{k1} & ca_{k2} & \cdots & ca_{kn} \\ a_{k+1,1} & a_{k+1,2} & \cdots & a_{k+1,n} \\ \vdots & \vdots & \ddots & \vdots \\ a_{n1} & a_{n2} & \cdots & a_{nn} \end{bmatrix}$$

We see that $\text{Det}(\mathbf{E}) = c$, $\text{Det}(\mathbf{E})\,\text{Det}(\mathbf{A}) = c\,\text{Det}(\mathbf{A})$, and $\text{Det}(\mathbf{EA}) = c\,\text{Det}(\mathbf{A}) = \text{Det}(\mathbf{E})\,\text{Det}(\mathbf{A})$. The remainder of the proof is left as General Exercise 39. ∎

EXAMPLE 5 Verify the assertion in Lemma 1 when

$$\mathbf{A} = \begin{bmatrix} 3 & 7 \\ 5 & -2 \end{bmatrix}, \qquad \mathbf{E} = \begin{bmatrix} 3 & 0 \\ 0 & 1 \end{bmatrix}$$

SOLUTION Evidently, we have

$$\mathbf{EA} = \begin{bmatrix} 9 & 21 \\ 5 & -2 \end{bmatrix}, \quad \text{Det}(\mathbf{A})\,\text{Det}(\mathbf{E}) = (-41)(3) = -123 = \text{Det}(\mathbf{EA}) \quad ∎$$

THEOREM 2

For square matrices of the same size, we have

$$\text{Det}(\mathbf{AB}) = \text{Det}(\mathbf{A})\,\text{Det}(\mathbf{B})$$

PROOF Case 1. Suppose that $\mathbf{A}$ is noninvertible. In this case, $\text{Det}(\mathbf{A}) = 0$ and $\text{Det}(\mathbf{A})\,\text{Det}(\mathbf{B}) = 0$. Thus, we have to prove that $\text{Det}(\mathbf{AB}) = 0$. By Theorem 5 in Section 4.1, it suffices to prove that $\mathbf{AB}$ is noninvertible. If $\mathbf{AB}$ is invertible, then we can write $(\mathbf{AB})(\mathbf{AB})^{-1} = \mathbf{I}$. Thus $\mathbf{B}(\mathbf{AB})^{-1}$ is a right inverse of $\mathbf{A}$. This contradicts Theorem 2 in Section 3.2.

Case 2. Suppose that $\mathbf{A}$ is invertible. In this case, a sequence of row operations on $\mathbf{A}$ will lead to $\mathbf{I}_n$. (Recall that this is how we usually calculate inverses.) Representing these row operations by elementary matrices, we can write $\mathbf{E}_k\mathbf{E}_{k-1}\cdots\mathbf{E}_2\mathbf{E}_1\mathbf{A} = \mathbf{I}$. The corresponding set of inverse row operations will lead from $\mathbf{I}$ to $\mathbf{A}$. We write this as $\mathbf{A} = \mathbf{E}_1^{-1}\mathbf{E}_2^{-1}\cdots\mathbf{E}_k^{-1}$. Consequently, by Lemma 1, we obtain

$$\begin{aligned}
\text{Det}(\mathbf{AB}) &= \text{Det}(\mathbf{E}_1^{-1}\mathbf{E}_2^{-1}\cdots\mathbf{E}_k^{-1}\mathbf{B}) \\
&= \text{Det}(\mathbf{E}_1^{-1})\,\text{Det}(\mathbf{E}_2^{-1}\cdots\mathbf{E}_k^{-1}\mathbf{B}) \\
&= \text{Det}(\mathbf{E}_1^{-1})\,\text{Det}(\mathbf{E}_2^{-1})\,\text{Det}(\mathbf{E}_3^{-1}\cdots\mathbf{E}_k^{-1}\mathbf{B}) \\
&= \text{Det}(\mathbf{E}_1^{-1}\mathbf{E}_2^{-1})\,\text{Det}(\mathbf{E}_3^{-1}\cdots\mathbf{E}_k^{-1}\mathbf{B}) \\
&\;\;\vdots \\
&= \text{Det}(\mathbf{E}_1^{-1}\mathbf{E}_2^{-1}\cdots\mathbf{E}_k^{-1})\,\text{Det}(\mathbf{B}) \\
&= \text{Det}(\mathbf{A})\,\text{Det}(\mathbf{B}) \quad ∎
\end{aligned}$$

EXAMPLE 6 Verify the assertion in Theorem 2 for this pair of matrices:

$$A = \begin{bmatrix} 1 & 0 & 2 \\ 3 & -1 & 4 \\ 1 & 5 & 1 \end{bmatrix}, \quad B = \begin{bmatrix} 2 & 2 & 3 \\ -1 & 1 & 3 \\ 5 & 1 & 4 \end{bmatrix}$$

SOLUTION　Using row operations, we have

$$\text{Det}(A) = \text{Det} \begin{bmatrix} 1 & 0 & 2 \\ 0 & -1 & -2 \\ 0 & 5 & -1 \end{bmatrix} = \text{Det} \begin{bmatrix} -1 & -2 \\ 5 & -1 \end{bmatrix} = 11$$

$$\text{Det}(B) = -\text{Det} \begin{bmatrix} -1 & 1 & 3 \\ 2 & 2 & 3 \\ 5 & 1 & 4 \end{bmatrix} = -\text{Det} \begin{bmatrix} -1 & 1 & 3 \\ 0 & 4 & 9 \\ 0 & 6 & 19 \end{bmatrix}$$

$$= \text{Det} \begin{bmatrix} 4 & 9 \\ 6 & 19 \end{bmatrix} = 22$$

$$\text{Det}(AB) = \text{Det} \begin{bmatrix} 12 & 4 & 11 \\ 27 & 9 & 22 \\ 2 & 8 & 22 \end{bmatrix} = \text{Det} \begin{bmatrix} 12 & 4 & 11 \\ 3 & 1 & 0 \\ -22 & 0 & 0 \end{bmatrix} = (11)(22) = 242 \quad ■$$

LEMMA 2

Recall that the minors of the matrix A are denoted by $M^{ij}(A)$. Then we have

$$M^{ij}(A^T) = [M^{ji}(A)]^T$$

PROOF　Write down the two matrices A and A^T side by side. In A, be sure to show row j and column i. In A^T, be sure to show column j and row i. These special rows and columns must be deleted to form $M^{ij}(A^T)$ and $M^{ji}(A)$. After that deletion, the matrices that remain should be transposes of each other. ■

EXAMPLE 7　Verify a nontrivial case of Lemma 2 using

$$A = \begin{bmatrix} 1 & 4 & 2 \\ 3 & -3 & 6 \\ 2 & 0 & 5 \end{bmatrix}, \quad A^T = \begin{bmatrix} 1 & 3 & 2 \\ 4 & -3 & 0 \\ 2 & 6 & 5 \end{bmatrix}$$

SOLUTION Use $i = 3$ and $j = 2$ in Lemma 2. Then

$$[\mathbf{M}^{23}(\mathbf{A})]^T = \begin{bmatrix} 1 & 2 \\ 4 & 0 \end{bmatrix}, \qquad \mathbf{M}^{32}(\mathbf{A}^T) = \begin{bmatrix} 1 & 2 \\ 4 & 0 \end{bmatrix} \qquad \blacksquare$$

THEOREM 3

The determinant of a matrix and the determinant of its transpose are equal:

$$\mathrm{Det}(\mathbf{A}) = \mathrm{Det}(\mathbf{A}^T)$$

PROOF The theorem is certainly correct for 1×1 matrices. Proceeding by induction, we assume that the theorem has been established for all $(n-1) \times (n-1)$ matrices. Let $\mathbf{A}$ be an $n \times n$ matrix. By Lemma 2 and the induction hypothesis, we have, for any k,

$$\mathrm{Det}(\mathbf{A}^T) = \sum_{j=1}^{n} (-1)^{k+j} (\mathbf{A}^T)_{jk} \, \mathrm{Det}[\mathbf{M}^{jk}(\mathbf{A}^T)]$$

$$= \sum_{j=1}^{n} (-1)^{k+j} \mathbf{A}_{kj} \, \mathrm{Det}[\mathbf{M}^{kj}(\mathbf{A})]^T$$

$$= \sum_{j=1}^{n} (-1)^{k+j} \mathbf{A}_{kj} \, \mathrm{Det}[\mathbf{M}^{kj}(\mathbf{A})]$$

$$= \mathrm{Det}(\mathbf{A}) \qquad \blacksquare$$

EXAMPLE 8 Illustrate Theorem 3 by computing $\mathrm{Det}(\mathbf{A})$ and $\mathrm{Det}(\mathbf{A}^T)$, when

$$\mathbf{A} = \begin{bmatrix} 2 & 1 & 2 \\ 4 & -1 & 3 \\ -6 & 2 & 2 \end{bmatrix}$$

SOLUTION Again, using row operations, we obtain

$$\mathrm{Det}(\mathbf{A}) = \mathrm{Det} \begin{bmatrix} 2 & 1 & 2 \\ 4 & -1 & 3 \\ -6 & 2 & 2 \end{bmatrix} = \mathrm{Det} \begin{bmatrix} 2 & 1 & 2 \\ 0 & -3 & -1 \\ 0 & 5 & 8 \end{bmatrix}$$

$$= (2)(-24 + 5) = -38$$

$$\mathrm{Det}(\mathbf{A}^T) = \mathrm{Det} \begin{bmatrix} 2 & 4 & -6 \\ 1 & -1 & 2 \\ 2 & 3 & 2 \end{bmatrix} = \mathrm{Det} \begin{bmatrix} 0 & 6 & -10 \\ 1 & -1 & 2 \\ 0 & 5 & -2 \end{bmatrix}$$

$$= (-1)(-12 + 50) = -38 \qquad \blacksquare$$

> ## THEOREM 4
>
> *The determinant of a matrix is a linear function of any one column (or row).*

PROOF We give the proof for the columns, and leave the other half to the exercises. The meaning of the first assertion in the theorem is this: Let $\mathbf{A}$ be an $n \times n$ matrix. Replace one column, say column j, by a variable vector, $\mathbf{x} = [x_1, x_2, \ldots, x_n]^T$. Call this new matrix $\mathbf{A}_j(\mathbf{x})$. Then $\mathbf{x} \mapsto \text{Det}[\mathbf{A}_j(\mathbf{x})]$ is a linear map from $\mathbb{R}^n$ to $\mathbb{R}$. To prove this, we expand $\text{Det}[\mathbf{A}_j(\mathbf{x} + \mathbf{y})]$ by cofactors using column j:

$$\text{Det}[\mathbf{A}_j(\mathbf{x} + \mathbf{y})] = \sum_{i=1}^{n}(x_i + y_i)C_{ij}$$

$$= \sum_{i=1}^{n}x_i C_{ij} + \sum_{i=1}^{n}y_i C_{ij}$$

$$= \text{Det}[\mathbf{A}_j(\mathbf{x})] + \text{Det}[\mathbf{A}_j(\mathbf{y})]$$

The other half of the linearity condition is similar and is left as General Exercise 40. ∎

> **EXAMPLE 9** Give a simple formula for $\text{Det}[\mathbf{A}_2(\mathbf{x})]$ when
>
> $$\mathbf{A} = \begin{bmatrix} 1 & 3 & 7 \\ 9 & 11 & -17 \\ 4 & 3 & 1 \end{bmatrix}$$

SOLUTION For column two, we obtain

$$\text{Det}[\mathbf{A}_2(\mathbf{x})] = \text{Det}\begin{bmatrix} 1 & x_1 & 7 \\ 9 & x_2 & -17 \\ 4 & x_3 & 1 \end{bmatrix}$$

$$= -x_1 \text{Det}\begin{bmatrix} 9 & -17 \\ 4 & 1 \end{bmatrix} + x_2 \text{Det}\begin{bmatrix} 1 & 7 \\ 4 & 1 \end{bmatrix}$$

$$- x_3 \text{Det}\begin{bmatrix} 1 & 7 \\ 9 & -17 \end{bmatrix}$$

$$= -x_1(77) + x_2(-27) - x_3(-80)$$

$$= -77x_1 - 27x_2 + 80x_3 \qquad \blacksquare$$

Cramer's Rule

Systems of linear equations can be solved by using determinants and Cramer's rule.[1]

THEOREM 5 Cramer's Rule

If A is invertible, then the solution to the system of equations Ax = b is given by these formulas:

$$x_j = \frac{\text{Det}[A_j(b)]}{\text{Det}(A)} \qquad (1 \le j \le n)$$

where $A_j(b)$ is matrix A with its jth column replaced by b.

PROOF Denote the columns of **A** by $a_1, a_2, \ldots, a_n$, and the columns of **I** by $e_1, e_2, \ldots, e_n$. Use $I_j(x)$ for the identity matrix with column j replaced by **x**. (We interpret **x** as a column vector.) Suppose that **Ax = b**. Then we have

$$A \cdot I_j(x) = A[e_1, e_2, \ldots, x, \ldots, e_n] = [Ae_1, Ae_2, \ldots, Ax, \ldots, Ae_n]$$
$$= [a_1, a_2, \ldots, b, \ldots, a_n] = A_j(b)$$

Taking determinants, we have $\text{Det}(A)\,\text{Det}[I_j(x)] = \text{Det}[A_j(b)]$. This confirms the equation to be proved because $\text{Det}[I_j(x)] = x_j$. ∎

EXAMPLE 10 Use Cramer's rule to find the value of x_2 in this linear system:

$$\begin{bmatrix} 1 & 4 & 2 \\ 3 & -3 & 6 \\ 2 & 0 & 5 \end{bmatrix} \begin{bmatrix} x_1 \\ x_2 \\ x_3 \end{bmatrix} = \begin{bmatrix} 3 \\ 5 \\ -4 \end{bmatrix}$$

---

[1] Gabriel Cramer (1704–1752) was only 18 years old when he was awarded a doctorate. The theorem to which his name is attached is only one of his many accomplishments.

SOLUTION Cramer's rule, with $j = 2$, leads to

$$x_2 = \text{Det} \begin{bmatrix} 1 & 3 & 2 \\ 3 & 5 & 6 \\ 2 & -4 & 5 \end{bmatrix} \div \text{Det} \begin{bmatrix} 1 & 4 & 2 \\ 3 & -3 & 6 \\ 2 & 0 & 5 \end{bmatrix} = (-4)/(-15) = 4/15$$

∎

Planes in $\mathbb{R}^3$

In three dimensions, the equation of a plane is

$$ax + by + cz = d$$

If three non-colinear points on a plane are given (x_1, y_1, z_1), (x_2, y_2, z_2), (x_3, y_3, z_3), then the coefficients a, b, c can be found by these determinants

$$a = \left(\frac{d}{D}\right) \text{Det} \begin{bmatrix} 1 & y_1 & z_1 \\ 1 & y_2 & z_2 \\ 1 & y_3 & z_3 \end{bmatrix}, \qquad b = \left(\frac{d}{D}\right) \text{Det} \begin{bmatrix} x_1 & 1 & z_1 \\ x_2 & 1 & z_2 \\ x_3 & 1 & z_3 \end{bmatrix}$$

$$c = \left(\frac{d}{D}\right) \text{Det} \begin{bmatrix} x_1 & y_1 & 1 \\ x_2 & y_2 & 1 \\ x_3 & y_3 & 1 \end{bmatrix}, \qquad D = \text{Det} \begin{bmatrix} x_1 & y_1 & z_1 \\ x_2 & y_2 & z_2 \\ x_3 & y_3 & z_3 \end{bmatrix} \neq 0$$

which are parametric in $d \neq 0$. Our remarks do not cover all cases. These equations can be found by using Cramer's rule.

Computing Inverses Using Determinants

The inverse of a matrix can be computed directly using determinants and Cramer's rule.

THEOREM 6

The inverse of an $n \times n$ invertible matrix $\mathbf{A}$ can be computed by the formulas

$$(\mathbf{A}^{-1})_{ij} = \frac{(-1)^{i+j}}{\text{Det}(\mathbf{A})} \text{Det}(\mathbf{M}^{ji}) = \frac{C_{ji}}{\text{Det}(\mathbf{A})}$$

Here, as usual, $\mathbf{M}^{ji}$ is the minor obtained from $\mathbf{A}$ by removing row j and column i and $C_{ji} = (-1)^{i+j} \text{Det}(\mathbf{M}^{ji})$ is the corresponding element of the cofactor matrix.

PROOF Fix an index j. We shall construct column j in $\mathbf{A}^{-1}$. Denote by $\mathbf{x}$ this column vector, which should satisfy the equation $\mathbf{Ax} = \mathbf{e}_j$, where $\mathbf{e}_j$ is the jth column in $\mathbf{I}$. By Cramer's rule, the components of the vector $\mathbf{x}$ are given by $x_i = \text{Det}[\mathbf{A}_i(\mathbf{e}_j)]/\text{Det}(\mathbf{A})$, where $\mathbf{A}_i(\mathbf{e}_j)$ is the matrix that arises when we replace column i in $\mathbf{A}$ by the column-vector $\mathbf{e}_j$. To evaluate $\text{Det}[\mathbf{A}_i(\mathbf{e}_j)]$, perform a column expansion using column i. Because column i contains the vector $\mathbf{e}_j$, only one term emerges from the column expansion:

$$\text{Det}[\mathbf{A}_i(\mathbf{e}_j)] = \sum_{k=1}^{n}(-1)^{i+k}(\mathbf{e}_j)_k\,\text{Det}(\mathbf{M}^{ki}) = (-1)^{i+j}\,\text{Det}(\mathbf{M}^{ji})$$

This calculation leads to

$$(\mathbf{A}^{-1})_{ij} = x_i = \frac{1}{\text{Det}(\mathbf{A})}\,\text{Det}[\mathbf{A}_i(\mathbf{e}_j)] = \frac{(-1)^{i+j}}{\text{Det}(\mathbf{A})}\,\text{Det}(\mathbf{M}^{ji}) \qquad \blacksquare$$

Recall the cofactors, $C_{ij} = (-1)^{i+j}\,\text{Det}(\mathbf{M}^{ij})$. By using cofactors, we obtain a simpler formula for the inverse:

$$\mathbf{A}^{-1} = \frac{1}{\text{Det}(\mathbf{A})}\mathbf{C}^T$$

where $\mathbf{C}$ is the cofactor matrix of $\mathbf{A}$, having generic elements C_{ij}. (The transpose of the matrix $\mathbf{C}$ is called the **classical adjoint matrix** of $\mathbf{A}$: $\text{Adj}(\mathbf{A}) = \mathbf{C}^T$. The term **adjugate** is also used.)

EXAMPLE 11 Using Theorem 6, compute the inverse of the matrix

$$\mathbf{A} = \begin{bmatrix} 5 & 7 & -2 \\ 3 & 2 & 1 \\ -2 & 4 & 6 \end{bmatrix}$$

SOLUTION This matrix has nine cofactors, of which a small sample are given here:

$$C_{11} = (-1)^{1+1}\,\text{Det}(\mathbf{M}^{11}) = \text{Det}\begin{bmatrix} 2 & 1 \\ 4 & 6 \end{bmatrix} = 8$$

$$C_{32} = (-1)^{3+2}\,\text{Det}(\mathbf{M}^{32}) = -\text{Det}\begin{bmatrix} 5 & -2 \\ 3 & 1 \end{bmatrix} = -11$$

The matrix containing the cofactors and the transpose of that matrix are

$$
\mathbf{C} = \begin{bmatrix} 8 & -20 & 16 \\ -50 & 26 & -34 \\ 11 & -11 & -11 \end{bmatrix}, \qquad \mathbf{C}^T = \begin{bmatrix} 8 & -50 & 11 \\ -20 & 26 & -11 \\ 16 & -34 & -11 \end{bmatrix}
$$

To finish this example, it is necessary to find the value of Det($\mathbf{A}$). An easy way to do this is to use the cofactor expansion and the matrices $\mathbf{A}$ and $\mathbf{C}$. If we choose the top row of $\mathbf{A}$, the corresponding cofactors will be the top row of $\mathbf{C}$. (See Theorem 1′.) Thus, we obtain

$$
\text{Det}(\mathbf{A}) = (5)(8) + (7)(-20) + (-2)(16) = -132
$$

We can verify this result with a direct calculation of the determinant of this 3×3 matrix:

$$
\mathbf{A}^{-1} = \frac{1}{\text{Det}(\mathbf{A})} \mathbf{C}^T = -\frac{1}{132} \begin{bmatrix} 8 & -50 & 11 \\ -20 & 26 & -11 \\ 16 & -34 & -11 \end{bmatrix}
$$

The work can be verified in an independent manner by computing

$$
\mathbf{A}\mathbf{C}^T = \begin{bmatrix} -132 & 0 & 0 \\ 0 & -132 & 0 \\ 0 & 0 & -132 \end{bmatrix} = -132 \begin{bmatrix} 1 & 0 & 0 \\ 0 & 1 & 0 \\ 0 & 0 & 1 \end{bmatrix} = \text{Det}(\mathbf{A})\mathbf{I}
$$

Notice that $\mathbf{A} \cdot \text{Adj}(\mathbf{A}) = \text{Det}(\mathbf{A})\mathbf{I}$.   ■

Another interesting relationship between a matrix and its matrix of cofactors is given in the following theorem.

THEOREM 7

If the elements of one row in a matrix are multiplied by the cofactors of a different row and added up, the result is 0. Thus, if the matrix in question is $\mathbf{A}$ and if $i \neq j$, then

$$
\sum_{k=1}^{n} a_{ik} C_{jk} = \sum_{k=1}^{n} a_{ik}(-1)^{j+k} \text{Det}(\mathbf{M}^{jk}) = 0
$$

A similar result is true for the columns of $\mathbf{A}$; it is left to General Exercise 31.

PROOF Fix a pair (i, j) with $i \neq j$, and form a matrix **B**, identical to **A**, except that the jth row of **B** is the ith row of **A**. Thus, **B** has two identical rows (namely rows i and j). Consequently, **B** is not invertible and has determinant 0. Notice that the minors of these two matrices obey this rule: $M^{jk}(A) = M^{jk}(B)$, because **A** and **B** differ only in their jth rows, and these minors do not involve the elements in row j. Carry out a calculation of Det(**B**) by a row expansion using row j:

$$0 = \text{Det}(\mathbf{B}) = \sum_{k=1}^{n} b_{jk}(-1)^{j+k} \text{Det}\left[\mathbf{M}^{jk}(\mathbf{B})\right] = \sum_{k=1}^{n} a_{ik}(-1)^{j+k} \text{Det}\left[\mathbf{M}^{jk}(\mathbf{A})\right] \blacksquare$$

EXAMPLE 12 Let $\mathbf{A} = \begin{bmatrix} 1 & 4 & 2 \\ 5 & -3 & 6 \\ 2 & 3 & 2 \end{bmatrix}$.

Multiply the entries in row two by the cofactors belonging to row one and add up the results, to see whether we get 0.

SOLUTION We have

$$5C_{11} - 3C_{12} + 6C_{13}$$

$$= 5(-1)^2 \text{Det}(\mathbf{M}^{11}) - 3(-1)^3 \text{Det}(\mathbf{M}^{12}) + 6(-1)^4 \text{Det}(\mathbf{M}^{13})$$

$$= 5 \text{Det} \begin{bmatrix} -3 & 6 \\ 3 & 2 \end{bmatrix} + 3 \text{Det} \begin{bmatrix} 5 & 6 \\ 2 & 2 \end{bmatrix} + 6 \text{Det} \begin{bmatrix} 5 & -3 \\ 2 & 3 \end{bmatrix}$$

$$= 5(-24) + 3(-2) + 6(21) = -120 - 6 + 126 = 0 \qquad \blacksquare$$

Vandermonde Matrix

The matrix that arises in the polynomial interpolation problem as described in Section 3.2, is called a **Vandermonde matrix**. If it involves n points, then its generic element is $v_{ij} = t_i^{j-1}$, where i and j run from 1 to n. It must be invertible, by Theorem 13, Section 3.2. As a matter of fact, the following theorem gives a formula for the determinant of the Vandermonde matrix.[2]

---

[2] Alexandre-Théophile Vandermonde (1735–1796) pursued a career as a violinist before turning to mathematical research and teaching. He was one of the first to study determinants as a worthwhile subject in itself.

THEOREM 8

The determinant of an $n \times n$ Vandermonde matrix $\mathbf{V}$ *having elements* t_i^{j-1} *is given by this formula:*

$$\mathrm{Det}(\mathbf{V}) = \prod_{1 \le j < i \le n} (t_i - t_j)$$

The $\prod$ symbol on the right signifies the product of all factors $(t_i - t_j)$, when $1 \le j < i \le n$. The formula shows at once that the determinant is *not* zero and thus that the matrix is invertible.

Consider the 4×4 Vandermonde matrix

$$\mathbf{V} = \begin{bmatrix} 1 & t_1 & t_1^2 & t_1^3 \\ 1 & t_2 & t_2^2 & t_2^3 \\ 1 & t_3 & t_3^2 & t_3^3 \\ 1 & t_4 & t_4^2 & t_4^3 \end{bmatrix}$$

The formula in Theorem 8 gives us

$$\prod_{1 \le j < i \le n} (t_i - t_j) = (t_2 - t_1)(t_3 - t_1)(t_4 - t_1)(t_3 - t_2)(t_4 - t_2)(t_4 - t_3)$$

$$= \prod_{j=1}^{n} \prod_{i=j+1}^{n} (t_i - t_j)$$

EXAMPLE 13 Find the determinant of this matrix by using the preceding formula, and verify the answer by using appropriate row operations to simplify the work.

$$\mathbf{V} = \begin{bmatrix} 1 & 5 & 25 & 125 \\ 1 & 4 & 16 & 64 \\ 1 & 3 & 9 & 27 \\ 1 & 2 & 4 & 8 \end{bmatrix}$$

SOLUTION We find

$$\mathrm{Det}(\mathbf{V}) = (t_2 - t_1)(t_3 - t_1)(t_4 - t_1)(t_3 - t_2)(t_4 - t_2)(t_4 - t_3)$$
$$= (-1)(-2)(-3)(-1)(-2)(-1) = 12$$

With row operations, we obtain the same result. ∎

Application: Coded Messages

A simple way to send a secret message is to assign a different integer to each letter in the alphabet and send the message as the resulting string of integers. For simplicity, we assign the positive integers $1, 2, 3, \ldots, 26$ to the letters of the alphabet A, B, C, ..., Z, in order, and use 27 for a space. Unfortunately, such a code is easy to break. However, matrix multiplication can be used to disguise the coded message further. We will use an integer matrix with determinant ± 1. It has an inverse with integer entries. (See General Exercise 35.) Using such a matrix to transform the message makes it more difficult to decipher. To construct a coding matrix with determinant ± 1, start with the identity matrix and apply swap and replacement operations with integer multipliers. For example, the message SEND HELP would be coded as $[19, 5, 14, 4, 27, 8, 5, 12, 16]$. Put the coded message into the columns of a matrix

$$\mathbf{K} = \begin{bmatrix} 19 & 4 & 5 \\ 5 & 27 & 12 \\ 14 & 8 & 16 \end{bmatrix}$$

Using the following integer coding matrix $\mathbf{N}$, which has $\text{Det}(\mathbf{N}) = -1$, we compute the product

$$\mathbf{NK} = \begin{bmatrix} 1 & 1 & 2 \\ 2 & 3 & 5 \\ 2 & 2 & 3 \end{bmatrix} \begin{bmatrix} 19 & 4 & 5 \\ 5 & 27 & 12 \\ 14 & 8 & 16 \end{bmatrix} = \begin{bmatrix} 42 & 47 & 49 \\ 103 & 129 & 126 \\ 70 & 86 & 82 \end{bmatrix} = \mathbf{S}$$

Then the coded message SEND HELP would be $[42, 103, 70, 47, 129, 86, 49, 126, 82]$. The message *received* can be decoded by multiplying by the inverse of the coding matrix

$$\mathbf{N}^{-1}\mathbf{S} = \begin{bmatrix} 1 & 1 & 2 \\ 2 & 3 & 5 \\ 2 & 2 & 3 \end{bmatrix} \begin{bmatrix} 19 & 4 & 5 \\ 5 & 27 & 12 \\ 14 & 8 & 16 \end{bmatrix} = \begin{bmatrix} 19 & 4 & 5 \\ 5 & 27 & 12 \\ 14 & 8 & 16 \end{bmatrix}$$

Once again, the coded message is contained in the columns of this matrix.

Mathematical Software

We can enter the data from Example 1 into the Maple system by using the following commands in Maple:

```
Maple
with(LinearAlgebra):
A := Matrix([[1,7,6],[-3,2,9],[11,-13,4]]);
M1 := Minor(A,2,1,output=matrix);
M2 := Minor(A,1,3,output=matrix);
```

Also, we can verify the results in Example 7 by using these commands in Maple:

```
Maple
with(LinearAlgebra):
A := Matrix([[1,4,2],[3,-3,6],[2,0,5]]);
Minor(Transpose(A),3,2,output=matrix);
Transpose(Minor(A),2,3,output=matrix);
```

We can use Maple to check our results from Example 11.

```
Maple
with(LinearAlgebra):
A := Matrix([[5,7,-2],[3,2,1],[-2,4,6]]);
B := Adjoint(A)/Determinant(A);
C := MatrixInverse(A);
```

Notice that Maple computes the *classical* adjoint.

Review of Determinant Notation and Properties

An $n \times n$ matrix has n^2 minors. Each minor is obtained from $\mathbf{A}$ by removing one row and one column from $\mathbf{A}$. Specifically, we denote by $\mathbf{M}^{ij}$ or $\mathbf{M}^{ij}(\mathbf{A})$ the minor resulting by the removal of the ith row and jth column from the matrix $\mathbf{A}$. The cofactor associated with the minor $\mathbf{M}^{ij}$ is defined to be the number $C_{ij} = (-1)^{i+j} \mathrm{Det}(\mathbf{M}^{ij})$. These cofactors are therefore simple real numbers obtained by evaluating the determinants of the minors and inserting signs $\pm$ as indicated. We can then put these cofactors into a matrix $\mathbf{C}$. An important relationship is $\mathbf{AC}^T = \mathrm{Det}(\mathbf{A})\mathbf{I}$. Thus, except for the factor $\mathrm{Det}(\mathbf{A})$, $\mathbf{C}^T$ is the inverse of $\mathbf{A}$.

SUMMARY 4.2

- A typical minor of $\mathbf{A}$ is $\mathbf{M}^{ij}$, obtained from $\mathbf{A}$ by removing row i and column j.

- A cofactor from $\mathbf{A}$ is a number $C_{ij} = (-1)^{i+j} \text{Det}(\mathbf{M}^{ij})$.

- $\text{Det}(\mathbf{A}) = \sum_{j=1}^{n}(-1)^{i+j}a_{ij}\text{Det}(\mathbf{M}^{ij}) = \sum_{j=1}^{n} a_{ij}C_{ij}$ (cofactor expansion of $\text{Det}(\mathbf{A})$ using row i.)

- $\text{Det}(\mathbf{A}) = \sum_{i=1}^{n}(-1)^{i+j}a_{ij}\text{Det}(\mathbf{M}^{ij}) = \sum_{i=1}^{n} a_{ij}C_{ij}$ (cofactor expansion of $\text{Det}(\mathbf{A})$ using column j.)

- $\text{Det}(\mathbf{AB}) = \text{Det}(\mathbf{A})\text{Det}(\mathbf{B})$, where $\mathbf{A}$ and $\mathbf{B}$ are arbitrary $n \times n$ matrices.

- $\mathbf{M}^{ij}(\mathbf{A}^T) = [\mathbf{M}^{ji}(\mathbf{A})]^T$

- $\text{Det}(\mathbf{A}) = \text{Det}(\mathbf{A}^T)$

- $\text{Det}(\mathbf{A})$ is a linear function of any one column (or row).

- (Cramer's rule.) The solution of $\mathbf{Ax} = \mathbf{b}$ is given by $x_j = \text{Det}[\mathbf{A}_j(\mathbf{b})]/\text{Det}(\mathbf{A})$, where $\mathbf{A}_j(\mathbf{b})$ is obtained from $\mathbf{A}$ by replacing the jth column by $\mathbf{b}$. (Here $\mathbf{A}$ is an $n \times n$ invertible matrix.)

- $(\mathbf{A}^{-1})_{ij} = [(-1)^{i+j}/\text{Det}(\mathbf{A})]\text{Det}(\mathbf{M}^{ji})$
  $\mathbf{A}^{-1} = (1/\text{Det}(\mathbf{A}))\mathbf{C}^T$
  (Here $\mathbf{A}$ is an $n \times n$ invertible matrix.)

- $\sum_{k=1}^{n} a_{ik}C_{jk} = \sum_{k=1}^{n} a_{ik}(-1)^{j+k}\text{Det}(\mathbf{M}^{jk}) = 0$ if $i \neq j$

(If the elements of a row in $\mathbf{A}$ are multiplied by the cofactors of a different row and added, the result is zero.) A similar result is true for columns instead of rows.

- The Vandermonde matrix of order n is denoted by $\mathbf{V}$ and has generic elements (t_i^{j-1}), where $1 \leq i, j \leq n$. The determinant of the nth order Vandermonde matrix is $\text{Det}(\mathbf{V}) = \prod_{1 \leq j < i \leq n}(t_i - t_j)$

- Summary of major properties:
  - $\text{Det}(\mathbf{A}^T) = \text{Det}(\mathbf{A})$
  - $\text{Det}(\mathbf{A}^{-1}) = 1/\text{Det}(\mathbf{A})$
  - $\text{Det}(\mathbf{AB}) = \text{Det}(\mathbf{A})\text{Det}(\mathbf{B})$
  - Usually, $\text{Det}(\mathbf{A} + \mathbf{B}) \neq \text{Det}(\mathbf{A}) + \text{Det}(\mathbf{B})$
  - $\text{Det}(c\mathbf{A}) = c^n\text{Det}(\mathbf{A})$ if $c \in \mathbb{R}$ and $\mathbf{A}$ is $n \times n$.
  - $\text{Det}(\mathbf{A}) = a_{11}a_{22}\cdots a_{nn}$ if $\mathbf{A}$ is upper or lower triangular.
  - $\text{Det}(\widetilde{\mathbf{A}}) = \text{Det}(\mathbf{A})$ if $\widetilde{\mathbf{A}}$ results from $\mathbf{A}$ by a row-replacement operation.
  - $\text{Det}(\widetilde{\mathbf{A}}) = -\text{Det}(\mathbf{A})$ if $\widetilde{\mathbf{A}}$ results from $\mathbf{A}$ by a row-swap operation.
  - $\text{Det}(\mathbf{A}) = 0$ if and only if the columns of $\mathbf{A}$ form a linearly dependent set. A similar result is true for the rows.
  - $\text{Det}(\mathbf{A}) = 0$ if and only if $\mathbf{A}$ is singular (not invertible).

KEY CONCEPTS 4.2

Minors, cofactors, row expansion, column expansion, Cramer's rule, Vandermonde matrix

GENERAL EXERCISES 4.2

1. Let $f(\mathbf{x}) = \text{Det} \begin{bmatrix} 3 & 7 & 2 \\ x_1 & x_2 & x_3 \\ 5 & 4 & -3 \end{bmatrix}$

Establish that f is a linear function of $\mathbf{x}$.

2. Use Cramer's rule to solve the system

$\mathbf{Ax} = \mathbf{b}$ if $\mathbf{A} = \begin{bmatrix} 3 & 2 & 7 \\ 1 & -4 & 1 \\ 4 & -1 & 3 \end{bmatrix}$ and

$\mathbf{b} = [1, 1, 1]^T$

3. In Example 4, which numbers are involved in computing Det($\mathbf{A}$) and which are not?

4. Let $\mathbf{A} = \begin{bmatrix} 1 & 3 & 2 \\ -2 & 1 & 4 \\ 3 & 2 & 1 \end{bmatrix}, \mathbf{B} = \begin{bmatrix} 2 & 3 & 1 \\ 4 & 1 & 1 \\ 6 & 2 & 2 \end{bmatrix}$

Verify the assertion in Theorem 2 for this pair of matrices.

5. Find the inverse of the 3×3 matrix

$\mathbf{A} = \begin{bmatrix} 1 & 2 & 3 \\ 4 & -1 & 2 \\ -3 & 1 & 1 \end{bmatrix}$ by the formula in

Theorem 6. Along the way you may wish to use the minors of $\mathbf{A}$, which are in the

matrix $\mathbf{M} = \begin{bmatrix} -3 & 10 & 1 \\ -1 & 10 & 7 \\ 7 & -10 & -9 \end{bmatrix}$

6. Use Cramer's rule to solve this system of equations:

$\begin{bmatrix} 3 & 2 & 7 \\ 1 & -4 & 1 \\ 4 & -1 & 3 \end{bmatrix} \begin{bmatrix} x_1 \\ x_2 \\ x_3 \end{bmatrix} = \begin{bmatrix} 7 \\ 5 \\ 3 \end{bmatrix}$

7. Use Cramer's rule to solve this system:

$\begin{bmatrix} 2 & -1 & 3 \\ 3 & 1 & -2 \\ 1 & -3 & 1 \end{bmatrix} \begin{bmatrix} x_1 \\ x_2 \\ x_3 \end{bmatrix} = \begin{bmatrix} 7 \\ 5 \\ 11 \end{bmatrix}$

8. Compute Det $\begin{bmatrix} -2 & 1 & 2 & 5 \\ 2 & 0 & 4 & 5 \\ 3 & 1 & 0 & 0 \\ 1 & 2 & 0 & 0 \end{bmatrix}$ by

cofactor expansions and by row operations.

9. Compute Det $\begin{bmatrix} 2 & 1 & 3 \\ 1 & 0 & 4 \\ -2 & 3 & 5 \end{bmatrix}$ by a

cofactor expansion using row three and then column two.

10. In $\begin{bmatrix} 1 & 3 & 5 \\ 2 & 0 & 1 \\ 1 & 6 & 7 \end{bmatrix}$, what are the minor $\mathbf{M}^{23}$, and

the cofactor C_{23}?

11. Compute Det $\begin{bmatrix} 5 & 4 & 2 & 1 \\ 2 & 3 & 1 & -2 \\ -5 & -7 & -3 & 9 \\ 1 & -2 & -1 & 4 \end{bmatrix}$

by doing the following steps (in order): row operations using the pivot position $(2, 3)$, cofactor expansion by the third column, and directly with a 3×3 matrix.

12. Let $\mathbf{A} = \begin{bmatrix} 1 & 1 & 3 \\ 5 & 4 & 2 \\ 1 & 3 & -4 \end{bmatrix}$

Verify Det($\mathbf{A}$) = Det($\mathbf{A}^T$)

13. Let $\mathbf{A} = \begin{bmatrix} 1 & 3 & 2 \\ -2 & 1 & 4 \\ 3 & 2 & 1 \end{bmatrix}, \mathbf{B} = \begin{bmatrix} 2 & 3 & 1 \\ 4 & 1 & 1 \\ 6 & 2 & 2 \end{bmatrix}$

Verify Det($\mathbf{AB}$) = Det($\mathbf{A}$) Det($\mathbf{B}$)

14. Use Theorem 6 to compute inverses of these three matrices:

a. $A = \begin{bmatrix} 1 & 0 & 4 \\ -2 & 3 & 1 \\ 0 & 1 & 3 \end{bmatrix}$

b. $B = \begin{bmatrix} 0 & 2 & 4 \\ 3 & 0 & 1 \\ 2 & 3 & -4 \end{bmatrix}$

c. $C = \begin{bmatrix} 1 & 1 & 2 \\ -1 & 3 & 4 \\ 5 & 6 & -2 \end{bmatrix}$

15. Using the cofactor matrices, compute the inverses of these matrices:

$$A = \begin{bmatrix} 1 & 2 & 1 \\ 2 & 5 & 3 \\ 2 & 3 & 2 \end{bmatrix}, B = \begin{bmatrix} 1 & 1 & 2 \\ 1 & 2 & 3 \\ 2 & 2 & 3 \end{bmatrix}$$

16. Let $A = \begin{bmatrix} 1 & 7 & 3 \\ 2 & 1 & 5 \\ 4 & 4 & 3 \end{bmatrix}$

Use Cramer's rule to compute the $(2, 3)$ entry of A^{-1}.

17. Let $A = \begin{bmatrix} 1 & 7 & 3 \\ 2 & 1 & 5 \\ 4 & 4 & 3 \end{bmatrix}$

Find the cofactor matrix C.

18. Use Cramer's rule to solve this system:

$$\begin{cases} 2x_1 - x_2 + 3x_3 = 7 \\ 3x_1 + x_2 - 2x_3 = 5 \\ x_1 - 3x_2 + x_3 = 11 \end{cases}$$

19. Let $A = \begin{bmatrix} 1 & 2 & 1 \\ 3 & -1 & 2 \\ 1 & 0 & 1 \end{bmatrix}$

Express the inverse matrix in terms of the cofactor matrix. Verify the results by carrying out the multiplication AC^T.

20. Using the direct method, compute

$$\text{Det} \begin{bmatrix} 1 & 4 & -2 \\ 0 & 3 & 1 \\ 5 & 1 & 2 \end{bmatrix}$$

21. Compute $\text{Det} \begin{bmatrix} -2 & 1 & 2 \\ 2 & 0 & 4 \\ 3 & 1 & 0 \end{bmatrix}$ by a direct method.

22. Consider $\begin{bmatrix} 4 & 7 & 8 & 0 & 0 & 5 \\ 3 & 6 & 9 & 0 & 7 & 3 \\ 2 & 0 & 0 & 0 & 0 & 0 \\ -2 & 4 & 2 & 4 & 5 & 2 \\ 1 & 5 & 3 & 0 & 0 & 0 \\ 1 & -3 & 4 & 0 & 0 & 0 \end{bmatrix}$

Compute the determinant of this matrix using cofactor expansions.

23. Compute the determinant of

$$\begin{bmatrix} 0 & 7 & 4 & 3 \\ 0 & 0 & 3 & 0 \\ 0 & 5 & 7 & 3 \\ 4 & 6 & -5 & -4 \end{bmatrix}$$

24. Compute the determinant of

$$\begin{bmatrix} 2 & 1 & 0 & 4 \\ 0 & -1 & 0 & 2 \\ 7 & -2 & 3 & 5 \\ 0 & 1 & 0 & -3 \end{bmatrix}$$

25. Compute the determinant of

$$\begin{bmatrix} 3 & 1 & 2 & 0 \\ 4 & 0 & 0 & 0 \\ 5 & 1 & -4 & 5 \\ 3 & 2 & 0 & 0 \end{bmatrix}$$

26. Let $A = \begin{bmatrix} 0 & 0 & 0 & 4 \\ 7 & 0 & 5 & 6 \\ 4 & 3 & 7 & -5 \\ 3 & 0 & 3 & -4 \end{bmatrix}$

Compute the determinant of A.

27. Explain why the determinant function is a linear function of any one row of the matrix on which it acts.

28. Let A be an $n \times n$ matrix, and fix two indices, p and q in the range from 1 to n. Replace a_{pq} by x. Now $Det(A)$ is a function of the real variable x. What is the derivative of this function of x?

29. Explain why $Det(A^T B) = Det(A) Det(B)$.

30. Let A be an $n \times n$ matrix, where n is odd. Establish that if A is **skew–symmetric** (meaning $A^T = -A$), then $Det(A) = 0$.

31. Establish the column version of Theorem 7.

32. The determinant of a general 3×3 matrix was computed in the text just before Theorem 2. Compute the same determinant by a column expansion using the third column and verify that the results are the same.

33. Let A be an arbitrary $n \times n$ invertible matrix. Explain why in each row there is an element that can be changed to make the matrix noninvertible.

34. Establish Theorem 7 by a direct calculation in this one case: Consider an arbitrary 3×3 matrix, and use the elements of row 2 and the cofactors belonging to row 3. Work out all the terms and see that they add up to 0.

35. Establish that if the entries in a square matrix A are integers and if $Det(A) = \pm 1$, then A^{-1} has the same two properties.

36. Explain why, for two $n \times n$ matrices, $Det(AB) = Det(BA)$. Also, explain why this does not imply that $AB = BA$.

37. Establish that for arbitrary $n \times n$ matrices, $Det(A_1 A_2 \ldots A_k) = Det(A_1) Det(A_2) \ldots Det(A_k)$ using induction.

38. Give an argument why, for elementary matrices,
$$Det(E_1 E_2 \ldots E_k) = Det(E_1) Det(E_2) \ldots Det(E_k).$$

39. Establish the one remaining case of Lemma 1.

40. Complete the proof of Theorem 4.

41. If we change one entry in a square matrix, will the determinant necessarily change? Give examples, and try to discover the theorems governing this question.

42. Establish that if A is invertible, then $Det(A^{-1}) = [Det(A)]^{-1}$. Find some other functions, f, for which $Det[f(A)] = f[Det(A)]$. To what extent are we limited to invertible matrices in this exploration?

43. Let A be an $n \times n$ matrix in which each entry has the form $A_{ij} = \alpha_{ij} t + \beta_{ij}$. What can you say about the determinant of A, as a function of t?

44. Find an example in which $Det(A + B) \neq Det(A) + Det(B)$. This will establish that Det is not linear. Reconcile this finding with Theorem 4.

45. Let A and B be $n \times n$ matrices such that $Det(A) = 17$ and $Det(B) = 2$. What are the numerical values of $Det(AB)$, $Det(BA^T)$, $Det(2A)$, $Det(A^{-1})$, and $Det(B^2)$?

46. Using determinants, prove that the product of two invertible $n \times n$ matrices is invertible.

47. Using Cramer's rule, derive the equation $ax + by + cz = d$ of a plane containing the three noncolinear points (x_1, y_1, z_1), (x_2, y_2, z_2), (x_3, y_3, z_3) in which the coefficients a, b, c, d can be found by determinants.

48. Is the formula $\prod_{i=2}^{n} \prod_{j=1}^{i} (t_i - t_j)$ correct for the Vandermonde determinant? Explain why or why not.

49. Suppose the vector $\mathbf{w} = (-1, 3, 7)$ is a linear combination of the vectors $\mathbf{u} = (4, 2, 7)$ and $\mathbf{v} = (3, 1, 4)$. When any one datum of this statement is altered, is it still true?

50. Determine which of the following inequalities are correct and which are incorrect:

a. $\text{Det} \begin{bmatrix} a & b \\ c & d \end{bmatrix} \le (|a| + |b|) \cdot (|c| + |d|)$

b. $\text{Det} \begin{bmatrix} a & b \\ c & d \end{bmatrix} \le \max\{|a|, |b|\} \cdot \max\{|c|, |d|\}$

c. $\text{Det} \begin{bmatrix} a & b \\ c & d \end{bmatrix} \le \sqrt{a^2 + b^2} \cdot \sqrt{c^2 + d^2}$

d. $\text{Det} \begin{bmatrix} a & b \\ c & d \end{bmatrix} \le 2C^2$ if $|a| \le C, |b| \le C,$ $|c| \le C,$ and $|d| \le C$

51. **a.** To fit the data points (t_0, y_0) and (t_1, y_1) with the polynomial $p(t) = a + bt$, set up a linear system of equations and solve it using Cramer's rule.

b. Repeat for the polymomial $p(t) = c + d(t - t_0)$. *Hint:* Use Cramer's rule to solve the 2×2 system in Section 2.2 on linear least squares.

52. (Continuation.) Find the polynomial of the form $p(t) = a + b(t - t_0) + c(t - t_0)(t - t_1)$ that fits the data points $(t_0, y_0), (t_1, y_1),$ and (t_2, y_2) using Cramer's rule.

53. Let $\mathbf{A} = \begin{bmatrix} 3 & 0 & 0 \\ -2 & 2 & 0 \\ 5 & 7 & 1 \end{bmatrix}$

$\mathbf{B} = \begin{bmatrix} -2 & 6 & 5 \\ 0 & 1 & 8 \\ 0 & 0 & 3 \end{bmatrix}$

and $\mathbf{C} = \mathbf{AB}$. What is the numerical value of $\text{Det}(\mathbf{C})$?

COMPUTER EXERCISES 4.2

1. In an $n \times n$ Vandermonde matrix $\mathbf{V}$, the (i, j)-element is $v_{ij} = t_i^{j-1}$. Use the symbol manipulation feature of Maple or Mathematica to obtain the general formula for the determinant of the 5×5 Vandermonde matrix. The $n \times n$ Vandermonde matrix is

$$\mathbf{V} = \begin{bmatrix} 1 & t_1 & t_1^2 & \cdots & t_1^{n-1} \\ 1 & t_2 & t_2^2 & \cdots & t_2^{n-1} \\ \vdots & \vdots & \vdots & \ddots & \vdots \\ 1 & t_n & t_n^2 & \cdots & t_n^{n-1} \end{bmatrix}$$

a. Devise a program for computing the determinant of an $n \times n$ Vandermonde matrix.

b. Use the Maple commands on page 288 to help you see a pattern in the factorization of the determinant of a Vandermonde matrix:

```
Maple
with(LinearAlgebra):
factor(Determinant(VandermondeMatrix([a,b])));
factor(Determinant(VandermondeMatrix([a,b,c])));
factor(Determinant(VandermondeMatrix([a,b,c,d])));
factor(Determinant(VandermondeMatrix([a,b,c,d,e])));
```

2. Using mathematical software, show that

$$\mathrm{Det} \begin{bmatrix} 1 & 1 & 1 & 1 & 1 \\ a & 1 & 1 & 1 & 1 \\ a & b & 1 & 1 & 1 \\ a & b & c & 1 & 1 \\ a & b & c & d & 1 \end{bmatrix}$$

$$= (1-a)(1-b)(1-c)(1-d)$$

Show that this result holds true for matrices of various sizes. What conditions guarantee that the matrix is noninvertible?

3. Compute $\mathbf{C}^{-1}$ where

$$\mathbf{C} = \begin{bmatrix} 0 & 1 & 1 & 1 & 1 \\ 1 & 0 & 1 & 1 & 1 \\ 1 & 1 & 0 & 1 & 1 \\ 1 & 1 & 1 & 0 & 1 \\ 1 & 1 & 1 & 1 & 0 \end{bmatrix}$$

4. Consider this $n \times n$ tridiagonal matrix

$$\mathbf{D}_n = \begin{bmatrix} 2 & -1 & 0 & 0 & \cdots & 0 \\ -1 & 2 & -1 & 0 & \cdots & 0 \\ 0 & -1 & 2 & -1 & \cdots & 0 \\ \vdots & & \ddots & \ddots & \ddots & \vdots \\ 0 & \cdots & \cdots & -1 & 2 & -1 \\ 0 & \cdots & \cdots & 0 & -1 & 2 \end{bmatrix}$$

Let $d_n = \mathrm{Det}(\mathbf{D}_n)$. For various values of n, verify numerically that $d_1 = 2$, $d_2 = 3$, and $d_n = 2d_{n-1} - d_{n-2}$ for $n \geq 3$. Moreover, we obtain $d_n = n + 1$.

5. Consider this $n \times n$ matrix

$$\mathbf{B}_n = \begin{bmatrix} a_1 + 1 & a_2 & a_3 & \cdots & a_{n-1} & a_n \\ a_1 & a_2 + 1 & a_3 & \cdots & a_{n-1} & a_n \\ a_1 & a_2 & a_3 + 1 & \cdots & a_{n-1} & a_n \\ \vdots & \vdots & \vdots & \ddots & \vdots & \vdots \\ a_1 & a_2 & a_3 & \cdots & a_{n-1} + 1 & a_n \\ a_1 & a_2 & a_3 & \cdots & a_{n-1} & a_n + 1 \end{bmatrix}$$

For various values of n, verify numerically that

$$\mathrm{Det}(\mathbf{B}_n) = 1 + a_1 + a_2 + \cdots + a_n$$

$$= 1 + \sum_{i=1}^{n} a_i$$

6. Compute $\mathrm{Det}(\mathbf{A}_3(\mathbf{b}))$ where

$$\mathbf{A} = \begin{bmatrix} 1 & 2 & 2 & 2 & 2 \\ 2 & 1 & 2 & 2 & 2 \\ 2 & 2 & 1 & 2 & 2 \\ 2 & 2 & 2 & 1 & 2 \\ 2 & 2 & 2 & 2 & 1 \end{bmatrix}, \quad \mathbf{b} = \begin{bmatrix} 1 \\ 1 \\ 1 \\ 1 \\ 1 \end{bmatrix}$$

Then use Cramer's Rule (Theorem 5) to compute x_3 for the linear system $\mathbf{A}\mathbf{x} = \mathbf{b}$.

CHAPTER FIVE

Vector Subspaces

5.1 COLUMN, ROW, AND NULL SPACES

> *The essence of mathematics is not to make simple things compli-cated, but to make complicated things simple.*
>
> —S. GUDDEN

> *Everything should be made as simple as possible, but not simpler.*
> —ALBERT EINSTEIN (1879–1955)

> *KISS Principle: Keep It Simple, Stupid!*

Introduction

In Example 5 of Section 2.4, we encountered a vector space that was con-tained in a larger vector space and shared with the larger space its def-initions of vector addition and scalar multiplication. This is a common phenomenon, deserving our special study.

DEFINITION

A subset in a vector space is a **subspace** *if it is nonempty and closed under the operations of adding vectors and multipling vectors by scalars.*

A subspace is nonempty by the definition of the term "*subspace.*" If an element **x** is in the subspace, then **0** is in the subspace because **0** = 0 · **x**. Thus, a subspace must contain the zero element of the larger space. We can state these requirements for a subspace U as follows:

1. If **x** and **y** are in U, then **x** + **y** is in U.
   (U is closed under vector addition.)
2. If **x** is in U and c is a scalar, then c**x** is in U.
   (U is closed under scalar multiplication.)
3. U is nonempty.

One sees immediately that under those circumstances, U is a vector space, too. All the axioms listed in Section 2.4 will be true for U. Most of these axioms are true automatically because U is a subset of V, and V obeys the axioms. The closure axioms are different in nature, however. Furthermore, we want to be sure that U is nonempty. The element that must be there is, of course, **0**.

We notice that two *extreme* cases of subspaces occur in any vector space V: namely, the subspace **0** and V itself. Any other subspace, U, must lie between these extremes: $\mathbf{0} \subseteq U \subseteq V$. The subspace of V that contains only the zero vector is denoted by **0** or {**0**}. We summarize all this in a formal statement:

> **THEOREM 1**
>
> *Let U be a subset of a vector space V. Suppose that U contains the zero element and is closed under vector addition and multiplication by scalars. Then U is a subspace of V.*

> **THEOREM 2**
>
> *The span of a nonempty set in a vector space is a subspace of that vector space.*

PROOF Let the vector space be V and let S be a nonempty subset of V. If **x** and **y** are two elements of Span(S), then each is a linear combination of elements of S. Simple algebra shows that the vector $\alpha\mathbf{x} + \beta\mathbf{y}$ will also be a linear combination of elements in S. This verifies the two closure axioms for Span(S). ∎

There are many examples of linear subspaces, such as subspaces of polynomials, subspaces of trigonometric polynomials, subspaces of matrices $\mathbb{R}^{m \times n}$, subspaces of upper or lower triangular matrices, and subspaces of the image or the kernel of some linear operators such as differentiation and integration.

EXAMPLE 1 In the vector space $\mathbb{P}$ consisting of all polynomials (of any degree), consider the four monomials p_0, p_1, p_2, and p_3, where $p_i(t) = t^i$. What is the span of the set $\{p_0, p_1, p_2, p_3\}$?

SOLUTION This is the vector space $\mathbb{P}_3$. It is a subspace of $\mathbb{P}$. (The zero-polynomial has no degree.) ∎

EXAMPLE 2 Illustrate Theorem 2.

SOLUTION The set of polynomials of the form $t \mapsto at + bt^2 + ct^5 + dt^7$ is a subspace of $\mathbb{P}$. ($\mathbb{P}$ is defined in Example 1.) ∎

EXAMPLE 3 Let s and t be variables that range independently over $\mathbb{R}$. Consider the vectors of the form

$$[s + 4t, \; 3s - t, \; 5s + t, \; 2t]^T$$

Is the set of all vectors of that form a subspace of $\mathbb{R}^4$?

SOLUTION We have no theorem that directly addresses such a problem. However, we may be able to see that the set in question is the linear span of a set of vectors. If so, Theorem 1 will apply. Observe that

$$\begin{bmatrix} s + 4t \\ 3s - t \\ 5s + t \\ 2t \end{bmatrix} = s \begin{bmatrix} 1 \\ 3 \\ 5 \\ 0 \end{bmatrix} + t \begin{bmatrix} 4 \\ -1 \\ 1 \\ 2 \end{bmatrix}$$

Hence, the set of vectors considered at the beginning is identical to the span of a pair of vectors:

$$\{[s + 4t, \; 3s - t, \; 5s + t, \; 2t]^T : s, t \in \mathbb{R}\} = \text{Span}\{[1, 3, 5, 0]^T, [4, -1, 1, 2]^T\}$$

This pair is obviously linearly independent, and, therefore, generates a two-dimensional subspace of $\mathbb{R}^4$. (The concept of dimension is taken up in Section 5.2.) ∎

From two subspaces X and Y in a vector space, we can construct two more subspaces, $X + Y$ and $X \cap Y$. (These arise in General Exercises 13–14.) The **vector sum**, $X + Y$, and the **intersection**, $X \cap Y$, are defined as follows:

$$X + Y = \{\mathbf{x} + \mathbf{y} : \mathbf{x} \in X \text{ and } \mathbf{y} \in Y\}$$
$$X \cap Y = \{\mathbf{x} : \mathbf{x} \in X \text{ and } \mathbf{x} \in Y\}$$

The union of two subspaces is usually *not* a subspace! (See General Exercises 15–16.)

THEOREM 3

The vector sum of two subspaces in a vector space is also a subspace.

PROOF Let X and Y be two subspaces of a vector space V. Then X and Y contain the zero element of V. Therefore, $X + Y$ contains the zero element of V. Is $X + Y$ closed under addition? Let $\mathbf{v}$ and $\widehat{\mathbf{v}}$ be two points in $X + Y$. Then $\mathbf{v} = \mathbf{x} + \mathbf{y}$ and $\widehat{\mathbf{v}} = \widehat{\mathbf{x}} + \widehat{\mathbf{y}}$ for suitable points $\mathbf{x}, \mathbf{y}, \widehat{\mathbf{x}}$, and $\widehat{\mathbf{y}}$. Thus, we obtain $\mathbf{v} + \widehat{\mathbf{v}} = \mathbf{x} + \mathbf{y} + \widehat{\mathbf{x}} + \widehat{\mathbf{y}} = (\mathbf{x} + \widehat{\mathbf{x}}) + (\mathbf{y} + \widehat{\mathbf{y}}) \in X + Y$. A similar equation shows that $X + Y$ is closed under scalar multiplication. ∎

EXAMPLE 4 In $\mathbb{R}^3$, let U be the subspace consisting of all scalar multiples of the vector $\mathbf{u} = (1, 3, -4)$. Let V be the subspace consisting of all scalar multiples of the vector $\mathbf{v} = (5, -1, 2)$. What is $U + V$?

SOLUTION The vector sum $U + V$ consists of all vectors expressible as $\alpha \mathbf{u} + \beta \mathbf{v}$. In other words, this is the span of the pair $\{\mathbf{u}, \mathbf{v}\}$. ∎

THEOREM 4

The intersection of two subspaces in a vector space is also a subspace of that vector space.

The proof is left as General Exercise 14.

EXAMPLE 5 Let U be the set of all vectors $\mathbf{u} = (u_1, u_2, u_3)$ in $\mathbb{R}^3$ such that $u_1 + 3u_2 - 4u_3 = 0$. Let V be the set of all vectors $\mathbf{v} = (v_1, v_2, v_3)$ such that $2v_1 + 3v_3 = 0$. What is $U \cap V$?

SOLUTION $U \cap V$ is the set of all $\mathbf{x}$ such that $x_1 + 3x_2 - 4x_3 = 0$ and $2x_1 + 3x_2 = 0$. Geometrically, we are considering two planes in $\mathbb{R}^3$ and their intersection, which in this case is a line in $\mathbb{R}^3$ through the origin. ■

Images and inverse images of subspaces by linear transformations are also subspaces.

DEFINITION

*If f is a mapping of a set X to a set Y, then for any set U in X, $f[U]$ is defined to be the set of **images** of points in U. In symbols, we have*

$$f[U] = \{f(x) \,:\, x \in U\}$$

THEOREM 5

If f is a linear transformation from a vector space X to a vector space Y, then for any subspace U in X, $f[U]$ is a subspace of Y.

PROOF Since U is a subspace, it contains $\mathbf{0}$. Since f is linear, $f(\mathbf{0}) = \mathbf{0}$. Thus, $\mathbf{0}$ is in $f[U]$. If $\mathbf{x}$ and $\mathbf{y}$ are in U, then so are $\alpha\mathbf{x} + \beta\mathbf{y}$. Hence, we have $f(\alpha\mathbf{x} + \beta\mathbf{y}) \in f[U]$. Since f is linear, $\alpha f(\mathbf{x}) + \beta f(\mathbf{y}) \in f[U]$. Thus, $f[U]$ is closed under addition and scalar multiplication. ■

EXAMPLE 6 This example illustrates Theorem 5 with specific vectors and subspaces. Define a linear transformation f from $\mathbb{R}^4$ to $\mathbb{R}^3$ by the equation
$$f(x_1, x_2, x_3, x_4) = (3x_1, \ 2x_2 - x_1, \ x_4)$$
Let W be the span of a set of two vectors in $\mathbb{R}^4$:

$$\mathbf{u} = (1, 2, 3, 0), \qquad \mathbf{v} = (0, 2, 3, 4)$$

Thus, f maps $\mathbb{R}^4$ into $\mathbb{R}^3$. Find a simple description of $f[W]$.

SOLUTION Using the formula for f we obtain

$$f(\mathbf{u}) = (3, 3, 0), \qquad f(\mathbf{v}) = (0, 4, 4)$$

It follows that

$$f[W] = f\big[\operatorname{Span}\{\mathbf{u}, \mathbf{v}\}\big] = \operatorname{Span}\big[\{f(\mathbf{u}), f(\mathbf{v})\}\big] = \operatorname{Span}\big\{(3, 3, 0), (0, 4, 4)\big\}$$
$$= \operatorname{Span}\big\{(1, 1, 0), (0, 1, 1)\big\}$$

The last expression defines a subspace, by Theorem 2.   ∎

DEFINITION

If f is a mapping from one set X to another set Y, then for any subset S in Y, $f^{-1}[S]$ is defined to be the set of all points in X that are mapped by f into S. This set is the **inverse image** *of S in X. In symbols, we have*

$$f^{-1}[S] = \{x \in X \,:\, f(x) \in S\}$$

See Figure 5.1 for an illustration of $f[U]$ and $f^{-1}[S]$. The notation $f^{-1}[S]$ involves a set S. Its use here does not imply that f is an invertible mapping.

THEOREM 6

If f is a linear transformation from a vector space X to a vector space Y, then for any subspace V in Y, $f^{-1}[V]$ is a subspace in X.

PROOF Since V is a subspace in Y, the zero vector is in V. Since f is a linear transformation and X is a vector space, $\mathbf{0} \in X$ and $f(\mathbf{0}) = \mathbf{0}$. Hence, we have $\mathbf{0} \in f^{-1}[V]$. If $\mathbf{x}$ and $\mathbf{y}$ are in $f^{-1}[V]$, then $\mathbf{x} = f(\mathbf{z})$ and $\mathbf{y} \in f(\mathbf{w})$, for some $\mathbf{z}$ and $\mathbf{w}$ in V. Thus, we obtain

$$\alpha \mathbf{x} + \beta \mathbf{y} = \alpha f(\mathbf{z}) + \beta f(\mathbf{w})$$
$$= f(\alpha \mathbf{z}) + f(\beta \mathbf{w})$$
$$= f(\alpha \mathbf{z} + \beta \mathbf{w})$$

This proves that $\alpha \mathbf{x} + \beta \mathbf{y} \in f^{-1}[V]$ and that $f^{-1}[V]$ is closed under vector addition and scalar multiplication. Hence, $f^{-1}[V]$ is a subspace of X.   ∎

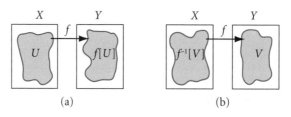

FIGURE 5.1 Sketch of (a) $f[U]$ and (b) $f^{-1}[V]$.

EXAMPLE 7 Continuing the previous example, let $V = \text{Span}\{\mathbf{z}, \mathbf{w}\}$, where $\mathbf{z} = (1, 1, 0)$ and $\mathbf{w} = (0, 1, 1)$. Describe the set $f^{-1}[V]$ in simple terms.

SOLUTION Letting $(3x_1, 2x_2 - x_1, x_4) = (1, 1, 0)$, we obtain $x_1 = \frac{1}{3}$, $x_2 = \frac{2}{3}$, and $x_4 = 0$. Hence, we have $f(\frac{1}{3}, \frac{2}{3}, x_3, 0) = (1, 1, 0)$. Similarly, letting $(3x_1, 2x_2 - x_1, x_4) = (0, 1, 1)$, we have $x_1 = 0$, $x_2 = \frac{1}{2}$, and $x_4 = 1$ and $f(0, \frac{1}{2}, x_3, 1) = (0, 1, 1)$. Consequently, we obtain

$$f^{-1}[V] = \text{Span}\left\{(\tfrac{1}{3}, \tfrac{2}{3}, s, 0), (0, \tfrac{1}{2}, s, 1)\right\} = \text{Span}\left\{(1, 2, s, 0), (0, 1, s, 2)\right\}$$

for any $s \in \mathbb{R}$. ∎

Linear Transformations

Recall the notion of null space (or kernel) of a linear transformation. It consists of all vectors that are mapped into $\mathbf{0}$ by the transformation.

THEOREM 7

The kernel of a linear transformation is a subspace in the domain of that transformation.

PROOF Let T be a linear transformation from one vector space V to another vector space W. The *kernel* or *null space* of T is the set

$$\text{Ker}(T) = \{\mathbf{v} \in V : T(\mathbf{v}) = \mathbf{0}\}$$

It is obviously a subset of V. It is a subspace of V because first, $\mathbf{0} \in \text{Ker}(T)$. Second, if $\mathbf{x}$ and $\mathbf{y}$ belong to the kernel of T, then $\alpha\mathbf{x} + \beta\mathbf{y}$ also belong to the kernel because

$$T(\alpha\mathbf{x} + \beta\mathbf{y}) = \alpha T(\mathbf{x}) + \beta T(\mathbf{y}) = \alpha\mathbf{0} + \beta\mathbf{0} = \mathbf{0} \qquad ∎$$

EXAMPLE 8 Let $(c_1, c_2, \ldots, c_n)$ be any fixed vector in $\mathbb{R}^n$. Define a linear transformation $T : \mathbb{R}^n \to \mathbb{R}$ by the following equation, in which $\mathbf{x} = (x_1, x_2, \ldots, x_n)$:

$$T(\mathbf{x}) = c_1 x_1 + c_2 x_2 + \ldots + c_n x_n$$

Describe the kernel of T.

SOLUTION It is the set of all vectors $\mathbf{x}$ that satisfy the equation

$$c_1 x_1 + c_2 x_2 + \ldots + c_n x_n = 0$$

If $n = 3$, we can visualize this set as a plane through the origin in $\mathbb{R}^3$. This is true if the vector $\mathbf{c}$ is not $\mathbf{0}$. If $\mathbf{c}$ is zero, the subspace here is all of $\mathbb{R}^3$. If $n = 2$, the set in question is a line through $\mathbf{0}$ if $\mathbf{c} \neq \mathbf{0}$. If $\mathbf{c} = \mathbf{0}$, the set is all of $\mathbb{R}^2$.   ■

EXAMPLE 9 Draw a sketch of the kernel of this linear map:
For $\mathbf{x} = (x_1, x_2)$, define $T(\mathbf{x}) = 3x_1 - 2x_2$.

SOLUTION The kernel of T consists of all vectors in $\mathbb{R}^2$ for which $3x_1 - 2x_2 = 0$. Thus, we have $x_2 = \frac{3}{2}x_1$. The graph of this equation is the straight line through the origin with slope $\frac{3}{2}$. The two points $(0, 0)$ and $(2, 3)$ determine this line. See Figure 5.2.   ■

EXAMPLE 10 We can obtain a subspace by this description:
Take all the vectors $\mathbf{x} = (x_1, x_2, x_3, x_4)$ in $\mathbb{R}^4$ that are linear combinations of $(3, 7, 4, 2)$ and $(2, 0, 1, 3)$, and in addition obey the equation

$$3x_1 + x_2 - 4x_3 + x_4 = 0$$

Here we are taking the intersection of two subspaces to get a third subspace.

SOLUTION We have

$$\mathbf{x} = a(3, 7, 4, 2) + b(2, 0, 1, 3) = (3a + 2b, 7a, 4a + b, 2a + 3b)$$

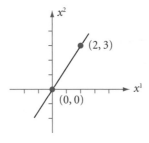

FIGURE 5.2 Kernel of $T(\mathbf{x}) = 3x_1 - 2x_2$.

We also set

$$0 = 3x_1 + x_2 - 4x_3 + x_4 = 3(3a + 2b) + 7a - 4(4a + b) + (2a + 3b) = 2a + 5b$$

Thus, we obtain $b = -\frac{2}{5}a$. In the first preceding equation, replace b by this multiple of a, and arrive at

$$\mathbf{x} = a(3 - \tfrac{4}{5}, \ 7, \ 4 - \tfrac{2}{5}, \ 2 - \tfrac{6}{5}) = \tfrac{1}{5}a(11, \ 35, \ 18, \ 4) = c(11, \ 35, \ 18, \ 4)$$

The solution is that the points described are all the scalar multiples of the vector $(11, 35, 18, 4)$. ∎

EXAMPLE 11 In $\mathbb{R}^4$, let U be the span of the single vector $\mathbf{u} = (3, 2, 4, 0)$. Let V be the kernel of the matrix

$$A = \begin{bmatrix} 2 & 1 & 3 & 4 \\ 1 & -1 & 2 & 1 \end{bmatrix}$$

Find a simple description of the subspace $U + V$ in $\mathbb{R}^4$.

SOLUTION To understand the kernel of A, we use the standard row-reduction process to get

$$\begin{bmatrix} 2 & 1 & 3 & 4 & 0 \\ 1 & -1 & 2 & 1 & 0 \end{bmatrix} \sim \begin{bmatrix} 1 & 0 & \frac{5}{3} & \frac{5}{3} & 0 \\ 0 & 1 & -\frac{1}{3} & \frac{2}{3} & 0 \end{bmatrix}$$

The general solution of the associated homogeneous system is given by

$$x_1 = -\tfrac{5}{3}x_3 - \tfrac{5}{3}x_4 \qquad \text{and} \qquad x_2 = \tfrac{1}{3}x_3 - \tfrac{2}{3}x_4$$

As usual, we write this as

$$\begin{bmatrix} x_1 \\ x_2 \\ x_3 \\ x_4 \end{bmatrix} = x_3 \begin{bmatrix} -\frac{5}{3} \\ \frac{1}{3} \\ 1 \\ 0 \end{bmatrix} + x_4 \begin{bmatrix} -\frac{5}{3} \\ -\frac{2}{3} \\ 0 \\ 1 \end{bmatrix} = s \begin{bmatrix} -5 \\ 1 \\ 3 \\ 0 \end{bmatrix} + t \begin{bmatrix} -5 \\ -2 \\ 0 \\ 3 \end{bmatrix}$$

where s and t are free parameters. Thus, the subspace in question is the span of a set of three vectors; namely,

$$U + V = \text{Span}\{(3, 2, 4, 0), \ (-5, 1, 3, 0), \ (-5, -2, 0, 3)\} \qquad ∎$$

THEOREM 8

The range of a linear transformation is a subspace of its co-domain.

PROOF If T is a linear transformation, say from U to V, then its range is

$$\text{Range}(T) = \left\{ T(\mathbf{x}) : \mathbf{x} \in U \right\}$$

The vector $\mathbf{0}$ of V is in $\text{Range}(T)$ because $\mathbf{0} = T(\mathbf{0})$. Thus, V is nonempty. If $\mathbf{x}$ and $\mathbf{y}$ are in $\text{Range}(T)$, then $\mathbf{x} = T(\mathbf{u})$ and $\mathbf{y} = T(\mathbf{w})$, for suitable $\mathbf{u}$ and $\mathbf{w}$ in U. It follows that

$$\alpha\mathbf{x} + \beta\mathbf{y} = \alpha T(\mathbf{u}) + \beta T(\mathbf{w}) = T(\alpha\mathbf{u} + \beta\mathbf{w})$$

This shows that $\text{Range}(T)$ is closed under vector addition and scalar multiplication. Notice that we used the linearity of T and the hypotheses that U and V are vector spaces. ■

EXAMPLE 12 Let $\mathbf{A} = \begin{bmatrix} 1 & -2 & -4 \\ 3 & 4 & 18 \\ 7 & -5 & -1 \end{bmatrix}$, and define T by the equation $T(\mathbf{x}) = \mathbf{A}\mathbf{x}$. Describe the range of T in simpler terms.

SOLUTION Every point in the range of T is of the form $\mathbf{A}\mathbf{x}$ for some $\mathbf{x}$. Thus, the range of T is the span of the columns of $\mathbf{A}$. If the columns of $\mathbf{A}$ form a linearly dependent set, then some columns are not needed in describing the range of T. A row reduction shows that

$$\mathbf{A} \sim \mathbf{B} = \begin{bmatrix} 1 & 0 & 2 \\ 0 & 1 & 3 \\ 0 & 0 & 0 \end{bmatrix}$$

Because the systems $\mathbf{A}\mathbf{x} = \mathbf{0}$ and $\mathbf{B}\mathbf{x} = \mathbf{0}$ have the same solutions, we can use columns 1 and 2 in $\mathbf{A}$ to span the range of T:

$$\text{Range}(T) = \text{Span} \left\{ \begin{bmatrix} 1 \\ 3 \\ 7 \end{bmatrix}, \begin{bmatrix} -2 \\ 4 \\ -5 \end{bmatrix} \right\}$$

Notice that we cannot use the columns of **B** to describe the range of **A** in this problem. Each of those columns has 0 as its last entry, whereas the range of **A** obviously contains some vectors that do not have that property. ∎

Revisiting Kernels and Null Spaces

Recall from your study of elementary algebra the important topic of solving equations. For example, you learned how to find the "*roots*" of an equation such as $7x^2 + 3x - 15 = 0$. In linear algebra, the corresponding topic is the finding of solutions of an equation of the form $L(\mathbf{x}) = \mathbf{v}$, where L is a linear transformation. Even the simple case, $L(\mathbf{x}) = \mathbf{0}$, requires some study. Several of the preceding chapters have dealt with the matrix version of that topic.

Recall that the terms **null space** and **kernel** are interchangable. They both refer to

$$\{\mathbf{x} \ : \ L(\mathbf{x}) = \mathbf{0}\} \qquad \text{or} \qquad \{\mathbf{x} \ : \ \mathbf{Ax} = \mathbf{0}\}$$

The first case is more general as it does not require a matrix, but only a linear mapping. We define the **null space** of a matrix **A** as the set

$$\text{Null }(\mathbf{A}) = \{\mathbf{x} : \mathbf{Ax} = \mathbf{0}\}$$

which we can find by solving $\mathbf{Ax} = \mathbf{0}$.

EXAMPLE 13 Let

$$C = \begin{bmatrix} 3 & -4 & 2 & 5 \\ 1 & 3 & 5 & 6 \end{bmatrix}$$

Are the two vectors, $\mathbf{u} = [3, 7, -8, 9]^T$ and $\mathbf{v} = [7, 1, 4, -5]^T$, in the null space of **C**? Are these vectors $\mathbf{w} = [5, -3]$ and $\mathbf{z} = [0, 0]$ in the left null space of $\mathbf{C}^T$?

SOLUTION The question is answered by computing $\mathbf{Cu} = [10, 38]^T$ and $\mathbf{Cv} = [0, 0]^T$. Thus, we obtain $\mathbf{v} \in \text{Null}(\mathbf{C})$, but $\mathbf{u} \notin \text{Null}(\mathbf{C})$. Also, we find that **w** is not in the null space of $\mathbf{C}^T$, but **z** is because $\mathbf{C}^T\mathbf{w} \neq \mathbf{0}$ and $\mathbf{C}^T\mathbf{z} = \mathbf{0}$. ∎

EXAMPLE 14 Use the matrix $\mathbf{C}$ in the preceding example, and find simple descriptions of the null spaces of $\mathbf{C}$ and $\mathbf{C}^T$.

SOLUTION The null space of $\mathbf{C}$ consists of all solutions to the equation $\mathbf{Cx} = \mathbf{0}$. One can use row reduction on the augmented matrix:

$$[\mathbf{C} \mid \mathbf{0}] = \begin{bmatrix} 3 & -4 & 2 & 5 & \mid & 0 \\ 1 & 3 & 5 & 6 & \mid & 0 \end{bmatrix} \sim \begin{bmatrix} 1 & 0 & 2 & 3 & \mid & 0 \\ 0 & 1 & 1 & 1 & \mid & 0 \end{bmatrix}$$

For the associated homogeneous system, the general solution is

$$x_1 = -2x_3 - 3x_4 \qquad \text{and} \qquad x_2 = -x_3 - x_4$$

This can be written as

$$\begin{bmatrix} x_1 \\ x_2 \\ x_3 \\ x_4 \end{bmatrix} = x_3 \begin{bmatrix} -2 \\ -1 \\ 1 \\ 0 \end{bmatrix} + x_4 \begin{bmatrix} -3 \\ -1 \\ 0 \\ 1 \end{bmatrix} = x_3 \mathbf{w} + x_4 \mathbf{z}$$

where x_3 and x_4 are free variables. Thus, the null space is the span of a pair of vectors; namely, $\mathbf{w} = [-2, -1, 1, 0]^T$ and $\mathbf{z} = [-3, -1, 0, 1]^T$.

With regard to the null space of $\mathbf{C}^T$, we obtain

$$[\mathbf{C}^T \mid \mathbf{0}] = \begin{bmatrix} 3 & 1 & \mid & 0 \\ -4 & 3 & \mid & 0 \\ 2 & 5 & \mid & 0 \\ 5 & 6 & \mid & 0 \end{bmatrix} \sim \begin{bmatrix} 1 & 0 & \mid & 0 \\ 0 & 1 & \mid & 0 \\ 0 & 0 & \mid & 0 \\ 0 & 0 & \mid & 0 \end{bmatrix}$$

Thus the null space of $\mathbf{C}^T$ is $\{\mathbf{0}\}$.   ∎

The Row Space and Column Space of a Matrix

Two important vector subspaces associated with a given matrix are its row space and its column space.

DEFINITION

The **row space** *of a matrix* $\mathbf{A}$ *is the span of the set of rows in* $\mathbf{A}$. *This is denoted by* Row($\mathbf{A}$).

The **column space** *of* $\mathbf{A}$ *is the span of the set of columns in* $\mathbf{A}$. *This is denoted by* Col($\mathbf{A}$).

Translating these definitions into alternative terminology, we can say that

$$\text{Row}(\mathbf{A}) = \text{Span}\{\mathbf{r}_1, \mathbf{r}_2, \ldots, \mathbf{r}_m\}$$
$$\text{Col}(\mathbf{A}) = \text{Span}\{\mathbf{a}_1, \mathbf{a}_2, \ldots, \mathbf{a}_n\}$$

Here we have assumed that $\mathbf{A}$ is an $m \times n$ matrix whose rows are $\mathbf{r}_i$ and whose columns are $\mathbf{a}_j$.

THEOREM 9

If two matrices are row equivalent to each other, then they have the same row space.

PROOF Let $\mathbf{A}$ be any matrix. We will prove that any single row operation performed on $\mathbf{A}$ to produce a new matrix $\widetilde{\mathbf{A}}$ has no effect on the row space of $\mathbf{A}$. Once this has been established, it will follow that a sequence of row operations performed on $\mathbf{A}$ will also have no effect on the row space. In carrying out this proof, it suffices to consider just the two row operations labeled as Types 1 and 2 in Section 1.1. Consider first the operation of adding to one row a multiple of another (**replacement**). There is no loss of generality in considering the operation $\mathbf{r}_1 \leftarrow \mathbf{r}_1 + \alpha \mathbf{r}_2$. This notation means *replace the first row by row 1 plus α times row 2*. Any element in the row space of $\widetilde{\mathbf{A}}$ has the form

$$c_1(\mathbf{r}_1 + \alpha \mathbf{r}_2) + c_2 \mathbf{r}_2 + \ldots + c_n \mathbf{r}_n$$

This is clearly a linear combination of rows $\mathbf{r}_1, \mathbf{r}_2, \ldots \mathbf{r}_n$. Hence, we have $\text{Row}(\widetilde{\mathbf{A}}) \subseteq \text{Row}(\mathbf{A})$. The reverse inclusion is also true because the row operation in question is reversible by another row operation of the same type (explicitly, use $\mathbf{r}_1 \leftarrow \mathbf{r}_1 - \alpha \mathbf{r}_2$). A similar analysis handles any row operation of Type 2 (**scale**, multiplying a row by a nonzero scalar). ∎

EXAMPLE 15 Find simple descriptions of the row space and column space of this matrix:

$$\mathbf{A} = \begin{bmatrix} 1 & 3 & 2 & 4 \\ 1 & 0 & 4 & -2 \\ 2 & 2 & 1 & 7 \\ 4 & 5 & 7 & 9 \end{bmatrix}$$

SOLUTION Because row operations generally simplify a matrix and the row space is not affected, we carry out a row reduction, arriving at

$$
\mathbf{A} \sim
\begin{bmatrix}
17 & 0 & 0 & 50 \\
0 & 17 & 0 & 20 \\
0 & 0 & 17 & -21 \\
0 & 0 & 0 & 0
\end{bmatrix}
= \mathbf{C} \sim
\begin{bmatrix}
1 & 0 & 0 & 50/17 \\
0 & 1 & 0 & 20/17 \\
0 & 0 & 1 & -21/17 \\
0 & 0 & 0 & 0
\end{bmatrix}
= \mathbf{B}
$$

The row space of $\mathbf{A}$ is the row space of $\mathbf{B}$, and it is spanned by the first three rows of $\mathbf{B}$. We can write

$$
\begin{aligned}
\mathrm{Row}(\mathbf{A}) &= \mathrm{Row}(\mathbf{B}) = \mathrm{Row}(\mathbf{C}) \\
&= \mathrm{Span}\{[17, 0, 0, 50], [0, 17, 0, 20], [0, 0, 17, -21]\}
\end{aligned}
$$

the latter being a more elegant description. The rows of an augmented matrix correspond to equations in a linear system. In the row-reduction process, each system has the same solution. Consequently, once the pivot positions are determined, we can use the corresponding rows in any of these coefficient matrices to span the row space! Conversely, the column space of $\mathbf{A}$ is given only by the span of the columns from $\mathbf{A}$ corresponding to the pivot positions. In this example, we obtain:

$$
\mathrm{Col}(\mathbf{A}) = \mathrm{Span}
\left\{
\begin{bmatrix} 1 \\ 1 \\ 2 \\ 4 \end{bmatrix},
\begin{bmatrix} 3 \\ 0 \\ 2 \\ 5 \end{bmatrix},
\begin{bmatrix} 2 \\ 4 \\ 1 \\ 7 \end{bmatrix}
\right\}
$$

Think about why this is so!   ∎

EXAMPLE 16 Do these matrices have the same row space and column space?

$$
\mathbf{G} =
\begin{bmatrix}
1 & 3 & 7 & 4 \\
2 & 5 & 1 & 5 \\
2 & 1 & 0 & 3
\end{bmatrix},
\qquad
\mathbf{H} =
\begin{bmatrix}
10 & 23 & 30 & 29 \\
1 & 4 & 20 & 7 \\
12 & 13 & 40 & 29
\end{bmatrix}
$$

SOLUTION We examine the reduced row echelon form of each matrix.

Calling upon mathematical software and changing some numbers into simple fractions, we arrive at

$$\mathbf{G} \sim \begin{bmatrix} 1 & 0 & 0 & 65/51 \\ 0 & 1 & 0 & 23/51 \\ 0 & 0 & 1 & 10/51 \end{bmatrix}, \qquad \mathbf{H} \sim \begin{bmatrix} 51 & 0 & 0 & 65 \\ 0 & 51 & 0 & 23 \\ 0 & 0 & 51 & 10 \end{bmatrix}$$

The answer is *yes!* The row spaces are the same, and the three rows of this last matrix form a simple set of three vectors that spans the common row space. The first three columns of $\mathbf{G}$ form a linearly independent set that spans $\text{Col}(\mathbf{G}) = \mathbb{R}^3$. Similarly, the first three columns of $\mathbf{H}$ span $\text{Col}(\mathbf{H}) = \mathbb{R}^3$. ∎

THEOREM 10

The column space of an $m \times n$ matrix $\mathbf{A}$ is the set $\{\mathbf{Ax} : \mathbf{x} \in \mathbb{R}^n\}$.

PROOF Recall that the product $\mathbf{Ax}$ is defined to be a linear combination of the columns of $\mathbf{A}$, in which the coefficients are the components of the vector $\mathbf{x}$. Thus, with the notation of the preceding paragraphs, we have

$$\mathbf{Ax} = \sum_{j=1}^{n} x_j \mathbf{a}_j$$

(We suppose that $\mathbf{A}$ has n columns, $\mathbf{a}_j$.) Because $\mathbf{x}$ runs over all possible vectors in $\mathbb{R}^n$, we get all the linear combinations of columns of $\mathbf{A}$. In other words, we get the span of the set of columns in $\mathbf{A}$. ∎

EXAMPLE 17 Is the vector $\mathbf{b} = [1, 20, 16]^T$ in the column space of the following matrix?

$$\mathbf{A} = \begin{bmatrix} 3 & -4 \\ 6 & 1 \\ 2 & 5 \end{bmatrix}$$

SOLUTION We are asking whether some vector $\mathbf{x}$ exists for which $\mathbf{Ax} = \mathbf{b}$. This question is answered by attempting to solve that equation. The appropriate augmented matrix is

$$[\mathbf{A} \mid \mathbf{b}] = \begin{bmatrix} 3 & -4 & \mid & 1 \\ 6 & 1 & \mid & 20 \\ 2 & 5 & \mid & 16 \end{bmatrix} \sim \begin{bmatrix} 1 & 0 & \mid & 3 \\ 0 & 1 & \mid & 2 \\ 0 & 0 & \mid & 0 \end{bmatrix}$$

Yes, the system is consistent and has the solution $\mathbf{x} = [3, 2]^T$. Checking with an independent verification, we have $3[3, 6, 2]^T + 2[-4, 1, 5]^T = [1, 20, 16]^T$. ∎

To describe the row space of a matrix in a similar manner, we observe that the row space of an $m \times n$ matrix $\mathbf{A}$ is the column space of the $n \times m$ matrix $\mathbf{A}^T$. One may object that vectors in the row space should be row vectors, whereas vectors in the column space should be column vectors. However, these distinctions come into play only when we insist on a vector being treated as a matrix. Then a column vector and a row vector must be interpreted differently. But we can take transposes to get rows from columns, and vice versa. Thus, we have

$$\text{Row}(\mathbf{A}) = \text{Col}(\mathbf{A}^T) = \left\{ \mathbf{A}^T \mathbf{x} : \mathbf{x} \in \mathbb{R}^m \right\}$$

> ### THEOREM 11
>
> *The row space of a matrix $\mathbf{A}$ is the column space of $\mathbf{A}^T$.*

EXAMPLE 18 Find simple descriptions of the row and column spaces of the matrix

$$\mathbf{A} = \begin{bmatrix} 3 & 5 & 11 & 5 & 8 \\ 0 & 0 & 2 & 1 & 5 \\ 6 & 10 & 22 & 10 & 20 \end{bmatrix}$$

SOLUTION Because elementary row operations do not change the row space (Theorem 7), we are free to carry out the row-reduction process, obtaining

$$\mathbf{A} = \begin{bmatrix} 3 & 5 & 11 & 5 & 8 \\ 0 & 0 & 2 & 1 & 5 \\ 6 & 10 & 22 & 10 & 20 \end{bmatrix} \sim \begin{bmatrix} 6 & 10 & 0 & -1 & 0 \\ 0 & 0 & 2 & 1 & 0 \\ 0 & 0 & 0 & 0 & 1 \end{bmatrix} = \mathbf{B}$$

The row space is the span of the set consisting of the row vectors in matrix $\mathbf{B}$:

$$\text{Row}(\mathbf{A}) = \text{Span}\left\{ [6, 10, 0, -1, 0], \ [0, 0, 2, 1, 0], \ [0, 0, 0, 0, 1] \right\}$$

The column space is spanned by columns 1, 3, and 5 of matrix $\mathbf{A}$:

$$\text{Col}(\mathbf{A}) = \text{Span}\left\{ \begin{bmatrix} 3 \\ 0 \\ 6 \end{bmatrix}, \begin{bmatrix} 11 \\ 2 \\ 22 \end{bmatrix}, \begin{bmatrix} 8 \\ 5 \\ 20 \end{bmatrix} \right\}$$

∎

EXAMPLE 19 Let

$$F = \begin{bmatrix} 1 & 3 & 2 & -5 \\ -2 & 3 & 6 & 7 \end{bmatrix}$$

Also, set $\mathbf{v} = [3, 12]^T$ and $\mathbf{w} = [-4, 15, 22, 11]$. Is $\mathbf{v}$ in the column space of $\mathbf{F}$? Is $\mathbf{w}$ in the row space of $\mathbf{F}$?

SOLUTION We see that the vector $\mathbf{v}$ is in the column space of $\mathbf{F}$ because

$$[\mathbf{F} \mid \mathbf{v}] = \begin{bmatrix} 1 & 3 & 2 & -5 & 3 \\ -2 & 3 & 6 & 7 & 12 \end{bmatrix} \sim \begin{bmatrix} 1 & 0 & -\frac{4}{3} & -4 & -3 \\ 0 & 1 & 10/9 & -\frac{1}{3} & 2 \end{bmatrix}$$

So we have $-3[1, -2]^T + 2[3, 3]^T = [3, 12]^T$.

To determine whether the vector $\mathbf{w}$ is in the row space of $\mathbf{F}$, we take the transpose of the matrix and vector.

$$[\mathbf{F}^T \mid \mathbf{w}^T] = \begin{bmatrix} 1 & -2 & -4 \\ 3 & 3 & 15 \\ 2 & 6 & 22 \\ -5 & 7 & 11 \end{bmatrix} \sim \begin{bmatrix} 1 & 0 & 2 \\ 0 & 1 & 3 \\ 0 & 0 & 0 \\ 0 & 0 & 0 \end{bmatrix}$$

The row reduction shows that column three is a linear combination of columns one and two. Checking, we find the relationship $2[1, 3, 2, -5] + 3[-2, 3, 6, 7] = [-4, 15, 22, 11]$. ∎

Here is another way to answer the question in Example 19 about a row space. The two rows of the matrix $\mathbf{F}$ form a linearly independent pair because neither row is a scalar multiple of the other. If we insert the vector $\mathbf{w}$ as a third row in this matrix, we can call the resulting matrix $\mathbf{G}$. In going from $\mathbf{F}$ to $\mathbf{G}$, the row space will either grow or remain unchanged. In the first case, the vector $\mathbf{w}$ is not in the row space of $\mathbf{F}$, but in the second case it is. We leave the details as General Exercise 38.

Caution

If $A \sim B$ and the matrix $\mathbf{B}$ is in reduced row echelon form, then $\text{Col}(\mathbf{A})$ is the span of the set of pivot columns in $\mathbf{A}$, whereas $\text{Row}(\mathbf{A})$ is the span of the set of pivot rows in $\mathbf{B}$.

SUMMARY 5.1

- A subset in a vector space V is a subspace of V if it is nonempty and closed under the operations of vector addition and scalar multiplication.

- Let U be a subset of a vector space V. Suppose that the zero element of V is in U, that U is closed under vector addition, and that U is closed under multiplication by scalars. Then U is a vector space—indeed, a subspace of V.

- The span of a nonempty set in a vector space is a subspace of that vector space.

- The kernel of a linear transformation is a subspace of the domain of that transformation.

- The vector sum of two subspaces in a vector space is also a subspace.

- The intersection of two subspaces in a vector space is also a subspace of that vector space.

- For any mapping $f : X \rightarrow Y$, we define $f[Q] = \{f(\mathbf{x}) : \mathbf{x} \in Q\}$ for any $Q \subseteq X$. We define $f^{-1}[S] = \{\mathbf{x} \in X : f(\mathbf{x}) \in S\}$ for any $S \subseteq Y$.

- The range of a linear transformation is a subspace of its co-domain.

- The null space of a matrix $\mathbf{A}$ is defined to be $\text{Null}(\mathbf{A}) = \{\mathbf{x} : \mathbf{A}\mathbf{x} = 0\}$.

- The row space of a matrix $\mathbf{A}$ is defined to be the span of the set of rows in $\mathbf{A}$. This is denoted by $\text{Row}(\mathbf{A})$. The column space of $\mathbf{A}$, denoted by $\text{Col}(\mathbf{A})$, is the span of the set of columns in $\mathbf{A}$.

- If two matrices are row equivalent to each other then they have the same row space.

- The column space of an $m \times n$ matrix $\mathbf{A}$ is the same as the set $\{\mathbf{A}\mathbf{x} : \mathbf{x} \in \mathbb{R}^n\}$.

- Different ways in which a subspace U can be created:

  - As the *span* of a given set of vectors: $U = \text{Span}(S)$

  - As the *kernel* of a linear transformation: $U = \text{Ker}(T)$

  - As the *range* of a linear transformation: $U = T(V)$

  - As the vector *sum* of two subspaces: $U = V + W$

  - As the *intersection* of two subspaces: $U = V \cap W$

  - As the *image* of a subspace via a linear transformation: $U = T(V)$

  - As the *inverse image* of a linear subspace by a linear transformation: $U = T^{-1}(V)$

KEY CONCEPTS 5.1

Subspaces, kernel, vector sum, vector intersection, images, inverse image, range, row space, column space, null space, creating subspaces, row equivalent

GENERAL EXERCISES 5.1

1. Determine whether the set of matrices having the form $\begin{bmatrix} a & -a \\ b & 0 \end{bmatrix}$ is a subspace of $\mathbb{R}^{2 \times 2}$.

2. All invertible $n \times n$ matrices have the same row space. What is it?

3. Verify that the set $\{p \in \mathbb{P} : p'(3) = 2p(5)\}$ is a subspace of the space $\mathbb{P}$ of all polynomials.

4. In $\mathbb{R}^5$, consider the set of all vectors that have the form $(3\alpha - 2\beta, \alpha + \beta + 2, -2\beta + \alpha, 5\alpha - \pi\beta, \alpha + \beta)$. Is this set a subspace of $\mathbb{R}^5$?

5. Is the set of all vectors of the form $(s^3 - t, 5s^3 + 2t, 0)$ a subspace of $\mathbb{R}^3$? Here s and t run over $\mathbb{R}$.

6. Determine whether the set of all 2×2 matrices of the form $\begin{bmatrix} a & b \\ b-a & c \end{bmatrix}$ is a subspace $\mathbb{R}^{2 \times 2}$.

7. In Example 1, we noted that $\mathbb{P}_3$ is a subspace of $\mathbb{P}$. Explain why we do *not* say that $\mathbb{R}^3$ is a subspace of $\mathbb{R}^4$.

8. Do these two matrices have the same row space?

$$\begin{bmatrix} 1 & 3 & -2 & 1 \\ 3 & 2 & 4 & -2 \\ 2 & -2 & 3 & 1 \end{bmatrix}, \quad \begin{bmatrix} 14 & 5 & 16 & -3 \\ 1 & -9 & -7 & 16 \\ 17 & -4 & 18 & 6 \end{bmatrix}$$

9. Define $T : \mathbb{R}^3 \to \mathbb{R}^2$ by $T(x_1, x_2, x_3) = (3x_1 - 2x_2 + 5x_3, 2x_1 + x_2 - x_3)$. Define $U = \{\mathbf{x} \in \mathbb{R}^3 : x_1 + x_2 - 2x_3 = 0\}$. Find a simple description of $T[U]$.

10. Describe in simple terms the row and column spaces of $\begin{bmatrix} 1 & 3 & 2 \\ -2 & 1 & 0 \\ -1 & 11 & 6 \end{bmatrix}$

11. What is the simplest example of two matrices that are row equivalent to each other yet have different column spaces?

12. The **trace** of a square matrix is the sum of its diagonal elements. Thus, if $\mathbf{A}$ is $n \times n$,

then $\text{Trace}(\mathbf{A}) = \sum_{i=1}^{n} \mathbf{A}_{ii}$. In the vector space $\mathbb{R}^{n \times n}$, consisting of all $n \times n$ matrices, is the set of matrices with zero trace a subspace?

13. Give an example and a geometric interpretation of two subspaces U and W of the vector space $\mathbb{R}^2$. Note that their sum is also a subspace of $\mathbb{R}^2$. The sum is defined by $U + W = \{\mathbf{u} + \mathbf{w} : \mathbf{u} \in U \text{ and } \mathbf{w} \in W\}$. This illustrates Theorem 3.

14. Establish that if U and W are subspaces of a vector space V, then their intersection is also a subspace of V. The intersection is defined by $U \cap W = \{\mathbf{x} : \mathbf{x} \in U \text{ and } \mathbf{x} \in W\}$. This is Theorem 4.

15. Is the union of two subspaces in a vector space always a subspace? Explain why or find a counterexample. The union is defined by $U \cup W = \{\mathbf{x} : \mathbf{x} \in U \text{ or } \mathbf{x} \in W\}$. (In mathematics, we always use the inclusive *or*. Thus, p or q is true when either p is true or q is true or both are true.)

16. (Continuation.) Find the exact conditions on subspaces U and W in a vector space V in order that their union be a subspace.

17. Let $\mathbb{P}$ be the vector space of all polynomials and let $\mathbb{P}_n$ be the vector space of all polynomials of degree at most n. Explain why $\mathbb{P}_n$ may be viewed as a subspace of $\mathbb{P}$. Let $\mathbb{Q}$ be the collection of polynomials with only even powers of t. Is $\mathbb{Q}$ a subspace of $\mathbb{P}$? (Assume that a constant polynomial $c \mapsto ct^0$ is an even power of t.)

18. Fix a set of vectors $\{\mathbf{u}_1, \mathbf{u}_2, \ldots, \mathbf{u}_n\}$ in some vector space. Explain why the set of n-tuples $(c_1, c_2, \ldots, c_n)$ such that $\sum_{i=1}^{n} c_i \mathbf{u}_i = 0$ is a subspace of $\mathbb{R}^n$.

19. Argue that if $A \sim B$, then each row of B is a linear combination of rows in A.

20. Give a proof of Theorem 1.

21. Show why the set of all noninvertible 2×2 matrices (with the usual operations) is *not* a vector space.

22. Explain why if $T : U \to V$ is a linear transformation, the inverse image of a subspace W in V via the transformation T is a subspace of U. The inverse image is defined by $T^{-1}[W] = \{x \in U : T(x) \in W\}$.
Caution: The notation $T^{-1}[W]$ does not insinuate that T has an inverse. Every map (invertible or not) has inverse images as defined in the preceding equation.

23. Establish that if T is a linear transformation, then the image of a subspace in the domain of T is a subspace in the co-domain of T.

24. Establish that the span of a set in a vector space is the smallest subspace containing that set.

25. In a vector space, suppose that $w = u + v$ and $z = u - v$. Explain why the spans of $\{u, v\}$ and $\{w, z\}$ are the same.

26. What conclusion can be drawn about a set S in a vector space if it satisfies the equation $\text{Span}(S) = S$?

27. Use the standard polynomials $p_0, p_1, \ldots,$ where $p_i(t) = t^i$. Then we write $\mathbb{P}_n = \text{Span}\{p_0, p_1, \ldots, p_n\}$. Let D denote the **differentiation operator**. Establish that $D(\mathbb{P}_n) = \mathbb{P}_{n-1}$. Also, show that $\text{Span}\{p_1, p_2, \ldots p_n\}$ is the same as the set $\{p \in \mathbb{P}_n : p(0) = 0\}$.

28. Explain why two matrices cannot have the same column space if one matrix has a row of zeros and the other does not.

29. If A and B are row equivalent to each other, does it follow that A and B have the same column space? (A proof or counterexample is needed.)

30. Explain why the ith row of AB is $x^T B$, where x^T is the ith row of A.

31. Establish that if u_i is the ith column of A and if v_i^T is the ith row of B, then $AB = \sum_{i=1}^{n} u_i v_i^T$. (Notice that if A is $m \times n$ and if B is $n \times p$, then each u_i is $m \times 1$ and each v_i^T is $1 \times p$. Hence each $u_i v_i^T$ is $m \times p$.)

32. If $f : X \to Y$ and if $U \subseteq X$, then we can create a new function g by the equation $g(x) = f(x)$ for all x in U. If U is a proper subset of X, then $g \neq f$ because they have different domains. The function g is called the **restriction** of f to U. The notation $g = f|U$ is often used. Suppose now that X and Y are linear spaces and L is a linear map from X to Y. Let U be a subspace of X. Is $L|U$ linear? What is its domain? Do L and $L|U$ have the same range? (Arguments and examples are needed.)

33. Can this be used as the definition of a *subspace*? A subspace in a vector space is a subset that is closed under the operations of vector addition and multiplication of a vector by an arbitrary scalar.

34. Let A and B be arbitrary matrices subject only to the condition that the product AB exist. Use the notation Col to indicate the column space of a matrix. Consider these two inclusion relations: $\text{Col}(AB) \subseteq \text{Col}(A)$

and $\text{Col}(\mathbf{AB}) \subseteq \text{Col}(\mathbf{B})$. Select the one of these that is always correct and prove it.

35. Give an example in which $\text{Col}(\mathbf{AB}) = \text{Col}(\mathbf{A})$. Give another example in which $\text{Col}(\mathbf{AB}) \neq \text{Col}(\mathbf{A})$. Finally, give an example in which $\text{Col}(\mathbf{AB}) = \text{Col}(\mathbf{B})$.

36. Establish that each row of $\mathbf{AB}$ is a linear combination of the rows of $\mathbf{B}$, and thus establish that $\text{Row}(\mathbf{AB}) \subseteq \text{Row}(\mathbf{B})$.

37. Let $\mathbb{R}^{\infty}$ be the space of all infinite sequences of the form $[x_1, x_2, \ldots]$, where the x_i are arbitrary real numbers. Let X be the subset of $\mathbb{R}^{\infty}$ consisting of all sequences that have only finitely many nonzero terms. Is X a subspace of $\mathbb{R}^{\infty}$?

38. In the last part of Example 19, use row reduction on the matrix $\mathbf{G}$ to show that $\mathbf{w}$ is in the row space of $\mathbf{F}$.

39. Use $\mathbf{A} = \begin{bmatrix} 5 & 0 & 0 & 10 \\ -1 & 0 & 1 & 4 \end{bmatrix}$ in this exercise.

  a. Find a set of vectors that spans the column space.

  b. Is $[-1, 1]^T$ in the column space?

  c. Find a set of vectors that spans the row space.

  d. Is $[-1, 1, 1, 4]$ in the row space?

  e. Find a set of vectors that spans the null space.

  f. Is $[-2, 1, -6, 1]^T$ in the null space?

40. Is the set of all vectors $\mathbf{x} = [x_1, x_2, x_3, x_4]^T$ that are linear combinations of vectors $[4, 2, 0, 1]^T$ and $[6, 3, -1, 2]^T$, and in addition satisfy the equation $x_1 = 2x_2$ a subspace of $\mathbb{R}^4$? Explain why or why not.

41. Write out the details in the solution to Example 2.

42. Write out the details in the alternative solution to Example 19.

43. Explain why $f[\text{Span}(W)] = \text{Span}(f[W])$ when f is linear. *Hint:* This is needed in Example 6.

44. Let $p_0, p_1, p_2, \ldots, p_n$ be polynomials such that the degree of p_k is k, for $0 \le k \le n$. Explain why $\{p_0, p_1, p_2, \ldots, p_n\}$ is a basis for $\mathbb{P}_n$.

45. Let $\mathbf{A} = \begin{bmatrix} 3 & 6 & 3 \\ 2 & 4 & 3 \\ 3 & 6 & 4 \end{bmatrix}$ and $\mathbf{B} = \begin{bmatrix} 3 & 6 \\ 2 & 4 \\ 3 & 6 \end{bmatrix}$

Find the null space of each of these matrices as well as for $\mathbf{A}^T$ and $\mathbf{B}^T$.

46. Let U consist of all vectors in $\mathbb{R}^3$ whose entries are equal. Explain why U is a subspace of $\mathbb{R}^3$ and describe U geometrically.

47. Let U consist of the span of all 2×2 matrices whose second row is zero. Let V be the span of those whose second column is zero. Explain why $U + W$ and $U \cap W$ are vector subspaces of $\mathbb{M}_{2 \times 2}$, which is the vector space of all 2×2 matrices.

48. Let W be the subspace of $\mathbb{R}^5$ spanned by these vectors: $\mathbf{u}_1 = (1, 2, 1, 3, 2)$, $\mathbf{u}_2 = (1, 3, 3, 5, 3)$, $\mathbf{u}_3 = (3, 8, 7, 13, 8)$, $\mathbf{u}_4 = (1, 4, 6, 9, 7)$, and $\mathbf{u}_5 = (5, 13, 13, 25, 19)$. Find a basis for W.

49. Consider

$$\mathbf{A} = \begin{bmatrix} 1 & 2 & 1 & 3 & 1 & 2 \\ 2 & 5 & 5 & 6 & 4 & 5 \\ 3 & 7 & 6 & 11 & 6 & 9 \\ 1 & 5 & 10 & 7 & 9 & 9 \\ 2 & 6 & 8 & 12 & 9 & 12 \end{bmatrix}$$

Find bases for the row space and the column space of $\mathbf{A}$.

50. Explain and discuss the following.

a. Span$\{v_1, v_2, v_3, v_4\}$ is a subspace of $\mathbb{R}^5$, where in the vector v_i the first i entries are one and all others are zero.

b. Let Span$\{1, p_1(x), p_2(x), \ldots, p_n(x)\}$ be a subspace of $\mathbb{P}_n$, where $p_i(x) = (x-1)^i$. Here $\mathbb{P}_n$ is the vector space of all polynomials of degree less than or equal to n.

c. Span$\{\mathbb{E}_{23}^{11}, \mathbb{E}_{23}^{12}, \mathbb{E}_{23}^{13}, \mathbb{E}_{23}^{21}, \mathbb{E}_{23}^{22}, \mathbb{E}_{23}^{23}\}$ is a subspace of $\mathbb{M}_{2\times3}$, where $\mathbb{E}_{23}^{ij}$ are 2×3 matrices with one in entry (i, j) and zero elsewhere. Here $\mathbb{M}_{m\times n}$ is the vector space of all $m \times n$ matrices for fixed m and n.

COMPUTER EXERCISES 5.1

1. Consider

$$A = \begin{bmatrix} 1 & 2 & 3 & 4 & 5 & 6 & 7 & 8 & 9 \\ 2 & 3 & 4 & 5 & 6 & 7 & 8 & 9 & 1 \\ 3 & 4 & 5 & 6 & 7 & 8 & 9 & 1 & 2 \\ 4 & 5 & 6 & 7 & 8 & 9 & 1 & 2 & 3 \\ 5 & 6 & 7 & 8 & 9 & 1 & 2 & 3 & 4 \end{bmatrix}$$

Find simple descriptions of the

a. null space
b. column space
c. row space

2. Consider the linear transformation $T(\mathbf{x}) = \mathbf{Bx}$ where

$$B = \begin{bmatrix} 1 & 1 & 1 \\ -1 & -1 & 0 \\ -2 & -2 & 0 \\ 0 & 0 & 1 \\ 1 & 1 & 2 \end{bmatrix}$$

Describe in simple terms the

a. kernel of T
b. range of T

3. Consider

$$C = \begin{bmatrix} 0 & 3 & 6 & 4 & 5 \\ 1 & 3 & 0 & 4 & 7 \\ 2 & 3 & 5 & 3 & 9 \end{bmatrix}$$

Find simple descriptions of the

a. null space
b. column space
c. row space

4. Consider

$$G = \begin{bmatrix} 0 & 3 & 6 \\ 1 & 3 & 0 \\ 2 & 3 & 5 \\ 1 & 4 & 5 \\ 2 & 4 & 7 \end{bmatrix}$$

Find simple descriptions of the

a. null space of G
b. null space of G^T

5.2 BASES AND DIMENSION

> *A thing is obvious mathematically after you see it.*
> —IN ROSE (ED.), *MATHEMATICAL MAXIMS AND MINIMS* [1988]

> *"Obvious" is the most dangerous word in mathematics.*
> —ERIC TEMPLE BELL (1883–1960)

> *Mathematics consists in proving the most obvious thing in the least obvious way.*
> —GEORGE POLYÁ (1889–1988)

Basis for a Vector Space

In this section, we explore the different ways in which the elements of a vector space can be represented by using various coordinate systems. The unifying concept is that of a *basis* for a vector space.

DEFINITION

In a vector space V, a linearly independent set that spans V is called a **basis** *for V.*

EXAMPLE 1 In $\mathbb{R}^n$, the set of **standard unit vectors** $\mathbf{e}_1, \mathbf{e}_2, \ldots, \mathbf{e}_n$ have components given by $(\mathbf{e}_i)_j = \delta_{ij}$. (Recall the Kronecker δ_{ij}, which is 1 if $i = j$ and 0 otherwise.) These vectors $\mathbf{e}_j$ are the column vectors in the $n \times n$ identity matrix $\mathbf{I}_n$. Is this set of vectors a basis for $\mathbb{R}^n$?

SOLUTION *Yes!* This set spans $\mathbb{R}^n$ and is linearly independent. Thus, it is a basis. It is often referred to as the *standard basis* for $\mathbb{R}^n$. ∎

EXAMPLE 2 Consider $\begin{bmatrix} 1 & 1 & 1 & 1 \\ 1 & 1 & 1 & 0 \\ 1 & 1 & 0 & 0 \\ 1 & 0 & 0 & 0 \end{bmatrix}$

Is the set of columns in this matrix a basis for $\mathbb{R}^4$?

SOLUTION We ask: *"Can we solve any equation* $\mathbf{A}\mathbf{x} = \mathbf{b}$*, when* $\mathbf{b}$ *is an arbitrary vector in* $\mathbb{R}^4$*?"* Yes; we start with the fourth equation, which gives $x_1 = b_4$ (and there is no other choice!). Then go to the third equation. It can be solved for x_2 because x_1 is known. Again, there is no other possible value for x_2. Continue upward through the set of four equations, noting that there are no choices to be made. The solution is uniquely determined by the vector **b**. Consequently, if **b** = **0**, then **x** = **0** necessarily follows. We have proved that the columns form a linearly independent set of vectors that spans $\mathbb{R}^4$. ∎

EXAMPLE 3 Let **u** = $(3, 2, -4)$, **v** = $(6, 7, 3)$, **w** = $(2, 1, 0)$. Do these vectors constitute a basis for $\mathbb{R}^3$?

SOLUTION Put the vectors into a matrix **A** as columns. We ask first whether the equation **Ax** = **0** has any nontrivial solutions. Row reduction leads to

$$\mathbf{A} = [\mathbf{u}, \ \mathbf{v}, \ \mathbf{w}] = \begin{bmatrix} 3 & 6 & 2 \\ 2 & 7 & 1 \\ -4 & 3 & 0 \end{bmatrix} \sim \begin{bmatrix} 1 & -1 & 1 \\ 0 & 1 & -4 \\ 0 & 0 & 35 \end{bmatrix}$$

(Here we omit the righthand-side zero-vector.) This shows that the equation **Ax** = **0** can be true only if **x** = **0**. Hence, the set of columns is linearly independent. The row reduction also shows that the equation **Ax** = **b** can be solved for any **b** in $\mathbb{R}^3$. Therefore, the set of columns spans $\mathbb{R}^3$ and provides a basis for $\mathbb{R}^3$. ∎

EXAMPLE 4 Let $\mathbf{A} = \begin{bmatrix} 1 & 4 & 3 \\ 2 & 7 & 5 \\ -13 & 8 & 21 \end{bmatrix}$

Do the columns of this matrix form a basis for $\mathbb{R}^3$?

SOLUTION We happen to notice that column three in this matrix is equal to column two minus column one. The set of columns is, therefore, a linearly dependent set. Hence, it is *not* a basis for anything. Of course, this set of columns spans a two-dimensional subspace in $\mathbb{R}^3$, but it is not a basis for that subspace because of its linear dependence. Columns one and two in this matrix, taken together, provide a basis for the column space of **A**. That is a two-dimensional subspace in $\mathbb{R}^3$. In other words, a plane in three-space passing through **0**. (In this discussion, the term *dimension* is being used in an informal way. Precise definitions are forthcoming.) ∎

The preceding four examples suggest that a general theorem can be proved.

> ## THEOREM 1
>
> *The set of columns of any $n \times n$ invertible matrix is a basis for $\mathbb{R}^n$.*

PROOF Let $\mathbf{A}$ be such a matrix. Do its columns span $\mathbb{R}^n$? Can we solve the system of equations $\mathbf{Ax} = \mathbf{b}$ for any $\mathbf{b}$ in $\mathbb{R}^n$? Yes, because $\mathbf{A}^{-1}\mathbf{b}$ solves the system. Thus, the columns of $\mathbf{A}$ span $\mathbb{R}^n$. What about the linear independence of the set of columns? Is there a nonzero solution to the equation $\mathbf{Ax} = \mathbf{0}$? No, because if $\mathbf{Ax} = \mathbf{0}$, then $\mathbf{x} = \mathbf{0}$, as is seen by multiplying by $\mathbf{A}^{-1}$. So the columns constitute a linearly independent set that spans $\mathbb{R}^n$. ∎

> **EXAMPLE 5** Do the first four Legendre polynomials, defined by
>
> $$P_0(t) = 1, \quad P_1(t) = t, \quad P_2(t) = \tfrac{1}{2}(3t^2 - 1), \quad P_3(t) = \tfrac{1}{2}(5t^3 - 3t)$$
>
> constitute a basis for $\mathbb{P}_3$?

SOLUTION Let us investigate the *spanning property* first. Given any polynomial q in $\mathbb{P}_3$, say $q(t) = a_0 + a_1 t + a_2 t^2 + a_3 t^3$, can we express it in terms of Legendre polynomials? In other words, can we solve this equation for c_0, c_1, c_2, c_3?

$$c_0 P_0 + c_1 P_1 + c_2 P_2 + c_3 P_3 = q$$

Putting in more detail, we have

$$c_0 + c_1 t + \tfrac{1}{2}c_2(3t^2 - 1) + \tfrac{1}{2}c_3(5t^3 - 3t) = a_0 + a_1 t + a_2 t^2 + a_3 t^3$$

Collecting similar terms on the left brings us to

$$(c_0 - \tfrac{1}{2}c_2) + (c_1 - \tfrac{3}{2}c_3)t + (\tfrac{3}{2}c_2)t^2 + (\tfrac{5}{2}c_3)t^3 = a_0 + a_1 t + a_2 t^2 + a_3 t^3$$

From elementary algebra, we recall that two polynomials are the same if and only if the coefficients of each power of t are the same. Thus, we must have $2c_0 - c_2 = 2a_0, 2c_1 - 3c_3 = 2a_1, 3c_2 = 2a_2$, and $5c_3 = 2a_3$. We can indeed solve for the coefficients c_i using these equations in reverse order. We arrive at $c_3 = \tfrac{2}{5}a_3, c_2 = \tfrac{2}{3}a_2, c_1 = \tfrac{3}{2}c_3 + a_1 = \tfrac{3}{5}a_3 + a_1$, and $c_0 = \tfrac{1}{2}c_2 + a_0 = \tfrac{1}{3}a_2 + a_0$. This proves that the span of the first four Legendre polynomials is $\mathbb{P}_3$. For the linear independence question, just substitute $p = 0$ in the preceding calculation to see that all c_j must be zero (because $a_0 = a_1 = a_2 = a_3 = 0$). ∎

THEOREM 2

If a basis is given for a vector space, then each vector in the space has a unique expression as a linear combination of elements in that basis.

PROOF This is established by Theorem 6 in Section 2.4. The reader can probably supply a proof without referring to Section 2.4. The proof relies on the concepts of *basis*, *spanning*, and *linear independence*. ∎

Let's look at a more provocative example, where the *vectors* involved are again polynomials.

EXAMPLE 6 Find a basis for the vector space spanned by the four polynomials p_1, p_2, p_3, p_4, where

$$p_1(t) = 2t^3 + 4, \quad p_2(t) = t^2 - 7, \quad p_3(t) = t^3 + t^2, \quad p_4(t) = 2t^2 + 2$$

SOLUTION The correspondence between these polynomials and their terms can be displayed as follows:

	p_1	p_2	p_3	p_4
t^0	4	-7	0	2
t^1	0	0	0	0
t^2	0	1	1	2
t^3	2	0	1	0

The four column vectors headed by p_1, p_2, p_3, and p_4 in this table need to be analyzed to uncover any linear dependencies. A row reduction is called for:

$$\begin{bmatrix} 4 & -7 & 0 & 2 \\ 0 & 1 & 1 & 2 \\ 2 & 0 & 1 & 0 \end{bmatrix} \sim \begin{bmatrix} 2 & 0 & 1 & 0 \\ 0 & 1 & 1 & 2 \\ 4 & -7 & 0 & 2 \end{bmatrix}$$

$$\sim \begin{bmatrix} 2 & 0 & 1 & 0 \\ 0 & 1 & 1 & 2 \\ 0 & -7 & -2 & 2 \end{bmatrix} \sim \begin{bmatrix} 2 & 0 & 1 & 0 \\ 0 & 1 & 1 & 2 \\ 0 & 0 & 5 & 16 \end{bmatrix}$$

(There is no need to keep the zero row.) The row-reduced form shows that column four is a linear combination of the first three columns. This is true of the beginning matrix as well as the final matrix. (The row-reduction process does not disturb linear dependencies among the columns of a matrix.) Hence, a basis is $\{p_1, p_2, p_3\}$. (We do not need p_4.) ∎

Another question for the same example: What nontrivial linear relationship exists among p_1, p_2, p_3, p_4? We continue the reduction process on the preceding matrix, ending at the reduced echelon form:

$$\begin{bmatrix} 2 & 0 & 1 & 0 \\ 0 & 1 & 1 & 2 \\ 0 & 0 & 5 & 16 \end{bmatrix} \sim \begin{bmatrix} 5 & 0 & 0 & -8 \\ 0 & 5 & 0 & -6 \\ 0 & 0 & 5 & 16 \end{bmatrix} \sim \begin{bmatrix} 1 & 0 & 0 & -\frac{8}{5} \\ 0 & 1 & 0 & -\frac{6}{5} \\ 0 & 0 & 1 & 3.2 \end{bmatrix}$$

We obtain $x_1 = \frac{8}{5}x_4$, $x_2 = \frac{6}{5}x_4$, and $x_3 = -3.2x_4$. Leting $x_4 = 5$, we have $\mathbf{x} = (8, 6, -16, 5)$. Therefore, we have $8p_1 + 6p_2 - 16p_3 + 5p_4 = 0$. This verifies that p_4 can be expressed in terms of the basis.

Coordinate Vector

We almost always restrict our attention to **ordered bases**, so that we can refer to their elements by subscript notation. For example, we might write

$$\mathcal{B} = \{\mathbf{u}_1, \mathbf{u}_2, \ldots, \mathbf{u}_n\}$$

If this is a basis for a vector space V, then each vector $\mathbf{x}$ in V can be written in one and only one way as $\mathbf{x} = \sum_{i=1}^{n} c_i \mathbf{u}_i$. The n-tuple $(c_1, c_2, \ldots, c_n)$ that arises in this way is called the **coordinate vector** of $\mathbf{x}$ associated with the (ordered) basis $\mathcal{B}$. It is denoted by $[\mathbf{x}]_{\mathcal{B}} = (c_1, c_2, \ldots, c_n)$.

> **EXAMPLE 7** What is the coordinate vector of $\mathbf{x}$ if $\mathbf{x} = (-15, 35, 2)$ and if the basis used is $\mathcal{B} = \{\mathbf{u}_1, \mathbf{u}_2, \mathbf{u}_3\}$ where $\mathbf{u}_1 = (1, 3, 2)$, $\mathbf{u}_2 = (-2, 4, -1)$, and $\mathbf{u}_3 = (2, -2, 0)$?

SOLUTION This problem can be turned into a system of linear equations that must be solved:

$$\begin{bmatrix} 1 & -2 & 2 \\ 3 & 4 & -2 \\ 2 & -1 & 0 \end{bmatrix} \begin{bmatrix} c_1 \\ c_2 \\ c_3 \end{bmatrix} = \begin{bmatrix} -15 \\ 35 \\ 2 \end{bmatrix}$$

Row reduction of the augmented matrix leads to $\mathbf{c} = (3, 4, -5)$, as shown next:

$$\begin{bmatrix} 1 & -2 & 2 & | & -15 \\ 3 & 4 & -2 & | & 35 \\ 2 & -1 & 0 & | & 2 \end{bmatrix} \sim \begin{bmatrix} 1 & 0 & 0 & | & 3 \\ 0 & 1 & 0 & | & 4 \\ 0 & 0 & 1 & | & -5 \end{bmatrix}$$

An independent verification consists in computing

$$c_1\mathbf{u}_1 + c_2\mathbf{u}_2 + c_3\mathbf{u}_3 = 3 \begin{bmatrix} 1 \\ 3 \\ 2 \end{bmatrix} + 4 \begin{bmatrix} -2 \\ 4 \\ -1 \end{bmatrix} - 5 \begin{bmatrix} 2 \\ -2 \\ 0 \end{bmatrix} = \begin{bmatrix} -15 \\ 35 \\ 2 \end{bmatrix} \quad \blacksquare$$

THEOREM 3

If $\mathcal{B}$ is a basis, then the mapping of $\mathbf{x}$ to its coordinate vector $[\mathbf{x}]_\mathcal{B}$ is linear.

PROOF There are two aspects of linearity, which we can combine and address as follows:

$$[\alpha\mathbf{x} + \beta\mathbf{y}]_\mathcal{B} = \alpha[\mathbf{x}]_\mathcal{B} + \beta[\mathbf{y}]_\mathcal{B}$$

where α and β are scalars. To see that this equation is true, let $[\mathbf{x}]_\mathcal{B} = (a_1, a_2, \ldots, a_n)$ and $[\mathbf{y}]_\mathcal{B} = (b_1, b_2, \ldots, b_n)$. Then these three equations follow:

$$\alpha\mathbf{x} = \sum_{i=1}^n \alpha a_i \mathbf{u}_i, \qquad \beta\mathbf{y} = \sum_{i=1}^n \beta b_i \mathbf{u}_i, \qquad \alpha\mathbf{x} + \beta\mathbf{y} = \sum_{i=1}^n (\alpha a_i + \beta b_i)\mathbf{u}_i$$

Hence, we obtain

$$\begin{aligned} [\alpha\mathbf{x} + \beta\mathbf{y}]_\mathcal{B} &= (\alpha a_1 + \beta b_1, \ \alpha a_2 + \beta b_2, \ \ldots, \ \alpha a_n + \beta b_n) \\ &= \alpha(a_1, a_2, \ldots, a_n) + \beta(b_1, b_2, \ldots, b_n) \\ &= \alpha[\mathbf{x}]_\mathcal{B} + \beta[\mathbf{y}]_\mathcal{B} \quad \blacksquare \end{aligned}$$

THEOREM 4

The mapping of a vector $\mathbf{x}$ to its coordinate vector $[\mathbf{x}]_\mathcal{B}$ (with respect to a given basis $\mathcal{B}$) is surjective (onto) and injective (one-to-one).

PROOF Is this map surjective (onto)? Is every vector in $\mathbb{R}^n$ an image of some $\mathbf{x}$ in V? Let $\mathbf{c}$ be any vector in $\mathbb{R}^n$, and write $\mathbf{c} = (c_1, c_2, \ldots, c_n)$. Is $\mathbf{c}$ equal to $[\mathbf{x}]_\mathcal{B}$ for some $\mathbf{x}$ in V? Of course: it is the image of the vector $\sum_{i=1}^n c_i \mathbf{u}_i$. (Refer to Section 2.3, Theorem 4.)

Is our mapping injective (one-to-one)? Because we know it to be linear, it suffices to verify that only the element $\mathbf{0}$ in V maps into the vector $\mathbf{0}$ of $\mathbb{R}^n$. But this is easy, because if $[\mathbf{x}]_\mathcal{B} = \mathbf{0}$, then $\mathbf{x} = \sum_{i=1}^n 0\mathbf{u}_i = \mathbf{0}$. (Refer to Section 2.3, Theorem 5.) $\quad \blacksquare$

Isomorphism and Equivalence Relations

In Theorem 4, we have observed a pair of vector spaces, V and $\mathbb{R}^n$, that are related by a linear, injective, and surjective map. Such a map is called an **isomorphism**, and the spaces are said to be **isomorphic** to each other. The relationship of two spaces being isomorphic to each other is an example of an equivalence relation.

DEFINITION

A relation (written as $\equiv$) is an **equivalence relation** *on a set X if it has these three properties for all* $\mathbf{x}$, $\mathbf{y}$, *and* $\mathbf{z}$ *in* X:

- $\mathbf{x} \equiv \mathbf{x}$ (**reflexive**)
- *If* $\mathbf{x} \equiv \mathbf{y}$, *then* $\mathbf{y} \equiv \mathbf{x}$ (**symmetric**)
- *If* $\mathbf{x} \equiv \mathbf{y}$ *and* $\mathbf{y} \equiv \mathbf{z}$, *then* $\mathbf{x} \equiv \mathbf{z}$ (**transitive**)

THEOREM 5

The relation of isomorphism between two vector spaces is an equivalence relation.

PROOF There are three properties to verify.

1. Is a vector space, V, isomorphic to itself? Yes, because the identity map $I : V \to V$ can serve as the isomorphism.
2. If V is isomorphic to W, does it follow that W is isomorphic to V? Yes, because if $T : V \to W$ is an isomorphism, then so is $T^{-1} : W \to V$.
3. If V is isomorphic to W and if W is isomorphic to U, is V isomorphic to U? Yes: Suppose that we are given the isomorphisms $T : V \to W$ and $S : W \to U$. To get from V to U, we use the composite mapping $S \circ T$, which has the desired properties. The reader should recall that $S \circ T$ is linear, injective, and surjective. ∎

We discuss equivalence relations briefly in Section 1.1 and will return to them again in Section 5.3.

EXAMPLE 8 Are the spaces $\mathbb{P}_n$ and $\mathbb{R}^{n+1}$ isomorphic to each other?

SOLUTION If we think that these spaces are isomorphic to each other, we must invent an isomorphism, and prove that it is one. Given a polynomial in $\mathbb{P}_n$, write it in detail—for example, in the standard form

$$p(t) = a_0 + a_1 t + a_2 t^2 + a_3 t^3 + \cdots + a_n t^n$$

With the polynomial p, we can now associate the vector in $\mathbb{R}^{n+1}$

$$\mathbf{a} = (a_0, a_1, a_2, \ldots, a_n)$$

This is another example of selecting a basis for a space and then identifying each element in the space with its coordinate vector relative to the basis. In this example, we associate $[p]_{\mathcal{B}}$ with p, and the isomorphism is the map $p \mapsto [p]_{\mathcal{B}}$. Theorems 3 and 4 indicate that we have an isomorphism. What basis for $\mathbb{P}_n$ are we using? It is the (ordered) set of functions $t \mapsto 1, t, t^2, \ldots, t^n$. ∎

EXAMPLE 9 Is $\mathbb{R}^{2 \times 2}$ isomorphic to $\mathbb{R}^4$?

SOLUTION Consider the map $L\left(\begin{bmatrix} a & b \\ c & d \end{bmatrix}\right) = \begin{bmatrix} a \\ b \\ c \\ d \end{bmatrix}$.

We get *all* of $\mathbb{R}^4$, and L is surjective. If $L\left(\begin{bmatrix} a & b \\ c & d \end{bmatrix}\right) = \begin{bmatrix} 0 \\ 0 \\ 0 \\ 0 \end{bmatrix}$,

then $\begin{bmatrix} a & b \\ c & d \end{bmatrix} = \begin{bmatrix} 0 & 0 \\ 0 & 0 \end{bmatrix}$. Hence, L is injective. Thus, L is one-to-one and maps $\mathbb{R}^{2 \times 2}$ onto $\mathbb{R}^4$. The mapping L is linear. Hence, it is an isomorphism. ∎

EXAMPLE 10 Is $\mathbb{P}_5$ isomorphic to $\mathbb{R}^{2 \times 3}$?

SOLUTION Let $p(t) = a_0 + a_1 t + a_2 t^2 + a_3 t^3 + a_4 t^4 + a_5 t^5$, so that p is a generic element of $\mathbb{P}_5$. The vector of coefficients $[a_0, a_1, a_2, a_3, a_4, a_5]^T$ is in $\mathbb{R}^6$, and the matrix $\begin{bmatrix} a_0 & a_1 & a_2 \\ a_3 & a_4 & a_5 \end{bmatrix}$ is in $\mathbb{R}^{2 \times 3}$. The mapping from p to the 2×3 matrix is linear, injective, and surjective. ∎

> ### THEOREM 6
>
> *If a vector space has a finite basis, then all of its bases have the same number of elements.*

PROOF I. Let $\mathcal{B}$ be a finite basis for the vector space being considered. Let $\mathcal{C}$ be another basis for the same space. Since $\mathcal{C}$ is linearly independent and $\mathcal{B}$ spans the space, the number of elements in $\mathcal{C}$ is not greater than the number of elements in $\mathcal{B}$, by Theorem 10 in Section 2.4. Conversely, since $\mathcal{B}$ is linearly independent and $\mathcal{C}$ spans the space, the number of elements in $\mathcal{B}$ is no greater than the number of elements in $\mathcal{C}$. ∎

We give another, slightly different, proof. This is by the *method of contradiction*, and it uses coordinate vectors.

PROOF II. Let n denote the number of elements in $\mathcal{B}$, and suppose that $\mathcal{C}$ has more than n elements. Select $\mathbf{v}_1, \mathbf{v}_2, \ldots, \mathbf{v}_m$ in $\mathcal{C}$, where $m > n$. We intend to show that $\mathcal{C}$ is linearly dependent, hence not a basis. Consider the equation

$$\sum_{i=1}^{m} c_i [\mathbf{v}_i]_{\mathcal{B}} = 0$$

Each $[\mathbf{v}_i]_{\mathcal{B}}$ is a column vector having n components, because n is the number of elements in $\mathcal{B}$. Thus, the preceding vector equation has n equations and m unknowns. By Corollary 2 in Section 1.3, this system has a nontrivial solution. By the linearity of the map $\mathbf{x} \mapsto [\mathbf{x}]_{\mathcal{B}}$, we have

$$\left[\sum_{i=1}^{m} c_i \mathbf{v}_i \right]_{\mathcal{B}} = 0 \quad \text{whence} \quad \sum_{i=1}^{m} c_i \mathbf{v}_i = 0$$

nontrivially. Thus, $\mathcal{C}$ is linearly dependent. ∎

Finite-Dimensional and Infinite-Dimensional Vector Spaces

A vector space having a finite basis is said to be **finite dimensional**. The number of elements in any basis for that space is called the **dimension** of the space. Thus, we can now say that $\mathbb{R}^n$ is n-dimensional, and $\mathbb{P}_n$ has dimension $n + 1$. In both cases, it suffices to exhibit a basis of the space and count the number of elements in it. In the same way, we find that the dimension of $\mathbb{R}^{m \times n}$ is mn. If V is a finite-dimensional vector space, we use $\mathrm{Dim}(V)$ for its dimension. Thus, for example, we have

$$\mathrm{Dim}(\mathbb{R}^m) = m$$

The space $\mathbb{R}^\infty$, defined in Section 2.4, Example 3, is not finite dimensional. This fact is most easily established by noting that the vectors $\mathbf{e}_n$ defined by $(\mathbf{e}_n)_i = \delta_{ni}$ are infinite in number yet form a linearly independent set. By Theorem 6 in Section 2.4, this vector space cannot have a finite spanning set. Naturally, we call such vector spaces **infinite dimensional**. Many such spaces are needed in applied mathematics and they have been assiduously studied for more than a century. Hilbert spaces and Banach spaces are examples. So are Sobolev spaces. (These topics are beyond the scope of this book.)

The linear space of all continuous real-valued functions on the interval $[0, 1]$ is infinite dimensional. To be convinced of this, it suffices to exhibit an infinite linearly independent family of continuous functions on $[0, 1]$. The set of monomials, $p_n(t) = t^n$, where $n = 0, 1, 2, \ldots$, serves this purpose.

DEFINITION

A vector space is **finite dimensional** *if it has a finite basis; in that event, its* **dimension** *is the number of elements in any basis.*

THEOREM 7

Two finite-dimensional vector spaces are isomorphic to each other if and only if their dimensions are the same.

PROOF First, suppose that the two spaces U and V are finite dimensional and have the same dimension. Select bases $\{\mathbf{u}_1, \mathbf{u}_2, \ldots, \mathbf{u}_n\}$ for U and $\{\mathbf{v}_1, \mathbf{v}_2, \ldots, \mathbf{v}_n\}$ for V. Define a linear transformation T by writing $T(\mathbf{u}_i) = \mathbf{v}_i$, for $i = 1, 2, \ldots, n$. Thus, the effect of T on any point

$$\mathbf{u} = \sum_{i=1}^{n} c_i \mathbf{u}_i$$

in U is like this:

$$T(\mathbf{u}) = T\left(\sum_{i=1}^{n} c_i \mathbf{u}_i\right) = \sum_{i=1}^{n} c_i T(\mathbf{u}_i) = \sum_{i=1}^{n} c_i \mathbf{v}_i$$

Is T injective? Suppose $T(\mathbf{u}) = \mathbf{0}$. Thus, we have

$$T\left(\sum_{i=1}^{n} c_i \mathbf{u}_i\right) = \mathbf{0} \quad \text{and} \quad \sum_{i=1}^{n} c_i T(\mathbf{u}_i) = \sum_{i=1}^{n} c_i \mathbf{v}_i = \mathbf{0}$$

Yes, T is one-to-one by Theorem 5 in Section 2.3 because it follows that $\sum_{i=1}^{n} |c_i| = 0$. Is T surjective? For any $\mathbf{v}$ in V, write

$$\mathbf{v} = \sum_{i=1}^{n} d_i \mathbf{v}_i = \sum_{i=1}^{n} d_i T(\mathbf{u}_i) = T\left(\sum_{i=1}^{n} d_i \mathbf{u}_i\right)$$

Yes, T is *onto* by Theorem 4 in Section 2.3.

For the other half of the proof, let T be an isomorphism from U onto V. Let $\{\mathbf{u}_1, \mathbf{u}_2, \ldots, \mathbf{u}_n\}$ be a basis for U. Then $\{T(\mathbf{u}_i) : 1 \le i \le n\}$ is a basis for V. This assertion has two parts, each requiring proof. First, is that set linearly independent? If

$$\sum_{i=1}^{n} c_i T(\mathbf{u}_i) = \mathbf{0}$$

then
$$T\left(\sum_{i=1}^{n} c_i \mathbf{u}_i\right) = \mathbf{0}$$

But, since T is an isomorphism, it is injective, and so $\sum_{i=1}^{n} c_i \mathbf{u}_i = \mathbf{0}$ and $\sum_{i=1}^{n} |c_i| = 0$ by the linear independence of $\{\mathbf{u}_i : 1 \le i \le n\}$. Second, does the set in question span V? If $\mathbf{v} \in V$, then $\mathbf{v} = T(\mathbf{u})$ for some $\mathbf{u} \in U$, because T is surjective. Then it follows that

$$\mathbf{v} = T(\mathbf{u}) = \sum_{i=1}^{n} c_i T(\mathbf{u}_i)$$ ∎

EXAMPLE 11 Are the spaces of matrices $\mathbb{R}^{m \times n}$ and $\mathbb{R}^{n \times m}$ isomorphic to one another? Are the spaces $\mathbb{R}^{3 \times 7}$ and $\mathbb{P}_{20}$ isomorphic to each other?

SOLUTION The two matrix spaces have dimension mn, and those two spaces are therefore isomorphic to each other by Theorem 7. The other pair of spaces is isomorphic for the same reason: they have the same dimension, namely 21. ∎

EXAMPLE 12 Consider the three vectors

$$\mathbf{u} = (2, 3, -1), \qquad \mathbf{v} = (13, 3, 4), \qquad \mathbf{w} = (-3, 1, -2)$$

What can be said of Span$\{\mathbf{u}, \mathbf{v}, \mathbf{w}\}$?

SOLUTION Notice that $\mathbf{v} = 2\mathbf{u} - 3\mathbf{w}$; namely, $(13, 3, 4) = 2(2, 3, -1) - 3(-3, 1, -2)$. Thus, if we have a generic element of the span, we can do some simplifying, like this:

$$a\mathbf{u} + b\mathbf{v} + c\mathbf{w} = a\mathbf{u} + b(2\mathbf{u} - 3\mathbf{w}) + c\mathbf{w} = a\mathbf{u} + 2b\mathbf{u} - 3b\mathbf{w} + c\mathbf{w}$$

$$= (a + 2b)\mathbf{u} + (c - 3b)\mathbf{w}$$

This equation shows that

$$\text{Span}\{\mathbf{u}, \mathbf{v}, \mathbf{w}\} = \text{Span}\{\mathbf{u}, \mathbf{w}\}$$

The vector $\mathbf{v}$, which was a linear combination of $\mathbf{u}$ and $\mathbf{w}$, is not needed in describing the span. This calculation can be done in greater generality, as shown in the next result. ∎

THEOREM 8

The span of a set is not affected by removing from the set one element that is a linear combination of other elements of that set.

PROOF Let S be the set, and suppose that

$$\mathbf{v}_0 = \sum_{i=1}^{n} c_i \mathbf{v}_i$$

where the vectors $\mathbf{v}_0, \mathbf{v}_1, \ldots, \mathbf{v}_n$ are distinct members of S. Let $\mathbf{w}$ be any vector in the span of S. Write

$$\mathbf{w} = \sum_{i=0}^{m} a_i \mathbf{v}_i$$

We may require that $m \geq n$, and we permit some coefficients a_i to be zero. Then rewrite

$$\mathbf{v}_0 = \sum_{i=1}^{m} c_i \mathbf{v}_i$$

where, if necessary, we have inserted some zero coefficients. Now we have

$$\mathbf{w} = a_0 \mathbf{v}_0 + \sum_{i=1}^{m} a_i \mathbf{v}_i = a_0 \sum_{i=1}^{m} c_i \mathbf{v}_i + \sum_{i=1}^{m} a_i \mathbf{v}_i = \sum_{i=1}^{m} (a_0 c_i + a_i) \mathbf{v}_i$$

This shows that $\mathbf{w}$ is representable as a linear combination of elements in S, but without using $\mathbf{v}_0$. ∎

> **THEOREM 9**
>
> *If a set of n vectors spans an n-dimensional vector space, then the set is a basis for that vector space.*

PROOF If S is the set and its span is V, we have only to prove that S is linearly independent. Suppose S is linearly dependent. Then some element of S is a linear combination of the other elements of S. Removing that element does not affect the span of the set, by Theorem 8. By Theorem 6 in Section 2.4, every linearly independent set in V can have at most $n - 1$ elements. Hence, we have $\text{Dim}(V) \leq n - 1$, a contradiction. ∎

> **THEOREM 10**
>
> *In an n-dimensional vector space, every linearly independent set of n vectors is a basis.*

PROOF If S is the given set and fails to be a basis, it must fail to span the vector space. Therefore, some point $\mathbf{x}$ in the vector space is not in the span of S. Now the set $S \cup \{\mathbf{x}\}$ is linearly independent, and the dimension of the space must be greater than n. ∎

> **EXAMPLE 13** Find a simple basis for the row space of
>
> $$\mathbf{A} = \begin{bmatrix} 3 & 9 & 4 & -4 & 3 & -24 \\ -1 & -3 & 0 & -2 & 6 & 9 \\ 2 & 6 & 6 & -11 & 9 & -17 \\ 4 & 12 & 10 & -17 & 18 & -32 \end{bmatrix}$$

SOLUTION Again, we use the principle that row-equivalent matrices have the same row space. The row-reduction process yields this reduced echelon matrix:

$$\mathbf{A} \sim \mathbf{B} = \begin{bmatrix} 1 & 3 & 0 & 2 & 0 & -7 \\ 0 & 0 & 1 & -\frac{5}{2} & 0 & -1 \\ 0 & 0 & 0 & 0 & 1 & \frac{1}{3} \\ 0 & 0 & 0 & 0 & 0 & 0 \end{bmatrix}$$

The dimension of the row space is three, and the first three rows of the matrix $\mathbf{B}$ provide a basis for the row space of $\mathbf{A}$. It may or may not be true that the first three rows in $\mathbf{A}$ give a basis for the row space. Do not assume that this is true. (The next example illustrates this.) ∎

EXAMPLE 14 Find a simple basis for the row space of

$$A = \begin{bmatrix} 12 & 2 & 46 \\ 6 & 1 & 23 \\ 3 & 1 & 14 \end{bmatrix}$$

SOLUTION A row reduction leads to

$$A \sim B = \begin{bmatrix} 1 & 0 & 3 \\ 0 & 1 & 5 \\ 0 & 0 & 0 \end{bmatrix}$$

The first two rows of **B** give a simple basis for the row space of **A**, but the first two rows of **A** do not span the row space. (Why?) ∎

THEOREM 11

If a set of n vectors spans a vector space, then the dimension of that space is at most n.

PROOF According to Theorem 6 in Section 2.4, a linearly independent set in a vector space can have no more elements than a spanning set. Hence, a basis can have no more elements than a spanning set. ∎

THEOREM 12

If a set of n vectors is linearly independent in a vector space, then the dimension of the space is at least n.

PROOF Let $\{v_1, v_2, \ldots, v_n\}$ be a linearly independent set in a vector space V. By Theorem 6 in Section 2.4, any spanning set in V must have at least n elements. Hence, any basis for V must have at least n elements, and $\text{Dim}(V) \geq n$. ∎

THEOREM 13

The pivot columns of a matrix form a basis for its column space.

PROOF Let $\mathbf{A}$ be any matrix, and denote by $\mathbf{B}$ its reduced echelon form. The pivot columns in $\mathbf{B}$ form a linearly independent set because no pivot column in $\mathbf{B}$ is a linear combination of columns that precede it. (Look at Example 15, which shows a typical case.) Because $\mathbf{A}$ is row equivalent to $\mathbf{B}$, the pivot columns of $\mathbf{A}$ form a linearly independent set. In this situation, remember that the vectors $\mathbf{x}$ that satisfy the equation $\mathbf{Ax} = \mathbf{0}$ are the same as the vectors that satisfy $\mathbf{Bx} = \mathbf{0}$. Each nonpivot column in $\mathbf{A}$ is a linear combination of the pivot columns of $\mathbf{A}$. The set of pivot columns in $\mathbf{A}$ spans $\mathrm{Col}(\mathbf{A})$ and is linearly independent. Hence, it is a basis for the column space of $\mathbf{A}$. ■

EXAMPLE 15 Find a basis for the column space of

$$\mathbf{A} = \begin{bmatrix} 1 & 2 & 4 & 7 & -2 \\ 3 & -1 & 5 & 7 & 3 \\ 2 & 0 & 4 & 6 & 0 \\ 2 & 4 & 8 & 14 & 1 \end{bmatrix}$$

SOLUTION The row-reduction process on $\mathbf{A}$ yields a matrix $\mathbf{B}$:

$$\mathbf{A} \sim \mathbf{B} = \begin{bmatrix} 1 & 0 & 2 & 3 & 0 \\ 0 & 1 & 1 & 2 & 0 \\ 0 & 0 & 0 & 0 & 1 \\ 0 & 0 & 0 & 0 & 0 \end{bmatrix}$$

Notice that in $\mathbf{B}$, columns three and four are linear combinations of columns one and two. Hence, the same is true for the columns of $\mathbf{A}$. We conclude that columns one, two, and five in $\mathbf{A}$ constitute a basis for $\mathrm{Col}(\mathbf{A})$. Observe that columns one, two, and five in $\mathbf{B}$ certainly do *not* constitute a basis for $\mathrm{Col}(\mathbf{A})$. (Why?) ■

EXAMPLE 16 Let $f(t) = 5t^3 + 7t^2 - 8t + 9$. Is the set $\{f, f', f'', f'''\}$ a basis for $\mathbb{P}_3$?

SOLUTION From $f(t) = 5t^3 + 7t^2 - 8t + 9$, we find

$$f'(t) = 15t^2 + 14t - 8, \qquad f''(t) = 30t + 14, \qquad f'''(t) = 30$$

Suppose $\sum_{i=0}^{3} c_i f^{(i)} = 0$; that is,

$$c_0(5t^3 + 7t^2 - 8t + 9) + c_1(15t^2 + 14t - 8) + c_2(30t + 14) + c_3(30) = 0$$

or

$$(5c_0)t^3 + (7c_0 + 15c_1)t^2 + (-8c_0 + 14c_1 + 30c_2)t + (9c_0 - 8c_1 + 14c_2 + 30c_3) = 0$$

Using the coefficients in these equations, we obtain this linear system

$$\begin{bmatrix} 5 & 0 & 0 & 0 \\ 7 & 15 & 0 & 0 \\ -8 & 14 & 30 & 0 \\ 9 & -8 & 14 & 30 \end{bmatrix} \begin{bmatrix} c_0 \\ c_1 \\ c_2 \\ c_3 \end{bmatrix} = \begin{bmatrix} 0 \\ 0 \\ 0 \\ 0 \end{bmatrix}$$

From forward substitution, we find that $c_0 = 0$, $c_1 = 0$, $c_2 = 0$, and $c_3 = 0$. The set is linearly independent. Therefore, the set is a basis for $\mathbb{P}_3$.   ∎

EXAMPLE 17 Is it possible to extend this pair of vectors to get a basis for $\mathbb{R}^4$?

$$\left\{ [1, 3, 1, 2]^T, [2, 9, 3, 6]^T \right\}$$

SOLUTION We create this matrix and transform it to reduced echelon form

$$\mathbf{A} = \begin{bmatrix} 1 & 2 & 1 & 0 & 0 & 0 \\ 3 & 9 & 0 & 1 & 0 & 0 \\ 1 & 3 & 0 & 0 & 1 & 0 \\ 2 & 6 & 0 & 0 & 0 & 1 \end{bmatrix} \sim \begin{bmatrix} 1 & 0 & 3 & 0 & 0 & -1 \\ 0 & 1 & -1 & 0 & 0 & \frac{1}{2} \\ 0 & 0 & 0 & 1 & 0 & -1 \\ 0 & 0 & 0 & 0 & 1 & \frac{1}{2} \end{bmatrix} = \mathbf{B}$$

Use columns one, two, four, and five in $\mathbf{A}$ as a basis. Notice that a linear combination of the first two columns gives the third column in $\mathbf{A}$. The columns of $\mathbf{B}$ do not solve the stated problem because $\mathbf{B}$ does not contain the two given vectors.   ∎

Linear Transformation of a Set

The next theorem asserts that the application of a linear transformation to a set cannot increase its dimension. Stated in other terms, the dimension of the range of a linear transformation cannot be greater than the dimension of its domain.

THEOREM 14

For any finite-dimensional vector space U, for any vector space V, and for any linear transformation L from U to V, we have

$$\text{Dim}\left(L(U)\right) \leq \text{Dim}(U)$$

PROOF Select a basis $\{\mathbf{u}_1, \mathbf{u}_2, \ldots, \mathbf{u}_n\}$ for U. If $\mathbf{x} \in U$, then $\mathbf{x} = \sum_{i=1}^{n} c_i \mathbf{u}_i$ for suitable c_i. Consequently, we have

$$L(\mathbf{x}) = \sum_{i=1}^{n} c_i L(\mathbf{u}_i) \in \text{Span}\left\{L(\mathbf{u}_1),\ L(\mathbf{u}_2), \ldots, L(\mathbf{u}_n)\right\}$$

Since $\mathbf{x}$ was arbitrary in U, we have

$$L[U] \subset \text{Span}\left\{L(\mathbf{u}_1),\ L(\mathbf{u}_2), \ldots, L(\mathbf{u}_n)\right\}$$

Hence, we obtain $\text{Dim}(L[U]) \leq n = \text{Dim}(U)$. ∎

THEOREM 15

Let $\{\mathbf{u}_1, \mathbf{u}_2, \ldots, \mathbf{u}_n\}$ be a basis for a vector space U. Let V be any vector space. A linear transformation $T : U \to V$ is completely determined by specifying the values of $\mathbf{v}_i = T(\mathbf{u}_i)$ for $i = 1, 2, \ldots, n$. These values, in turn, can be any vectors whatsoever in V.

PROOF If $\mathbf{x}$ is any point in U, its coordinates $(a_1, a_2, \ldots, a_n)$ with respect to the basis $\{\mathbf{u}_1, \mathbf{u}_2, \ldots, \mathbf{u}_n\}$ are uniquely determined. Then

$$\mathbf{x} = \sum_{i=1}^{n} a_i \mathbf{u}_i$$

Since T is linear,

$$T(\mathbf{x}) = \sum_{i=1}^{n} a_i T(\mathbf{u}_i) = \sum_{i=1}^{n} a_i \mathbf{v}_i \qquad ∎$$

A common pitfall is to assume that a linear transformation can be defined by specifying $T(\mathbf{w}_i) = \mathbf{y}_i$ for $i = 1, 2, \ldots, n$ for *any* vectors $\mathbf{w}_i$ and $\mathbf{y}_i$. Theorem 15 does not say that, does it?

EXAMPLE 18 Find the linear transformation T such that $T(\mathbf{u}_i) = \mathbf{v}_i$, where

$$\mathbf{u}_1 = (1, 3, 2), \qquad \mathbf{v}_1 = (2, -3, 1)$$

$$\mathbf{u}_2 = (2, 1, 1), \qquad \mathbf{v}_2 = (-1, 4, 1)$$

$$\mathbf{u}_3 = (4, 7, 5), \qquad \mathbf{v}_3 = (3, -2, 1)$$

SOLUTION We are tempted to use Theorem 15. But the set $\{\mathbf{u}_1, \mathbf{u}_2, \mathbf{u}_3\}$ is linearly dependent, in as much as

$$2\mathbf{u}_1 + \mathbf{u}_2 - \mathbf{u}_3 = \mathbf{0}$$

If T is a linear transformation, then

$$T(2\mathbf{u}_1 + \mathbf{u}_2 - \mathbf{u}_3) = \mathbf{0}$$

Putting in the details, we eventually get a contradiction:

$$2T(\mathbf{u}_1) + T(\mathbf{u}_2) - T(\mathbf{u}_3) = \mathbf{0}$$

$$2\mathbf{v}_1 + \mathbf{v}_2 - \mathbf{v}_3 = \mathbf{0}$$

$$2(2, -3, 1) + (-1, 4, 1) - (3, -2, 1) = \mathbf{0}$$

$$(0, 0, 2) = \mathbf{0}$$

Thus there is no such linear transformation. ∎

THEOREM 16

In a finite-dimensional vector space, any linearly independent set can be expanded to create a basis.

PROOF Let $\{\mathbf{u}_1, \mathbf{u}_2, \ldots, \mathbf{u}_k\}$ be linearly independent. If $\{\mathbf{u}_1, \mathbf{u}_2, \ldots, \mathbf{u}_k\}$ spans the given vector space V, then it is a basis (because it is linearly independent and spans V). If $\{\mathbf{u}_1, \mathbf{u}_2, \ldots, \mathbf{u}_k\}$ does not span V, find $\mathbf{u}_{k+1}$ that is in V but not in Span$\{\mathbf{u}_1, \mathbf{u}_2, \ldots, \mathbf{u}_k\}$. Then $\{\mathbf{u}_1, \mathbf{u}_2, \ldots, \mathbf{u}_k, \mathbf{u}_{k+1}\}$ is linearly independent. Repeat this process. See Theorem 11 in Section 2.4. ∎

Dimensions of Various Subspaces

If T is a linear transformation defined on one vector space U and taking values in another vector space V, we already have two spaces related to T: its **domain** (U) and its **co-domain** (V). Then we can go on to define the kernel of T by writing

$$\text{Ker}(T) = \{\mathbf{u} \in U \;:\; T(\mathbf{u}) = \mathbf{0}\}$$

This is also called the **null space** of $\mathbf{T}$ and is denoted by Null($\mathbf{T}$). The **nullity** of a linear transformation T is the dimension of its null space. The kernel of T consists of all the vectors in U that are mapped by T into $\mathbf{0}$. Next, we can define the **range** of T,

$$\text{Range}(T) = \big\{T(\mathbf{u}) \;:\; \mathbf{u} \in U\big\}$$

This is the set of all images created by T when it acts on all the vectors in U. The kernel of T is a subspace of U, and the range of T is a subspace of V. These assertions have been proved in Section 5.1. (See Theorems 7 and 8 in Section 5.1.) The preceding definitions are valid without restricting U or V to be finite dimensional. However, if U is finite dimensional, we can establish some relations among the dimensions of these spaces. (All of these dimension numbers are nonnegative integers.) Begin by observing that if U is of dimension n, then the range of T cannot be of dimension greater than n. (Recall Theorem 14, which asserts that one cannot increase the dimension of a set by applying a linear transformation to it.)

The theorem we will prove can be illustrated in a simple case where the dimensions are easily discerned.

EXAMPLE 19 Consider a linear transformation $T : \mathbb{R}^5 \to \mathbb{R}^6$ defined by the following matrix $\mathbf{A}$. Recall that this terminology means $T(\mathbf{x}) = \mathbf{Ax}$. We show also the reduced row echelon form of $\mathbf{A}$ as the matrix $\tilde{\mathbf{A}}$.

$$\mathbf{A} = \begin{bmatrix} 1 & 2 & 0 & 5 & 0 \\ 2 & 4 & 1 & 13 & 1 \\ 2 & 4 & 0 & 10 & 1 \\ 4 & 8 & 3 & 29 & 0 \\ 1 & 2 & 0 & 5 & 1 \\ 10 & 20 & 7 & 71 & 3 \end{bmatrix} \sim \begin{bmatrix} 1 & 2 & 0 & 5 & 0 \\ 0 & 0 & 1 & 3 & 0 \\ 0 & 0 & 0 & 0 & 1 \\ 0 & 0 & 0 & 0 & 0 \\ 0 & 0 & 0 & 0 & 0 \\ 0 & 0 & 0 & 0 & 0 \end{bmatrix} = \tilde{\mathbf{A}}$$

What are the dimensions of the various subspaces associated with $\mathbf{A}$?

SOLUTION The dimension of the kernel of $\mathbf{A}$ will be two because there are two free variables in $\widetilde{\mathbf{A}}$. Columns one, three, and five in $\widetilde{\mathbf{A}}$ form a linearly independent set because they are pivot columns. Therefore, columns one, three, and five in $\mathbf{A}$ form a linearly independent set. These columns in $\mathbf{A}$, (*not* in $\widetilde{\mathbf{A}}$) give a basis for the range of T. Hence, the dimension of the range of T is 3. The sum of the dimensions of the range and kernel is 5, which is also the dimension of the domain of T. ∎

The preceding calculation should generalize to any matrix. Here is the general case, for an arbitrary matrix.

EXAMPLE 20 Let $\mathbf{A}$ be an $m \times n$ matrix. Let $\widetilde{\mathbf{A}}$ be its reduced echelon form. Suppose there are k pivots in $\widetilde{\mathbf{A}}$ or $\mathbf{A}$. How are the dimensions of the column space and null space related?

SOLUTION There will be k columns having pivots and $n - k$ columns without pivots. Each column without a pivot adds to the dimension of the null space because those columns correspond to free variables. Hence, we have

$$\text{Dim}\left(\text{Null}(\mathbf{A})\right) = n - k$$

The columns of $\mathbf{A}$ that have pivot positions form a basis for the range of $\mathbf{A}$. Hence, the equation $(n - k) + k = n$ allows us to conclude that

$$\text{Dim}\left(\text{Col}(\mathbf{A})\right) + \text{Dim}\left(\text{Null}(\mathbf{A})\right) = n$$

This theoretical result can be formulated without referring to matrices as follows. ∎

THEOREM 17

If T is a linear transformation whose domain is an n-dimensional vector space, then
$$\text{Dim}\left(\text{Ker}(T)\right) + \text{Dim}\left(\text{Range}(T)\right) = n$$

PROOF Select a basis $\{\mathbf{y}_1, \mathbf{y}_2, \ldots \mathbf{y}_m\}$ for the range of T. Select a basis $\{\mathbf{x}_1, \mathbf{x}_2, \ldots, \mathbf{x}_k\}$ for the kernel of T. Select vectors $\mathbf{u}_i$ so that $T(\mathbf{u}_i) = \mathbf{y}_i$ for $i = 1, 2, \ldots, m$. Define

$$\mathcal{B} = \{\mathbf{x}_1, \mathbf{x}_2, \ldots, \mathbf{x}_k, \mathbf{u}_1, \mathbf{u}_2, \ldots, \mathbf{u}_m\}$$

It is to be proved that $\mathcal{B}$ is a basis for the domain of T. Let the domain of T be the space U. First, we will prove that $\mathcal{B}$ spans U. Let $\mathbf{v} \in U$. Then $T(\mathbf{v})$ is in the range of T. Therefore, we obtain

$$T(\mathbf{v}) = \sum_{i=1}^{m} c_i \mathbf{y}_i = \sum_{i=1}^{m} c_i T(\mathbf{u}_i) = T\left(\sum_{i=1}^{m} c_i \mathbf{u}_i\right)$$

This shows that

$$T\left(\mathbf{v} - \sum_{i=1}^{m} c_i \mathbf{u}_i\right) = \mathbf{0}$$

and that $\mathbf{v} - \sum_{i=1}^{m} c_i \mathbf{u}_i$ is in the kernel of T. From this it follows that $\mathbf{v} - \sum_{i=1}^{m} c_i \mathbf{u}_i = \sum_{j=1}^{k} d_j \mathbf{x}_j$, whence $\mathbf{v} = \sum_{i=1}^{m} c_i \mathbf{u}_i + \sum_{j=1}^{k} d_j \mathbf{x}_j$. Next, we prove that $\mathcal{B}$ is linearly independent. Suppose that $\sum_{i=1}^{k} a_i \mathbf{x}_i + \sum_{j=1}^{m} b_j \mathbf{u}_j = \mathbf{0}$. Then

$$\mathbf{0} = T\left(\sum_{i=1}^{k} a_i \mathbf{x}_i + \sum_{j=1}^{m} b_j \mathbf{u}_j\right) = \sum_{i=1}^{k} a_i T(\mathbf{x}_i) + \sum_{j=1}^{m} b_j \mathbf{y}_j = \sum_{j=1}^{m} b_j \mathbf{y}_j$$

Hence, all b_j are zero. Then $\sum_{i=1}^{k} a_i \mathbf{x}_i = \mathbf{0}$ and all a_i are zero. Conclusion:

$$\mathrm{Dim}(U) = k + m = \mathrm{Dim}\left(\mathrm{Ker}(T)\right) + \mathrm{Dim}\left(\mathrm{Range}(T)\right) \qquad \blacksquare$$

The purely matrix form of Theorem 17 can be stated as follows.

THEOREM 18 Rank–Nullity Theorem
For any matrix, the number of columns equals the dimension of the column space plus the dimension of the null space.

Think of this theorem as stating that each column of a matrix contributes either to the dimension of the null space or the dimension of the column space. Ultimately, this depends on the very simple observation that each column either has a pivot position or it does not. The columns with pivots add to the dimension of the column space, whereas the columns without pivots add to the dimension of the null space (because they correspond to free variables in solving the homogeneous problem $\mathbf{A}\mathbf{x} = \mathbf{0}$). In other words, for an $m \times n$ matrix, we have

$$\mathrm{Rank}(\mathbf{A}) + \mathrm{Nullity}(\mathbf{A}) = n$$

The next theorem is an important one in linear algebra. It may even be surprising, because it draws a connection between the row space and the column space of a matrix. In general, these are subspaces in completely different vector spaces, $\mathbb{R}^n$ and $\mathbb{R}^m$, if the matrix in question is $m \times n$.

THEOREM 19

The row space and column space of a matrix have the same dimension:

$$\text{Dim}\left(\text{Row}(\mathbf{A})\right) = \text{Dim}\left(\text{Col}(\mathbf{A})\right)$$

PROOF Let $\mathbf{A}$ be an $m \times n$ matrix of rank r. Let $\widetilde{\mathbf{A}}$ denote its reduced row echelon form. Then r is the number of nonzero rows in $\widetilde{\mathbf{A}}$, by the definition of the term *rank*. Each nonzero row in $\widetilde{\mathbf{A}}$ must have a pivot element, and these rows form a linearly independent set. Hence, they constitute a basis for the row space of $\mathbf{A}$. Thus, r is the dimension of the row space of $\mathbf{A}$. Because r rows have pivots, r columns have pivots. The remaining $n - r$ columns do not have pivots. Each of these columns gives rise to a free variable, and the dimension of the null space of $\mathbf{A}$ equals this number $n - r$. The column space of $\mathbf{A}$ has dimension r, because the pivotal columns (of $\mathbf{A}$) constitute a basis for the column space.

Another way to see that the column space has dimension r is to use Theorem 17. The dimension of the domain of $\mathbf{A}$ (interpreted as a linear transformation in the usual fashion $\mathbf{x} \mapsto \mathbf{Ax}$) is n. The kernel of $\mathbf{A}$ has dimension $n - r$. Therefore, the range of $\mathbf{A}$ is of dimension r. The range of $\mathbf{A}$ is its column space. ∎

In some books and technical papers, you will encounter the terminology **row rank** and **column rank** of a matrix. These are the dimension of the row space and the dimension of the column space, respectively. Thus, Theorem 18 can be stated in the form "*The row rank and column rank of a matrix are equal.*"

EXAMPLE 21 Is there an example of a linear transformation T from $\mathbb{R}^{17}$ to $\mathbb{R}^{32}$ whose range has dimension seven and whose kernel has dimension nine?

SOLUTION Before attempting to construct such an example, we test the equation

$$\text{Dim}\left(\text{Ker}(T)\right) + \text{Dim}\left(\text{Range}(T)\right) = \text{Dim}\left(\text{Domain}(T)\right)$$

The information given would lead to $9 + 7 = 17$, which is incorrect. No such linear transformation exists. ∎

EXAMPLE 22 Use Theorem 17 to prove that $\text{Dim}\left(T[U]\right) \leq \text{Dim}(U)$. Here T is a linear transformation defined on a vector space U.

SOLUTION In this inequality, let U be the domain of T. The notation $T[U]$ represents the image of the vector space U under the linear transformation T. Hence, it is the range of T. By Theorem 17, the inequality follows at once, because the kernel of T cannot have a negative dimension. ∎

EXAMPLE 23 Let $\mathbf{A} = \begin{bmatrix} 1 & 2 & 3 & 2 & 3 \\ 3 & 6 & 9 & 2 & 3 \\ 2 & 4 & 6 & 2 & 3 \\ 1 & 2 & 3 & 2 & 3 \end{bmatrix}$

Find bases for each of the spaces $\text{Null}(\mathbf{A})$, $\text{Col}(\mathbf{A})$, and $\text{Row}(\mathbf{A})$.

SOLUTION We find this row equivalence:

$$\mathbf{A} = \begin{bmatrix} 1 & 2 & 3 & 2 & 3 \\ 3 & 6 & 9 & 2 & 3 \\ 2 & 4 & 6 & 2 & 3 \\ 1 & 2 & 3 & 2 & 3 \end{bmatrix} \sim \begin{bmatrix} 1 & 2 & 3 & 0 & 0 \\ 0 & 0 & 0 & 1 & \frac{3}{2} \\ 0 & 0 & 0 & 0 & 0 \\ 0 & 0 & 0 & 0 & 0 \end{bmatrix} = \mathbf{B}$$

The reduced echelon form reveals that in solving the equation $\mathbf{A}\mathbf{x} = \mathbf{0}$ we obtain $x_1 + 2x_2 + 3x_3 = 0$ and $x_4 + \frac{3}{2}x_5 = 0$, or

$$\begin{bmatrix} x_1 \\ x_2 \\ x_3 \\ x_4 \\ x_5 \end{bmatrix} = x_2 \begin{bmatrix} -2 \\ 1 \\ 0 \\ 0 \\ 0 \end{bmatrix} + x_3 \begin{bmatrix} -3 \\ 0 \\ 1 \\ 0 \\ 0 \end{bmatrix} + x_5 \begin{bmatrix} 0 \\ 0 \\ 0 \\ -\frac{3}{2} \\ 1 \end{bmatrix} = r\mathbf{u} + s\mathbf{v} + t\mathbf{w}$$

A basis for the null space is $\{\mathbf{u}, \mathbf{v}, \mathbf{w}\}$. Therefore, we obtain $\text{Dim}(\text{Null}(\mathbf{A})) = 3$. A basis for the column space of $\mathbf{A}$ is given by columns one and four of $\mathbf{A}$ or

$$\text{Col}(\mathbf{A}) = \text{Span} \left\{[1, 3, 2, 1]^T, [2, 2, 2, 2]^T\right\}$$

Consequently, we obtain $\text{Dim}(\text{Col}(\mathbf{A})) = 2$. A basis for the row space of $\mathbf{A}$ is given by rows one and two of $\mathbf{B}$ or

$$\text{Row}(\mathbf{B}) = \text{Span}\left\{[1, 2, 3, 0, 0],\ [0, 0, 0, 1, \tfrac{3}{2}]\right\}$$

Finally, as a check, we have $\text{Dim}\left(\text{Col}(\mathbf{A})\right) + \text{Dim}\left(\text{Null}(\mathbf{A})\right) = 2 + 3 = 5 = n$, where $\mathbf{A}$ is $m \times n = 4 \times 5$. ∎

Caution

If $\mathbf{A} \sim \mathbf{B}$, where $\mathbf{B}$ is the reduced echelon form of $\mathbf{A}$, then select pivot columns from $\mathbf{A}$ for the basis of $\text{Col}(\mathbf{A})$, but select pivot rows from $\mathbf{B}$ for $\text{Row}(\mathbf{A})$. Remember that row operations preserve the linear dependence among the columns but not among the rows!

SUMMARY 5.2

- In a vector space V, a linearly independent set that spans V is a basis for V.
- The set of columns of any $n \times n$ invertible matrix is a basis for $\mathbb{R}^n$.
- If a basis is given for a vector space, then each vector in the space has a unique expression as a linear combination of elements in that basis.
- The mapping of $\mathbf{x}$ to its coordinate vector $[\mathbf{x}]_{\mathcal{B}}$ (as described previously) is linear.
- The mapping $\mathbf{x} \mapsto [\mathbf{x}]_{\mathcal{B}}$ defined previously is injective (*one-to-one*) and surjective (*onto*) from V to $\mathbb{R}^n$.
- The relation of isomorphism between two vector spaces is an equivalence relation.
- If a vector space has a finite basis, then all of its bases have the same number of elements.
- A vector space is finite dimensional if it has a finite basis; in that event, its dimension is the number of elements in any basis.
- Two finite-dimensional vector spaces are isomorphic to each other if and only if their dimensions are the same.

- The span of a set is not affected by removing from the set one element that is a linear combination of other elements of that set.
- If a set of n vectors spans an n-dimensional vector space, then the set is a basis for that vector space.
- In an n-dimensional vector space every linearly independent set of n vectors is a basis.
- If a set of n vectors spans a vector space, then the dimension of that space is at most n.
- If a set of n vectors is linearly independent in a vector space V, then the dimension of V is at least n.
- The pivot columns of a matrix form a basis for its column space.
- For any finite-dimensional vector space U and any linear transformation $L : U \to V$, we have $\text{Dim}\left(L[U]\right) \le \text{Dim}(U)$.
- Let $\{\mathbf{u}_1, \mathbf{u}_2, \ldots, \mathbf{u}_n\}$ be a basis for a vector space U. Let V be any vector space. A linear transformation $T : U \to V$ is completely determined by specifying the values

of $T(\mathbf{u}_i)$ for $i = 1, 2, \ldots, n$. These values, in turn, can be any vectors whatsoever in V.

- In a finite-dimensional vector space, any linearly independent set can be expanded to get a basis.

- If T is a linear transformation whose domain is an n-dimensional vector space, then
$$\text{Dim}\left(\text{Ker}(T)\right) + \text{Dim}\left(\text{Range}(T)\right) = n$$

- (Rank–Nullity Theorem.) For any matrix, the number of columns, n, equals the dimension of the column space plus the dimension of the null space:
$$\text{Dim}\left(\text{Col}(\mathbf{A})\right) + \text{Dim}\left(\text{Null}(\mathbf{A})\right) = n$$

- The row space and column space of a matrix have the same dimension:
$$\text{Dim}\left(\text{Row}(\mathbf{A})\right) = \text{Dim}\left(\text{Col}(\mathbf{A})\right)$$

- **Caution:** If $\mathbf{A} \sim \mathbf{B}$, (the latter being the reduced echelon form) then select pivot columns from $\mathbf{A}$ as the basis vectors of $\text{Col}(\mathbf{A})$ but select pivot rows from $\mathbf{B}$ for $\text{Row}(\mathbf{A})$.

KEY CONCEPTS 5.2

Basis for a vector space, standard basis, coordinate vector of an element in a vector space $[\mathbf{x}]_{\mathcal{B}}$, isomorphism, isomorphic, equivalence relations, dimension of a vector space, finite dimensional and infinite dimensional, kernel, domain, range, co-domain, dimensions of various subspaces, column space, row space, null space, rank, nullity

GENERAL EXERCISES 5.2

1. Consider
$$\begin{bmatrix} 1 & 3 & 2 & 7 & 5 \\ 2 & 0 & 1 & 4 & 4 \\ 6 & 2 & 3 & -1 & -6 \\ -3 & 1 & 0 & 12 & 15 \end{bmatrix}$$

Do the rows of this matrix form a basis for a subspace in $\mathbb{R}^5$?

2. Let $\mathcal{B} = \{(1, 2, 3), (2, 2, 2), (3, -3, 1)\}$ and $\mathbf{x} = (3, 2, -1)$. What is $[\mathbf{x}]_{\mathcal{B}}$?

3. Express the polynomial p defined by $p(t) = 3(t-2) + 4(3-5t)^2 - t^3$ in terms of Legendre polynomials.

4. Express the polynomial defined by $p = 3P_0 - 4P_1 + 2P_2 - P_3$ in standard form. (Here, P_k is the kth degree Legendre polynomial defined in Example 5.)

5. Find the coordinates of $\mathbf{x} = (-5, 1, 2)$ with respect to the basis consisting of $\mathbf{u}_1 = (1, 3, 2), \mathbf{u}_2 = (2, 1, 4)$, and $\mathbf{u}_3 = (1, 0, 6)$.

6. Find the dimension of the span of four polynomials whose definitions are $p_1(t) = 4 + 3t + 2t^2 + t^3$, $p_2(t) = 5 + 4t + 3t^2 + 2t^3$, $p_3(t) = 6 + 5t + 4t^2 + 3t^3$, $p_4(t) = 7 + 6t + 5t^2 + 4t^3$.

7. Let $\mathbf{A} = \begin{bmatrix} 3 & 9 & -12 & 1 \\ 1 & 3 & -4 & 0 \\ 2 & 6 & -8 & 1 \\ 3 & 9 & -12 & 0 \end{bmatrix}$

Find a basis for the null space of $\mathbf{A}$. Check your work by an independent calculation.

8. Determine whether $(1, 5, 6)$ is in the row space of this matrix $\begin{bmatrix} 1 & 2 & 3 \\ 3 & 6 & 9 \\ 1 & 3 & 4 \end{bmatrix}$

9. Explain why the map $L : \mathbb{P}_n \to \mathbb{P}_n$ defined by the equation $L(p) = p + p'$ is an isomorphism. You may assume that L is linear. (Here p' is the derivative of p.)

10. Define polynomials $p_1(t) = 1 - 2t - t^2$, $p_2(t) = t + t^2 + t^3$, $p_3(t) = 1 - t + t^3$, and $p_4(t) = 3 + 4t + t^2 + 4t^3$. Let S be the set of these four functions. Find a subset of S that is a basis for the span of S.

11. Suppose $\mathbf{A}$ is 5×6 and all solutions of $\mathbf{Ax} = \mathbf{0}$ are multiples of one nonzero vector. Will the equation $\mathbf{Ax} = \mathbf{b}$ be solvable for all $\mathbf{b}$?

12. Consider $S = \{[1, 3, 2, 0]^T, [-2, 0, 6, 7]^T, [0, 6, 10, 7]^T, [2, 10, -3, 1]^T\}$. Find a basis for $\text{Span}(S)$.

13. Suppose that $\mathbf{A}$ is a 6×8 matrix, and the system $\mathbf{Ax} = \mathbf{b}$ has a solution with two free variables. Are there inconsistent cases?

14. Let p be a polynomial of degree n, $p(t) = \sum_{i=0}^{n} a_i t^i$ with $a_n \neq 0$. Is the set $\{p, p', p'', \ldots, p^{(n)}\}$ necessarily a basis for $\mathbb{P}_n$? Explain why or give a counterexample. (Related to Example 16.)

15. Define four polynomials as follows: $p_1(t) = t^3 - 2t^2 + 1$, $p_2(t) = 2t^2 + t$, $p_3(t) = 2t^3 + 3t + 2$, $p_4(t) = 4t^2 + 4t$. Find a subset of $\{p_1, p_2, p_3, p_4\}$ that is a basis for the span of this set of four polynomials. Explain.

16. Consider the four vectors $\mathbf{u} = (1, 1, 1, 1)$, $\mathbf{v} = (1, 1, 1, -1)$, $\mathbf{w} = (1, 1, -1, -1)$, and $\mathbf{x} = (1, -1, -1, -1)$. Verify that $\{\mathbf{u}, \mathbf{v}, \mathbf{w}, \mathbf{x}\}$ is a basis for $\mathbb{R}^4$. Explain what you are doing.

17. Determine whether $\begin{bmatrix} 1 & 5 & 6 \end{bmatrix}$ is in the row space of this matrix: $\begin{bmatrix} 1 & 2 & 3 \\ 3 & 5 & 9 \\ 1 & 3 & 4 \end{bmatrix}$

18. Let $L : \mathbb{P}_4 \to \mathbb{P}_2$ be defined by $L(p) = p''$. What are $\text{Ker}(L)$, $\text{Range}(L)$, and $\text{Dim}\big(\text{Domain}(L)\big)$?

19. Let $\mathbf{u} = (1, 1)$, $\mathbf{v} = (3, -2)$, $\mathbf{w} = (4, 3)$, and $\mathbf{y} = (-3, 4)$. Let L be a linear map from $\mathbb{R}^2$ to $\mathbb{R}^2$ such that $L(\mathbf{u}) = \mathbf{w}$ and $L(\mathbf{v}) = \mathbf{y}$. If $\mathbf{z} = (7, 5)$, what is $L(\mathbf{z})$?

20. Let $\mathbf{A} = \begin{bmatrix} 0 & 0 & 0 & 0 \\ 3 & 2 & 1 & 2 \\ 2 & -1 & 1 & 3 \\ 0 & 7 & -1 & -5 \\ 2 & 1 & 1 & 1 \end{bmatrix}$

Find a basis for the row space of $\mathbf{A}$.

21. Establish that the Chebyshev polynomials $p_0(t) = 1$, $p_1(t) = t$, $p_2(t) = 2t^2 - 1$, $p_3(t) = 4t^3 - 3t$ form a basis for $\mathbb{P}_3$.

22. Let $\mathbf{A} = \begin{bmatrix} 1 & 2 & 1 & 0 & -1 \\ 2 & -1 & 0 & 1 & 3 \\ 1 & -3 & -1 & 1 & 4 \\ 2 & 9 & 4 & -1 & -7 \end{bmatrix}$

Find a simple basis for the column space of $\mathbf{A}$. Explain why $\{\mathbf{x} : \mathbf{Ax} = \mathbf{0}\}$ is a subspace of $\mathbb{R}^5$, and compute its dimension.

23. Use determinants to show that each of these is a basis for $\mathbb{R}^3$:
  a. $[1, 0, 0]^T$, $[0, 1, 0]^T$, $[0, 0, 1]^T$
  b. $[1, 1, 1]^T$, $[1, 1, 0]^T$, $[1, 0, 0]^T$
  c. $[2, 1, 3]^T$, $[5, 2, 1]^T$, $[3, 7, 2]^T$

24. (Continuation.) Consider $[1, 3, 2]^T$, $[4, 2, 3]^T$, $[-1, 1, 0]^T$, $[3, 0, 1]$. Find a basis for $\mathbb{R}^3$ from among these vectors. Use determinants to verify your results.

25. Consider $\{[-1, 1, 0]^T, [4, 2, 3]^T, [1, 3, 2]^T, [4, 6, 5]^T\}$. What is a basis for the space spanned by this set?

26. Extend this set $\{[1,2,3]^T, [0,2,3]^T\}$ to a basis for $\mathbb{R}^3$.

27. Let $\mathbf{A} = \begin{bmatrix} 2 & -2 & 3 & 5 & 5 \\ -1 & 1 & 4 & 14 & -8 \\ 4 & -4 & -2 & -14 & 18 \\ 3 & -3 & -1 & -9 & 13 \end{bmatrix}$

Find a simple basis for the row space of $\mathbf{A}$.

28. Construct a simple basis for the span of the set of four vectors given here: $(1,0,3,0)$, $(3,1,11,0)$, $(5,1,17,1)$, and $(9,2,31,1)$.

29. Let $\mathbf{A} = \begin{bmatrix} 1 & 3 & 2 \\ 2 & 3 & 1 \\ 3 & 5 & 2 \\ 1 & 2 & 1 \end{bmatrix}$

Find bases for the range and the kernel of $\mathbf{A}$. Find values for $\text{Dim}\left(\text{Ker}(\mathbf{A})\right)$, $\text{Dim}\left(\text{Range}(\mathbf{A})\right)$, $\text{Dim}\left(\text{Domain}(\mathbf{A})\right)$.

30. Explain why a line through the origin in a vector space is a subspace of dimension one. Recall the representation of a line as discussed in Sections 2.1–2.2.

31. Establish that if S is a linearly independent set of n elements in a vector space, then $\text{Dim}\left(\text{Span}(S)\right) = n$. *Every linearly independent set is a basis for something.*

32. Affirm that the rank of a matrix equals the dimension of its row space. (The term *rank* was defined in Section 1.3.)

33. Establish that if T is a linearly independent set in an n-dimensional vector space, and if S spans that vector space, then $\#(T) \le n \le \#(S)$. The notation $\#(T)$ denotes the number of elements in the set T.

34. Verify that the dimension of $\mathbb{R}^{m \times n}$ is mn. What is the dimension of the subspace of $\mathbb{R}^{n \times n}$ consisting of symmetric matrices? An argument is required.

35. What is the dimension of the space of $n \times n$ matrices having zero trace? (The trace of an $n \times n$ matrix $\mathbf{A}$ is $\sum_{i=1}^{n} \mathbf{A}_{ii}$.) Justify your answer.

36. What is the dimension of the space of polynomials having degree at most n and taking the value 0 at the point 0? An argument is required. Also explain why this space really is a vector subspace.

37. Let $\mathbf{B}$ be an $n \times n$ noninvertible matrix. Let V be the set of all $n \times n$ matrices $\mathbf{A}$ such that $\mathbf{BA} = \mathbf{0}$. Is V a vector space? If so, what is its dimension?

38. Apropos Theorem 12: Give an example of a map from $\mathbb{R}$ onto $\mathbb{R}^2$. Thus, if we admit nonlinear maps, the dimension of the range can be greater than the dimension of the domain.

39. Explain why the composition of two isomorphisms is an isomorphism.

40. Select a nonzero vector $\mathbf{v}$ in $\mathbb{R}^n$, and define a subspace by the equation $U = \{\mathbf{x} \in \mathbb{R}^n : \mathbf{v}^T \mathbf{x} = 0\}$. (Here we interpret $\mathbf{v}^T$ as a matrix with one row and $\mathbf{x}$ as a matrix with one column, so that the matrix product $\mathbf{v}^T \mathbf{x}$ makes sense.) What is the dimension of U?

41. What is the dimension of the set of all polynomials that can be written in the form $p(t) = at^9 + bt^{17} + ct^7$?

42. Criticize this argument: We have three vectors, $\mathbf{u}_1 = (1,3,2)$, $\mathbf{u}_2 = (-2,4,5)$, and $\mathbf{u}_3 = (-1,7,7)$. We notice that $\mathbf{u}_3$ is a linear combination of $\mathbf{u}_1$ and $\mathbf{u}_2$. Furthermore, $\mathbf{u}_2$ is a linear combination of $\mathbf{u}_1$ and $\mathbf{u}_3$. Using Theorem 8 twice, we conclude that both $\mathbf{u}_2$ and $\mathbf{u}_3$ can be removed from the set $\{\mathbf{u}_1, \mathbf{u}_2, \mathbf{u}_3\}$ without affecting the span of that set.

43. Establish that the vector space of all sequences of real numbers having only a finite number of nonzero terms has the basis $\{\mathbf{u}_1, \mathbf{u}_2, \ldots, \mathbf{u}_k, \ldots\}$, where $\mathbf{u}_1 = (1, 0, 0, \ldots)$, $\mathbf{u}_2 = (0, 1, 0, \ldots)$, and so on. (There are infinitely many vectors $\mathbf{u}_k$.)

44. In Theorem 14, why do we not prove that $\text{Dim}\left(L(V)\right) = \text{Dim}(V)$?

45. Establish that if the rows of an $n \times n$ matrix span $\mathbb{R}^n$, then the same is true of the columns.

46. Is the row space of a matrix isomorphic to its column space?

47. Argue that if the rows of an $n \times n$ matrix form a linearly independent set, then the columns span $\mathbb{R}^n$.

48. Finish the proof of Theorem 3.

49. Justify that $\left[\sum_{i=1}^{n} a_i\mathbf{u}_i\right]_{\mathcal{B}} = (a_1, a_2, \ldots, a_n)$ if $\mathcal{B} = \{\mathbf{u}_1, \mathbf{u}_2, \ldots, \mathbf{u}_n\}$ is a basis.

50. In the set $\mathbb{Z}$ consisting of all integers, define $n \equiv m$ to signify that $n - m$ is divisible by 7. Explain why this defines an equivalence relation.

51. Substantiate that if a vector space V contains a linearly independent set of k vectors and if another set of k vectors spans V, then $\text{Dim}(V) = k$.

52. An isomorphism preserves dimensions. More precisely, if there is an isomorphism between two finite dimensional vector spaces, then the spaces have the same dimension. Justify this result.

53. An isomorphism maps every linearly independent set into a linearly independent set. It also maps a linearly dependent set into a linearly dependent set. Justify these assertions.

54. Explain why every subset of a linearly independent set is linearly independent.

55. Establish that a basis for a finite-dimensional vector space is any maximal linearly independent set. (Such a set is linearly independent *and* is not a proper subset of another linearly independent set.)

56. Explain why a basis for a finite-dimensional vector space is any minimal spanning set. (Such a set spans the given vector space but it contains no smaller spanning set. A *spanning set* is any set that spans the given vector space.)

57. Verify that if the columns of an $m \times n$ matrix form a basis for $\mathbb{R}^m$, then $n = m$.

58. Is there a linear map from $\mathbb{R}^2$ onto $\mathbb{R}^3$? Why or why not?

59. Confirm that $\mathbb{R}^{n \times m}$ is isomorphic to $\mathbb{R}^{p \times q}$ if and only if the product nm equals the product pq.

60. Verify that if V and W are finite-dimensional vector spaces that are not isomorphic to each other, then their dimensions are not the same.

61. Let $p_j(t) = t^j$ for $j = 0, 1, 2, \ldots$. Explain why any set of such functions is linearly independent.

62. Let U be a subspace of a vector space X. Define $\mathbf{x} \equiv \mathbf{y}$ to mean $\mathbf{x} - \mathbf{y} \in U$. Is this an equivalence relation on X?

63. Use the notation $\#S$ to signify the number of elements of a set S. Explain why a finite set S in a vector space must obey the inequality Dim $\big(\text{Span}(S)\big) \leq \#S$.

64. Let $\mathbf{A}$ be an $m \times n$ matrix, and $\mathbf{B}$ be an $n \times k$ matrix. Establish that if $k > m$, then the columns of $\mathbf{AB}$ form a linearly dependent set.

65. Consider the vector space consisting of all real-valued sequences, $(\ldots, x_{-2}, x_{-1}, x_0, x_1, x_2, \ldots)$. The algebraic operations are the standard ones. This space is infinite-dimensional, and we see at once that it is the same as the set of all real-valued functions defined on $\mathbb{Z}$ (the set of all integers, positive, negative, and zero). Find the dimension of the subspace consisting of all sequences that obey the rule $x_n = 3x_{n-1} - x_{n-2}$ for all n.

66. Let S be a subset of a vector space, V. Suppose that S spans V, and that there is at least one vector in V that has a unique expression as a linear combination of the vectors in S. Is S necessarily a basis for V?

67. Let Q be a set of vectors that spans a vector space X. If some vector in X can be expressed in two different ways as a linear combination of vectors in Q, does it follow that every vector in X has this property?

68. Let $\mathbf{A}$ be an $m \times n$ matrix. Let $\mathbf{B}$ be an $n \times k$ matrix. Let $k > n$. Is it a justifiable conclusion that the columns of $\mathbf{AB}$ form a linearly dependent set?

69. Let $\mathbf{A}$ be a 10×7 matrix of rank five. What are the dimensions of the domain of $\mathbf{A}$, the kernel of $\mathbf{A}$, the row space of $\mathbf{A}$, the column space of $\mathbf{A}$, and the range of $\mathbf{A}$? Explain how you arrive at each answer, demonstrating along the way that you know the definitions of the terms and some applicable theorems.

70. Explain why the rows of an $n \times n$ invertible matrix provide a basis for $\mathbb{R}^n$.

71. Establish that the columns of an $n \times n$ invertible matrix form a basis for $\mathbb{R}^n$. Your proof should use directly the definitions of *basis* and *invertible*. It should demonstrate that you know the meanings of these terms. (Do *not* simply quote part of the Invertible Matrix Theorem.)

72. In a certain vector space W, we have two sets of vectors $V = \{\mathbf{v}_1, \mathbf{v}_2, \ldots, \mathbf{v}_7\}$, and $U = \{\mathbf{u}_1, \mathbf{u}_2, \ldots, \mathbf{u}_8\}$. One of these sets spans W and the other is linearly independent. Which set has which property? What is the dimension of W? Explain fully.

73. State the theorem that relates the dimensions of certain vector spaces connected with a linear transformation. This theorem concerns an arbitrary linear transformation from a finite-dimensional vector space into another vector space (which need not be finite dimensional). Illustrate the theorem by using the polynomials $p_i(t) = t^i$ and a linear transformation L defined by $L(ap_4 + bp_3 + cp_2 + dp_1 + ep_0) = cp_2 + dp_1 + ep_0$. Find the dimensions of the three important subspaces connected with such a transformation. You should compute the dimensions of those three linear spaces. The transformation L maps $\mathbb{P}_5$ to $\mathbb{P}_3$.

74. Let $\{\mathbf{y}_1, \mathbf{y}_2, \ldots, \mathbf{y}_n\}$ be a basis for the range of a linear transformation T. Let $\mathbf{x}_i$ be points in the domain of T such that $T(\mathbf{x}_i) = \mathbf{y}_i$ for $1 \leq i \leq n$. Establish that $\{\mathbf{x}_1, \mathbf{x}_2, \ldots, \mathbf{x}_n\}$ is linearly independent. If possible, improve this result by weakening the hypothesis on $\{\mathbf{y}_1, \mathbf{y}_2, \ldots, \mathbf{y}_n\}$.

75. We call a linear transformation *robust* if it maps no nonzero vectors to zero. Explain why the domain and range of a robust transformation have equal dimensions.

76. Establish that for any $m \times n$ matrix $\mathbf{A}$, the rank of $\mathbf{A}$ is n minus the dimension of the kernel of $\mathbf{A}$.

77. Let T be the operator $T(f) = f + f' + f''$, being applied in the polynomial space $\mathbb{P}_7$. What are the dimensions of the domain, the range, and the null space of T? What polynomials are in the kernel?

78. Let $\mathbf{A}$ be a 12×15 matrix. Suppose that the equation $\mathbf{Ax} = \mathbf{b}$ has a solution for every $\mathbf{b} \in \mathbb{R}^{12}$. What are the dimensions of the domain, the range, and the kernel of $\mathbf{A}$?

79. Explain why a matrix and its transpose have the same rank.

80. In detail, explain the last questions in Examples 14 and 15. Verify them. Find other examples to illustrate this concept.

81. Give an example of a matrix in reduced row echelon form for which the co-domain, domain, range, and kernel have dimensions 8, 7, 4, and 3, respectively.

82. Let $\mathbf{A}$ be an 9×6 matrix whose kernel is two dimensional. What are the dimensions of the row space, the column space, and the range?

83. Confirm that the rank of an $m \times n$ matrix cannot be greater than the lesser of m and n. In symbols, $\mathrm{Rank}(\mathbf{A}) \leq \min\{m, n\}$.

84. Let $\mathbf{A}$ be any matrix and $\widetilde{\mathbf{A}}$ its reduced row echelon form. Explain why $\mathbf{Ax} = \mathbf{0}$ is true if and only if $\widetilde{\mathbf{A}}\mathbf{x} = \mathbf{0}$.

85. Is there a 2×2 matrix whose powers span $\mathbb{R}^{2 \times 2}$?

86. Confirm, for an arbitrary $m \times n$ matrix $\mathbf{A}$, that the rank of $\mathbf{A}$ is n if and only if the kernel of $\mathbf{A}$ contains only the vector $\mathbf{0}$.

87. Establish that the kernel of a matrix and the column space of its transpose can have only $\mathbf{0}$ in common.

88. (Continuation.) Let $\mathbf{A}$ be an $m \times n$ matrix. Explain the logical equivalence of these two conditions:
   a. The equation $\mathbf{Ax} = \mathbf{b}$ is consistent for all $\mathbf{b}$ in $\mathbb{R}^m$.
   b. The kernel of $\mathbf{A}^T$ contains only $\mathbf{0}$.

89. Verify that if integers k, n, m satisfy the inequality $0 \leq k \leq n \leq m$, then there is an $m \times n$ matrix $\mathbf{A}$ such that $\mathrm{Dim}\big(\mathrm{Ker}(\mathbf{A})\big) = n - k$ and $\mathrm{Dim}\big(\mathrm{Range}(\mathbf{A})\big) = k$.

90. Explain why the following inequality holds for any linear transformation T: $\mathrm{Dim}\big(\mathrm{Range}(T)\big) \leq \mathrm{Dim}\big(\mathrm{Domain}(T)\big)$.

COMPUTER EXERCISES 5.2

1. Let

$$\mathbf{A} = \begin{bmatrix} 0 & 2 & 0 & 4 & 1 & 3 & 0 \\ 0 & 2 & 1 & 4 & 0 & 3 & 0 \\ 1 & 2 & 0 & 4 & 0 & 3 & 1 \\ 1 & 2 & 0 & 4 & 0 & 3 & 0 \end{bmatrix}$$

Numerically compute the following:
   a. Null space
   b. Column space
   c. Row space

2. Consider this $n \times n$ matrix

$$\begin{bmatrix} m & 1 & 1 & \cdots & 1 & 1 & 1 \\ 1 & m & 1 & \cdots & 1 & 1 & 1 \\ 1 & 1 & m & \cdots & 1 & 1 & 1 \\ \vdots & \vdots & \vdots & \ddots & \vdots & \vdots & \vdots \\ 1 & 1 & 1 & \cdots & m & 1 & 1 \\ 1 & 1 & 1 & \cdots & 1 & m & 1 \\ 1 & 1 & 1 & \cdots & 1 & 1 & m \end{bmatrix}$$

Determine numerically for what values of m this matrix is noninvertible.

3. Consider this matrix

$$\begin{bmatrix} -1 & 1 & 1 & -2 & 1 & -2 \\ -1 & 0 & 3 & -4 & 2 & -5 \\ -1 & 0 & 3 & -5 & 3 & -6 \\ -1 & 0 & 3 & -6 & 4 & -7 \\ -1 & 0 & 3 & -6 & 4 & -7 \end{bmatrix}$$

Numerically compute the following:

a. Null space
b. Column space
c. Row space

4. Consider this augmented matrix:

$$[\mathbf{A} \,|\, \mathbf{b}] = \begin{bmatrix} -1 & 1 & 1 & -2 & 1 & | & -2 \\ -1 & 0 & 3 & -4 & 2 & | & -5 \\ -1 & 0 & 3 & -5 & 3 & | & -6 \\ -1 & 0 & 3 & -6 & 4 & | & -7 \\ -1 & 0 & 3 & -6 & 4 & | & -7 \end{bmatrix}$$

a. Determine numerically the general solution.
b. Find the basis for the column space of $\mathbf{A}$ and its dimension.
c. Is the righthand vector $\mathbf{b}$ in the column space?

5. Consider the matrices

$$\mathbf{A} = \begin{bmatrix} 1 & -1 & 2 \\ 2 & 3 & -1 \end{bmatrix}$$

$$\mathbf{B} = \begin{bmatrix} 1 & -2 & -1 \\ 3 & -4 & 5 \end{bmatrix}$$

$$\mathbf{C} = \begin{bmatrix} 2 & -1 & 10 \\ 3 & -5 & 1 \end{bmatrix}$$

Determine numerically which of these matrices have the same row space.

5.3 COORDINATE SYSTEMS

The more you know, the less sure you are.

—VOLTAIRE (1694–1778)

You know that I write slowly. This is chiefly because I am never satisfied until I have said as much as possible in a few words, and writing briefly takes far more time than writing at length.

—KARL FRIEDRICH GAUSS (1777–1855)

Coordinate Vectors

In Section 5.2, we considered an arbitrary n-dimensional vector space V and a prescribed ordered basis for V: $\mathcal{B} = \{\mathbf{u}_1, \mathbf{u}_2, \ldots, \mathbf{u}_n\}$. Every vector $\mathbf{x}$ in V has a uniquely determined set of coefficients $(a_1, a_2, \ldots, a_n)$ such that $\mathbf{x} = \sum_{i=1}^{n} a_i \mathbf{u}_i$. This set of coefficients is denoted by $[\mathbf{x}]_\mathcal{B}$ and is called the **coordinate vector** of $\mathbf{x}$ relative to the basis $\mathcal{B}$, or the **coordinatization** of $\mathbf{x}$ associated with $\mathcal{B}$. It is a vector in $\mathbb{R}^n$. Thus, the equation $[\mathbf{x}]_\mathcal{B} = \mathbf{a}$ is true if and only if

$$\mathbf{x} = \sum_{i=1}^{n} a_i \mathbf{u}_i$$

Here, of course, we have written $\mathbf{a} = (a_1, a_2, \ldots, a_n)$. One must think of $\mathcal{B}$ as an *ordered* set $\{\mathbf{u}_1, \mathbf{u}_2, \ldots, \mathbf{u}_n\}$ so that the coefficients a_i will match with the vectors $\mathbf{u}_i$.

EXAMPLE 1 Let $\mathbf{v}_1 = \begin{bmatrix} 1 \\ 3 \\ 1 \end{bmatrix}$, $\mathbf{v}_2 = \begin{bmatrix} 2 \\ 1 \\ 4 \end{bmatrix}$, $\mathbf{v}_3 = \begin{bmatrix} 3 \\ -2 \\ 3 \end{bmatrix}$.

What is the coordinate vector of $\mathbf{w} = \begin{bmatrix} 3 \\ 17 \\ 13 \end{bmatrix}$

with respect to the basis $\mathcal{B} = \{\mathbf{v}_1, \mathbf{v}_2, \mathbf{v}_3\}$?

SOLUTION We place the vectors in an augmented matrix and find its reduced echelon form:

$$\begin{bmatrix} \mathbf{v}_1 & \mathbf{v}_2 & \mathbf{v}_3 & | & \mathbf{w} \end{bmatrix} = \begin{bmatrix} 1 & 2 & 3 & | & 3 \\ 3 & 1 & -2 & | & 17 \\ 1 & 4 & 3 & | & 13 \end{bmatrix}$$

$$\sim \begin{bmatrix} 1 & 0 & 0 & | & 2 \\ 0 & 1 & 0 & | & 5 \\ 0 & 0 & 1 & | & -3 \end{bmatrix}$$

The result is $[\mathbf{w}]_\mathcal{B} = \begin{bmatrix} 2 \\ 5 \\ -3 \end{bmatrix}$, and one can verify it by computing $2\mathbf{v}_1 + 5\mathbf{v}_2 - 3\mathbf{v}_3 = \mathbf{w}$. ∎

Changing Coordinates

Now there is the question of how to change coordinates from one basis to another. Thus, two bases will be active in a single vector space, and we shall seek a method to go back and forth between the two types of coordinatization. Suppose, then, that the vector space V is finite-dimensional, and that two bases, $\mathcal{B}$ and $\mathcal{C}$, have been prescribed. How can $[\mathbf{x}]_{\mathcal{C}}$ be computed from $[\mathbf{x}]_{\mathcal{B}}$? Here is the appropriate analysis.

First, we must name the elements in the first basis. $\mathcal{B} = \{\mathbf{u}_1, \mathbf{u}_2, \ldots, \mathbf{u}_n\}$. Let $\mathbf{x}$ be any element of V, and let $[\mathbf{x}]_{\mathcal{B}} = (a_1, a_2, \ldots, a_n)$. Then $\mathbf{x} = \sum_{i=1}^{n} a_i \mathbf{u}_i$ and

$$[\mathbf{x}]_{\mathcal{C}} = \left[\sum_{i=1}^{n} a_i \mathbf{u}_i\right]_{\mathcal{C}} = \sum_{i=1}^{n} a_i [\mathbf{u}_i]_{\mathcal{C}}$$

This equation used crucially the linear property of coordinatization (Theorem 3 in Section 5.2). Now recall that a linear combination of vectors can be written as a matrix times a column vector. The columns of the matrix are the vectors figuring in the linear combination, whereas the entries of the single additional column vector are the coefficients in this linear combination. Accordingly, we construct a matrix $\mathbf{P}$ whose columns are $[\mathbf{u}_i]_{\mathcal{C}}$, and write

$$[\mathbf{x}]_{\mathcal{C}} = \mathbf{P} \begin{bmatrix} a_1 \\ a_2 \\ \vdots \\ a_n \end{bmatrix} = \mathbf{P}[\mathbf{x}]_{\mathcal{B}}$$

where

$$\mathbf{P} = \begin{bmatrix} [\mathbf{u}_1]_{\mathcal{C}} & [\mathbf{u}_2]_{\mathcal{C}} & \cdots & [\mathbf{u}_n]_{\mathcal{C}} \end{bmatrix}$$

The matrix $\mathbf{P}$ is called the **transition matrix** that goes from basis $\mathcal{B}$ to basis $\mathcal{C}$. This discussion has proved the following theorem.

THEOREM 1

Let V be a finite-dimensional vector space in which two ordered bases $\mathcal{B}$ and $\mathcal{C}$ have been prescribed. Let $\mathbf{P}$ be the matrix whose columns are the $\mathcal{C}$-coordinates of the basis vectors in $\mathcal{B}$. Then we have $[\mathbf{x}]_{\mathcal{C}} = \mathbf{P}[\mathbf{x}]_{\mathcal{B}}$ for all $\mathbf{x}$ in V.

This theorem can be illustrated with the following diagram

$$\mathcal{B} = \{\mathbf{u}_1, \mathbf{u}_2, \ldots, \mathbf{u}_n\} \xrightarrow{\mathbf{P}} \mathcal{C} = \{\mathbf{v}_1, \mathbf{v}_2, \ldots, \mathbf{v}_n\}$$

in which the transition matrix $\mathbf{P}$ goes from basis $\mathcal{B}$ to basis $\mathcal{C}$.

EXAMPLE 2 Let two bases for $\mathbb{R}^2$ be prescribed as follows:

$$\mathcal{B} = \{\mathbf{u}_1, \mathbf{u}_2\}, \qquad \mathcal{C} = \{\mathbf{v}_1, \mathbf{v}_2\}$$

where

$$\mathbf{u}_1 = \begin{bmatrix} 1 \\ 1 \end{bmatrix}, \qquad \mathbf{u}_2 = \begin{bmatrix} -1 \\ 1 \end{bmatrix}, \qquad \mathbf{v}_1 = \begin{bmatrix} 2 \\ 4 \end{bmatrix}, \qquad \mathbf{v}_2 = \begin{bmatrix} 1 \\ 3 \end{bmatrix}$$

Find the transition matrix $\mathbf{P}$ that is needed for converting $\mathcal{B}$-coordinates to $\mathcal{C}$-coordinates.

SOLUTION By Theorem 1, the columns of $\mathbf{P}$ should be $[\mathbf{u}_1]_\mathcal{C}$ and $[\mathbf{u}_2]_\mathcal{C}$. Concentrating on the first column of $\mathbf{P}$, we ask for the coefficients x_1 and x_2 such that

$$\mathbf{u}_1 = x_1\mathbf{v}_1 + x_2\mathbf{v}_2 \qquad \text{or} \qquad x_1\begin{bmatrix} 2 \\ 4 \end{bmatrix} + x_2\begin{bmatrix} 1 \\ 3 \end{bmatrix} = \begin{bmatrix} 1 \\ 1 \end{bmatrix}$$

The augmented matrix for this problem is $\begin{bmatrix} 2 & 1 & | & 1 \\ 4 & 3 & | & 1 \end{bmatrix}$. The solution

is $x_1 = 1$ and $x_2 = -1$. Hence, column one in $\mathbf{P}$ is $\mathbf{x} = \begin{bmatrix} 1 \\ -1 \end{bmatrix}$.

Going on to column two in $\mathbf{P}$, we must find the coefficients y_1 and y_2 such that

$$\mathbf{u}_2 = y_1\mathbf{v}_1 + y_2\mathbf{v}_2 \qquad \text{or} \qquad y_1\begin{bmatrix} 2 \\ 4 \end{bmatrix} + y_2\begin{bmatrix} 1 \\ 3 \end{bmatrix} = \begin{bmatrix} -1 \\ 1 \end{bmatrix}$$

The augmented matrix for this system of equations is $\begin{bmatrix} 2 & 1 & | & -1 \\ 4 & 3 & | & 1 \end{bmatrix}$.

The solution is $y_1 = -2$ and $y_2 = 3$. Hence, column two in $\mathbf{P}$ is

$\mathbf{y} = \begin{bmatrix} -2 \\ 3 \end{bmatrix}$. It follows that the $\mathbf{P}$-matrix is $\mathbf{P} = \begin{bmatrix} \mathbf{x} & \mathbf{y} \end{bmatrix} = \begin{bmatrix} 1 & -2 \\ -1 & 3 \end{bmatrix}$.

A more efficient approach is to note that there are two systems of equations to solve and that they share a coefficient matrix, $\mathbf{A}$. The two systems of equations are $\mathbf{Ax} = \mathbf{u}_1$ and $\mathbf{Ay} = \mathbf{u}_2$. One should think of this problem as

$$\mathbf{A}[\mathbf{x}\ \mathbf{y}] = [\mathbf{u}_1\ \mathbf{u}_2]$$

For the augmented matrix, the steps in row reduction are like this:

$$\left[\ \mathbf{A}\ \middle|\ \mathbf{u}_1\ \ \mathbf{u}_2\ \right] = \left[\ \mathbf{v}_1\ \ \mathbf{v}_2\ \middle|\ \mathbf{u}_1\ \ \mathbf{u}_2\ \right]$$

$$= \left[\begin{array}{cc|cc} 2 & 1 & 1 & -1 \\ 4 & 3 & 1 & 1 \end{array}\right] \sim \left[\begin{array}{cc|cc} 2 & 1 & 1 & -1 \\ 0 & 1 & -1 & 3 \end{array}\right]$$

$$\sim \left[\begin{array}{cc|cc} 2 & 0 & 2 & -4 \\ 0 & 1 & -1 & 3 \end{array}\right] \sim \left[\begin{array}{cc|cc} 1 & 0 & 1 & -2 \\ 0 & 1 & -1 & 3 \end{array}\right]$$

$$= \left[\ \mathbf{I}\ \middle|\ \mathbf{x}\ \ \mathbf{y}\ \right] = \left[\ \mathbf{I}\ \middle|\ \mathbf{P}\ \right]$$

One obtains the same $\mathbf{P}$ as found previously. ∎

EXAMPLE 3 For $\mathbb{P}_3$, let us use two bases, where the first consists of Chebyshev[1] polynomials:

$\mathcal{B}:\quad T_0(x) = 1,\quad T_1(x) = x,\quad T_2(x) = 2x^2 - 1,\quad T_3(x) = 4x^3 - 3x$

For the second basis, the simple monomials are used. This basis is called the *natural* basis or the *standard* basis:

$\mathcal{C}:\quad p_0(x) = 1,\quad p_1(x) = x,\quad p_2(x) = x^2,\quad p_3(x) = x^3$

What is the matrix needed for transforming $\mathcal{B}$ coordinates to $\mathcal{C}$ coordinates?

SOLUTION Call this matrix $\mathbf{P}$. The columns of $\mathbf{P}$ should be $[T_i]_{\mathcal{C}}$, where $0 \le i \le 3$. We observe that

$$T_0 = p_0,\qquad T_1 = p_1,\qquad T_2 = 2p_2 - p_0,\qquad T_3 = 4p_3 - 3p_1$$

---

[1] Pafnuty Lvovich Chebyshev (1821–1894) founded a strong school of mathematics in St. Petersburg, Russia. In number theory, his name is attached to the theorem that asserts the existence of at least one prime number between n and $2n - 2$, for $n = 4, 5, \ldots$. He made outstanding contributions to approximation theory and statistics.

Hence, we obtain

$$
\mathbf{P} = \begin{bmatrix} 1 & 0 & -1 & 0 \\ 0 & 1 & 0 & -3 \\ 0 & 0 & 2 & 0 \\ 0 & 0 & 0 & 4 \end{bmatrix}
$$

∎

EXAMPLE 4 In the same situation as Example 3, what is the transition matrix for coordinates $\mathcal{C}$ to coordinates $\mathcal{B}$?

SOLUTION Let's not call it $\mathbf{P}$ because that will confuse the two examples. Use instead $\mathbf{Q}$. The columns of $\mathbf{Q}$ should be $[p_0]_\mathcal{B}$, $[p_1]_\mathcal{B}$, $[p_2]_\mathcal{B}$, and $[p_3]_\mathcal{B}$. Consider first p_0. How can it be expressed as a linear combination of T_0, T_1, T_2, T_3? Call the unknown coefficients z_0, z_1, z_2, z_3. We want $z_0 T_0 + z_1 T_1 + z_2 T_2 + z_3 T_3 = p_0$. Representing the polynomials just by their coefficients, we must have

$$
z_0 \begin{bmatrix} 1 \\ 0 \\ 0 \\ 0 \end{bmatrix} + z_1 \begin{bmatrix} 0 \\ 1 \\ 0 \\ 0 \end{bmatrix} + z_2 \begin{bmatrix} -1 \\ 0 \\ 2 \\ 0 \end{bmatrix} + z_3 \begin{bmatrix} 0 \\ -3 \\ 0 \\ 4 \end{bmatrix} = \begin{bmatrix} 1 \\ 0 \\ 0 \\ 0 \end{bmatrix}
$$

This is a system of equations that can be solved by row reduction. Each T_i must be treated in the same way, and the coefficient matrix remains the same for each. Hence, we should solve the system whose augmented matrix is

$$
\left[\begin{array}{cccc|cccc} 1 & 0 & -1 & 0 & 1 & 0 & 0 & 0 \\ 0 & 1 & 0 & -3 & 0 & 1 & 0 & 0 \\ 0 & 0 & 2 & 0 & 0 & 0 & 1 & 0 \\ 0 & 0 & 0 & 4 & 0 & 0 & 0 & 1 \end{array} \right] = [\mathbf{P} \mid \mathbf{I}]
$$

It is clear that we are finding the inverse of $\mathbf{P}$ by this calculation! But now it is obvious from Theorem 1 that the inverse of $\mathbf{P}$ is indeed what is wanted: From $[\mathbf{x}]_\mathcal{C} = \mathbf{P}[\mathbf{x}]_\mathcal{B}$, we have immediately $[\mathbf{x}]_\mathcal{B} = \mathbf{P}^{-1}[\mathbf{x}]_\mathcal{C}$. Thus, we obtain $\mathbf{Q} = \mathbf{P}^{-1}$. For the record, the answer is

$$
\mathbf{P}^{-1} = \mathbf{Q} = \begin{bmatrix} 1 & 0 & \frac{1}{2} & 0 \\ 0 & 1 & 0 & \frac{3}{4} \\ 0 & 0 & \frac{1}{2} & 0 \\ 0 & 0 & 0 & \frac{1}{4} \end{bmatrix}
$$

∎

Examples 3 and 4 can be illustrated with the following diagram:

$$\mathcal{B} = \{T_0, T_1, T_2, T_3\} \overset{\mathbf{P}}{\rightarrow} \mathcal{C} = \{p_0, p_1, p_2, p_3\}$$
$$\mathcal{B} = \{T_0, T_1, T_2, T_3\} \overset{\mathbf{P^{-1}}}{\leftarrow} \mathcal{C} = \{p_0, p_1, p_2, p_3\}$$

Here the transition matrix $\mathbf{P}$ goes from basis $\mathcal{B}$ to basis $\mathcal{C}$, and its inverse $\mathbf{P}^{-1}$ goes from basis $\mathcal{C}$ back to basis $\mathcal{B}$.

Linear Transformations

Theorem 3 in Section 2.3 asserted that a linear transformation from $\mathbb{R}^n$ to $\mathbb{R}^m$ is completely determined by the images of the standard basis elements $\mathbf{e}_i$ in $\mathbb{R}^n$. A more general result along these lines is possible; it is the same as Theorem 15 in Section 5.2 except that there the basis is finite.

THEOREM 2

A linear transformation from one vector space to another is completely determined by the images of any basis in the domain. These images can, in turn, be completely arbitrary.

PROOF We employ the usual notation, $L : X \rightarrow Y$, to name the map, the domain, and the co-domain. Let $\mathcal{B}$ be a basis for the domain. (It need not be finite.) If $\mathbf{x}$ is any point in X, then for a suitable finite set of elements $\mathbf{u}_i$ in $\mathcal{B}$ and accompanying scalars a_i, we have $\mathbf{x} = \sum_{i=1}^{n} a_i \mathbf{u}_i$. By linearity, it follows that

$$L(\mathbf{x}) = \sum_{i=1}^{n} a_i L(\mathbf{u}_i) \qquad \blacksquare$$

EXAMPLE 5 Is there a linear transformation that maps these vectors $\mathbf{u}_1 = (2, 3)$ to $\mathbf{v}_1 = (-1, 5)$ and $\mathbf{u}_2 = (4, 1)$ to $\mathbf{v}_2 = (3, 0)$?

SOLUTION We only have to be sure that the two vectors $\mathbf{u}_1$ and $\mathbf{u}_2$ form a linearly independent set. That they do so is obvious because neither is a multiple of the other. These two vectors form a basis for their span. Notice that $\mathbf{AU} = \mathbf{V}$ where

$$\mathbf{A} = \begin{bmatrix} 1 & -1 \\ -\frac{1}{2} & 2 \end{bmatrix}, \qquad \mathbf{U} = \begin{bmatrix} 2 & 4 \\ 3 & 1 \end{bmatrix}, \qquad \mathbf{V} = \begin{bmatrix} -1 & 3 \\ 5 & 0 \end{bmatrix} \qquad \blacksquare$$

EXAMPLE 6 Find a linear transformation that maps $(1, 3, -7)$ to $(2, 0, 2)$ and $(3, 2, 1)$ to $(-1, 1, 5)$ and $(-3, 5, -23)$ to $(8, -2, 5)$.

SOLUTION If the linear transformation is given by a 3×3 matrix $\mathbf{X}$, then we want

$$\mathbf{XB} = \mathbf{X} \begin{bmatrix} 1 & 3 & -3 \\ 3 & 2 & 5 \\ -7 & 1 & -23 \end{bmatrix} = \begin{bmatrix} 2 & -1 & 8 \\ 0 & 1 & -2 \\ 2 & 5 & -5 \end{bmatrix} = \mathbf{C}$$

The equation $\mathbf{XB} = \mathbf{C}$ is not the sort of problem we have become skilled at solving. Take transposes to get a familiar problem, $\mathbf{B}^T \mathbf{X}^T = \mathbf{C}^T$. The accompanying augmented matrix is

$$\begin{bmatrix} \mathbf{B}^T & | & \mathbf{C}^T \end{bmatrix} = \begin{bmatrix} 1 & 3 & -7 & 2 & 0 & 2 \\ 3 & 2 & 1 & -1 & 1 & 5 \\ -3 & 5 & -23 & 8 & -2 & -5 \end{bmatrix}$$

$$\sim \begin{bmatrix} 1 & 0 & 2.4 & -1 & 0.42 & 0 \\ 0 & 1 & -3.1 & 1 & -0.14 & 0 \\ 0 & 0 & 0 & 0 & 0 & 1 \end{bmatrix}$$

The system is obviously inconsistent. There is a linear relation among the columns of $\mathbf{B}$, and any linear transformation must preserve that. In other words, the third vector is not free. The following theorem is pertinent. ∎

THEOREM 3

If $\sum_{i=1}^{n} a_i \mathbf{u}_i = \mathbf{0}$ and if L is linear, then $\sum_{i=1}^{n} a_i L(\mathbf{u}_i) = \mathbf{0}$.

PROOF We have at once

$$\mathbf{0} = L(\mathbf{0}) = L\left(\sum_{i=1}^{n} a_i \mathbf{u}_i \right) = \sum_{i=1}^{n} a_i L(\mathbf{u}_i)$$ ∎

A linear transformation mapping a finite-dimensional vector space into another vector space (not necessarily finite-dimensional) can be described by a matrix, after bases have been chosen for the domain and range of the transformation. Suppose that the transformation is $L : X \rightarrow Y$ and that bases have been chosen: $\mathcal{B} = \{\mathbf{u}_1, \mathbf{u}_2, \ldots, \mathbf{u}_n\}$ for X, and $\mathcal{C} = \{\mathbf{v}_1, \mathbf{v}_2, \ldots, \mathbf{v}_m\}$ for the range of L. By Theorem 11 (in Section 5.2), we know that $m \leq n$.

For any $\mathbf{x}$ in X, we can write $\mathbf{x} = \sum_{i=1}^{n} \alpha \mathbf{u}_i$. The scalars α_i are the coordinates of $\mathbf{x}$ with respect to the basis $\mathcal{B}$. Because L is linear, we can proceed to the equation $L(\mathbf{x}) = \sum_{i=1}^{n} \alpha_i L(\mathbf{u}_i)$. Now take the $\mathcal{C}$-coordinates of both sides, using the fact that this process is also linear (Theorem 3 of Section 5.2). The result is $[L(\mathbf{x})]_{\mathcal{C}} = \sum_{i=1}^{n} \alpha_i [L(\mathbf{u}_i)]_{\mathcal{C}}$. Form a matrix $\mathbf{A}$ by putting $[L(\mathbf{u}_i)]_{\mathcal{C}}$ as the columns of $\mathbf{A}$. The preceding equation is then

$$[L(\mathbf{x})]_{\mathcal{C}} = \mathbf{A}[\mathbf{x}]_{\mathcal{B}}$$

because the α_i are the entries in the coordinate vector for $\mathbf{x}$. We have established the following result.

THEOREM 4

Let L be a linear transformation from a finite-dimensional vector space X into a vector space Y. Let $\mathcal{B}$ and $\mathcal{C}$ be ordered bases for X and $L(X)$, respectively. Let $\mathbf{A}$ be the matrix whose columns are the vectors $[L(\mathbf{u}_i)]_{\mathcal{C}}$, where $\mathbf{u}_i$ are the ordered elements of $\mathcal{B}$. Then for all $\mathbf{x}$ in X, the equation $[L(\mathbf{x})]_{\mathcal{C}} = \mathbf{A}[\mathbf{x}]_{\mathcal{B}}$ is valid.

EXAMPLE 7 Let $\mathbf{X} = \mathbb{P}_4$, and use the basis of Chebyshev polynomials: $T_0(x) = 1$, $T_1(x) = x$, $T_2(x) = 2x^2 - 1$, $T_3(x) = 4x^3 - 3x$, and $T_4(x) = 8x^4 - 8x^2 + 1$. Let L be the differentiation operator that maps $\mathbb{P}_4$ onto $\mathbb{P}_3$. Just for variety, let us use the standard basis for $\mathbb{P}_3$, $p_i(x) = x^i$. What is the matrix $\mathbf{A}$ in this case?

SOLUTION To answer this, we must find $L(p_i)$ and $[L(p_i)]_{\mathcal{C}}$. We have $L(p_0) = 0$, $L(p_1) = 1$, $L(p_2) = 4x$, $L(p_3) = 12x^2 - 3$, $L(p_4) = 32x^3 - 16x$. Therefore, the matrix sought is

$$\mathbf{A} = \begin{bmatrix} 0 & 1 & 0 & -3 & 0 \\ 0 & 0 & 4 & 0 & -16 \\ 0 & 0 & 0 & 12 & 0 \\ 0 & 0 & 0 & 0 & 32 \end{bmatrix}$$

If we desire to know the effect of applying L to a typical random polynomial $\mathbf{x} = 3T_0 - 2T_1 + 7T_2 - 4T_3 + 6T_4$, the answer is $\mathbf{A}[\mathbf{x}]_{\mathcal{C}}$, which is computed from the matrix $\mathbf{A}$:

$$
\begin{bmatrix}
0 & 1 & 0 & -3 & 0 \\
0 & 0 & 4 & 0 & -16 \\
0 & 0 & 0 & 12 & 0 \\
0 & 0 & 0 & 0 & 32
\end{bmatrix}
\begin{bmatrix}
3 \\ -2 \\ 7 \\ -4 \\ 6
\end{bmatrix}
=
\begin{bmatrix}
10 \\ -68 \\ -48 \\ 192
\end{bmatrix}
$$

In other words, the resulting polynomial is $x \mapsto 10 - 68x - 48x^2 + 192x^3$. Checking with an independent verification:

$2T_0 - 2T_1 + 7T_2 - 4T_3 + 6T_4$

$\qquad = 3 - 2x + 7(2x^2 - 1) - 4(4x^3 - 3x) + 6(8x^4 - 8x^2 + 1)$

$\qquad = 2 + 10x - 34x^2 - 16x^3 + 48x^4$

The derivative is $10 - 68x - 48x^2 + 192x^3$. ■

EXAMPLE 8 Consider the linear transformation $L : \mathbb{R}^{2\times2} \to \mathbb{R}^3$ defined by

$$
L\left(\begin{bmatrix} a & c \\ b & d \end{bmatrix}\right) = \begin{bmatrix} a - b + 2d \\ 2a + c - d \\ -2b + d \end{bmatrix}
$$

Suppose that the basis for $\mathbb{R}^{2\times2}$ is chosen to be

$$
\mathcal{B} = \{\mathbf{M}_1, \ \mathbf{M}_2, \ \mathbf{M}_3, \ \mathbf{M}_4\}
$$

where

$$
\mathbf{M}_1 = \begin{bmatrix} 2 & 5 \\ 2 & 1 \end{bmatrix}, \qquad \mathbf{M}_2 = \begin{bmatrix} -2 & -2 \\ 0 & 1 \end{bmatrix}
$$

$$
\mathbf{M}_3 = \begin{bmatrix} -3 & -4 \\ 1 & 2 \end{bmatrix}, \qquad \mathbf{M}_4 = \begin{bmatrix} -1 & -3 \\ 0 & 1 \end{bmatrix}
$$

and the basis for $\mathbb{R}^3$ is

$$
\mathcal{C} = \{\mathbf{w}_1, \mathbf{w}_2, \mathbf{w}_3,\}
$$

where

$$
\mathbf{w}_1 = \begin{bmatrix} 7 \\ 0 \\ -3 \end{bmatrix}, \qquad \mathbf{w}_2 = \begin{bmatrix} 2 \\ -1 \\ -2 \end{bmatrix}, \qquad \mathbf{w}_3 = \begin{bmatrix} -2 \\ 0 \\ 1 \end{bmatrix}
$$

Find the matrix $\mathbf{A}$ such that $[L(\mathbf{v})]_\mathcal{C} = \mathbf{A}[\mathbf{v}]_\mathcal{B}$.

SOLUTION The transformation is

$$
L\left(\begin{bmatrix} a & c \\ b & d \end{bmatrix}\right) = \begin{bmatrix} a - b + 2d \\ 2a + c - d \\ -2b + d \end{bmatrix} = \begin{bmatrix} 1 & -1 & 0 & 2 \\ 2 & 0 & 1 & -1 \\ 0 & -2 & 0 & 1 \end{bmatrix} \begin{bmatrix} a \\ b \\ c \\ d \end{bmatrix}
$$

Associate with each matrix $\mathbf{M}_i = \begin{bmatrix} a_i & c_i \\ b_i & d_i \end{bmatrix}$ in the basis $\mathcal{B}$ the vector

$$
\mathbf{v}_i = \begin{bmatrix} a_i \\ b_i \\ c_i \\ d_i \end{bmatrix} ; \text{namely,}
$$

$$
\mathbf{v}_1 = \begin{bmatrix} 2 \\ 2 \\ 5 \\ 1 \end{bmatrix}, \quad \mathbf{v}_2 = \begin{bmatrix} -2 \\ 0 \\ -2 \\ 1 \end{bmatrix}, \quad \mathbf{v}_3 = \begin{bmatrix} -3 \\ 1 \\ -4 \\ 2 \end{bmatrix}, \quad \mathbf{v}_4 = \begin{bmatrix} -1 \\ 0 \\ -3 \\ 1 \end{bmatrix}
$$

We seek the matrix $\mathbf{A}$ such that $[L(\mathbf{v})]_\mathcal{C} = \mathbf{A}[\mathbf{v}]_\mathcal{B}$. Matrix $\mathbf{A}$ has columns $[L(\mathbf{v}_i)]_\mathcal{C}$ for $i = 1, 2, 3, 4$. We find

$$
L(\mathbf{v}_1) = \begin{bmatrix} 2 \\ 8 \\ -3 \end{bmatrix}, \quad L(\mathbf{v}_2) = \begin{bmatrix} 0 \\ -7 \\ 1 \end{bmatrix}
$$

$$
L(\mathbf{v}_3) = \begin{bmatrix} 0 \\ -12 \\ 0 \end{bmatrix}, \quad L(\mathbf{v}_4) = \begin{bmatrix} 1 \\ -6 \\ 1 \end{bmatrix}
$$

The matrix $\mathbf{A}$ is therefore

$$
\mathbf{A} = \begin{bmatrix} 2 & 0 & 0 & 1 \\ 8 & -7 & -12 & -6 \\ -3 & 1 & 0 & 1 \end{bmatrix} \qquad ∎
$$

Mapping a Vector Space into Itself

The final question that we want to address pertains to the case of a linear transformation mapping a vector space into itself. We want to understand the various matrices that can represent our mapping when the bases are changed. Let $L : X \to X$ be a linear map, where X is an n-dimensional vector space. Let X be given two bases, $\mathcal{B}$ and $\mathcal{C}$. We already know how to get the matrix for L when we restrict our attention to $\mathcal{B}$, using it twice, once for the domain and once for the co-domain. Specifically, we form the matrix $\mathbf{A}$ whose columns are $[L(\mathbf{u}_i)]_{\mathcal{B}}$, where $\mathbf{u}_i$ are the vectors in $\mathcal{B}$. Having done so, we have, by Theorem 4,

$$[L(\mathbf{x})]_{\mathcal{B}} = \mathbf{A}[\mathbf{x}]_{\mathcal{B}}$$

In the same way, there is a matrix $\mathbf{B}$ such that for all $\mathbf{x}$,

$$[L(\mathbf{x})]_{\mathcal{C}} = \mathbf{B}[\mathbf{x}]_{\mathcal{C}}$$

Finally, another matrix $\mathbf{P}$ enters as the transition matrix in Theorem 1. Thus, $\mathbf{P}$ does the following task:

$$[\mathbf{x}]_{\mathcal{C}} = \mathbf{P}[\mathbf{x}]_{\mathcal{B}}$$

for all $\mathbf{x}$. Putting this all together, we have

$$\mathbf{B}[\mathbf{x}]_{\mathcal{C}} = [L(\mathbf{x})]_{\mathcal{C}} = \mathbf{P}[L(\mathbf{x})]_{\mathcal{B}} = \mathbf{P}\mathbf{A}[\mathbf{x}]_{\mathcal{B}} = \mathbf{P}\mathbf{A}\mathbf{P}^{-1}[\mathbf{x}]_{\mathcal{C}}$$

Because $[\mathbf{x}]_{\mathcal{C}}$ can be any vector in $\mathbb{R}^n$, we conclude that $\mathbf{B} = \mathbf{P}\mathbf{A}\mathbf{P}^{-1}$. This is the relationship we sought to uncover. The result is the following theorem.

> **THEOREM 5**
>
> *If $n \times n$ matrices $\mathbf{A}$ and $\mathbf{B}$ represent the same linear transformation, but with respect to two bases, then for an appropriate invertible matrix $\mathbf{P}$, we have $\mathbf{B} = \mathbf{P}\mathbf{A}\mathbf{P}^{-1}$.*

Similar Matrices

The relationship between $\mathbf{A}$ and $\mathbf{B}$ in Theorem 5 is called **similarity**. Thus, as often happens, a common English word receives a highly technical meaning in mathematics.

> **DEFINITION**
>
> *Let $\mathbf{A}$ and $\mathbf{B}$ be two $n \times n$ matrices. We say that $\mathbf{A}$ is **similar** to $\mathbf{B}$ if there is an invertible matrix $\mathbf{P}$ such that $\mathbf{A} = \mathbf{P}\mathbf{B}\mathbf{P}^{-1}$.*

Similarity is an equivalence relation, as discussed in Section 5.2. Thus, if we express "$\mathbf{A}$ *is similar to* $\mathbf{B}$" by $\mathbf{A} \simeq \mathbf{B}$, we have these properties:

1. $A \simeq A$ (**reflexive**)

2. If $A \simeq B$, then $B \simeq A$ (**symmetric**)

3. If $A \simeq B$ and $B \simeq C$, then $A \simeq C$ (**transitive**)

EXAMPLE 9 Do these two matrices represent the same linear transformation with respect to different bases? In other words, are these two matrices similar to each other?

$$A = \begin{bmatrix} 2 & 0 & -1 \\ -9 & 29 & 23 \\ 12 & -36 & -29 \end{bmatrix}, \qquad B = \begin{bmatrix} -1 & 0 & 0 \\ 0 & 1 & 0 \\ 0 & 0 & 2 \end{bmatrix}$$

SOLUTION The answer is *Yes*, because the matrices

$$P = \begin{bmatrix} 1 & 2 & 3 \\ -2 & -1 & 1 \\ 3 & 2 & 0 \end{bmatrix}, \qquad P^{-1} = \begin{bmatrix} -2 & 6 & 5 \\ 3 & -9 & -7 \\ -1 & 4 & 3 \end{bmatrix}$$

have the property $A = PBP^{-1}$. How did we find P? See the next example. ∎

EXAMPLE 10 Let $A = \begin{bmatrix} 0 & \frac{3}{2} & 6 \\ \frac{2}{3} & 3 & 14 \\ -\frac{4}{3} & 1 & 1 \end{bmatrix}$, $B = \begin{bmatrix} 3 & 1 & 2 \\ 1 & 0 & 1 \\ 2 & 2 & 1 \end{bmatrix}$.

Are these matrices similar to each other?

SOLUTION Given $n \times n$ matrices A and B, finding a matrix P such that

$$A = PBP^{-1}$$

can be a complicated problem. We would need to solve the homogeneous system

$$AP - PB = 0$$

involving n^2 equations in n^2 unknowns. Letting

$$P = \begin{bmatrix} a & b & c \\ u & v & w \\ x & y & z \end{bmatrix}$$

and using the particular matrices $\mathbf{A}$ and $\mathbf{B}$ given previously, we find that the equation $\mathbf{AP} - \mathbf{PB} = \mathbf{0}$ leads to these nine linear equations: $-3a - b - 2c + \frac{3}{2}u + 6x = 0$, $-a - 2c + \frac{3}{2}v + 6y = 0$, $-2a - b - c + \frac{3}{2}w + 6z = 0$, $\frac{2}{3}a - v - 2w + 14x = 0$, $\frac{2}{3}b - u + 3v - 2w + 14y = 0$, $\frac{2}{3}c - 2u - v + 2w + 14z = 0$, $-\frac{4}{3}a + u - 2x - y - 2z = 0$, $-\frac{4}{3}b + v - x + y - 2z = 0$, and $-\frac{4}{3}c + w - 2x - y = 0$. Setting up the augmented matrix and transforming it to reduced echelon form, we have

$$
\begin{bmatrix}
-3 & -1 & -2 & \frac{3}{2} & 0 & 0 & 6 & 0 & 0 \\
-1 & 0 & -2 & 0 & \frac{3}{2} & 0 & 0 & 6 & 0 \\
-2 & -1 & -1 & 0 & 0 & \frac{3}{2} & 0 & 0 & 6 \\
\frac{2}{3} & 0 & 0 & 0 & -1 & -2 & 14 & 0 & 0 \\
0 & \frac{2}{3} & 0 & -1 & 3 & -2 & 0 & 14 & 0 \\
0 & 0 & \frac{2}{3} & -2 & -1 & 2 & 0 & 0 & 14 \\
-\frac{4}{3} & 0 & 0 & 1 & 0 & 0 & -2 & -1 & -2 \\
0 & -\frac{4}{3} & 0 & 0 & 1 & 0 & -1 & 1 & -2 \\
0 & 0 & -\frac{4}{3} & 0 & 0 & 1 & -2 & 1 & 0
\end{bmatrix}
$$

$$
\sim
\begin{bmatrix}
1 & 0 & 0 & 0 & 0 & 0 & -\frac{3}{2} & 0 & -3 \\
0 & 1 & 0 & 0 & 0 & 0 & -\frac{5}{2} & 0 & 0 \\
0 & 0 & 1 & 0 & 0 & 0 & -\frac{5}{2} & 0 & 0 \\
0 & 0 & 0 & 1 & 0 & 0 & -4 & 0 & -6 \\
0 & 0 & 0 & 0 & 1 & 0 & -\frac{13}{3} & 0 & -2 \\
0 & 0 & 0 & 0 & 0 & 1 & -\frac{16}{3} & 0 & 0 \\
0 & 0 & 0 & 0 & 0 & 0 & 0 & 1 & 0 \\
0 & 0 & 0 & 0 & 0 & 0 & 0 & 0 & 0 \\
0 & 0 & 0 & 0 & 0 & 0 & 0 & 0 & 0
\end{bmatrix}
$$

So we obtain the general solution $a = \frac{3}{2}x + 3z$, $b = \frac{5}{2}x$, $c = \frac{5}{2}x$, $u = 4x + 6z$, $v = 13/3x + 2z$, $w = 16/3x$, and $y = 0$. Yes, $\mathbf{A}$ and $\mathbf{B}$ are similar! For example, letting $x = 0$ and $z = 1$, we obtain

$$
\mathbf{P} = \begin{bmatrix} 3 & 0 & 0 \\ 6 & 2 & 0 \\ 0 & 0 & 1 \end{bmatrix}, \qquad
\mathbf{P}^{-1} = \begin{bmatrix} \frac{1}{3} & 0 & 0 \\ -1 & \frac{1}{2} & 0 \\ 0 & 0 & 1 \end{bmatrix}
$$

Checking by simple matrix multiplication verifies that $\mathbf{A} = \mathbf{PBP}^{-1}$. ∎

THEOREM 6

If $\mathbf{A} = \mathbf{PQP}^{-1}$ and if the columns of $\mathbf{P}$ are taken as the basis $\mathcal{B}$, then $[\mathbf{Ax}]_{\mathcal{B}} = \mathbf{Q}[\mathbf{x}]_{\mathcal{B}}$. In other words, $\mathbf{Q}$ is the matrix for $\mathbf{x} \mapsto \mathbf{Ax}$ when the chosen basis is $\mathcal{B}$.

PROOF Observe that for all $\mathbf{x}$, $\mathbf{P}[\mathbf{x}]_\mathcal{B} = \mathbf{x}$. It follows that $[\mathbf{x}]_\mathcal{B} = \mathbf{P}^{-1}\mathbf{x}$ and that $[\mathbf{Ax}]_\mathcal{B} = \mathbf{P}^{-1}\mathbf{Ax}$. From $\mathbf{A} = \mathbf{PQP}^{-1}$, we obtain $\mathbf{P}^{-1}\mathbf{A} = \mathbf{QP}^{-1}$ and finally we obtain $[\mathbf{Ax}]_\mathcal{B} = \mathbf{QP}^{-1}\mathbf{x} = \mathbf{Q}[\mathbf{x}]_\mathcal{B}$. ∎

More on Equivalence Relations

This is a topic that arises naturally in many diverse situations. The basic idea is very simple: We divide a set of elements into disjoint subclasses, and think of the entities of each subclass as being related in some way. You have seen many examples of this, such as dividing people into the two classes, male and female, or dividing the set of all polynomials into distinct classes depending on their degree. A division of the set of all integers into even or odd integers is another familiar example. A generalization of this idea is as follows. We divide each integer by a fixed number n and classify the integers by the remainder that results in this division. It must be 0 or 1 or 2 or ... or $(n - 1)$. This is call **modulo(n) arithmetic**.

Now, having divided a given set $\mathbf{X}$ into mutually disjoint subclasses, let us concentrate on the relationship between elements of the same subclass. We say that two elements, $\mathbf{x}$ and $\mathbf{y}$, of the same subclass are **equivalent** to each other, and write $\mathbf{x} \equiv \mathbf{y}$. Then we can prove these properties of the $\equiv$ relation:

1. For each $\mathbf{x}$ in X, $\mathbf{x} \equiv \mathbf{x}$.	(**reflexive**)
2. If $\mathbf{x} \equiv \mathbf{y}$, then $\mathbf{y} \equiv \mathbf{x}$.	(**symmetric**)
3. If $\mathbf{x} \equiv \mathbf{y}$ and $\mathbf{y} \equiv z$, then $\mathbf{x} \equiv z$.	(**transitive**)

Any relation that has these three properties is called an **equivalence relation**. Some equivalence relations have special names. For example, in plane geometry, we speak of *similar triangles* and *congruent triangles*. These concepts involve equivalence relations.

The tables can be turned. Suppose that an equivalence relation $\equiv$ on a set X is given. With each element x in X, we associate an *equivalence class* $\{ y : y \equiv x \}$. These equivalence classes are mutually disjoint. To prove this, let $E(x)$ denote the equivalence class containing x, as just defined. Consider a different equivalence class, say $E(u)$. Can they have an element in common? Suppose y belongs to both. Then $y \equiv x$ and $y \equiv u$. By the preceding axioms, it follows that $x \equiv u$, and so $E(x) = E(u)$. The last equation is demonstrated by taking an arbitrary element z in $E(x)$. Then it follows that $z \equiv x$, $z \equiv u$, and $z \in E(u)$. Hence, we have $E(x) \subseteq E(u)$. By symmetry in the argument, $E(u) \subseteq E(x)$. Because $E(u) = E(x)$, we have contradicted the choice of $E(u)$ as an equivalence class different from $E(x)$. The reader is urged to go over this argument carefully to reach a thorough understanding of the concepts involved. We do not claim that it is easy.

Among matrices there are several useful equivalence relations. For example one can define the *row equivalence* of two matrices to mean that they have the same reduced row echelon form. This special equivalence relation is denoted in this book by $\sim$. For example, we have

$$\begin{bmatrix} 1 & 2 & 3 \\ 4 & 5 & 6 \\ 10 & 8 & 7 \end{bmatrix} \sim \begin{bmatrix} 17 & 31 & -11 \\ 2 & 5 & 47 \\ -19 & 3 & 13 \end{bmatrix}$$

because both of these matrices are row equivalent to $\mathbf{I}_3$.

As mentioned previously, another equivalence relation discussed here is that of similarity, written $\mathbf{A} \simeq \mathbf{B}$. These two equivalence relations, $\sim$ and $\simeq$, are completely different from each other!

Further Examples

The following five examples illustrate the new concepts introduced in this section.

EXAMPLE 11 Let $g_1(t) = 3t^2+t-1$, $g_2(t) = t^2-t+2$, $g_3(t) = 2t^2+3t+1$. Do these polynomials provide a basis for $\mathbb{P}_2$? If so, what is the coordinate vector for the polynomial p defined by $p(t) = 11t^2 + 17t - 4$?

SOLUTION We put the coefficients of the polynomials in an augmented matrix and find its reduced echelon form:

$$\begin{bmatrix} -1 & 2 & 1 & | & -4 \\ 1 & -1 & 3 & | & 17 \\ 3 & 1 & 2 & | & 11 \end{bmatrix} \sim \begin{bmatrix} 1 & 0 & 0 & | & 2 \\ 0 & 1 & 0 & | & -3 \\ 0 & 0 & 1 & | & 4 \end{bmatrix}$$

Yes, $\mathcal{C} = \{g_1, g_2, g_3\}$ is a basis for $\mathbb{P}_2$. Furthermore, $[p]_\mathcal{C} = \begin{bmatrix} 2 \\ -3 \\ 4 \end{bmatrix}$.

We can verify these results with a separate calculation:
$2g_1 - 3g_2 + 4g_3 = 2(3t^2 + t - 1) - 3(t^2 - t + 2) + 4(2t^2 + 3t + 1)$
$= 11t^2 + 17t - 4$. ∎

EXAMPLE 12 Use the transition matrix of Example 3 to express the polynomial $p = 3T_0 - 7T_1 + 5T_2 + 4T_3$ in terms of the natural (*standard*) basis for $\mathbb{P}_3$. Here T_0, T_1, T_2, T_3 are Chebyshev polynomials as in Example 7.

SOLUTION The polynomial p has coordinate vector $[p]_\mathcal{B} = (3, -7, 5, 4)$. Hence, we have

$$[p]_\mathcal{C} = \mathbf{P}[p]_\mathcal{B} = \begin{bmatrix} 1 & 0 & -1 & 0 \\ 0 & 1 & 0 & -3 \\ 0 & 0 & 2 & 0 \\ 0 & 0 & 0 & 4 \end{bmatrix} \begin{bmatrix} 3 \\ -7 \\ 5 \\ 4 \end{bmatrix} = \begin{bmatrix} -2 \\ -19 \\ 10 \\ 16 \end{bmatrix}$$

The result is $p(t) = -2 - 19t + 10t^2 + 16t^3$. An independent calculation will verify the results:

$$3T_0 - 7T_1 + 5T_2 + 4T_3 = 3 - 7x + 5(2x^2 - 1) + 4(4x^3 - 3x)$$
$$= -2 - 19x + 10xy^2 + 16x^3.$$ ∎

EXAMPLE 13 First, define matrices

$$\mathbf{A}_1 = \begin{bmatrix} 2 & 2 \\ 2 & 3 \end{bmatrix}, \qquad \mathbf{A}_2 = \begin{bmatrix} 3 & 3 \\ 2 & 0 \end{bmatrix}, \qquad \mathbf{A}_3 = \begin{bmatrix} 0 & 3 \\ 2 & 2 \end{bmatrix}$$

Then use coordinate mappings to determine whether the set $\{\mathbf{A}_1, \mathbf{A}_2, \mathbf{A}_3\}$ is linearly independent.

SOLUTION One basis for $\mathbb{R}^{2\times2}$ is

$$\mathbf{B}_1 = \begin{bmatrix} 1 & 0 \\ 0 & 0 \end{bmatrix}, \qquad \mathbf{B}_2 = \begin{bmatrix} 0 & 1 \\ 0 & 0 \end{bmatrix}$$

$$\mathbf{B}_3 = \begin{bmatrix} 0 & 0 \\ 1 & 0 \end{bmatrix}, \qquad \mathbf{B}_4 = \begin{bmatrix} 0 & 0 \\ 0 & 1 \end{bmatrix}$$

It is easy to express the **A** matrices in terms of the **B** matrices:

$$\mathbf{A}_1 = 2\mathbf{B}_1 + 2\mathbf{B}_2 + 2\mathbf{B}_3 + 3\mathbf{B}_4, \quad \mathbf{A}_2 = 3\mathbf{B}_1 + 3\mathbf{B}_2 + 2\mathbf{B}_3, \quad \mathbf{A}_3 = 3\mathbf{B}_2 + 2\mathbf{B}_3 + 2\mathbf{B}_4$$

The question is whether there is a linear dependence of the form

$$\alpha\mathbf{A}_1 + \beta\mathbf{A}_2 + \gamma\mathbf{A}_3 = \mathbf{0}$$

Therefore, we want to find the solutions of the equation

$$\alpha \begin{bmatrix} 2 \\ 2 \\ 2 \\ 3 \end{bmatrix} + \beta \begin{bmatrix} 3 \\ 3 \\ 2 \\ 0 \end{bmatrix} + \gamma \begin{bmatrix} 0 \\ 3 \\ 2 \\ 2 \end{bmatrix} = \begin{bmatrix} 2 & 3 & 0 \\ 2 & 3 & 3 \\ 2 & 2 & 2 \\ 3 & 0 & 2 \end{bmatrix} \begin{bmatrix} \alpha \\ \beta \\ \gamma \end{bmatrix} = \begin{bmatrix} 0 \\ 0 \\ 0 \\ 0 \end{bmatrix}$$

The row-reduction process applied to the coefficient matrix yields

$$\begin{bmatrix} 2 & 3 & 0 \\ 2 & 3 & 3 \\ 2 & 2 & 2 \\ 3 & 0 & 2 \end{bmatrix} \sim \begin{bmatrix} 1 & 0 & 0 \\ 0 & 1 & 0 \\ 0 & 0 & 1 \\ 0 & 0 & 0 \end{bmatrix}$$

This proves that the homogeneous problem has no solution other than the zero vector, and $\{\mathbf{A}_1, \mathbf{A}_2, \mathbf{A}_3\}$ is linearly independent. ∎

EXAMPLE 14 Two bases for $\mathbb{R}^3$ are given:

$$\mathcal{B} = \{\mathbf{b}_1, \ \mathbf{b}_2, \ \mathbf{b}_3,\} \quad \text{and} \quad \mathcal{C} = \{\mathbf{c}_1, \ \mathbf{c}_2, \ \mathbf{c}_3\}$$

where

$$\mathbf{b}_1 = \begin{bmatrix} 5 \\ -1 \\ 3 \end{bmatrix}, \quad \mathbf{b}_2 = \begin{bmatrix} 25 \\ -7 \\ 12 \end{bmatrix}, \quad \mathbf{b}_3 = \begin{bmatrix} 34 \\ -16 \\ 17 \end{bmatrix}$$

and

$$\mathbf{c}_1 = \begin{bmatrix} 1 \\ -3 \\ 2 \end{bmatrix}, \quad \mathbf{c}_2 = \begin{bmatrix} -2 \\ 0 \\ 1 \end{bmatrix}, \quad \mathbf{c}_3 = \begin{bmatrix} 3 \\ 1 \\ 0 \end{bmatrix}$$

What is the **P**-matrix that enables us to convert from the $\mathcal{B}$-coordinates to the $\mathcal{C}$-coordinates?

SOLUTION The **P**-matrix should have this property:

$$[\mathbf{x}]_{\mathcal{C}} = \mathbf{P}[\mathbf{x}]_{\mathcal{B}}$$

for all $\mathbf{x} \in \mathbb{R}^3$. Our theory tells us that the columns of **P** should be $[\mathbf{b}_1]_{\mathcal{C}}, [\mathbf{b}_2]_{\mathcal{C}}, [\mathbf{b}_3]_{\mathcal{C}}$. To start, we must represent $\mathbf{b}_1$ in terms of the $\mathcal{C}$-basis. Thus, we need to solve the equation $x_1 \mathbf{c}_1 + x_2 \mathbf{c}_2 + x_3 \mathbf{c}_3 = \mathbf{b}_1$ or

$$\begin{bmatrix} \mathbf{c}_1 & \mathbf{c}_2 & \mathbf{c}_3 \end{bmatrix} \mathbf{x} = \begin{bmatrix} 1 & -2 & 3 \\ -3 & 0 & 1 \\ 2 & 1 & 0 \end{bmatrix} \begin{bmatrix} x_1 \\ x_2 \\ x_3 \end{bmatrix} = \begin{bmatrix} 5 \\ -1 \\ 3 \end{bmatrix} = \mathbf{b}_1$$

The vector $\mathbf{x} = [x_1, x_2, x_3]^T$ (when it has been computed) will be the first column in the matrix $\mathbf{P}$. Similar remarks pertain to the vectors $\mathbf{b}_2$ and $\mathbf{b}_3$. The computations can be done simultaneously:

$$\begin{bmatrix} \mathbf{c}_1 & \mathbf{c}_2 & \mathbf{c}_3 & | & \mathbf{b}_1 & \mathbf{b}_2 & \mathbf{b}_3 \end{bmatrix} = \begin{bmatrix} 1 & -2 & 3 & | & 5 & 25 & 34 \\ -3 & 0 & 1 & | & -1 & -7 & -16 \\ 2 & 1 & 0 & | & 3 & 12 & 17 \end{bmatrix}$$

Row reduction leads to

$$\begin{bmatrix} 1 & 0 & 0 & | & 1 & 5 & 7 \\ 0 & 1 & 0 & | & 1 & 2 & 3 \\ 0 & 0 & 1 & | & 2 & 8 & 11 \end{bmatrix} = \begin{bmatrix} \mathbf{I} & | & \mathbf{P} \end{bmatrix}$$

One can test the equation

$$\mathbf{P}[\mathbf{x}]_{\mathcal{B}} = [\mathbf{x}]_{\mathcal{C}}$$

For example let $\mathbf{x} = [5, -1, 3]^T$. It is known that $\mathbf{x} = 1\mathbf{b}_1 + 0\mathbf{b}_2 + 0\mathbf{b}_3$. Therefore, we have $[\mathbf{x}]_{\mathcal{B}} = [1, 0, 0]^T$. We obtain $\mathbf{P}[\mathbf{x}]_{\mathcal{B}} = \mathbf{P}[1, 0, 0]^T = [1, 1, 2]^T = [\mathbf{x}]_{\mathcal{C}}$. Further testing obtains $1\mathbf{c}_1 + 1\mathbf{c}_2 + 2\mathbf{c}_3 = [1, -3, 2]^T + [-2, 0, 1]^T + 2[3, 1, 0]^T = [5, -1, 3]^T$. ∎

In the solution to Example 14, we follow this path of reasoning: $[\mathbf{C} | \mathbf{B}] \sim [\mathbf{EC} | \mathbf{EB}] = [\mathbf{I} | \mathbf{P}]$ by row reduction. Thus, we have $\mathbf{EC} = \mathbf{I}$, $\mathbf{EB} = \mathbf{P}, \mathbf{E} = \mathbf{C}^{-1}$, and $\mathbf{P} = \mathbf{EB} = \mathbf{C}^{-1}\mathbf{B}$.

EXAMPLE 15 The first few Chebyshev polynomials are defined by the equations $T_0(t) = 1, T_1(t) = t, T_2(t) = 2t^2 - 1, T_3(t) = 4t^3 - 3t$. Express the polynomial $p(x) = 5x^3 - 2x^2 + 3x - 1$ as a linear combination of these Chebyshev polynomials.

SOLUTION We want to find the coefficients in the equation

$$p(x) = a_0 T_0 + a_1 T_1 + a_2 T_2 + a_3 T_3$$

The correct coefficients will solve this linear system:

$$
\begin{bmatrix}
1 & 0 & -1 & 0 \\
0 & 1 & 0 & -3 \\
0 & 0 & 2 & 0 \\
0 & 0 & 0 & 4
\end{bmatrix}
\begin{bmatrix}
a_0 \\ a_1 \\ a_2 \\ a_3
\end{bmatrix}
=
\begin{bmatrix}
-1 \\ 3 \\ -2 \\ 5
\end{bmatrix}
$$

The augmented matrix and its row-reduced form are

$$
\left[\begin{array}{cccc|c}
1 & 0 & -1 & 0 & -1 \\
0 & 1 & 0 & -3 & 3 \\
0 & 0 & 2 & 0 & -2 \\
0 & 0 & 0 & 4 & 4
\end{array}\right]
\sim
\left[\begin{array}{cccc|c}
1 & 0 & 0 & 0 & -2 \\
0 & 1 & 0 & 0 & 27/4 \\
0 & 0 & 1 & 0 & -1 \\
0 & 0 & 0 & 1 & \frac{5}{4}
\end{array}\right]
$$

Verifying independently, we find

$$
-2T_0(x) + (27/4)T_1(x) - T_2(x) + \tfrac{5}{4}T_3(x)
$$
$$
= -2 + 27/4x - (2x^2 - 1) + \tfrac{5}{4}(4x^3 - 3x) = -1 + 3x - 2x^2 + 5x^3 = p(x)
$$

■

SUMMARY 5.3

- Every vector $\mathbf{x}$ in an n-dimensional vector space V, with an ordered basis $\mathcal{B} = \{\mathbf{u}_1, \mathbf{u}_2, \ldots, \mathbf{u}_n\}$, can be written $\mathbf{x} = \sum_{i=1}^{n} a_i\mathbf{u}_i$, where the coefficients $(a_1, a_2, \ldots, a_n)$ are uniquely determined. Here $[\mathbf{x}]_{\mathcal{B}} = (a_1, a_2, \ldots, a_n)$ is the **coordinate vector** of $\mathbf{x}$ relative to the basis $\mathcal{B}$.

- The **transition matrix** $\mathbf{P}$ from a basis $\mathcal{B} = \{\mathbf{u}_1, \mathbf{u}_2, \ldots, \mathbf{u}_n\}$ to a basis $\mathcal{C} = \{\mathbf{v}_1, \mathbf{v}_2, \ldots, \mathbf{v}_n\}$ has the form $[\mathbf{x}]_{\mathcal{C}} = \mathbf{P}[\mathbf{x}]_{\mathcal{B}}$, where $\mathbf{P} = [[\mathbf{u}_1]_{\mathcal{C}}, [\mathbf{u}_2]_{\mathcal{C}}, \ldots, [\mathbf{u}_n]_{\mathcal{C}}]$.

- The **transition matrix** $\mathbf{P}^{-1}$ from a basis $\mathcal{C} = \{\mathbf{v}_1, \mathbf{v}_2, \ldots, \mathbf{v}_n\}$ to a basis $\mathcal{B} = \{\mathbf{u}_1, \mathbf{u}_2, \ldots, \mathbf{u}_n\}$ has the form $[\mathbf{x}]_{\mathcal{B}} = \mathbf{P}^{-1}[\mathbf{x}]_{\mathcal{C}}$.

- An **equivalence relationship** $\equiv$ on any set has these properties:

  - For each $\mathbf{x}$ in $X, \mathbf{x} \equiv \mathbf{x}$. (**reflexive**).
  - If $\mathbf{x} \equiv \mathbf{y}$, then $\mathbf{y} \equiv \mathbf{x}$. (**symmetric**).
  - If $\mathbf{x} \equiv \mathbf{y}$ and $\mathbf{y} \equiv \mathbf{z}$, then $\mathbf{x} \equiv \mathbf{z}$. (**transitive**).

- $\mathbf{A}$ is **similar** to $\mathbf{B}$ if $\mathbf{B} = \mathbf{PAP}^{-1}$ for some invertible matrix $\mathbf{P}$.

- Similarity is an equivalence relationship.

- Theorems:

  - Let V be a finite-dimensional vector space in which two ordered bases $\mathcal{B}$ and $\mathcal{C}$ have been prescribed. Let $\mathbf{P}$ be the matrix whose columns are the $\mathcal{C}$-coordinates of the basis vectors in $\mathcal{B}$. Then for all $\mathbf{x}$ in V, we have $[\mathbf{x}]_{\mathcal{C}} = \mathbf{P}[\mathbf{x}]_{\mathcal{B}}$.

- A linear transformation from one vector space to another is completely determined by the images of any basis for the domain. These images can, in turn, be completely arbitrary.

- If $\sum_{i=1}^{n} a_i\mathbf{u}_i = \mathbf{0}$ and if L is linear, then $\sum_{i=1}^{n} a_iL(\mathbf{u}_i) = \mathbf{0}$.

- Let L be a linear transformation from a finite-dimensional vector space X into a vector space Y. Let $\mathcal{B}$ and $\mathcal{C}$ be ordered bases for X and $L(X)$, respectively. Let $\mathbf{A}$ be the matrix whose columns are the vectors $[L(\mathbf{u}_i)]_{\mathcal{C}}$, where $\mathbf{u}_i$ are the ordered elements of $\mathcal{B}$. Then for all $\mathbf{x}$ in X, we obtain $[L(\mathbf{x})]_\mathcal{C} = \mathbf{A}[\mathbf{x}]_\mathcal{B}$.

- If $n \times n$ matrices $\mathbf{A}$ and $\mathbf{B}$ represent the same linear transformation, but with respect to two bases, then for an appropriate invertible matrix $\mathbf{P}$, we have $\mathbf{B} = \mathbf{PAP}^{-1}$.

- If $\mathbf{A} = \mathbf{PQP}^{-1}$ and if the columns of $\mathbf{P}$ are taken as the basis $\mathcal{B}$, then $[\mathbf{Ax}]_\mathcal{B} = \mathbf{Q}[\mathbf{x}]_\mathcal{B}$. Here, $\mathbf{Q}$ is the matrix for $\mathbf{x} \mapsto \mathbf{Ax}$ when the chosen basis is $\mathcal{B}$.

KEY CONCEPTS 5.3

Coordinate vector, coordination, bases, changing coordinates, transition matrix, Chebyshev polynomials, natural/standard basis, linear transformation, finite/infinite dimensional vector spaces, mapping a vector space into itself, similar matrices, similarity, equivalence relation, reflexive, symmetric, transition

GENERAL EXERCISES 5.3

1. Find the transition matrix from basis $\mathcal{B}$ to basis $\mathcal{C}$ when $\mathcal{B}$ has elements $\mathbf{u}_1 = (-2, -2)$ and $\mathbf{u}_2 = (-3, -2)$ and $\mathcal{C}$ has $\mathbf{v}_1 = (1, 2)$ and $\mathbf{v}_2 = (3, 4)$. Use the matrix to compute $[\mathbf{x}]_\mathcal{C}$ when $[\mathbf{x}]_\mathcal{B} = (5, 3)$. What is $\mathbf{x}$ in its usual form, that is, as an element of $\mathbb{R}^2$?

2. Let $\{\mathbf{e}_1, \mathbf{e}_2\}$ be the standard basis for $\mathbb{R}^2$. Form a new basis by selecting a positive angle θ and rotating $\mathbf{e}_1$ and $\mathbf{e}_2$ counterclockwise through the angle θ. Find the explicit form of the new basis elements by using trigonometry.

3. (Continuation.) Find the transition matrix for converting standard coordinates to coordinates in the rotated system.

4. The first four Chebyshev polynomials are given in Example 3. You will need also the fifth: $T_4(t) = 9t^4 - 7t^2 + 1$. The first five Legendre[2] polynomials are $P_0, P_1, \ldots, P_4$, where $P_0(t) = 1$, $P_1(t) = t$, $P_2(t) = (3t^2 - 1)/2$, $P_3(t) = (5t^3 - 3t)/2$, and $P_4(t) = (35t^4 - 30t^2 + 3)/8$. Find the transition matrix for converting polynomials in $\mathbb{P}^4$ from combinations of Chebyshev polynomials to combinations of Legendre polynomials.

5. The Chebyshev polynomials (discussed in Example 3) obey the following relations: $T_0(x) = 1$, $T_1(x) = x$, and, in general, $T_{k+1}(x) = 2xT_k(x) - T_{k-1}(x)$. Here, $k = 1, 2, 3, \ldots$. (We say that these polynomials are defined *recursively*.) Find the transition matrix that converts a linear combination of $T_0, T_1, \ldots, T_5$ into a polynomial in standard form.

6. Define these six vectors: $\mathbf{u}_1 = (1, 4, 7, 2)$, $\mathbf{u}_2 = (3, 1, 1, 5)$, $\mathbf{u}_3 = (-3, 10, 19, -4)$, $\mathbf{v}_1 = (7, 1, 6, 5)$, $\mathbf{v}_2 = (-2, 1, 3, 1)$, and $\mathbf{v}_3 = (11, 3, 2, 9)$. Find a linear transformation L such that $L(\mathbf{u}_i) = \mathbf{v}_i$ for $1 \leq i \leq 3$.

7. Let one basis, $\mathcal{B}$, for $\mathbb{R}^3$ have the vectors $(2, 1, 7), (3, 2, 1)$, and $(-2, 0, 1)$. Let another basis, $\mathcal{C}$, have vectors $(1, 3, 2)$, $(-4, 0, 5)$, and $(2, -1, 3)$. Find the transition matrix $\mathbf{P}$ that allows us to calculate $[x]_\mathcal{C}$ from $[x]_\mathcal{B}$.

---

[2] Adrien Marie Legendre (1752–1833) worked on number theory and differential equations, one of which is named after him: $(1 - x^2)f''(x) - 2f' + n(n-1)f(x) = 0$. The Legendre polynomials satisfy this differential equation. Legendre published a popular geometry textbook in 1794.

8. Let $\mathcal{B} = \{\mathbf{u}_1, \mathbf{u}_2\}$, $\mathbf{u}_1 = [3, 1]^T$, $\mathbf{u}_2 = [2, 5]^T$, and $\mathbf{x} = [-8, -33]^T$. Compute $[\mathbf{x}]_{\mathcal{B}}$.

9. The matrices $\mathbf{A} = \begin{bmatrix} 3 & 4 \\ 2 & 5 \end{bmatrix}$ and $\mathbf{B} = \begin{bmatrix} 1 & -1 \\ 7 & 0 \end{bmatrix}$ are similar. Solve for $\mathbf{P}$ in the equation $\mathbf{A} = \mathbf{P}^{-1}\mathbf{B}\mathbf{P}$.

10. In $\mathbb{R}^3$, consider the bases

$$\mathcal{B} = \{\mathbf{b}_1, \mathbf{b}_2, \mathbf{b}_3\} = \left\{ \begin{bmatrix} 1 \\ 3 \\ 2 \end{bmatrix}, \begin{bmatrix} -2 \\ 1 \\ 2 \end{bmatrix}, \begin{bmatrix} 1 \\ 1 \\ 1 \end{bmatrix} \right\}$$

$$\mathcal{C} = \{\mathbf{c}_1, \mathbf{c}_2, \mathbf{c}_3\} = \left\{ \begin{bmatrix} 1 \\ 2 \\ 3 \end{bmatrix}, \begin{bmatrix} 1 \\ 4 \\ 6 \end{bmatrix}, \begin{bmatrix} 2 \\ 7 \\ 11 \end{bmatrix} \right\}$$

What is the matrix $\mathbf{P}$ needed to change $\mathcal{B}$-coordinates to $\mathcal{C}$-coordinates?

11. Let $\mathcal{D} = \{\mathbf{d}_1, \mathbf{d}_2, \mathbf{d}_3\}$ and $\mathcal{F} = \{\mathbf{f}_1, \mathbf{f}_2, \mathbf{f}_3\}$ be bases for V. Suppose $\mathbf{f}_1 = 2\mathbf{d}_1 - \mathbf{d}_2 + \mathbf{d}_3$, $\mathbf{f}_2 = 3\mathbf{d}_2 + \mathbf{d}_3$, and $\mathbf{f}_3 = -3\mathbf{d}_1 + 2\mathbf{d}_3$. Find the matrix for changing $\mathcal{F}$ coordinates to $\mathcal{D}$ coordinates. Find $[\mathbf{x}]_{\mathcal{D}}$ if $\mathbf{x} = \mathbf{f}_1 - 2\mathbf{f}_2 + 2\mathbf{f}_3$.

12. Let $T : V \rightarrow W$, where V has basis $\mathcal{B} = \{\mathbf{b}_1, \mathbf{b}_2\}$, W has basis $\mathcal{C} = \{\mathbf{c}_1, \mathbf{c}_2, \mathbf{c}_3\}$, $T(\mathbf{b}_1) = 3\mathbf{c}_1 + 5\mathbf{c}_2 - 7\mathbf{c}_3$, and $T(\mathbf{b}_2) = 2\mathbf{c}_1 - \mathbf{c}_2 + 4\mathbf{c}_3$. Find $\mathbf{M}$ so that $[T(\mathbf{x})]_{\mathcal{C}} = \mathbf{M}[\mathbf{x}]_{\mathcal{B}}$. If $[\mathbf{x}]_{\mathcal{B}} = \begin{bmatrix} 7 \\ -2 \end{bmatrix}$, then what is $[T(\mathbf{x})]_{\mathcal{C}}$?

13. Let $\mathbf{u}_1(t) = 1$, $\mathbf{u}_2(t) = \sin t$, and $\mathbf{u}_3(t) = \cos t$. These constitute a basis for a three-dimensional vector space. If $T(\mathbf{u}_1) = \mathbf{0}$, $T(\mathbf{u}_2) = \mathbf{u}_3$, and $T(\mathbf{u}_3) = -\mathbf{u}_2$, what is the matrix for T? If $\mathbf{x} = 3\mathbf{u}_1 - 2\mathbf{u}_2 + 7\mathbf{u}_3$, then what is $T(\mathbf{x})$? Use the matrix to find this.

14. Let $L : \mathbb{R}^4 \rightarrow \mathbb{R}^2$ be defined by

$$L[a, b, c, d] = \begin{bmatrix} 5a + b - 2c + 8d \\ 3a - 5b + c - 2d \end{bmatrix}$$

Use for $\mathcal{B}$ the standard basis in $\mathbb{R}^4$ and for $\mathcal{C}$ the standard basis in $\mathbb{R}^2$. What is the matrix that does this task?

15. Define a **linear transformation** L from $\mathbb{P}_2$ to $\mathbb{R}^2$ as follows: If $p(t) = a_3 t^2 + a_2 t + a_1$, then $L(p) = (a_3 + a_2, 2a_1 - a_3)$. Let the standard bases be prescribed: $\mathcal{B}$ for $\mathbb{P}_2$: $q_2(t) = t^2, q_1(t) = t, q_0(t) = 1$; $\mathcal{C}$ for $\mathbb{R}^2$: $\mathbf{v}_1 = (1, 0), \mathbf{v}_2 = (0, 1)$. What is the matrix for L with respect to these two ordered bases? In other words, what is $\mathbf{A}_{\mathcal{B}\mathcal{C}}$? Use the matrix found to compute $L(p)$ when $p(t) = 4t^2 - t + 1$. Also verify your result by a direct calculation using the definition of L.

16. Let V be a vector space with ordered basis $\mathcal{D} = \{\mathbf{d}_1, \mathbf{d}_2\}$, and let W be a vector space with ordered basis $\mathcal{B} = \{\mathbf{b}_1, \mathbf{b}_2\}$. Let T be a linear transformation from V to W such that $T(\mathbf{d}_1) = -3\mathbf{b}_2 + 2\mathbf{b}_1$ and $T(\mathbf{d}_2) = 5\mathbf{b}_2 - 4\mathbf{b}_1$. What is the matrix for T relative to these two ordered bases?

17. Let $\mathcal{B} = \{\mathbf{u}_1, \mathbf{u}_2\}$, where $\mathbf{u}_1 = (1, 3), \mathbf{u}_2 = (-2, 1)$. Let $\mathcal{C} = \{\mathbf{v}_1, \mathbf{v}_2\}$, where $\mathbf{v}_1 = (2, 7), \mathbf{v}_2 = (1, 2)$. Define a linear transformation T such that $T(\mathbf{u}_1) = 2\mathbf{v}_1 - 3\mathbf{v}_2$ and $T(\mathbf{u}_2) = \mathbf{v}_1 + 4\mathbf{v}_2$. Find the matrix $\mathbf{A}$ such that $[T(\mathbf{x})]_{\mathcal{C}} = \mathbf{A}[\mathbf{x}]_{\mathcal{B}}$.

18. Let $\mathcal{B} = \{\mathbf{u}_1, \mathbf{u}_2\}$, where $\mathbf{u}_1 = (3, 1), \mathbf{u}_2 = (2, 5)$, and $\mathbf{x} = (-8, -33)$. Compute $[\mathbf{x}]_{\mathcal{B}}$.

19. Verify that $\begin{bmatrix} 2 & 1 \\ 0 & 2 \end{bmatrix} \simeq \begin{bmatrix} 2 & \alpha \\ 0 & 2 \end{bmatrix}$ if $\alpha \neq 0$.

(Here $\simeq$ is the similarity equivalence relation.)

20. Establish that if $\mathbf{P}$ is the transition matrix for changing $\mathcal{B}$-coordinates to $\mathcal{C}$-coordinates, and if $\mathbf{Q}$ does the same for $\mathcal{C}$ to $\mathcal{D}$, then $\mathbf{Q}\mathbf{P}$ is the transition matrix for $\mathcal{B}$ to $\mathcal{D}$.

21. Give an argument for this assertion. If $\mathcal{B}$ is a basis for some vector space and $\mathcal{B} = \{\mathbf{u}_1, \mathbf{u}_2, \ldots, \mathbf{u}_n\}$, then for each i, the co-ordinate vector for $\mathbf{u}_i$ is $\mathbf{e}_i$ (the ith standard unit vector in $\mathbb{R}^n$). Thus, in symbols, $[\mathbf{u}_i]_{\mathcal{B}} = \mathbf{e}_i$.

22. Let $\mathcal{B}$ be a basis for a finite-dimensional vector space, V. Establish that if S is a linearly dependent set in V then the set $\{[\mathbf{x}]_{\mathcal{B}} : \mathbf{x} \in S\}$ is also linearly dependent. Establish a stronger result, if you can.

23. Let $\{\mathbf{u}_1, \mathbf{u}_2, \ldots, \mathbf{u}_n\}$ be a basis for $\mathbb{R}^n$. Inter-pret these vectors as columns of a matrix $\mathbf{A}$, and call this basis $\mathcal{B}$. In terms of $\mathbf{A}$, what are the transition matrices for converting standard coordinates to $\mathcal{B}$-coordinates and vice versa?

24. Substantiate that two similar matrices have the same determinant. Is the converse true?

25. Affirm that $\simeq$ is an equivalence relation. (Recall that $\simeq$ means **similarity**.)

26. Find a basis for the set of all 2×2 matrices that commute with the matrix $\begin{bmatrix} 0 & 2 \\ 3 & 1 \end{bmatrix}$

27. Explain that if one of $\mathbf{A}$ and $\mathbf{B}$ is invertible, then $\mathbf{AB} \simeq \mathbf{BA}$. Is the invertibility hypoth-esis necessary? (An example or theorem is needed.)

28. If two matrices are similar (in the technical meaning of this word), does it follow that they are row equivalent?

29. Show that every 2×2 matrix is similar to its transpose.

30. Verify by an independent calculation the correctness of the final assertion in Exam-ple 5.

31. Does the matrix equation $\mathbf{AP} = \mathbf{PB}$ imply that $\mathbf{A}$ is similar to $\mathbf{B}$?

32. Explain why transition matrices are always invertible.

33. Explain why $\mathbf{Px} = \mathbf{0}$ implies that $\mathbf{x} = \mathbf{0}$. Here, $\mathbf{P}$ is a transition matrix,

34. Give an example of matrices such that $\mathbf{AP} = \mathbf{PB}$ but $\mathbf{P} \neq \mathbf{0}$ and $\mathbf{A}$ is not similar to $\mathbf{B}$.

35. Find necessary and sufficient conditions on general 2×2 matrices $\mathbf{A}$ and $\mathbf{B}$ so that they are similar.

36. For this exercise only, we say that two sets of vectors in a vector space are equiv-alent to each other if each is in the span of the other. Is this an equivalence relation?

37. The **Fibonacci sequence** is $F = (1, 1, 2, 3, 5, 8, 13, 21, \ldots)$. Its definition is $x_1 = 1$, $x_2 = 1$, and (for $n \geq 3$) $x_n = x_{n-1} + x_{n-2}$. Its name is from a mathemati-cian Fibonacci (born around 1175, died around 1250), who needed it in his study of the breeding of rabbits. Explain why any se-quence that satisfies the recurrence relation $x_n = x_{n-1} + x_{n-2}$ is a linear combination of F and $G = (0, 1, 2, 3, 5, 8, 13, 21, \ldots)$.

38. Derive $\mathbf{P}$ and $\mathbf{P}^{-1}$ in Example 9. Verify in both of these examples that $\mathbf{A} = \mathbf{PBP}^{-1}$.

39. Are these equivalence relations? Explain.
   a. Set inclusion: $A \subseteq B$
   b. Similar matrices: $\mathbf{A} = \mathbf{PBP}^{-1}$

40. Find the coordinate vector for the given vector relative to the basis shown.

   a. $x = (8, 3)$

      $A = \{a_1, a_2\} = \{(1, 0), (2, 1)\}$

   b. $y = (-2, 1)$

      $B = \{b_1, b_2\}$

         $= \{(1, -3), (2, -5)\}$

   c. $z = (1, 3, 2)$

      $C = \{c_1, c_2, c_3\}$

         $= \{(1, 1, 0), (0, 1, 1), (1, 0, 1)\}$

41. Find the transition matrix Q for transforming the C-coordinates to the B-coordinates.

   a. $B = \{b_1, b_2, b_3, b_4\}$

         $= \{1, t, t^2, t^3\}$

      $C = \{c_1, c_2, c_3, c_4\}$

         $= \{1, 1 + t, 1 + t + t^2, 1 + t$

            $+ t^2 + t^3\}$

   b. $B = \{b_1, b_2, b_3\} = \{x^2, x, 1\}$

      $C = \{c_1, c_2, c_3\}$

         $= \{x^2 - x + 1, 3x^2 + 1, 2x^2 + x - 2\}$

42. (Continuation.) Using the previous results, find the transition matrix P used in converting from the B-coordinates back to the C-coordinates.

43. In $\mathbb{R}^2$, are these equivalence relations? Explain.

   a. aRb means lines a and b have a point in common.

   b. aSb means lines a and b are parallel or coincident.

44. Find the transition matrix Q used in converting from the C-coordinates to the B-coordinates,

   a. $B = \{u_1, u_2\} = \{(1, 0), (0, 1)\}$

      $C = \{v_1, v_2\} = \{(2, 1), (-1, 1)\}$

   b. $B = \{u_1, u_2\} = \{(1, 2), (3, 5)\}$

      $C = \{v_1, v_2\} = \{(1, -1), (1, -2)\}$

   c. $B = \{u_1, u_2, u_3\}$

         $= \{(6, 3, 3), (4, -1, 3), (5, 5, 2)\}$

      $C = \{v_1, v_2, v_3\}$

         $= \{(2, 0, 1), (1, 2, 0), (1, 1, 1)\}$

45. (Continuation.) Using the previous results, find the transition matrix P used in converting from the B-coordinates back to the C-coordinates.

46. Find the matrix representation for the linear mapping given by $T(A) = AX$ where $X = \begin{bmatrix} a & b \\ c & d \end{bmatrix}$ with regard to the standard basis

$$\mathcal{E} = \{E_{11}, E_{12}, E_{21}, E_{22}\}$$

where E_{ij} is the 2×2 matrix with 1 in the (i, j) location and 0's elsewhere.

47. Find the coordinate vector for the given vector relative to the basis.

   a. $u = (1, 3, 2)$

      $C = \{c_1, c_2, c_3\}$

         $= \{(1, 1, 0), (0, 1, 1), (1, 0, 1)\}$

   b. $v = (3, -1, 2)$

      $D = \{d_1, d_2, d_3\}$

         $= \{(0, 1, 1), (0, 1, 2), (1, 0, 1)\}$

48. Find the coordinate vectors for the given vector relative to these bases.

   a. $u = 5x^2 + 2x - 3$

      $B = \{x^2, x, 1\}$

   b. $v = 5x^2 + 2x - 3$

      $C = \{x^2 - x + 1, 3x^2 - 1, 2x^2 + x - 2\}$

49. Verify that these are similar matrices. Explain.

a. $\begin{bmatrix} -5 & -12 \\ 2 & 5 \end{bmatrix} \simeq \begin{bmatrix} 1 & 0 \\ 0 & -1 \end{bmatrix}$

b. $\begin{bmatrix} 0 & 2 \\ 3 & -1 \end{bmatrix} \simeq \begin{bmatrix} -30 & -48 \\ 18 & 29 \end{bmatrix}$

c. $\begin{bmatrix} 1 & 3 & 2 \\ 1 & 0 & -4 \\ 0 & 1 & 3 \end{bmatrix} \simeq \begin{bmatrix} 4 & 1 & 0 \\ -7 & 0 & 1 \\ 9 & 3 & 0 \end{bmatrix}$

50. Establish the following.

   a. If $\mathbf{D}$ is an invertible diagonal matrix, then $\mathbf{D}$ is similar to itself.

   b. If $\mathbf{A}$ is similar to $\mathbf{B}$, then $\mathbf{B}$ is similar to $\mathbf{A}$, and $\mathbf{B}^{-1}$ is similar to $\mathbf{A}^{-1}$.

   c. If $\mathbf{A}$ is similar to $\mathbf{B}$, then $\mathbf{A}^k$ is similar to $\mathbf{B}^k$.

51. Are these linear mappings? Explain.

   a. $F(x, y) = (x, x + y)$

   b. $G(x, y) = |x - y|$

   c. $H(x, y, z) = (y, x + 1, y + z)$

   d. $L(x, y, z) = (2x, -3y, 4z)$

52. Let $F(x, y) = (3x + y, 2x - 5y)$ be a linear mapping from $\mathbb{R}^2$ to $\mathbb{R}^2$ relative to the basis $S = \{\mathbf{u}_1, \mathbf{u}_2\} = \{(1, -2), (3, -4)\}$.

   a. For an arbitrary vector (a, b), find the coordinate vector $[(a, b)]_S$.

   b. Write $F(\mathbf{u}_1)$ and $F(\mathbf{u}_2)$ as a linear combination of the basis vectors.

   c. Find $\mathbf{F}$, which is the matrix representation of F in the basis.

53. **a.** Let $(a, b) \in \mathbb{R}^2$ be an arbitrary vector. Find the coordinate vector $[(a, b)]_T$ with respect to the basis
$$T = \{\mathbf{u}_1, \mathbf{u}_2\} = \{(1, 3), (3, 8)\}$$

   b. Write the basis vectors of $S = \{\mathbf{v}_1, \mathbf{v}_2\} = \{(1, -2), (3, -4)\}$ as a linear combination of the basis vectors of T.

   c. Find the change of basis matrix $\mathbf{P}$ from basis T to S.

   d. Find the change of basis matrix $\mathbf{Q}$ from basis T to S.

   e. Show that $\mathbf{PQ} = \mathbf{I}$.

54. Let $T(\mathbf{v}) = \mathbf{Av}$ where $\mathbf{A} = \begin{bmatrix} 2 & 1 \\ 4 & 3 \end{bmatrix}$. Find the matrix representation relative to these bases.

   a. $\mathcal{V} = \{\mathbf{v}_1, \mathbf{v}_2\} = \{(3, 1), (5, 2)\}$

   b. $\mathcal{E} = \{\mathbf{e}_1, \mathbf{e}_2\} = \{(1, 0), (0, 1)\}$

55. Determine if these are similar matrices.

   a. $\begin{bmatrix} 4 & 6 \\ 3 & 4 \end{bmatrix} \simeq \begin{bmatrix} -2 & -3 \\ 8 & 10 \end{bmatrix}$

   b. $\begin{bmatrix} \cos\theta & -\sin\theta \\ \sin\theta & \cos\theta \end{bmatrix} \simeq \begin{bmatrix} e^{-i\theta} & 0 \\ 0 & e^{i\theta} \end{bmatrix}$

   Hint: $e^{i\theta} = \cos\theta + i\sin\theta$

COMPUTER EXERCISES 5.3

Use mathematical software such as Maple to solve these problems numerically.

1. Is this set of polynomials linearly independent?
$$\{p(x) = 2x^3 + 4x^2 + 9x + 5,$$
$$q(x) = x^3 - x^2 + 8x + 2,$$
$$r(x) = x^3 - 3x^2 + 5x + x\}$$

2. Is this set of matrices linearly independent?

$$\{\mathbf{A} = \begin{bmatrix} 1 & -3 \\ -4 & 0 \end{bmatrix},$$

$$\mathbf{B} = \begin{bmatrix} 3 & -1 \\ 2 & 2 \end{bmatrix},$$

$$\mathbf{C} = \begin{bmatrix} 1 & 2 \\ 3 & 1 \end{bmatrix}\}$$

3. Is this set of polynomials linearly independent?

$$\{u(t) = 3t^3 + 8t^2 - 8t + 7,$$
$$v(t) = t^3 + 6t^2 - t + 4,$$
$$w(t) = t^3 + 4t^2 - 2t + 1\}$$

4. Is this set of matrices linearly independent?

$$\{R = \begin{bmatrix} 1 & 1 \\ 0 & 0 \end{bmatrix},$$

$$S = \begin{bmatrix} 1 & 0 \\ 0 & 1 \end{bmatrix},$$

$$T = \begin{bmatrix} 1 & 1 \\ 1 & 1 \end{bmatrix}\}$$

5. Compute the rank of this matrix.

$$F = \begin{bmatrix} 1 & 3 & 1 & -2 & -3 \\ 1 & 4 & 3 & -1 & -4 \\ 2 & 3 & -4 & -7 & -3 \\ 3 & 8 & 1 & -7 & -8 \end{bmatrix}$$

CHAPTER SIX

Eigensystems

6.1 EIGENVALUES AND EIGENVECTORS

The purpose of computing is insight, not numbers.
— RICHARD W. HAMMING (1915–1998)

Numbers are intellectual witnesses that belong only to mankind.
— HONORE DE BALZAC (1799–1850)

Introduction

Sometimes, the effect of multiplying a vector by a matrix is to produce simply a scalar multiple of that vector. For example,

$$\begin{bmatrix} 3 & 2 \\ -4 & 9 \end{bmatrix} \begin{bmatrix} 1 \\ 1 \end{bmatrix} = \begin{bmatrix} 5 \\ 5 \end{bmatrix} = 5 \begin{bmatrix} 1 \\ 1 \end{bmatrix}$$

We often want to know to what extent this phenomenon occurs for a given matrix. That is, we ask for all instances of the equation $\mathbf{Ax} = \lambda\mathbf{x}$, where $\mathbf{A}$ is given but we know neither the vector $\mathbf{x}$ nor the scalar λ. Because there are two unknowns on the right multiplied together, this is *not* a linear problem! If λ were fixed, and we only wanted to find all the appropriate $\mathbf{x}$-vectors, it would be a standard linear problem. Even easier is the case when $\mathbf{x}$ is known, and we only want to determine λ. Each of these is a linear problem. But it is nonlinear if $\mathbf{x}$ and λ are both unknown. Therefore, we must expect to encounter some obstacles in solving this problem.

367

Eigenvectors and Eigenvalues

First, we notice that the vector $\mathbf{0}$ always solves the equation $\mathbf{Ax} = \lambda\mathbf{x}$. We rule out that case as uninteresting. The properly posed question is therefore: For what scalars λ is there a *nontrivial* solution ($\mathbf{x} \neq \mathbf{0}$) to the equation

$$\mathbf{Ax} = \lambda\mathbf{x}$$

where the matrix $\mathbf{A}$ is given and the vector $\mathbf{x}$ is sought? This is known as the **eigenvalue problem** for the given matrix, $\mathbf{A}$.

DEFINITION

*Let $\mathbf{A}$ be any square matrix, real or complex. A number λ is an **eigenvalue** of $\mathbf{A}$ if the equation*

$$\mathbf{Ax} = \lambda\mathbf{x}$$

*is true for some nonzero vector $\mathbf{x}$. (Here λ is allowed to be a real or complex number.) The vector $\mathbf{x}$ is an **eigenvector** associated with the eigenvalue λ. The eigenvector may also be complex.*

Arithmetic with complex numbers cannot be avoided in the subject of eigenvalues. That is why the preceding definition explicitly allows for complex numbers to enter our calculations. See Appendix B for a summary of complex arithmetic.

In matrix theory, some terminology has changed over time. For example, *eigenvalues* have been known as *characteristic values*, *latent roots*, and *eigenwerte*. As recently as 1949, the eminent mathematician D. E. Littlewood was calling an eigenvalue a "*latent root*," an eigenvector a "*pole*," and the inverse of a matrix its "*reciprocal*," as well as using $x_1^2 + x_2^2 + \cdots + x_n^2$ as the "*norm*" of the vector $\mathbf{x} = (x_1, x_2, \ldots, x_n)$, which is not the standard definition given in Section 7.1. (See *The Skeleton Key of Mathematics* by Littlewood [2002].) The German word "*eigen*" has several different English translations, such as *proper*, *peculiar*, and *strange*. Many German idioms contain this word.

EXAMPLE 1 Illustrate the use of these new terms *eigenvalue* and *eigenvector* by using the equation above.

SOLUTION That equation has the form $\mathbf{Ax} = \lambda\mathbf{x}$, where

$$\mathbf{A} = \begin{bmatrix} 3 & 2 \\ -4 & 9 \end{bmatrix}, \qquad \mathbf{x} = \begin{bmatrix} 1 \\ 1 \end{bmatrix}, \qquad \lambda = 5$$

Thus, 5 is an eigenvalue of the given matrix $\mathbf{A}$, and a corresponding eigenvector is $[1, 1]^T$. ∎

We must be prepared for complex eigenvalues and eigenvectors, even if $\mathbf{A}$ is real. Notice that if λ is an eigenvalue of $\mathbf{A}$ and if $\mathbf{x}$ and $\mathbf{y}$ are eigenvectors of $\mathbf{A}$ associated with λ, then $\mathbf{x} + \mathbf{y}$ (if not $\mathbf{0}$) is also an eigenvector associated with λ. A similar observation applies to $c\,\mathbf{x}$ for any nonzero scalar c. (Verify this!) Hence for fixed $\mathbf{A}$ and λ, the set $\{\mathbf{x} \ : \ \mathbf{A}\mathbf{x} = \lambda\mathbf{x}\}$ is a subspace. It will often be the subspace containing only the vector, $\mathbf{0}$, as happens when λ is not an eigenvalue of $\mathbf{A}$.

Using Determinants in Finding Eigenvalues

If λ is an eigenvalue of $\mathbf{A}$, then the equation $\mathbf{A}\mathbf{x} = \lambda\mathbf{x}$ has a nontrivial solution. Consequently, the equation $(\mathbf{A} - \lambda\mathbf{I})\mathbf{x} = \mathbf{0}$ has a nontrivial solution, and the matrix $\mathbf{A} - \lambda\mathbf{I}$ is noninvertible. Hence, we have $\text{Det}(\mathbf{A} - \lambda\mathbf{I}) = 0$. Because the argument is reversible, we have this result:

THEOREM 1

A scalar λ is an eigenvalue of a matrix $\mathbf{A}$ if and only if $\text{Det}(\mathbf{A} - \lambda\mathbf{I}) = 0$.

The effect of this theorem is to turn the intractable nonlinear problem (with unknowns λ and $\mathbf{x}$) into a problem involving only λ. Conceptually this is very important, although finding the correct values of λ may still be numerically challenging.

The equation $\text{Det}(\mathbf{A}\mathbf{x} - \lambda\mathbf{I}) = 0$ is called the **characteristic equation** of $\mathbf{A}$. It is the equation from which we can compute the eigenvalues of $\mathbf{A}$. The function $\lambda \mapsto \text{Det}(\mathbf{A} - \lambda\mathbf{I})$ is the **characteristic polynomial** of $\mathbf{A}$.

EXAMPLE 2 What are the characteristic equation and the eigenvalues of the matrix $\mathbf{A} = \begin{bmatrix} 2 & 1 \\ 4 & -1 \end{bmatrix}$? For each eigenvalue, find an eigenvector.

SOLUTION To answer this, we set up the matrix equation

$$\text{Det}(\mathbf{A} - \lambda\mathbf{I}) = \text{Det} \begin{bmatrix} 2 - \lambda & 1 \\ 4 & -1 - \lambda \end{bmatrix} = 0$$

Next, we expand the determinant and obtain:

$$(2 - \lambda)(-1 - \lambda) - 4 = 0$$

This leads to $\lambda^2 - \lambda - 6 = 0$, or $(\lambda - 3)(\lambda + 2) = 0$. Hence, the given matrix **A** has precisely two eigenvalues: 3 and -2. Along the way one gets the characteristic polynomial, $\lambda \mapsto \lambda^2 - \lambda - 6$.

Flushed with success we pursue the eigenvectors. First, let $\lambda = 3$. The homogeneous equation that we must solve is then

$$(\mathbf{A} - 3\mathbf{I})\mathbf{x} = \mathbf{0} \qquad \text{or} \qquad \begin{bmatrix} -1 & 1 \\ 4 & -4 \end{bmatrix} \begin{bmatrix} x_1 \\ x_2 \end{bmatrix} = \begin{bmatrix} 0 \\ 0 \end{bmatrix}$$

Notice that the matrix in this subproblem is noninvertible. We expect that, because λ has been chosen to satisfy the equation $\text{Det}(\mathbf{A} - \lambda\mathbf{I}) = \mathbf{0}$, and that is the same as making $\mathbf{A} - \lambda\mathbf{I}$ noninvertible. (Recall Theorem 5 in Section 4.1.) Row reduction of the augmented matrix leads to

$$\begin{bmatrix} -1 & 1 & | & 0 \\ 4 & -4 & | & 0 \end{bmatrix} \sim \begin{bmatrix} 1 & -1 & | & 0 \\ 0 & 0 & | & 0 \end{bmatrix}$$

We obtain $x_1 - x_2 = 0$. The simplest eigenvector for the eigenvalue 3 is therefore $\mathbf{u} = (1, 1)$.*

In the same way, we let $\lambda = -2$ and search for a vector in the null space of the matrix

$$\mathbf{A} - 2\mathbf{I} = \begin{bmatrix} 4 & 1 \\ 4 & 1 \end{bmatrix} \sim \begin{bmatrix} 4 & 1 \\ 0 & 0 \end{bmatrix}$$

We obtain $4x_1 + x_2 = 0$. A convenient eigenvector is then $\mathbf{v} = (1, -4)$. We now know that $\mathbf{Au} = 3\mathbf{u}$ and $\mathbf{Av} = -2\mathbf{v}$. Can there be any other eigenvalues for this matrix? No, any eigenvalue would necessarily satisfy the equations we arrived at, and there were no more than two eigenvalues. ∎

EXAMPLE 3 Find all the eigenvalues and at least one eigenvector of the matrix

$$\mathbf{A} = \begin{bmatrix} 3 & 0 & 0 & 13 \\ -25 & 7 & 11 & -6 \\ 18 & 0 & 1 & 5 \\ 0 & 0 & 0 & -2 \end{bmatrix}$$

SOLUTION The characteristic equation is, by definition,

$$\text{Det} \begin{bmatrix} 3-\lambda & 0 & 0 & 13 \\ -25 & 7-\lambda & 11 & -6 \\ 18 & 0 & 1-\lambda & 5 \\ 0 & 0 & 0 & -2-\lambda \end{bmatrix} = 0$$

---

* Here and elsewhere, $\mathbf{u} = (u_1, u_2)$ means the column vector $\mathbf{u} = [u_1, u_2]^T$.

The steps involved in getting the values of λ are as follows:

$$(7 - \lambda) \, \text{Det} \begin{bmatrix} 3 - \lambda & 0 & 13 \\ 18 & 1 - \lambda & 5 \\ 0 & 0 & -2 - \lambda \end{bmatrix}$$

$$= (7 - \lambda)(1 - \lambda) \, \text{Det} \begin{bmatrix} 3 - \lambda & 13 \\ 0 & -2 - \lambda \end{bmatrix}$$

$$= (7 - \lambda)(1 - \lambda)(3 - \lambda)(-2 - \lambda) = 0$$

Our work indicates that the eigenvalues are $-2, 1, 3,$ and 7. (Typical eigenvalues in the real world are *not* integers!) Now, select an eigenvalue, say 3, and search for a corresponding eigenvector. We must solve $(\mathbf{A} - 3\mathbf{I})\mathbf{x} = \mathbf{0}$ with a nonzero vector $\mathbf{x}$. This is a homogeneous problem whose augmented matrix is

$$[\mathbf{A} - 3\mathbf{I} \mid \mathbf{0}] = \begin{bmatrix} 0 & 0 & 0 & 13 & \bigm| & 0 \\ -25 & 4 & 11 & -6 & \bigm| & 0 \\ 18 & 0 & -2 & 5 & \bigm| & 0 \\ 0 & 0 & 0 & -5 & \bigm| & 0 \end{bmatrix}$$

The solution is $\mathbf{u} = (2, -37, 18, 0)$. This can be verified in a separate calculation. Namely, $\mathbf{Au} = (6, -111, 54, 0) = 3\mathbf{u}$, and this indeed is 3 times $\mathbf{u}$. ∎

Linear Transformations

The concepts of eigenvalues and eigenvectors carry over to linear transformations.

> **DEFINITION**
>
> *Let L be a linear operator mapping a vector space V into itself. If*
>
> $$L(\mathbf{v}) = \lambda \mathbf{v} \qquad \text{and} \qquad \mathbf{v} \neq \mathbf{0}$$
>
> *then we call λ an **eigenvalue** and $\mathbf{v}$ an **eigenvector** of the linear operator L.*

For example, let $L(\mathbf{v}) = \mathbf{v}''$, where V is the vector space of twice-differentiable functions on the symmetric interval $[-\pi, \pi]$. Since $d(\sin t)/dt = \cos t$ and $d(\cos t)/dt = -\sin t$, -1 is an eigenvalue and the *sine* function is an eigenvector.

In Example 2, the pertinent conclusions can be conveyed as follows:

$$\mathbf{A}[\mathbf{u} \; \mathbf{v}] = [\mathbf{u} \; \mathbf{v}] \text{Diag}(\lambda_1, \lambda_2)$$

$$\mathbf{AP} = \begin{bmatrix} 2 & 1 \\ 4 & -1 \end{bmatrix} \begin{bmatrix} 1 & 1 \\ 1 & -4 \end{bmatrix} = \begin{bmatrix} 1 & 1 \\ 1 & -4 \end{bmatrix} \begin{bmatrix} 3 & 0 \\ 0 & -2 \end{bmatrix} = \mathbf{PD}$$

The matrix containing two eigenvectors as columns is denoted by $\mathbf{P}$, and the diagonal matrix on the right is denoted by $\mathbf{D}$. Thus, we have $\mathbf{AP} = \mathbf{PD}$. In this case, one can take another step and write $\mathbf{P}^{-1}\mathbf{AP} = \mathbf{D}$ or $\mathbf{A} = \mathbf{PDP}^{-1}$. This is possible because $\mathbf{P}$ in this case is invertible. For some other examples, this step is not permitted because $\mathbf{P}$ is singular. In the terminology of Section 5.3, the equation $\mathbf{A} = \mathbf{PDP}^{-1}$ indicates that our original matrix $\mathbf{A}$ is **similar** to the diagonal matrix $\mathbf{D}$. Remember that *similar* is a highly technical term in matrix theory, meaning in this context that $\mathbf{A} = \mathbf{PDP}^{-1}$.

DEFINITION

A matrix $\mathbf{A}$ *is* **diagonalizable** *if there is a diagonal matrix* $\mathbf{D}$ *and an invertible matrix* $\mathbf{P}$ *such that*

$$\mathbf{P}^{-1}\mathbf{AP} = \mathbf{D} \qquad \text{or} \qquad \mathbf{A} = \mathbf{PDP}^{-1}$$

We repeat that there is a profound difference between the two equations $\mathbf{AP} = \mathbf{PD}$ and $\mathbf{P}^{-1}\mathbf{AP} = \mathbf{D}$. The latter equation requires that $\mathbf{P}$ be invertible. Although we can always arrive at an equation of the form $\mathbf{AP} = \mathbf{PD}$, we may be unable to proceed to $\mathbf{P}^{-1}\mathbf{AP} = \mathbf{D}$. Obviously, the invertibility (nonsingularity) of the matrix $\mathbf{P}$ is the issue. (Notice that the equation $\mathbf{AP} = \mathbf{PD}$ is always solvable, since $\mathbf{P} = \mathbf{0}$ is one solution. Therefore, this case must be ruled out.) The next example illustrates what may happen.

EXAMPLE 4 Find the eigenvalues and eigenvectors of the matrix

$$\mathbf{A} = \begin{bmatrix} 1 & 1 \\ 0 & 1 \end{bmatrix}$$

Carry out the diagonalization process, if possible.

SOLUTION To find the eigenvalues, write

$$\text{Det} \begin{bmatrix} 1 - \lambda & 1 \\ 0 & 1 - \lambda \end{bmatrix} = (\lambda - 1)^2 = 0$$

There is only one eigenvalue—namely, $\lambda = 1$. When we look for an eigenvector, we solve the homogeneous equation having augmented matrix

$$\left[\begin{array}{cc|c} 0 & 1 & 0 \\ 0 & 0 & 0 \end{array}\right]$$

One possible eigenvector is $(1, 0)$. All others are multiples of this one. We can, of course, write $\mathbf{AP} = \mathbf{PD}$, which looks like this:

$$\mathbf{AP} = \left[\begin{array}{cc} 1 & 1 \\ 0 & 1 \end{array}\right]\left[\begin{array}{cc} 1 & 1 \\ 0 & 0 \end{array}\right] = \left[\begin{array}{cc} 1 & 1 \\ 0 & 0 \end{array}\right]\left[\begin{array}{cc} 1 & 0 \\ 0 & 1 \end{array}\right] = \mathbf{PD}$$

The matrix $\mathbf{P}$ is noninvertible and $\mathbf{A}$ is not diagonalizable! We do not have enough eigenvectors to provide a basis for $\mathbb{R}^2$. ∎

In computing the eigenvalues of a matrix, we may encounter repeated roots (*multiple roots*) in the characteristic polynomial. As mentioned previously, the multiplicity of an eigenvalue as a root of the characteristic polynomial is called its **algebraic multiplicity**. The dimension of the accompanying eigenspace is termed the **geometric multiplicity** of that eigenvalue. These can differ, but the algebraic multiplicity is always greater than or equal to the geometric multiplicity. Example 4 exhibits this phenomenon in a 2×2 matrix. The eigenvalue 1 has *algebraic multiplicity* two, because the factor $(\lambda - 1)$ occurred twice in the characteristic polynomial $\lambda \mapsto \text{Det}(\mathbf{A} - \lambda \mathbf{I})$. Also, we say that the eigenvalue has *geometric multiplicity* one, since it leads to a one-dimensional subspace of eigenvectors.

For a formal definition of these two concepts, let $\mathbf{A}$ be an $n \times n$ matrix. Let its characteristic polynomial be

$$\text{Det}(\mathbf{A} - \lambda \mathbf{I}) = \pm(\lambda - \lambda_1)(\lambda - \lambda_2)(\lambda - \lambda_3)\cdots(\lambda - \lambda_n)$$

In this equation some factors can be repeated; that is, they can occur more than once. The complex numbers λ_j are the eigenvalues of $\mathbf{A}$. If a certain factor $\lambda - \lambda_j$ occurs k_j times in the factorization, the corresponding eigenvalue has **algebraic multiplicity** k_j. If the corresponding subspace has dimension ℓ_j, then we say that the eigenvalue has **geometric multiplicity** ℓ_j. This means that ℓ_j is the dimension of $\text{Null}(\mathbf{A} - \lambda_j \mathbf{I})$. The following theorem is given without proof; it asserts that $k_j \geq \ell_j$.

THEOREM 2

The geometric multiplicity of an eigenvalue cannot exceed its algebraic multiplicity.

Distinct Eigenvalues

When all the eigenvalues of a matrix are different, the associated eigenvectors form a linearly independent set, according to the following theorem.

THEOREM 3

Any set of eigenvectors corresponding to distinct eigenvalues of a matrix is linearly independent.

PROOF Let $\mathbf{A}\mathbf{x}^{(i)} = \lambda_i \mathbf{x}^{(i)}$ for $1 \leq i \leq k$. Assume that the vectors $\mathbf{x}^{(i)}$ are nonzero and that the eigenvalues λ_i are all different from each other. We hope to establish that the set $\{\mathbf{x}^{(1)}, \mathbf{x}^{(2)}, \ldots, \mathbf{x}^{(k)}\}$ is linearly independent. Our proof will be by the method of contradiction. Suppose that $\{\mathbf{x}^{(1)}, \mathbf{x}^{(2)}, \ldots, \mathbf{x}^{(k)}\}$ is linearly dependent. By Theorem 9 in Section 2.4, some element of this ordered set is a linear combination of the preceding elements in the set. Let j be the first integer such that $\mathbf{x}^{(j+1)}$ is a linear combination of $\mathbf{x}^{(1)}, \mathbf{x}^{(2)}, \ldots, \mathbf{x}^{(j)}$. Then the set $\{\mathbf{x}^{(1)}, \mathbf{x}^{(2)}, \ldots, \mathbf{x}^{(j)}\}$ is linearly independent, and $j < k$. We carry out the following steps:

$$\sum_{i=1}^{j} c_i \mathbf{x}^{(i)} = \mathbf{x}^{(j+1)}$$

$$\sum_{i=1}^{j} c_i \mathbf{A}\mathbf{x}^{(i)} = \mathbf{A}\mathbf{x}^{(j+1)}$$

$$\sum_{i=1}^{j} c_i \lambda_i \mathbf{x}^{(i)} = \lambda_{j+1}\mathbf{x}^{(j+1)} = \lambda_{j+1} \sum_{i=1}^{j} c_i \mathbf{x}^{(i)}$$

$$\sum_{i=1}^{j} c_i (\lambda_i - \lambda_{j+1}) \mathbf{x}^{(i)} = 0$$

By the linear independence of $\{\mathbf{x}^{(1)}, \mathbf{x}^{(2)}, \ldots, \mathbf{x}^{(j)}\}$, the coefficients $c_i(\lambda_i - \lambda_{j+1})$ are all zero. Because the eigenvalues are distinct, we conclude that all c_i are zero. Hence, by a preceding equation, $\mathbf{x}^{(j+1)} = 0$. This cannot be true. (All the eigenvectors are nonzero.) This contradiction concludes the proof. ∎

Another, slightly different, proof can be given as follows.

PROOF Let $\mathbf{A}\mathbf{x}^{(i)} = \lambda_i \mathbf{x}^{(i)}$, for $1 \leq i \leq k$, where all $\mathbf{x}^{(i)}$ are nonzero and the eigenvalues are distinct. Let r be the largest index for which

$\{\mathbf{x}^{(1)}, \mathbf{x}^{(2)}, \ldots, \mathbf{x}^{(r)}\}$ is linearly independent. If $r = k$, we are finished. Assume, therefore, that $r < k$. Then $\{\mathbf{x}^{(1)}, \mathbf{x}^{(2)}, \ldots, \mathbf{x}^{(r+1)}\}$ is linearly dependent. It follows that $\mathbf{x}^{(r+1)}$ is in the span of $\{\mathbf{x}^{(1)}, \mathbf{x}^{(2)}, \ldots, \mathbf{x}^{(r)}\}$. Thus, we obtain $\mathbf{x}^{(r+1)} = \sum_{i=1}^{r} a_i \mathbf{x}^{(i)}$, where the coefficients a_i are not all zero because $\mathbf{x}^{(r+1)} \neq \mathbf{0}$. Now compute a few things:

$$\lambda_{r+1}\mathbf{x}^{(r+1)} = \mathbf{A}\mathbf{x}^{(r+1)} = \mathbf{A}\sum_{i=1}^{r} a_i \mathbf{x}^{(i)} = \sum_{i=1}^{r} a_i \mathbf{A}\mathbf{x}^{(i)} = \sum_{i=1}^{r} a_i \lambda_i \mathbf{x}^{(i)}$$

$$\lambda_{r+1}\mathbf{x}^{(r+1)} = \lambda_{r+1}\sum_{i=1}^{r} a_i \mathbf{x}^{(i)} = \sum_{i=1}^{r} \lambda_{r+1} a_i \mathbf{x}^{(i)}$$

By subtraction between the two preceding equations, we get $\mathbf{0} = \sum_{i=1}^{r} a_i(\lambda_i - \lambda_{r+1})\mathbf{x}^{(i)}$. None of the factors $\lambda_i - \lambda_{r+1}$ are zero, by hypothesis. Also, the coefficients a_i are not all zero. We have here a contradiction of the linear independence of $\{\mathbf{x}^{(1)}, \mathbf{x}^{(2)}, \ldots, \mathbf{x}^{(r)}\}$. ∎

The preceding theorem is valid if we define eigenvalues and eigenvectors for any linear transformation, T. No finite-dimensionality assumption is required. A slight rewording of the proof would be needed: At first, the equation $T(\mathbf{x}^{(\alpha)}) = \lambda_\alpha \mathbf{x}^{(\alpha)}$ would be assumed, where α runs over some index set. If the set of $\mathbf{x}^{(\alpha)}$ is linearly dependent, we can then pass to a finite linearly dependent set, as in the proof of Theorem 3.

Because of its practical importance, we state an immediate consequence of Theorem 3:

COROLLARY 1

If an $n \times n$ matrix $\mathbf{A}$ has n distinct eigenvalues, then $\mathbf{A}$ is diagonalizable.

Bases of Eigenvectors

If an $n \times n$ matrix $\mathbf{A}$ is given, there *may* or *may not* exist a basis for $\mathbb{R}^n$ consisting of eigenvectors of $\mathbf{A}$. For example, we *can* find a basis for $\mathbb{R}^2$ consisting of eigenvectors of the matrix $\mathbf{A} = \begin{bmatrix} 3 & 5 \\ 0 & 7 \end{bmatrix}$. Because $\mathbf{A}$ is triangular, the eigenvalues are $\lambda_1 = 3$ and $\lambda_2 = 7$. The eigenvalue–eigenvector pairs are $\lambda_1 = 3, \begin{bmatrix} 1 \\ 0 \end{bmatrix}$ and $\lambda_2 = 7, \begin{bmatrix} 5 \\ 4 \end{bmatrix}$. Hence, for $\mathbf{A} = \begin{bmatrix} 3 & 5 \\ 0 & 7 \end{bmatrix}$, there *is* a basis for $\mathbb{R}^2$ made up of eigenvectors: $\left\{ \begin{bmatrix} 1 \\ 0 \end{bmatrix}, \begin{bmatrix} 5 \\ 4 \end{bmatrix} \right\}$.

However, for $\mathbf{B} = \begin{bmatrix} 0 & 1 \\ 0 & 0 \end{bmatrix}$, 0 is the only eigenvalue (a repeated root in the characteristic polynomial $\lambda \mapsto \lambda^2$). There is only a one-dimensional eigenspace—namely, the span of $\begin{bmatrix} 1 \\ 0 \end{bmatrix}$.

Finally, for $\mathbf{C} = \begin{bmatrix} 0 & 0 \\ 0 & 0 \end{bmatrix}$, 0 is again the only eigenvalue (repeated), but there are two free variables in this case. Here there *is* a basis for $\mathbb{R}^2$ consisting of eigenvectors of $\mathbf{C}$—namely, $\begin{bmatrix} 1 \\ 0 \end{bmatrix}$ and $\begin{bmatrix} 0 \\ 1 \end{bmatrix}$.

COROLLARY 2

If an $n \times n$ real matrix $\mathbf{A}$ has n distinct real eigenvalues, then the corresponding eigenvectors of $\mathbf{A}$ constitute a basis for $\mathbb{R}^n$.

If $\mathbf{A}$ has repeated eigenvalues, no conclusion can be drawn without further analysis!

Application: Powers of a Matrix

The factorization of a matrix

$$\mathbf{A} = \mathbf{PDP}^{-1}$$

discussed previously can be used to compute efficiently the powers of $\mathbf{A}$. Let us see how this works by starting with $\mathbf{A}^2$. Assume as known the equation $\mathbf{A} = \mathbf{PDP}^{-1}$. Multiply both sides of this equation on the left by $\mathbf{A}$, getting

$$\mathbf{A}^2 = \mathbf{AA} = (\mathbf{PDP}^{-1})(\mathbf{PDP}^{-1}) = \mathbf{PD}(\mathbf{P}^{-1}\mathbf{P})\mathbf{DP}^{-1} = \mathbf{PDIDP}^{-1} = \mathbf{PD}^2\mathbf{P}^{-1}$$

(Notice how the associative law for matrix multiplication enters this calculation.) This argument can be repeated, with the result that

$$\mathbf{A}^k = \mathbf{PD}^k\mathbf{P}^{-1} \qquad (k \geq 1)$$

This is easy to compute, because the powers of a diagonal matrix are also diagonal, and the diagonal entries are powers of the diagonal elements in $\mathbf{D}$. In algebraic form, this is

$$\mathbf{D}^k = [\text{Diag}(d_1, d_2, \ldots, d_n)]^k = \text{Diag}(d_1^k, d_2^k, \ldots, d_n^k)$$

$$= \begin{bmatrix} d_1 & 0 & 0 & \cdots & 0 \\ 0 & d_2 & 0 & \cdots & 0 \\ \vdots & \vdots & \vdots & \vdots & \vdots \\ 0 & 0 & 0 & \cdots & d_n \end{bmatrix}^k = \begin{bmatrix} d_1^k & 0 & 0 & \cdots & 0 \\ 0 & d_2^k & 0 & \cdots & 0 \\ \vdots & \vdots & \vdots & \vdots & \vdots \\ 0 & 0 & 0 & 0 & d_n^k \end{bmatrix}$$

EXAMPLE 5 Using the matrices from Example 2, compute $\mathbf{A}^k$.

SOLUTION From the equation $\mathbf{A} = \mathbf{PDP}^{-1}$, we get $\mathbf{A}^k = \mathbf{PD}^k\mathbf{P}^{-1}$ or

$$\begin{bmatrix} 2 & 1 \\ 4 & -1 \end{bmatrix}^k = \begin{bmatrix} 1 & 1 \\ 1 & -4 \end{bmatrix} \begin{bmatrix} 3^k & 0 \\ 0 & (-2)^k \end{bmatrix} \begin{bmatrix} \frac{4}{5} & \frac{1}{5} \\ \frac{1}{5} & -\frac{1}{5} \end{bmatrix}$$

$$= \begin{bmatrix} \frac{1}{5}[4(3^k) + (-2)^k] & \frac{1}{5}[(3^k) - (-2)^k] \\ \frac{1}{5}[4(3^k) - 4(-2)^k] & \frac{1}{5}[(3^k) + 4(-2)^k] \end{bmatrix}$$

We obtain a quick verification by letting $k = 1$. ∎

Characteristic Equation and Characteristic Polynomial

In the general case of an $n \times n$ matrix, the function p defined by

$$p(\lambda) = \text{Det}(\mathbf{A} - \lambda\mathbf{I})$$

will be a polynomial of degree n in the variable λ. As we know, it is called the **characteristic polynomial** of $\mathbf{A}$. When we set it equal to zero, the result is called the **characteristic equation**. This polynomial may have n different roots, or there may be repeated roots. Repeated roots occur when the characteristic polynomial has factors of the form $(\lambda - \lambda_j)^k$, where $k > 1$. The characteristic polynomial may have complex roots, as we now illustrate.

EXAMPLE 6 Let $\mathbf{A} = \begin{bmatrix} 1 & 1 \\ -2 & 3 \end{bmatrix}$

Find the eigenvalues of $\mathbf{A}$ by using the characteristic polynomial.

SOLUTION Proceeding as before, we get

$$p(\lambda) = \text{Det}(\mathbf{A} - \lambda\mathbf{I}) = \text{Det}\begin{bmatrix} 1 - \lambda & 1 \\ -2 & 3 - \lambda \end{bmatrix} = (1-\lambda)(3-\lambda) + 2 = \lambda^2 - 4\lambda + 5$$

The roots of this polynomial are computed with the quadratic formula. Recall that the quadratic equation $a\lambda^2 + b\lambda + c = 0$ has roots

$$\lambda = \frac{-b \pm \sqrt{b^2 - 4ac}}{2a}$$

For the case at hand, we obtain

$$\lambda = \frac{4 \pm \sqrt{16 - (4)(5)}}{2} = 2 \pm \sqrt{-1} = 2 \pm i$$

Notice that a real matrix may have complex eigenvalues.   ∎

Diagonalization Involving Complex Numbers

The diagonalization of matrices involving complex eigenvalues and eigenvectors is illustrated in the next example.

EXAMPLE 7 What is the diagonalization of the matrix in Example 6?

$$\mathbf{A} = \begin{bmatrix} 1 & 1 \\ -2 & 3 \end{bmatrix}$$

SOLUTION In principle, this is worked out in the way we have already illustrated in the real case. But now we must undertake some arithmetic calculations with complex numbers. (See Appendix B.) Start with the eigenvalue $\lambda_1 = 2 + i$. For this value of λ, we have

$$\mathbf{A} - \lambda_1\mathbf{I} = \begin{bmatrix} 1 & 1 \\ -2 & 3 \end{bmatrix} - (2+i)\begin{bmatrix} 1 & 0 \\ 0 & 1 \end{bmatrix} = \begin{bmatrix} -1-i & 1 \\ -2 & 1-i \end{bmatrix}$$

If our work is correct, this matrix will be noninvertible. Looking at the second column, we conclude that row 2 should be $(1 - i)$ times row 1. We verify this:

$$(1 - i)(\text{row } 1) = (1 - i)[-1 - i, 1] = [(1 - i)(-1 - i), 1 - i]$$
$$= [-(1 - i^2), 1 - i] = [-2, 1 - i] = (\text{row } 2)$$

Hence, suitable row operations (which we need not carry out!) will lead us to

$$\mathbf{A} - \lambda_1\mathbf{I} = \begin{bmatrix} -1-i & 1 \\ -2 & 1-i \end{bmatrix} \sim \begin{bmatrix} 1+i & -1 \\ 0 & 0 \end{bmatrix}$$

Let $\mathbf{x}$ be a nonzero solution to this homogeneous system of equations. Let x_2 be the free variable, so that we obtain $(1+i)x_1 = x_2$. Letting $x_2 = 1+i$, we get $x_1 = 1$, and a convenient eigenvector is then $[1, 1+i]^T$. Proceeding in the same way with the eigenvalue $\lambda_2 = 2 - i$, we arrive at the matrix

$$\mathbf{A} - \lambda_2\mathbf{I} = \begin{bmatrix} -1+i & 1 \\ -2 & 1+i \end{bmatrix}$$

Again, this must be noninvertible. Just for amusement, we check this non-invertibility property by computing the determinant, which should be zero:

$$(-1+i)(1+i) + 2 = -(1-i)(1+i) + 2 = -(1-i^2) + 2 = -2 + 2 = 0$$

Hence, suitable row operations (which again need not be performed) will reduce the matrix:

$$\mathbf{A} - \lambda_2\mathbf{I} = \begin{bmatrix} -1+i & 1 \\ -2 & 1+i \end{bmatrix} \sim \begin{bmatrix} 1-i & -1 \\ 0 & 0 \end{bmatrix}$$

We obtain $x_2 = (1-i)x_1$, where x_1 is a free variable. By letting $x_1 = 1$, we obtain a convenient solution to the homogeneous system of equations $\mathbf{x} = [1, 1-i]^T$. The $\mathbf{P}$-matrix is then

$$\mathbf{P} = \begin{bmatrix} 1 & 1 \\ 1+i & 1-i \end{bmatrix}$$

Computing the inverse of $\mathbf{P}$ in the usual way leads to

$$\mathbf{P}^{-1} = \tfrac{1}{2}\begin{bmatrix} 1+i & -i \\ 1-i & i \end{bmatrix}$$

The diagonalization equation, $\mathbf{A} = \mathbf{P}\mathbf{D}\mathbf{P}^{-1}$, is then

$$\begin{bmatrix} 1 & 1 \\ -2 & 3 \end{bmatrix} = \begin{bmatrix} 1 & 1 \\ 1+i & 1-i \end{bmatrix} \begin{bmatrix} 2+i & 0 \\ 0 & 2-i \end{bmatrix} \begin{bmatrix} \tfrac{1}{2}(1+i) & -\tfrac{i}{2} \\ \tfrac{1}{2}(1-i) & \tfrac{i}{2} \end{bmatrix} \blacksquare$$

Notice in Examples 6 and 7 that both the eigenvalues and the eigenvectors occur as complex conjugate pairs. The reason for this is that the matrix $\mathbf{A}$ is real but the eigenvalues and eigenvectors are complex. When $\mathbf{A}$ is real and $\mathbf{A}\mathbf{x} = \lambda\mathbf{x}$, it follows that $\mathbf{A}\overline{\mathbf{x}} = \overline{\lambda}\overline{\mathbf{x}}$.

THEOREM 4

If $\mathbf{A}$ is a real matrix and has a complex eigenvalue λ, then the conjugate $(\overline{\lambda})$ is also an eigenvalue. Thus, we have $\mathbf{A}\mathbf{x} = \lambda\mathbf{x}$ and $\mathbf{A}\overline{\mathbf{x}} = \overline{\lambda}\overline{\mathbf{x}}$.

Of course, if a real matrix does not have any complex eigenvalues, then there are no complex conjugate pairs!

Application: Dynamical Systems

Here we return to the computation of $\mathbf{A}^k\mathbf{x}$ for high values of k. (We shall see in Examples 8 and 9 how this problem arises in simple dynamical systems.) Start with an eigenvector and eigenvalue for $\mathbf{A}$, say

$$\mathbf{A}\mathbf{v} = \lambda\mathbf{v}$$

Then

$$\mathbf{A}^2\mathbf{v} = \mathbf{A}(\mathbf{A}\mathbf{v}) = \mathbf{A}(\lambda\mathbf{v}) = \lambda(\mathbf{A}\mathbf{v}) = \lambda(\lambda\mathbf{v}) = \lambda^2\mathbf{v}$$

We can repeat this process to obtain $\mathbf{A}^3\mathbf{v} = \lambda^3\mathbf{v}$, and so on. In general, we have

$$\mathbf{A}^k\mathbf{v} = \lambda^k\mathbf{v} \qquad \text{for} \qquad k = 1, 2, 3, \ldots$$

Thus, for any positive integer k, it is easy to compute $\mathbf{A}^k\mathbf{v}$ when $\mathbf{v}$ is an eigenvector of $\mathbf{A}$. If $\mathbf{A}$ is an $n \times n$ matrix, and if we have a basis for $\mathbb{R}^n$ consisting of eigenvectors of $\mathbf{A}$, then we can compute $\mathbf{A}^k\mathbf{x}$ for any $\mathbf{x}$ with relative ease, as follows. Let $\{\mathbf{v}^{(1)}, \mathbf{v}^{(2)}, \ldots, \mathbf{v}^{(n)}\}$ be a basis for $\mathbb{R}^n$ consisting of eigenvectors of $\mathbf{A}$. Suppose explicitly that

$$\mathbf{A}\mathbf{v}^{(j)} = \lambda_j\mathbf{v}^{(j)} \qquad \text{for} \qquad j = 1, 2, \ldots, n$$

Given any $\mathbf{x}$ in $\mathbb{R}^n$, first express $\mathbf{x}$ in terms of the basis: $\mathbf{x} = \sum_{j=1}^{n} c_j\mathbf{v}^{(j)}$. Then multiply by $\mathbf{A}^k$, getting

$$\mathbf{A}^k\mathbf{x} = \mathbf{A}^k\sum_{j=1}^{n} c_j\mathbf{v}^{(j)} = \sum_{j=1}^{n} c_j\mathbf{A}^k\mathbf{v}^{(j)} = \sum_{j=1}^{n} c_j\lambda_j^k\mathbf{v}^{(j)}$$

No powers of $\mathbf{A}$ need be computed, and that advantage is worth noting for future applications. (See Example 8.)

If the eigenvalues and eigenvectors of **A** are not available, the powers of **A** can be computed in the usual way. Remember that multiplying two $n \times n$ matrices involves roughly n^3 arithmetic operations, and therefore computing $\mathbf{A}^k$ is costly if carried out simply by repeated multiplications. For studying the long-term behavior of a dynamical system, one can compute $\mathbf{A}^{2^k}$ efficiently by repeated squaring:

$$\mathbf{A}^2 = \mathbf{A}\mathbf{A}, \qquad \mathbf{A}^4 = \mathbf{A}^2\mathbf{A}^2, \qquad \mathbf{A}^8 = \mathbf{A}^4\mathbf{A}^4, \quad \text{etc.}$$

EXAMPLE 8 A dynamical system is modeled by the equation $\mathbf{x}^{(k)} = \mathbf{A}\mathbf{x}^{(k-1)}$, where

$$\mathbf{A} = \begin{bmatrix} 0.6 & 0.3 \\ -0.2 & 1.2 \end{bmatrix}$$

For any starting point, $\mathbf{x}^{(0)}$, this equation defines an infinite sequence of vectors in $\mathbb{R}^2$. What is the long-term behavior of this system if the starting vector is $\mathbf{x}^{(0)} = (6.97, 3.87)$?

SOLUTION To answer this, we can use mathematical software and brute force by repeatedly carrying out matrix–vector multiplications $\mathbf{x}^{(k)} = \mathbf{A}\mathbf{x}^{(k-1)}$ starting with $\mathbf{x}^{(0)} = (6.97, 3.87)$. We plot these points in Figure 6.1 using the symbol $*$.

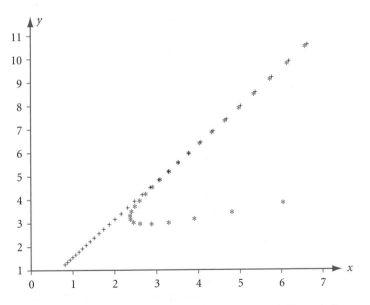

FIGURE 6.1 A trajectory of the dynamical system in Example 8.

The next step is to express the starting vector $\mathbf{x}^{(0)} = (6.97, 3.87)$ as a linear combination of the eigenvectors. First, using mathematical software, we obtain the matrix $\mathbf{D}$ of eigenvalues and matrix $\mathbf{V}$ of eigenvectors:

$$\mathbf{V} = \begin{bmatrix} -0.9211 & -0.5354 \\ -0.3893 & -0.8446 \end{bmatrix}, \qquad \mathbf{D} = \begin{bmatrix} 0.7268 & 0 \\ 0 & 1.0732 \end{bmatrix}$$

Because the eigenvectors are the columns of $\mathbf{V}$, we must solve $\mathbf{Vc} = \mathbf{x}^{(0)}$ for $\mathbf{c}$. Because both $\mathbf{V}$ and $\mathbf{c}$ are known, this 2×2 system can be solved directly or by using mathematical software to find $\mathbf{c} = (-6.6981, -1.4947)$. Hence, we obtain

$$\mathbf{x}^{(0)} = -6.6981\, \mathbf{v}^{(1)} - 1.4947\, \mathbf{v}^{(2)}$$

Next, we derive the formulas suitable for computing any $\mathbf{x}^{(k)}$ in an efficient manner.

$$\begin{aligned}
\mathbf{x}^{(k)} &= \mathbf{A}^k \mathbf{x}^{(0)} \\
&= \mathbf{A}^k (-6.6981\, \mathbf{v}^{(1)} - 1.4947\, \mathbf{v}^{(2)}) \\
&= -6.6981\, \mathbf{A}^k \, \mathbf{v}^{(1)} - 1.4947\, \mathbf{A}^k \, \mathbf{v}^{(2)} \\
&= -6.6981\, \lambda_1^k \, \mathbf{v}^{(1)} - 1.4947\, \lambda_2^k \, \mathbf{v}^{(2)} \\
&= -6.6981\, (0.7268)^k \, \mathbf{v}^{(1)} - 1.4947\, (1.0732)^k \, \mathbf{v}^{(2)} \\
&\approx -1.4947\, (1.0732)^k \, \mathbf{v}^{(2)} \\
&= -1.4947\, (1.0732)^k \, (-0.5354, -0.8446) \\
&= (1.0732)^k (0.80026, 1.26242)
\end{aligned}$$

The vectors $\mathbf{x}^{(k)}$ eventually settle down in the direction of $\mathbf{v}^{(2)}$, because the component in the direction of $\mathbf{v}^{(1)}$ is diminishing quickly. The magnitude of the vectors is increasing because of the factor $(1.0732)^k$.

For comparison purposes, Figure 6.1 shows two sets of points: the repeated matrix–vector multiplication (marked with the $*$ symbol) and the dominant-term approximation (marked with the $+$ symbol). The latter appears to lie on a straight line because the dominant eigenvalue term overwhelms any effect from the other eigenvalues.

Where the points appear to lie on a straight line, it is because the term involving the dominant eigenvalue overwhelms the effect of the other eigenvalue. ∎

EXAMPLE 9 A dynamical system is given in the form $\mathbf{x}^{(1)} = \mathbf{A}\mathbf{x}^{(0)}$, $\mathbf{x}^{(2)} = \mathbf{A}\mathbf{x}^{(1)}$, etc., where

$$\mathbf{A} = \begin{bmatrix} 1.08 & -0.229 \\ 0.229 & 1.08 \end{bmatrix} \quad \text{and} \quad \mathbf{x}^{(0)} = \begin{bmatrix} 1.2 \\ 0.3 \end{bmatrix}$$

Discover the long-term behavior of this system.

SOLUTION We use mathematical software to generate the sequence $\mathbf{x}^{(1)}$, $\mathbf{x}^{(2)}$, and so on, up to $\mathbf{x}^{(30)}$. We use commands similar to those in Example 8 except that $\mathbf{A}$ and $\mathbf{x}^{(0)}$ are different.

In an example like this, it is easy to continue this sequence of vectors $\mathbf{x}^{(k)}$. Figure 6.2 shows that the points are following an orbit that spirals around the origin at ever-increasing distances. To find a dominant-term approximation for the successive points, we would follow a procedure similar to the one in Example 8. Again, we can use mathematical software to compute the eigenvalues and eigenvectors of $\mathbf{A}$. The eigenvalues are $1.08 \pm 0.229i$, and the matrix of eigenvectors can be taken to be $\begin{bmatrix} 1 & 1 \\ i & -i \end{bmatrix}$. ■

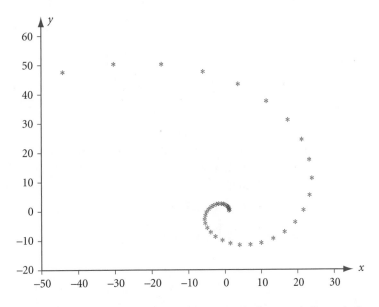

FIGURE 6.2 A spiral trajectory of the dynamical system in Example 9.

Further Dynamical Systems in $\mathbb{R}^2$

We describe the geometrical behavior of some simple dynamical systems having the form

$$\mathbf{x}^{(k)} = \mathbf{A}\mathbf{x}^{(k-1)} \qquad (k = 0, 1, 2, \ldots)$$

If the points on one trajectory approach a point, that point is called an **attractor** or a **sink**. If the points on a trajectory move away from a point, then such a point is a **repeller** or a **source**. If points on some trajectory move toward a point and then move away from the point, that point is a **saddle point**. Also, a trajectory can **cycle** by repeating with a certain period, or **spiral** about a point (inward or outward).

Suppose that $\mathbf{A}$ is the simple matrix

$$\mathbf{A} = \begin{bmatrix} a & 0 \\ 0 & b \end{bmatrix}$$

The eigenvalue–eigenvector pairs are $a, (1, 0)$ and $b, (0, 1)$. We have

$$\mathbf{x}^{(k)} = \mathbf{A}^k \mathbf{x}^{(0)} = \mathbf{A}^k (c_1 \mathbf{v}^{(1)} + c_2 \mathbf{v}^{(2)})$$

$$= c_1 \mathbf{A}^k \mathbf{v}^{(1)} + c_2 \mathbf{A}^k \mathbf{v}^{(2)}$$

$$= c_1 \lambda^k \mathbf{v}^{(1)} + c_2 \lambda^k \mathbf{v}^{(2)}$$

$$= c_1 a^k \begin{bmatrix} 1 \\ 0 \end{bmatrix} + c_2 b^k \begin{bmatrix} 0 \\ 1 \end{bmatrix}$$

$$= \begin{bmatrix} c_1 a^k \\ c_2 b^k \end{bmatrix}$$

Depending on the values of the elements a, b of the matrix $\mathbf{A}$ and of the starting vector $\mathbf{x}^{(0)} = (c_1, c_2)$, we obtain various behaviors for the dynamical system.

If $\mathbf{x}^{(0)} = (1, 1)$, then $\mathbf{x}^{(k)} = (a^k, b^k)$. If $0 < a < 1$ and $0 < b < 1$, then $\mathbf{x}^{(k)} \to (0, 0)$ and the origin is an *attractor*. If $0 < a < 1$ and $b > 1$, then $\mathbf{x}^{(k)} \to (0, \infty)$, and the origin is a *repeller* along the x_2-axis. If $a > 1$ and $0 < b < 1$, then $\mathbf{x}^{(k)} \to (\infty, 0)$, and the origin is a *repeller* along the x_1-axis. If $a > 1$ and $b > 1$, then $\mathbf{x}^{(k)} \to (\infty, \infty)$, and the origin is a *repeller*.

Analysis of a Dynamical System

Here are the steps in analyzing a discrete dynamical system, such as

$$\mathbf{x}^{(k)} = \mathbf{A}\mathbf{x}^{(k-1)} \qquad \text{for} \qquad k = 0, 1, 2, \ldots$$

where $\mathbf{A}$ and $\mathbf{x}^{(0)}$ are prescribed, and $\mathbf{A}$ is $n \times n$.

1. Observe that

$$\mathbf{x}^{(k)} = \mathbf{A}\mathbf{x}^{(k-1)} = \mathbf{A}^2\mathbf{x}^{(k-2)} = \cdots = \mathbf{A}^k\mathbf{x}^{(0)}$$

2. Find the eigenvalues $\lambda_1, \lambda_2, \ldots, \lambda_n$ of $\mathbf{A}$.
3. Order the eigenvalues according to magnitude:
   $|\lambda_1| > |\lambda_2| \geq \cdots \geq |\lambda_n|$.
4. Find eigenvectors, one for each eigenvalue, say $\mathbf{v}^{(1)}, \mathbf{v}^{(2)}, \ldots, \mathbf{v}^{(n)}$. Thus, we have

$$\mathbf{A}\mathbf{v}^{(j)} = \lambda_j \mathbf{v}^{(j)}$$

5. To proceed, we must assume that the initial vector, $\mathbf{x}^{(0)}$, is in the span of the set of eigenvectors. Thus, we write

$$\mathbf{x}^{(0)} = \sum_{j=1}^{n} c_j \mathbf{v}^{(j)}$$

for suitable coefficients c_j. Compute these coefficients.
6. Now there is a simple formula for computing the vectors $\mathbf{x}^{(k)}$ in the dynamical system:

$$\mathbf{x}^{(k)} = \mathbf{A}^k\mathbf{x}^{(0)} = \mathbf{A}^k \sum_{j=1}^{n} c_j \mathbf{v}^{(j)} = \sum_{j=1}^{n} c_j \mathbf{A}^k \mathbf{v}^{(j)} = \sum_{j=1}^{n} c_j \lambda_j^k \mathbf{v}^{(j)}$$

7. For isolating the dominant term in the preceding equation, we can write

$$\mathbf{x}^{(k)} = \lambda_1^k \sum_{j=1}^{n} c_j (\lambda_j/\lambda_1)^k \mathbf{v}^{(j)} \approx \lambda_1^k c_1 \mathbf{v}^{(1)}$$

This equation shows the influence of λ_1 on the sequence of vectors $\mathbf{x}^{(0)}, \mathbf{x}^{(1)}, \mathbf{x}^{(2)}, \ldots$. The factors (λ_j/λ_1) are all at most 1 in absolute value.

Application: Economic Models

Here, we consider an application of linear algebra in the science of economics. It is not unreasonable to expect linear algebra to play a role in analyzing the economy, since, at first glance, causes and effects seem to have a linear relation. For example, if the input to an industry is doubled, one expects the output to double as well.

The economic model discussed here assumes that the economy of a country is divided into a number of *industries.* The term **industry** is interpreted broadly: not only do we have the coal industry and the steel industry, but we also have the medical industry, the education industry, the government industry, and so forth. The word **sectors** can also be used in this context.

Suppose we have n of these industries. A realistic model may require a rather large value of n—say $n = 500$ or $10,000$. Each industry produces in one year a product having a certain total value. Let us say that the value of the output from industry i, in one year, is x_i dollars. We choose the letter x_i, because it will be an unknown quantity. Why should the value of the year's output from an industry be unknown? It is because the value is determined by the price at which the output is sold, and that price is, to some extent, free to be manipulated by the various industries in the model.

The ith industry will consume a certain fraction of each industry's output in a year, including its own output. For example, the electric industry (whose product is electricity) will consume some outputs from the coal industry, the natural gas industry, and others. It will probably consume some of its own product.

In general, let a_{ij} be the fraction of the total yearly production from industry j used by industry i in one year. Thus, in row i of the matrix $\mathbf{A}$ being constructed, the needs of industry i are listed. These numbers represent the fractions of each industry's yearly output that are needed by industry i in carrying out its role in the economy.

We assume that all the yearly output from each industry is bought by the industries being considered in our model. Hence, each column sum, $\sum_{i=1}^{n} a_{ij}$, must equal 1. That sum accounts for the entire output of industry j going back to the various industries in our model.

Conversely, the row sums are restricted only by nonnegativity.

EXAMPLE 10 Explain this concrete case involving three industries, called E, C, and S (perhaps standing for electricity, coal, and steel).

	E	C	S
E	0.1	0.5	0.5
C	0.3	0.0	0.4
S	0.6	0.5	0.1

SOLUTION This table shows that the output of industry E goes one-tenth to E, three-tenths to C, and six-tenths to S, totaling 1.0. It also shows that industry E needs one-tenth of the output of industry E, five-tenths of the output from industry C, and five-tenths of the output from industry S. Each row and column have similar interpretations. Notice that the column sums are all unity, reflecting the fact that *everything has to go somewhere.*

∎

In the general model, industry i purchases fractions $a_{i1}, a_{i2}, \ldots, a_{in}$ of the outputs from industries $1, 2, \ldots, n$. Because x_j is the value of the total output from industry j, the total yearly cost incurred by industry i in one year is

$$x_1 a_{i1} + x_2 a_{i2} + \cdots + x_n a_{in} = \sum_{j=1}^{n} a_{ij} x_j = \text{cost of inputs to the } i\text{th industry}$$

In a state of equilibrium, the earnings of each industry would equal its expenditures. This means that $x_i = \sum_{j=1}^{n} a_{ij} x_j$. This is a system of equations having the form $\mathbf{x} = \mathbf{A}\mathbf{x}$. Hence, $\mathbf{A}$ must have 1 as one of its eigenvalues! (There is no interest in the solution $\mathbf{x} = \mathbf{0}$.)

THEOREM 5

If each column of a square matrix $\mathbf{A}$ sums to 1, then 1 is an eigenvalue of $\mathbf{A}$, and the system $\mathbf{A}\mathbf{x} = \mathbf{x}$ has a nontrivial solution.

PROOF The hypothesis can be stated in the form $\sum_{i=1}^{n} a_{ij} = 1$ for all j. This means that the vector $\mathbf{u} = (1, 1, \ldots, 1)$ is an eigenvector of $\mathbf{A}^T$ corresponding to the eigenvalue 1 of $\mathbf{A}^T$. But, by General Exercise 32, $\mathbf{A}$ and $\mathbf{A}^T$ have the same eigenvalues. Hence, 1 is an eigenvalue of $\mathbf{A}$. ∎

In the language of economics, Theorem 1 asserts that there exists a set of prices for the commodities such that each industry has an income that equals its costs. These special prices are called **equilibrium prices**. Because this equilibrium price vector $\mathbf{x}$ is an eigenvector of the matrix $\mathbf{A}$, all multiples of it are also equilibrium price vectors. Thus, if each industry decides to double its prices, there will still be equilibrium. In this discussion, we think of x_i as the *price*, or the declared value of the yearly output of an industry.

EXAMPLE 11 The problem $\mathbf{x} = \mathbf{A}\mathbf{x}$ is to be solved nontrivially, using the matrix from Example 10.

SOLUTION We have

$$\mathbf{I} - \mathbf{A} = \begin{bmatrix} 1 & 0 & 0 \\ 0 & 1 & 0 \\ 0 & 0 & 1 \end{bmatrix} - \begin{bmatrix} 0.1 & 0.5 & 0.5 \\ 0.3 & 0 & 0.4 \\ 0.6 & 0.5 & 0.1 \end{bmatrix}$$

$$= \begin{bmatrix} 0.9 & -0.5 & -0.5 \\ -0.3 & 1.0 & -0.4 \\ -0.6 & -0.5 & 0.9 \end{bmatrix} \sim \begin{bmatrix} 15 & 0 & -14 \\ 0 & 25 & -17 \\ 0 & 0 & 0 \end{bmatrix}$$

We can use the integer solution $[70, 51, 75]^T$ for the homogeneous system of equations. ∎

Application: Systems of Linear Differential Equations

Let $\mathbf{x} = [x_1, x_2, \ldots, x_n]$. Each x_i is interpreted as a function of a variable t. Often, t is *time* in a concrete problem. The notation x_i' signifies the derivative of x_i with respect to t, and it is therefore a rate of change in the variable x_i. Naturally, we write

$$\mathbf{x}' = [x_1', x_2', \ldots, x_n'] \qquad \text{or} \qquad \mathbf{x}'(t) = [x_1'(t), x_2'(t), \ldots, x_n'(t)]$$

Now let $\mathbf{A}$ be an $n \times n$ matrix. A system of n linear differential equations with n variables then can be written simply as $\mathbf{x}' = \mathbf{A}\mathbf{x}$. The ith equation in this system is

$$x_i' = a_{i1}x_1 + a_{i2}x_2 + \cdots + a_{in}x_n$$

For a down-to-earth numerical example, we can take

$$\begin{cases} x_1' = 3x_1 + 2x_2 \\ x_2' = 7x_1 - 2x_2 \end{cases} \qquad \text{or} \qquad \mathbf{x}' = \mathbf{A}\mathbf{x}, \qquad \mathbf{A} = \begin{bmatrix} 3 & 2 \\ 7 & -2 \end{bmatrix}$$

Here we see two simple differential equations, but neither can be solved alone, since the righthand sides involve both x_1 and x_2. Does it help to know that the eigenvalues of $\mathbf{A}$ are 5 and -4? At first glance there seems to be no connection between the system of first-order differential equations and the eigenvalues of $\mathbf{A}$. However, we shall use the eigenvalues of $\mathbf{A}$ to produce solutions of the system of differential equations. Even at this stage we can enunciate a general theorem—that is, one that is valid for any n.

THEOREM 6

If $\mathbf{A}\mathbf{u} = \lambda\mathbf{u}$, *then the function* $t \mapsto e^{\lambda t}\mathbf{u}$ *is a solution of the differential equation* $\mathbf{x}' = \mathbf{A}\mathbf{x}$.

PROOF The proof is trivial if $\mathbf{u} = 0$. In the other case, $\mathbf{u} \neq 0$ and we have a genuine eigenvalue and eigenvector. In the differential equation $\mathbf{x}' = \mathbf{A}\mathbf{x}$, we replace $\mathbf{x}$ by $e^{\lambda t}\mathbf{u}$, getting $\mathbf{A}(e^{\lambda t}\mathbf{u}) = \lambda e^{\lambda t}\mathbf{u}$. Now cancel the positive scalar $e^{\lambda t}$, leaving $\mathbf{A}\mathbf{u} = \lambda\mathbf{u}$, which is true. ∎

Returning to the 2×2 numerical example, we find that the eigenvalues of $\mathbf{A}$ are 5 and -4. Corresponding eigenvectors are $\mathbf{v} = [1, 1]$ and $\mathbf{w} = [2, -7]$. Therefore, the functions $t \mapsto e^{5t}\mathbf{v}$ and $t \mapsto e^{-4t}\mathbf{w}$ are solutions of the previous 2×2 system. By linearity, a more general solution is

$$t \mapsto (\alpha e^{5t}\mathbf{v} + \beta e^{-4t}\mathbf{w})$$

What recourse do we have for solving $\mathbf{x}' = \mathbf{A}\mathbf{x}$ when $\mathbf{A}$ is diagonalizable? This term refers to the factorization $\mathbf{A} = \mathbf{P}^{-1}\mathbf{D}\mathbf{P}$, in which $\mathbf{D}$ is a diagonal matrix. The differential equation is now $\mathbf{x}' = \mathbf{P}^{-1}\mathbf{D}\mathbf{P}\mathbf{x}$ or $\mathbf{P}\mathbf{x}' = \mathbf{D}\mathbf{P}\mathbf{x}$. Letting $\mathbf{y} = \mathbf{P}\mathbf{x}$, we have a familiar sort of problem: $\mathbf{y}' = \mathbf{D}\mathbf{y}$. After solving this problem for $\mathbf{y}$, we get $\mathbf{x}$ from the equation $\mathbf{x} = \mathbf{P}^{-1}\mathbf{y}$.

There is much more to the theory, and the reader can pursue this topic in books such as Sadun [2001], Nobel and Daniel [1988], and Mirsky [1990].

Epilogue: Eigensystems without Determinants

Many professional mathematicians advocate a separation of the subject of eigenvalues from determinant theory. This view of the subject is backed by the knowledge gained from numerical analysis, where determinants are almost *never* used. (They are not competitive with other algorithms in speed and accuracy.) Here, we give one example showing how determinants can be banished from eigenvalue theory. The following theorem and proof are adapted from the paper by Axler [1995], cited in the references.

THEOREM 7

Every linear operator defined on and taking values in a finite-dimensional vector space has at least one eigenvalue.

PROOF Let L be such an operator, and let $\mathbf{u}$ be a nonzero vector in the domain of L. The sequence $\{\mathbf{u}, L(\mathbf{u}), L^2(\mathbf{u}), \ldots\}$ cannot be linearly independent because the space involved here is finite dimensional. Thus, there must exist a nontrivial relationship $\sum_{i=0}^{n} a_i L^i(\mathbf{u}) = \mathbf{0}$. Let p be the polynomial defined by the equation $p(t) = \sum_{i=0}^{n} a_i t^i$. Thus, we have $p(L)\mathbf{u} = 0$. By the Fundamental Theorem of Algebra (Appendix B), this polynomial has roots (zeros) in the complex plane. (This is a deep theorem, and it would not be true if we insisted on real-valued roots.) These roots lead to a factored form for p; namely, $p(t) = c(t - r_1)(t - r_2) \cdots (t - r_n)$. Now we have

$$0 = p(L)\mathbf{u} = c(L - r_1 I)(L - r_2 I) \cdots (L - r_n I)\mathbf{u}$$

If each factor $L - r_i I$ were injective, $p(L)\mathbf{u}$ could not be $\mathbf{0}$, since we started with the nonzero vector $\mathbf{u}$. Thus, for some value of i, $L - r_i I$ is not injective, and we have a nonzero solution to the equation $L(\mathbf{x}) - r_i \mathbf{x} = \mathbf{0}$. ∎

EXAMPLE 12 The process used in the preceding proof can be illustrated in a concrete case. Let the linear transformation be $L(\mathbf{x}) = \mathbf{Ax}$, where

$$\mathbf{A} = \begin{bmatrix} 1 & 3 & 7 \\ -2 & 1 & 4 \\ 0 & 5 & -3 \end{bmatrix}$$

Find an eigenvalue of $\mathbf{A}$ by the method used in the proof of Theorem 7.

SOLUTION We select any nonzero vector for $\mathbf{u}$, such as $\mathbf{u} = [1, 2, 3]^T$. With the help of mathematical software, we find that

$$\mathbf{u} = \begin{bmatrix} 1 \\ 2 \\ 3 \end{bmatrix}, \qquad \mathbf{Au} = \begin{bmatrix} 28 \\ 12 \\ 1 \end{bmatrix}$$

$$\mathbf{A}^2\mathbf{u} = \begin{bmatrix} 71 \\ -40 \\ 57 \end{bmatrix}, \qquad \mathbf{A}^3\mathbf{u} = \begin{bmatrix} 350 \\ 46 \\ -371 \end{bmatrix}$$

We need go no further because the four points listed form a linearly dependent set. (Why?) Create a matrix $\mathbf{B}$ whose columns are the four vectors already constructed. By row reduction we can find a linear combination of the columns of $\mathbf{B}$ that is zero, in a nontrivial manner:

$$\mathbf{B} = \begin{bmatrix} 1 & 28 & 71 & 350 \\ 2 & 12 & -40 & 46 \\ 3 & 1 & 57 & -371 \end{bmatrix} \sim \begin{bmatrix} 1 & 0 & 0 & -111 \\ 0 & 1 & 0 & 19 \\ 0 & 0 & 1 & -1 \end{bmatrix} = \mathbf{C}$$

The systems $\mathbf{Bx} = \mathbf{0}$ and $\mathbf{Cx} = \mathbf{0}$ have the same solutions, and we can set the free variable equal to 1, thereby getting $\mathbf{x} = [111, -19, 1, 1]^T$. At this stage we know that

$$111\mathbf{u} - 19\mathbf{Au} + \mathbf{A}^2\mathbf{u} + \mathbf{A}^3\mathbf{u} = \mathbf{0}$$

Defining the polynomial p by the equation $p(t) = 111 - 19t + t^2 + t^3$, we can write $p(\mathbf{A})\mathbf{u} = \mathbf{0}$. Now we need the roots of the polynomial p. In MATLAB, one defines a polynomial by giving its coefficients: $p = [1, 1, -19, 111]$. Then the roots are obtained with the command $r = \text{roots}(p)$. They are

$$r_1 = -6.5223, \qquad r_2 = 2.7612 + 3.0650i, \qquad r_3 = 2.7612 - 3.0650i$$

Then

$$p(t) = (t - r_1)(t - r_2)(t - r_3), \qquad p(\mathbf{A})\mathbf{u} = (\mathbf{A} - r_1\mathbf{I})(\mathbf{A} - r_2\mathbf{I})(\mathbf{A} - r_3\mathbf{I})\mathbf{u}$$

In this example, it turns out that each factor $\mathbf{A} - r_k\mathbf{I}$ is not injective, and r_1, r_2, and r_3 are all eigenvalues of $\mathbf{A}$. ■

Mathematical Software

In Example 1, we can use mathematical software to verify the results. Here is a summary of the basic commands available in three well-known mathematical software systems:

MATLAB	Maple	Mathematica
A = [3,2;-4,9] eig(A)	with(LinearAlgebra): A := Matrix([[3,2], [-4,9]]); Eigenvalues(A);	A = {{3,2}, {-4,9}} Eigenvalues[A]

We find that the matrix has two eigenvalues, 5 and 7.

In Example 2, we can check our results using mathematical software.

MATLAB	Maple	Mathematica
A = [2,1;4,-1] [P,D] = eig(A)	with(LinearAlgebra): A := Matrix([[2,1], [4,-1]]); (DD,P) := Eigenvectors(A);	A = {{2,1},{4,-1}} {DD,P} = Eigensystem[A]

MATLAB returns the eigenvalues 3 and -2 with eigenvectors $(0.7071, 0.7071)$ and $(-0.2425, 0.9701)$, These are correct because they are scalar multiples of the simple eigenvectors $\mathbf{u} = (1, 1)$ and $\mathbf{v} = (1, -4)$ already known. In Maple and Mathematica, we obtain similar results but the eigenvector corresponding to the eigenvalue -2 has been scaled to $(-\frac{1}{4}, 1)$ and $(-1, 4)$, respectively.

In Example 4, matrix **A** is diagonalizable and can be written in the form $\mathbf{A} = \mathbf{PDP}^{-1}$. As we noted previously, this equation contains an abundance of information! To begin with, the columns of **P** are eigenvectors of **A**. Furthermore, the diagonal elements of the diagonal matrix **D** are the eigenvalues of **A**. All of this additional information about **A** is hard-won, although mathematical software seems to do it effortlessly. Here are the procedures to get eigenvalues and eigenvectors for a matrix using common mathematical software packages:

MATLAB	Maple	Mathematica
A = [1,1;0,1] [P, D] = eig(A) A*P - P*D	with(LinearAlgebra): A:= Matrix([[1,1],[0,1]]); (DD,P) := Eigenvectors(A); A.P = P.DiagonalMatrix(DD);	A = {{1,1},{0,1}} {DD,P} = Eigensystem[A] A.P = P.DiagonalMatrix[DD]

MATLAB returns the repeated eigenvalue 1 with eigenvectors $(1,0)$ and $(-1,0)$. The equation $\mathbf{AP} = \mathbf{PD}$ is essentially correct, but there is some roundoff error of magnitude 10^{-15}. Results from Maple and Mathematica are similar.

In Example 6, we can find the roots of the characteristic polynomial using either MATLAB, Maple, or Mathematica:

MATLAB	Maple
A = [1,1;-2,3] roots(poly(A))	with(LinearAlgebra): A := Matrix([[1,1],[-2,3]]); Solve(CharacteristicPolynomial(A,x));

Mathematica
A = {{1,1},{-2,3}} NSolve[CharacteristicPolynomial[A,x]==0,x]

The results from all of these mathematical software systems are the same.

Again, in Example 7, we can verify the results using mathematical software:

MATLAB	Maple
A = [1,1;-2,3] [P, D] = eig(A) A - P*D*inv(P)	with(LinearAlgebra): A := Matrix([[1.,1.],[-2.,3.]]); (DD,P) := Eigenvectors(A); A = P.DiagonalMatrix(DD[1..2],2,2). MatrixInverse(P);

```
Mathematica
A = {{1,1},{-2,3}}
{DD, P} = Eigensystem[A]
A == P.DiagonalMatrix[DD].Inverse[P]
```

MATLAB returns eigenvalues $2 \pm i$ and eigenvectors $(0.40821 \pm 0.40821i,$ $0.8165)$, which can be normalized to agree with the preceding results. With Maple and Mathematica we obtain similar results within roundoff error.

In Example 8, we have used the following MATLAB commands:

```
MATLAB
A = [0.6,0.3;-0.2,1.2];
x = [6.07;3.87];
hold on
plot(x(1),x(2),'*')
for k=1:30
   x = A*x;
   plot(x(1),x(2),'*')
end
hold off
```

These points are plotted in Figure 6.1 and marked with the symbol $*$. We can also apply the analysis that leads to the dominant term. The following MATLAB code computes the points $\mathbf{x}^{(j)}$ and plots them in Figure 6.1 using the + symbol.

```
MATLAB
hold on
y = [0.80026;1.262421];
plot(y(1),y(2),'+')
for k=1:30
   y = 1.0732.^k*[0.80026; 1.26242];
   plot(y(1),y(2),'+')
end
hold off
```

In Example 8, we can compute the long-term behavior of these points in a more efficient way. First, we compute eigenvalues and eigenvectors for the matrix **A**. The commands are as follows:

MATLAB	Maple

`A = [0.6,0.3;`
 `    -0.2,1.2]`
 `[V,D] = eig(A)`
 `A*V - V*D`
 	
 `c = V\[6.97;3.87]`
	`with(LinearAlgebra):`
	 `A := Matrix([[0.6,0.3],`
	 `             [-0.2,1.2]]);`
	 `(DD,P) := Eigenvectors(A);`
	 `Norm(MatrixInverse(P).A.P -`
	 `     DiagonalMatrix(DD),2);`
	 `c := Solve[V,Vector([6.97,3.87]);`

Mathematica

`A = {{0.6,0.3},{-0.2,1.2}}`
 `{DD,P} = Eigensystem[A]`
 `A.P == P.DiagonalMatrix[DD]`
 `x = Solve(V,{{6.97,3.87}})`

We display the matrices $\mathbf{V}$ and $\mathbf{D}$ from MATLAB; they turn out to be

$$\mathbf{V} = \begin{bmatrix} -0.9211 & -0.5354 \\ -0.3893 & -0.8446 \end{bmatrix}, \qquad \mathbf{D} = \begin{bmatrix} 0.7268 & 0 \\ 0 & 1.0732 \end{bmatrix}$$

Remember, MATLAB rounds numbers for display. The full-precision values are stored, however, and can be seen by using the command `format long`. The columns of the matrix $\mathbf{V}$ are normalized eigenvectors of $\mathbf{A}$, and the diagonal elements of $\mathbf{D}$ are the eigenvalues, listed in the same order as the eigenvectors. (A vector is **normalized** by dividing it by its length, and the length of a vector $\mathbf{x}$ in $\mathbb{R}^n$ is $\sqrt{x_1^2 + x_2^2 + \cdots + x_n^2}$. These matters are explained in Section 7.1.) We can verify the work by computing $\mathbf{AV} - \mathbf{VD}$, which should be the $\mathbf{0}$-matrix. In this example with MATLAB, this matrix turns out to be

$$\begin{bmatrix} 0 & 0 \\ 0.55 \times 10^{-16} & 0 \end{bmatrix}$$

and that is "close enough for government work."

SUMMARY 6.1

- Let $\mathbf{A}$ be an $n \times n$ matrix. A number λ is an **eigenvalue** of $\mathbf{A}$ if the equation $\mathbf{Ax} = \lambda\mathbf{x}$ is true for some nonzero vector $\mathbf{x}$. Then the vector $\mathbf{x}$ is an **eigenvector** associated with the eigenvalue λ. (Here the eigenvalue and eigenvector may be complex.)

- The polynomial $p(\lambda) = \text{Det}(\mathbf{A} - \lambda\mathbf{I})$ of degree n in the variable λ is the **characteristic polynomial** of $\mathbf{A}$, and $\text{Det}(\mathbf{A} - \lambda\mathbf{I}) = 0$ is the **characteristic equation**.

- The characteristic polynomial has the form $\text{Det}(\mathbf{A} - \lambda\mathbf{I}) = \pm(\lambda - \lambda_1)^{k_1}(\lambda - \lambda_2)^{k_2}$ $(\lambda - \lambda_3)^{k_3} \cdots (\lambda - \lambda_m)^{k_m}$, where $k_1 + k_2 + k_3 + \cdots + k_m = n$. Each eigenvalue λ_j has **algebraic multiplicity** k_j and if $\text{Dim(Null}(\mathbf{A} - \lambda_j\mathbf{I})) = \ell_j$, then the **geometric multiplicity** is ℓ_j. Moreover, $k_j \geq \ell_j$.

- If $\mathbf{A} = \mathbf{PDP}^{-1}$, then $\mathbf{A}^k = \mathbf{PD}^k\mathbf{P}^{-1}$ and this is true for any invertible $\mathbf{P}$. If $\mathbf{D}$ is a diagonal matrix, then $\mathbf{A}^k = \mathbf{P}\cdot\text{Diag}(d_1^k, d_2^k, \ldots, d_n^k)\mathbf{P}^{-1}$ for $k \geq 1$.

- Suppose that the dynamical system $\mathbf{x}^{(k)} = \mathbf{A}\mathbf{x}^{(k-1)}$ has eigenvalues of $\mathbf{A}$ that are ordered $|\lambda_1| > |\lambda_2| \geq \cdots \geq |\lambda_n|$ and eigenvectors $\mathbf{v}^{(1)}, \mathbf{v}^{(2)}, \ldots, \mathbf{v}^{(n)}$. Express the given initial vector as $\mathbf{x}^{(0)} = \sum_{j=1}^{n} c_j\mathbf{v}^{(j)}$ for suitable coefficients c_j. Then the dominant term is $\mathbf{x}^{(k)} \approx \lambda_1^k c_1\mathbf{v}^{(1)}$.

- Theorems and Corollaries:

  - If $L(\mathbf{v}) = \lambda\mathbf{v}$ and $\mathbf{v} \neq \mathbf{0}$, then λ is an **eigenvalue** and $\mathbf{v}$ is an **eigenvector** of the linear operator defined on any vector space.

- $\mathbf{A}$ is **diagonalizable** if the equation $\mathbf{P}^{-1}\mathbf{A}\mathbf{P} = \mathbf{D}$ is true, where $\mathbf{D}$ is a diagonal matrix and $\mathbf{P}$ is an invertible matrix.

- A scalar λ is an eigenvalue of a matrix $\mathbf{A}$ if and only if $\text{Det}(\mathbf{A} - \lambda\mathbf{I}) = 0$.

- The geometric multiplicity of an eigenvalue cannot exceed its algebraic multiplicity.

- Any set of eigenvectors corresponding to distinct eigenvalues of a matrix is linearly independent.

- If an $n \times n$ matrix has n distinct eigenvalues, then it is diagonalizable.

- Every polynomial of degree 1 or more has a root in the complex field.

- If each column of a square matrix $\mathbf{A}$ sums to 1, then 1 is an eigenvalue of $\mathbf{A}$, and the system $\mathbf{A}\mathbf{x} = \mathbf{x}$ has a nontrivial solution.

- Every linear operator defined on and taking values in a finite-dimensional vector space has at least one eigenvalue.

KEY CONCEPTS 6.1

Eigenvalues and eigenvectors for matrices and linear transformations, diagonalizable, basis of eigenvectors, powers of a matrix, characteristic polynomial, characteristic equation, dynamical systems, economic models, eigensystems without determinants

GENERAL EXERCISES 6.1

1. What are the eigenvalues of these matrices:

$$\mathbf{A} = \begin{bmatrix} 1.3 & 0 \\ 2.8 & -7.4 \end{bmatrix}, \qquad \mathbf{B} = \begin{bmatrix} 3 & 2 \\ -4 & 9 \end{bmatrix}$$

Show the calculations for the matrix $\mathbf{B}$ and, if possible, check your answers using MATLAB or similar software.

2. Carry out the diagonalization of matrices:

$$\begin{bmatrix} 7 & 2 \\ 0 & 5 \end{bmatrix}, \begin{bmatrix} 11 & -6 \\ 4 & 1 \end{bmatrix}$$

3. Let $\mathbf{A} = \begin{bmatrix} a & b & c \\ a & b & c \\ a & b & c \end{bmatrix}$

Find the eigenvalues and eigenvectors of **A**. (Assume $abc \neq 0$.)

4. Let $\mathbf{A} = \begin{bmatrix} 0 & 1 \\ -1 & 0 \end{bmatrix}$

Find the eigenvalue–eigenvector pairs. Explore the geometric effect of letting $\mathbf{x}^{(k)} = \mathbf{Ax}^{(k-1)}$ and $k = 0, 1, 2, \ldots$.

5. The matrix $\mathbf{A} = \begin{bmatrix} 0 & -1 \\ 1 & 0 \end{bmatrix}$

represents a rotation of $90°$ counterclockwise. Obviously, **Ax** is never a simple multiple of **x** (if $\mathbf{x} \neq \mathbf{0}$), because the vectors **x** and **Ax** point in different directions. However, the equation $\mathbf{Ax} = \lambda\mathbf{x}$ must have nontrivial solutions because the characteristic polynomial must have roots. Explain this apparent contradiction. Also, find the diagonalization of **A** and check your work.

6. Find the diagonalization of $\begin{bmatrix} 1 & 0 \\ 0 & 0 \end{bmatrix}$
Check your work.

7. Show why the conjugate of $a + ib$ is not necessarily $a - ib$. (Give some examples.)

8. Find the eigenvalues of $\begin{bmatrix} 1 & 2 & 1 \\ 0 & 1 & 0 \\ 1 & 3 & 1 \end{bmatrix}$

9. Find a basis for $\mathbb{R}^2$ consisting of eigenvectors of the matrix $\begin{bmatrix} 11 & -6 \\ 4 & 1 \end{bmatrix}$

10. Criticize this argument: With a sequence of row operations, we establish that

$$\mathbf{A} = \begin{bmatrix} 1 & 3 & 5 \\ 4 & -9 & 2 \\ 3 & 6 & -5 \end{bmatrix} \sim \begin{bmatrix} 1 & 3 & 5 \\ 0 & 3 & 20 \\ 0 & 0 & -122 \end{bmatrix} = \mathbf{B}$$

Therefore, the eigenvalues of **A** are 1, 3, and -122.

11. By examining the column sums, prove that 3 is an eigenvalue of the matrix

$$\begin{bmatrix} 7 & 2 & 5 & -3 \\ 5 & -6 & -8 & 9 \\ -1 & 3 & 5 & 12 \\ -8 & 4 & 1 & -15 \end{bmatrix}$$

12. (Practice with complex numbers.) Find the real and imaginary parts of these complex numbers:
 a. $(5 + 2i)(3 - 4i)$
 b. $(5 + 2i)/(3 - 4i)$
 c. $(5 + 2i)/|3 - 4i|$
 d. $(3 - 5i)(2 + i)/(4 + 3i)$

13. Find all the values of λ for which this equation has nontrivial solutions:

$$\begin{bmatrix} 2 & 3 \\ 1 & 4 \end{bmatrix} \begin{bmatrix} x_1 \\ x_2 \end{bmatrix} = \lambda \begin{bmatrix} 1 & -2 \\ 3 & -1 \end{bmatrix} \begin{bmatrix} x_1 \\ x_2 \end{bmatrix}$$

14. For what values of α will the matrix

$$\mathbf{A} = \begin{bmatrix} 3 & \alpha \\ 7 & 2 \end{bmatrix}$$

have real eigenvalues?

15. Find the eigenvalues of $\mathbf{A} = \begin{bmatrix} 3 & 2 \\ -4 & 9 \end{bmatrix}$

For each eigenvalue, find an eigenvector. Find a nonzero matrix **P** such that $\mathbf{AP} = \mathbf{PD}$, where **D** is a diagonal matrix. Give **D** explicitly. Verify your work with independent calculations.

16. Let $\mathbf{A} = \begin{bmatrix} 2 & 4 & 6 \\ 0 & -3 & 5 \\ 0 & 0 & 1 \end{bmatrix}$

Find all the eigenvalues of **A**. For each eigenvalue, find an eigenvector.

17. Let $A = \begin{bmatrix} 2 & 1 \\ 4 & -1 \end{bmatrix}$

Find the characteristic polynomial. Find the eigenvalues. Find an eigenvector for each eigenvalue. Find a matrix P such that $P^{-1}AP$ is a diagonal matrix.

18. Let $A = \begin{bmatrix} 1 & 1 & -1 \\ 0 & 1 & 1 \\ 0 & 0 & 1 \end{bmatrix}$

Find the eigenvalues of A. Find the eigenspaces for A. Give the algebraic and geometric multiplicities of the eigenvalues. Find matrices P and D that make the equation $AP = PD$ nontrivially true. Here D should be a 3×3 diagonal matrix, with the eigenvalues on its diagonal. Also, P should be a 3×3 matrix. *Note:* We are not claiming that P is invertible. There can be many answers for P.

19. In Example 3, find an eigenvector corresponding to the eigenvalue 7.

20. (Continuation.) What is the diagonalization of the matrix in the preceding exercise?

21. Let $A = \begin{bmatrix} a & b \\ -b & a \end{bmatrix}$

where a and b are arbitrary real numbers.
a. Find the eigenvalue–eigenvector pairs.
b. Show that $\begin{bmatrix} a & b \\ -b & a \end{bmatrix} = \begin{bmatrix} r & 0 \\ 0 & r \end{bmatrix} \times$

$\begin{bmatrix} \cos\theta & -\sin\theta \\ \sin\theta & \cos\theta \end{bmatrix}$ where $r = \sqrt{a^2 + b^2}$

and θ is the angle in the triangle at the origin in the complex plane with sides a, b, and hypotenuse r.
c. Discuss the geometric effect of the iteration $x^{(k)} = Ax^{(k-1)}$. Suppose that $0 < a < 1, 0 < b < 1$, and $a^2 + b^2 = 1$.
d. Verify numerically your conclusions using $a = 0.6$ and $b = 0.8$.

22. Let $A = \begin{bmatrix} a & b \\ c & d \end{bmatrix}$, where a, b, c, and d are arbitrary real numbers.
a. Establish that the eigenvalues of the matrix A are $\frac{1}{2}(a+d) \pm \sqrt{[\frac{1}{2}(a - d)]^2 + bc}$.
b. Find the necessary and sufficient condition on the matrix A in order that it have two different eigenvalues. Find the eigenvectors.
c. Explain why A has only one eigenvalue if $(a - d)^2 = -4bc$. What are the eigenvectors in this case? *Hint:* You may use mathematical software such as Maple to do the symbolic manipulations.

23. (Continuation.)
a. Find the necessary and sufficient conditions on the matrix $\begin{bmatrix} a & b \\ c & d \end{bmatrix}$ in order that its eigenvalues be real. Assume that the entries in the matrix are real numbers.
b. Explain why the eigenvalues of A are real if $bc \geq 0$.
c. Using your results, establish that a symmetric 2×2 matrix must have real eigenvalues.

24. (Continuation.) Find the necessary and sufficient conditions on the matrix $\begin{bmatrix} a & b \\ c & d \end{bmatrix}$ in order that its eigenvalues be complex. (Remember that the numbers a, b, c, and d are real.)

25. (Continuation.)
a. Find the necessary and sufficient condition on the matrix $\begin{bmatrix} a & b \\ c & d \end{bmatrix}$ in order that it *not* be diagonalizable.
b. Explain what happens when $a = d$ and $bc \neq 0$.
c. Give an example of a nondiagonalizable 2×2 matrix.

26. Assume that A is a 2×2 matrix whose elements are integers. Find the necessary and sufficient conditions on the entries of the matrix in order that the eigenvalues of A be integers.

27. Find all the 2×2 matrices that have 1 as one of their eigenvalues.

28. Find all the 2×2 matrices that have eigenvalues 3 and 7.

29. Explain why a square matrix is invertible if and only if 0 is not one of its eigenvalues.

30. Let A be an $n \times n$ matrix whose rows all sum to the same value. What can you say about eigenvalues and eigenvectors of A?

31. Establish that if A is invertible and if λ is an eigenvalue of A, then λ^{-1} is an eigenvalue of A^{-1}.

32. Argue that the eigenvalues of a matrix A are also eigenvalues of A^T. Do A and A^T have the same characteristic polynomial? (Explain why or give a counterexample.)

33. Verify that similar matrices (as defined in Section 5.3.) have the same eigenvalues.

34. Substantiate that if an $n \times n$ matrix has n different eigenvalues, then the matrix is diagonalizable.

35. Consider a dynamical system where

$$\mathbf{x}^{(r)} = A\mathbf{x}^{(r-1)} \quad \text{and} \quad A = \begin{bmatrix} \frac{1}{2} & \frac{1}{3} \\ \frac{1}{2} & \frac{2}{3} \end{bmatrix}$$

Explain why the sum of the two components of $\mathbf{x}^{(r)}$ is a constant; that is, is independent of r. (Here each $\mathbf{x}^{(r)}$ is a vector in $\mathbb{R}^2$.)

36. Let A be an upper triangular matrix whose diagonal elements are all different from one

another. Establish that if B is invertible, then BAB^{-1} is diagonalizable.

37. Explain why the hypothesis $\{\mathbf{u}_1, \mathbf{u}_2, \ldots, \mathbf{u}_n\}$ is a linearly independent set in $\mathbb{R}^n$ leads to the conclusion that there is an $n \times n$ invertible matrix A having the vectors $\mathbf{u}_i$ as eigenvectors. Can the eigenvalues also be assigned freely?

38. All eigenvalues are real for a **symmetric** matrix $(A = A^T)$. Explain why and give some numerical examples.

39. All eigenvalues are purely imaginary or zero for a **skew-symmetric** matrix $(A = -A^T)$. Explain why and give some examples.

40. Argue that if the $n \times n$ matrix A has n different real eigenvalues, then there is a basis for $\mathbb{R}^n$ consisting of eigenvectors of A.

41. For each of these 2×2 matrices, find its eigenvalues and eigenvectors as well as the algebraic multiplicity of the eigenvalues and geometric multiplicity of the eigenvectors.

a. $\begin{bmatrix} 4 & 1 \\ 1 & 4 \end{bmatrix}$ **b.** $\begin{bmatrix} 4 & 1 \\ 1 & 2 \end{bmatrix}$ **c.** $\begin{bmatrix} 3 & 1 \\ -1 & 2 \end{bmatrix}$

d. $\begin{bmatrix} 4 & 1 \\ -1 & 2 \end{bmatrix}$ **e.** $\begin{bmatrix} 2 & 2 \\ 2 & -1 \end{bmatrix}$

f. $\begin{bmatrix} 0 & 1 \\ -1 & 0 \end{bmatrix}$ **g.** $\begin{bmatrix} 1 & 1 \\ 0 & 1 \end{bmatrix}$

42. There are sixteen 2×2 matrices with elements from the set $\{0, 1\}$. How many of these matrices do not have eigenvalues from the same set? What are these matrices and their eigenvalues?

43. Let λ be an eigenvalue of an $n \times n$ matrix A, and define $V = \{\mathbf{x} : A\mathbf{x} = \lambda\mathbf{x}\}$. This set is called an **eigenspace** of A. Establish in detail that V is a subspace of $\mathbb{R}^n$. Is each vector in V an eigenvector of A?

44. **a.** In Figure 6.1, some of the points arising from the dynamical system in Example 8 are shown. How can we plot a curve instead of a finite set of points in this example? There may be many answers. Find the best one.

 b. In Figure 6.2, some of the points arising from the dynamical system in Example 9 are shown. How can we plot a curve instead of a finite set of points in this example? There may be many answers. Find the best one.

45. If possible, find a 2×2 matrix for which the corresponding dynamical system has these three initial points: $\mathbf{x}^{(0)} = (1,1)$, $\mathbf{x}^{(1)} = (3,0)$, and $\mathbf{x}^{(2)} = (3,-3)$.

46. Find the eigenvalues and eigenvectors of these matrices:

 a. $\begin{bmatrix} 33 & 16 & 72 \\ -24 & -10 & -57 \\ -8 & -4 & -17 \end{bmatrix}$

 b. $\begin{bmatrix} 8 & -1 & -5 \\ -4 & 4 & -2 \\ 18 & -5 & -7 \end{bmatrix}$

 c. $\begin{bmatrix} 4 & 1 & 1 \\ 2 & 4 & 1 \\ 0 & 1 & 4 \end{bmatrix}$

 d. $\begin{bmatrix} 2 & -i & 0 \\ i & 2 & 0 \\ 0 & 0 & 3 \end{bmatrix}$

47. Establish that the eigenvalues of a triangular matrix are the numbers on its diagonal. Explain why the eigenvalues of a matrix cannot be computed by first using row-reduction techniques to get a triangular matrix, and then reading off the diagonal elements.

48. Explain why $(\mathrm{Diag}[d_1, d_2, \ldots, d_n])^k = \mathrm{Diag}(d_1^k, d_2^k, \ldots, d_n^k)$. Here, $\mathrm{Diag}(d_1, d_2,$ $\ldots, d_n)$ is the diagonal matrix having the given numbers on its diagonal.

49. Let $\mathbf{A} = \begin{bmatrix} a & b+c \\ b-c & -a \end{bmatrix}$ for $a^2 + b^2 \geq c^2$. Show that the eigenvalues of $\mathbf{A}$ are real.

50. Let $\mathbf{A} = \begin{bmatrix} a & b \\ 0 & c \end{bmatrix}$ where $a \neq c$ and $b \neq 0$. Determine $\mathbf{A}^k$ and verify for $k = 1, 2, 3, 4$.

51. If $\mathbf{A}^2 = \mathbf{0}$, what are the eigenvalues of $\mathbf{A}$? Assume $\mathbf{A} \neq \mathbf{0}$.

52. Affirm that if $\mathbf{AP} = \mathbf{PD}$, where $\mathbf{D}$ is a diagonal matrix, and $\mathbf{P}$ is invertible, then the columns of $\mathbf{P}$ are eigenvectors of $\mathbf{A}$ and the diagonal elements of $\mathbf{D}$ are eigenvalues of $\mathbf{A}$.

53. Confirm that if $\mathbf{A}$ is any $n \times n$ matrix, then its characteristic polynomial has as its term of highest degree $\pm \lambda^n$.

54. Let $\{\mathbf{u}_1, \mathbf{u}_2\}$ be a linearly independent pair of vectors in $\mathbb{R}^2$. Let arbitrary real numbers λ_1 and λ_2 be given. Does there necessarily exist a 2×2 matrix $\mathbf{A}$ such that $\mathbf{Au}_i = \lambda_i \mathbf{u}_i$ for $i = 1, 2$?

55. The **generalized eigenvalue problem** asks for the values of λ for which $\mathbf{A} - \lambda \mathbf{B}$ is noninvertible. To begin with, prove that $\mathbf{AV} = \mathbf{BVD}$ where $\mathbf{V}$ is a full matrix with columns containing the generalized eigenvectors and $\mathbf{D}$ is a diagonal matrix containing the generalized eigenvalues on the main diagonal. Recall that for the eigenvalue exercise $\mathbf{AV} = \mathbf{VD}$. (For a fixed pair of matrices, the set $\{\mathbf{A} - \lambda \mathbf{B} : \lambda \in \mathbb{C}\}$ is called a **pencil** of matrices.)

56. (Continuation.) If $\mathbf{B}$ is invertible, then the generalized eigenvalue problem described in the preceding exercise can be turned into the equivalent ordinary eigenvalue exercise:

Det($\mathbf{B}^{-1}\mathbf{A} - \lambda\mathbf{I}$) = 0. Verify this assertion. Use this procedure to find the eigenvalues λ that make $\mathbf{A} - \lambda\mathbf{B}$ noninvertible, when $\mathbf{A} = \begin{bmatrix} 2 & 3 \\ 1 & 5 \end{bmatrix}$ and $\mathbf{B} = \begin{bmatrix} 1 & 2 \\ 3 & 7 \end{bmatrix}$.

57. (Continuation.) Solve the generalized eigenvalue problem $\mathbf{A} = \lambda\mathbf{B}$, where

$$\mathbf{A} = \begin{bmatrix} 1 & 3 \\ 2 & 7 \end{bmatrix}, \qquad \mathbf{B} = \begin{bmatrix} 1 & 1 \\ 1 & 1 \end{bmatrix}$$

Here, one expects only to make $\mathbf{A} - \lambda\mathbf{B}$ noninvertible. Repeat for $\mathbf{B} = \begin{bmatrix} 1 & 2 \\ 1 & 1 \end{bmatrix}$.

58. In Example 8, we discarded a term that eventually became negligible. This was the term $-6.6981(.7268)^k$. The retained (dominant) term was $-1.4947(1.0732)^k$. For what values of k will the *dominant* term be greater than 10 times the *negligible* term?

59. Let $\mathbf{A}$ and $\mathbf{B}$ be diagonal matrices of the same size. Find a simple formula for $\mathbf{AB}$ and its inverse, if it has one.

60. (Micro-Research Project.) In a dynamical system of the type considered in this section, we have an equation $\mathbf{x}^{(k)} = \mathbf{A}\mathbf{x}^{(k-1)}$ and a starting value $\mathbf{x}^{(0)}$. How can we estimate the sensitivity of $\mathbf{x}^{(200)}$ to small changes in $\mathbf{x}^{(0)}$?

61. In the dynamical systems of Examples 8 and 9, explain why successive points become farther apart as they recede from the origin.

62. Explain why or find a counterexample: Every 2×2 matrix is similar to its transpose.

63. Refer to Example 8 and Figure 6.1. Does the curve have two asymptotes? If so, find the formulas for the asymptotes and explain the phenomenon.

64. Let $\mathbf{A}$ be an $n \times n$ matrix, all of whose rows are the same. Find two eigenvalues of $\mathbf{A}$.

65. (Micro-Research Project.) Can a two-dimensional dynamical system produce points lying on a circle? Answer the same question for an ellipse, a parabola, and a hyperbola.

66. Explain why if $\mathbf{A} = \mathbf{PBP}^{-1}$, then $\mathbf{A}^k = \mathbf{PB}^k\mathbf{P}^{-1}$ for $k = 1, 2, 3, \ldots$.

67. We know that every square matrix has an eigenvalue and a corresponding nonzero eigenvector. The simplest example of a linear transformation that has no eigenvalues is outlined here. Let $\mathbb{P}$ be the vector space of all polynomials. Define a linear transformation L by the equation $(Lp)(t) = tp(t)$. Complete the argument by assuming that L has an eigenvalue and a corresponding nonzero eigenvector.

68. All eigenvalues have absolute value one for an **orthogonal** matrix ($\mathbf{AA}^T = \mathbf{I}$). Explain why and give some examples.

69. Let $\mathbf{A}$ be an $n \times n$ matrix, and let $\mathbf{u} = [1, 1, 1, \ldots, 1]^T$. If $\mathbf{A}^T\mathbf{u} = \mathbf{u}$, then the equation $\mathbf{A}\mathbf{x} = \mathbf{x}$ has a nontrivial solution. Explain.

70. Let $\mathbf{A} = \begin{bmatrix} 0 & 0 & 1 \\ 0 & 2 & 0 \\ 0 & 0 & 2 \end{bmatrix}$

Find all the eigenvalues. Find all the eigenvectors. What are the algebraic and geometric multiplicities? Find the matrices such that $\mathbf{AP} = \mathbf{PD}$, where $\mathbf{D}$ is a diagonal matrix. Determine the matrices in the diagonalization factorization $\mathbf{A} = \mathbf{PDP}^{-1}$. Give a general formula for $\mathbf{A}^k$.

71. In Example 7, find the diagonalization equation $\mathbf{A} = \mathbf{PDP}^{-1}$, when the order of the entries in the diagonal matrix $\mathbf{D}$ is reversed.

72. Suppose $\mathbf{x}$ and $\mathbf{y}$ are eigenvectors of $\mathbf{A}$ associated with λ. Explain why all $a\mathbf{x} + b\mathbf{y}$, for scalars a and b except when $a\mathbf{x} + b\mathbf{y} = \mathbf{0}$, are eigenvectors associated with λ.

73. All eigenvalues are positive for a **positive definite** matrix: $\mathbf{x}^T \mathbf{A} \mathbf{x} > 0$ for all $\mathbf{x} \ne \mathbf{0}$. Explain why and give some numerical examples.

74. Show that the **Fibonacci sequence** $x_0 = 0$, $x_1 = 1$, and $x_{k+1} = x_k + x_{k-1}$ for $k \ge 1$ can be written as

$$\begin{bmatrix} x_{k+1} \\ x_k \end{bmatrix} = \mathbf{A} \begin{bmatrix} x_k \\ x_{k-1} \end{bmatrix} = \mathbf{A}^k \begin{bmatrix} x_1 \\ x_0 \end{bmatrix}$$

where $\mathbf{A} = \begin{bmatrix} 1 & 1 \\ 1 & 0 \end{bmatrix}$.

Determine $\mathbf{A}^k$.

75. In Example 5, verify the results for $k = 2$ and $k = 3$.

76. In Example 7, find an expression for $\mathbf{A}^k$ as a single matrix and verify that it is correct for $k = 1$ and 2.

77. Let $\mathbf{A}$ be an $n \times n$ real matrix, where n is odd. Explain why $\mathbf{A}$ must have at least one real eigenvalue.

COMPUTER EXERCISES 6.1

1. Let $\mathbf{A} = \begin{bmatrix} 0 & 1 & 0 \\ 0 & 0 & 1 \\ 4 & -17 & 8 \end{bmatrix}$, $\mathbf{B} = \begin{bmatrix} 6 & 5 & -5 \\ 2 & 6 & -2 \\ 2 & 5 & -1 \end{bmatrix}$

$$\mathbf{C} = \begin{bmatrix} 6 & 8 & -6 \\ 3 & -2 & -3 \\ 1 & 2 & -1 \end{bmatrix}$$

For each of these three matrices, determine the eigenvalues and eigenvectors. Find the diagonal matrix containing the eigenvalues. Find the matrix whose columns are unit eigenvectors. Verify the equation $\mathbf{A} = \mathbf{P}\mathbf{D}\mathbf{P}^{-1}$ and the two similar equations for $\mathbf{B}$. In verifying the answers (if you are using MAT-LAB) first use the command `format long` to get the values to 15-digit accuracy from that point on. Use the eigenvalue test to determine whether any of these three matrices are noninvertible. How do you account for the fact that the eigenvalues are integers? Is that phenomenon to be expected whenever the given matrix has integer entries?

2. Using the matrix $\mathbf{A} = \begin{bmatrix} 0.4 & 0.4 & -0.5 \\ 0.4 & 0.5 & 0.6 \\ -1.1 & -0.3 & -0.8 \end{bmatrix}$

study the dynamical system defined by the equations $\mathbf{x}^{(0)} = (1, 1, 1)$ and $\mathbf{x}^{(k+1)} = \mathbf{A}\mathbf{x}^{(k)}$.

a. Are the successive points tending to lie on a line? If so, what is the equation of the line (in $\mathbb{R}^3$)? You will need the eigenvalues and eigenvectors of $\mathbf{A}$ to answer these questions.

b. Can this dynamical system be run *backward* to get $\mathbf{x}^{(-1)}$, $\mathbf{x}^{(-2)}$ and so forth? If so, give the formula by which we compute $\mathbf{x}^{(i-1)}$ from $\mathbf{x}^{(i)}$, where $i = 0, -1, -2, \ldots$.

c. Also run the system backward to find out the long-term behavior in the negative direction. Probably 30 or 40 steps will suggest the ultimate behavior in both cases (forward and backward).

3. Write programs in MATLAB, Maple, and Mathematica for Example 5 involving powers of a matrix. Verify the results by carrying out the calculation with matrix–matrix multiplications and by using the formula derived.

4. Write programs for Example 9 and compare the results of (1) repeated matrix–vector multiplication and of (2) an analysis involving just the dominant term. Plot the results.

5. Use Maple or Mathematica to compute the eigenvalues of a general matrix $\begin{bmatrix} a & b \\ c & d \end{bmatrix}$.

   Show that these results agree with your direct algebraic calculations done by hand.

6. (Continuation.) Use Maple or Mathematica to compute the eigenvectors of a general matrix $\begin{bmatrix} a & b \\ c & d \end{bmatrix}$ of order two. Show that these results agree with your direct algebraic calculations done by hand.

7. Political affiliations change from one generation to the next. We consider three parties, labeled $D, R,$ and S (not necessarily Democrats, Republicans, and Socialists). The following table shows how party membership shifts from one generation to the next. For example, membership in S goes 40% to S, 10% to R, and 50% to D in one generation. What is the long-term outlook for the three parties?

	D	R	S
D	0.8	0.2	0.5
R	0.1	0.8	0.1
S	0.1	0.0	0.4

8. Carry out the complete analysis of the dynamical system in Example 9. Find the eigenvalues and eigenvectors. Express the terms in the dynamical system in the manner of Example 8. When does $\mathbf{x}^{(k)}$ enter the fourth quadrant? Solve this analytically, if possible (that is, not experimentally). This is rather difficult. You may find the Euler equation helpful: $x + iy = re^{i\theta}$, where $r = |x + iy|$ and $e^{i\theta} = \cos\theta + i\sin\theta$.

9. Suppose that a dynamical system is governed by the equation $\mathbf{x}^{(k)} = \mathbf{Ax}^{(k-1)}$, for $k = 1, 2, \ldots$, where the parameter k is often measured in discrete units (seconds, hours, days, years, etc.). Usually, we are interested in letting $t \to \infty$. However, if $\mathbf{A}$ is invertible, we can let t go to $-\infty$. The equation $\mathbf{x}^{(k)} = \mathbf{Ax}^{(k-1)}$ becomes $\mathbf{x}^{(k-1)} = \mathbf{A}^{-1}\mathbf{x}^{(k)}$, and we let $k = 0, -1, -2, \ldots$. This simple idea opens the possibility of seeing into the past. Carry out this strategy on the system in Example 9.

10. Compute the eigenvalues and eigenvectors of these matrices:

   a. $\begin{bmatrix} 5 & 4 & 1 & 1 \\ 4 & 5 & 1 & 1 \\ 1 & 1 & 4 & 2 \\ 1 & 1 & 2 & 4 \end{bmatrix}$
   b. $\begin{bmatrix} 6 & 4 & 4 & 1 \\ 4 & 6 & 1 & 4 \\ 4 & 1 & 6 & 4 \\ 1 & 4 & 4 & 6 \end{bmatrix}$

   c. $\begin{bmatrix} -2 & 2 & 2 & 2 \\ -3 & 3 & 2 & 2 \\ -2 & 0 & 4 & 2 \\ -1 & 0 & 0 & 5 \end{bmatrix}$
   d. $\begin{bmatrix} 6 & -3 & 4 & 1 \\ 4 & 3 & 3 & 0 \\ 4 & -2 & 3 & 1 \\ 4 & 2 & 3 & 1 \end{bmatrix}$

   e. $\begin{bmatrix} 4 & -5 & 0 & 3 \\ 0 & 4 & -3 & -5 \\ 5 & -3 & 4 & 0 \\ 3 & 0 & 5 & 41 \end{bmatrix}$

11. For the following matrices $\mathbf{A}$, plot the trajectories for the dynamical systems $\mathbf{x}^{(k+1)} = \mathbf{A}\mathbf{x}^{(k)}$ using a variety of different values for the starting point $\mathbf{x}^{(0)}$:

a. $\begin{bmatrix} 4 & 1 \\ 1 & 4 \end{bmatrix}$ b. $\begin{bmatrix} 0.65 & -0.15 \\ -0.15 & 0.65 \end{bmatrix}$

c. $\begin{bmatrix} 1 & 0.5 \\ 1 & 0.5 \end{bmatrix}$

12. For the following matrices $\mathbf{A}$, plot the trajectories for the dynamical systems $\mathbf{x}^{(k+1)} = \mathbf{A}\mathbf{x}^{(k)}$ starting with $\mathbf{x}^{(0)} = (4, 4)$:

a. $\begin{bmatrix} 0.2 & -1.2 \\ 0.6 & 1.4 \end{bmatrix}$ b. $\begin{bmatrix} 0.5 & -0.5 \\ 0.5 & 0.5 \end{bmatrix}$

If a real 2×2 matrix $\mathbf{A}$ has complex eigenvalues $\lambda = a \pm bi$, then the trajectories of the dynamical system $\mathbf{x}^{(k+1)} = \mathbf{A}\mathbf{x}^{(k)}$ lie on a closed orbit if $|\lambda| = 1$ ($\mathbf{0}$ is an orbital center); spiral inward if $|\lambda| < 1$ ($\mathbf{0}$ is a spiral attractor); spiral outward if $|\lambda| > 1$ ($\mathbf{0}$ is a spiral repeller).

13. Using mathematical software in Example 12, find all the eigenvalues and eigenvectors to show that they agree with the results given in the text. What are the algebraic and geometric multiplicities?

14. By repeating Examples 8 and 9 on your computer system, reproduce the plots in Figures 6.1 and 6.2.

CHAPTER SEVEN

Inner-Product Vector Spaces

7.1 INNER-PRODUCT SPACES

> " *All the efforts of nature are only mathematical results of a small number of immutable laws.*
>
> —PIERRE-SIMON LAPLACE (1749–1827)

> *The latest authors, like the most ancient, strove to subordinate the phenomena to the laws of mathematics.* "
>
> —ISAAC NEWTON (1642–1727)

Inner-Product Spaces and Their Properties

An **inner-product space** is a linear space endowed with an additional algebraic operation, called an **inner product**. In many situations where an inner product is called for, the scalars must be allowed to be arbitrary **complex** numbers. Because every real number is at the same time a complex number, it is best to state the postulates in the complex case, as this will cover both. If a and b are real numbers, then the **conjugate** of $a + ib$ is $a - ib$. The notation is as follows: If $z = a + ib$, then $\bar{z} = a - ib$. (Further information about complex numbers will be found in Appendix B.) The formal definition of an inner-product space is

> **DEFINITION**
>
> An **inner product** on a vector space is an operation that creates, from any two vectors $\mathbf{x}$ and $\mathbf{y}$, a scalar, denoted by $\langle \mathbf{x}, \mathbf{y} \rangle$. These postulates must be fulfilled:
>
> 1. $\langle \mathbf{x}, \mathbf{x} \rangle > 0$ whenever $\mathbf{x} \neq \mathbf{0}$.
> 2. $\langle \mathbf{x}, \mathbf{y} \rangle = \overline{\langle \mathbf{y}, \mathbf{x} \rangle}$
> $\langle \mathbf{x}, \mathbf{y} \rangle = \langle \mathbf{y}, \mathbf{x} \rangle$, in the real case.
> 3. $\langle \mathbf{x} + \mathbf{y}, \mathbf{z} \rangle = \langle \mathbf{x}, \mathbf{z} \rangle + \langle \mathbf{y}, \mathbf{z} \rangle$
> 4. $\langle \alpha \mathbf{x}, \mathbf{y} \rangle = \alpha \langle \mathbf{x}, \mathbf{y} \rangle$, for any scalar α.

Many consequences flow from the postulates (or *axioms*) given in the preceding inner-product definition. Here are some immediate properties, the proofs of which are relegated to the problems:

> **INNER PRODUCT PROPERTIES**
>
> a. $\langle \mathbf{x}, \alpha \mathbf{y} \rangle = \overline{\alpha} \langle \mathbf{x}, \mathbf{y} \rangle$
> $\langle \mathbf{x}, \alpha \mathbf{y} \rangle = \alpha \langle \mathbf{x}, \mathbf{y} \rangle$, in the real case.
> b. $\langle \mathbf{x}, \mathbf{y} + \mathbf{z} \rangle = \langle \mathbf{x}, \mathbf{y} \rangle + \langle \mathbf{x}, \mathbf{z} \rangle$
> c. $\langle \sum_{i=1}^{m} \mathbf{x}_i, \mathbf{y} \rangle = \sum_{i=1}^{m} \langle \mathbf{x}_i, \mathbf{y} \rangle$
> d. $\langle \mathbf{x}, \sum_{i=1}^{m} \mathbf{y}_i \rangle = \sum_{i=1}^{m} \langle \mathbf{x}, \mathbf{y}_i \rangle$
> e. $\langle \mathbf{x} + \mathbf{y}, \mathbf{x} + \mathbf{y} \rangle = \langle \mathbf{x}, \mathbf{x} \rangle + \langle \mathbf{x}, \mathbf{y} \rangle + \langle \mathbf{y}, \mathbf{x} \rangle + \langle \mathbf{y}, \mathbf{y} \rangle$
> $\langle \mathbf{x} + \mathbf{y}, \mathbf{x} + \mathbf{y} \rangle = \langle \mathbf{x}, \mathbf{x} \rangle + 2\langle \mathbf{x}, \mathbf{y} \rangle + \langle \mathbf{y}, \mathbf{y} \rangle$, in the real case.

In $\mathbb{R}^n$, the usual or *standard* inner product is defined by

$$\langle \mathbf{x}, \mathbf{y} \rangle = x_1 y_1 + x_2 y_2 + \cdots + x_n y_n$$

EXAMPLE 1 Let $\mathbf{x} = (3, -1, 4)$ and $\mathbf{y} = (2, 5, 2)$. What is the value of the standard inner product?

SOLUTION Then $\langle \mathbf{x}, \mathbf{y} \rangle = (3)(2) + (-1)(5) + (4)(2) = 9$. ∎

In $\mathbb{R}^n$, we regard vectors as *column* vectors, and therefore, $\mathbf{x}^T$ is a **row vector** or a $1 \times n$ matrix. Likewise, $\mathbf{y}$ is a **column vector** or an $n \times 1$ matrix. The matrix product $\mathbf{x}^T \mathbf{y}$ is a 1×1 matrix, which we identify (as always) with a simple scalar. The inner product can be written in several different ways:

$$\langle \mathbf{x}, \mathbf{y} \rangle = \mathbf{x}^T \mathbf{y} = \mathbf{y}^T \mathbf{x} = \sum_{j=1}^{n} x_j y_j$$

The **standard inner product** on $\mathbb{R}^n$ is known also as the **dot product**, and we have written it as $\mathbf{x} \cdot \mathbf{y}$. The matrix product $\mathbf{x}^T \mathbf{y}$ is then the matrix form for the dot product $\mathbf{x} \cdot \mathbf{y}$. We refrain from using the dot product notation for other inner products, hoping to avoid confusion.

EXAMPLE 2 The vector space $\mathbb{C}^n$ is similar to the space $\mathbb{R}^n$, but the components of the vectors and the scalars are allowed to be complex numbers. Discuss the standard inner product in $\mathbb{C}^n$ defined by the equation

$$\langle \mathbf{x}, \mathbf{y} \rangle = x_1 \overline{y}_1 + x_2 \overline{y}_2 + \cdots + x_n \overline{y}_n$$

SOLUTION The inner product can be written in different ways:

$$\langle \mathbf{x}, \mathbf{y} \rangle = \mathbf{y}^H \mathbf{x} = \sum_{j=1}^{n} x_j \overline{y}_j$$

For the column vector $\mathbf{y} = [y_1, y_2, \ldots, y_n]^T$, the complex conjugate transpose is the row vector $\mathbf{y}^H = [\overline{y}_1, \overline{y}_2, \ldots, \overline{y}_n]$. The superscript H calls for the conjugate transpose of a matrix. (It honors the mathematician Charles Hermite; see footnote at the beginning of Chapter 8.) The Hermitian $\mathbf{A}^H$ becomes the simple transpose $\mathbf{A}^T$ for a matrix in the real case. ■

In the space $\mathbb{R}^n$, it is conventional to use the inner product given in Example 1, unless we are warned otherwise. In this setting, the inner product is the dot product, and because it is real-valued we need not concern ourselves with the complex conjugate that enters the second postulate and other calculations.

EXAMPLE 3 In $\mathbb{R}^n$, we can select a vector $\mathbf{w} = (w_1, w_2, \ldots, w_n)$ whose components are all positive, and use it to define another inner product:

$$\langle \mathbf{x}, \mathbf{y} \rangle_{\mathbf{w}} = w_1 x_1 y_1 + w_2 x_2 y_2 + \cdots + w_n x_n y_n$$

SOLUTION The positive numbers w_j are called the **weights** in this context. Why do we specify *positive constants* w_i? It is to ensure the positivity of $\langle \mathbf{x}, \mathbf{x} \rangle$, as demanded in the first postulate. (Details are left to General Exercise 2.) In some applications, it is helpful to use this more versatile type of inner product. ∎

EXAMPLE 4 What is a simple inner product in the vector space $C[a, b]$?

SOLUTION Here $C[a, b]$ consists of all the continuous real-valued functions defined on the closed interval $[a, b]$. An inner product often used in this space is

$$\langle f, g \rangle = \int_a^b f(t)g(t)\, dt$$

∎

The Norm in an Inner-Product Space

If an inner product has been specified in a vector space, we can introduce a concept of **magnitude** (or **length** or **norm**) of a vector. It is defined by the equation

$$\|\mathbf{x}\| = \sqrt{\langle \mathbf{x}, \mathbf{x} \rangle}$$

When the standard inner product is being used in the space $\mathbb{R}^n$, this norm is called the **Euclidean norm**. We can easily convince ourselves that when $n = 2$, and the Euclidean norm is used, we get the familiar length of the line segment joining the point $\mathbf{x} = (x_1, x_2)$ to the origin: $\|\mathbf{x}\| = \sqrt{x_1^2 + x_2^2}$. The Pythagorean law asserts that the square of this line segment equals the sum of the squares of the two legs of the right triangle. These legs are x_1 and x_2. Figure 7.1 illustrates the Euclidean norm in $\mathbb{R}^2$. This explanation depends on our using the standard inner product for $\mathbb{R}^2$, as given in Example 1.

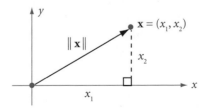

FIGURE 7.1 Euclidean norm.

Most of the examples in Chapter 7 involve the Euclidean inner product. The scalar field is usually $\mathbb{R}$ and the vectors are in $\mathbb{R}^n$. Hence, throughout this chapter, it is understood that we are using the Euclidean norm.

If one uses a more general inner product, we will have a more general concept of length or magnitude of a vector. Such generalizations are, in fact, quite important.

EXAMPLE 5a Illustrate some of the algebraic operations using these vectors: $\mathbf{x} = (3, 2, -4)$, $\mathbf{y} = (2, -5, 6)$, and $\mathbf{z} = (-2, 1, 4)$.

SOLUTION From the data, we obtain $\langle \mathbf{x}, \mathbf{y} \rangle = -28$, $\langle 3\mathbf{x}, \mathbf{y} \rangle = -84$, $\langle \mathbf{x} + \mathbf{y}, \mathbf{z} \rangle = -5$, $\|\mathbf{x}\| = \sqrt{29}$, and $\mathbf{x}/\|\mathbf{x}\| = (1/\sqrt{29})(3, 2, -4)$. ∎

A vector $\mathbf{x}$ is **normalized** by dividing it by its length $\|\mathbf{x}\|$. Here and elsewhere it is understood that $\mathbf{x}/\|\mathbf{x}\|$ means $(1/\|\mathbf{x}\|) \cdot \mathbf{x}$.

EXAMPLE 5b Continue with some examples of vectors with complex entries. Such as $\mathbf{u} = (2 + i, 4 - 3i)$ and $\mathbf{v} = (5 - 2i, 2 + 3i)$.

SOLUTION The calculations lead to $\langle \mathbf{u}, \mathbf{v} \rangle = 7 + 9i$, $\langle \mathbf{v}, \mathbf{u} \rangle = 7 - 9i$, $\langle \mathbf{u}, \mathbf{u} \rangle = 30$, $\langle \mathbf{v}, \mathbf{v} \rangle = 42$, and $\langle \mathbf{u} + \mathbf{v}, \mathbf{u} + \mathbf{v} \rangle = 86$. ∎

EXAMPLE 5c Suppose the inner product being used is weighted like this: $\langle \mathbf{x}, \mathbf{y} \rangle_\mathbf{w} = 3x_1 y_1 + 2x_2 y_2$ where $\mathbf{w} = (3, 2)$.
Let $\mathbf{x} = (5, 2)$ and $\mathbf{y} = (-1, 4)$.

SOLUTION Then $\langle \mathbf{x}, \mathbf{y} \rangle_\mathbf{w} = (3)(5)(-1) + (2)(2)(4) = 1$. ∎

Some crucial properties of the norm in an inner-product space are collected in the next theorem.[1] We are careful to use only the four postulates in

---

[1] Here is some brief information about the mathematicians whose names are attached to the mathematical entities in Theorem 1. Pythagoras of Samos (569–475 BC) is often regarded as the first pure mathematician. He is credited with the Pythagorean law for right triangles. Augustin Louis Cauchy (1789–1857) was a genius who left an immense outpouring of high-quality mathematics. Hermann Amandus Schwarz (1843–1921) was a professor of mathematics in Germany. Sometimes the name of Viktor Yakovlevich Bunyakovsky (1804–1889) is also attached to the Cauchy–Schwarz inequality because Bunyakovsky published a paper containing a version of this inequality 25 years before Schwarz.

the definition of an inner-product space and the immediate consequences of the postulates. Proceeding in this way ensures that our results are valid for all inner-product spaces, not only the familiar ones. We also profit by having greater efficiency in proofs.

THEOREM 1

In any inner-product space, the norm has these properties:

a. $\|\mathbf{x}\| > 0$ *for every nonzero vector* $\mathbf{x}$

b. $\|\alpha\mathbf{x}\| = |\alpha|\|\mathbf{x}\|$

c. $|\langle \mathbf{x}, \mathbf{y} \rangle| \leq \|\mathbf{x}\| \cdot \|\mathbf{y}\|$ **(Cauchy–Schwarz Inequality)**

d. $\|\mathbf{x} + \mathbf{y}\| \leq \|\mathbf{x}\| + \|\mathbf{y}\|$ **(Triangle Inequality)**

e. *If* $\langle \mathbf{x}, \mathbf{y} \rangle = 0$, *then* $\|\mathbf{x} + \mathbf{y}\|^2 = \|\mathbf{x}\|^2 + \|\mathbf{y}\|^2$ **(Pythagorean Theorem)**

f. $\|\mathbf{x} + \mathbf{y}\|^2 = \|\mathbf{x}\|^2 + \langle \mathbf{x}, \mathbf{y} \rangle + \langle \mathbf{y}, \mathbf{x} \rangle + \|\mathbf{y}\|^2$

$\|\mathbf{x} + \mathbf{y}\|^2 = \|\mathbf{x}\|^2 + 2\langle \mathbf{x}, \mathbf{y} \rangle + \|\mathbf{y}\|^2$, *in the real case*

PROOF Only **c** and **d** pose any difficulty. For these parts, we give the proof only in the case of the real scalar field. To prove the Cauchy–Schwarz inequality in **c**, observe that it is obviously correct when $\mathbf{y} = \mathbf{0}$. Hence, assume that $\mathbf{y}$ is not zero, and let $\mathbf{u} = \mathbf{y}/\|\mathbf{y}\|$ and $t = \langle \mathbf{x}, \mathbf{u} \rangle$. Then $\|\mathbf{u}\| = 1$ (easily verified), and

$$0 \leq \langle \mathbf{x} - t\mathbf{u}, \mathbf{x} - t\mathbf{u} \rangle = \langle \mathbf{x}, \mathbf{x} \rangle - 2t\langle \mathbf{x}, \mathbf{u} \rangle + t^2 \langle \mathbf{u}, \mathbf{u} \rangle$$
$$= \|\mathbf{x}\|^2 - 2t^2 + t^2 = \|\mathbf{x}\|^2 - t^2 = \|\mathbf{x}\|^2 - \langle \mathbf{x}, \mathbf{u} \rangle^2$$

This leads to $|\langle \mathbf{x}, \mathbf{u} \rangle| \leq \|\mathbf{x}\|$ and then to $|\langle \mathbf{x}, \mathbf{y}/\|\mathbf{y}\| \rangle| \leq \|\mathbf{x}\|$ and the Cauchy–Schwarz inequality in **c**. To prove the triangle inequality in **d**, we use the Cauchy–Schwarz inequality in **c** as follows:

$$\|\mathbf{x} + \mathbf{y}\|^2 = \langle \mathbf{x} + \mathbf{y}, \mathbf{x} + \mathbf{y} \rangle = \langle \mathbf{x}, \mathbf{x} \rangle + 2\langle \mathbf{x}, \mathbf{y} \rangle + \langle \mathbf{y}, \mathbf{y} \rangle$$
$$\leq \|\mathbf{x}\|^2 + 2\|\mathbf{x}\| \cdot \|\mathbf{y}\| + \|\mathbf{y}\|^2 = \left[\|\mathbf{x}\| + \|\mathbf{y}\| \right]^2$$

Figure 7.2 illustrates the triangle inequality in $\mathbb{R}^2$.   ∎

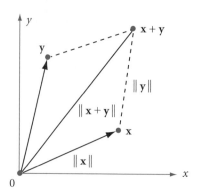

FIGURE 7.2 Triangle inequality.

To illustrate the power of Theorem 1, consider Example 4 with two continuous functions u and v. Theorem 1 applies in this situation, and we get this inequality, valid for any two continuous functions:

$$\left| \int_a^b f(t)g(t)dt \right| \le \sqrt{\int_a^b |f(t)|^2 dt} \sqrt{\int_a^b |g(t)|^2 dt}$$

Distance Function

In an inner-product space, the distance between two vectors $\mathbf{x}$ and $\mathbf{y}$ is defined to be $\|\mathbf{x} - \mathbf{y}\|$. This is the length of the line segment joining the vectors $\mathbf{x}$ and $\mathbf{y}$.

In mathematics, there is almost universal agreement that any **distance function**, d, should obey these rules:

- $d(\mathbf{x}, \mathbf{y}) > 0$ if $\mathbf{x} \ne \mathbf{y}$
- $d(\mathbf{x}, \mathbf{y}) = d(\mathbf{y}, \mathbf{x})$
- $d(\mathbf{x}, \mathbf{y}) \le d(\mathbf{x}, \mathbf{z}) + d(\mathbf{z}, \mathbf{y})$

Needless to say, our distance function, $d(\mathbf{x}, \mathbf{y}) = \|\mathbf{x} - \mathbf{y}\|$, does conform to these rules. (The reader is invited to verify this.)

A vector of norm 1 in an inner-product space is called a **unit vector**. If $\mathbf{x}$ is a nonzero vector, one creates a unit vector pointing in the same direction as $\mathbf{x}$ by writing $\mathbf{y} = \mathbf{x}/\|\mathbf{x}\|$. This process is called **normalization**. Once again, we remind the reader that throughout it is understood that $\mathbf{x}/\|\mathbf{x}\|$ means $(1/\|\mathbf{x}\|) \cdot \mathbf{x}$. That the new vector is of norm 1 follows from the computation $\|\mathbf{y}\| = \|(1/\|\mathbf{x}\|) \cdot \mathbf{x}\| = (1/\|\mathbf{x}\|) \cdot \|\mathbf{x}\| = 1$.

Mutually Orthogonal Vectors

> **DEFINITION**
>
> *Two vectors* $\mathbf{x}$ *and* $\mathbf{y}$ *are* **mutually orthogonal** *if* $\langle \mathbf{x}, \mathbf{y} \rangle = 0$.

The expression is often shortened by saying *the vectors are orthogonal*, but that is misleading, since orthogonality is a property of *sets* of vectors. The Pythagorean law in Theorem 1 concerns orthogonal pairs of vectors.

A set of vectors is said to be **orthogonal** if $\langle \mathbf{x}, \mathbf{y} \rangle = 0$ for every pair of different vectors $\mathbf{x}, \mathbf{y}$ chosen from the given set. For example, to prove that a set $\{\mathbf{x}, \mathbf{y}, \mathbf{z}, \mathbf{w}\}$ is orthogonal, we must verify $\langle \mathbf{x}, \mathbf{y} \rangle = 0$, $\langle \mathbf{x}, \mathbf{z} \rangle = 0$, $\langle \mathbf{x}, \mathbf{w} \rangle = 0$, $\langle \mathbf{y}, \mathbf{z} \rangle = 0$, $\langle \mathbf{y}, \mathbf{w} \rangle = 0$, and $\langle \mathbf{z}, \mathbf{w} \rangle = 0$. The other cases such as $\langle \mathbf{w}, \mathbf{y} \rangle = 0$ follow from Postulate **2** in the definition of an inner product at the beginning of this section.

If $\mathbf{x}$ is orthogonal to $\mathbf{y}$, we often write $\mathbf{x} \perp \mathbf{y}$. If $\mathbf{x}$ is orthogonal to all the vectors in a set U, we write $\mathbf{x} \perp U$. If every member of one set V is orthogonal to every member of another set U, we write $V \perp U$. (Thus, the meaning of the symbol $\perp$ shifts slightly, depending on the context.) To illustrate these matters, let $\mathbf{x} = (5, 7)$ and $\mathbf{y} = (-14, 10)$. Then $\mathbf{x} \perp \mathbf{y}$ because $\langle \mathbf{x}, \mathbf{y} \rangle = (5)(-14) + (7)(10) = 0$. Also, we obtain $\|\mathbf{x} + \mathbf{y}\|^2 = \|(-9, 17)\|^2 = 81 + 289 = 370$, while $\|\mathbf{x}\|^2 + \|\mathbf{y}\|^2 = 25 + 49 + 196 + 100 = 370$, in accordance with the Pythagorean law: $\|\mathbf{x} + \mathbf{y}\|^2 = \|\mathbf{x}\|^2 + \|\mathbf{y}\|^2$, if $\mathbf{x} \perp \mathbf{y}$.

> **THEOREM 2**
>
> *Every orthogonal set of nonzero vectors in an inner-product space is linearly independent.*

PROOF Let U be an orthogonal set of nonzero vectors in an inner-product space. Suppose that this set is linearly dependent. Then there exists a nontrivial equation $\sum_{i=1}^{n} c_i \mathbf{u}_i = \mathbf{0}$, where each $\mathbf{u}_i \in U$. We can assume that all the coefficients c_i are nonzero. Take the inner product of this vector with any of the $\mathbf{u}_j$, getting

$$\mathbf{0} = \left\langle \sum_{i=1}^{n} c_i \mathbf{u}_i, \ \mathbf{u}_j \right\rangle = \sum_{i=1}^{n} c_i \langle \mathbf{u}_i, \ \mathbf{u}_j \rangle = c_j \langle \mathbf{u}_j, \ \mathbf{u}_j \rangle$$

Since $\mathbf{u}_j$ is not zero, we conclude that $c_j = 0$, contrary to our hypotheses. ∎

> ### THEOREM 3
>
> *In $\mathbb{R}^n$, there does not exist an orthogonal set of $n + 1$ nonzero vectors.*

PROOF Since $\mathbb{R}^n$ is of dimension n, it cannot contain a linearly independent set of $n + 1$ vectors. (Recall Theorem 12 in Section 5.2.)   ∎

> ### THEOREM 4
>
> *Any orthogonal set of n nonzero vectors in $\mathbb{R}^n$ is a basis for $\mathbb{R}^n$.*

An orthogonal set of vectors, each of length 1, is said to be **orthonormal**.

> **EXAMPLE 6** Find an orthogonal basis and an orthonormal basis for $\mathbb{R}^3$ starting with these two vectors: $\mathbf{u} = (3, 5, 1)$ and $\mathbf{v} = (2, -2, 4)$.

SOLUTION We seek a nonzero vector $\mathbf{x} = (x_1, x_2, x_3)$ such that $\langle \mathbf{u}, \mathbf{x} \rangle = 0$ and $\langle \mathbf{v}, \mathbf{x} \rangle = 0$. This leads to the homogeneous system of equations

$$\begin{cases} 3x_1 + 5x_2 + x_3 = 0 \\ 2x_1 - 2x_2 + 4x_3 = 0 \end{cases}$$

Its coefficient matrix can be reduced to echelon form:

$$\begin{bmatrix} 3 & 5 & 1 \\ 2 & -2 & 4 \end{bmatrix} \sim \begin{bmatrix} 1 & 0 & 11/8 \\ 0 & 1 & -\frac{5}{8} \end{bmatrix}$$

The solution is $x_1 = (-11/8)x_3$ and $x_2 = \frac{5}{8}x_3$, where x_3 is a free variable. Use $x_3 = 8, x_2 = 5, x_1 = -11$ to form the vector $\mathbf{x} = (-11, 5, 8)$. The three columns of this matrix form an orthogonal basis for $\mathbb{R}^3$:

$$\mathbf{A} = \begin{bmatrix} \mathbf{u} & \mathbf{v} & \mathbf{x} \end{bmatrix} = \begin{bmatrix} 3 & 2 & -11 \\ 5 & -2 & 5 \\ 1 & 4 & 8 \end{bmatrix}$$

Before normalization, we have

$$\mathbf{A}^T\mathbf{A} = \begin{bmatrix} 3 & 5 & 1 \\ 2 & -2 & 4 \\ -11 & 5 & 8 \end{bmatrix} \begin{bmatrix} 3 & 2 & -11 \\ 5 & -2 & 5 \\ 1 & 4 & 8 \end{bmatrix} = \begin{bmatrix} 35 & 0 & 0 \\ 0 & 24 & 0 \\ 0 & 0 & 210 \end{bmatrix} = \mathbf{D}$$

We find $\|\mathbf{u}\|^2 = 35$, $\|\mathbf{v}\|^2 = 24$, and $\|\mathbf{x}\|^2 = 210$. The columns of the next matrix form an orthonormal basis for $\mathbb{R}^3$:

$$\mathbf{B} = \begin{bmatrix} 3/\sqrt{35} & 2/\sqrt{24} & -11/\sqrt{210} \\ 5/\sqrt{35} & -2/\sqrt{24} & 5/\sqrt{210} \\ 1/\sqrt{35} & 4/\sqrt{24} & 8/\sqrt{210} \end{bmatrix} = \mathbf{A}\mathbf{D}^{-\frac{1}{2}}$$

To verify the calculation, one notes that

$$\mathbf{A}^T\mathbf{A} = \mathbf{D} = \mathbf{D}^{\frac{1}{2}}\mathbf{D}^{\frac{1}{2}}, \quad \mathbf{D}^{-\frac{1}{2}}\mathbf{A}^T\mathbf{A}\mathbf{D}^{-\frac{1}{2}} = (\mathbf{A}\mathbf{D}^{-\frac{1}{2}})^T(\mathbf{A}\mathbf{D}^{-\frac{1}{2}}) = \mathbf{I}, \quad \mathbf{B}^T\mathbf{B} = \mathbf{I}$$

■

Orthogonal Projection

One vector $\mathbf{x}$ can be projected orthogonally onto another vector $\mathbf{y}$, provided that $\mathbf{y}$ is not zero. The idea is that the **projection** of $\mathbf{x}$ onto $\mathbf{y}$ should be a scalar multiple of $\mathbf{y}$, say $\alpha\mathbf{y}$, such that $\mathbf{x} - \alpha\mathbf{y}$ is orthogonal to $\mathbf{y}$. (See Figure 7.3.) What is the correct value of α? We want $\langle \mathbf{x} - \alpha\mathbf{y}, \mathbf{y} \rangle = 0$, or $\langle \mathbf{x}, \mathbf{y} \rangle - \alpha \langle \mathbf{y}, \mathbf{y} \rangle = 0$, and the proper value for α is then $\alpha = \langle \mathbf{x}, \mathbf{y} \rangle / \langle \mathbf{y}, \mathbf{y} \rangle$. To summarize, we have proved the following theorem.

THEOREM 5

In any inner-product space, the orthogonal projection of a vector $\mathbf{x}$ onto a nonzero vector $\mathbf{y}$ is the point

$$\mathbf{p} = \frac{\langle \mathbf{x}, \mathbf{y} \rangle}{\langle \mathbf{y}, \mathbf{y} \rangle} \mathbf{y}$$

It has the property that $\mathbf{x} - \mathbf{p}$ is orthogonal to $\mathbf{y}$. Thus, $\mathbf{x}$ is split into an orthogonal pair of vectors in the equation $\mathbf{x} = \mathbf{p} + (\mathbf{x} - \mathbf{p})$.

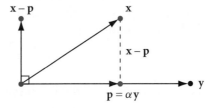

FIGURE 7.3 Orthogonal projection of **x** onto **y**.

If one forgets the formula for an orthogonal projection, it can easily be reconstructed as follows: The orthogonal projection of the vector $\mathbf{x}$ onto the vector $\mathbf{y}$ is a vector $\mathbf{p}$ that is determined by two properties (as in Figure 7.3):

- $\mathbf{p} = \alpha\mathbf{y}$ for some scalar α
- $(\mathbf{x} - \mathbf{p}) \perp \mathbf{y}$

Therefore, we obtain $\langle \mathbf{x}, \mathbf{y} \rangle = \langle \alpha\mathbf{y}, \mathbf{y} \rangle = \alpha\langle \mathbf{y}, \mathbf{y} \rangle$ and $\alpha = \langle \mathbf{x}, \mathbf{y} \rangle / \langle \mathbf{y}, \mathbf{y} \rangle$. The orthogonal projection of $\mathbf{x}$ onto $\mathbf{y}$ is given by the formula in Theorem 5.

Notice that our concept of projecting $\mathbf{x}$ onto $\mathbf{y}$ does not depend on the magnitude of the vector $\mathbf{y}$. Therefore, one must expect the formula to be insensitive to changes in the length of $\mathbf{y}$. Actually the formula can be remembered more easily as

$$\mathbf{p} = \langle \mathbf{x}, \mathbf{v} \rangle \, \mathbf{v}$$

where $\mathbf{v}$ is the normalized $\mathbf{y}$; that is, $\mathbf{v} = \mathbf{y} / \|\mathbf{y}\|$.

EXAMPLE 7 In $\mathbb{R}^3$, use the standard inner product. What is the orthogonal projection of the vector $\mathbf{x} = (1, 2, 3)$ onto the vector $\mathbf{y} = (3, 2, 1)$?

SOLUTION To answer this, we need to compute the projection as follows:

$$\mathbf{p} = \frac{\langle \mathbf{x}, \mathbf{y} \rangle}{\langle \mathbf{y}, \mathbf{y} \rangle} \, \mathbf{y} = (10/14)\mathbf{y} = \tfrac{5}{7}(3, 2, 1)$$

One can also verify that $\mathbf{x} - \mathbf{p}$ is orthogonal to $\mathbf{y}$. Thus, we obtain

$$\langle \mathbf{x} - \mathbf{p}, \mathbf{y} \rangle = (\mathbf{x} - \mathbf{p})^T\mathbf{y} = \left\langle \tfrac{4}{7}(-2, 1, 4), (3, 2, 1) \right\rangle = 0 \qquad \blacksquare$$

The calculation of an orthogonal projection can be carried out in several different ways. For example, we can begin with the point $\mathbf{z}$ that is to be projected and the matrix $\mathbf{U}$ whose columns are the vectors $\mathbf{u}_i$ previously used. The point $\mathbf{p}$ that we seek is a linear combination of the columns of $\mathbf{U}$ and is therefore of the form

$$\mathbf{p} = \mathbf{U}\mathbf{c}$$

for some unknown vector $\mathbf{c}$ in $\mathbb{R}^n$. The orthogonality condition is that $\mathbf{z} - \mathbf{p}$ should be orthogonal to all the columns of $\mathbf{U}$, or in other terms,

$$(\mathbf{z} - \mathbf{p})^T\mathbf{U} = \mathbf{0}$$

Since $\mathbf{p} = \mathbf{Uc}$, this last equation becomes

$$(\mathbf{z} - \mathbf{Uc})^T\mathbf{U} = \mathbf{0}, \qquad \mathbf{U}^T(\mathbf{z} - \mathbf{Uc}) = \mathbf{0}, \qquad \mathbf{U}^T\mathbf{Uc} = \mathbf{U}^T\mathbf{z}$$

This equation will reappear in Section 7.2, where we shall see how a well-chosen basis of W will lead to $\mathbf{U}^T\mathbf{U} = \mathbf{I}$. In this case, we obtain $\mathbf{c} = \mathbf{U}^T\mathbf{z}$. No solving of linear equations is called for, but some effort must be expended to get the matrix $\mathbf{U}$ in the first place. Theorem 9 in this section has another formalism for solving this important problem.

Angle between Vectors

A concept of angle between vectors can be introduced in any inner-product space. If $\mathbf{x}$ and $\mathbf{y}$ are two nonzero vectors, we define the angle θ between them by the formula

$$\cos\theta = \frac{\langle \mathbf{x}, \mathbf{y} \rangle}{\|\mathbf{x}\| \cdot \|\mathbf{y}\|}$$

It is understood that the angle θ should be chosen in the closed interval $[0, \pi]$. To justify this definition, observe first that if $\mathbf{x}$ and $\mathbf{y}$ are mutually orthogonal, then the angle θ between them should be $\frac{\pi}{2}$, and our formula is correct. Next, we can see whether the formula is correct in $\mathbb{R}^2$ when we use the standard inner product. Here we must call upon the **Law of Cosines** from trigonometry, which asserts that in a triangle having sides a, b, c, and opposing angles A, B, C, the formula

$$c^2 = a^2 + b^2 - 2ab \cdot \cos C$$

holds. Create a triangle having sides $\mathbf{x}$, $\mathbf{y}$, and $\mathbf{x} - \mathbf{y}$. Then in the Law of Cosines let $C = \theta$, $a = \|\mathbf{x}\|$, $b = \|\mathbf{y}\|$, and $c = \|\mathbf{x} - \mathbf{y}\|$. This produces the equation

$$\|\mathbf{x} - \mathbf{y}\|^2 = \|\mathbf{x}\|^2 + \|\mathbf{y}\|^2 - 2\|\mathbf{x}\| \cdot \|\mathbf{y}\| \cos\theta$$

Hence, we obtain

$$\|\mathbf{x}\|^2 - 2\langle \mathbf{x}, \mathbf{y} \rangle + \|\mathbf{y}\|^2 = \|\mathbf{x}\|^2 + \|\mathbf{y}\|^2 - 2\|\mathbf{x}\| \cdot \|\mathbf{y}\| \cos\theta$$

by using Theorem 1, Part **f**. When this equation is simplified, we arrive at the equation adopted previously as the definition of $\cos\theta$. One last point: How do we know that the righthand term in the definition of $\cos\theta$ is in the interval $[-1, 1]$? The Cauchy–Schwarz inequality assures us that this is true!

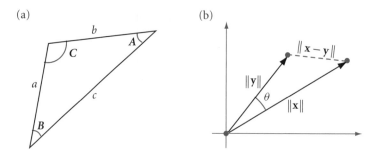

FIGURE 7.4 Law of cosines.

Figure 7.4 illustrates the Law of Cosines in elementary trigonometry. For the special case $\theta = 90° = \frac{\pi}{2}$, the Law of Cosines becomes

$$\left\|\mathbf{x} - \mathbf{y}\right\|^2 = \left\|\mathbf{x}\right\|^2 + \left\|\mathbf{y}\right\|^2$$

which is the condition of orthogonality for vectors $\mathbf{x}$ and $\mathbf{y}$.

EXAMPLE 8 What is the angle between the two vectors $\mathbf{x} = (1, 3, 7)$ and $\mathbf{y} = (-3, 2, 4)$?

SOLUTION The formula for $\cos \theta$ requires these ingredients: $\langle \mathbf{x}, \mathbf{y} \rangle = 31$, $\left\|\mathbf{x}\right\| = \sqrt{59}$, and $\left\|\mathbf{y}\right\| = \sqrt{29}$. Hence, we obtain

$$\cos \theta = \frac{\langle \mathbf{x}, \mathbf{y} \rangle}{\left\|\mathbf{x}\right\| \cdot \left\|\mathbf{y}\right\|} = \frac{31}{\sqrt{1711}} \approx 0.74944$$

and $\theta \approx 0.72358$ radians, or $41.458°$ (degrees). ■

Orthogonal Complements

DEFINITION

The **orthogonal complement** *of a set S in an inner-product space is the set*

$$\{ \mathbf{x} : \mathbf{x} \perp \mathbf{s} \quad \text{for all } \mathbf{s} \in S \} \qquad \text{or} \qquad \{ \mathbf{x} : \mathbf{x} \perp S \}$$

The orthogonal complement of S is denoted by $S^{\perp}$.

EXAMPLE 9 Is the vector $\mathbf{x} = (3, 7, 1)$ in the orthogonal complement of the pair of vectors $\{\mathbf{u}, \mathbf{v}\}$, where $\mathbf{u} = (5, -2, -1)$ and $\mathbf{v} = (2, -3, 15)$?

SOLUTION To be in the orthogonal complement of $\{\mathbf{u}, \mathbf{v}\}$, the vector $\mathbf{x}$ must be orthogonal to $\mathbf{u}$ and $\mathbf{v}$. The orthogonality is confirmed as follows:

$$\langle \mathbf{x}, \mathbf{u} \rangle = (3)(5) + (7)(-2) + (1)(-1) = 15 - 14 - 1 = 0$$

$$\langle \mathbf{x}, \mathbf{v} \rangle = (3)(2) + (7)(-3) + (1)(15) = 6 - 21 + 15 = 0 \qquad \blacksquare$$

THEOREM 6

In an inner-product space, the orthogonal complement of a set is the same as the orthogonal complement of its span.

EXAMPLE 10 Let $\mathbf{U}$ be the subspace of $\mathbb{R}^3$ whose basis consists of the two vectors $\mathbf{u} = (5, -2, -1)$ and $\mathbf{v} = (2, -3, 15)$ as in Example 9. Is the vector $\mathbf{x}$ (from that example) in the orthogonal complement of $\mathbf{U}$?

SOLUTION Because we already know that $\mathbf{x}$ is orthogonal to $\mathbf{u}$ and $\mathbf{v}$, Theorem 5 allows us to conclude that $\mathbf{x}$ is orthogonal to the span of $\mathbf{u}$ and $\mathbf{v}$, which is $\mathbf{U}$. $\qquad \blacksquare$

THEOREM 7

In an inner-product space, the orthogonal complement of any subset is a subspace.

EXAMPLE 11 Let $\mathbf{x} = (3, 7, 1)$. What is the dimension of the orthogonal complement of this single vector?

SOLUTION Our work in Example 10 shows that $\mathbf{u}$ and $\mathbf{v}$ are in this orthogonal complement. Neither $\mathbf{u}$ nor $\mathbf{v}$ is a multiple of the other. Hence, together they span a two-dimensional subspace of $\mathbb{R}^3$, which is the orthogonal complement of the single vector $\mathbf{x}$. $\qquad \blacksquare$

THEOREM 8

The null space of a matrix is the orthogonal complement of its row space:

$$\text{Null}(\mathbf{A}) = [\text{Row}(\mathbf{A})]^{\perp}$$

PROOF In the following list of assertions, we have logical equivalences:

1. $\mathbf{x}$ is in the null space of $\mathbf{A}$
2. $\mathbf{Ax} = \mathbf{0}$
3. $\mathbf{x}$ is orthogonal to the rows of $\mathbf{A}$
4. $\mathbf{x}$ is in $[\text{Row}(\mathbf{A})]^{\perp}$

Consequently, the null space of a matrix and the orthogonal complement of its row space are the same. ∎

EXAMPLE 12 Find a basis for the null space of the matrix

$$\mathbf{A} = \begin{bmatrix} 5 & -2 & -1 \\ 2 & -3 & 15 \end{bmatrix}$$

and verify that the null space is the orthogonal complement of the row space.

SOLUTION To find a simple description of the null space of this matrix, we use row reduction to solve the homogeneous equation $\mathbf{Ax} = \mathbf{0}$:

$$\begin{bmatrix} 5 & -2 & -1 \\ 2 & -3 & 15 \end{bmatrix} \sim \begin{bmatrix} 1 & 4 & -31 \\ 2 & -3 & 15 \end{bmatrix} \sim \begin{bmatrix} 1 & 4 & -31 \\ 0 & -11 & 77 \end{bmatrix}$$

$$\sim \begin{bmatrix} 1 & 4 & -31 \\ 0 & 1 & -7 \end{bmatrix} \sim \begin{bmatrix} 1 & 0 & -3 \\ 0 & 1 & -7 \end{bmatrix}$$

Thus, we obtain $x_1 - 3x_3 = 0$ and $x_2 - 7x_3 = 0$. The general solution is then

$$\begin{bmatrix} x_1 \\ x_2 \\ x_3 \end{bmatrix} = x_3 \begin{bmatrix} 3 \\ 7 \\ 1 \end{bmatrix}$$

The null space is one-dimensional and is spanned by the vector $(3, 7, 1)$. Our work in Example 9 has shown that the orthogonal complement of the row space is generated by this same vector. ∎

Examples of orthogonal sets appear among the exercises at the end of this section. Here is a very important example, lying at the heart of Fourier analysis.

EXAMPLE 13 In the space $C[0, \pi]$ (consisting of all continuous functions on the interval $[0, \pi]$), we consider the sequence of functions w_k, defined by $w_k(t) = \cos(kt)$. (Here, $k = 0, 1, 2, \ldots$.) Is this an orthogonal sequence?

SOLUTION To find out, we must compute the inner products $\langle w_k, w_j \rangle$, where $k \neq j$. Refer to Example 4 for the appropriate inner product. Using standard trigonometric formulas, we obtain

$$\langle w_k, w_j \rangle = \int_0^\pi [\cos(kt) \cos(jt)]\, dt$$

$$= \frac{1}{2} \int_0^\pi \left[\cos[(k+j)t] + \cos[(k-j)t] \right] dt$$

$$= \frac{1}{2} \left[\frac{\sin[(k+j)t]}{k+j} \right]_0^\pi + \frac{1}{2} \left[\frac{\sin[(k-j)t]}{k-j} \right]_0^\pi = 0$$

Here we have used these trigonometric identities:

$$\sin(x \pm y) = \sin x \cos y \pm \cos x \sin y$$

$$\cos x + \cos y = 2 \cos[\tfrac{1}{2}(x+y)] \cos[\tfrac{1}{2}(x-y)]$$ ∎

Orthonormal Bases

Recall the definition of **orthonormality**: A set of vectors in an inner-product space is orthonormal if it is an orthogonal set and each vector in it has length 1. To create an orthonormal set from an orthogonal set of nonzero vectors, simply *normalize* each element. Thus, every $\mathbf{u}_i$ in the given set is replaced by $\mathbf{u}_i/\|\mathbf{u}_i\|$. The condition of orthonormality is succinctly expressed by the equation $\langle \mathbf{u}_i, \mathbf{u}_j \rangle = \delta_{ij}$. (Recall that the Kronecker delta function has the properties $\delta_{ij} = 0$ when $i \neq j$ and $\delta_{ii} = 1$.)

THEOREM 9

If $\{\mathbf{u}_1, \mathbf{u}_2, \ldots, \mathbf{u}_n\}$ is an orthonormal set in an inner-product space, then each $\mathbf{x}$ in the span of that set is expressible as

$$\mathbf{x} = \sum_{i=1}^n \langle \mathbf{x}, \mathbf{u}_i \rangle \mathbf{u}_i$$

PROOF Since $\mathbf{x}$ is in the span of the orthonormal set, there exist scalars c_i such that $\mathbf{x} = \sum_{i=1}^{n} c_i \mathbf{u}_i$. Take the inner product on both sides of this equation with the vector $\mathbf{u}_j$, getting

$$\langle \mathbf{x}, \mathbf{u}_j \rangle = \left\langle \sum_{i=1}^{n} c_i \mathbf{u}_i, \mathbf{u}_j \right\rangle = \sum_{i=1}^{n} c_i \langle \mathbf{u}_i, \mathbf{u}_j \rangle = \sum_{i=1}^{n} c_i \delta_{ij} = c_j \qquad \blacksquare$$

Suppose we are working in the space $\mathbb{R}^n$, and we have an orthonormal set of n vectors, $\mathbf{u}_1, \mathbf{u}_2, \ldots, \mathbf{u}_n$. Put them into a matrix $\mathbf{U}$ as columns. The resulting matrix is square, and this property is crucial. The orthonormality now gives us the equation $\mathbf{U}^T \mathbf{U} = \mathbf{I}$. Such a matrix $\mathbf{U}$ is said to be **orthogonal**.[2] It is obviously invertible as $\mathbf{U}^T$ is its inverse. Also, we obtain $\mathbf{U}\mathbf{U}^T = \mathbf{I}$ by Theorem 4 in Section 3.2. (That theorem asserts that for square matrices, the equation $\mathbf{BA} = \mathbf{I}$ implies the equation $\mathbf{AB} = \mathbf{I}$.) It follows, from the equation $\mathbf{U}\mathbf{U}^T = \mathbf{I}$, that the rows of $\mathbf{U}$ also form an orthonormal set of vectors! This is an impressive bit of magic. We summarize these results as follows.

DEFINITION

A *real matrix* $\mathbf{U}$ *is* **orthogonal** *if* $\mathbf{U}\mathbf{U}^T = \mathbf{U}^T\mathbf{U} = \mathbf{I}$.
A *complex matrix* $\mathbf{U}$ *is* **unitary** *if* $\mathbf{U}\mathbf{U}^H = \mathbf{U}^H\mathbf{U} = \mathbf{I}$.

EXAMPLE 14 Here is a matrix whose columns form an orthogonal set:

$$\mathbf{A} = \begin{bmatrix} 4 & 1 & 1 \\ 1 & 2 & -2 \\ -2 & 3 & 1 \end{bmatrix}$$

By normalizing each column, produce an orthogonal matrix, $\mathbf{U}$. Find further interesting relations and the inverse of $\mathbf{U}$.

---

[2] The standard nomenclature here is faulty: One would expect the matrix to be called **orthonormal**, but that is not done. A complex square matrix satisfying $\mathbf{A}^H \mathbf{A} = \mathbf{I}$ is said to be **unitary**.

SOLUTION Each column must be divided by its norm, and this produces

$$\mathbf{U} = \begin{bmatrix} 4/\sqrt{21} & 1/\sqrt{14} & 1/\sqrt{6} \\ 1/\sqrt{21} & 2/\sqrt{14} & -2/\sqrt{6} \\ -2/\sqrt{21} & 3/\sqrt{14} & 1/\sqrt{6} \end{bmatrix}$$

The columns of $\mathbf{U}$ form an orthonormal base for $\mathbb{R}^3$. The rows do the same. The matrix $\mathbf{U}$ is obtained by multiplying $\mathbf{A}$ on the right by a diagonal matrix

$$\mathbf{D}^{-\frac{1}{2}} = \begin{bmatrix} 1/\sqrt{21} & 0 & 0 \\ 0 & 1/\sqrt{14} & 0 \\ 0 & 0 & 1/\sqrt{6} \end{bmatrix}$$

Thus, we obtain $\mathbf{U} = \mathbf{AD}^{-\frac{1}{2}}$. Furthermore, we have

$$\mathbf{I} = \mathbf{U}^T\mathbf{U} = (\mathbf{AD}^{-\frac{1}{2}})^T(\mathbf{AD}^{-\frac{1}{2}}) = \mathbf{D}^{-\frac{1}{2}}\mathbf{A}^T\mathbf{AD}^{-\frac{1}{2}}$$

$$\mathbf{I} = \mathbf{UU}^T = (\mathbf{AD}^{-\frac{1}{2}})(\mathbf{AD}^{-\frac{1}{2}})^T = \mathbf{AD}^{-\frac{1}{2}}\mathbf{D}^{-\frac{1}{2}}\mathbf{A}^T = \mathbf{ADA}^T \qquad \blacksquare$$

The next theorem is a variant of Theorem 2.

THEOREM 10

Every orthonormal set is linearly independent.

The proof is left as General Exercise 33c for the reader.

THEOREM 11

Let $\{\mathbf{u}_1, \mathbf{u}_2, \ldots, \mathbf{u}_n\}$ be an orthonormal basis for a subspace U in an inner-product space. The orthogonal projection of any $\mathbf{x}$ onto U is the point

$$\mathbf{p} = \sum_{i=1}^{n} \langle \mathbf{x}, \mathbf{u}_i \rangle \mathbf{u}_i$$

PROOF It is obvious that the point $\mathbf{p}$ defined by the summation is in $\mathbf{U}$. What remains is the question of whether $\mathbf{x} - \mathbf{p} \perp \mathbf{U}$. For this, one can verify that $\mathbf{x} - \mathbf{p} \perp \mathbf{u}_j$ for all j. The appropriate calculation is

$$\langle \mathbf{x} - \mathbf{p}, \mathbf{u}_j \rangle = \langle \mathbf{x}, \mathbf{u}_j \rangle - \langle \mathbf{p}, \mathbf{u}_j \rangle = \langle \mathbf{x}, \mathbf{u}_j \rangle - \sum_{i=1}^{n} \langle \mathbf{x}, \mathbf{u}_i \rangle \langle \mathbf{u}_i, \mathbf{u}_j \rangle$$

$$= \langle \mathbf{x}, \mathbf{u}_j \rangle - \langle \mathbf{x}, \mathbf{u}_j \rangle = 0 \qquad \blacksquare$$

THEOREM 12

In order that a vector be orthogonal to a subspace (in an inner-product space), it is sufficient that the vector be orthogonal to each member of a set that spans the subspace.

The proof is left as General Exercise 33d.

Subspaces in Inner-Product Spaces

DEFINITION

*A vector space X is the **direct sum** of two subspaces, U and V,*

$$if\ X = U + V\ and\ U \cap V = \{0\}$$

THEOREM 13

If U is a subspace in an n-dimensional inner-product space, then

$$\text{Dim}(U) + \text{Dim}(U^{\perp}) = n$$

PROOF Let $\{\mathbf{u}_1, \mathbf{u}_2, \ldots, \mathbf{u}_k\}$ be an orthonormal base for U. Let $\{\mathbf{v}_1, \mathbf{v}_2, \ldots, \mathbf{v}_p\}$ be an orthonormal base for $U^{\perp}$. Because each $\mathbf{v}_j$ belongs to $U^{\perp}$ and each $\mathbf{u}_i$ belongs to U, we conclude that $\langle \mathbf{v}_j, \mathbf{u}_i \rangle = 0$ for all i and j. Hence, we conclude that the set

$$\{\mathbf{u}_1, \mathbf{u}_2, \ldots, \mathbf{u}_k, \mathbf{v}_1, \mathbf{v}_2, \ldots, \mathbf{v}_p\}$$

is orthonormal and is a base for the space containing U and $U^{\perp}$. Thus, we have $k + p = n$. ∎

THEOREM 14

Let U and V be subsets of an inner-product space. If $U \subseteq V$, then $V^{\perp} \subseteq U^{\perp}$.

PROOF Assume the hypothesis $U \subseteq V$ and let $\mathbf{x} \in V^{\perp}$. Then $\langle \mathbf{x}, \mathbf{v} \rangle = 0$ for all $\mathbf{v} \in V$. Since $U \subseteq V$, we have $\langle \mathbf{x}, \mathbf{u} \rangle = 0$ for all $\mathbf{u} \in U^{\perp}$. ∎

> ### THEOREM 15
>
> *If U is a finite-dimensional subspace in an inner-product space X, then X is the direct sum of U and $U^\perp$:*
>
> $$X = U + U^\perp, \qquad U \cap U^\perp = \{\mathbf{0}\}, \qquad U = U^{\perp\perp}$$

PROOF Let $\{\mathbf{u}_1, \mathbf{u}_2, \ldots, \mathbf{u}_n\}$ be an orthonormal base for U.

Step 1. The orthogonal projection onto U is the linear transformation P defined by $P(\mathbf{x}) = \sum_{i=1}^n \langle \mathbf{x}, \mathbf{u}_i \rangle \mathbf{u}_i$. For any $\mathbf{x} \in X$, we have $\mathbf{x} - P(\mathbf{x}) \perp U$ because

$$\langle \mathbf{x} - P(\mathbf{x}), \mathbf{u}_j \rangle = \langle \mathbf{x}, \mathbf{u}_j \rangle - \left\langle \sum_{i=1}^n \langle \mathbf{x}, \mathbf{u}_i \rangle \mathbf{u}_i, \mathbf{u}_j \right\rangle$$

$$= \langle \mathbf{x}, \mathbf{u}_j \rangle - \sum_{i=1}^n \langle \mathbf{x}, \mathbf{u}_i \rangle \langle \mathbf{u}_i, \mathbf{u}_j \rangle = 0$$

The equation $\mathbf{x} = P(\mathbf{x}) + [\mathbf{x} - P(\mathbf{x})]$ proves that $X = U + U^\perp$.

Step 2. The only vector in $U \cap U^\perp$ is the zero vector because if $\mathbf{x} \in U \cap U^\perp$, then $\mathbf{x} \perp \mathbf{x}$.

Step 3. To prove $U \subseteq U^{\perp\perp}$, let $\mathbf{u} \in U$. Since $X = U + U^\perp$, we have $\mathbf{u} \perp U^\perp$ and $\mathbf{u} \in U^{\perp\perp}$. In other words, if $\mathbf{u} \in U$ and $\mathbf{v} \in U^\perp$, then $\mathbf{v} \perp U$ and $\mathbf{v} \perp \mathbf{u}$ so $\mathbf{u} \in U^{\perp\perp}$.

Step 4. To prove $U^{\perp\perp} \subseteq U$, let $\mathbf{x} \in U^{\perp\perp}$. Then $\mathbf{x} \perp U^\perp$. Since $X = U + U^\perp$, we conclude that $\mathbf{x} \in U$. ∎

Here are some examples of direct sums.

a. Even functions ($f(t) = f(-t)$) and odd functions ($f(t) = -f(-t)$) both form subspaces. Any function f is a direct sum of an even function and an odd function since $f = f_E + f_O$ where $f_E(t) = \frac{1}{2}[f(t) + f(-t)]$ and $f_O(t) = \frac{1}{2}[f(t) - f(-t)]$.

b. The space of all $n \times n$ real matrices is the direct sum of the subspaces of all symmetric matrices ($\mathbf{A}^T = \mathbf{A}$) and of all skew symmetric matrices ($\mathbf{A}^T = -\mathbf{A}$) because $\mathbf{A} = \frac{1}{2}(\mathbf{A} + \mathbf{A}^T) + \frac{1}{2}(\mathbf{A} - \mathbf{A}^T)$.

Application: Work and Forces

We can describe work and forces in terms of the notation and concepts discussed in this section. Let the vector $\mathbf{f}$ be the force exerted on an object, let the vector $\mathbf{d}$ be the displacement caused by the force, and let θ be the angle between **f** and **d**. For example, suppose we are pulling a heavy load on a dolly with a constant force so that it moves horizontally along the ground. The work done in moving the dolly through a distance **d** is given by the distance moved multiplied by the magnitude of the component of the force in the direction of motion. This is illustrated in Figure 7.5. The component of **f** in the direction **d** is $\|\mathbf{f}\| \cos \theta$. By the definition, the work accomplished is

$$W = \|\mathbf{f}\| \cdot \|\mathbf{d}\| \cos \theta = \langle \mathbf{f}, \mathbf{d} \rangle$$

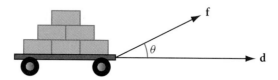

FIGURE 7.5 Force and displacement.

Application: Collision

The Law of Cosines can be applied to determine the final location of a ball after a glancing collision with a wall, as shown in Figure 7.6. Let $\mathbf{u} = (u_1, u_2)$ be the initial position, $\mathbf{v} = (v_1, v_2)$ be the final position, and $\mathbf{u} - \mathbf{v}$ be the change in position as a result of the collision. From the Law of Cosines, it follows that the magnitude of the change in position is:

$$\|\mathbf{u} - \mathbf{v}\|^2 = \|\mathbf{u}\|^2 + \|\mathbf{v}\|^2 - 2\|\mathbf{u}\| \cdot \|\mathbf{v}\| \cos \varphi$$

From this expression, we can obtain

$$\langle \mathbf{u}, \mathbf{v} \rangle = \|\mathbf{u}\| \cdot \|\mathbf{v}\| \cos \varphi$$

which gives a connection between the inner product of the vectors **u** and **v** and the angle φ between them.

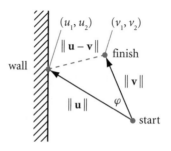

FIGURE 7.6 Glancing ball.

SUMMARY 7.1

- For any vector $\mathbf{x}$ in $\mathbb{C}^n$, if $\mathbf{x} = (x_1, x_2, \ldots, x_n)$, we write $\bar{\mathbf{x}} = (\bar{x}_1, \bar{x}_2, \ldots, \bar{x}_n)$.
  An **inner product** $\langle \mathbf{x}, \mathbf{y} \rangle$ has these properties:

  - $\langle \mathbf{x}, \mathbf{x} \rangle > 0$ whenever $\mathbf{x} \neq \mathbf{0}$

  - $\langle \mathbf{x}, \mathbf{y} \rangle = \overline{\langle \mathbf{y}, \mathbf{x} \rangle}$

  - $\langle \mathbf{x} + \mathbf{y}, \mathbf{z} \rangle = \langle \mathbf{x}, \mathbf{z} \rangle + \langle \mathbf{y}, \mathbf{z} \rangle$

  - $\langle \alpha\mathbf{x}, \mathbf{y} \rangle = \alpha \langle \mathbf{x}, \mathbf{y} \rangle$ for any scalar α

  - $\langle \mathbf{x}, \alpha\mathbf{y} \rangle = \bar{\alpha} \langle \mathbf{x}, \mathbf{y} \rangle$

  - $\langle \mathbf{x}, \mathbf{y} + \mathbf{z} \rangle = \langle \mathbf{x}, \mathbf{y} \rangle + \langle \mathbf{x}, \mathbf{z} \rangle$

  - $\langle \sum_{j=1}^{m} \mathbf{x}_j, \mathbf{y} \rangle = \sum_{j=1}^{m} \langle \mathbf{x}_j, \mathbf{y} \rangle$

  - $\langle \mathbf{x}, \sum_{j=1}^{m} \mathbf{y}_j \rangle = \sum_{j=1}^{m} \langle \mathbf{x}, \mathbf{y}_j \rangle$

  - $\langle \mathbf{x} + \mathbf{y}, \mathbf{x} + \mathbf{y} \rangle = \langle \mathbf{x}, \mathbf{x} \rangle + \langle \mathbf{x}, \mathbf{y} \rangle + \langle \mathbf{y}, \mathbf{x} \rangle + \langle \mathbf{y}, \mathbf{y} \rangle$

- $\langle \mathbf{x}, \mathbf{y} \rangle = \mathbf{x}^T \bar{\mathbf{y}} = \mathbf{y}^T \mathbf{x} = \sum_{j=1}^{n} x_j y_j$
  (**Standard inner product** in $\mathbb{R}^n$)

- $\langle \mathbf{x}, \mathbf{y} \rangle = \bar{\mathbf{y}}^T \mathbf{x} = \sum_{j=1}^{n} x_j \bar{y}_j$
  (**Standard inner product** in $\mathbb{C}^n$)

- $\langle \mathbf{x}, \mathbf{y} \rangle_{\mathbf{w}} = \sum_{j=1}^{n} w_j x_j y_j$
  (**Weighted inner product** in $\mathbb{R}^n$)

- $\langle \mathbf{x}, \mathbf{y} \rangle = \int_a^b \mathbf{x}(t) \overline{\mathbf{y}(t)} \, dt$
  (**Usual inner product** in $C[a, b]$)

- **Distance function** d:

  - $d(\mathbf{x}, \mathbf{y}) > 0$ if $\mathbf{x} \neq \mathbf{y}$

  - $d(\mathbf{x}, \mathbf{y}) = d(\mathbf{y}, \mathbf{x})$

  - $d(\mathbf{x}, \mathbf{y}) \leq d(\mathbf{x}, \mathbf{z}) + d(\mathbf{z}, \mathbf{y})$

- $\mathbf{x} \perp \mathbf{y}$ means that $\langle \mathbf{x}, \mathbf{y} \rangle = 0$. In words, $\mathbf{x}$ and $\mathbf{y}$ are mutually orthogonal.

- $V \perp U$ means that $\langle \mathbf{v}, \mathbf{u} \rangle = 0$ for all $\mathbf{v}$ in V and $\mathbf{u}$ in U.

- $\cos\theta = \langle \mathbf{x}, \mathbf{y} \rangle / \|\mathbf{x}\| \cdot \|\mathbf{y}\|$
  (**Angle between vectors**)

- $c^2 = a^2 + b^2 - 2ab \cdot \cos C$ (**Law of Cosines**)

- $\|\mathbf{x}\|^2 - 2\langle \mathbf{x}, \mathbf{y} \rangle + \|\mathbf{y}\|^2 = \|\mathbf{x}\|^2 + \|\mathbf{y}\|^2 - 2\|\mathbf{x}\| \cdot \|\mathbf{y}\| \cos\theta$

- $\|\mathbf{x} - \mathbf{y}\|^2 = \|\mathbf{x}\|^2 + \|\mathbf{y}\|^2$ when $\theta = \frac{\pi}{2}$

- $\{\mathbf{x} : \mathbf{x} \perp \mathbf{s} \text{ for all } \mathbf{s} \in S\} = \{\mathbf{x} : \mathbf{x} \perp S\} = S^\perp$
  (orthogonal complement of S)

- **Direct sum** of subspaces U and V: $X = U + V$ and $U \cap V = \{\mathbf{0}\}$. It is an orthogonal direct sum if $U \perp V$.

- Inner-Product Theorems:

  - In an inner-product space, the norm has these properties:
    $\|\mathbf{x}\| > 0$ for every nonzero vector $\mathbf{x}$
    $\|\alpha\mathbf{x}\| = |\alpha| \|\mathbf{x}\|$
    $|\langle \mathbf{x}, \mathbf{y} \rangle| \leq \|\mathbf{x}\| \cdot \|\mathbf{y}\|$
    (**Cauchy–Schwarz inequality**)
    $\|\mathbf{x} + \mathbf{y}\| \leq \|\mathbf{x}\| + \|\mathbf{y}\|$
    (**Triangle inequality**)
    If $\langle \mathbf{x}, \mathbf{y} \rangle = 0$, then $\|\mathbf{x} + \mathbf{y}\|^2 = \|\mathbf{x}\|^2 + \|\mathbf{y}\|^2$
    (**Pythagorean Law**)
    $\|\mathbf{x} + \mathbf{y}\|^2 = \|\mathbf{x}\|^2 + \langle \mathbf{x}, \mathbf{y} \rangle + \langle \mathbf{y}, \mathbf{x} \rangle + \|\mathbf{y}\|^2$
    $\|\mathbf{x} + \mathbf{y}\|^2 = \|\mathbf{x}\|^2 + 2\langle \mathbf{x}, \mathbf{y} \rangle + \|\mathbf{y}\|^2$
    if the scalar field is $\mathbb{R}$

  - Every orthogonal set of nonzero vectors in an inner-product space is linearly independent.

  - In $\mathbb{R}^n$, there does not exist an orthogonal set of $n + 1$ nonzero vectors.

  - In any inner-product space, the orthogonal projection of a vector $\mathbf{x}$ onto a nonzero vector $\mathbf{y}$ is the point $\mathbf{p} = [\langle \mathbf{x}, \mathbf{y} \rangle / \langle \mathbf{y}, \mathbf{y} \rangle] \mathbf{y}$. It has the property that $\mathbf{x} - \mathbf{p}$ is orthogonal to $\mathbf{y}$. Thus, $\mathbf{x}$ is split into an orthogonal pair of vectors in the equation:
    $$\mathbf{x} = \mathbf{p} + (\mathbf{x} - \mathbf{p})$$

  - In an inner-product space, the orthogonal complement of a set is the same as the orthogonal complement of its span.

  - In an inner-product space, the orthogonal complement of any subset is a subspace.

  - The null space of a matrix is the orthogonal complement of its row space.

- If $\{\mathbf{u}_1, \mathbf{u}_2, \ldots, \mathbf{u}_n\}$ is an orthonormal set in an inner-product space, then each $\mathbf{x}$ in the span of that set is expressible as $\mathbf{x} = \sum_{i=1}^{n} \langle \mathbf{x}, \mathbf{u}_i \rangle \mathbf{u}_i$.
- Every orthonormal set is linearly independent.
- Let $\{\mathbf{u}_1, \mathbf{u}_2, \ldots \mathbf{u}_n\}$ be an orthonormal basis for a subspace U in an inner-product space. The orthogonal projection of any $\mathbf{x}$ onto U is $\sum_{i=1}^{n} \langle \mathbf{x}, \mathbf{u}_i \rangle \mathbf{u}_i$.

- $\text{Dim}(U) + \text{Dim}(U^\perp) = n$, where U is a subspace in an n-dimensional inner-product space.
- If $U \subseteq V$, then $V^\perp \subseteq U^\perp$, where U and V are subsets of an inner-product space.
- $U^{\perp\perp} = U, X = U + U^\perp, U \cap U^\perp = \{\mathbf{0}\}$, where U is a subspace in a finite-dimensional inner-product space X.

KEY CONCEPTS 7.1

Inner products, dot product, Hermitian, magnitude or length or norm of a vector, distance function, unit vector, normalization, mutually orthogonal vectors, orthogonality, Cauchy–Schwarz inequality, triangle inequality, Pythagorean law, orthonormal sets of vectors, orthogonal projection, angles between vectors, Law of Cosines, orthogonal complement of a set, orthogonal vectors, orthonormal vectors, orthogonal direct sums of subspaces

GENERAL EXERCISES 7.1

1. In $\mathbb{R}^2$, show that the pair of vectors $\{(a, b), (b, -a)\}$ is orthogonal.

2. Calculate the orthogonal projection of $(1, 3, 7)$ onto $(2, -1, 3)$.

3. Show that the vectors $\mathbf{u} = (2, 3, -5)$, $\mathbf{v} = (2, 2, 2)$, and $\mathbf{w} = (-8, 7, 1)$ form an orthogonal triple; that is, each vector is orthogonal to the other two. Find the projection of $\mathbf{x} = (4, -3, 5)$ onto each of the vectors $\mathbf{u}$, $\mathbf{v}$, and $\mathbf{w}$. Verify that the three projections add up to give $\mathbf{x}$. Thus the vector $\mathbf{x}$ will be dissected into three mutually orthogonal pieces or *components*.

4. Verify that these three vectors form an orthogonal set: $\mathbf{u}_1 = (3, 3, 1), \mathbf{u}_2 = (2, -1, -3)$, and $\mathbf{u}_3 = (8, -11, 9)$. Place

these as rows in a matrix $\mathbf{A}$. Normalize the rows to get a matrix $\mathbf{B}$. Verify that the columns of $\mathbf{B}$ form an orthonormal set.

5. Explain why, in an inner-product space, the equations $\|\mathbf{x} - \mathbf{y}\| = 0$ and $\mathbf{x} = \mathbf{y}$ are equivalent.

6. Let $\mathbf{x} = (1, 3, 7)$ and $\mathbf{y} = (-4, 2, 1)$. Compute $\|\mathbf{x}\|, \langle \mathbf{x}, \mathbf{y} \rangle, \|\mathbf{y}\|$, and $\|\mathbf{x} + \mathbf{y}\|$. Verify in this example that the Cauchy–Schwarz inequality and the triangle inequality are true.

7. Explain why it is not true that every orthogonal set is linearly independent.

8. In an inner-product space, let $\|\mathbf{x}\| = 7$ and $\|\mathbf{y}\| = 9$. How large can $\|\mathbf{x} - \mathbf{y}\|$ be?

9. Determine whether this mapping is linear: $\mathbf{x} \mapsto (\langle \mathbf{x}, \mathbf{v} \rangle / \langle \mathbf{v}, \mathbf{v} \rangle) \mathbf{v}$. (Here $\mathbf{v}$ is any fixed vector, and the setting can be any inner-product space.)

10. Consider the matrix $\mathbf{A} = \begin{bmatrix} 4 & 3 & 1 & 2 \\ 3 & 2 & 0 & 2 \\ -2 & 3 & 1 & 2 \end{bmatrix}$ find a simple description of the null space and the orthogonal complement of the row space.

11. Answer these questions and draw simple sketches to illustrate them:
   a. If $\mathbf{x} \perp \mathbf{y}$, does it follow that for all scalars α and β, we have $\alpha \mathbf{x} \perp \beta \mathbf{y}$?
   b. If $\mathbf{x} \perp \mathbf{y}$ and $\mathbf{y} \perp \mathbf{z}$, does it follow that $\mathbf{x} \perp \mathbf{z}$?
   c. If $\mathbf{x} \perp \mathbf{y}$ and $\mathbf{x} \perp \mathbf{z}$, does it follow that $\alpha \mathbf{x} \perp (\beta \mathbf{y} + \gamma \mathbf{z})$ for all scalars α, β, and γ?

12. Explain this apparent contradiction: For the vector $\mathbf{x} = (7 + 3i, -3 + 7i)$, we find that $\|\mathbf{x}\|^2 = \langle \mathbf{x}, \mathbf{x} \rangle = x_1^2 + x_2^2 = (7 + 3i)^2 + (-3 + 7i)^2 = (49 + 42i - 9) + (9 - 42i - 40) = 0$.

13. Find the orthogonal projection of $\mathbf{x} = (2, 2, 3)$ onto the subspace of $\mathbb{R}^3$ spanned by the two vectors $\mathbf{u} = (4, 1, -2)$ and $\mathbf{v} = (1, -2, 1)$. Note that $\mathbf{u} \perp \mathbf{v}$. Check your work by verifying independently the two properties that the projection should have.

14. Let $\mathbf{u}$ and $\mathbf{v}$ be nonzero vectors in an inner-product space. Under what conditions will we have $\|\mathbf{u}\|^2 = \langle \mathbf{u}, \mathbf{v} \rangle = \|\mathbf{v}\|^2$? A condition on the angle between $\mathbf{u}$ and $\mathbf{v}$ would be satisfactory.

15. In an inner-product space, if $\langle \mathbf{x}, \mathbf{y} \rangle = 0$, then $\|\mathbf{x} + \mathbf{y}\|^2 = \|\mathbf{x}\|^2 + \|\mathbf{y}\|^2$. Establish this and determine whether the converse is also true.

16. Let $\mathbf{u} = (1, 2, 3)$, $\mathbf{v} = (2, -1, 1)$, and $\mathbf{w} = (3, 1, 0)$. Verify that if $\langle \mathbf{x}, \mathbf{u} \rangle = \langle \mathbf{x}, \mathbf{v} \rangle = \langle \mathbf{x}, \mathbf{w} \rangle = 0$, then $\mathbf{x} = \mathbf{0}$.

17. Let $f(t) = \|(1, 3, t, 7)\|$. What is the minimum value of this function of t? Answer the same question for the function g defined by $g(t) = \|(1 + t, 3, 7 - t)\|$.

18. If $\langle \mathbf{x}, \mathbf{y} \rangle > 0$, then the angle between $\mathbf{x}$ and $\mathbf{y}$ is acute. Explain.

19. If $\langle \mathbf{x}, \mathbf{y} \rangle = 0$ for all $\mathbf{y}$, then $\mathbf{x} = \mathbf{0}$. Explain why or why not.

20. Let $\mathbf{x} \neq \mathbf{0}$ and $\mathbf{y} \neq \mathbf{0}$. Does it follow that $\text{Proj}_{\mathbf{x}} \mathbf{y} = \mathbf{0}$ if and only if $\text{Proj}_{\mathbf{y}} \mathbf{x} = \mathbf{0}$? Explain why or why not.

21. Refer to Example 3 and explain why if any one of the numbers w_i is less than or equal to 0, then the formula will not produce a genuine inner product.

22. Write out in detail the proofs of the consequences **a**–**e** from the four postulates in the inner-product definition.

23. Establish the validity of
   a. Theorem 4   **b.** Theorem 5

24. In any inner-product space, if $\|\mathbf{x}\|^2 = \|\mathbf{y}\|^2 = \langle \mathbf{x}, \mathbf{y} \rangle$, then $\mathbf{x} = \mathbf{y}$. Explain why or why not.

25. What is the maximum value of $3x - 7y + 4z$ when x, y, and z are real numbers subject only to the constraint $x^2 + y^2 + z^2 = 1$?

26. Explain why the hypotheses $\|\mathbf{x} + \mathbf{y}\| = \|\mathbf{y}\|$ and $\mathbf{x} \neq \mathbf{0}$ lead to $\langle \mathbf{x}, \mathbf{y} \rangle < 0$.

27. Explain why, for any matrix $\mathbf{A}$, $[\text{Col}(\mathbf{A})]^{\perp} = \text{Null}(\mathbf{A}^T)$

28. Establish the validity of
   a. Theorem 6   **b.** Theorem 7

29. Explain why, in any inner-product space, if $\langle \mathbf{x}, \mathbf{y} \rangle = \|\mathbf{x}\| \cdot \|\mathbf{y}\| \neq 0$, then $\mathbf{y}$ is a positive multiple of $\mathbf{x}$.

30. What is the simplest decomposition of $\mathbf{x} = (x_1, x_2, x_3)$ into three mutually orthogonal components? (Several answers are reasonable.)

31. Let $\mathbf{A}$ be a square matrix whose columns form an orthogonal set. If we normalize each row of $\mathbf{A}$, we obtain a new matrix, $\mathbf{B}$. Is $\mathbf{B}$ an orthogonal matrix? What interesting properties does $\mathbf{B}$ have? Answer the same questions for the matrix $\mathbf{C}$ obtained by normalizing the columns of $\mathbf{A}$.

32. Refer to Example 13, and determine the orthonormal set that arises by normalizing the functions $\mathbf{w}_k$.

33. Establish the validity of the following:
   a. Theorem 1, Parts **a, b, e,** and **f**
   b. Theorem 10
   c. Theorem 12

34. Establish these computational rules when the scalar field is taken to be $\mathbb{R}$:
   a. $\langle \mathbf{x} + \mathbf{y}, \mathbf{u} + \mathbf{v} \rangle =$
      $\langle \mathbf{x}, \mathbf{u} \rangle + \langle \mathbf{x}, \mathbf{v} \rangle + \langle \mathbf{y}, \mathbf{u} \rangle + \langle \mathbf{y}, \mathbf{v} \rangle$
   b. $\|\mathbf{x} + \mathbf{y}\|^2 = \|\mathbf{x}\|^2 + 2\langle \mathbf{x}, \mathbf{y} \rangle + \|\mathbf{y}\|^2$
   c. $\|\mathbf{x} + \mathbf{y}\|^2 + \|\mathbf{x} - \mathbf{y}\|^2 = 2\|\mathbf{x}\|^2 + 2\|\mathbf{y}\|^2$

   (The equality in Part **c** is known as the **Parallelogram Law**.)

35. Establish that if $\langle \mathbf{x}, \mathbf{y} \rangle$ is real, then we have
$$\|\mathbf{x} + i\mathbf{y}\|^2 = \|i\mathbf{x} + \mathbf{y}\|^2 = \|\mathbf{x}\|^2 + \|\mathbf{y}\|^2$$

36. Explain why, for any matrix, the kernel is the orthogonal complement of the row space.

37. In $\mathbb{R}^n$, let $\mathbf{x}$ and $\mathbf{y}$ be vectors written as $n \times 1$ matrices. Explain that with the standard inner product, we have $\langle \mathbf{x}, \mathbf{y} \rangle = \mathbf{x}^T \mathbf{y} = \mathbf{y}^T \mathbf{x}$. Establish that in $\mathbb{C}^n$, the corresponding equation is $\langle \mathbf{x}, \mathbf{y} \rangle = \mathbf{y}^H \mathbf{x} = (\mathbf{x}^H \mathbf{y})^H$. The notation $\mathbf{x}^H$ indicates the conjugate transpose of the column vector $\mathbf{x}$. That is, each component of the vector is replaced by its complex conjugate, and the vector is written as a row vector.

38. Refer to Theorem 5. Can you explain why $\mathbf{p}$ is the multiple of $\mathbf{y}$ closest to $\mathbf{x}$? Suggestion: Use the methods of calculus to minimize the expression $\|\mathbf{x} - t\mathbf{y}\|^2$.

39. Let S and Q be subsets in $\mathbb{R}^n$, and suppose that $S \subseteq Q$. Explain why $Q^\perp \subseteq S^\perp$.

40. Compute the work done by a force of 40 newtons acting in the direction specified by $(3, 5, 7)$ and causing a displacement of 25 units in the direction $(1, 1, 6)$.

41. Let T be a linear transformation defined by $T(\mathbf{x}) = \sum_{i=1}^{n} \langle \mathbf{x}, \mathbf{v}_i \rangle \mathbf{u}_i$ in an inner-product space. Find a nonzero solution of the equation $T(\mathbf{x}) = \mathbf{x}$. You may need to make some judicious assumptions.

42. Establish this identity in any real inner-product space:
$$\|\mathbf{x} - \mathbf{y}\|^2 = \|\mathbf{x}\|^2 - \|\mathbf{y}\|^2 + 2\langle \mathbf{y} - \mathbf{x}, \mathbf{y} \rangle$$

43. Explain why the inequality
$$|\langle \mathbf{x}, \mathbf{y} \rangle| \leq \lambda \|\mathbf{x}\|^2 + [1/(4\lambda)] \|\mathbf{y}\|^2$$
holds for any $\lambda > 0$.

44. Find all solutions to $\langle \mathbf{x}, \mathbf{a} \rangle \mathbf{c} = \mathbf{b}$, where $\mathbf{a}, \mathbf{b},$ and $\mathbf{c}$ are fixed vectors in an inner-product space. If necessary, take the scalar field to be $\mathbb{R}$. There may be several special cases to consider separately.

45. Find the solutions of the equation $\mathbf{x} + \langle \mathbf{x}, \mathbf{a} \rangle \mathbf{c} = \mathbf{b}$. There may be a number of special cases to consider. If necessary, assume the scalar field to be $\mathbb{R}$.

46. Solve the equation $T(\mathbf{x}) = \mathbf{b}$ using the transformation T in General Exercise 41.

47. In the space $C[-1, 1]$, use the inner product given in Example 4. Show why the Legendre polynomials form an orthogonal set: $P_0(t) = 1$, $P_1(t) = t$, and $P_2(t) = (3t^2 - 1)/2$.

48. In an inner-product space, $\|\mathbf{x}\| = \|\mathbf{y}\| = 1$. Consider the triangle with vertices $\mathbf{0}$, $\mathbf{x}$, and $\mathbf{y}$. How big can its area be?

49. Explain why, in $\mathbb{R}^2$, if $\mathbf{x} \neq \mathbf{0}$ and $\mathbf{x} \perp \mathbf{y}$, then $\mathbf{y} = \mathbf{Ax}$ for some matrix of the form

$$\mathbf{A} = \begin{bmatrix} 0 & c \\ -c & 0 \end{bmatrix}$$

50. Establish, in any inner-product space, that if $1 = \|\mathbf{x}\| < \|\mathbf{y}\|$, then $\|\mathbf{x} - \widehat{\mathbf{y}}\| < \|\mathbf{x} - \mathbf{y}\|$, where $\widehat{\mathbf{y}} = \mathbf{y}/\|\mathbf{y}\|$.

51. Show why the orthogonal projection of $\mathbf{x}$ onto a nonzero $\mathbf{y}$ is a vector whose length is no greater than the length of $\mathbf{x}$.

52. Explain why if $\|\mathbf{x}\| = \|\mathbf{y}\|$, then the projection of $\mathbf{x}$ onto $\mathbf{y}$ has the same length as the projection of $\mathbf{y}$ onto $\mathbf{x}$.

53. Write out in detail all parts of Theorem 1.

54. Establish, in any inner-product space, that the Cauchy–Schwarz inequality becomes an equality if and only if $\|\mathbf{y}\|\mathbf{x} = \pm\|\mathbf{x}\|\mathbf{y}$.

55. Establish that the distance from a point $\mathbf{x}$ to a line through $\mathbf{0}$ and $\mathbf{v}$ is $\|\mathbf{x} - [\langle \mathbf{x}, \mathbf{v} \rangle / \langle \mathbf{v}, \mathbf{v} \rangle] \mathbf{v}\|$. When we assume $\|\mathbf{v}\| = 1$, the distance is equal to $\|\mathbf{x} - \langle \mathbf{x}, \mathbf{v} \rangle \mathbf{v}\|$.

56. Use techniques of calculus to find the value of t for which $\|\mathbf{x} - t\mathbf{v}\|$ is a minimum. Draw a connection between this problem and the formula for an orthogonal projection in Theorem 3.

57. A linear transformation $P : X \to X$ is called a **projection** if $P^2 = P$. Determine whether this map is a projection: $T(x_1, x_2) = (0, x_2 - x_1)$.

58. Is the distance from the point $(3, 7, -5)$ to the line through 0 and $(2, 3, 4)$ equal to $(1/29)\sqrt{52142}$?

59. Establish the uniqueness of the orthogonal projection of a point onto a subspace in an inner-product space. A careful statement of this fact is as follows. If V is a subspace in an inner-product space, and if $\mathbf{x}$ is any point of that space, then there is at most one point $\mathbf{p}$ in V such that $\mathbf{x} - \mathbf{p} \perp V$.

60. Criticize this argument or provide a correct version: If $\{\mathbf{u}_1, \mathbf{u}_2, \ldots, \mathbf{u}_n\}$ is orthonormal then every $\mathbf{x}$ satisfies $\mathbf{x} = \sum_{i=1}^{n} \langle \mathbf{x}, \mathbf{u}_i \rangle \mathbf{u}_i$ because we can start with $\mathbf{x} = \sum_{i=1}^{n} c_i \mathbf{u}_i$ and then take the inner product on both sides with $\mathbf{u}_j$, getting $c_j = \langle \mathbf{x}, \mathbf{u}_j \rangle$.

61. Use the formula for the angle between two vectors to verify the formula for the orthogonal projection as given in Theorem 5.

62. The vector space $C[a, b]$ has as its elements all continuous real-valued functions defined on the closed interval $[a, b]$. In this space, define an **inner product** with the formula $\langle f, g \rangle = \int_a^b f(t)g(t)w(t)\, dt$, where w is a fixed member of $C[a, b]$. What must be assumed of w in order that this equation define an inner product? Establish the Postulates 1–4 for this inner-product definition.

63. What are the postulates (*axioms*) for an inner-product space? (Assume the scalars are real numbers.) Use the axioms to prove the **Pythagorean Law**: If $\mathbf{x} \perp \mathbf{y}$, then $\|\mathbf{x} + \mathbf{y}\|^2 = \|\mathbf{x}\|^2 + \|\mathbf{y}\|^2$. Is this true only for the case of real scalars, or is it also valid when the scalars are complex?

64. Explain why, in any two-dimensional inner-product space, if $\mathbf{u} \neq \mathbf{0}$, $\mathbf{v} \neq \mathbf{0}$, and $\langle \mathbf{u}, \mathbf{v} \rangle = \mathbf{0}$, then $\{\mathbf{u}, \mathbf{v}\}$ is a basis for the space. What is the analogous theorem for a three-dimensional space?

65. State and prove a slightly different version of Theorem 10, where the hypothesis is changed to *orthogonal* instead of *orthonormal*.

66. Try to establish the Cauchy–Schwarz inequality by an argument based on the quadratic function q defined by $q(t) = \|\mathbf{x} - t\mathbf{u}\|^2$, where $\mathbf{u}$ is a unit vector. Find the minimum point, t_0, and note that $q(t_0) \geq 0$, etc.

67. Find the exact conditions on $\mathbf{x}$ and $\mathbf{y}$ so that this inequality becomes correct: $\|\mathbf{x} + \mathbf{y}\|^2 \leq \|\mathbf{x}\|^2 + \|\mathbf{y}\|^2$. Also show, by example, that it is not always valid.

68. Explore this approach to the orthogonal projection of one point onto another: To project $\mathbf{x}$ onto $\mathbf{y}$, we look for a point $\alpha\mathbf{y}$ as close as possible to $\mathbf{x}$. This leads to a simple minimization problem that can be solved with elementary calculus.

69. Let $\mathbf{x}$ and $\mathbf{y}$ be vectors in $\mathbb{R}^n$, all of whose components are ± 1. Explain why if $\mathbf{x}$ is orthogonal to $\mathbf{y}$, then n is even.

70. Let $\mathbf{u}$ be a unit vector in $\mathbb{R}^n$. Establish that if $-1 \leq c \leq 1$, then there is a unit vector $\mathbf{x}$ such that $\langle \mathbf{x}, \mathbf{u} \rangle = c$.

71. Establish that if the angle between $\mathbf{x}$ and $\mathbf{y}$ is acute, then $\|\mathbf{x} + \mathbf{y}\|^2 \geq \|\mathbf{x}\|^2 + \|\mathbf{y}\|^2$.

72. Explain why the projection of $\mathbf{u}$ onto $\mathbf{v}$ has norm at most $\|\mathbf{u}\|$.

73. Let $\mathbf{u}$ and $\mathbf{v}$ be vectors in an inner-product space, neither a multiple of the other. Define operators P and Q by $P(\mathbf{x}) = [\langle \mathbf{x}, \mathbf{u} \rangle / \langle \mathbf{u}, \mathbf{u} \rangle]\mathbf{u}$ and $Q(\mathbf{x}) = [\langle \mathbf{x}, \mathbf{v} \rangle / \langle \mathbf{v}, \mathbf{v} \rangle]\mathbf{v}$. Explain why the sequence of vectors $\mathbf{x}$, $P(\mathbf{x})$, $(QP)(\mathbf{x})$, $(PQP)(\mathbf{x})$, and so on converges to $\mathbf{0}$.

74. Give an example of points $\mathbf{u}$ and $\mathbf{v}$ in $\mathbb{R}^2$ such that $\mathbf{u} \neq \mathbf{0}$, $\mathbf{v} \neq \mathbf{0}$, $\mathbf{u} \neq \mathbf{v}$, and $\langle \mathbf{u}, \mathbf{v} \rangle = \|\mathbf{v}\|^2$. For a fixed $\mathbf{v}$, describe these two sets: $\{\mathbf{x} : \langle \mathbf{x}, \mathbf{v} \rangle = \|\mathbf{v}\|\}$ and $\{\mathbf{x} : \langle \mathbf{x}, \mathbf{v} \rangle = \|\mathbf{x}\|^2\}$.

75. Fill in the details of the proof of the Cauchy–Schwarz inequality along these lines: We have $\langle (\mathbf{x} + t\mathbf{y}), (\mathbf{x} + t\mathbf{y}) \rangle \geq 0$ and $at^2 + bt + c \geq 0$, where $a = \langle \mathbf{y}, \mathbf{y} \rangle$, $b = 2\langle \mathbf{x}, \mathbf{y} \rangle$, and $c = \langle \mathbf{x}, \mathbf{x} \rangle$. This quadratic function corresponds to a parabola that is on or above the t-axis when $b^2 - 4ac \leq 0$. A sketch may be helpful.

76. Let $\mathbf{x} \neq \mathbf{0}$, $\mathbf{y} \neq \mathbf{0}$. We have $\langle \mathbf{x}, \mathbf{y} \rangle \geq 0$ if and only if $\|\mathbf{x} + \mathbf{y}\| > \|\mathbf{y}\|$. Explain.

77. Explain why or find a counterexample: Let $\mathbf{x} \neq \mathbf{0}$. We have $\langle \mathbf{x}, \mathbf{y} \rangle = 0$ if and only if $\|\mathbf{x} + \mathbf{y}\| > \|\mathbf{y}\|$.

78. Let $\mathbf{x} \neq \mathbf{0}$ and $\mathbf{y} \neq \mathbf{0}$. Explain why we have $\langle \mathbf{x}, \mathbf{y} \rangle = \|\mathbf{x}\| \|\mathbf{y}\|$ if and only if $\mathbf{y} = c\mathbf{x}$ with $c > 0$.

79. Establish that $\left| \|\mathbf{x}\| - \|\mathbf{y}\| \right| \leq \|\mathbf{x} - \mathbf{y}\|$.

80. Establish that $\|\mathbf{x}\| \leq \sum_{i=1}^{n} |x_i|$ for $\mathbf{x} \in \mathbb{R}^n$.

81. Establish that
$$\|a\mathbf{x} + b\mathbf{y}\| \leq \max\{|a|, |b|\}(\|\mathbf{x}\| + \|\mathbf{y}\|).$$

82. Explain why if $\mathbf{v}_1, \mathbf{v}_2, \ldots, \mathbf{v}_n$ are mutually orthogonal and nonzero, then $\mathbf{v}_1$ is not expressible as $\sum_{i=2}^{n} c_i \mathbf{v}_i$.

83. If the orthogonal projection of $\mathbf{y}$ onto $\mathbf{x}$ is $\mathbf{x}$, does it follow that either $\|\mathbf{x}\| = 1$ or $\mathbf{x} \cdot \mathbf{y} = 1$? Explain why or why not.

84. Establish, in any inner-product space, that if $\mathbf{x}$ and $\mathbf{y}$ are two points such that $\|\mathbf{x}\| \leq \|\mathbf{x} + \alpha\mathbf{y}\|$ for all scalars α, then $\langle x, y \rangle = 0$.

85. Let $\mathbf{u}$ and $\mathbf{v}$ be two unit vectors in an inner-product space. Assume that for every $\mathbf{x}$ in the space, $\langle \mathbf{u}, \mathbf{x} \rangle \langle \mathbf{v}, \mathbf{x} \rangle \leq 0$. Establish that $\mathbf{u} = \mathbf{v}$.

86. Critique this argument: Every orthonormal set is linearly independent and every orthonormal set is orthogonal. Therefore, every orthogonal set is linearly independent.

87. Find the point on the parabola $y = x^2$ that is as close as possible to the point $(9, 8)$. This problem illustrates how simple problems concerning linear functions become quite difficult when any nonlinearity is introduced.

88. Define an inner product in $\mathbb{R}^3$ by the equation $\langle \mathbf{x}, \mathbf{y} \rangle_o = 3x_1 y_1 + 5x_2 y_2 + 2x_3 y_3$. (The subscript o is there to remind us that this is *not* the *standard* inner product.) What is the largest value that $\langle \mathbf{x}, \mathbf{y} \rangle_o$ can attain when $\mathbf{y}$ is a fixed vector and $\mathbf{x}$ is a free vector constrained only by $\langle \mathbf{x}, \mathbf{x} \rangle_o \leq 7$.

89. Compute the work done by a force of 40 newtons acting in the direction specified by $(3, 5, 7)$ and causing a displacement of 25 units in the direction $(1, 1, 6)$.

90. Derive the last formula in the collision application.

COMPUTER EXERCISES 7.1

1. Let $\mathbf{P} = \begin{bmatrix} \frac{1}{3} & \frac{2}{3} & \frac{2}{3} \\ \frac{2}{3} & x & y \\ \frac{2}{3} & y & z \end{bmatrix}$

Use mathematical software such as Maple to find all symmetric orthogonal matrices $\mathbf{P}$ whose first row and first column are as shown.

2. Consider
$$\left\{ \begin{bmatrix} \frac{1}{2} & -\frac{1}{2} \\ \frac{1}{2} & \frac{1}{2} \end{bmatrix}, \begin{bmatrix} \frac{1}{2} & \frac{1}{2} \\ -\frac{1}{2} & \frac{1}{2} \end{bmatrix}, \begin{bmatrix} 0 & \frac{1}{2}\sqrt{2} \\ \frac{1}{2}\sqrt{2} & 0 \end{bmatrix}, \begin{bmatrix} \frac{1}{2}\sqrt{2} & 0 \\ 0 & -\frac{1}{2}\sqrt{2} \end{bmatrix} \right\}$$

Determine numerically if this set is an orthonormal basis for $\mathbb{R}^{2\times 2}$.
Hint: Use the dot-product
$$\langle \mathbf{A}, \mathbf{B} \rangle = \mathrm{Trace}(\mathbf{B}^T \mathbf{A})$$

3. Consider
$$\left\{ \begin{bmatrix} 1 \\ 1 \\ 1 \\ 0 \end{bmatrix}, \begin{bmatrix} 1 \\ -1 \\ 0 \\ 2 \end{bmatrix}, \begin{bmatrix} -1 \\ -1 \\ 2 \\ 0 \end{bmatrix} \right\}$$

Establish numerically if the vectors in this set are mutually orthogonal in $\mathbb{R}^4$

4. Let

$$U = \text{Span}\left\{ \begin{bmatrix} 1 \\ 3 \\ 4 \\ -3 \end{bmatrix}, \begin{bmatrix} 2 \\ 3 \\ 2 \\ -3 \end{bmatrix}, \begin{bmatrix} 1 \\ 2 \\ 2 \\ -2 \end{bmatrix} \right\}$$

$$V = \text{Span}\left\{ \begin{bmatrix} 2 \\ 3 \\ 3 \\ -1 \end{bmatrix}, \begin{bmatrix} 1 \\ 2 \\ 3 \\ 0 \end{bmatrix}, \begin{bmatrix} 1 \\ 1 \\ 0 \\ -1 \end{bmatrix} \right\}$$

Numerically compute the basis for $U + V$ and its dimension.

7.2 ORTHOGONALITY

> *The art of doing mathematics is finding that special case that contains all the germs of generality.*
>
> —DAVID HILBERT (1862–1943)

> *... the source of all great mathematics is the special case. It is frequent in mathematics that every instance of a concept of seemingly great generality is in essence the same as a small and concrete special case.*
>
> —PAUL R. HALMOS (1916–2006)

Introduction

In the preceding section, we discussed the orthogonal projection of a point onto a finite-dimensional subspace V in an inner-product space. Our procedure utilized an orthonormal basis for V. How do we know that V has an orthonormal basis? We can produce one by imitating the method of finding an (unrestricted) basis for a subspace. Start by selecting any nonzero vector $\mathbf{v}_1$ in V. If V contains a nonzero vector $\mathbf{v}_2$ that is orthogonal to $\mathbf{v}_1$, put it in the basis. Then, if V contains a nonzero vector $\mathbf{v}_3$ that is orthogonal to $\mathbf{v}_1$ and $\mathbf{v}_2$, put it in the basis. Proceed in this way. (See Figure 7.7.) The chosen points $\mathbf{v}_1, \mathbf{v}_2, \ldots$ will be mutually orthogonal. The generated set is also linearly independent, by Theorem 2 in Section 7.1. Thus, if V is n dimensional, the selection process certainly must stop after n steps. We will show that the algorithm generates an orthogonal set of vectors. The resulting set is, therefore, orthogonal, and each vector in the set is nonzero. If each vector $\mathbf{v}_i$ is now normalized, the result is an **orthonormal** base for V. Normalizing a vector $\mathbf{v}$ means replacing $\mathbf{v}$ by $\mathbf{v}/\|\mathbf{v}\|$. The norm of a vector is derived from the inner product: $\|\mathbf{x}\| = \sqrt{\langle \mathbf{x}, \mathbf{x} \rangle}$. Throughout this section the norm used is the familiar Euclidean norm.

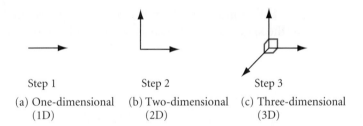

Step 1   Step 2   Step 3

(a) One-dimensional   (b) Two-dimensional   (c) Three-dimensional
(1D)   (2D)   (3D)

FIGURE 7.7   Mutually orthogonal vectors.

The Gram–Schmidt Process

A concrete realization of a process similar to the one just described is the Gram–Schmidt[3] process. It operates in any finite-dimensional inner-product space and produces an orthonormal base by modifying a given non-orthogonal base. In this algorithm, the vectors are normalized as we go along. Suppose that a basis $\{\mathbf{v}_1, \mathbf{v}_2, \ldots\}$ is available for an inner-product space. For the first step, we define $\mathbf{w}_1$ to be the normalized version of $\mathbf{v}_1$; that is, $\mathbf{w}_1 = \mathbf{v}_1 / \|\mathbf{v}_1\|$. For an inductive definition, suppose that we have constructed an orthonormal system $\mathbf{w}_1, \mathbf{w}_2, \ldots, \mathbf{w}_{k-1}$ whose span is the same as $\mathrm{Span}\{\mathbf{v}_1, \mathbf{v}_2, \ldots, \mathbf{v}_{k-1}\}$. To get $\mathbf{w}_k$, subtract from $\mathbf{v}_k$ its projection on the span of $\{\mathbf{w}_1, \mathbf{w}_2, \ldots, \mathbf{w}_{k-1}\}$, and then normalize it. The formula for this process is

$$\mathbf{w}_k = \frac{\mathbf{v}_k - \sum_{j=1}^{k-1} \langle \mathbf{v}_k, \mathbf{w}_j \rangle \mathbf{w}_j}{\left\| \mathbf{v}_k - \sum_{j=1}^{k-1} \langle \mathbf{v}_k, \mathbf{w}_j \rangle \mathbf{w}_j \right\|} \qquad (k = 2, 3, \ldots)$$

The new basis has the property that for each k,

$$\mathrm{Span}\{\mathbf{w}_1, \mathbf{w}_2, \ldots, \mathbf{w}_k\} = \mathrm{Span}\{\mathbf{v}_1, \mathbf{v}_2, \ldots, \mathbf{v}_k\}$$

---

[3] A footnote in Section 7.2 gives information about Jorge Pedersen Gram. Another important figure in linear algebra was Erhard Schmidt (1876–1959), who served as a professor of mathematics in several German universities. He became an expert in the eigenfunctions that arise in the study of integral equations and partial differential equations, and he was one of the first to make use of infinite-dimensional vector spaces in his work. He introduced the notation $\| \cdot \|$ for the magnitude of a vector, and $\langle \mathbf{x}, \mathbf{y} \rangle$ for the inner product. He proved the Pythagorean Theorem in abstract inner-product spaces and many other results in this subject while it was in its infancy and new to almost all mathematicians. In a 1907 paper, Schmidt described what is now called the Gram–Schmidt process. (It should be noted that Pierre-Simon Laplace was familiar with this process before either Gram or Schmidt.)

If that inductive definition is hard to grasp, look at the first few steps in the algorithm. The vectors $\mathbf{w}_1$, $\mathbf{w}_2$, and $\mathbf{w}_3$ are calculated as shown:

$$\mathbf{w}_1 = \frac{\mathbf{v}_1}{\|\mathbf{v}_1\|}$$

$$\mathbf{w}_2 = \frac{\mathbf{v}_2 - \langle \mathbf{v}_2, \mathbf{w}_1 \rangle \mathbf{w}_1}{\|\mathbf{v}_2 - \langle \mathbf{v}_2, \mathbf{w}_1 \rangle \mathbf{w}_1\|}$$

$$\mathbf{w}_3 = \frac{\mathbf{v}_3 - \langle \mathbf{v}_3, \mathbf{w}_1 \rangle \mathbf{w}_1 - \langle \mathbf{v}_3, \mathbf{w}_2 \rangle \mathbf{w}_2}{\|\mathbf{v}_3 - \langle \mathbf{v}_3, \mathbf{w}_1 \rangle \mathbf{w}_1 - \langle \mathbf{v}_3, \mathbf{w}_2 \rangle \mathbf{w}_2\|}$$

and so on. If unnormalized vectors $\widetilde{\mathbf{w}}_k$ are introduced, the process can be described in another set of equations:

$$\widetilde{\mathbf{w}}_1 = \mathbf{v}_1, \qquad\qquad\qquad \mathbf{w}_1 = \frac{\widetilde{\mathbf{w}}_1}{\|\widetilde{\mathbf{w}}_1\|}$$

$$\widetilde{\mathbf{w}}_2 = \mathbf{v}_2 - \langle \mathbf{v}_2, \mathbf{w}_1 \rangle \mathbf{w}_1, \qquad \mathbf{w}_2 = \frac{\widetilde{\mathbf{w}}_2}{\|\widetilde{\mathbf{w}}_2\|}$$

$$\widetilde{\mathbf{w}}_3 = \mathbf{v}_3 - \langle \mathbf{v}_3, \mathbf{w}_1 \rangle \mathbf{w}_1 - \langle \mathbf{v}_3, \mathbf{w}_2 \rangle \mathbf{w}_2, \qquad \mathbf{w}_3 = \frac{\widetilde{\mathbf{w}}_3}{\|\widetilde{\mathbf{w}}_3\|}$$

and so on.

EXAMPLE 1 Apply the Gram–Schmidt process to the three vectors $\mathbf{v}_1 = (1, 0, 3)$, $\mathbf{v}_2 = (2, 2, 0)$, and $\mathbf{v}_3 = (3, 1, 2)$, in the order given.

SOLUTION In the first step, $\widetilde{\mathbf{w}}_1 = \mathbf{v}_1 = (1, 0, 3)$. The first normalized vector is

$$\mathbf{w}_1 = \widetilde{\mathbf{w}}_1 / \|\widetilde{\mathbf{w}}_1\| = \left(1/\sqrt{10}\right)(1, 0, 3)$$

The next vector (in unnormalized form) is

$$\widetilde{\mathbf{w}}_2 = \mathbf{v}_2 - \langle \mathbf{v}_2, \mathbf{w}_1 \rangle \mathbf{w}_1 = (2, 2, 0) - \left(2/\sqrt{10}\right) \left[\left(1/\sqrt{10}\right)(1, 0, 3)\right] = \tfrac{1}{5}(9, 10, -3)$$

Note that at this stage there is a vector $\widetilde{\mathbf{w}}_2$ that is orthogonal to $\widetilde{\mathbf{w}}_1$, by Theorem 2 in Section 7.1. The numerical verification in this problem yields $\langle \widetilde{\mathbf{w}}_1, \widetilde{\mathbf{w}}_2 \rangle = (1, 0, 3) \cdot \tfrac{1}{5}(9, 10, -3) = 0$. Normalization now produces

$$\|\widetilde{\mathbf{w}}_2\| = (\tfrac{1}{5})\sqrt{190}, \qquad \mathbf{w}_2 = \widetilde{\mathbf{w}}_2 / \|\widetilde{\mathbf{w}}_2\| = \left(1/\sqrt{190}\right)(9, 10, -3)$$

We obtain $\langle \mathbf{w}_1, \mathbf{w}_2 \rangle = 0$.

The next step subtracts from $\mathbf{v}_3$ its orthogonal projection onto the span of $\mathbf{w}_1$ and $\mathbf{w}_2$, creating

$$\begin{aligned}
\widetilde{\mathbf{w}}_3 &= \mathbf{v}_3 - \langle \mathbf{v}_3, \mathbf{w}_1 \rangle \mathbf{w}_1 - \langle \mathbf{v}_3, \mathbf{w}_2 \rangle \mathbf{w}_2 \\
&= (3, 1, 2) - (9/\sqrt{10})[(1/\sqrt{10})(1, 0, 3)] - (31/\sqrt{190})[(1/\sqrt{190})(9, 10, -3)] \\
&= (3, 1, 2) - (9/10)(1, 0, 3) - (31/190)(9, 10, -3) \\
&= (4/19)(3, -3, -1)
\end{aligned}$$

Normalizing produces

$$\left\| \mathbf{w}_3 \right\|_2 = 4/\sqrt{19}, \qquad \mathbf{w}_3 = \widetilde{\mathbf{w}}_3 / \left\| \widetilde{\mathbf{w}}_3 \right\| = (1/\sqrt{19})(3, -3, -1)$$

As usual, it is gratifying to make a few checks to see that the results are correct. For example, $\langle \mathbf{w}_1, \mathbf{w}_3 \rangle = [(1/\sqrt{10})(1, 0, 3)] \cdot [(1/\sqrt{19})(3, -3, -1)] = 0$. Also, we obtain $\langle \mathbf{w}_2, \mathbf{w}_3 \rangle = [(1/\sqrt{190})(9, 10, -3)] \cdot [(1/\sqrt{19})(3, -3, -1)] = 0$. $\blacksquare$

Before leaving this example, let us review what we have done. First, we have created an orthonormal basis for $\mathbb{R}^3$. But we could have immediately written down $\{\mathbf{e}_1, \mathbf{e}_2, \mathbf{e}_3\}$ as an orthonormal basis. The special feature possessed by the basis created in the example is that $\mathbf{w}_1$ is a multiple of $\mathbf{v}_1$, $\mathbf{w}_2$ is in the span of $\{\mathbf{v}_1, \mathbf{v}_2\}$, and, finally, $\mathbf{w}_3$ is in the span of $\{\mathbf{v}_1, \mathbf{v}_2, \mathbf{v}_3\}$. (This last fact is not surprising, since the three $\mathbf{v}_i$-vectors span $\mathbb{R}^3$.)

THEOREM 1

Every finite-dimensional inner-product space has an orthonormal basis.

PROOF The Gram–Schmidt process can be applied to any basis. $\blacksquare$

Unnormalized Gram–Schmidt Algorithm

A somewhat different algorithm can be used to produce an orthogonal basis for a given subspace. (Such a basis can be normalized, term-by-term, to yield an orthonormal base if one is needed.) This algorithm has the distinct advantage of bypassing the extraction of square roots.

The basic idea is the same: Subtract from a given vector its orthogonal projection onto a subspace. Formally, start with a basis $\{\mathbf{v}_1, \mathbf{v}_2, \ldots\}$. The newly created vectors will be denoted by $\mathbf{z}_1, \mathbf{z}_2, \ldots$ so that confusion with the earlier algorithm is avoided. Thus, define $\mathbf{z}_1 = \mathbf{v}_1$ to start. Next, construct $\mathbf{z}_2$ by subtracting from $\mathbf{v}_2$ its projection on $\mathbf{z}_1$. The result is

$$\mathbf{z}_2 = \mathbf{v}_2 - \frac{\langle \mathbf{v}_2, \mathbf{z}_1 \rangle}{\langle \mathbf{z}_1, \mathbf{z}_1 \rangle} \mathbf{z}_1$$

In the next step, we subtract from $\mathbf{v}_3$ its projections onto $\mathbf{z}_1$ and $\mathbf{z}_2$:

$$\mathbf{z}_3 = \mathbf{v}_3 - \frac{\langle \mathbf{v}_3, \mathbf{z}_1 \rangle}{\langle \mathbf{z}_1, \mathbf{z}_1 \rangle} \mathbf{z}_1 - \frac{\langle \mathbf{v}_3, \mathbf{z}_2 \rangle}{\langle \mathbf{z}_2, \mathbf{z}_2 \rangle} \mathbf{z}_2$$

Step k does this:

$$\mathbf{z}_k = \mathbf{v}_k - \sum_{j=1}^{k-1} \frac{\langle \mathbf{v}_k, \mathbf{z}_j \rangle}{\langle \mathbf{z}_j, \mathbf{z}_j \rangle} \mathbf{z}_j$$

The main difference between the algorithms for the $\mathbf{w}_k$ and $\mathbf{z}_k$ is that the vectors $\mathbf{w}_k$ are normalized after each step, whereas the $\mathbf{z}_k$ are not. Hence, they remain unnormalized! Avoiding the calculation of square roots is another advantage.

EXAMPLE 2 Find an orthogonal basis for $\mathbb{R}^3$ by applying the *unnormalized* version of the Gram–Schmidt process to the three vectors $\mathbf{v}_1 = (1, 0, 3)$, $\mathbf{v}_2 = (2, 2, 0)$, and $\mathbf{v}_3 = (3, 1, 2)$ from Example 1.

SOLUTION In Step 1, we simply set $\mathbf{z}_1 = \mathbf{v}_1 = (1, 0, 3)$. In Step 2, we obtain $\mathbf{z}_2$ by subtracting from $\mathbf{v}_2$ its orthogonal projection on $\mathbf{z}_1$. Thus, we have

$$\mathbf{z}_2 = \mathbf{v}_2 - \frac{\langle \mathbf{v}_2, \mathbf{z}_1 \rangle}{\langle \mathbf{z}_1, \mathbf{z}_1 \rangle} \mathbf{z}_1 = (2, 2, 0) - \tfrac{1}{5}(1, 0, 3) = \tfrac{1}{5}(9, 10, -3)$$

This calculation gives us a scalar multiple of $(9, 10, -3)$. In this version of the Gram–Schmidt process, we do not hesitate to use convenient multiples of the vectors, since we seek only the orthogonality and not the normality of the vectors. Hence, we can simply let $\mathbf{z}_2 = (9, 10, -3)$. In Step 3, we subtract from $\mathbf{v}_3$ its orthogonal projections on $\mathbf{z}_1$ and $\mathbf{z}_2$:

$$\mathbf{z}_3 = \mathbf{v}_3 - \frac{\langle \mathbf{v}_3, \mathbf{z}_1 \rangle}{\langle \mathbf{z}_1, \mathbf{z}_1 \rangle} \mathbf{z}_1 - \frac{\langle \mathbf{v}_3, \mathbf{z}_2 \rangle}{\langle \mathbf{z}_2, \mathbf{z}_2 \rangle} \mathbf{z}_2 = \mathbf{v}_3 - (9/10)\mathbf{z}_1 - (31/190)\mathbf{z}_2 = (3, 1, 2)$$

$$- (9/10)(1, 0, 3) - (31/190)(9, 10, -3) = (1/190)(120, -120, -40)$$

Again, ignoring scalar multiples, we get $\mathbf{z}_3 = (-3, 3, 1)$. A skeptical reader can check the orthogonality, $\langle \mathbf{z}_1, \mathbf{z}_2 \rangle = 0$, $\langle \mathbf{z}_1, \mathbf{z}_3 \rangle = 0$, and $\langle \mathbf{z}_2, \mathbf{z}_3 \rangle = 0$. ∎

For hand calculations, it is easier to construct an orthonormal basis by first constructing an orthogonal basis and then normalizing the vectors all at once at the end. From the preceding work, we see that the unnormalized vectors can be $\mathbf{z}_1 = (1, 0, 3)$, $\mathbf{z}_2 = (9, 10, -3)$, and $\mathbf{z}_3 = (-3, 3, 1)$.

Their norms are $\left\|\mathbf{z}_1\right\|_2 = \sqrt{10}$, $\left\|\mathbf{z}_2\right\|_2 = \sqrt{190}$, and $\left\|\mathbf{z}_3\right\|_2 = \sqrt{19}$. The normalized vectors are then $\tilde{\mathbf{z}}_1 = \left(1/\sqrt{10}\right)(1, 0, 3)$, $\tilde{\mathbf{z}}_2 = \left(1/\sqrt{190}\right)(9, 10, -3)$, and $\tilde{\mathbf{z}}_3 = \left(1/\sqrt{19}\right)(-3, 3, 1)$. These agree with the results in Example 1.

Modified Gram–Schmidt Process

By Theorem 5 in Section 7.1, the orthogonal projection of $\mathbf{x}$ onto $\mathbf{y}$ is defined to be

$$\text{proj}_{\mathbf{y}}\,\mathbf{x} = \left[\frac{\langle \mathbf{x}, \mathbf{y} \rangle}{\langle \mathbf{y}, \mathbf{y} \rangle}\right]\mathbf{y}$$

This projects the vector $\mathbf{x}$ orthogonally onto the vector $\mathbf{y}$. The Gram–Schmidt process starting with vectors $\mathbf{v}_1, \mathbf{v}_2, \ldots$ can be written as

$$\mathbf{z}_1 = \mathbf{v}_1, \qquad\qquad\qquad \mathbf{q}_1 = \frac{\mathbf{z}_1}{\left\|\mathbf{z}_1\right\|}$$

$$\mathbf{z}_2 = \mathbf{v}_2 - \text{proj}_{\mathbf{z}_1}\,\mathbf{v}_2, \qquad\qquad \mathbf{q}_2 = \frac{\mathbf{z}_2}{\left\|\mathbf{z}_2\right\|}$$

$$\mathbf{z}_3 = \mathbf{v}_3 - \text{proj}_{\mathbf{z}_1}\,\mathbf{v}_3 - \text{proj}_{\mathbf{z}_2}\,\mathbf{v}_3, \qquad \mathbf{q}_3 = \frac{\mathbf{z}_3}{\left\|\mathbf{z}_3\right\|}$$

In general, the kth step is

$$\mathbf{z}_k = \mathbf{v}_k - \sum_{j=1}^{k-1} \text{proj}_{\mathbf{z}_j}\,\mathbf{v}_k, \qquad \mathbf{q}_k = \frac{\mathbf{z}_k}{\left\|\mathbf{z}_k\right\|}$$

Here $\{\mathbf{z}_1, \mathbf{z}_2, \mathbf{z}_3, \ldots, \mathbf{z}_k\}$ is an orthogonal set and $\{\mathbf{q}_1, \mathbf{q}_2, \mathbf{q}_3, \ldots, \mathbf{q}_k\}$ is an orthonormal set. When implemented on a computer, the Gram–Schmidt process is numerically unstable because the vectors $\mathbf{z}_k$ may not be exactly orthogonal due to roundoff errors. By a minor modification, the Gram–Schmidt process can be stabilized. Instead of computing the vectors $\mathbf{u}_k$ as above, we can compute them as

$$\mathbf{z}_k \leftarrow \mathbf{v}_k - \text{proj}_{\mathbf{z}_1}\,\mathbf{v}_2$$

$$\mathbf{z}_k \leftarrow \mathbf{z}_k - \text{proj}_{\mathbf{z}_2}\,\mathbf{z}_k$$

$$\mathbf{z}_k \leftarrow \mathbf{z}_k - \text{proj}_{\mathbf{z}_3}\,\mathbf{z}_k$$

$$\vdots$$

$$\mathbf{z}_k \leftarrow \mathbf{z}_k - \text{proj}_{\mathbf{z}_{k-2}}\,\mathbf{z}_k$$

$$\mathbf{z}_k \leftarrow \mathbf{z}_k - \text{proj}_{\mathbf{z}_{k-1}}\,\mathbf{z}_k$$

where $\mathbf{z}_k$ is overwritten repeatedly. In exact arithmetic, this computation gives the same results as the original form above. However, it produces smaller errors in finite-precision computer arithmetic. A computer algorithm for the modified Gram–Schmidt process follows:

$$
\begin{aligned}
&\textbf{for } j = 1 \textbf{ to } n \\
&\quad \textbf{for } i = 1 \textbf{ to } j - 1 \\
&\qquad s \leftarrow \langle \mathbf{v}_j, \mathbf{v}_i \rangle \\
&\qquad \mathbf{v}_j \leftarrow \mathbf{v}_j - s\mathbf{v}_i \\
&\quad \textbf{end for} \\
&\quad \mathbf{v}_j \leftarrow \mathbf{v}_j / \|\mathbf{v}_j\| \\
&\textbf{end for}
\end{aligned}
$$

Here the vectors $\mathbf{v}_1, \mathbf{v}_2, \ldots, \mathbf{v}_n$ are replaced by an orthonormal set of vectors that spans the same subspace. The i-loop removes components in the $\mathbf{v}_i$ direction followed by normalization of the vector.

EXAMPLE 3 Apply the Modified Gram–Schmidt process to the three vectors $\mathbf{v}_1 = (1, 0, 3)$, $\mathbf{v}_2 = (2, 2, 0)$, and $\mathbf{v}_3 = (3, 1, 2)$ from Example 1, finding an orthonormal basis for $\mathbb{R}^3$.

SOLUTION In Step 1 $(j = 1)$, we set $\mathbf{v}_1 \leftarrow \mathbf{v}_1 / \|\mathbf{v}_1\| = (1, 0, 3)/\sqrt{10} \approx (0.31623, 0, 0.94868)$. In Step 2 $(j = 2, i = 1)$, we obtain $s \leftarrow \langle \mathbf{v}_2, \mathbf{v}_1 \rangle = (2, 2, 0) \cdot (1, 0, 3)/\sqrt{10} = \sqrt{10}/5$ and $\mathbf{v}_2 \leftarrow \mathbf{v}_2 - s\mathbf{v}_1 = (2, 2, 0) - \frac{1}{5}(1, 0, 3) = (\frac{9}{5}, 2, -\frac{3}{5})$, $\|\mathbf{v}_2\| = \frac{1}{5}\sqrt{(190)}$, $\mathbf{v}_2 \leftarrow \mathbf{v}_2 / \|\mathbf{v}_2\| = (9, 10, -3)/\sqrt{190} \approx (0.65293, 0.72548, -0.21764)$. In Step 3 $(j = 3, i = 1, 2)$, we obtain $s \leftarrow \langle \mathbf{v}_3, \mathbf{v}_1 \rangle = (3, 1, 2) \cdot (1, 0, 3)/\sqrt{10} = 9/\sqrt{10}$, $\mathbf{v}_3 \leftarrow \mathbf{v}_3 - s\mathbf{v}_1 = (3, 1, 2) - 0.9(1, 0, 3) = (2.1, 1, -0.7)$, and $s \leftarrow \langle \mathbf{v}_3, \mathbf{v}_2 \rangle = (2.1, 1, -0.7) \cdot (9, 10, -3)/\sqrt{190} = (31/190)$, $\mathbf{v}_3 \leftarrow \mathbf{v}_3 - s\mathbf{v}_2 = (2.1, 1, -0.7) - (31/\sqrt{190})(9, 10, -3) = (12, -12, -4)/19$, and $\|\mathbf{v}_3\| = 4/\sqrt{19} \approx 1.48748$, $\mathbf{v}_3 \leftarrow \mathbf{v}_3 / \|\mathbf{v}_3\| = (3, 2, -1)/\sqrt{19} \approx (0.68825, -0.68825, -0.22942)$. Verify the orthonormality. ∎

EXAMPLE 4 Find a basis for $[\text{Col}(\mathbf{A})]^\perp$, where $\mathbf{A}$ is the matrix

$$
\begin{bmatrix}
3 & -1 \\
1 & 2 \\
-1 & 5
\end{bmatrix}
$$

SOLUTION The question is posed in $\mathbb{R}^3$, and the columns of $\mathbf{A}$ form a basis for a two-dimensional subspace in $\mathbb{R}^3$. Thus, we need only one vector

x that is orthogonal to the two columns in **A**. Using $\mathbf{A}^T\mathbf{x} = \mathbf{0}$, the conditions on the vector **x** are

$$\begin{cases} 3x_1 + x_2 - x_3 = 0 \\ -1x_1 + 2x_2 + 5x_3 = 0 \end{cases}$$

With familiar row manipulations, we arrive at

$$\begin{bmatrix} 3 & 1 & -1 \\ -1 & 2 & 5 \end{bmatrix} \sim \begin{bmatrix} 1 & 0 & -1 \\ 0 & 1 & 2 \end{bmatrix}$$

We can use the solution $\mathbf{x} = [1, -2, 1]^T$, obtained by setting the free variable x_3 equal to 1. ∎

THEOREM 2

A point **p** *is the orthogonal projection of a point* **x** *onto a subspace* V *(in an inner-product space) if and only if* **p** *is the point in* V *closest to* **x**.

PROOF Let **v** be any point in V other than **p**, and assume that **p** is the orthogonal projection of **x** on V. Then $\mathbf{p} - \mathbf{v}$ is in V, and is orthogonal to $\mathbf{x} - \mathbf{p}$. The Pythagorean law then is applicable:

$$\left\|\mathbf{x} - \mathbf{v}\right\|^2 = \left\|(\mathbf{x} - \mathbf{p}) + (\mathbf{p} - \mathbf{v})\right\|^2 = \left\|\mathbf{x} - \mathbf{p}\right\|^2 + \left\|\mathbf{p} - \mathbf{v}\right\|^2 > \left\|\mathbf{x} - \mathbf{p}\right\|^2$$

Conversely, suppose that **p** is a point in V as close as possible to **x**. Then for any $\mathbf{v} \in V$ and for any positive λ, we have

$$\begin{aligned} 0 &\le \left\|\mathbf{x} - \mathbf{p} - \lambda\mathbf{v}\right\|^2 - \left\|\mathbf{x} - \mathbf{p}\right\|^2 \\ &= \left\|\mathbf{x} - \mathbf{p}\right\|^2 - 2\langle\mathbf{x} - \mathbf{p}, \lambda\mathbf{v}\rangle + \lambda^2\left\|\mathbf{v}\right\|^2 - \left\|\mathbf{x} - \mathbf{p}\right\|^2 \\ &= \lambda\left[-2\langle\mathbf{x} - \mathbf{p}, \mathbf{v}\rangle + \lambda\left\|\mathbf{v}\right\|^2\right] \end{aligned}$$

It follows that $2\langle\mathbf{x} - \mathbf{p}, \mathbf{v}\rangle \le \lambda\left\|\mathbf{v}\right\|^2$. Let λ converge downward to 0, to conclude that $\langle\mathbf{x} - \mathbf{p}, \mathbf{v}\rangle \le 0$. Because **v** was arbitrary in V, we can replace it by $-\mathbf{v}$ and conclude that $\langle\mathbf{x} - \mathbf{p}, \mathbf{v}\rangle \ge 0$. Hence, we have $\mathbf{x} - \mathbf{p} \perp \mathbf{v}$ and $\mathbf{x} - \mathbf{p} \perp V$. ∎

Linear Least-Squares Solution

Many important applications depend on Theorem 2. For example, we might wish to replace a complicated function by a simpler one, giving up a small amount of precision. Among the functions offered as possible replacements, we may want to select the one closest to the original function. Under the right circumstances, an orthogonal projection can be employed. We shall discuss at greater length the problem of finding the *best approximate solution* to a system of equations $\mathbf{Ax} = \mathbf{b}$ when the system itself is inconsistent.

Suppose that we have an inconsistent system

$$\mathbf{Ax} = \mathbf{b}$$

where $\mathbf{A}$ is an $m \times n$ matrix. Often, the number of equations, m, is much greater than the number of unknowns, n. This occurs when there is a mass of data available for determining a small number of parameters in a linear model. The data would typically be inconsistent, and our task is to invent a way of computing the parameters despite this inconsistency. Inconsistency means that the vector $\mathbf{b}$ is not in the span of the set of columns of $\mathbf{A}$. The strategy is to replace $\mathbf{b}$ with the point $\mathbf{p}$ in the column space of $\mathbf{A}$ that is nearest to $\mathbf{b}$. Then solve the system

$$\mathbf{Ax} = \mathbf{p}$$

The point $\mathbf{p}$ should be the orthogonal projection of $\mathbf{b}$ onto the column space of $\mathbf{A}$. This point is characterized by two properties:

$$\mathbf{p} \in \mathrm{Col}(\mathbf{A}), \qquad \mathbf{b} - \mathbf{p} \perp \mathrm{Col}(\mathbf{A})$$

Figure 7.8 illustrates the situation just described. These requirements are met if

$$\mathbf{p} = \mathbf{Ax} \quad \text{for some } \mathbf{x}, \qquad \mathbf{A}^T(\mathbf{b} - \mathbf{p}) = \mathbf{0}$$

This last equation asserts that the rows of $\mathbf{A}^T$ are orthogonal to the column vector $\mathbf{b} - \mathbf{p}$. Putting all this together, we have

$$\mathbf{A}^T(\mathbf{b} - \mathbf{Ax}) = \mathbf{0} \qquad \text{or} \qquad \mathbf{A}^T\mathbf{Ax} = \mathbf{A}^T\mathbf{b}$$

The equations in this system are called the **normal equations** for the original problem, $\mathbf{Ax} = \mathbf{b}$. (The term *normal* is used in mathematics to mean perpendicular.) The resulting solution vector $\mathbf{x}$ is called the **least-squares solution** of the original problem. That terminology will be explained later. To summarize our findings in this paragraph, we state the relevant theorem.

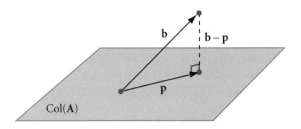

FIGURE 7.8 Properties of least squares.

> **THEOREM 3**
>
> *If a system of equations $\mathbf{Ax} = \mathbf{b}$ is inconsistent, then the system of normal equations,*
>
> $$\mathbf{A}^T\mathbf{Ax} = \mathbf{A}^T\mathbf{b}$$
>
> *can be solved for $\mathbf{x}$, and with this choice of $\mathbf{x}$, the point $\mathbf{Ax}$ will be as close to $\mathbf{b}$ as possible.*

EXAMPLE 5 Find the least-squares solution of the system

$$\begin{cases} x_1 + 4x_2 = 1 \\ 2x_1 + 5x_2 = 1 \\ 3x_1 + 6x_2 = -2 \end{cases}$$

SOLUTION The system is inconsistent, and we therefore pass directly to the *normal equations*:

$$\mathbf{A}^T\mathbf{Ax} = \mathbf{A}^T\mathbf{b}$$

$$\begin{bmatrix} 1 & 2 & 3 \\ 4 & 5 & 6 \end{bmatrix} \begin{bmatrix} 1 & 4 \\ 2 & 5 \\ 3 & 6 \end{bmatrix} \begin{bmatrix} x_1 \\ x_2 \end{bmatrix} = \begin{bmatrix} 1 & 2 & 3 \\ 4 & 5 & 6 \end{bmatrix} \begin{bmatrix} 1 \\ 1 \\ -2 \end{bmatrix}$$

Carrying out the multiplications leads to the *normal equations*

$$\begin{bmatrix} 14 & 32 \\ 32 & 77 \end{bmatrix} \begin{bmatrix} x_1 \\ x_2 \end{bmatrix} = \begin{bmatrix} -3 \\ -3 \end{bmatrix}$$

As we know, this should be consistent. Its solution is $\mathbf{x} = [-\frac{5}{2}, 1]^T$. ∎

It is interesting to note that the normal equations for a least-squares matrix problem are automatically consistent, although the matrix $\mathbf{A}^T\mathbf{A}$ may be noninvertible! Expressed in another way, the matrix $\mathbf{A}^T\mathbf{A}$ may undergo a loss in rank, yet the system of normal equations will be consistent. This is because the expression

$$\min_{\mathbf{x} \neq \mathbf{0}} \|\mathbf{Ax} - \mathbf{b}\|$$

has a minimum point that must solve the normal equations. An example showing a loss of rank follows.

EXAMPLE 6 Find the least-squares solution of this overdetermined system:

$$\begin{cases} 1x_1 + 3x_2 + 2x_3 = 1 \\ 2x_1 + x_2 - 5x_3 = 2 \\ x_1 + 13x_2 + 20x_3 = -3 \\ 5x_1 + 10x_2 + x_3 = 4 \end{cases}$$

SOLUTION The matrices involved are

$$A = \begin{bmatrix} 1 & 3 & 2 \\ 2 & 1 & -5 \\ 1 & 13 & 20 \\ 5 & 10 & 1 \end{bmatrix}, \qquad b = \begin{bmatrix} 1 \\ 2 \\ -3 \\ 4 \end{bmatrix}$$

To get the least-squares solution of the system $Ax = b$, we solve the normal equations

$$A^T A x = A^T b$$

After carrying out those matrix multiplications, we arrive at

$$\begin{bmatrix} 31 & 68 & 17 \\ 68 & 279 & 271 \\ 17 & 271 & 430 \end{bmatrix} \begin{bmatrix} x_1 \\ x_2 \\ x_3 \end{bmatrix} = \begin{bmatrix} 22 \\ 6 \\ -64 \end{bmatrix}$$

$$\left[\begin{array}{ccc|c} 31 & 68 & 17 & 22 \\ 68 & 279 & 271 & 6 \\ 17 & 271 & 430 & -64 \end{array} \right] \sim \left[\begin{array}{ccc|c} 1 & 0 & -17/5 & 205/144 \\ 0 & 1 & \frac{9}{5} & -69/212 \\ 0 & 0 & 0 & 0 \end{array} \right]$$

Thus, we have $x_1 = (17/5)x_3 + 205/144$ and $x_2 = -\frac{9}{5}x_3 - 69/212$, where there is a free variable and there are multiple least-squares solutions. Taking $x_3 = 0$, we get one solution $x = [205/144, -69/212, 0]^T \approx [1.4236, -0.3255, 0]^T$, which has the norm value 0.7960. If MATLAB is applied to the original system of equations, however, we would get the approximate solution of least norm, which is $[0, 1172/2737, -479/1144]^T \approx [0, 0.4282, -0.4187]^T$ and the norm value is also 0.7960. This example has a coefficient matrix of rank two. ∎

There is another way of being led to an approximate solution of an overdetermined system $Ax = b$. It is a direct frontal attack on the problem of minimizing the expression $\|Ax - b\|$ by choice of the vector x. This will give us immediately the x for which Ax is as close as possible to b. Notice that the minimum point of $\|Ax - b\|$ is the same as the minimum point of $\left\| \mathbf{Ax} - \mathbf{b} \right\|^2$, since the squaring function is increasing on the positive half of the real line. To find the minimum of $\left\| \mathbf{Ax} - \mathbf{b} \right\|^2$, we use the standard technique from calculus; namely, we set the partial derivatives equal to zero and solve the resulting equations.

$$0 = \frac{\partial}{\partial x_k} \left\| \mathbf{Ax} - \mathbf{b} \right\|^2 = \frac{\partial}{\partial x_k} \sum_{i=1}^{m} (\mathbf{Ax} - \mathbf{b})_i^2$$

$$= \frac{\partial}{\partial x_k} \sum_{i=1}^{m} \left(\sum_{j=1}^{n} a_{ij} x_j - b_i \right)^2 = \sum_{i=1}^{m} 2 \left(\sum_{j=1}^{n} a_{ij} x_j - b_i \right) a_{ik}$$

The result is the system

$$\sum_{i=1}^{m} \sum_{j=1}^{n} a_{ij} a_{ik} x_j = \sum_{i=1}^{m} a_{ik} b_i, \qquad (k = 1, 2, \ldots, n)$$

This is the same as

$$\mathbf{A}^T \mathbf{Ax} = \mathbf{A}^T \mathbf{b}$$

which is the **normal equation**.

A separate argument shows that the expression we are minimizing has no local maxima, and the point obtained by solving the normal equations really is a minimum point. To verify this, let $\widehat{\mathbf{x}}$ be the solution of the normal equations. Then

$$\mathbf{A}^T (\mathbf{A}\widehat{\mathbf{x}} - \mathbf{b}) = 0$$

Let $\mathbf{v}$ be any vector. Then we have

$$\langle \mathbf{A}\widehat{\mathbf{x}} - \mathbf{b}, \mathbf{Av} \rangle = \langle \mathbf{A}^T (\mathbf{A}\widehat{\mathbf{x}} - \mathbf{b}), \mathbf{v} \rangle = 0$$

Hence, we have

$$\left\| \mathbf{A}(\widehat{\mathbf{x}} + \mathbf{v}) - \mathbf{b} \right\|^2 = \left\| (\mathbf{A}\widehat{\mathbf{x}} - \mathbf{b}) + \mathbf{Av} \right\|^2$$

$$= \left\| \mathbf{A}\widehat{\mathbf{x}} - \mathbf{b} \right\|^2 + 2 \langle \mathbf{A}\widehat{\mathbf{x}} - \mathbf{b}, \mathbf{Av} \rangle + \left\| \mathbf{Av} \right\|^2$$

$$= \left\| \mathbf{A}\widehat{\mathbf{x}} - \mathbf{b} \right\|^2 + \left\| \mathbf{Av} \right\|^2$$

$$\geq \left\| \mathbf{A}\widehat{\mathbf{x}} - \mathbf{b} \right\|^2$$

This shows that if we change $\widehat{\mathbf{x}}$ by adding to it any vector $\mathbf{v}$, then the value of $\left\| \mathbf{A}\widehat{\mathbf{x}} - \mathbf{b} \right\|$ can only increase, and $\widehat{\mathbf{x}}$ is a minimum point.

EXAMPLE 7 A common use of the linear least-squares procedure is in finding the best fit of n data points by a straight line. We seek the equation of a line in the form $y = mx + b$, and the given data points are (x_i, y_i), where $1 \leq i \leq n$. Suppose we are given this table of values:

x_i	1.0	2.0	2.5	3.0
y_i	3.7	4.1	4.3	5.0

How do we proceed?

SOLUTION With four data points (x_i, y_i), we have four equations of the form

$$\begin{cases} mx_1 + b = y_1 \\ mx_2 + b = y_2 \\ mx_3 + b = y_3 \\ mx_4 + b = y_4 \end{cases}$$

This can be written as

$$\mathbf{Aw = y}: \qquad \begin{bmatrix} x_1 & 1 \\ x_2 & 1 \\ x_3 & 1 \\ x_4 & 1 \end{bmatrix} \begin{bmatrix} m \\ b \end{bmatrix} = \begin{bmatrix} y_1 \\ y_2 \\ y_3 \\ y_4 \end{bmatrix}$$

After we insert the given data values, we compute the normal equations

$$\mathbf{A}^T\mathbf{Aw = A}^T\mathbf{y}: \qquad \begin{bmatrix} 1 & 2 & 2.5 & 3 \\ 1 & 1 & 1 & 1 \end{bmatrix} \begin{bmatrix} 1 & 1 \\ 2 & 1 \\ 2.5 & 1 \\ 3 & 1 \end{bmatrix} \begin{bmatrix} m \\ b \end{bmatrix}$$

$$= \begin{bmatrix} 1 & 2 & 2.5 & 3 \\ 1 & 1 & 1 & 1 \end{bmatrix} \begin{bmatrix} 3.7 \\ 4.1 \\ 4.3 \\ 5.0 \end{bmatrix}$$

The augmented matrix for the normal equations is

$$\begin{bmatrix} 20.25 & 8.5 & | & 17.10 \\ 8.5 & 4.0 & | & 37.65 \end{bmatrix} \sim \begin{bmatrix} 1 & 0 & | & 0.6 \\ 0 & 1 & | & 3.0 \end{bmatrix}$$

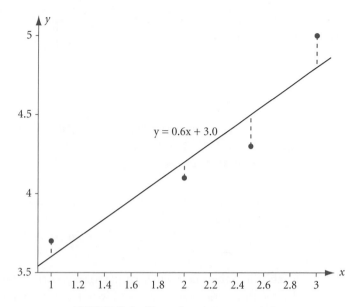

FIGURE 7.9 Linear least-squares solution.

Solving this system, we find that the line $y = 0.6x + 3.0$ is the best fit in the least-squares sense. (The sum of the squares of the vertical distances from the data points to the line is minimized.) Figure 7.9 illustrates this situation. ∎

Gram Matrix

At the beginning of this section, we saw how to create an orthonormal basis for an inner-product space. This turned out to be useful in projecting a point orthogonally onto a subspace. There are other ways to proceed, however, and the labor of concocting an orthonormal base can be avoided, to some extent.

If U is a subspace and $\mathbf{x}$ is a point in the containing space, then the orthogonal projection of $\mathbf{x}$ onto U is a point $\mathbf{p}$ having the following properties:

$$\mathbf{p} \in U, \qquad \mathbf{x} - \mathbf{p} \perp U$$

This definition does not refer to an orthonormal or orthogonal base for the subspace. See Figure 7.10 for the mental picture that guides us here.

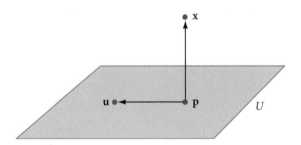

FIGURE 7.10 The point **p** is the orthogonal projection of **x** on U.

If $\{\mathbf{u}_1, \mathbf{u}_2, \ldots, \mathbf{u}_n\}$ is an orthonormal basis for the subspace U, then

$$\mathbf{p} = \sum_{i=1}^{n} \langle \mathbf{x}, \mathbf{u}_i \rangle \mathbf{u}_i$$

Establishing that the given formula for **p** is correct can be done by verifying the two Properties 1 and 2 above. First, it is clear that $\mathbf{p} \in U$, by its definition. That $\mathbf{x} - \mathbf{p} \perp U$ can be seen by computing $\langle \mathbf{x} - \mathbf{p}, \mathbf{u}_j \rangle$ for $1 \le j \le n$. The appropriate calculation is

$$\langle \mathbf{x} - \mathbf{p}, \mathbf{u}_j \rangle = \langle \mathbf{x}, \mathbf{u}_j \rangle - \langle \mathbf{p}, \mathbf{u}_j \rangle = \langle \mathbf{x}, \mathbf{u}_j \rangle - \left\langle \sum_{i=1}^{n} \langle \mathbf{x}, \mathbf{u}_i \rangle \mathbf{u}_i, \ \mathbf{u}_j \right\rangle$$

$$= \langle \mathbf{x}, \mathbf{u}_j \rangle - \sum_{i=1}^{n} \langle \mathbf{x}, \mathbf{u}_i \rangle \langle \mathbf{u}_i, \mathbf{u}_i \rangle = \langle \mathbf{x}, \mathbf{u}_j \rangle - \langle \mathbf{x}, \mathbf{u}_j \rangle = 0$$

If $\{\mathbf{w}_1, \mathbf{w}_2, \ldots, \mathbf{w}_n\}$ is an orthogonal basis for U, then

$$\mathbf{p} = \sum_{i=1}^{n} \left[\frac{\langle \mathbf{x}, \mathbf{w}_i \rangle}{\langle \mathbf{w}_i, \mathbf{w}_i \rangle} \right] \mathbf{w}_i$$

From each of these formulas, one can see that the map $\mathbf{x} \mapsto \mathbf{p}$ is linear.

If we have a basis $\{\mathbf{v}_1, \mathbf{v}_2, \ldots, \mathbf{v}_n\}$ for U that has no other special properties, we set $\mathbf{p} = \sum_{i=1}^{n} c_i \mathbf{v}_i$ and impose the orthogonality condition:

$$(\mathbf{x} - \mathbf{p}) \perp U \qquad \text{or} \qquad \left(\mathbf{x} - \sum_{i=1}^{n} c_i \mathbf{v}_i \right) \perp \mathbf{v}_j \qquad (1 \le j \le n)$$

This leads to a system of linear equations to be solved:

$$\sum_{i=1}^{n} c_i \langle \mathbf{v}_i, \mathbf{v}_j \rangle = \langle \mathbf{x}, \mathbf{v}_j \rangle \quad (1 \le j \le n)$$

This is a system of n linear equations in the n unknowns c_j. The coefficient matrix for this system is an $n \times n$ **Gram**[4] **matrix** $\mathbf{G} = (g_{ij})$ with elements $g_{ij} = \langle \mathbf{v}_i, \mathbf{v}_j \rangle$ for $1 \leq i, j \leq n$.

The right-hand side in our system of equations is a vector $\mathbf{b} = (b_i)$ whose components are the numbers $b_j = \langle \mathbf{x}, \mathbf{v}_j \rangle$. The system of linear equations can now be written in the familiar form

$$\mathbf{Gc} = \mathbf{b}$$

The coefficient matrix $\mathbf{G}$ is always invertible, according to the next theorem. Thus, the solution of our problem is unique and given explicitly by $\mathbf{c} = \mathbf{G}^{-1}\mathbf{b}$. The system above often arises in applications.

Consider an overdetermined system

$$\mathbf{Ax} = \mathbf{b}$$

where $\mathbf{A}$ is $m \times n$. In ordinary least squares, we minimize $\|\mathbf{Ax} - \mathbf{b}\|^2$ by solving the normal equations

$$\mathbf{A}^T \mathbf{A} \widehat{\mathbf{x}} = \mathbf{A}^T \mathbf{b}$$

When the columns of $\mathbf{A}$ form a linearly independent set, $\mathbf{A}^T\mathbf{A}$ is invertible and the best approximate solution is $\widehat{\mathbf{x}} = (\mathbf{A}^T\mathbf{A})^{-1}\mathbf{A}^T\mathbf{b}$.

In applying the least-squares strategy to a practical problem, it can happen that the function values reported from experiments are not of uniform reliability. A flexible approach would be to assign *weights* to the data points. If $x_1 < x_2 < \cdots < x_m$ are nodes and $y_1, y_2, \ldots, y_m$ are the measured quantities at the n nodes, then let $w_1, w_2, \ldots, w_m$ be the positive weights that indicate in a coarse manner the variations in precision of the data. For the best approximate solution of the new system

$$\mathbf{WAx} = \mathbf{Wb}$$

we compute the solution $\widehat{\mathbf{x}}_w$ of the **weighted normal equations**

$$(\mathbf{WA})^T (\mathbf{WA}) \widehat{\mathbf{x}}_w = (\mathbf{WA})^T \mathbf{Wb}$$

---

[4] Jorgen Pedersen Gram (1850–1916) is best remembered for the Gram–Schmidt orthogonalization process. He was not the first to use this method, for it seems to have been known to Laplace and Cauchy. Gram published his first important mathematical paper while still a university student! Rather than teaching mathematics at a university he became a research mathematician employed by an insurance company. He published papers, gave lectures, and won awards for his mathematical research. At the age of 65, Gram was killed after being struck by a bicycle.

Here $\mathbf{W}$ is an invertible matrix related to the weights w_i. We have a new inner product $\langle \mathbf{x}, \mathbf{y} \rangle_w = (\mathbf{W}\mathbf{y})^T (\mathbf{W}\mathbf{x})$ and new length $\|\mathbf{x}\|_w = \|\mathbf{W}\mathbf{x}\|$. For example, in the case of the weighted linear least squares, the function to be minimized is $\sum_{i=1}^{m} w_i^2 (a + bx_i - y_i)^2$ and we might use a diagonal weight matrix such as $\mathbf{W} = \mathrm{Diag}(w_i^{\frac{1}{2}})$.

THEOREM 4

If a set $\{\mathbf{u}_1, \mathbf{u}_2, \ldots, \mathbf{u}_n\}$ is linearly independent (in an inner-product space), then the corresponding Gram matrix with elements $\langle \mathbf{u}_i, \mathbf{u}_j \rangle$ is nonsingular.

PROOF To prove the invertibility of the Gram matrix, one can show that its rows form a linearly independent set. (It will follow that the rank of the Gram matrix is n, and that the matrix is invertible. Recall the Invertible Matrix Theorem, in Section 3.2.) Suppose that the rows of the Gram matrix satisfy an equation

$$\sum_{i=1}^{n} c_i \langle \mathbf{u}_i, \mathbf{u}_j \rangle = 0 \qquad (1 \le j \le n)$$

In quick succession, we conclude that

$$\left\langle \sum_{i=1}^{n} c_i \mathbf{u}_i, \ \sum_{j=1}^{n} c_j \mathbf{u}_j \right\rangle = 0, \qquad \left\| \sum_{i=1}^{n} c_i \mathbf{u}_i \right\|^2 = 0, \qquad \sum_{i=1}^{n} c_i \mathbf{u}_i = 0$$

Since the set $\{\mathbf{u}_1, \mathbf{u}_2, \ldots, \mathbf{u}_n\}$ is assumed to be linearly independent, the coefficients c_i must be 0. ∎

EXAMPLE 8 Let U be the subspace in $\mathbb{R}^3$ spanned by the two vectors $\mathbf{v}_1 = (2, 11, -5)$ and $\mathbf{v}_2 = (6, 2, 1)$. Find the orthogonal projection of the point $\mathbf{z} = (-9, 50, -27)$ onto U. Carry out some independent verifications of the work.

SOLUTION The projection point can be denoted by $\mathbf{p}$. Because $\mathbf{p}$ must be a point in the subspace, we can write

$$\mathbf{p} = c_1 \mathbf{v}_1 + c_2 \mathbf{v}_2$$

The orthogonality conditions are $z - p \perp v_1$ and $z - p \perp v_2$. This leads to $\langle p, v_i \rangle = \langle z, v_i \rangle$ for $i = 1, 2$. Substituting the form of p, we have

$$\langle c_1 v_1 + c_2 v_2, v_i \rangle = \langle z, v_i \rangle \qquad (i = 1, 2)$$

The following pair of equations is equivalent to the preceding pair:

$$\begin{cases} c_1 \langle v_1, v_1 \rangle + c_2 \langle v_2, v_1 \rangle = \langle z, v_1 \rangle \\ c_1 \langle v_1, v_2 \rangle + c_2 \langle v_2, v_2 \rangle = \langle z, v_2 \rangle \end{cases}$$

Carrying out the calculations, we arrive at

$$\begin{cases} 150c_1 + 29c_2 = 667 \\ 29c_1 + 41c_2 = 19 \end{cases}$$

MATLAB gives the answer to this as $c_1 \approx 5.0473$ and $c_2 \approx -3.1066$. We have solved a least-squares problem in this example. Namely, we have found the "*least-squares solution*" to the inconsistent system

$$\mathbf{Ac} = \mathbf{b}: \qquad \begin{bmatrix} 2 & 6 \\ 11 & 2 \\ -5 & 1 \end{bmatrix} \begin{bmatrix} c_1 \\ c_2 \end{bmatrix} = \begin{bmatrix} -9 \\ 50 \\ -27 \end{bmatrix}$$

What is the projection p? It is $c_1 v_1 + c_2 v_2$, which turns out to be approximately

$$p = \begin{bmatrix} -8.5451 \\ 49.3069 \\ -23.3430 \end{bmatrix}$$

How close is z to its projected point? This question is answered by computing

$$\| z - p \| \approx \left\| \begin{bmatrix} -0.4549 \\ 0.6931 \\ 1.3430 \end{bmatrix} \right\| \approx 1.5783$$

For one verification of the work, we can test whether $z - p$ is orthogonal to the subspace spanned by v_1 and v_2. Put the vectors v_1 and v_2 into a matrix A as columns, $A = [v_1, v_2]$, and premultiply this matrix by the row vector $z - p$.

$$(z - p)\mathbf{A} = \begin{bmatrix} -0.4549, & 0.6931, & 1.3430 \end{bmatrix} \begin{bmatrix} 2 & 6 \\ 11 & 2 \\ -5 & 1 \end{bmatrix}$$

$$= 10^{-13} \begin{bmatrix} 0.0711, & -0.4263 \end{bmatrix}$$

Recall that MATLAB will solve inconsistent systems of linear equations in the least-squares sense with the simple command $\texttt{x = A\b}$. For the problem in this example, the result (after entering the matrix $\mathbf{A}$ and vector $\mathbf{b}$) is

$$\mathbf{x} = \begin{bmatrix} 5.0473 \\ -3.1066 \end{bmatrix} \qquad \blacksquare$$

Distance from a Point to a Hyperplane

A typical problem in higher-dimensional geometry is the calculation of the distance from a point to a hyperplane in $\mathbb{R}^n$. In $\mathbb{R}^2$, a hyperplane is a line. In $\mathbb{R}^3$, it will be a plane, and so on. A hyperplane through the origin in $\mathbb{R}^n$ will be a subspace of dimension $n - 1$. It is a special case that frequently arises in applications.

EXAMPLE 9 How can we compute the distance from a given point to a hyperplane passing through $\mathbf{p}$ in $\mathbb{R}^n$?

SOLUTION A hyperplane in $\mathbb{R}^n$ through the point $\mathbf{p}$ is a set of the form

$$H = \big\{ \mathbf{x} : \langle \mathbf{x} - \mathbf{p}, \mathbf{u} \rangle = 0 \big\}$$

where $\mathbf{u}$ is a nonzero vector in $\mathbb{R}^n$. In this example, we let $\mathbf{p} = \mathbf{0}$ if the hyperplane contains the vector $\mathbf{0}$.

Let $\mathbf{y}$ be any point in $\mathbb{R}^n$. We seek a formula for the distance from $\mathbf{y}$ to H. Let $\mathbf{x}$ be a point on H as close as possible to the point $\mathbf{y}$. Thus, $\mathbf{x}$ will be the orthogonal projection of $\mathbf{y}$ onto H. (Recall Theorem 2.) The distance sought will then be $\|\mathbf{x} - \mathbf{y}\|$. The point $\mathbf{x}$ is characterized by two properties:

$$\mathbf{x} \in H, \qquad \mathbf{y} - \mathbf{x} \perp H$$

These two conditions lead to $\langle \mathbf{x} - \mathbf{p}, \mathbf{u} \rangle = 0$ and $\mathbf{y} - \mathbf{x} = \lambda \mathbf{u}$. Putting these two properties together, we have $\langle \mathbf{y} - \mathbf{p} - \lambda \mathbf{u}, \mathbf{u} \rangle = 0$. Thus, we obtain $\langle \mathbf{y} - \mathbf{p}, \mathbf{u} \rangle = \lambda \langle \mathbf{u}, \mathbf{u} \rangle$ and $\lambda = \langle \mathbf{y} - \mathbf{p}, \mathbf{u} \rangle / \langle \mathbf{u}, \mathbf{u} \rangle$. Hence, we have

$$\mathbf{x} = \mathbf{y} - \lambda \mathbf{u} = \mathbf{y} - \left[\frac{\langle \mathbf{y} - \mathbf{p}, \mathbf{u} \rangle}{\langle \mathbf{u}, \mathbf{u} \rangle} \right] \mathbf{u}$$

The formula for the distance from $\mathbf{x}$ to $\mathbf{y}$ is

$$\|\mathbf{x} - \mathbf{y}\| = \left\| \left[\frac{\langle \mathbf{y} - \mathbf{p}, \mathbf{u} \rangle}{\langle \mathbf{u}, \mathbf{u} \rangle} \right] \mathbf{u} \right\| = \frac{|\langle \mathbf{y} - \mathbf{p}, \mathbf{u} \rangle|}{\|\mathbf{u}\|} \qquad \blacksquare$$

Suppose **a** and **p** are vectors in $\mathbb{R}^n$ with $\mathbf{a} \neq \mathbf{0}$. The set of all vectors **x** in $\mathbb{R}^n$ satisfying the equation

$$\langle \mathbf{x} - \mathbf{p}, \mathbf{a} \rangle = 0$$

is a hyperplane through the point **p**. This is a **normal equation** for the hyperplane with a **normal vector a**. Let $\mathbf{a} = (a_1, a_2, \ldots, a_n), \mathbf{p} = (p_1, p_2, \ldots, p_n)$, and $\mathbf{x} = (x_1, x_2, \ldots, x_n)$. We may write the normal equation for the hyperplane as

$$a_1 x_1 + a_2 x_2 + \cdots + a_n x_n = d$$

where $d = \langle \mathbf{p}, \mathbf{a} \rangle$.

EXAMPLE 10 Compute a point in the plane $x + 2y + z = 4$ in $\mathbb{R}^3$ closest to the given point $\mathbf{y} = (2, 3, 3)$ and find the distance between these two points.

SOLUTION We have $\mathbf{a} = (1, 2, 1)$, $\mathbf{p} = (1, 1, 1)$, and $d = \langle \mathbf{p}, \mathbf{a} \rangle = 4$. From Example 9, the closest point in the plane to **y** is

$$\mathbf{x} = \mathbf{y} - \left[\frac{\langle \mathbf{y} - \mathbf{p}, \mathbf{a} \rangle}{\langle \mathbf{a}, \mathbf{a} \rangle} \right] \mathbf{a} = \begin{bmatrix} 2 \\ 3 \\ 3 \end{bmatrix} - \frac{7}{6} \begin{bmatrix} 1 \\ 2 \\ 1 \end{bmatrix} = \frac{1}{6} \begin{bmatrix} 5 \\ 4 \\ 11 \end{bmatrix}$$

and the distance between **y** and **x** is $|\langle \mathbf{y} - \mathbf{p}, \mathbf{a} \rangle| / \|\mathbf{a}\| = 7/\sqrt{6}$. ∎

Mathematical Software

Here are the MATLAB, Maple, and Mathematica commands to carry out the calculations in Example 5. We show the input typed by the user but not the output.

MATLAB	Maple
format long	with(LinearAlgebra):
A = [1,4;2,5;3,6]	A := Matrix([[1,4],[2,5],[3,6]]);
b = [1;1;-2]	b := Vector([1,1,-1]);
E = A'*A	E := Transpose(A).A;
f = A'*b	f := Transpose(A).b;
G = [E f]	G := Matrix(E,f);
rref(G)	ReducedRowEchelonForm(G);

```
Mathematica
A = {{1,4},{2,5},{3,6}}
b = {1,1,-2}
E = Transpose[A].A
f = Transpose[A].b
G = {E, f}
RowReduce[G]
```

The results from the three software systems are similar.

What alternative is available in these mathematical software packages for the problem in Example 5? After entering the data $\mathbf{A}$ and $\mathbf{b}$ of the problem, one can use the MATLAB command x=A\b. MATLAB will use its own preferred method for solving the equation in the least-squares sense. One can use commands in Maple or in Mathematica to solve least-squares problems directly.

SUMMARY 7.2

- **Gram–Schmidt process**, starting with a basis $\{\mathbf{v}_1, \mathbf{v}_2, \dots, \mathbf{v}_n\}$:

  $\mathbf{w}_1 = \mathbf{v}_1 / \|\mathbf{v}_1\|$

  $\mathbf{w}_2 = \left[\mathbf{v}_2 - \langle \mathbf{v}_2, \mathbf{w}_1\rangle \mathbf{w}_1\right] / \|\mathbf{v}_2 - \langle \mathbf{v}_2, \mathbf{w}_1\rangle \mathbf{w}_1\|$

  $\mathbf{w}_3 = \left[\mathbf{v}_3 - \langle \mathbf{v}_3, \mathbf{w}_1\rangle \mathbf{w}_1 - \langle \mathbf{v}_3, \mathbf{w}_2\rangle \mathbf{w}_2\right] /$

  $\qquad \|\mathbf{v}_3 - \langle \mathbf{v}_3, \mathbf{w}_1\rangle \mathbf{w}_1 - \langle \mathbf{v}_3, \mathbf{w}_2\rangle \mathbf{w}_2\|$

  $\mathbf{w}_k = \left[\mathbf{v}_k - \sum_{j=1}^{k-1} \langle \mathbf{v}_k, \mathbf{w}_j\rangle \mathbf{w}_j\right] /$

  $\qquad \|\mathbf{v}_k - \sum_{j=1}^{k-1} \langle \mathbf{v}_k, \mathbf{w}_j\rangle \mathbf{w}_j\|$

  $\text{Span}\{\mathbf{v}_1, \mathbf{v}_2, \dots, \mathbf{v}_k\} = \text{Span}\{\mathbf{w}_1, \mathbf{w}_2, \dots, \mathbf{w}_k\}$, for all k, and $\{\mathbf{w}_1, \dots, \mathbf{w}_n\}$ is an orthonormal base.

- **Gram–Schmidt process**, starting with a basis $\{\mathbf{v}_1, \mathbf{v}_2, \dots, \mathbf{v}_n\}$:

  $\widetilde{\mathbf{w}}_1 = \mathbf{v}_1$, and $\mathbf{w}_1 = \widetilde{\mathbf{w}} / \|\widetilde{\mathbf{w}}_1\|$

  $\widetilde{\mathbf{w}}_2 = \mathbf{v}_2 - \langle \mathbf{v}_2, \mathbf{w}_1\rangle \mathbf{w}_1$, and $\mathbf{w}_2 = \widetilde{\mathbf{w}} / \|\widetilde{\mathbf{w}}_2\|$

  $\widetilde{\mathbf{w}}_3 = \mathbf{v}_3 - \langle \mathbf{v}_3, \mathbf{w}_1\rangle \mathbf{w}_1 - \langle \mathbf{v}_3, \mathbf{w}_2\rangle \mathbf{w}_2$

  $\mathbf{w}_3 = \widetilde{\mathbf{w}}_3 / \|\widetilde{\mathbf{w}}_3\|$, and in general

  $\widetilde{\mathbf{w}}_k = \mathbf{v}_k - \sum_{j=1}^{k-1} \langle \mathbf{v}_k, \mathbf{w}_j\rangle \mathbf{w}_j$ and $\mathbf{w}_k = \widetilde{\mathbf{w}}_k / \|\widetilde{\mathbf{w}}_k\|$

  Here $\widetilde{\mathbf{w}}_k$ are unnormalized vectors.

  $\text{Span}\{\mathbf{v}_1, \mathbf{v}_2, \dots, \mathbf{v}_n\} = \text{Span}\{\mathbf{w}_1, \mathbf{w}_2, \dots, \mathbf{w}_n\}$, the latter being an orthonormal basis.

- **Unnormalized Gram–Schmidt algorithm**, starting with a basis $\{\mathbf{v}_1, \mathbf{v}_2, \dots, \mathbf{v}_n\}$:

  $\mathbf{z}_1 = \mathbf{v}_1$

  $\mathbf{z}_2 = \mathbf{v}_2 - \left[\langle \mathbf{v}_2, \mathbf{z}_1\rangle / \langle \mathbf{z}_1, \mathbf{z}_1\rangle\right] \mathbf{z}_1$

  $\mathbf{z}_3 = \mathbf{v}_3 - \left[\langle \mathbf{v}_3, \mathbf{z}_1\rangle / \langle \mathbf{z}_1, \mathbf{z}_1\rangle\right] \mathbf{z}_1 -$

  $\qquad \langle \mathbf{v}_3, \mathbf{z}_2\rangle / \langle \mathbf{z}_2, \mathbf{z}_2\rangle\right] \mathbf{z}_2$

  In general, $\mathbf{z}_k = \mathbf{v}_k - \sum_{j=1}^{k-1} \left[\langle \mathbf{v}_k, \mathbf{z}_j\rangle / \langle \mathbf{z}_j, \mathbf{z}_j\rangle\right] \mathbf{z}_j$

  $\text{Span}\{\mathbf{v}_1, \mathbf{v}_2, \dots, \mathbf{v}_k\} = \text{Span}\{\mathbf{z}_1, \mathbf{z}_2, \dots, \mathbf{z}_k\}$, the latter being an orthogonal basis.

- Linear least squares: These three pairs of assertions are logically equivalent:

  $\left\{ \mathbf{p} \in \text{Col}(\mathbf{A}) \text{ and } \mathbf{b} - \mathbf{p} \perp \text{Col}(\mathbf{A}) \right\}$

  $\left\{ \mathbf{p} = \mathbf{A}\mathbf{x} \text{ for some } \mathbf{x} \text{ and } \mathbf{A}^T(\mathbf{b} - \mathbf{p}) = \mathbf{0} \right\}$

  $\left\{ \mathbf{A}^T\mathbf{A}\mathbf{x} = \mathbf{A}^T\mathbf{b} \right\}$

- The vector $\mathbf{x}$ that gives the minimum value of the function $\mathbf{x} \mapsto \|\mathbf{A}\mathbf{x} - \mathbf{b}\|$ is the $\mathbf{x}$ that satisfies the equation $\mathbf{A}^T\mathbf{A}\mathbf{x} = \mathbf{A}^T\mathbf{b}$. (This is the normal equation.)

- The orthogonal projection of a point $\mathbf{x}$ onto a subspace U is the point $\mathbf{p} \in U$ such that $\mathbf{x} - \mathbf{p} \perp U$.

- The orthogonal projection of a point $\mathbf{x}$ onto a subspace possessing an orthonormal basis $\{\mathbf{u}_1, \mathbf{u}_2, \ldots, \mathbf{u}_n\}$ is $\sum_{i=1}^{n} \langle \mathbf{x}, \mathbf{u}_i \rangle \mathbf{u}_i$.
- The orthogonal projection of a point $\mathbf{x}$ onto a subspace possessing an orthogonal basis $\{\mathbf{u}_i\}$ is $\mathbf{p} = \sum_{i=1}^{n} \left[\langle \mathbf{x}, \mathbf{u}_i \rangle / \langle \mathbf{u}_i, \mathbf{u}_i \rangle \right] \mathbf{u}_i$.
- If a set $\{\mathbf{u}_1, \mathbf{u}_2, \ldots, \mathbf{u}_n\}$ is linearly independent (in an inner-product space), then the corresponding Gram matrix with elements $\langle \mathbf{u}_i, \mathbf{u}_j \rangle$ is invertible.

- Orthogonality theorems:
  - Every finite-dimensional inner-product space has an orthonormal basis.
  - A point $\mathbf{p}$ is the orthogonal projection of a point $\mathbf{x}$ onto a subspace V if and only if $\mathbf{p}$ is the point in V closest to $\mathbf{x}$.
  - If the equation $\mathbf{A}\mathbf{x} = \mathbf{b}$ is inconsistent, we get an approximate solution by solving $\mathbf{A}^T\mathbf{A}\mathbf{x} = \mathbf{A}^T\mathbf{b}$ for $\mathbf{x}$. The point $\mathbf{A}\mathbf{x}$, with this $\mathbf{x}$, is as close to $\mathbf{b}$ as possible.

KEY CONCEPTS 7.2

Gram-Schmidt process (normalized or unnormalized), orthonormal bases, orthogonal projections, least-squares solutions of equations, normal equations, approximate solutions of overdetermined systems, Gram matrix, distance from a point to a hyperplane

GENERAL EXERCISES 7.2

1. Consider
$$\begin{bmatrix} 1 & -3 \\ 2 & 6 \\ 7 & -3 \\ 3 & 4 \end{bmatrix} \begin{bmatrix} x_1 \\ x_2 \end{bmatrix} = \begin{bmatrix} 1 \\ 3 \\ 2 \\ 1 \end{bmatrix}$$

Find the least-squares solution of this system. Take advantage of the fact that the two columns in the coefficient matrix are mutually orthogonal.

2. Find an orthogonal basis for the polynomial space $\mathbb{P}_2$, using the inner product $\langle p, q \rangle = \int_{-1}^{1} p(t)q(t)\, dt$. Apply the Gram–Schmidt algorithm to the ordered basis $\{1, t, t^2\}$.

3. Let $\mathbf{A} = \begin{bmatrix} 1 & 3 \\ 2 & -1 \\ 4 & 0 \end{bmatrix}$, $\mathbf{b} = \begin{bmatrix} -2 \\ 7 \\ 13 \end{bmatrix}$

Find the least-squares solution of the system $\mathbf{A}\mathbf{x} = \mathbf{b}$. Proceed by finding first the projection $\mathbf{p}$ of $\mathbf{b}$ onto $\mathrm{Col}(\mathbf{A})$ and then solving the consistent system $\mathbf{A}\mathbf{x} = \mathbf{p}$.

4. Find the least-squares solution of the system $\begin{bmatrix} 3 & 2 \\ 2 & 3 \\ 1 & 2 \end{bmatrix} \begin{bmatrix} x \\ y \end{bmatrix} = \begin{bmatrix} 3 \\ 0 \\ 1 \end{bmatrix}$

5. In the space $\mathbb{R}^4$, let V be the two-dimensional subspace spanned by the set $\{(1, 2, 7, 3), (-3, 6, -3, 4)\}$. Find the point in V closest to $(11, -2, 1, 2)$. Note that the basis given for V is orthogonal.

6. Apply the Gram–Schmidt process to this (ordered) set of three vectors: $\mathbf{v}_1 = (0, 0, 1, 0)$, $\mathbf{v}_2 = (1, 0, 1, 0)$, and $\mathbf{v}_3 = (1, 1, 1, 1)$. The result should be an orthonormal set.

7. Apply the Gram–Schmidt process to this set of three vectors: $\mathbf{u}_1 = (1, 2, 1)$, $\mathbf{u}_2 = (1, 0, 1)$, and $\mathbf{u}_3 = (3, 1, 0)$.

8. For what values of α and β will the vector $(\alpha, 1, \beta)$ be orthogonal to $(2, 0, 1)$ and $(-1, 1, 2)$?

9. Find an orthonormal base for the row space

of $\begin{bmatrix} 1 & 1 & -1 & 1 \\ 1 & -1 & 1 & 1 \\ 1 & 2 & 0 & 1 \end{bmatrix}$

10. Find the best approximation of the vector $(1, 3, 2)$ by a linear combination of the two vectors $(1, 1, 4)$ and $(4, 0, -1)$. Observe that the latter pair is orthogonal.

11. Find the distance from the point $(1, 0, 2)$ to the subspace spanned by $(1, 2, -1)$ and $(3, -2, -1)$. The latter pair is orthogonal.

12. In Example 7, what is the projection of the point $(1, 2, -3, 4)$ on the column space of the given 4×3 matrix? Is this projected point a unique linear combination of the columns?

13. Find the orthogonal projection of the vector $(5, 3, -7)$ onto the subspace spanned by $(1, 3, 3)$ and $(2, -2, 1)$.

14. Apply the Gram–Schmidt process to the three vectors: $\mathbf{u}_1 = (-1, 1, -1, 3)$, $\mathbf{u}_2 = (0, 0, 0, 4)$, and $\mathbf{u}_3 = (-1, 4, -1, 6)$. Do all the normalization of vectors at the end of the process.

15. A function f defined on $\mathbb{R}$ is said to be **even** if $f(x) = f(-x)$ for all x. It is said to be **odd** if $f(x) = -f(-x)$ for all x. Use the inner product $\langle f, g \rangle = \int_{-1}^{1} f(x)g(x)\, dx$ and prove that any even function is orthogonal to any odd function. Assume that the functions in question are continuous and real valued.

16. Consider the system $\begin{cases} x_1 + 2x_2 = 3 \\ 3x_1 + 2x_2 = 5 \\ 4x_1 + 5x_2 = 2 \end{cases}$

Investigate the validity of the following solution to the system of equations. First we write the system in matrix form as

$$\begin{bmatrix} 1 & 2 \\ 3 & 2 \\ 4 & 5 \end{bmatrix} \begin{bmatrix} x_1 \\ x_2 \end{bmatrix} = \begin{bmatrix} 3 \\ 5 \\ 2 \end{bmatrix}$$

Next, we multiply both sides of the equation by the matrix $\begin{bmatrix} 1 & 3 & 4 \\ 2 & 2 & 5 \end{bmatrix}$ getting the

equation $\begin{bmatrix} 26 & 28 \\ 28 & 33 \end{bmatrix} \begin{bmatrix} x_1 \\ x_2 \end{bmatrix} = \begin{bmatrix} 26 \\ 26 \end{bmatrix}$

The row-reduction process leads then to the

solution $\begin{bmatrix} 65/37 \\ -26/37 \end{bmatrix}$

17. Do you expect these two systems to have the same least-squares solutions?

$$\begin{cases} 3x - 2y = -5 \\ 2x + 3y = 4 \\ x - 2y = 3 \end{cases} \quad \text{and} \quad \begin{cases} 3x - 2y = -5 \\ 4x + 6y = 8 \\ x - 2y = 3 \end{cases}$$

Explain why or why not.

18. If we wish to find the least-squares solution

of $\begin{bmatrix} 3 & 2 \\ 5 & 1 \\ 4 & 2 \\ 3 & 2 \end{bmatrix} \begin{bmatrix} x_1 \\ x_2 \end{bmatrix} = \begin{bmatrix} 6 \\ 7 \\ 5 \\ 6 \end{bmatrix}$ can we begin by

dropping the fourth equation? Explain.

19. Find the orthogonal projection of the point $(1, 2)$ onto the line generated by the points $(3, 1)$ and $(0, 0)$.

20. Determine the orthogonal projection of $\mathbf{x} = (2, 2, 3)$ onto the subspace of $\mathbb{R}^3$ spanned by the two vectors $\mathbf{u} = (4, 1, -2)$

and $\mathbf{v} = (1, -2, 1)$. Notice that $\mathbf{u} \perp \mathbf{v}$. Check your work by verifying independently the two properties that the projection should have.

21. If an orthogonal basis is to be created from a given basis $\{\mathbf{v}_1, \mathbf{v}_2, \ldots, \mathbf{v}_n\}$, we can use this algorithm: $\mathbf{u}_1 = \mathbf{v}_1$ and

$$\mathbf{u}_{k+1} = \mathbf{v}_{k+1} - \sum_{j=1}^{k} \left[\frac{\langle \mathbf{v}_{k+1}, \mathbf{u}_j \rangle}{\langle \mathbf{u}_j, \mathbf{u}_j \rangle} \right] \mathbf{u}_j$$

for $k = 1, 2, \ldots$

Establish that the $\mathbf{u}$-vectors form an orthogonal set. The advantage of this algorithm is that no square roots are needed.

22. (Continuation.) Find an orthogonal basis by using the algorithm in General Exercise 1 on the three vectors $[1, 1, 0]$, $[0, 1, 1]$, and $[1, 0, 1]$. Give your solution in terms of vectors whose components are integers. Is it always possible to find an integer-valued orthogonal system by applying the algorithm in the previous problem to an integer-valued basis?

23. In fitting a table of empirical data $\{(t_i, y_i)\}$, it is usually assumed that the values of the independent variable t are given exactly, whereas the dependent variable y may be infected with observational errors. Assume that the data are to be represented by a function of the form $y = \alpha t + \beta$. The errors at each data point t_i are then $\alpha t_i + \beta - y_i$, and we want to minimize the sum of their squares, $\sum_{i=1}^{m} (\alpha t_i + \beta - y_i)^2$. Find the formulas for α and β. Your formula will involve these quantities:

$$\sum_{i=1}^{m} t_i^2, \quad \sum_{i=1}^{m} t_i, \quad \sum_{i=1}^{m} t_i y_i, \quad \sum_{i=1}^{m} y_i$$

24. Let $\mathbf{A}$ be a matrix whose columns form an orthogonal set. What is the formula for the projection of a vector $\mathbf{b}$ onto the column space of $\mathbf{A}$? Use your formula to give a

simple solution to the least-squares problem $\mathbf{A}\mathbf{x} = \mathbf{b}$ in this special case. (An explicit formula for the vector $\mathbf{x}$ is desired.) Notice that the example in General Exercise 2 is of the type considered here.

25. Use the unnormalized Gram–Schmidt process applied to the ordered basis $\mathbf{x}_1 = [1, 2, 0]^T$, $\mathbf{x}_2 = [2, -1, 1]^T$, and $\mathbf{x}_3 = [-2, 6, 4]^T$ to obtain an orthogonal basis $\{\mathbf{v}_1, \mathbf{v}_2, \mathbf{v}_3\}$. Express the vector $\mathbf{y} = (3, 2, 1)$ as a sum $\mathbf{u} + \mathbf{v}$, where $\mathbf{u}$ is in the span of $\{\mathbf{x}_1, \mathbf{x}_2\}$ and $\mathbf{v}$ is orthogonal to the span of $\{\mathbf{x}_1, \mathbf{x}_2\}$. What are $\mathbf{u}$ and $\mathbf{v}$?

26. Explain why $\|\mathbf{x}+\mathbf{y}\|^2 = \|\mathbf{x}\|^2 + \|\mathbf{y}\|^2$ if $\mathbf{x} \perp \mathbf{y}$. What can you say about the converse? The scalar field should be $\mathbb{C}$ in this problem.

27. Confirm that the null space of $\mathbf{A}^T$ is the orthogonal complement of the column space of $\mathbf{A}$.

28. Affirm the parallelogram law for any inner-product space:
$$\|\mathbf{x} + \mathbf{y}\|^2 + \|\mathbf{x} - \mathbf{y}\|^2 = 2\|\mathbf{x}\|^2 + 2\|\mathbf{y}\|^2$$

29. For any set S in an inner-product space, establish that $S^\perp$ and $\mathrm{Span}(S)$ have only one element in common. What is that one element?

30. Explain why for a specific vector $\mathbf{z}$ and finite-dimensional subspace W in an inner-product space, there is one and only one point $\mathbf{p} \in W$ such that $\mathbf{z} - \mathbf{p} \perp W$.

31. Establish that if $\mathbf{x}$ is orthogonal to $\mathbf{u}$, then every multiple of $\mathbf{x}$ is orthogonal to every multiple of $\mathbf{u}$.

32. Argue that if $\mathbf{u}$ is a unit vector in an inner-product space, then the projection of $\mathbf{x}$ onto $\mathbf{u}$ is $\langle \mathbf{x}, \mathbf{u} \rangle \mathbf{u}$.

33. Affirm that the map $\mathbf{x} \mapsto \mathbf{p}$, where $\mathbf{p}$ is the orthogonal projection of $\mathbf{x}$ onto a finite-dimensional subspace in an inner-product space, is a linear map.

34. Let $\{\mathbf{u}_1, \mathbf{u}_2, \ldots, \mathbf{u}_n\}$ be an orthogonal set of nonzero vectors in an inner-product space. Can you verify that if $\mathbf{x} = \sum_{i=1}^{n} c_i \mathbf{u}_i$, then $\mathbf{x}$ is orthogonal to each vector $\mathbf{u}_j$, and $c_j = \langle \mathbf{x}, \mathbf{u}_j \rangle / \langle \mathbf{u}_j, \mathbf{u}_j \rangle$, for $j = 1, 2, \ldots, n$?

35. In an inner-product space, establish that if $\mathbf{x}$ is orthogonal to $\mathbf{u}_1, \mathbf{u}_2, \ldots, \mathbf{u}_n$, then $\mathbf{x}$ is orthogonal to $\text{Span}\{\mathbf{u}_1, \mathbf{u}_2, \ldots, \mathbf{u}_n\}$. Is the converse true?

36. Let $\{\mathbf{v}_1, \mathbf{v}_2, \ldots, \mathbf{v}_n\}$ be an orthonormal set in an m-dimensional inner-product space, where $m > n$. Explain why there is a vector $\mathbf{v}_{n+1}$ in the space such that $\{\mathbf{v}_1, \mathbf{v}_2, \ldots, \mathbf{v}_{n+1}\}$ is orthonormal.

37. Give an argument that for any matrix $\mathbf{A}$, the matrices $\mathbf{A}^T\mathbf{A}$ and $\mathbf{A}\mathbf{A}^T$ are symmetric.

38. A square matrix $\mathbf{A}$ is said to be **nonnegative definite** if $\mathbf{x}^T\mathbf{A}\mathbf{x} \geq 0$ for all vectors $\mathbf{x}$. Establish that for every matrix $\mathbf{A}$, $\mathbf{A}^T\mathbf{A}$ is nonnegative definite.

39. (Continuation.) Affirm that the eigenvalues of a nonnegative definite matrix are nonnegative.

40. Establish that the least-squares solution $\mathbf{x}$ of the system $\mathbf{A}\mathbf{x} = \mathbf{b}$ has the property $\mathbf{b}^T\mathbf{A}\mathbf{x} \geq 0$.

41. If $\mathbf{A}$ is an $m \times n$ matrix of rank m, then $\mathbf{A}\mathbf{A}^T$ is invertible. Explain why or why not.

42. Find all the matrices $\mathbf{A}$ that have the property $\mathbf{A}^{-1} = \mathbf{A}^T$.

43. Confirm that if $\mathbf{A}^T\mathbf{A}\mathbf{x} = \mathbf{0}$, then $\mathbf{A}\mathbf{x} = \mathbf{0}$.

44. Explain why the matrices $\mathbf{A}^T\mathbf{A}$ and $\mathbf{A}$ have the same kernel.

45. Let $\{\mathbf{w}_1, \mathbf{w}_2, \cdots, \mathbf{w}_n\}$ be an orthonormal base for $\mathbb{R}^n$. Let $V = \text{Span}\{\mathbf{w}_1, \mathbf{w}_2, \ldots, \mathbf{w}_r\}$, where $r < n$. What is a convenient orthonormal basis for $V^\perp$?

46. Let V be a subspace in an inner-product space, and let $\mathbf{v} \in V$. If $\mathbf{w} \perp \mathbf{v}$, does it follow that $\mathbf{w} \perp V$? A proof or example is needed.

47. **a.** Fix a unit vector $\mathbf{v}$ in an inner-product space, and define a linear transformation T by the formula $T(\mathbf{x}) = 2\langle \mathbf{x}, \mathbf{v} \rangle \mathbf{v} - \mathbf{x}$. Draw a picture to see what T does to a given vector $\mathbf{x}$. This transformation is an example of a *reflection*. Affirm that $T^2 = \mathbf{I}$.

  b. Fix a vector $\mathbf{v}$ and define $T(\mathbf{x}) = \langle \mathbf{x}, \mathbf{v} \rangle \mathbf{v} - 2\mathbf{x}$. Under what conditions will T have a nonzero fixed point?

48. Use the ideas in Example 9 to derive this formula for the distance between two hyperplanes: $|\beta - \alpha| / \|\mathbf{u}\|$. The two hyperplanes have equations $\langle \mathbf{u}, \mathbf{x} \rangle = \alpha$ and $\langle \mathbf{u}, \mathbf{y} \rangle = \beta$.

49. Find a 3×3 matrix $\mathbf{A}$ whose columns are mutually orthogonal and whose first two columns are $[3, 1, 2]^T$ and $[2, 0, -3]^T$. Then produce a new matrix $\mathbf{B}$ from $\mathbf{A}$ by dividing each column of $\mathbf{A}$ by its norm. Compute $\mathbf{B}\mathbf{B}^T$ and $\mathbf{B}^T\mathbf{B}$. Are the results what you expect? Explain.

50. Every subspace of $\mathbb{R}^n$ is the kernel of a matrix. Explain why or why not.

51. Verify the orthogonality property of the vectors in Example 2.

52. Find all the 2×2 matrices, $\mathbf{A}$, such that $\mathbf{A}^T\mathbf{A} = \mathbf{I}$.

53. Let U be a subset of an inner-product space X. Assume that $\text{Span}(U) = X$. Explain why the equation $\langle \mathbf{x}, \mathbf{u} \rangle = \langle \mathbf{y}, \mathbf{u} \rangle$ for all $\mathbf{u} \in U$ implies that $\mathbf{x} = \mathbf{y}$.

54. Find an orthonormal base for the space $\mathbb{P}_2$ when the inner-product is defined by the equations $\langle p, g \rangle = \int_{-1}^{1} p(t)g(t)t^2 \, dt$.

55. For any $n \times n$ matrix $\mathbf{A} = (a_{ij})$, establish these inequalities:
a. $|\text{Det}(\mathbf{A})| \le \prod_{i=1}^{n} \sum_{j=1}^{n} |a_{ij}|$

b. $|\text{Det}(\mathbf{A})| \le \prod_{i=1}^{n} \left(\sum_{j=1}^{n} |a_{ij}|^2 \right)^{\frac{1}{2}}$
(Hadamard's inequality)

56. If $\mathbf{x}^T \mathbf{A} \mathbf{x} = 0$ for some nonzero $\mathbf{x}$, does it follow that $\mathbf{A}$ is noninvertible? Explain why or why not.

COMPUTER EXERCISES 7.2

1. The purpose of this problem is to assist the reader in learning the MATLAB commands and their use in solving a typical matrix problem. Start with two column vectors, $[4, 3, 2]^T$ and $[1, 2, -5]^T$. Observe that this is an orthogonal pair. Create the matrix $\mathbf{A}$ with the two given vectors as columns. To find a third vector orthogonal to the two we already have, set up the appropriate homogeneous equations and solve them with the `rref` command. Create the 3×3 matrix using all three column vectors. Check that the columns of $\mathbf{A}$ are mutually orthogonal by computing $\mathbf{A}^T \mathbf{A}$ and being sure it is a diagonal matrix. Next, normalize the columns of $\mathbf{A}$. Now one should have $\mathbf{A}\mathbf{A}^T = \mathbf{I}$ and $\mathbf{A}^T \mathbf{A} = \mathbf{I}$.

2. Redo Example 8 with $\mathbf{z} = (1, 3, 5)$.

3. Find a column vector that can be appended to the matrix $\mathbf{A} = \begin{bmatrix} 4 & 2 \\ 1 & 1 \\ 3 & -3 \end{bmatrix}$, making the set of three columns orthogonal.

4. Using the method of linear least squares, find the linear equation $y = ax + b$ that best fits the data shown between x and y. Plot the graph of the line as well as the original data points.

x	5	7	10	12	14	20	25
y	677	996	2294	2025	3683	3778	4921

5. Find a quadratic equation of the form $y = a + bx + cx^2$ that best fits this data in the least squares sense. Plot the data and the resulting curve.

x	0	25	50	75	100	125	150
y	20	80	185	190	200	155	25

Hint: See Computer Exercise 51 in Section 2.2.

CHAPTER EIGHT

Additional Topics

8.1 HERMITIAN MATRICES AND THE SPECTRAL THEOREM

> *It is clear that the chief end of mathematical study must be to make the student think.*
>
> —JOHN WESLEY YOUNG (1880–1932)

> *Cogito Ergo Sum. I think, therefore I am.*
>
> —RENÉ DESCARTES (1596–1650)

Introduction

In Section 5.3 on coordinate systems, we saw instances of a linear transformation that could be put into a simple form—at the expense of changing the basis of the underlying space. Later, in Section 5.3, we saw how to find a basis so that a given linear transformation would be represented by a diagonal matrix. This turned out to be possible in some cases, but not in all. Here we take up this theme again, and shall find that for a symmetric real matrix or Hermitian complex matrix, the situation is favorable: The diagonalization is always possible, and the new basis is orthonormal. Further, the eigenvalues of matrices that are symmetric and Hermitian are necessarily real. It is easy to see why these theorems are regarded as pinnacles in the

458

theory of linear algebra. The main results are two versions of the *Spectral Theorems* (Theorems 6–7).

Hermitian Matrices and Self-Adjoint Mappings

We proceed in as general a way as possible. Thus, we consider arbitrary inner-product spaces and linear maps rather than limiting ourselves to $\mathbb{R}^n$ and matrices. Furthermore, it is better to consider inner-product spaces in which the scalars are allowed to be complex numbers. The first important example of this is the space $\mathbb{C}^n$, whose elements are vectors with n complex components. The **standard inner product** in $\mathbb{C}^n$ is

$$\langle \mathbf{x}, \mathbf{y} \rangle = \mathbf{y}^H \mathbf{x} = \begin{bmatrix} \overline{y}_1, \overline{y}_2, \cdots, \overline{y}_n \end{bmatrix} \begin{bmatrix} x_1 \\ x_2 \\ \vdots \\ x_n \end{bmatrix} = x_1 \overline{y}_1 + x_2 \overline{y}_2 + \cdots + x_n \overline{y}_n = \sum_{i=1}^{n} x_i \overline{y}_i$$

If a vector $\mathbf{y}$ in $\mathbb{C}^n$ is regarded as an $n \times 1$ matrix, (i.e., a column vector), then $\mathbf{y}^H$ is the corresponding row vector, with the conjugates of the components in place of the original components.

> ## DEFINITION
>
> *A matrix* $\mathbf{A}$ *is* **Hermitian**[1] *if* $\mathbf{A} = \mathbf{A}^H$. *Equivalently, the entries in* $\mathbf{A}$ *must obey this rule:* $a_{ij} = \overline{a}_{ji}$.

> **EXAMPLE 1** Show Hermitian matrices containing some imaginary numbers.

---

[1] Charles Hermite (1822–1901) is remembered for many accomplishments in mathematics, one of the first being his establishing in 1873 that the number e is transcendental. (That means that e is not the root of any polynomial whose coefficients are integers.) Later, Hermite's name was attached to an orthogonal system of polynomials that eventually played a crucial role in quantum mechanics. Hermite had a defect in his right foot and walked with some difficulty. After a year in the university, he was refused the right to continue his studies because of this disability! Although this unfair decision was reversed after pressure from some important people, Hermite decided he did not want to stay under the strict conditions imposed on him. Four years later, in recognition of his brilliant research, he was appointed to a faculty position at the same institution that had tried to prevent him from continuing his studies.

SOLUTION

$$A = \begin{bmatrix} 3 & 2-i \\ 2+i & 7 \end{bmatrix} \quad \text{and} \quad B = \begin{bmatrix} 5 & 2-i & 1+i\sqrt{5} \\ 2+i & 4 & 3i \\ 1-i\sqrt{5} & -3i & -1 \end{bmatrix} \quad \blacksquare$$

EXAMPLE 2 Illustrate some algebraic operations in $\mathbb{C}^n$.

SOLUTION Notice that

$$[3+2i, \, 4-i]^H = \begin{bmatrix} 3-2i \\ 4+i \end{bmatrix}$$

$$\begin{bmatrix} 3+2i & -1+i \\ 2-5i & 6+i \end{bmatrix}^H = \begin{bmatrix} 3-2i & 2+5i \\ -1-i & 6-i \end{bmatrix}$$

Moreover, for two given vectors

$$\mathbf{x} = \begin{bmatrix} 2+3i \\ 5-4i \end{bmatrix}, \qquad \mathbf{y} = \begin{bmatrix} 3-2i \\ 1+i \end{bmatrix}$$

we find that

$$\langle \mathbf{x}, \mathbf{y} \rangle = 1+4i, \qquad \langle \mathbf{y}, \mathbf{x} \rangle = 1-4i, \qquad \langle \mathbf{x}, \mathbf{x} \rangle = 54, \qquad \langle \mathbf{y}, \mathbf{y} \rangle = 15 \quad \blacksquare$$

A real matrix is Hermitian if and only if it is symmetric. This fact is obvious from the definition of "Hermitian." Thus, we have generalized the concept of *symmetric* to *Hermitian*. However, there is this twist that after taking the transpose, we also take the complex conjugates of all entries. In an inner-product space with *complex* vectors, we must remember that

$$\langle \mathbf{x}, \mathbf{y} \rangle = \mathbf{y}^H \mathbf{x} = \overline{\mathbf{y}}^T \mathbf{x} = \mathbf{x}^T \overline{\mathbf{y}} = \overline{\mathbf{x}^T \, \overline{\mathbf{y}}} = \overline{\mathbf{x}^H \mathbf{y}} = \overline{\langle \mathbf{y}, \mathbf{x} \rangle}$$

and

$$\langle \mathbf{x}, \alpha\mathbf{y} \rangle = (\alpha\mathbf{y})^H \mathbf{x} = \overline{\alpha}\, \overline{\mathbf{y}}^T \mathbf{x} = \overline{\alpha}\, \mathbf{y}^H \mathbf{x} = \overline{\alpha}\, \langle \mathbf{x}, \mathbf{y} \rangle$$

A consequence is that for a square matrix A, we have

$$\langle A\mathbf{x}, \mathbf{y} \rangle = \mathbf{y}^H A\mathbf{x} = (A^H \mathbf{y})^H \mathbf{x} = \langle \mathbf{x}, A^H \mathbf{y} \rangle$$

In this setting, we often use the notation $\mathbf{A}^H$ to denote the *conjugate transpose* of a matrix $\mathbf{A}$. It is obtained by taking the transpose of $\mathbf{A}$ and then replacing each entry by its complex conjugate. The order of these two steps can be reversed. In symbols: $\mathbf{A}^H = \overline{\mathbf{A}^T}$, or $\mathbf{A}^H = (\overline{\mathbf{A}})^T$. (If $z = \alpha + i\beta$, where α and β are real, then the **conjugate** of the complex number z is $\overline{z} = \alpha - i\beta$.) In matrix notation, we have

$$\langle \mathbf{x}, \mathbf{y} \rangle = \mathbf{y}^H \mathbf{x}$$
$$\langle \mathbf{A}\mathbf{x}, \mathbf{y} \rangle = \mathbf{y}^H \mathbf{A}\mathbf{x}$$

Then, we have

$$\langle \mathbf{x}, \mathbf{A}^H \mathbf{y} \rangle = (\mathbf{A}^H \mathbf{y})^H \mathbf{x} = \mathbf{y}^H (\mathbf{A}^H)^H \mathbf{x} = \mathbf{y}^H \mathbf{A}\mathbf{x} = \langle \mathbf{A}\mathbf{x}, \mathbf{y} \rangle$$

Notice that if we use the inner-product notation, it is inconsequential whether the vectors are regarded as rows or columns. For the matrix version of the inner product, however, a vector $\mathbf{y}$ *must* be treated as a column vector, and then $\mathbf{y}^H$ is a row vector. The so-called **outer product** of two n-component vectors is also defined; it is

$$\mathbf{x}\mathbf{y}^H = \begin{bmatrix} x_1 \\ x_2 \\ x_3 \\ \vdots \\ x_n \end{bmatrix} \begin{bmatrix} \overline{y}_1 & \overline{y}_2 & \overline{y}_3 & \cdots & \overline{y}_n \end{bmatrix}$$

$$= \begin{bmatrix} x_1\overline{y}_1 & x_1\overline{y}_2 & x_1\overline{y}_3 & \cdots & x_1\overline{y}_n \\ x_2\overline{y}_1 & x_2\overline{y}_2 & x_2\overline{y}_3 & \cdots & x_2\overline{y}_n \\ x_3\overline{y}_1 & x_3\overline{y}_2 & x_3\overline{y}_3 & \cdots & x_3\overline{y}_n \\ \vdots & \vdots & \vdots & \ddots & \vdots \\ x_n\overline{y}_1 & x_n\overline{y}_2 & x_n\overline{y}_3 & \cdots & x_n\overline{y}_n \end{bmatrix}$$

and is an $n \times n$ matrix of rank at most one.

Self-Adjoint Mapping

The concept of a matrix being Hermitian corresponds to the concept of a self-adjoint mapping in an inner-product space.

DEFINITION

A mapping f from an inner-product space into itself is **self-adjoint** *if the equation*

$$\langle f(\mathbf{x}), \mathbf{y} \rangle = \langle \mathbf{x}, f(\mathbf{y}) \rangle$$

is true for all vectors $\mathbf{x}$ *and* $\mathbf{y}$. *In this definition, the scalar field for the inner-product space may be either* $\mathbb{C}$ *or* $\mathbb{R}$.

THEOREM 1

Every self-adjoint mapping is linear.

PROOF Let f be self-adjoint. For given vectors $\mathbf{x}$ and $\mathbf{y}$, we have, for an arbitrary vector $\mathbf{z}$

$$
\begin{aligned}
\langle f(\alpha \mathbf{x} + \beta \mathbf{y}), \mathbf{z} \rangle &= \langle \alpha \mathbf{x} + \beta \mathbf{y}, f(\mathbf{z}) \rangle \\
&= \alpha \langle \mathbf{x}, f(\mathbf{z}) \rangle + \beta \langle \mathbf{y}, f(\mathbf{z}) \rangle \\
&= \alpha \langle f(\mathbf{x}), \mathbf{z} \rangle + \beta \langle f(\mathbf{y}), \mathbf{z} \rangle \\
&= \langle \alpha f(\mathbf{x}) + \beta f(\mathbf{y}), \mathbf{z} \rangle
\end{aligned}
$$

Since $\mathbf{z}$ is arbitrary, we conclude that $f(\alpha \mathbf{x} + \beta \mathbf{y}) = \alpha f(\mathbf{x}) + \beta f(\mathbf{y})$ and f is linear. ∎

THEOREM 2

All eigenvalues of a self-adjoint operator are real.

PROOF Let L be a self-adjoint operator on some inner-product space, and suppose that

$$L(\mathbf{x}) = \lambda \mathbf{x}$$

for some nonzero $\mathbf{x}$. (At this stage in the proof, the eigenvalue λ is allowed to be complex.) We have

$$\lambda \langle \mathbf{x}, \mathbf{x} \rangle = \langle \lambda \mathbf{x}, \mathbf{x} \rangle = \langle L(\mathbf{x}), \mathbf{x} \rangle = \langle \mathbf{x}, L(\mathbf{x}) \rangle = \langle \mathbf{x}, \lambda \mathbf{x} \rangle = \overline{\lambda} \langle \mathbf{x}, \mathbf{x} \rangle$$

Cancelling the nonzero term $\langle \mathbf{x}, \mathbf{x} \rangle$ leaves the equation $\lambda = \overline{\lambda}$; whence, λ must be real. ∎

THEOREM 3

Any two eigenvectors of a self-adjoint operator, if associated with different eigenvalues, are mutually orthogonal.

PROOF Let L be a self-adjoint transformation on an inner-product space. Suppose that

$$L(\mathbf{x}) = \lambda\mathbf{x}, \qquad L(\mathbf{y}) = \mu\mathbf{y}$$

It may be assumed that $\mathbf{x}$ and $\mathbf{y}$ are nonzero vectors and that $\lambda \neq \mu$. By straightforward calculation, it is revealed that

$$\lambda\langle\mathbf{x}, \mathbf{y}\rangle = \langle\lambda\mathbf{x}, \mathbf{y}\rangle = \langle L(\mathbf{x}), \mathbf{y}\rangle = \langle\mathbf{x}, L(\mathbf{y})\rangle = \langle\mathbf{x}, \mu\mathbf{y}\rangle = \overline{\mu}\langle\mathbf{x}, \mathbf{y}\rangle = \mu\langle\mathbf{x}, \mathbf{y}\rangle$$

(Notice that Theorem 2 allowed us to write $\overline{\mu} = \mu$.) Since $\lambda \neq \mu$, it must be that $\langle\mathbf{x}, \mathbf{y}\rangle = 0$. ∎

THEOREM 4

Consider $\mathbb{C}^n$ with its standard inner product, and let $\mathbf{A}$ be an $n \times n$ matrix. Define a linear transformation L by writing $L(\mathbf{x}) = \mathbf{A}\mathbf{x}$, where $\mathbf{x} \in \mathbb{C}^n$. The operator L is self-adjoint if and only if $\mathbf{A}$ is Hermitian.

PROOF We have the following equivalences, where, for brevity, we have omitted all the universal quantifiers "*for all* $\mathbf{x}$" and "*for all* $\mathbf{y}$." The symbol $\Leftrightarrow$ means "*if and only if*."

$$L \text{ is self-adjoint} \quad \Leftrightarrow \quad \langle L(\mathbf{x}), \mathbf{y}\rangle = \langle\mathbf{x}, L(\mathbf{y})\rangle \quad \Leftrightarrow \quad \langle\mathbf{A}\mathbf{x}, \mathbf{y}\rangle = \langle\mathbf{x}, \mathbf{A}\mathbf{y}\rangle$$

$$\Leftrightarrow \quad \mathbf{y}^H\mathbf{A}\mathbf{x} = (\mathbf{A}\mathbf{y})^H\mathbf{x} \quad \Leftrightarrow \quad \mathbf{y}^H\mathbf{A}\mathbf{x} = \mathbf{y}^H\mathbf{A}^H\mathbf{x} \quad \Leftrightarrow \quad \mathbf{A} = \mathbf{A}^H \qquad ∎$$

COROLLARY 1

All eigenvalues of a Hermitian matrix are real.

PROOF If $\mathbf{A}$ is Hermitian and $L(\mathbf{x}) = \mathbf{A}\mathbf{x}$, then L is self-adjoint, by Theorem 4. The eigenvalues of $\mathbf{A}$ and L are the same and are real, by Theorem 2. ∎

COROLLARY 2

All eigenvalues of a symmetric real matrix are real.

THEOREM 5

Let L be a linear transformation on a finite-dimensional vector space. If $\mathbf{A}$ is the matrix for L relative to a given basis, then the eigenvalues of L and $\mathbf{A}$ are the same.

PROOF Recall Theorem 4 in Section 5.3. According to it, if the basis in question is $\mathcal{B}$, then the matrix $\mathbf{A}$ plays this role:

$$[L(\mathbf{x})]_{\mathcal{B}} = \mathbf{A}[\mathbf{x}]_{\mathcal{B}}$$

for every point $\mathbf{x}$ in the space. Now we have these equivalences:

$$L(\mathbf{x}) = \lambda\mathbf{x} \quad \Leftrightarrow \quad [L(\mathbf{x})]_{\mathcal{B}} = \lambda[\mathbf{x}]_{\mathcal{B}} \quad \Leftrightarrow \quad \mathbf{A}[\mathbf{x}]_{\mathcal{B}} = \lambda[\mathbf{x}]_{\mathcal{B}}$$

∎

EXAMPLE 3 Find the eigenvalues of the first matrix in Example 1.

SOLUTION Since $\mathbf{A} = \begin{bmatrix} 3 & 2-i \\ 2+i & 7 \end{bmatrix}$ and is Hermitian, its eigenvalues must be real. To calculate the eigenvalues, we find the roots of the characteristic polynomial:

$$\mathrm{Det}\begin{bmatrix} 3-\lambda & 2-i \\ 2+i & 7-\lambda \end{bmatrix} = (3-\lambda)(7-\lambda) - (2-i)(2+i)$$
$$= \lambda^2 - 10\lambda + 16 = (\lambda - 8)(\lambda - 2)$$

Hence, the eigenvalues are 2 and 8. ∎

EXAMPLE 4 Find the eigenvectors for the matrix $\mathbf{A}$ in the preceding example. Are they mutually orthogonal?

SOLUTION Taking $\lambda = 2$, we find an eigenvector $\mathbf{u} = \begin{bmatrix} -2+i \\ 1 \end{bmatrix}$. Taking $\lambda = 8$, we find an eigenvector $\mathbf{v} = \begin{bmatrix} 2-i \\ 5 \end{bmatrix}$. We have $\langle\mathbf{u},\mathbf{v}\rangle = \mathbf{v}^H\mathbf{u} = (2+i)(-2+i) + (5)(1) = -5 + 5 = 0$. Yes, they are! ∎

The Spectral Theorem

Now we present a version of the *spectral theorem* that is as brief as possible. The reader will see at once how it reveals the structure of any self-adjoint operator. Endearing properties of the eigenvalues and eigenspaces will result easily from the spectral theorem.

> **THEOREM 6   Spectral Theorem**
> *Let L be a self-adjoint operator on an n-dimensional inner-product space,*
> *X. (The scalar field for X can be either $\mathbb{R}$ or $\mathbb{C}$.) Then there exists an*
> *orthonormal basis $\{\mathbf{u}_1, \mathbf{u}_2, \ldots, \mathbf{u}_n\}$ for X and a set of n real numbers λ_j such*
> *that*
> $$L(\mathbf{x}) = \sum_{i=1}^{n} \lambda_i \langle \mathbf{x}, \mathbf{u}_i \rangle \mathbf{u}_i \qquad (\mathbf{x} \in X)$$

PROOF   We begin by noticing that if the preceding displayed equation is true, then we can infer that

$$L(\mathbf{u}_j) = \sum_{i=1}^{n} \lambda_i \langle \mathbf{u}_j, \mathbf{u}_i \rangle \mathbf{u}_i = \sum_{i=1}^{n} \lambda_j \delta_{ij} \mathbf{u}_i = \lambda_j \mathbf{u}_j$$

Thus, the vectors $\mathbf{u}_j$ are eigenvectors of L, and the numbers λ_j are corresponding eigenvalues of L. Select an eigenvalue, say λ_1, together with an eigenvector $\mathbf{u}_1$, which we may assume to be a unit vector. By Theorem 2, λ_1 is real. This is Step 1 in an n-step process.

Let us describe a typical step, say Step k, where $k < n$. Suppose that we have chosen eigenvalues $\lambda_1, \lambda_2, \ldots, \lambda_k$ (not necessarily distinct). Suppose also that we have an orthonormal set $\{\mathbf{u}_1, \mathbf{u}_2, \ldots, \mathbf{u}_k\}$, where

$$L(\mathbf{u}_j) = \lambda_j \mathbf{u}_j \qquad (1 \le j \le k)$$

Let V be the span of $\{\mathbf{u}_1, \mathbf{u}_2, \ldots, \mathbf{u}_k\}$. We restrict the operator L to $V^{\perp}$, and discover that $V^{\perp}$ is *invariant under L.* This means that $L(V^{\perp}) \subseteq V^{\perp}$. Indeed, if $\mathbf{x} \in V^{\perp}$, then, for j in the range $1 \le j \le k$, we have

$$\langle L(\mathbf{x}), \mathbf{u}_j \rangle = \langle \mathbf{x}, L(\mathbf{u}_j) \rangle = \langle \mathbf{x}, \lambda_j \mathbf{u}_j \rangle = \lambda_j \langle \mathbf{x}, \mathbf{u}_j \rangle = 0$$

This establishes that

$$L(\mathbf{x}) \perp \mathbf{u}_j, \qquad L(\mathbf{x}) \perp V, \qquad L(\mathbf{x}) \in V^{\perp}$$

The restricted operator is written $L|V^{\perp}$. (It is the same as L, except for its having a smaller domain, namely $V^{\perp}$.) It is also self-adjoint. Hence, we can find an eigenvalue λ_{k+1} and an eigenvector $\mathbf{u}_{k+1}$ (taken to be of unit length) in $V^{\perp}$. We know that $\mathbf{u}_{k+1}$ is orthogonal to $\{\mathbf{u}_1, \mathbf{u}_2, \ldots, \mathbf{u}_k\}$. This process is repeated until we have found n eigenvalues and n eigenvectors. (The eigenvalues are not necessarily distinct. For a given eigenvalue, there may be several mutually orthogonal eigenvectors.) When the process stops, we have an orthonormal base for the space. Hence for each $\mathbf{x}$, we have

$$\mathbf{x} = \sum_{i=1}^{n} \langle \mathbf{x}, \mathbf{u}_i \rangle \mathbf{u}_i$$

whence we obtain

$$L(\mathbf{x}) = \sum_{i=1}^{n} \langle \mathbf{x}, \mathbf{u}_i \rangle L(\mathbf{u}_i) = \sum_{i=1}^{n} \lambda_i \langle \mathbf{x}, \mathbf{u}_i \rangle \mathbf{u}_i \qquad \blacksquare$$

Unitary and Orthogonal Matrices

> **DEFINITION**
>
> A square matrix $\mathbf{U}$ is **unitary** if $\mathbf{U}^H \mathbf{U} = \mathbf{I}$. If $\mathbf{U}$ is real and $\mathbf{U}^T \mathbf{U} = \mathbf{I}$, the term **orthogonal** is used.

The matrix version of the Spectral Theorem is as follows. It uses the concept of a unitary matrix.

> **THEOREM 7   Spectral Theorem: Matrix Version**
>
> Let $\mathbf{A}$ be a Hermitian matrix. Then there exists a unitary matrix $\mathbf{U}$ such that
>
> $$\mathbf{U}^H \mathbf{A} \mathbf{U} = \mathbf{D}$$
>
> where $\mathbf{D}$ is a diagonal matrix having all the eigenvalues of $\mathbf{A}$ on its diagonal. The columns of $\mathbf{U}$ are unit eigenvectors of $\mathbf{A}$.

PROOF Define an operator L on $\mathbb{C}^n$ by the equation

$$L(\mathbf{x}) = \mathbf{A}\mathbf{x}$$

By Theorem 4, L is self-adjoint. Hence, by the Spectral Theorem, we have

$$\mathbf{A}\mathbf{x} = L(\mathbf{x}) = \sum_{k=1}^{n} \lambda_k \langle \mathbf{x}, \mathbf{u}_k \rangle \mathbf{u}_k = \sum_{k=1}^{n} \lambda_k \left[(\mathbf{u}_k)^H \mathbf{x} \right] \mathbf{u}_k = \sum_{k=1}^{n} \lambda_k \left(\sum_{j=1}^{n} x_j (\overline{\mathbf{u}}_k)_j \right) \mathbf{u}_k$$

We use the notation $(\mathbf{u}_k)_j$ to represent the jth component of the vector $\mathbf{u}_k$. Let $\mathbf{U}$ be the matrix whose columns are $\mathbf{u}_1, \mathbf{u}_2, \ldots, \mathbf{u}_n$. Then $(\mathbf{u}_k)_i = \mathbf{U}_{ik}$. From the preceding equation, we then can write

$$\sum_{j=1}^{n} \mathbf{A}_{ij} x_j = (\mathbf{A}\mathbf{x})_i = \sum_{k=1}^{n} \lambda_k \sum_{j=1}^{n} x_j \overline{\mathbf{U}}_{jk} (\mathbf{u}_k)_i = \sum_{j=1}^{n} \sum_{k=1}^{n} \lambda_k (\mathbf{U}^H)_{kj} \mathbf{U}_{ik} x_j$$

Consequently, we have

$$\mathbf{A}_{ij} = \sum_{k=1}^{n} \lambda_k \mathbf{U}_{ik} (\mathbf{U}^H)_{kj}$$

Now let $\mathbf{D}$ be the diagonal matrix with the eigenvalues on its diagonal, $\mathbf{D}_{\nu\mu} = \lambda_\nu \delta_{\nu\mu}$. Then we can continue the preceding calculation of $\mathbf{A}$ as follows:

$$\mathbf{A}_{ij} = \sum_{\nu=1}^{n} \sum_{\mu=1}^{n} \mathbf{U}_{i\nu} \mathbf{D}_{\nu\mu} (\mathbf{U}^H)_{\mu j}$$

This equation means that $\mathbf{A} = \mathbf{U}\mathbf{D}\mathbf{U}^H$. ∎

EXAMPLE 5 Find the eigenvalues of the matrix $\mathbf{A} = \begin{bmatrix} 10 & 3i \\ -3i & 2 \end{bmatrix}$ and the unitary matrix $\mathbf{U}$ such the $\mathbf{U}^H \mathbf{A} \mathbf{U} = \mathbf{D}$.

SOLUTION Being Hermitian, $\mathbf{A}$ must have real eigenvalues. In the usual way, we find these to be 11 and 1. The corresponding eigenvectors are $\mathbf{v} = [3i, 1]^T$ and $\mathbf{w} = [1, 3i]^T$. This pair is orthogonal:

$$\langle \mathbf{v}, \mathbf{w} \rangle = v_1 \overline{w}_1 + v_2 \overline{w}_2 = (3i)(1) + (1)(-3i) = 0$$

We can normalize these two eigenvectors and put them as columns in a matrix, $\mathbf{U}$. Thus, we have $\mathbf{U} = \left(1/\sqrt{10} \right) \begin{bmatrix} 3i & 1 \\ 1 & 3i \end{bmatrix}$. The matrix $\mathbf{U}$ satisfies $\mathbf{U}\mathbf{U}^H = \mathbf{I}$ and $\mathbf{U}^{-1} = \mathbf{U}^H$. ∎

THEOREM 8

Let $\mathbf{A}$ be a real, symmetric matrix. Then there is an orthogonal matrix $\mathbf{U}$ such that

$$\mathbf{U}^T\mathbf{A}\mathbf{U} = \mathbf{D}$$

where $\mathbf{D}$ is a diagonal matrix having the eigenvalues of $\mathbf{A}$ on its diagonal. The columns of $\mathbf{U}$ are the corresponding eigenvectors of unit length.

PROOF This is the special case of Theorem 7 that arises when $\mathbf{A}$ is real. ■

Theorem 7 is often summarized by saying that a Hermitian matrix is unitarily similar to a diagonal matrix. Recall that **similar** means the relationship $\mathbf{A} = \mathbf{S}\mathbf{B}\mathbf{S}^{-1}$ holds for two matrices, $\mathbf{A}$ and $\mathbf{B}$. When we say **unitarily similar**, we mean that $\mathbf{A} = \mathbf{U}\mathbf{B}\mathbf{U}^{-1}$, where $\mathbf{U}$ is a unitary matrix. Thus, we have $\mathbf{U}\mathbf{U}^H = \mathbf{I}$ and the inverse of $\mathbf{U}$ is $\mathbf{U}^H$; that is, $\mathbf{U}^{-1} = \mathbf{U}^H$.

DEFINITION

Two matrices $\mathbf{A}$ and $\mathbf{B}$ are **unitarily similar** to each other if there is a unitary matrix $\mathbf{U}$ such that $\mathbf{A} = \mathbf{U}\mathbf{B}\mathbf{U}^H$.

Likewise, Theorem 8 is summarized by saying that a symmetric, real matrix is orthogonally similar to a diagonal matrix. In this case, $\mathbf{U}$ should be an orthogonal matrix: $\mathbf{U}\mathbf{U}^T = \mathbf{I}$. Theorems 6, 7, and 8 are often regarded as high points in the subject of linear algebra.

EXAMPLE 6 Compute the diagonalization of the matrix

$$\mathbf{A} = \begin{bmatrix} 3 & 4i \\ -4i & 3 \end{bmatrix}$$

SOLUTION The characteristic polynomial is $(3 - \lambda)^2 + 16i^2 = \lambda^2 - 6\lambda - 7$ and its roots (eigenvalues of $\mathbf{A}$) are 7 and -1. Taking the eigenvalue 7, we find a corresponding eigenvector by solving the equation $(\mathbf{A} - 7\,\mathbf{I})\mathbf{x} = 0$. A convenient solution (not normalized) is $[i, 1]^T$. The normalized eigenvector is then $[c\,i, c]^T$, where $c = \frac{1}{2}\sqrt{2}$. We treat the second eigenvalue, -1, in the same way and find the normalized eigenvector $[c\,i, -c]^T$. These normalized eigenvectors go into a matrix $\mathbf{U}$ as columns, keeping the correct order. Thus, we obtain $\mathbf{U} = \begin{bmatrix} c\,i & c\,i \\ c & -c \end{bmatrix}$. To verify all the important conclusions, we calculate

$$\mathbf{AU} = \begin{bmatrix} 3 & 4i \\ -4i & 3 \end{bmatrix} \begin{bmatrix} c\,i & c\,i \\ c & -c \end{bmatrix} = \begin{bmatrix} 7c\,i & -c\,i \\ 7c & c \end{bmatrix}$$

$$\mathbf{U}^H\mathbf{AU} = \begin{bmatrix} -c\,i & c \\ -c\,i & -c \end{bmatrix} \mathbf{AU} = \begin{bmatrix} -c\,i & c \\ -c\,i & -c \end{bmatrix} \begin{bmatrix} 7c\,i & -c\,i \\ 7c & c \end{bmatrix}$$

$$= \begin{bmatrix} 14c^2 & 0 \\ 0 & -2c^2 \end{bmatrix} = \begin{bmatrix} 7 & 0 \\ 0 & -1 \end{bmatrix} = \mathbf{D}$$

We can view this as a factorization of the matrix $\mathbf{A}$:

$$\mathbf{A} = \mathbf{UDU}^H$$

Recall that this equation allows us to compute $\mathbf{A}^k$ efficiently:

$$\mathbf{A}^k = \mathbf{UD}^k\mathbf{U}^H$$

The powers of the diagonal matrix $\mathbf{D}$ are $\mathbf{D}^k = \begin{bmatrix} 7^k & 0 \\ 0 & (-1)^k \end{bmatrix}$. ∎

The Cayley–Hamilton Theorem

We turn now to a remarkable theorem, named after Arthur Cayley[2] and Sir William Rowan Hamilton[3]. One of the important mathematicians of the nineteenth century was Arthur Cayley. In an 1858 paper entitled *A Memoir on the Theory of Matrices,* he wrote the following passage: "*I obtain the remarkable theorem that any matrix whatever satisfies an algebraic equation*

---

[2] Arthur Cayley (1821–1895) published three mathematical papers while still an undergraduate! He spent 14 years working as a lawyer so that he could retire and pursue mathematics. He published about 250 mathematical papers. If anybody should be credited as the creator of matrix theory, Cayley might be the best choice. He also was a learned man in other diffuse areas, such as law, literature, and architecture. He showed that our present-day definition of matrix multiplication is the correct one from the standpoint of linear operator theory (preserving the composition of two linear maps). He also studied quadratic forms and matrix inverses. Cayley's accomplishments must be appreciated in the context of nineteenth-century mathematics. Many mathematicians shrank from participating in matrix theory, principally because the product of matrices is not commutative! In their eyes, that *flaw* in the theory discredited the entire subject.

[3] William Rowan Hamilton (1805–1865) was a genius, able to read Latin, Greek, and Hebrew by the age of 5. At Trinity College, Dublin, he excelled in mathematics and physics, especially the subject of optics, to which he made substantial contributions. In mathematics, he is remembered chiefly as the inventor of quaternions, which are four-dimensional vectors with special noncommutative rules for multiplication.

of its own order, the coefficient of the highest power being unity... and the last coefficient being in fact the determinant." The formal statement of this result is Theorem 9.

THEOREM 9 Cayley–Hamilton Theorem

Every square matrix satisfies its own characteristic equation.

To understand this, recall first that the characteristic equation of a matrix **A** is

$$p(\lambda) \equiv \text{Det}(\mathbf{A} - \lambda \mathbf{I}) = 0$$

The lefthand side of this equation is a polynomial in the variable λ, and we have named it p. The zeros of the polynomial p are the eigenvalues of **A**. (Recall that one conceptually appealing algorithm for finding the eigenvalues of **A** is to compute the roots of the characteristic equation—that is, to solve for the zeros of p.)

In the characteristic polynomial, we are free to replace λ with any mathematical entity for which $p(...)$ is meaningful. Thus, we can contemplate $p(\mathbf{A})$, because the positive integer powers of **A** are well defined, and we can agree on using $\mathbf{A}^0 = \mathbf{I}$. The Cayley–Hamilton Theorem asserts that

$$p(\mathbf{A}) = \mathbf{0}$$

EXAMPLE 7 Let $\mathbf{A} = \begin{bmatrix} 11 & -6i \\ 4i & 1 \end{bmatrix}$

Verify that **A** obeys the Cayley–Hamilton Theorem.

SOLUTION We find the characteristic polynomial of **A**:

$$\text{Det} \begin{bmatrix} 11 - \lambda & -6i \\ 4i & 1 - \lambda \end{bmatrix} = (11 - \lambda)(1 - \lambda) - 24 = \lambda^2 - 12\lambda - 13$$

Substituting **A** for λ produces

$$\mathbf{A}^2 - 12\,\mathbf{A} - 13\,\mathbf{I} = \begin{bmatrix} 145 & -72i \\ 48i & 25 \end{bmatrix} - \begin{bmatrix} -132 & +72i \\ -48i & -12 \end{bmatrix} + \begin{bmatrix} -13 & 0 \\ 0 & -13 \end{bmatrix}$$

$$= \begin{bmatrix} 0 & 0 \\ 0 & 0 \end{bmatrix}$$

In this same example, we could use the factored form of the characteristic polynomial, which is given by $p(\lambda) = (\lambda + 1)(\lambda - 13)$. Substituting A for λ, we get

$$(\mathbf{A} - 13\,\mathbf{I})(\mathbf{A} + \mathbf{I}) = \begin{bmatrix} -2 & -6i \\ 4i & -12 \end{bmatrix} \begin{bmatrix} -12 & -6i \\ 4i & 2 \end{bmatrix} = \begin{bmatrix} 0 & 0 \\ 0 & 0 \end{bmatrix} \quad \blacksquare$$

When $\mathbf{A}$ is assumed to be a diagonalizable matrix, a proof of the Cayley–Hamilton Theorem is not difficult! Here are the steps. Let

$$\mathbf{A} = \mathbf{P}\mathbf{D}\mathbf{P}^{-1}$$

As we saw in Section 6.1,

$$\mathbf{A}^k = \mathbf{P}\mathbf{D}^k\mathbf{P}^{-1}$$

for all k, and we have $q(\mathbf{A}) = \mathbf{P}\,q(\mathbf{D})\,\mathbf{P}^{-1}$, for any polynomial q. If $\mathbf{D} = \text{Diag}(\lambda_1, \lambda_2, \ldots, \lambda_n)$, then $\mathbf{D}^k = \text{Diag}(\lambda_1^k, \lambda_2^k, \ldots, \lambda_n^k)$, and

$$q(\mathbf{D}) = \text{Diag}(q(\lambda_1), q(\lambda_2), \ldots, q(\lambda_n))$$

In particular, if q is the characteristic polynomial, then $q(\mathbf{D}) = \mathbf{0}$, and $q(\mathbf{A}) = \mathbf{0}$.

EXAMPLE 8 Let $\mathbf{A} = \begin{bmatrix} 5 & 2 & 1 \\ 1 & 1 & 7 \\ 3 & 0 & 11 \end{bmatrix}$

Find the coefficients in the equation $\mathbf{A}^3 = c_0\mathbf{I} + c_1\mathbf{A} + c_2\mathbf{A}^2$.

SOLUTION Let the characteristic polynomial of $\mathbf{A}$ be p. Then

$$p(\lambda) = \text{Det} \begin{bmatrix} 5 - \lambda & 2 & 1 \\ 1 & 1 - \lambda & 7 \\ 3 & 0 & 11 - \lambda \end{bmatrix} = -\lambda^3 + 17\lambda^2 - 66\lambda + 72$$

By the Cayley–Hamilton Theorem, $p(\mathbf{A}) = -\mathbf{A}^3 + 17\mathbf{A}^2 - 66\mathbf{A} + 72\mathbf{I} = \mathbf{0}$, and the coefficients asked for are $c_0 = 72$, $c_1 = -66$, and $c_2 = 17$. $\blacksquare$

EXAMPLE 9 Use the information in the preceding example to obtain $\mathbf{A}^{-1}$ in terms of the powers of $\mathbf{A}$.

SOLUTION From Example 8, we have

$$\mathbf{I} = 1/72(\mathbf{A}^3 - 17\mathbf{A}^2 + 66\mathbf{A}) = 1/72\mathbf{A}(\mathbf{A}^2 - 17\mathbf{A} + 66\mathbf{I})$$

Hence, we obtain

$$\mathbf{A}^{-1} = (1/72)\,(\mathbf{A}^2 - 17\mathbf{A} + 66\mathbf{I})$$ ■

EXAMPLE 10 Use the Cayley–Hamilton Theorem to find the span of a set $\{\mathbf{I}, \mathbf{A}, \mathbf{A}^2, \mathbf{A}^3, \ldots\}$, where $\mathbf{A}$ is an arbitrary $n \times n$ matrix.

SOLUTION The Cayley–Hamilton Theorem asserts that

$$c_0\mathbf{I} + c_1\mathbf{A} + c_2\mathbf{A}^2 + \cdots + c_{n-1}\mathbf{A}^{n-1} + \mathbf{A}^n = \mathbf{0}$$

if the coefficients c_j are taken from the characteristic polynomial of $\mathbf{A}$. It follows that $\mathbf{A}^n$ is in the span of $\{\mathbf{I}, \mathbf{A}, \mathbf{A}^2, \ldots, \mathbf{A}^{n-1}\}$. Multiply the preceding equation by $\mathbf{A}$ to see that $\mathbf{A}^{n+1}$ is also in the same span. All the powers $\mathbf{A}^k$ for $k \geq n$ are in the span of $\{\mathbf{I}, \mathbf{A}, \mathbf{A}^2, \ldots, \mathbf{A}^{n-1}\}$:

$$\text{Span}\{\mathbf{I}, \mathbf{A}, \mathbf{A}^2, \ldots, \mathbf{A}^{n-1}\} = \text{Span}\{\mathbf{I}, \mathbf{A}, \mathbf{A}^2, \ldots, \mathbf{A}^{n-1}, \mathbf{A}^n, \mathbf{A}^{n+1}, \ldots\}$$

The powers of an $n \times n$ matrix span a vector space of dimension at most n in the n^2-dimensional space $\mathbb{R}^{n \times n}$.   ■

Quadratic Forms

A quadratic form is any homogeneous polynomial of degree two in any number of variables. In this situation, **homogeneous** means that all the terms are of degree two. For example, the expression $7x_1x_2 + 3x_2x_4$ is homogeneous, but the expression $x_1 - 3x_1x_2$ is not. The square of the distance between two points in an inner-product space is a quadratic form. Quadratic forms were introduced by Hermite, and 70 years later they turned out to be essential in the theory of quantum mechanics! The formal definition follows.

DEFINITION

A **quadratic form** is a function of this type: $\mathbf{x} \mapsto \mathbf{x}^T \mathbf{A} \mathbf{x}$. Here, $\mathbf{A}$ is an arbitrary $n \times n$ matrix and $\mathbf{x}$ is a vector variable in $\mathbb{R}^n$.

EXAMPLE 11 What is the explicit formula for the quadratic form that arises from the matrix $\mathbf{A} = \begin{bmatrix} 6 & 3 \\ 1 & 4 \end{bmatrix}$?

SOLUTION The quadratic form, according to the definition, is

$$\mathbf{x}^T\mathbf{Ax} = [x_1 \ x_2] \begin{bmatrix} 6 & 3 \\ 1 & 4 \end{bmatrix} \begin{bmatrix} x_1 \\ x_2 \end{bmatrix} = [x_1 \ x_2] \begin{bmatrix} 6x_1 + 3x_2 \\ x_1 + 4x_2 \end{bmatrix}$$

$$= x_1(6x_1 + 3x_2) + x_2(x_1 + 4x_2) = 6x_1^2 + 4x_1x_2 + 4x_2^2$$

The resulting function is a polynomial of degree two in the variables x_1 and x_2, and each term is of degree two. We can think of it as a quadratic polynomial in two variables $ax^2 + 2bxy + cy^2$. ∎

This example illustrates the fact that a quadratic form induced by an $n \times n$ nonzero matrix is a polynomial of degree two in the n variables, $x_1, x_2, \ldots, x_n$.

If a quadratic form is presented as $\mathbf{x} \mapsto \mathbf{x}^T\mathbf{Bx}$, where $\mathbf{B}$ is a real matrix, then there is no loss of generality in requiring $\mathbf{B}$ to be symmetric. The next theorem restates and proves this.

THEOREM 10

If $\mathbf{A}$ and $\mathbf{B}$ are $n \times n$ real matrices connected by the relation

$$\mathbf{B} = \tfrac{1}{2}(\mathbf{A} + \mathbf{A}^T)$$

then the corresponding quadratic forms of $\mathbf{A}$ and $\mathbf{B}$ are identical, and $\mathbf{B}$ is symmetric.

PROOF Since $\mathbf{x}^T\mathbf{Ax}$ is a 1×1 matrix (a real number), we have $\mathbf{x}^T\mathbf{A}^T\mathbf{x} = (\mathbf{x}^T\mathbf{Ax})^T$. Consequently, we obtain

$$\mathbf{x}^T\mathbf{Bx} = \tfrac{1}{2}\mathbf{x}^T(\mathbf{A} + \mathbf{A}^T)\mathbf{x} = \tfrac{1}{2}\mathbf{x}^T\mathbf{Ax} + \tfrac{1}{2}\mathbf{x}^T\mathbf{A}^T\mathbf{x}$$

$$= \tfrac{1}{2}\mathbf{x}^T\mathbf{Ax} + \tfrac{1}{2}(\mathbf{x}^T\mathbf{Ax})^T = \tfrac{1}{2}\mathbf{x}^T\mathbf{Ax} + \tfrac{1}{2}\mathbf{x}^T\mathbf{Ax} = \mathbf{x}^T\mathbf{Ax}$$

The quadratic forms $\mathbf{x}^T\mathbf{Ax}$ and $\mathbf{x}^T\mathbf{Bx}$ are the same but the matrices $\mathbf{A}$ and $\mathbf{B}$ are not, unless $\mathbf{A}$ is already symmetric. ∎

EXAMPLE 12 Illustrate Theorem 10 using the matrix in Example 11 by finding the symmetric matrix $\mathbf{B}$ that has the same quadratic form as the nonsymmetric matrix $\mathbf{A}$.

SOLUTION With $\mathbf{A} = \begin{bmatrix} 6 & 3 \\ 1 & 4 \end{bmatrix}$, we obtain

$$\mathbf{B} = \tfrac{1}{2}(\mathbf{A} + \mathbf{A}^T) = \tfrac{1}{2}\begin{bmatrix} 6 & 3 \\ 1 & 4 \end{bmatrix} + \tfrac{1}{2}\begin{bmatrix} 6 & 1 \\ 3 & 4 \end{bmatrix} = \begin{bmatrix} 6 & 2 \\ 2 & 4 \end{bmatrix}$$

The quadratic forms are

$$\mathbf{x}^T\mathbf{A}\mathbf{x} = [x_1 \; x_2]\begin{bmatrix} 6 & 3 \\ 1 & 4 \end{bmatrix}\begin{bmatrix} x_1 \\ x_2 \end{bmatrix} = 6x_1^2 + 4x_1x_2 + 4x_2^2$$

$$\mathbf{x}^T\mathbf{B}\mathbf{x} = [x_1 \; x_2]\begin{bmatrix} 6 & 2 \\ 2 & 4 \end{bmatrix}\begin{bmatrix} x_1 \\ x_2 \end{bmatrix} = 6x_1^2 + 4x_1x_2 + 4x_2^2$$

We have obtained equivalent quadratic forms! ∎

EXAMPLE 13 Let $\mathbf{D} = \begin{bmatrix} 5 + \sqrt{5} & 0 \\ 0 & 5 - \sqrt{5} \end{bmatrix}$

What is the quadratic form induced by this matrix? By a change of variables, show that the diagonal matrix $\mathbf{D}$ has the same quadratic form, with a different vector, as the quadratic form of the symmetric matrix $\mathbf{B}$ and nonsymmetric matrix $\mathbf{A}$; namely,

$$\mathbf{y}^T\mathbf{D}\mathbf{y} = \mathbf{x}^T\mathbf{B}\mathbf{x} = \mathbf{x}^T\mathbf{A}\mathbf{x}$$

SOLUTION Again, it is a straightforward calculation based on the definition:

$$\mathbf{y}^T\mathbf{D}\mathbf{y} = [y_1, y_2]\begin{bmatrix} 5 + \sqrt{5} & 0 \\ 0 & 5 - \sqrt{5} \end{bmatrix}\begin{bmatrix} y_1 \\ y_2 \end{bmatrix} = (5 + \sqrt{5})y_1^2 + (5 - \sqrt{5})y_2^2$$

We see a quadratic form in especially simple garb. There are no *cross-product* terms, such as y_1y_2. If we are willing to make a change of variables, such simple forms can always be found! We leave it to the reader to show that

$$\mathbf{x}^T\mathbf{B}\mathbf{x} = \mathbf{y}^T\mathbf{D}\mathbf{y}$$

$$6x_1^2 + 4x_1x_2 + 4x_2^2 = (5 + \sqrt{5})y_1^2 + (5 - \sqrt{5})y_2^2$$

by such a change of variables (see General Exercise 67). ∎

Now we explain the important technique of simplifying a quadratic form arising from a symmetric matrix. Recall from Theorem 8 that every real and symmetric matrix $\mathbf{A}$ can be diagonalized by an orthogonal matrix:

$$\mathbf{U}^T\mathbf{A}\mathbf{U} = \mathbf{D}$$

where $\mathbf{D}$ is diagonal and $\mathbf{U}$ is an orthogonal matrix, $\mathbf{U}\mathbf{U}^T = \mathbf{I}$.

EXAMPLE 14 Let $\mathbf{B} = \begin{bmatrix} 3 & 2 \\ 2 & 3 \end{bmatrix}$

Diagonalize the quadratic form that arises from this matrix.

SOLUTION The quadratic form involved here is

$$\mathbf{x}^T\mathbf{B}\mathbf{x} = [x_1\ x_2] \begin{bmatrix} 3 & 2 \\ 2 & 3 \end{bmatrix} \begin{bmatrix} x_1 \\ x_2 \end{bmatrix} = 3x_1^2 + 4x_1x_2 + 3x_2^2$$

With a change of variables, we will obtain a diagonal matrix for the quadratic form. We find that the eigenvalues of $\mathbf{B}$ are 5 and 1. Furthermore, the corresponding eigenvectors, in normalized form, are the columns of the matrix

$$\mathbf{U} = \tfrac{1}{\sqrt{2}} \begin{bmatrix} -1 & 1 \\ 1 & 1 \end{bmatrix}$$

The matrix $\mathbf{U}$ is orthogonal: $\mathbf{U}\mathbf{U}^T = \mathbf{I}$. The change of variable needed here is $\mathbf{x} = \mathbf{U}\mathbf{y}$; namely, $x_1 = \tfrac{1}{2}\sqrt{2}(-y_1 + y_2)$ and $x_2 = \tfrac{1}{2}\sqrt{2}(y_1 + y_2)$. The effect on the quadratic form is then

$$\mathbf{x}^T\mathbf{B}\mathbf{x} = (\mathbf{U}\mathbf{y})^T\mathbf{B}(\mathbf{U}\mathbf{y}) = \mathbf{y}^T\mathbf{U}^T\mathbf{B}\mathbf{U}\mathbf{y} = \mathbf{y}^T\mathbf{D}\mathbf{y}$$
$$3x_1^2 + 4x_1x_2 + 3x_2^2 = y_1^2 + 5y_2^2$$

In this equation, $\mathbf{D}$ is the diagonal matrix having the eigenvalues 1 and 5 on its diagonal. ∎

EXAMPLE 15 What is the quadratic form induced by the matrix

$$\mathbf{D} = \begin{bmatrix} 3 & 0 \\ 0 & 7 \end{bmatrix}$$

Set the quadratic polynomial equal to a positive constant, such as 42. What is the corresponding geometric figure?

SOLUTION   Clearly, we obtain:

$$\mathbf{x}^T\mathbf{D}\mathbf{x} = [x_1 \ x_2] \begin{bmatrix} 3 & 0 \\ 0 & 7 \end{bmatrix} \begin{bmatrix} x_1 \\ x_2 \end{bmatrix} = [x_1 \ x_2] \begin{bmatrix} 3x_1 \\ 7x_2 \end{bmatrix} = 3x_1^2 + 7x_2^2$$

Setting $3x_1^2 + 7x_2^2 = 42$ and dividing all terms by 42 produces the *standard form* of an ellipse, $x_1^2/14 + x_2^2/6 = 1$. We shall have an *ellipse* for the graph. The semimajor axis is $\sqrt{14}$, and the semiminor axis is $\sqrt{6}$. See Figure 8.1.   ∎

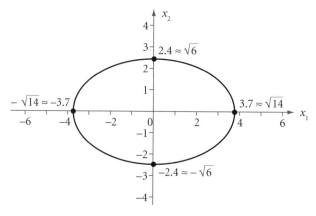

FIGURE 8.1   Ellipse $3x_1^2 + 7x_2^2 = 42$.

EXAMPLE 16   If the term $7x_2^2$ is changed to $-7x_2^2$ in the equation of Example 15, what geometric figure emerges?

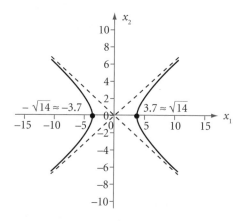

FIGURE 8.2   Hyperbola $3x_1^2 - 7x_2^2 = 42$.

SOLUTION The equation now is $3x_1^2 - 7x_2^2 = 42$. If a large value is assigned to x_1, we can satisfy the equation by choosing an appropriate large value for x_2. The figure is therefore not bounded. In fact it is a *hyperbola*, having symmetry with respect to both the horizontal and vertical axes, and symmetry with respect to the origin. See Figure 8.2. ∎

Application: World Wide Web Searching

Beginning in 1998, Web search engines such as Google started to use the Web hyperlink structure to improve their search analysis. This is known as *link analysis* and it is based on fundamental linear algebra concepts. This approach resulted in remarkably accurate Web searches and it revolutionized the design of search engines!

Jon Kleinberg from Cornell University developed the **HITS algorithm**, which is a link analysis algorithm, during his postdoctoral studies at IBM Almaden. Dr. Kleinberg noticed a pattern among Web pages: *Good hubs point to good authorities.*

Pages with many outlinks serve as *hubs* or portal pages, whereas other pages with many inlinks are *authorities* on topics. All pages in the query list and all pages pointing to or from the query page are in a neighborhood set of the query list, which may contain anywhere from hundreds to hundreds of thousands of Web pages. The hub and authority scores are refined iteratively until they *converge* to stationary values. Let **h** and **a** be column vectors of hub and authority scores and let **L** be the **adjacency matrix** for the neighborhood set. It can be shown that

$$\mathbf{a}^{(k)} = \mathbf{L}^T \mathbf{h}^{(k-1)}, \qquad \mathbf{h}^{(k)} = \mathbf{L} \mathbf{a}^{(k)}$$

and, therefore, we obtain

$$\mathbf{a}^{(k)} = \mathbf{L}^T \mathbf{L} \mathbf{a}^{(k-1)}, \qquad \mathbf{h}^{(k)} = \mathbf{L} \mathbf{L}^T \mathbf{h}^{(k-1)}$$

Here $\mathbf{L}^T \mathbf{L}$ is the hub matrix and $\mathbf{L} \mathbf{L}^T$ is the authority matrix. Both of these are semidefinite matrices. From these equations, we see that the HITS algorithm is the *power method* applied to these matrices. It solves the eigenvalue problems

$$\mathbf{L}^T \mathbf{L} \mathbf{a} = \lambda_1 \mathbf{a}, \qquad \mathbf{L} \mathbf{L}^T \mathbf{h} = \lambda_1 \mathbf{h}$$

where λ_1 is the largest eigenvalue of $\mathbf{L}^T \mathbf{L}$ (or $\mathbf{L} \mathbf{L}^T$). The vectors **a** and **h** are the corresponding eigenvectors. There are many other important issues such as convergence, existence, uniqueness, and numerical computation of these scores that we leave to the reader to explore. Variations of the HITS algorithm have been suggested, each having advantages and disadvantages.

Google is based on the **PageRank algorithm**. It is a link–analysis algorithm, similar to the HITS algorithm, and involves a recursive scheme and the power method. Sergey Brin and Larry Page conceived of the PageRank algorithm and the Google system when they were computer science graduate students at Stanford University in 1998. Their original idea was that "*a page is important if it is pointed to by other important pages.*" The importance of a page is its PageRank score and it is determined by summing the Page-Ranks of all pages that point to this page. When an important page points to several pages, its weight (PageRank) is distributed proportionally.

The page rank is a measure of the importance of each Web page relative to all other Web pages in the World Wide Web. For each Web page, a ballot is taken among all Web pages with a hyperlink to that page counting as a vote for it. So the PageRank of a page is defined recursively and it depends on the number of *incoming links* and their PageRank metric. (The word "PageRank" is a pun on Larry Page's name.) Although the exact details are a trade secret, the PageRank in the Google ToolBar goes from 0 to 10 using a logarithmic scale. Web pages have a high rank if they have many Web pages linked to them with high rank. Votes cast by important pages weigh more heavily and help to make other pages important.

For example, consider a small universe of four Web pages A, B, C, and D. If all these pages link to page A, then the PageRank of A, denoted $\text{pr}(A)$, would be the sum of the PageRank of pages B, C, and D:

$$\text{pr}(A) = \text{pr}(B) + \text{pr}(C) + \text{pr}(D)$$

In addition, suppose that page B links to page C, whereas page D links to pages A, B, and C. Each page splits its vote over several pages. Page B gives half a vote to each of pages A and C. Page C gives one vote to page A. Page D gives one-third of a vote to pages A, B, and C. (See the sketch in Figure 8.3.) For each page that is linked to page A, its PageRank is divided

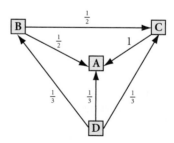

FIGURE 8.3   Web page links.

by the total number of links from that page:

$$\text{pr}(A) = \tfrac{1}{2}\text{pr}(B) + \text{pr}(C) + \tfrac{1}{3}\text{pr}(D)$$

The actual PageRank formula incorporates two more considerations and is given by

$$\text{pr}(A) = q \left[\frac{\text{pr}(B)}{\ell(B)} + \frac{\text{pr}(C)}{\ell(C)} + \frac{\text{pr}(D)}{\ell(D)} \right] + (1 - q)$$

For each Web page linked to A, its PageRank is divided by the number of links that come from that page. For example, $\ell(B)$ is the number of incoming links to page A from page B. The factor q (usually 0.85) is used to scale down indirect votes because direct votes are trusted more than indirect votes. To start off, all pages are given a small authority of $1 - q = 0.15$, resulting in the property that the average PageRank of all pages will be 1. Google is constantly recalculating the PageRank of Web pages. When a PageRank tends to stabilize, it is used by a search engine. For additional details, see the online encyclopedia Wikipedia: `en.wikipedia.org/wiki/PageRank`

Let $\mathbf{y}^{(k)^T}$ be the PageRank row vector at the kth iteration, which can be written as

$$\mathbf{y}^{(k+1)^T} = \mathbf{y}^{(k)^T} \mathbf{H}$$

where $\mathbf{H}$ is a row-normalized hyperlink matrix; that is, h_{ij} is nonzero if there is a link from page i to page j and is zero otherwise. Because of convergence problems such as cycling or dependence on the starting vector, the PageRank concept was revised to involve the power method applied to a primitive stochastic iteration matrix that always converges to the PageRank vector $\mathbf{y}^T$ (a unique stationary distribution vector) independent of the starting vector. The asymptotic rate of convergence is governed by the magnitude of the subdominant eigenvalue λ_2 of the transitive probability matrix. Google also uses nonmathematical *metrics* when responding to a query and is secretive about exactly what it does, but at the heart of the software is the PageRank algorithm.

Google's algorithm for the ranking of Web pages uses a *popularity contest* style of search. Other iterative ranking methods predate Google's PageRank algorithm by over seventy years. In his 1941 paper, Nobel Prize winning economist Wassily Leontief discussed an iterative method for ranking industries. In 1965, sociologist Charles Hubbell published an iterative method for ranking people. Computer scientist Jon Kleinberg developed the Hypertext Induced Topic algorithm for optimizing Web information and it was referenced by Google's Sergey Brin and Larry Page. It has been said that "*PageRank stands on the shoulders of giants!*". For additional details on Web searching see Langville and Meyer [2005–2006] as well as Berry and Browne [2005].

Mathematical Software

Here is an example of using mathematical software to find the diagonalization of a complex matrix. Define the matrix

$$\mathbf{A} = \begin{bmatrix} 5 & 2-i & 1+i\sqrt{5} \\ 2+i & 4 & 3 \\ 1-i\sqrt{5} & 3 & -1 \end{bmatrix}$$

How does one use mathematical software to find the diagonalization of this matrix?

We can use MATLAB, Maple, or Mathematica as follows:

```
MATLAB
A = [5, 2-i,1+i*sqrt(5);2+i,4,3;1-i*sqrt(5),3,-1]
[P,D] = eig(A)
P'*A*P
```

```
Maple
with(LinearAlgebra):
A := Matrix([[5, 2-I,1+I*sqrt(5)], [2+I,4,3],
      [1-I*sqrt(5),3,-1]]);
(DD,P) := Eigenvectors(A);
Transpose[P].A.P;
```

```
Mathematica
A = {{5, 2-I,1+I*Sqrt[5]},{2+I,4,3},{1-I*Sqrt[5]I,3,-1}}
{DD,P} = Eigensystem[A]
Transpose[P].A.P
```

The computer programs respond with $\mathbf{A}$ and the two new matrices $\mathbf{P}$ and $\mathbf{D}$. The matrix equation $\mathbf{P}^H\mathbf{AP} = \mathbf{D}$ should be true, and that is readily verified.

SUMMARY 8.1

- $\langle \mathbf{x}, \mathbf{y} \rangle = \mathbf{y}^H\mathbf{x} = \sum_{j=1}^{n} x_j \overline{y}_j$
  This is the **Standard Inner Product** for $\mathbb{C}^n$.

- $\mathbf{xy}^H$ is an $n \times n$ matrix of rank at most one (called the **Outer Product**)

- $\langle \mathbf{x}, \mathbf{y} \rangle = \overline{\langle \mathbf{y}, \mathbf{x} \rangle}$, $\langle \mathbf{x}, \alpha \mathbf{y} \rangle = \overline{\alpha}\langle \mathbf{x}, \mathbf{y} \rangle$, and $\langle \mathbf{Ax}, \mathbf{y} \rangle = \langle \mathbf{x}, \mathbf{A}^H\mathbf{y} \rangle$

- $\langle \mathbf{x}, \mathbf{y} \rangle = \mathbf{y}^H\mathbf{x}$ and $\langle \mathbf{Ax}, \mathbf{y} \rangle = \mathbf{y}^H\mathbf{Ax}$

- $\mathbf{A} = \mathbf{A}^H$ is the condition for a **Hermitian matrix**.

- A mapping f is **self-adjoint** if and only if $\langle f(\mathbf{x}), \mathbf{y} \rangle = \langle \mathbf{x}, f(\mathbf{y}) \rangle$.

- $\mathbf{U}^H\mathbf{U} = \mathbf{I}$ is the condition for a square matrix to be **unitary**.
- $\mathbf{U}^T\mathbf{U} = \mathbf{I}$ is the condition for a square real matrix to be **orthogonal**.
- $\mathbf{x} \mapsto \mathbf{x}^T\mathbf{A}\mathbf{x}$ is the definition of a **quadratic** form in $\mathbb{R}^n$.
- Theorems and Corollaries:
  - Every self-adjoint mapping is linear.
  - All eigenvalues of a self-adjoint operator are real.
  - Any two eigenvectors of a self-adjoint operator are mutually orthogonal if they are associated with different eigenvalues.
  - Eigenvalues of L and $\mathbf{A}$ are the same if $\mathbf{A}$ is the matrix for a linear transformation L relative to a given basis.
  - The operator L is self-adjoint if its matrix satisfies $\mathbf{A} = \mathbf{A}^H$.
  - **Spectral Theorem**: Let L be a self-adjoint operator on an n-dimensional inner-product space, X. Then there exists an orthonormal basis $\{\mathbf{u}_1, \mathbf{u}_2, \ldots, \mathbf{u}_n\}$ for

X and a set of n real numbers λ_j such that $L(\mathbf{u}_j) = \lambda_j\mathbf{u}_j$, for $1 \leq j \leq n$, and $L(\mathbf{x}) = \sum_{j=1}^{n} \lambda_j\langle\mathbf{x}, \mathbf{u}_j\rangle\mathbf{u}_j$ for all $\mathbf{x} \in X$.

- **Spectral Theorem, Matrix Version**: If $\mathbf{A}^H = \mathbf{A}$, then $\mathbf{U}^H\mathbf{A}\mathbf{U} = \mathbf{D}$, where $\mathbf{D}$ is a diagonal matrix having the eigenvalues of $\mathbf{A}$ on its diagonal, and $\mathbf{U}$ is a unitary matrix whose columns are eigenvectors of $\mathbf{A}$.
- If $\mathbf{A}^H = \mathbf{A}$, then all eigenvalues of $\mathbf{A}$ are real.
- If $\mathbf{A}$ is real and $\mathbf{A}^T = \mathbf{A}$, then all eigenvalues of $\mathbf{A}$ are real.
- If $\mathbf{A}$ is a real, symmetric matrix, then $\mathbf{U}^T\mathbf{A}\mathbf{U} = \mathbf{D}$, where $\mathbf{D}$ is diagonal and $\mathbf{U}$ is orthogonal. The eigenvalues of $\mathbf{A}$ are on the diagonal of $\mathbf{D}$.
- **Cayley–Hamilton Theorem**: If p is the characteristic polynomial of a matrix $\mathbf{A}$, then $p(\mathbf{A}) = \mathbf{0}$.
- If $\mathbf{B} = \frac{1}{2}(\mathbf{A} + \mathbf{A}^T)$, then the quadratic forms of $\mathbf{A}$ and $\mathbf{B}$ are identical and $\mathbf{B}$ is symmetric.

KEY CONCEPTS 8.1

Complex inner products, conjugate transpose, symmetric matrices, Hermitian matrices, self-adjoint transformations, spectral theorem, orthogonal matrices, unitary matrices, unitarily similar, diagonalization, Cayley–Hamilton Theorem, quadratic form, Web searching

GENERAL EXERCISES 8.1

1. Find the diagonalization of $\mathbf{A} = \begin{bmatrix} 2 & 3 \\ 3 & 2 \end{bmatrix}$

2. Let $\mathbf{u} = 3^{-\frac{1}{2}}(1, 1, 1)$, $\mathbf{v} = 6^{-\frac{1}{2}}(-1, 2, -1)$, and $\mathbf{w} = 2^{-\frac{1}{2}}(-1, 0, 1)$. Verify that $\{\mathbf{u}, \mathbf{v}, \mathbf{w}\}$ is orthonormal.

3. Find the eigenvalues of this matrix:
$$\begin{bmatrix} 2 - 3i & 1 - i \\ -2 + i & 1 - 2i \end{bmatrix}$$

4. Find the successive powers of the matrices $\mathbf{A}$ and $\mathbf{B}$ given here, and discuss the negative integer powers also.

a. $A = \begin{bmatrix} 0 & 1 & 0 & 0 & 0 \\ 0 & 0 & 1 & 0 & 0 \\ 0 & 0 & 0 & 1 & 0 \\ 0 & 0 & 0 & 0 & 1 \\ 0 & 0 & 0 & 0 & 0 \end{bmatrix}$

b. $B = \begin{bmatrix} b_1 & 0 & 0 & 0 & 0 \\ 0 & b_2 & 0 & 0 & 0 \\ 0 & 0 & b_3 & 0 & 0 \\ 0 & 0 & 0 & b_4 & 0 \\ 0 & 0 & 0 & 0 & b_5 \end{bmatrix}$

5. Can every orthogonal 2×2 matrix be put into the form $\begin{bmatrix} \cos\theta & \sin\theta \\ \sin\theta & -\cos\theta \end{bmatrix}$ for a suitable value of θ?

6. Carry out the diagonalization of this self-adjoint matrix: $\begin{bmatrix} 2 & 1+i \\ 1-i & 3 \end{bmatrix}$
For encouragement along the way, we offer these guideposts: The eigenvalues are 1 and 4. The eigenvectors, before normalization, can be $(1 + i, -1)$ and $(1 + i, 2)$. For those eigenvectors the normalizing divisors are $\sqrt{6}$ and $\sqrt{3}$.

7. Let $A = \begin{bmatrix} 2 & 1+i & 2+2i \\ 1-i & 3 & 5-i \\ 2-2i & 5+i & -5 \end{bmatrix}$
Without doing any calculation, conclude that the eigenvalues of this matrix are real.

8. Let $A = \begin{bmatrix} 3 & 1 \\ 0 & 7 \end{bmatrix}$
Explain the meaning of the assertion that A obeys the Cayley–Hamilton theorem. Verify the assertion.

9. Explain why the diagonal of a Hermitian matrix is real.

10. What are the graphs of $x^T A x = c$, when $A = \begin{bmatrix} 1 & -1 \\ -1 & 1 \end{bmatrix}$ and c is given different values such as $c = -1, 0, 1$?

11. Find all the matrices (symmetric and otherwise) that correspond to this quadratic form: $2x_1^2 + 6x_1 x_2 + 5x_2^2$.

12. (Continuation.) With a change of variables, express the quadratic form in the preceding problem without the cross-product terms. Along the way, you should find eigenvalues $\frac{1}{2}(7 \pm 3\sqrt{5})$.

13. Let $A = \begin{bmatrix} c & -c \\ -c & c \end{bmatrix}$ where c is any real number. What are the eigenvalues and eigenvectors of A in terms of c?

14. What is A if its corresponding quadratic form is $2x_1^2 + 6x_1 x_2 + 7x_2^2$?

15. Find all the matrices A such that $\begin{bmatrix} \alpha & 0 \\ 0 & \beta \end{bmatrix} = \frac{1}{2}(A + A^T)$ where $\alpha \neq 0$ and $\beta \neq 0$.

16. Establish as true or find a counterexample to the assertion that the nonzero eigenvalues of a skew symmetric matrix are pure imaginary. The matrix A is **skew symmetric** if $A^T = -A$.

17. Explain why an arbitrary square complex matrix is the sum of a Hermitian matrix and a skew Hermitian matrix. A matrix S is **skew-Hermitian** if $S^H = -S$.

18. In the space $\mathbb{P}$ consisting of all polynomials, use the inner product $\langle x, y \rangle = \int_{-1}^{1} x(t)\overline{y(t)}\, dt$. Select a polynomial p, and define a linear operator $L : \mathbb{P} \to \mathbb{P}$ by the formula $L(x) = px$. (This is the simple product of two polynomials: $t \mapsto p(t)x(t)$.) Find the operator L^* that has this property: $\langle L(x), y \rangle = \langle x, L^*(y) \rangle$. Assume that the polynomials can have complex coefficients, and that the variable t can be complex. The operator L^* is called the **adjoint** of L and is a generalization of A^H for matrices to operators for possible infinite-dimensional situations.

19. Establish that for any $m \times n$ real matrix $\mathbf{A}$, if $\mathbf{Ax} = \mathbf{0}$ for all $\mathbf{x}$ in $\mathbb{R}^n$, then $\mathbf{A} = \mathbf{0}$.

20. Explain why the columns of a square matrix must form an orthonormal set if the rows do so. Cite the crucial theorem from an earlier point in the book that justifies this.

21. (Continuation.) If the matrix in the preceding problem is not square, can we draw the same conclusion? A counterexample or a theorem is needed.

22. Establish directly that the eigenvalues of a real symmetric matrix are real and that the eigenvectors associated with different eigenvalues are mutually orthogonal.

23. A unitarily diagonalizable matrix $\mathbf{A}$ commutes with $\mathbf{A}^H$, which is called a **normal matrix**. Establish this fact.

24. If $\mathbf{A}$ is unitary, explain why $\mathbf{A}^{-1}$ and $\mathbf{A}^H$ are also unitary. Is the product of two unitary matrices unitary? (Theorems or examples are needed.) Does the sum of two unitary matrices have any interesting properties? What useful property does the matrix $\frac{1}{2}(\mathbf{A} + \mathbf{A}^H)$ have?

25. Explain why $\langle \mathbf{Ax}, \mathbf{y} \rangle = \langle \mathbf{x}, \mathbf{A}^H \mathbf{y} \rangle$, using only the definition of $\mathbf{A}^H$ and the properties of an inner product.

26. Explain why an invertible matrix $\mathbf{A}$ satisfies the equation $(\mathbf{A}^H)^{-1} = (\mathbf{A}^{-1})^H$.

27. Let L be a self-adjoint operator on a finite-dimensional inner-product space (the scalar field being $\mathbb{R}$). Let $\mathbf{A}$ be the matrix for L with respect to some basis. Is $\mathbf{A}$ necessarily symmetric? A theorem or example is needed.

28. Affirm that if we introduce a new inner product on $\mathbb{C}^2$ by the equation $\langle \mathbf{x}, \mathbf{y} \rangle_D = 2x_1 y_1 + 3x_2 y_2$, then the transformation T defined by $T(x_1, x_2) = (2x_1 + 6x_2, 4x_1 + 7x_2)$ is self-adjoint with respect to the new inner product.

29. Let $\mathbf{A}$ be an $n \times n$ Hermitian matrix. Explain why this hypothesis leads to the conclusion that the map $\mathbf{x} \mapsto \mathbf{Ax}$ is self-adjoint on the space $\mathbb{C}^n$, with its standard inner product.

30. Let V be an inner-product space. Establish that the family of all self-adjoint operators from V to V is closed under addition and under multiplication by *real* scalars.

31. (Continuation.) In detail, establish that $F(\mathbf{x}) = \sum_{i=1}^m f_i$ is self-adjoint and $G(\mathbf{x}) = \sum_{i=1}^m r_i g_i$ is self-adjoint, where $f_1, f_2, \ldots, f_n$ and $g_1, g_2, \ldots, g_n$ are self-adjoint mappings from an inner-product space into itself and $r_1, r_2, \ldots, r_m$ are real numbers.

32. Explain why the map $\mathbf{x} \mapsto \langle \mathbf{x}, \mathbf{u} \rangle \mathbf{v}$ is self-adjoint if and only if $\mathbf{u} = \mathbf{v}$.

33. Let $\{\mathbf{v}_1, \mathbf{v}_2, \ldots, \mathbf{v}_m\}$ be points in an inner-product space. Define $T(\mathbf{x}) = \sum_{i=1}^n r_i \langle \mathbf{x}, \mathbf{v}_i \rangle \mathbf{v}_i$, where $r_1, r_2, \ldots r_n$, are real numbers. Establish that T is self-adjoint.

34. In an inner-product space, let $\{\mathbf{u}_1, \mathbf{u}_2, \ldots, \mathbf{u}_n\}$ and $\{\mathbf{v}_1, \mathbf{v}_2, \ldots, \mathbf{v}_n\}$ be linearly independent sets. Define a linear operator $L(\mathbf{x}) = \sum_{i=1}^n \langle \mathbf{x}, \mathbf{u}_i \rangle \mathbf{v}_i$. Show that if $\mathbf{u}_i = \mathbf{v}_i$ for $1 \leq i \leq n$, then L is self-adjoint.

35. Let f be a mapping from $\mathbb{R}^n$ into $\mathbb{R}^n$ such that $\langle f(\mathbf{x}), \mathbf{y} \rangle = \langle \mathbf{x}, f(\mathbf{y}) \rangle$ for all $\mathbf{x}$ and $\mathbf{y}$. Establish that there is a symmetric matrix $\mathbf{A}$ such that $f(\mathbf{x}) = \mathbf{Ax}$ for all $\mathbf{x}$ and $\mathbf{y}$.

36. Establish that if the matrix $\mathbf{A}$ is Hermitian, then the quadratic form $\mathbf{x} \mapsto \langle \mathbf{Ax}, \mathbf{x} \rangle$ is real valued. Assume that $\mathbf{x}$ ranges over the space $\mathbb{C}^n$.

37. Affirm that if the real matrix $\mathbf{A}$ is orthogonally diagonalizable, then $\mathbf{A}$ must be symmetric. The hypothesis means that for some orthogonal matrix $\mathbf{U}$, the matrix $\mathbf{UAU}^T$ is a diagonal matrix.

38. (Continuation.) Use the preceding problem and Theorem 8 to establish that if $\mathbf{A}$ and $\mathbf{B}$ are orthogonally diagonalizable, then the same is true of $\mathbf{A} + \mathbf{B}$.

39. Use the Cayley–Hamilton Theorem to get a formula for $\mathbf{A}^{-1}$ when $\mathbf{A}$ is invertible. Be sure to use somewhere in your work the hypothesis that $\mathbf{A}$ is invertible! You can assume that the characteristic equation of $\mathbf{A}$ is known.

40. Use the matrix $\mathbf{A}$ in Example 6. Find a formula for $\mathbf{A}^k$ by diagonalizing $\mathbf{A}$. (Refer to Section 6.1), in particular, for this technique. The eigenvalues are 7 and 5. Then compute $\mathbf{A}^2$ by using the Cayley–Hamilton Theorem.

41. Explain why, for any $n \times n$ matrix $\mathbf{A}$, $\mathbf{A}^n$ is in the span of $\{\mathbf{I}, \mathbf{A}, \mathbf{A}^2, \ldots, \mathbf{A}^{n-1}\}$.

42. Let q be any polynomial and $\mathbf{A}$ any square matrix. Verify that if λ is an eigenvalue of $\mathbf{A}$, then $q(\lambda)$ is an eigenvalue of $q(\mathbf{A})$.

43. Establish that if $n > 1$, then there is no $n \times n$ matrix whose powers span $\mathbb{R}^{n \times n}$.

44. Find out what $n \times n$ matrices have the property that $\mathbf{A}^n \in \mathrm{Span}\{\mathbf{A}, \mathbf{A}^2, \ldots, \mathbf{A}^{n-1}\}$.

45. Explain, using induction, that for any $n \times n$ matrix $\mathbf{A}$, $\mathbf{A}^k \in \mathrm{Span}\{\mathbf{A}^0, \mathbf{A}^1, \ldots, \mathbf{A}^{n-1}\}$ for all nonnegative integers k.

46. Fix an $n \times n$ matrix $\mathbf{A}$, and let matrices $\mathbf{B}$ and $\mathbf{C}$ be in the span of $\{\mathbf{A}^0, \mathbf{A}^1, \mathbf{A}^2, \ldots\}$. Affirm that $\mathbf{BC} = \mathbf{CB}$.

47. Critique this proof of the Cayley–Hamilton Theorem. Let $p(\lambda) = \mathrm{Det}(\mathbf{A} - \lambda \mathbf{I})$. Therefore, we have $p(\mathbf{A}) = \mathrm{Det}(\mathbf{A} - \mathbf{AI}) = \mathrm{Det}(\mathbf{0}) = 0$. (See Barbeau [2000].)

48. Let $\mathbf{A}$ be a square matrix having possibly complex elements. Confirm that the diagonal of $\mathbf{AA}^H$ has only nonnegative real numbers. Also, establish that $\mathbf{AA}^H$ is Hermitian.

49. Explain why every 2×2 real matrix is similar to its transpose. Is this true for 2×2 complex matrices?

50. **a.** Find a 2×2 matrix, $\mathbf{A}$, for which the dimension of $\mathrm{Span}\{\mathbf{I}, \mathbf{A}\}$ is two. Then find other 2×2 matrices for which this dimension is zero and one, respectively.
b. Find a 3×3 matrix, $\mathbf{A}$, for which the dimension of $\mathrm{Span}\{\mathbf{I}, \mathbf{A}, \mathbf{A}^2\}$ is three. Then find other 3×3 matrices for which this dimension is zero, one, and two, respectively.
c. Can you create examples of this sort for arbitrary n?

51. For $\mathbf{z} = (z_1, z_2, \ldots, z_n) \in \mathbb{C}$, we use the notation $\bar{\mathbf{z}} = (\bar{z}_1, \bar{z}_2, \ldots, \bar{z}_n)$. Is the mapping $\mathbf{z} \to \bar{\mathbf{z}}$ linear?

52. If $\mathbf{A}$ is an $n \times n$ matrix, define $S(\mathbf{A})$ to be the span of the nonnegative powers of $\mathbf{A}$. This will be a subspace of $\mathbb{R}^{n \times n}$. How many matrices $\mathbf{A}, \mathbf{B}, \mathbf{C} \ldots$ are needed so that $S(\mathbf{A}), S(\mathbf{B}), S(\mathbf{C}) \ldots$ spans $\mathbb{R}^{n \times n}$?

53. Explain that if $\mathbf{A}$ is an $m \times n$ matrix of rank m, then $\mathbf{AA}^T$ is invertible.

54. (Continuation.) Confirm that if $\mathbf{A}$ is an $m \times n$ matrix of rank m and $\mathbf{B}$ is an $n \times n$ invertible matrix, then $\mathbf{ABB}^T\mathbf{A}^T$ is invertible. What about the case $\mathbf{ABA}^T$?

55. Let $\mathbf{A}$ be a real matrix that is similar to a real diagonal matrix. Does it follow that $\mathbf{A}$ is symmetric? An example or theorem is needed.

56. Confirm that if $\langle \mathbf{u}, \mathbf{x} \rangle = \langle \mathbf{v}, \mathbf{x} \rangle$ for all $\mathbf{x} \in \mathbb{C}^n$, then $\mathbf{u} = \mathbf{v}$.

57. In modern terms, explain the quote by Cayley on pp. 469–470. For additional information, see Midonic [1965, p. 198].

58. Can you find a 3×3 matrix whose powers span $\mathbb{R}^{3 \times 3}$?

59. Let $\mathbf{A}$ and $\mathbf{B}$ be $n \times n$ matrices such that $2\mathbf{A} = \mathbf{B} + \mathbf{B}^T$. Does it follow that $2\mathbf{B} = \mathbf{A} + \mathbf{A}^T$?

60. Using the matrix in Example 6, find the general expression for $\mathbf{A}^k$ as a single matrix.

61. We are given $\mathbf{A} = \begin{bmatrix} 1 & 2 & 3 \\ 4 & 5 & 6 \\ 7 & 8 & 9 \end{bmatrix}$

Find the explicit formula for the quadratic form induced by the matrix and show that it is a polynomial of degree two in three variables $x_1, x_2,$ and x_3.

62. Can the quadratic forms of the following matrices be expressed without the $x_1 x_2$ term using changes of variables? Explain why or why not.

a. $\begin{bmatrix} 3 & 5 \\ 2 & -1 \end{bmatrix}$ **b.** $\begin{bmatrix} 3 & 6 \\ 1 & 4 \end{bmatrix}$ **c.** $\begin{bmatrix} -5 & -2 \\ 2 & -5 \end{bmatrix}$

63. Redo Example 13 using a direct change of variables.

64. Let $C[-1, 1]$ be the vector space of all continuous real-valued functions defined on the closed interval $[-1, 1]$. Use the inner product $\langle u, v \rangle = \int_{-1}^{1} u(t)v(t)dt$. Define T on $C[-1, 1]$ by the equation $T(u) = wu$, where w is a fixed function in $C[-1, 1]$. Is T self-adjoint?

65. (Continuation.) Use the definitions in the preceding problem, except that $[T(u)](x) = (x^2 + y^2)u(y)$. Is T self-adjoint?

66. (Continuation.) Let V stand for the vector space of all polynomials p such that $p(1) = p(-1) = 0$. Use the inner-product of General Exercise 64. Establish that V is an inner-product space. Find out whether the operation of differentiation is self-adjoint. Define L as $L(u) = u' = du/dt$.

67. Find the change of variables in Examples 12 and 13 so that $\mathbf{x}^T\mathbf{Ax} = \mathbf{x}^T\mathbf{Bx} = \mathbf{y}^T\mathbf{Dy}$.

68. (**Adjacency Matrix.**) In the figure, each of $n = 15$ nodes represents a Web page and a directed edge from node i to node j means that page i contains a Web link to page j. Construct the adjacency matrix $\mathbf{A}$, which is an $n \times n$ matrix whose (i, j) entry is 1 if there is a link from node i to node j.

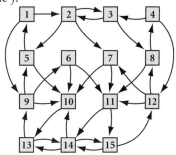

COMPUTER EXERCISES 8.1

1. For any matrix of the form $\mathbf{A} = \begin{bmatrix} a & b \\ b & c \end{bmatrix}$, use mathematical software to find formulas for the factors in the equation $\mathbf{A} = \mathbf{Q}\mathbf{D}\mathbf{Q}^T$, where $\mathbf{Q}\mathbf{Q}^T = \mathbf{I}_2$ and $\mathbf{D}$ is diagonal.

2. Determine numerically whether these matrices are Hermitian or skew-Hermitian:

a. $\mathbf{A} = \begin{bmatrix} 5i & -4+i & 3-2i \\ 4+i & 2i & -2+i \\ -3-2i & 2+i & -6i \end{bmatrix}$

b. $\mathbf{B} = \begin{bmatrix} 3 & 1-2i & 4+7i \\ 1+2i & -4 & -2i \\ 4-7i & 2i & 2 \end{bmatrix}$

3. Determine numerically whether these matrices are orthogonal or unitary:

a. $\mathbf{C} = \begin{bmatrix} \frac{1}{2} & -\frac{i}{2} & -\frac{1}{2}+\frac{i}{2} \\ \frac{i}{2} & \frac{1}{2} & \frac{1}{2}+\frac{i}{2} \\ \frac{1}{2}+\frac{i}{2} & -\frac{1}{2}+\frac{i}{2} & 0 \end{bmatrix}$

b. $\mathbf{D} = \begin{bmatrix} \frac{1}{9} & \frac{8}{9} & -\frac{4}{9} \\ \frac{4}{9} & -\frac{4}{9} & -\frac{7}{9} \\ \frac{8}{9} & \frac{1}{9} & \frac{4}{9} \end{bmatrix}$

4. Let $\begin{bmatrix} a & \frac{2}{3} & \frac{2}{3} \\ \frac{2}{3} & \frac{1}{3} & b \\ x & y & z \end{bmatrix}$

Determine numerically the values of a, b, and c so that this matrix is orthogonal.

5. Let $\begin{bmatrix} a & b \\ c & d \end{bmatrix}$

Determine the general form of a real 2×2 orthogonal matrix.

6. Let $\begin{bmatrix} a & b \\ c & d \end{bmatrix}$

Establish that a general 2×2 *normal* matrix is either symmetric or the sum of a scalar matrix and a skew-symmetric matrix. *Note:* A matrix $\mathbf{A}$ is a **normal** matrix if $\mathbf{A}\mathbf{A}^T = \mathbf{A}^T\mathbf{A}$ when $\mathbf{A}$ is real or if $\mathbf{A}\mathbf{A}^H = \mathbf{A}^H\mathbf{A}$ when $\mathbf{A}$ is complex.

7. (Continuation.) Establish that each of these are normal matrices:

a. real symmetric matrices

b. real orthogonal matrices

c. complex Hermitian matrices

d. complex unitary matrices

8. Consider

$$\mathbf{A} = \begin{bmatrix} 5 & 1 & 2 & 0 & 4 \\ 1 & 4 & 2 & 1 & 3 \\ 2 & 2 & 5 & 4 & 0 \\ 0 & 1 & 4 & 1 & 3 \\ 4 & 3 & 0 & 3 & 4 \end{bmatrix}$$

For the matrix $\mathbf{A}$ shown, compute its eigenvalues and eigenvectors as well as the unitary matrix $\mathbf{U}$ such that $\mathbf{U}^H\mathbf{A}\mathbf{U} = \mathbf{D}$.

9. Consider

$$\mathbf{B} = \begin{bmatrix} 1 & 1 & 1 & 1 & 1 \\ -1 & -1 & 0 & 0 & 1 \\ -2 & -2 & 0 & 0 & 1 \\ 0 & 0 & 1 & 1 & 3 \\ 1 & 1 & 2 & 2 & 4 \end{bmatrix}$$

Compute the diagonalization of matrix $\mathbf{B}$.

10. Consider

$$\mathbf{C} = \begin{bmatrix} 0 & 3 & 6 & 4 & 5i \\ 1 & 3i & 0 & 4 & 7 \\ 2 & 3 & 5 & 3 & 9i \\ 1 & 4i & 5 & 9 & 3 \\ 2 & 8 & 5 & 2 & i \end{bmatrix}$$

Verify that the matrix $\mathbf{C}$ obeys the Cayley-Hamilton Theorem.

11. Consider

$$F = \begin{bmatrix} 0 & 3i & 6 & 4 & 2 \\ 1 & 3 & 0 & 4i & 1 \\ 2 & 3i & 5 & 3 & 4 \\ 1 & 4 & 5 & 9 & 2i \\ 2i & 4 & 7 & i & 0 \end{bmatrix}$$

For the given matrix, compute the symmetric matrix $G = \frac{1}{2}(F + R^H)$. Show that the corresponding quadratic forms are equivalent.

8.2 MATRIX FACTORIZATIONS AND BLOCK MATRICES

Young men should prove theorems, old men should write books.
— GODFREY H. HARDY (1877–1947)

Mathematics is like checkers in being suitable for the young, not too difficult, amusing, and without peril to the state.
— PLATO (ca. 429–347 BC)

Introduction

Whenever we encounter an equation $A = BC$, we have a factorization of the matrix A into a product of the two matrices B and C. In some cases, we have more factors, as in this familiar case encountered in Section 6.1: $A = PDP^{-1}$. There is also a QR-factorization $A = QR$, and a singular-value decomposition (SVD), $A = PDQ$. These and other factorizations all have their uses, and we begin with a simple one, the LU factorization, that is often a part of the row-reduction process.

Permutation Matrix

First, however, the concept of a **permutation matrix** is needed. Such a matrix is obtained by reordering the rows of an identity matrix, I. For example, here is a typical permutation matrix:

$$P = \begin{bmatrix} 0 & 0 & 1 & 0 \\ 0 & 0 & 0 & 1 \\ 1 & 0 & 0 & 0 \\ 0 & 1 & 0 & 0 \end{bmatrix}$$

Notice that the product PA is the matrix A with its rows reordered in exactly the same way as the rows of I were reordered to produce P. For example, we obtain

$$PA = \begin{bmatrix} 0 & 0 & 1 & 0 \\ 0 & 0 & 0 & 1 \\ 1 & 0 & 0 & 0 \\ 0 & 1 & 0 & 0 \end{bmatrix} \begin{bmatrix} 1 & 4 & 2 & -3 \\ 3 & -2 & 8 & 9 \\ 5 & 0 & 3 & 1 \\ 9 & 8 & 7 & 6 \end{bmatrix} = \begin{bmatrix} 5 & 0 & 3 & 1 \\ 9 & 8 & 7 & 6 \\ 1 & 4 & 2 & -3 \\ 3 & -2 & 8 & 9 \end{bmatrix}$$

The mapping from the original ordering to the reordering is $(1, 2, 3, 4) \mapsto (3, 4, 1, 2)$.

EXAMPLE 1 What permutation matrix can be applied to the matrix

$$A = \begin{bmatrix} 0 & 3 & 5 \\ 1 & 2 & -7 \\ 0 & 0 & 4 \end{bmatrix}$$

in order to get an echelon form?

SOLUTION We can interchange the first row and the second row. An equivalent procedure would be to multiply A on the left with this permutation matrix:

$$P = \begin{bmatrix} 0 & 1 & 0 \\ 1 & 0 & 0 \\ 0 & 0 & 1 \end{bmatrix}$$

Consequently, we obtain

$$PA = \begin{bmatrix} 0 & 1 & 0 \\ 1 & 0 & 0 \\ 0 & 0 & 1 \end{bmatrix} \begin{bmatrix} 0 & 3 & 5 \\ 1 & 2 & -7 \\ 0 & 0 & 4 \end{bmatrix} = \begin{bmatrix} 1 & 2 & -7 \\ 0 & 3 & 5 \\ 0 & 0 & 4 \end{bmatrix} \quad \blacksquare$$

Well-written software for solving systems of linear equations chooses the pivot elements carefully to ameliorate the bad effects of round-off error. The natural order of the rows gives way to an ordering that pays attention to the magnitudes of the elements available as pivots. In general, we choose *strong* pivot elements to avoid growth of round-off errors.

LU-Factorization

The *LU*-factorization of a matrix **A** is usually written as **A** = **LU**. In this equation, **L** is a lower triangular matrix having 1's on its diagonal. (One says that **L** is *unit* lower triangular.) The factor **U** is upper triangular. Here is an example of the equation **A** = **LU**:

$$
\mathbf{A} = \begin{bmatrix} 2 & 2 & 3 \\ 6 & 7 & 14 \\ 14 & 10 & -5 \end{bmatrix} = \begin{bmatrix} 1 & 0 & 0 \\ 3 & 1 & 0 \\ 7 & -4 & 1 \end{bmatrix} \begin{bmatrix} 2 & 2 & 3 \\ 0 & 1 & 5 \\ 0 & 0 & -6 \end{bmatrix} = \mathbf{LU}
$$

The usefulness of the *LU*-factorization depends on the following thought process. Suppose that we wish to solve a linear system

$$\mathbf{Ax} = \mathbf{b}$$

If a factorization **A** = **LU** is available, it can be put to immediate use. Because **A** = **LU**, the problem can be written as **LUx** = **b**. Introduce a new variable **y** = **Ux**. Solve these two linear systems (in the order indicated):

$$\begin{cases} \mathbf{Ly} = \mathbf{b} \\ \mathbf{Ux} = \mathbf{y} \end{cases}$$

The solution sought is **x** because

$$\mathbf{Ax} = \mathbf{LUx} = \mathbf{Ly} = \mathbf{b}$$

We illustrate with the preceding matrix.

EXAMPLE 2 Solve the following system of equations, making use of the *LU*-factorization of the preceding coefficient matrix.

$$\begin{cases} 2x_1 + 2x_2 + 3x_3 = 13 \\ 6x_1 + 7x_2 + 14x_3 = 60 \\ 14x_1 + 10x_2 - 5x_3 = -23 \end{cases}$$

SOLUTION There are two systems to be solved, each having a triangular coefficient matrix. First, we have

$$\mathbf{Ly} = \mathbf{b}: \qquad \begin{bmatrix} 1 & 0 & 0 \\ 3 & 1 & 0 \\ 7 & -4 & 1 \end{bmatrix} \begin{bmatrix} y_1 \\ y_2 \\ y_3 \end{bmatrix} = \begin{bmatrix} 13 \\ 60 \\ -23 \end{bmatrix}$$

We solve this using *forward substitution*, getting $\mathbf{y} = [13, 21, -30]^T$. Then we solve the second system:

$$\mathbf{Ux = y}: \quad \begin{bmatrix} 2 & 2 & 3 \\ 0 & 1 & 5 \\ 0 & 0 & -6 \end{bmatrix} \begin{bmatrix} x_1 \\ x_2 \\ x_3 \end{bmatrix} = \begin{bmatrix} 13 \\ 21 \\ -30 \end{bmatrix}$$

The solution of this system is $\mathbf{x} = [3, -4, 5]^T$, which we obtained by using *back substitution*. ∎

If we carry out Gaussian elimination on this matrix

$$\mathbf{A} = \begin{bmatrix} 2 & 2 & 3 \\ 6 & 7 & 14 \\ 14 & 10 & -5 \end{bmatrix}$$

we shall see that the LU-factorization is a by-product. By using elementary matrices, one can turn $\mathbf{A}$ into an upper triangular matrix. Here are the steps:

$$\mathbf{E_1 A} = \begin{bmatrix} 1 & 0 & 0 \\ -3 & 1 & 0 \\ 0 & 0 & 1 \end{bmatrix} \begin{bmatrix} 2 & 2 & 3 \\ 6 & 7 & 14 \\ 14 & 10 & -5 \end{bmatrix} = \begin{bmatrix} 2 & 2 & 3 \\ 0 & 1 & 5 \\ 14 & 10 & -5 \end{bmatrix}$$

$$\mathbf{E_2(E_1 A)} = \begin{bmatrix} 1 & 0 & 0 \\ 0 & 1 & 0 \\ -7 & 0 & 1 \end{bmatrix} \begin{bmatrix} 2 & 2 & 3 \\ 0 & 1 & 5 \\ 14 & 10 & -5 \end{bmatrix} = \begin{bmatrix} 2 & 2 & 3 \\ 0 & 1 & 5 \\ 0 & -4 & -26 \end{bmatrix}$$

$$\mathbf{E_3(E_2 E_1 A)} = \begin{bmatrix} 1 & 0 & 0 \\ 0 & 1 & 0 \\ 0 & 4 & 1 \end{bmatrix} \begin{bmatrix} 2 & 2 & 3 \\ 0 & 1 & 5 \\ 0 & -4 & -26 \end{bmatrix} = \begin{bmatrix} 2 & 2 & 3 \\ 0 & 1 & 5 \\ 0 & 0 & -6 \end{bmatrix}$$

We have established the equation

$$\mathbf{E_3 E_2 E_1 A} = \begin{bmatrix} 2 & 2 & 3 \\ 0 & 1 & 5 \\ 0 & 0 & -6 \end{bmatrix} = \mathbf{U}$$

Consequently, we have also

$$\mathbf{A = E_1^{-1} E_2^{-1} E_3^{-1} U}$$

The inverses of these elementary matrices are obtained by simply changing the signs of the multipliers. Hence, we find

$$\mathbf{E}_1^{-1}\mathbf{E}_2^{-1}\mathbf{E}_3^{-1} = \begin{bmatrix} 1 & 0 & 0 \\ 3 & 1 & 0 \\ 0 & 0 & 1 \end{bmatrix} \begin{bmatrix} 1 & 0 & 0 \\ 0 & 1 & 0 \\ 7 & 0 & 1 \end{bmatrix} \begin{bmatrix} 1 & 0 & 0 \\ 0 & 1 & 0 \\ 0 & -4 & 1 \end{bmatrix}$$

$$= \begin{bmatrix} 1 & 0 & 0 \\ 3 & 1 & 0 \\ 7 & -4 & 1 \end{bmatrix} = \mathbf{L}$$

Notice how the entries in $\mathbf{L}$ are built up directly from entries in the elementary matrices $\mathbf{E}_i^{-1}$, which are the negative values of the multipliers.

To illustrate how the LU-factors of a matrix can be found in a more direct way, let us use the same matrix $\mathbf{A}$ as in Example 2. Start with $\mathbf{A}$ as given, and $\mathbf{L}$ in skeleton form:

$$\mathbf{A} = \begin{bmatrix} 2 & 2 & 3 \\ 6 & 7 & 14 \\ 14 & 10 & -5 \end{bmatrix}, \qquad \mathbf{L} = \begin{bmatrix} 1 & 0 & 0 \\ ? & 1 & 0 \\ ? & ? & 1 \end{bmatrix}$$

We carry out some row-reductive steps, adding a multiple of one row to another row. The negative of the multipliers are stored in the matrix $\mathbf{L}$.

$$\widetilde{\mathbf{A}} = \begin{bmatrix} 2 & 2 & 3 \\ 0 & 1 & 5 \\ 0 & -4 & -26 \end{bmatrix}, \qquad \mathbf{L} = \begin{bmatrix} 1 & 0 & 0 \\ 3 & 1 & 0 \\ 7 & ? & 1 \end{bmatrix}$$

$$\mathbf{U} \doteq \begin{bmatrix} 2 & 2 & 3 \\ 0 & 1 & 5 \\ 0 & 0 & -6 \end{bmatrix}, \qquad \mathbf{L} = \begin{bmatrix} 1 & 0 & 0 \\ 3 & 1 & 0 \\ 7 & -4 & 1 \end{bmatrix}$$

At the end of the process, $\mathbf{U}$ is on the left and $\mathbf{L}$ is on the right. It is not necessary to continue writing the emerging matrix $\mathbf{L}$; we are only filling in the *holes*, so to speak.

Also, notice that when we store the negative of the multipliers in $\mathbf{L}$, we assume that a multiple of one row in $\mathbf{A}$ is *added* to a row below it. One sees that the negative multipliers to be stored in $\mathbf{L}$ are plus or minus the quotients of elements from $\mathbf{A}$ or $\widetilde{\mathbf{A}}$. Thus, in the previous example, the first multiplier is $-\frac{6}{2} = -3$, and 3 is stored. The next multiplier is $-14/2 = -7$, and we store 7. This is a simple way to describe the negative of multipliers that are being used to fill in the gaps in $\mathbf{L}$. Also, we see that the negative of the multipliers are stored in the matrix **L** in positions corresponding to the locations in the other matrix where zero entries are being created. (For nonsquare matrices, the matrix **L** may need to be extended with columns from the identity matrix. See Example 4.)

It is easy to give examples where there is *no LU*-factorization. Here is one:

$$\mathbf{A} = \begin{bmatrix} 0 & 1 \\ 1 & 0 \end{bmatrix}$$

If we set up the equation **A** = **LU**, it looks like this:

$$\begin{bmatrix} 0 & 1 \\ 1 & 0 \end{bmatrix} = \begin{bmatrix} 1 & 0 \\ a & 1 \end{bmatrix} \begin{bmatrix} b & c \\ 0 & d \end{bmatrix} = \begin{bmatrix} b & c \\ ab & ac + d \end{bmatrix}$$

Then we must have $b = 0$ and $ab = 1$, which are incompatible demands. However, it is obvious that by *interchanging* or *swapping* the rows of **A** in this example, we obtain a (trivial) example of an *LU*-factorization:

$$\begin{bmatrix} 1 & 0 \\ 0 & 1 \end{bmatrix} = \begin{bmatrix} 1 & 0 \\ 0 & 1 \end{bmatrix} \begin{bmatrix} 1 & 0 \\ 0 & 1 \end{bmatrix}$$

To handle a wider class of matrices in the *LU*-factorization process, we can permit permutations of the rows. This means changing the order of the rows. This is done automatically in most mathematical software packages. To permute the rows of a matrix, we can premultiply the matrix by a **permutation matrix**, as explained previously.

EXAMPLE 3 Illustrate permuting the rows of a matrix **A** by multiplication on the left by a permutation matrix **P**. Then find the *LU*-factorization of **PA**.

SOLUTION

$$\mathbf{PA} = \begin{bmatrix} 0 & 0 & 1 \\ 0 & 1 & 0 \\ 1 & 0 & 0 \end{bmatrix} \begin{bmatrix} 3 & 2 & 1 \\ 4 & 1 & 5 \\ 6 & 3 & 4 \end{bmatrix} = \begin{bmatrix} 6 & 3 & 4 \\ 4 & 1 & 5 \\ 3 & 2 & 1 \end{bmatrix}$$

$$= \begin{bmatrix} 1 & 0 & 0 \\ \frac{2}{3} & 1 & 0 \\ \frac{1}{2} & -\frac{1}{2} & 1 \end{bmatrix} \begin{bmatrix} 6 & 3 & 4 \\ 0 & -1 & \frac{7}{3} \\ 0 & 0 & \frac{1}{6} \end{bmatrix} = \mathbf{LU}$$

These are the matrices that MATLAB returns as the *LU*-factorization of the matrix **A**. ∎

EXAMPLE 4 The LU-factorization is also available when the matrix $\mathbf{A}$ is not square. The principles remain unchanged. For example, how can one compute the LU-factorization of the following matrix?

$$\mathbf{A} = \begin{bmatrix} 3 & -1 & 4 \\ 6 & -3 & 10 \\ -9 & 5 & -11 \\ -3 & 0 & -7 \\ 6 & -4 & 2 \end{bmatrix}$$

SOLUTION The basic steps are as in the square case. In the first step we use row operations to create zeros in column one of $\mathbf{A}$, and store the negatives of the multipliers in the first column of the square matrix $\mathbf{L}$ in the order in which they were used to zero out entries:

$$\mathbf{A} \sim \begin{bmatrix} 3 & -1 & 4 \\ 0 & -1 & 2 \\ 0 & 2 & 1 \\ 0 & -1 & -3 \\ 0 & -2 & -6 \end{bmatrix}, \qquad \mathbf{L} = \begin{bmatrix} 1 & 0 & 0 & 0 & 0 \\ 2 & 1 & 0 & 0 & 0 \\ -3 & ? & 1 & 0 & 0 \\ -1 & ? & ? & 1 & 0 \\ 2 & ? & ? & 0 & 1 \end{bmatrix}$$

In the next step, two further row operations are applied and the negatives of the multipliers are stored in the second column of matrix $\mathbf{L}$:

$$\mathbf{A} \sim \begin{bmatrix} 3 & -1 & 4 \\ 0 & -1 & 2 \\ 0 & 0 & 5 \\ 0 & 0 & -5 \\ 0 & 0 & -10 \end{bmatrix}, \qquad \mathbf{L} = \begin{bmatrix} 1 & 0 & 0 & 0 & 0 \\ 2 & 1 & 0 & 0 & 0 \\ -3 & -2 & 1 & 0 & 0 \\ -1 & 1 & ? & 1 & 0 \\ -2 & 2 & ? & 0 & 1 \end{bmatrix}$$

The final step brings us to

$$\mathbf{U} = \begin{bmatrix} 3 & -1 & 4 \\ 0 & -1 & 2 \\ 0 & 0 & 5 \\ 0 & 0 & 0 \\ 0 & 0 & 0 \end{bmatrix}, \qquad \mathbf{L} = \begin{bmatrix} 1 & 0 & 0 & 0 & 0 \\ 2 & 1 & 0 & 0 & 0 \\ -3 & -2 & 1 & 0 & 0 \\ -1 & 1 & -1 & 1 & 0 \\ -2 & 2 & -1 & 0 & 1 \end{bmatrix}$$

The conclusion is that

$$\mathbf{A} = \mathbf{LU}$$

$$
\begin{bmatrix}
3 & -1 & 4 \\
6 & -3 & 10 \\
-9 & 5 & -11 \\
-3 & 0 & -7 \\
6 & -4 & 2
\end{bmatrix}
=
\begin{bmatrix}
1 & 0 & 0 & 0 & 0 \\
2 & 1 & 0 & 0 & 0 \\
-3 & -2 & 1 & 0 & 0 \\
-1 & 1 & -1 & 1 & 0 \\
-2 & 2 & -1 & 0 & 1
\end{bmatrix}
\begin{bmatrix}
3 & -1 & 4 \\
0 & -1 & 2 \\
0 & 0 & 5 \\
0 & 0 & 0 \\
0 & 0 & 0
\end{bmatrix}
$$

Here the entries in the matrix **L** below the main diagonal are the negative values of the multipliers used. They are located in the positions where zeros are being created. For nonsquare matrices **A**, the lower triangular matrix **L** is square and it may need to be extended with columns from the identity matrix. It is usually clear what needs to be done. If **A** is $m \times n$, then **L** is $m \times m$ and **U** is $m \times n$.   ■

LL^T-Factorization: Cholesky Factorization

The LU-factorization of a symmetric matrix **A** into the product of a lower and an upper triangular matrix has a special form. Namely, we can write $\mathbf{A} = \mathbf{LL}^T$. This may involve complex numbers, even if the matrix **A** is real. But the special case where **A** is a real, symmetric, and positive definite matrix involves only real numbers. This is called the **Cholesky factorization**.[4] It is important because symmetric, positive definite matrices arise frequently in applications.

> **DEFINITION**
>
> *A real matrix* **A** *is* **positive definite** *if* $\mathbf{x}^T\mathbf{A}\mathbf{x} > 0$ *whenever the vector* **x** *is different from* 0. *The corresponding definition for a complex matrix is that* $\mathbf{x}^H\mathbf{A}\mathbf{x} > 0$ *for all nonzero* **x** *(real or complex).*

There are various tests for positive definiteness, such as the positivity of the determinants of the principal minors. This is called the **Sylvester criterion**. The principal minors are the square submatrices lying in the upper left-hand corner of **A**. An $n \times n$ matrix has n of these principal minors. All the eigenvalues of a positive definite matrix are positive. Further consequences of positive definiteness are explored in General Exercises 21–25. Every positive definite matrix is invertible (nonsingular), and its inverse is positive definite.

---

[4] The LL^T-factorization is named after Major Andre-Louis Cholesky (1875–1918), who was a geodesist in the French army.

A property weaker than positive definite is **nonnegative definite**, which means that $\mathbf{x}^H \mathbf{A} \mathbf{x} \geq 0$ for all $\mathbf{x}$.

EXAMPLE 5 Use the test alluded to above to determine whether the real symmetric matrix

$$\mathbf{A} = \begin{bmatrix} 4 & 10 & 14 \\ 10 & 41 & 59 \\ 14 & 59 & 94 \end{bmatrix}$$

is positive definite.

SOLUTION The determinants of the principal minors are

$$\text{Det}[4] = 4 > 0, \qquad \text{Det} \begin{bmatrix} 4 & 10 \\ 10 & 41 \end{bmatrix} = 64 > 0$$

$$\text{Det} \begin{bmatrix} 4 & 10 & 14 \\ 10 & 41 & 59 \\ 14 & 59 & 94 \end{bmatrix} = 576 > 0$$

Because these are positive numbers, the matrix $\mathbf{A}$ is positive definite. ∎

EXAMPLE 6 Find the Cholesky factorization of this real, symmetric, and positive definite matrix:

$$\mathbf{A} = \begin{bmatrix} 4 & 10 & 14 \\ 10 & 41 & 59 \\ 14 & 59 & 94 \end{bmatrix}$$

SOLUTION We want to find the values of $r, s, t, u, v,$ and w in the equation

$$\mathbf{A} = \begin{bmatrix} 4 & 10 & 14 \\ 10 & 41 & 59 \\ 14 & 59 & 94 \end{bmatrix} = \begin{bmatrix} r & 0 & 0 \\ s & t & 0 \\ u & v & w \end{bmatrix} \begin{bmatrix} r & s & u \\ 0 & t & v \\ 0 & 0 & w \end{bmatrix} = \mathbf{L}\mathbf{L}^T$$

One can solve for the unknowns in a systematic way:

1. $4 = r^2$ and $r = 2$
2. $10 = rs = 2s$ and $s = 5$
3. $14 = ru = 2u$ and $u = 7$
4. $41 = s^2 + t^2 = 25 + t^2$ and $t = 4$

5. $59 = su + tv = 35 + 4v$ and $v = 6$

6. $94 = u^2 + v^2 + w^2 = 49 + 36 + w^2$ and $w = 3$

Finally, one can verify the factorization as shown here:

$$\mathbf{A} = \begin{bmatrix} 4 & 10 & 14 \\ 10 & 41 & 59 \\ 14 & 59 & 94 \end{bmatrix} = \begin{bmatrix} 2 & 0 & 0 \\ 5 & 4 & 0 \\ 7 & 6 & 3 \end{bmatrix} \begin{bmatrix} 2 & 5 & 7 \\ 0 & 4 & 6 \\ 0 & 0 & 3 \end{bmatrix} = \mathbf{LL}^T \qquad \blacksquare$$

EXAMPLE 7 Find the Cholesky factorization for an arbitrary 4×4, symmetric, positive definite matrix.

SOLUTION We have to solve for the unknowns, called ℓ_{ij}, in this equation:

$$\begin{bmatrix} a_{11} & a_{12} & a_{13} & a_{14} \\ & a_{22} & a_{23} & a_{24} \\ & & a_{33} & a_{34} \\ & & & a_{44} \end{bmatrix}$$

$$= \begin{bmatrix} \ell_{11} & 0 & 0 & 0 \\ \ell_{21} & \ell_{22} & 0 & 0 \\ \ell_{31} & \ell_{32} & \ell_{33} & 0 \\ \ell_{41} & \ell_{42} & \ell_{43} & \ell_{44} \end{bmatrix} \begin{bmatrix} \ell_{11} & \ell_{21} & \ell_{31} & \ell_{41} \\ 0 & \ell_{22} & \ell_{32} & \ell_{42} \\ 0 & 0 & \ell_{33} & \ell_{43} \\ 0 & 0 & 0 & \ell_{44} \end{bmatrix}$$

The equations look like this:

$$\begin{aligned}
a_{11} &= \ell_{11}^2, & a_{23} &= \ell_{21}\ell_{31} + \ell_{22}\ell_{32} \\
a_{12} &= \ell_{11}\ell_{21}, & a_{24} &= \ell_{21}\ell_{41} + \ell_{22}\ell_{42} \\
a_{13} &= \ell_{11}\ell_{31}, & a_{33} &= \ell_{31}^2 + \ell_{32}^2 + \ell_{33}^2 \\
a_{14} &= \ell_{11}\ell_{41}, & a_{34} &= \ell_{32}\ell_{41} + \ell_{32}\ell_{42} + \ell_{33}\ell_{43} \\
a_{22} &= \ell_{21}^2 + \ell_{22}^2, & a_{44} &= \ell_{41}^2 + \ell_{42}^2 + \ell_{43}^2 + \ell_{44}^2
\end{aligned}$$

The solutions are as follows:

$$\begin{aligned}
\ell_{11} &= \sqrt{a_{11}}, & \ell_{32} &= (a_{23} - \ell_{21}\ell_{31})/\ell_{22} \\
\ell_{21} &= a_{12}/\ell_{11}, & \ell_{42} &= (a_{24} - \ell_{21}\ell_{41})/\ell_{22} \\
\ell_{31} &= a_{13}/\ell_{11}, & \ell_{33} &= \sqrt{a_{33} - \ell_{31}^2 - \ell_{32}^2} \\
\ell_{41} &= a_{14}/\ell_{11}, & \ell_{43} &= (a_{34} - \ell_{32}\ell_{41} - \ell_{32}\ell_{42})/\ell_{33} \\
\ell_{22} &= \sqrt{a_{22} - \ell_{21}^2}, & \ell_{44} &= \sqrt{a_{44} - \ell_{41}^2 - \ell_{42}^2 - \ell_{43}^2}
\end{aligned}$$

This can be simplified, as shown in the following pseudocode. $\qquad \blacksquare$

A computer code for the Cholesky factorization can be quite succinct. Start with a symmetric matrix $\mathbf{A}$, assumed to be positive definite and of dimensions $n \times n$. We want $\mathbf{A} = \mathbf{L}\mathbf{L}^T$. Let the elements of $\mathbf{L}$ be denoted by ℓ_{ij}. Then $\ell_{ij} = 0$ for $j > i$. The pseudocode for the **Cholesky factorization algorithm** is

> **for** $k = 1$ **to** n
> $$\ell_{kk} \leftarrow \left(a_{kk} - \sum_{s=1}^{k-1} \ell_{ks}^2 \right)^{\frac{1}{2}}$$
> **for** $i = k + 1$ **to** n
> $$\ell_{ik} \leftarrow \left(a_{ik} - \sum_{s=1}^{k-1} \ell_{is}\ell_{ks} \right) \Big/ \ell_{kk}$$
> **end for**
> **end for**

In this abbreviated code, the convention is adopted that a summation is defined to be zero if the beginning index exceeds the ending index. Thus, we have $\sum_{i=1}^{0} x_i = 0$, for example. The requirement that the matrix $\mathbf{A}$ be symmetric positive definite guarantees that the second line of the pseudocode involves only the square roots of positive numbers according to the following theorem.

THEOREM 1

Every real symmetric positive definite matrix can be factored as $\mathbf{L}\mathbf{L}^T$, where $\mathbf{L}$ is a real lower triangular matrix having positive elements on its diagonal.

PROOF Consult Golub and van Loan [1996, p. 87]. ∎

LDL^T-Factorization

Under certain conditions, we can find the Cholesky LL^T-factorization directly from the LU-factorization. Suppose we are given a real, symmetric, positive definite matrix $\mathbf{A}$ and its factorization $\mathbf{A} = \mathbf{LU}$, where $\mathbf{L}$ is a unit lower triangular matrix and $\mathbf{U}$ is an upper triangular matrix with positive diagonal elements. We can extract the diagonal elements from the matrix $\mathbf{U}$ and use them to form a diagonal matrix $\mathbf{D}$ so that $\mathbf{A} = \mathbf{LDL}^T$. Then we can split the matrix so that $\mathbf{D} = \mathbf{D}^{\frac{1}{2}}\mathbf{D}^{\frac{1}{2}}$ because the diagonal matrix $\mathbf{D}$ has positive diagonal elements. We obtain

$$\mathbf{A} = (\mathbf{LD}^{\frac{1}{2}})(\mathbf{D}^{\frac{1}{2}}\mathbf{L}^T) = \widehat{\mathbf{L}}\widehat{\mathbf{L}}^T$$

where $\widehat{\mathbf{L}} = \mathbf{LD}^{\frac{1}{2}}$.

EXAMPLE 8 Carry out the process just described on the following matrix:

$$\mathbf{A} = \begin{bmatrix} 6 & 3 & 4 \\ 4 & 1 & 5 \\ 3 & 2 & 1 \end{bmatrix}$$

SOLUTION First, we carry out the forward elimination phase of Gaussian elimination

$$\mathbf{A} = \begin{bmatrix} 6 & 3 & 4 \\ 3 & 2 & 1 \\ 4 & 1 & 5 \end{bmatrix} \sim \begin{bmatrix} 6 & 3 & 4 \\ 0 & \frac{1}{2} & -1 \\ 0 & -1 & \frac{7}{3} \end{bmatrix} \sim \begin{bmatrix} 6 & 3 & 4 \\ 0 & \frac{1}{2} & -1 \\ 0 & 0 & \frac{1}{3} \end{bmatrix}$$

and

$$\mathbf{A} = \begin{bmatrix} 6 & 3 & 4 \\ 3 & 2 & 1 \\ 4 & 1 & 5 \end{bmatrix} = \begin{bmatrix} 1 & 0 & 0 \\ \frac{1}{2} & 1 & 0 \\ \frac{2}{3} & -2 & 1 \end{bmatrix} \begin{bmatrix} 6 & 3 & 4 \\ 0 & \frac{1}{2} & -1 \\ 0 & 0 & \frac{1}{3} \end{bmatrix} = \mathbf{LU}$$

$$= \begin{bmatrix} 1 & 0 & 0 \\ \frac{1}{2} & 1 & 0 \\ \frac{2}{3} & -2 & 1 \end{bmatrix} \begin{bmatrix} 6 & 0 & 0 \\ 0 & \frac{1}{2} & 0 \\ 0 & 0 & \frac{1}{3} \end{bmatrix} \begin{bmatrix} 1 & \frac{1}{2} & \frac{2}{3} \\ 0 & 1 & -2 \\ 0 & 0 & 1 \end{bmatrix} = \mathbf{LDL}^T$$

$$= \begin{bmatrix} 1 & 0 & 0 \\ \frac{1}{2} & 1 & 0 \\ \frac{2}{3} & -2 & 1 \end{bmatrix} \begin{bmatrix} \sqrt{6} & 0 & 0 \\ 0 & (\frac{1}{2})\sqrt{2} & 0 \\ 0 & 0 & (\frac{1}{3})\sqrt{3} \end{bmatrix}$$

$$\begin{bmatrix} \sqrt{6} & 0 & 0 \\ 0 & (\frac{1}{2})\sqrt{2} & 0 \\ 0 & 0 & (\frac{1}{3})\sqrt{3} \end{bmatrix} \begin{bmatrix} 1 & \frac{1}{2} & \frac{2}{3} \\ 0 & 1 & -2 \\ 0 & 0 & 1 \end{bmatrix}$$

$$= \begin{bmatrix} \sqrt{6} & 0 & 0 \\ (\frac{1}{2})\sqrt{6} & (\frac{1}{2})\sqrt{2} & 0 \\ (\frac{2}{3})\sqrt{6} & -\sqrt{2} & (\frac{1}{3})\sqrt{3} \end{bmatrix} \begin{bmatrix} \sqrt{6} & (\frac{1}{2})\sqrt{6} & (\frac{2}{3})\sqrt{6} \\ 0 & (\frac{1}{2})\sqrt{2} & -\sqrt{2} \\ 0 & 0 & (\frac{1}{3})\sqrt{3} \end{bmatrix} = \widehat{\mathbf{L}}\widehat{\mathbf{L}}^T$$

Evidently, we have found the LU-factorization, the LDL^T-factorization, and the LL^T-factorization. ∎

QR-Factorization

Another type of matrix factoring is called the *QR*-factorization. Here we include nonsquare matrices in the discussion because this factoring is useful in solving overdetermined systems of linear equations, where a least-squares solution is sought. In the factorization $\mathbf{A} = \mathbf{QR}$, we allow $\mathbf{A}$ and $\mathbf{Q}$ to be $m \times n$. The columns of the matrix $\mathbf{Q}$ form an orthonormal set in $\mathbb{R}^m$ so that $\mathbf{Q}^T\mathbf{Q} = \mathbf{I}$. Therefore, we suppose that $m \geq n$. The factor $\mathbf{R}$ is an $n \times n$, upper triangular, invertible matrix. How is this factorization employed to get the least-squares solution of a system of equations $\mathbf{Ax} = \mathbf{b}$? Simply solve the system $\mathbf{Rx} = \mathbf{Q}^T\mathbf{b}$ for the vector $\mathbf{x}$.

THEOREM 2

The vector $\mathbf{x}$ is the least-squares solution of the system of equations $\mathbf{Ax} = \mathbf{b}$ under the circumstances just described.

PROOF We compute

$$\begin{aligned}
\mathbf{A}^T(\mathbf{Ax} - \mathbf{b}) &= (\mathbf{QR})^T(\mathbf{QRx} - \mathbf{b}) \\
&= \mathbf{R}^T(\mathbf{Q}^T\mathbf{Q})\mathbf{Rx} - \mathbf{R}^T\mathbf{Q}^T\mathbf{b} \\
&= \mathbf{R}^T(\mathbf{Rx}) - \mathbf{R}^T\mathbf{Q}^T\mathbf{b} \\
&= \mathbf{R}^T(\mathbf{Q}^T\mathbf{b}) - \mathbf{R}^T\mathbf{Q}^T\mathbf{b} = \mathbf{0}
\end{aligned}$$

Consequently, the normal equations are satisfied:

$$\mathbf{A}^T\mathbf{Ax} = \mathbf{A}^T\mathbf{b} \qquad \blacksquare$$

In MATLAB, the command `[Q,R]=qr(A)` produces the desired factors. Actually, in MATLAB, if we want to solve a system only in the least-squares sense, the command `x=A\b` automatically does this. Behind the scenes, the *QR*-factorization is being employed. There are analogous commands in Maple and Mathematica. The theorem governing this factorization follows.

THEOREM 3

Any $m \times n$ matrix of rank n has a QR-factorization, where $\mathbf{Q}$ is an $m \times n$ matrix, $\mathbf{Q}^T\mathbf{Q} = \mathbf{I}_n$, and $\mathbf{R}$ is an $n \times n$, upper triangular, and invertible matrix.

PROOF Let $\mathbf{A}$ be an $m \times n$ matrix of rank n. Since $\mathbf{A}$ has rank n, $m \geq n$. Let the columns of $\mathbf{A}$ be $\mathbf{a}_1, \mathbf{a}_2, \ldots, \mathbf{a}_n$. By applying the Gram–Schmidt process to this set of columns, we obtain an orthonormal set of vectors $\{\mathbf{q}_1, \mathbf{q}_2, \ldots, \mathbf{q}_n\}$ such that, for each value of i,

$$\text{Span}\{\mathbf{a}_1, \mathbf{a}_2, \ldots, \mathbf{a}_i\} = \text{Span}\{\mathbf{q}_1, \mathbf{q}_2, \ldots, \mathbf{q}_i\}$$

Hence for suitable coefficients, r_{ij}, we have

$$\mathbf{a}_i = \sum_{j=1}^{i} r_{ji} \mathbf{q}_j$$

for $i = 1, 2, \ldots, n$. Create an $n \times n$ matrix $\mathbf{R}$ whose generic elements are r_{ij}. Supply zero values for r_{ij} when $i > j$. This matrix satisfies the equation

$$\mathbf{A} = \mathbf{QR}$$

Since $\mathbf{A}$ has rank n, $\mathbf{R}$ is invertible. The orthonormality of the set of columns in $\mathbf{Q}$ is expressed by the equation $\mathbf{Q}^T \mathbf{Q} = \mathbf{I}$. ∎

EXAMPLE 9 Find the QR-factorization of this matrix: $\mathbf{A} = \begin{bmatrix} 1 & 2 \\ 1 & 0 \\ 1 & 2 \\ 1 & 0 \end{bmatrix}$

SOLUTION The proof of Theorem 3 can be followed, step-by-step. To start the Gram–Schmidt process, normalize the first column $\mathbf{a}_1$. This produces $\mathbf{q}_1 = [\frac{1}{2}, \frac{1}{2}, \frac{1}{2}, \frac{1}{2}]^T$ as column one in $\mathbf{Q}$. Next, we subtract from the second column $\mathbf{a}_2$ its projection on $\mathbf{q}_1$. This yields a temporary vector

$$\mathbf{v} = \mathbf{a}_2 - \langle \mathbf{a}_2, \mathbf{q}_1 \rangle \mathbf{q}_1 = \mathbf{a}_2 - 2\mathbf{q}_1 = [1, -1, 1, -1]^T$$

Normalizing $\mathbf{v}$ yields $\mathbf{q}_2 = [\frac{1}{2}, -\frac{1}{2}, \frac{1}{2}, -\frac{1}{2}]^T$. Note that our work gives us $\mathbf{a}_1 = 2\mathbf{q}_1$ and $\mathbf{a}_2 = 2\mathbf{q}_1 + 2\mathbf{q}_2$. These equations imply that $r_{11} = r_{12} = r_{22} = 2$ and $r_{22} = 0$. Hence, we obtain

$$\mathbf{A} = \begin{bmatrix} 1 & 2 \\ 1 & 0 \\ 1 & 2 \\ 1 & 0 \end{bmatrix} = \begin{bmatrix} \frac{1}{2} & \frac{1}{2} \\ \frac{1}{2} & -\frac{1}{2} \\ \frac{1}{2} & \frac{1}{2} \\ \frac{1}{2} & -\frac{1}{2} \end{bmatrix} \begin{bmatrix} 2 & 2 \\ 0 & 2 \end{bmatrix} = [\mathbf{q}_1 \ \mathbf{q}_2]\, \mathbf{R} = \mathbf{QR}$$ ∎

Singular-Value Decomposition (SVD)

This section is devoted to the *singular-value decomposition* of an arbitrary matrix. It has many uses, among them a way of producing a reliable estimate of the rank of a matrix. The main theorem is easily stated:

> **THEOREM 4**
>
> *Any matrix can be put into the factored form* $\mathbf{PDQ}$*, where* $\mathbf{P}$ *and* $\mathbf{Q}$ *are unitary and* $\mathbf{D}$ *is diagonal.*

PROOF Let $\mathbf{A}$ be $m \times n$. Since unitary matrices are square, $\mathbf{P}$ must be $m \times m$, and $\mathbf{Q}$ must be $n \times n$. Consequently, $\mathbf{D}$ must be $m \times n$. All entries above and below the diagonal of $\mathbf{D}$ are to be zero. The matrix $\mathbf{A}^H\mathbf{A}$ is Hermitian because $(\mathbf{A}^H\mathbf{A})^H = \mathbf{A}^H(\mathbf{A}^H)^H = \mathbf{A}^H\mathbf{A}$. Furthermore, $\mathbf{A}^H\mathbf{A}$ is nonnegative definite because

$$\mathbf{x}^H(\mathbf{A}^H\mathbf{A})\mathbf{x} = (\mathbf{x}^H\mathbf{A}^H)(\mathbf{A}\mathbf{x}) = (\mathbf{A}\mathbf{x})^H(\mathbf{A}\mathbf{x}) = \left\|\mathbf{A}\mathbf{x}\right\|_2^2 \geq 0$$

By Corollary 2 in Section 8.1, the eigenvalues of $\mathbf{A}^H\mathbf{A}$ are real and nonnegative; we can therefore denote them by σ_i^2. The labeling can be such that

$$\sigma_1^2 \geq \sigma_2^2 \geq \cdots \geq \sigma_r^2 > 0 = \sigma_{r+1}^2 = \cdots = \sigma_n^2$$

The numbers σ_i are called the **singular values**[5] of $\mathbf{A}$. If this notation is used, σ_r is the last nonzero singular value. The rank of $\mathbf{A}^H\mathbf{A}$ is r, because $n - r$ of the eigenvalues are zero. (Why?)

By the Spectral Theorem (Section 8.1), there is an orthonormal set of eigenvectors $\mathbf{u}_i$ for $\mathbf{A}^H\mathbf{A}$. Hence, we obtain

$$\mathbf{A}^H\mathbf{A}\mathbf{u}_i = \sigma_i^2\mathbf{u}_i$$

and $\mathbf{u}_i^H\mathbf{u}_j = \delta_{ij}$ for $1 \leq i \leq n$. Now we have

$$\left\|\mathbf{A}\mathbf{u}_i\right\|_2^2 = (\mathbf{A}\mathbf{u}_i)^H(\mathbf{A}\mathbf{u}_i) = \mathbf{u}_i^H\mathbf{A}^H\mathbf{A}\mathbf{u}_i = \mathbf{u}_i^H(\mathbf{A}^H\mathbf{A}\mathbf{u}_i) = \mathbf{u}_i^H\sigma_i^2\mathbf{u}_i$$

$$= \sigma_i^2\mathbf{u}_i^H\mathbf{u}_i = \sigma_i^2$$

---

[5] Gene Howard Golub (1932–2007) created, along with William Kahan, an algorithm for computing the singular value decomposition. The SVD algorithm is used in a wide variety of applications such as signal processing, data analysis, and Internet search engines. Because of its versatility, this algorithm has been called the *Swiss Army Knife* of numerical computations. See Golub and Kahan [1965]. Gene was fondly known as *Professor SVD*.

Consequently, we obtain $\mathbf{Au}_i = \mathbf{0}$ for $i = r + 1, r + 2, \ldots, n$. We have

$$r = \text{Rank}(\mathbf{A}^H\mathbf{A}) \leq \min\{\text{Rank}(\mathbf{A}^H), \text{Rank}(\mathbf{A})\} = \text{Rank}(\mathbf{A}) \leq \min\{m, n\}$$

Define $\mathbf{Q}$ to have rows $\mathbf{u}_1^H, \mathbf{u}_2^H, \ldots, \mathbf{u}_n^H$. Then $\mathbf{Q}$ is $n \times n$ and is unitary: $\mathbf{QQ}^H = \mathbf{Q}^H\mathbf{Q} = \mathbf{I}_n$. Define $\mathbf{v}_i = \sigma_i^{-1}\mathbf{Au}_i$ for $1 \leq i \leq r$. These vectors have the property

$$\mathbf{v}_i^H\mathbf{v}_j = (\sigma_i^{-1}\mathbf{Au}_i)^H(\sigma_j^{-1}\mathbf{Au}_j) = (\sigma_i^{-1}\sigma_j^{-1})\mathbf{u}_i^H\mathbf{A}^H\mathbf{Au}_j$$

$$= (\sigma_i\sigma_j)^{-1}\mathbf{u}_i^H\sigma_j^2\mathbf{u}_j = \delta_{ij}$$

Select additional vectors $\mathbf{v}_i$ so that $\{\mathbf{v}_1, \mathbf{v}_2, \ldots, \mathbf{v}_n\}$ is an orthonormal base for $\mathbb{C}^n$. Let $\mathbf{P}$ be the matrix whose columns are $\mathbf{v}_i$ for $1 \leq i \leq m$. Let $\mathbf{D}$ be the $m \times n$ diagonal matrix having σ_i on its diagonal. Now we have

$$\mathbf{A} = \mathbf{PDQ}$$

because

$$(\mathbf{P}^H\mathbf{AQ}^H)_{ij} = \mathbf{v}_i^H\mathbf{Au}_j = \mathbf{v}_i^H\sigma_j\mathbf{v}_j = \sigma_j\delta_{ij} = \mathbf{D}_{ij} \qquad\blacksquare$$

Schur Decomposition

In 1909, Schur[6] established that any square matrix can be reduced to a triangular matrix using a unitary similarity transformation.

> **THEOREM 5   Schur Decomposition**
> *If $\mathbf{A} \in \mathbb{C}^{n \times n}$, then there exists a unitary matrix $\mathbf{U} \in \mathbb{C}^{n \times n}$ so that $\mathbf{A} = \mathbf{UTU}^H$ and the eigenvalues of $\mathbf{A}$ appear on the diagonal of the upper triangular matrix $\mathbf{T} \in C^{n \times n}$.*

The proof of Schur's Theorem can be found in Kincaid and Cheney [2002] and Golub and Van Loan [1996]. As a special case, every Hermitian matrix is unitarily similar to a diagonal matrix.

---

[6] Issai Schur (1875–1941) was born in Germany and died on his birthday in the city of Tel Aviv. He was the victim of Nazi persecution in the 1930s and emigrated to what is now Israel. He was a student of the famous mathematician Frobenius in Berlin. Schur was a superb expositor and attracted as many as 400 students to his lectures in huge auditoriums! He ascended through the academic ranks at the University of Berlin to Professor in 1919.

COROLLARY 1

If $A \in \mathbb{C}^{n \times n}$ is a Hermitian matrix, then there exists a Hermitian matrix $U \in \mathbb{C}^{n \times n}$ so that $A = UDU^H$, where the eigenvalues appear in the diagonal matrix D.

Another special case is that every square real matrix is similar to a triangular matrix.

COROLLARY 2

If $A \in \mathbb{R}^{n \times n}$, then there exists an invertible matrix $S \in \mathbb{R}^{n \times n}$ and a triangular matrix T such that $A = STS^{-1}$ and the eigenvalues of A appear on the diagonal of T.

EXAMPLE 10 We illustrate Schur's theorem by finding the decomposition of

$$A = \begin{bmatrix} 5 & -2 \\ 7 & -4 \end{bmatrix}$$

SOLUTION The characteristic equation $\text{Det}(A - \lambda I) = \lambda^2 - \lambda - 6 = 0$ gives us the eigenvalues -2 and 3. By solving $A - \lambda I = 0$ with each of these eigenvalues, the corresponding eigenvectors are $v_1 = [-1, 1]^T$ and $v_2 = [7, -2]^T$. Using the Gram–Schmidt orthogonalization process, we obtain $u_1 = v_1$ and $u_2 = v_2 - [v_2^H u_1 / u_1^H u_1] u_1 = [1, 1]^T$. After normalizing these vectors, we obtain the unitary matrix

$$U = \frac{1}{\sqrt{2}} \begin{bmatrix} -1 & 1 \\ 1 & 1 \end{bmatrix}$$

which has the property $UU^H = I$. Finally, we obtain the Schur decomposition

$$U^H A U = \begin{bmatrix} 3 & -9 \\ 0 & -2 \end{bmatrix} = T$$

which is an upper triangular matrix having the eigenvalues on the diagonal. ∎

Partitioned Matrices

Frequently, it is convenient to partition matrices into submatrices (called *blocks*) and to carry out matrix operations using the blocks. Let $\mathcal{A}$, $\mathcal{B}$, and $\mathcal{C}$ be $m \times k$, $k \times \ell$, and $\ell \times n$ matrices, respectively, partitioned into the submatrices $\mathbf{A}_{ij}$, $\mathbf{B}_{ij}$, and $\mathbf{C}_{ij}$ of appropriate dimensions.

$$\mathcal{A} = \begin{bmatrix} \mathbf{A}_{11} & \mathbf{A}_{12} & \cdots & \mathbf{A}_{1k} \\ \mathbf{A}_{21} & \mathbf{A}_{22} & \cdots & \mathbf{A}_{2k} \\ \vdots & \vdots & \ddots & \vdots \\ \mathbf{A}_{m1} & \mathbf{A}_{m2} & \cdots & \mathbf{A}_{mk} \end{bmatrix}$$

$$\mathcal{B} = \begin{bmatrix} \mathbf{B}_{11} & \mathbf{B}_{12} & \cdots & \mathbf{B}_{1\ell} \\ \mathbf{B}_{21} & \mathbf{B}_{22} & \cdots & \mathbf{B}_{2\ell} \\ \vdots & \vdots & \ddots & \vdots \\ \mathbf{B}_{k1} & \mathbf{B}_{k2} & \cdots & \mathbf{B}_{k\ell} \end{bmatrix}$$

$$\mathcal{C} = \begin{bmatrix} \mathbf{C}_{11} & \mathbf{C}_{12} & \cdots & \mathbf{C}_{1n} \\ \mathbf{C}_{21} & \mathbf{C}_{22} & \cdots & \mathbf{C}_{2n} \\ \vdots & \vdots & \ddots & \vdots \\ \mathbf{C}_{\ell1} & \mathbf{C}_{\ell2} & \cdots & \mathbf{C}_{\ell n} \end{bmatrix}$$

If $\mathcal{A}$ and $\mathcal{B}$ have the same dimensions and the same partitioning, then multiplication by the scalar α is

$$\mathcal{C} = \alpha \mathcal{A}$$

which is done block-by-block as $\mathbf{C}_{ij} = \alpha \mathbf{A}_{ij}$. Matrix addition and subtraction are similarly done blockwise:

$$\mathcal{C} = \mathcal{A} \pm \mathcal{B}$$

where $\mathbf{C}_{ij} = \mathbf{A}_{ij} \pm \mathbf{B}_{ij}$ for each block.

To establish the general results for multiplication of partitioned matrices, we compute products as if the blocks were scalars. We assume that the partitioning has been done in such a way that each product $\mathbf{A}_{is}\mathbf{B}_{sj}$ can be formed. Then

$$\mathcal{C} = \mathcal{A}\mathcal{B}$$

where

$$C_{ij} = \sum_{s=1}^{n} A_{is} B_{sj}$$

The column partitions of $\mathcal{A}$ match the row partitions of $\mathcal{B}$ so that $\mathcal{A}$ and $\mathcal{B}$ are conformable for block multiplication.

An example of a 2×3 block partitioned matrix is

$$
\mathcal{A} = \left[\begin{array}{c|ccc|cc}
10 & -19 & 17 & -12 & 4 & 1 \\
9 & -18 & 17 & -12 & 4 & 1 \\
8 & -16 & 15 & -11 & 4 & 1 \\
6 & -12 & 12 & -10 & 4 & 1 \\
\hline
4 & 8 & 8 & -6 & 1 & 2 \\
2 & -4 & 4 & -3 & 1 & 0
\end{array} \right]
$$

$$
= \left[\begin{array}{c|c|c}
\mathbf{A}_{11} & \mathbf{A}_{12} & \mathbf{A}_{13} \\
\hline
\mathbf{A}_{21} & \mathbf{A}_{22} & \mathbf{A}_{23}
\end{array} \right]
$$

Consider this 3×2 block partitioned matrix:

$$
\mathcal{B} = \left[\begin{array}{ccc|cc}
1 & 0 & 1 & -1 & 1 \\
\hline
0 & -1 & 1 & -1 & 0 \\
1 & -1 & 0 & 0 & 1 \\
0 & -1 & 1 & -1 & 0 \\
\hline
1 & -1 & 0 & 0 & 1 \\
0 & -1 & 1 & -1 & 1
\end{array} \right]
= \left[\begin{array}{c|c}
\mathbf{B}_{11} & \mathbf{B}_{12} \\
\hline
\mathbf{B}_{21} & \mathbf{B}_{22} \\
\hline
\mathbf{B}_{31} & \mathbf{B}_{32}
\end{array} \right]
$$

The block product $\mathcal{C} = \mathcal{AB}$ exists and turns out to be

$$
\mathcal{C} = \mathcal{AB} = \left[\begin{array}{ccc|cc}
31 & 9 & -20 & 20 & 32 \\
30 & 8 & -20 & 20 & 31 \\
27 & 7 & -18 & 18 & 28 \\
22 & 5 & -15 & 15 & 23 \\
\hline
13 & -13 & 8 & -8 & 15 \\
7 & 2 & -5 & 5 & 7
\end{array} \right]
$$

As one sample of the calculation, we show how $\mathbf{C}_{22}$ is computed:

$$\mathbf{C}_{22} = \mathbf{A}_{21}\mathbf{B}_{12} + \mathbf{A}_{22}\mathbf{B}_{22} + \mathbf{A}_{23}\mathbf{B}_{32}$$

$$= \begin{bmatrix} 4 \\ 2 \end{bmatrix} \begin{bmatrix} -1 & 1 \end{bmatrix} + \begin{bmatrix} 8 & 8 & -6 \\ -4 & 4 & -3 \end{bmatrix} \begin{bmatrix} -1 & 0 \\ 0 & 1 \\ -1 & 0 \end{bmatrix}$$

$$+ \begin{bmatrix} 1 & 2 \\ 1 & 0 \end{bmatrix} \begin{bmatrix} 0 & 1 \\ -1 & 1 \end{bmatrix}$$

$$= \begin{bmatrix} -4 & 4 \\ -2 & 2 \end{bmatrix} + \begin{bmatrix} -2 & 8 \\ 7 & 4 \end{bmatrix} + \begin{bmatrix} -2 & 3 \\ 0 & 1 \end{bmatrix}$$

$$= \begin{bmatrix} -8 & 15 \\ 5 & 7 \end{bmatrix}$$

One must be careful not to disturb the order of the submatrices in the multiplications.

Solving a System Having a 2 × 2 Block Matrix

Consider the 2 × 2 block linear system

$$\begin{bmatrix} \mathbf{A} & \mathbf{B} \\ \mathbf{B}^T & \mathbf{0} \end{bmatrix} \begin{bmatrix} \mathbf{x} \\ \mathbf{y} \end{bmatrix} = \begin{bmatrix} \mathbf{b} \\ \mathbf{c} \end{bmatrix}$$

where $\mathbf{A}$ is an $n \times n$ real, symmetric, invertible matrix. The matrix $\mathbf{B}$ is an $n \times m$ real matrix of full rank. We assume that $\mathbf{A}$ is positive definite on the nullspace of $\mathbf{B}$. The vectors $\mathbf{x}$ and $\mathbf{b}$ are column vectors in $\mathbb{R}^n$, whereas $\mathbf{y}$ and $\mathbf{c}$ are column vectors in $\mathbb{R}^m$.

In a typical application, the coefficient matrix is a 2 × 2 block matrix whose dimension in the usual sense is $(n + m) \times (n + m)$. It is a symmetric and nonnegative definite matrix (having nonnegative eigenvalues).

Suppose that the solution of the preceding block linear system is sought. If $\mathbf{B}$ is invertible ($n = m$), the solution can be found immediately:

$$\begin{cases} \mathbf{x} = \mathbf{B}^{-T}\mathbf{c} \\ \mathbf{y} = \mathbf{B}^{-1}(\mathbf{b} - \mathbf{A}\mathbf{x}) \end{cases}$$

The notation $\mathbf{B}^{-T}$ means $(\mathbf{B}^{-1})^T$.

If the matrix $\mathbf{B}$ is *not* invertible, Gaussian elimination can be used on the augmented matrix of the original 2 × 2 block system. Supposing that $\mathbf{A}$ is invertible, then

$$\left[\begin{array}{cc|c} \mathbf{A} & \mathbf{B} & \mathbf{b} \\ \mathbf{B}^T & \mathbf{0} & \mathbf{c} \end{array} \right] \sim \left[\begin{array}{cc|c} \mathbf{A} & \mathbf{B} & \mathbf{b} \\ \mathbf{0} & -\mathbf{B}^T\mathbf{A}^{-1}\mathbf{B} & \mathbf{c} - \mathbf{B}^T\mathbf{A}^{-1}\mathbf{b} \end{array} \right]$$

Here we multiply the first system by $-\mathbf{B}^T\mathbf{A}^{-1}$ and add it to the second system of equations. Consequently, we obtain the **reduced system**

$$(\mathbf{B}^T\mathbf{A}^{-1}\mathbf{B})\mathbf{y} = (\mathbf{B}^T\mathbf{A}^{-1}\mathbf{b} - \mathbf{c})$$

leading to the solution

$$\begin{cases} \mathbf{y} = (\mathbf{B}^T\mathbf{A}^{-1}\mathbf{B})^{-1}(\mathbf{B}^T\mathbf{A}^{-1}\mathbf{b} - \mathbf{c}) \\ \mathbf{x} = \mathbf{A}^{-1}(\mathbf{b} - \mathbf{B}\mathbf{y}) \end{cases}$$

provided that $\mathbf{B}^T\mathbf{A}^{-1}\mathbf{B}$ is invertible.

The LU-factorization of the block matrix is

$$\begin{bmatrix} \mathbf{A} & \mathbf{B} \\ \mathbf{B}^T & \mathbf{0} \end{bmatrix} = \begin{bmatrix} \mathbf{I} & \mathbf{0} \\ \mathbf{B}^T\mathbf{A}^{-1} & \mathbf{I} \end{bmatrix} \begin{bmatrix} \mathbf{A} & \mathbf{B} \\ \mathbf{0} & -\mathbf{B}^T\mathbf{A}^{-1}\mathbf{B} \end{bmatrix}$$

The LDL^T-factorization is

$$\begin{bmatrix} \mathbf{A} & \mathbf{B} \\ \mathbf{B}^T & \mathbf{0} \end{bmatrix} = \begin{bmatrix} \mathbf{I} & \mathbf{0} \\ \mathbf{B}^T\mathbf{A}^{-1} & \mathbf{L}_2 \end{bmatrix} \begin{bmatrix} \mathbf{A} & \mathbf{0} \\ \mathbf{0} & -\mathbf{D}_2 \end{bmatrix} \begin{bmatrix} \mathbf{I} & \mathbf{A}^{-1}\mathbf{B} \\ \mathbf{0} & \mathbf{L}_2^T \end{bmatrix}$$

where the factorization of the reduced system is $\mathbf{B}^T\mathbf{A}^{-1}\mathbf{B} = \mathbf{L}_2\mathbf{D}_2\mathbf{L}_2^T$. These factorizations, and others, such as in General Exercises 33–35, can be useful in reducing the number of floating-point operations when the conjugate gradient method is being used. See Dollar and Wathen [2006].

Applications in which one wishes to solve matrix systems of this form are, for example, **convex quadratic programming problems**, where we seek to minimize

$$f(\mathbf{x}) = \tfrac{1}{2}\mathbf{x}^T\mathbf{A}\mathbf{x} + \mathbf{s}^T\mathbf{x} \quad \text{subject to the constraint} \quad \mathbf{B}\mathbf{x} = \mathbf{t}$$

and saddle-point problems involving finite-element approximations of variational problems.

Suppose we want to solve the block linear system

$$\begin{bmatrix} \mathbf{A} & \mathbf{B} \\ \mathbf{C} & \mathbf{D} \end{bmatrix} \begin{bmatrix} \mathbf{x} \\ \mathbf{y} \end{bmatrix} = \begin{bmatrix} \mathbf{b} \\ \mathbf{c} \end{bmatrix}$$

where dimensions of the submatrices are as follows: $\mathbf{A}$ is $n \times n$, $\mathbf{B}$ is $n \times m$, $\mathbf{C}$ is $m \times n$, and $\mathbf{D}$ is $m \times m$. The column vectors $\mathbf{x}$ and $\mathbf{b}$ have n components and $\mathbf{y}$ and $\mathbf{c}$ have m entries. It is assumed that $\mathbf{A}$ is invertible.

We are given the block coefficient matrix with the assumptions mentioned previously as well as the right-hand side vectors **b** and **c**. Using the reduction process with the augmented block matrices, we have

$$\left[\begin{array}{cc|c} \mathbf{A} & \mathbf{B} & \mathbf{b} \\ \mathbf{C} & \mathbf{D} & \mathbf{c} \end{array}\right] \sim \left[\begin{array}{cc|c} \mathbf{A} & \mathbf{B} & \mathbf{b} \\ \mathbf{0} & \mathbf{D} - \mathbf{CA}^{-1}\mathbf{B} & \mathbf{c} - \mathbf{CA}^{-1}\mathbf{b} \end{array}\right]$$

Rather than solving the $(n + m) \times (n + m)$ system, we can work with the following smaller $m \times m$ and $n \times n$ systems

$$\begin{cases} (\mathbf{D} - \mathbf{CA}^{-1}\mathbf{B})\mathbf{y} = \mathbf{c} - \mathbf{CA}^{-1}\mathbf{b} \\ \mathbf{Ax} = \mathbf{b} - \mathbf{By} \end{cases}$$

Here we assume that we can find the inverses of **A** and $\mathbf{D} - \mathbf{CA}^{-1}\mathbf{B}$. The latter is called the **Schur complement of A**. In practice, the matrix **A** should be well conditioned to obtain accurate results. Similarly, one could solve for **x** first and then **y** as in General Exercise 53.

Inverting a 2 × 2 Block Matrix

Here we derive a formula for the inverse of a 2×2 block matrix of the form

$$\left[\begin{array}{cc} \mathbf{A} & \mathbf{B} \\ \mathbf{C} & \mathbf{D} \end{array}\right]$$

The dimensions of the submatrices are as follows: **A** is $n \times n$, **B** is $n \times m$, **C** is $m \times n$, and **D** is $m \times m$. It is assumed that **A** and **D** are invertible.

Using the reduction process with block matrices, we have

$$\left[\begin{array}{cc} \mathbf{A} & \mathbf{B} \\ \mathbf{C} & \mathbf{D} \end{array}\right] \sim \left[\begin{array}{cc} \mathbf{A} & \mathbf{B} \\ \mathbf{0} & \mathbf{D} - \mathbf{CA}^{-1}\mathbf{B} \end{array}\right] \mapsto \left[\begin{array}{cc} \mathbf{A} & \mathbf{0} \\ \mathbf{0} & \mathbf{D} - \mathbf{CA}^{-1}\mathbf{B} \end{array}\right]$$

The term $\mathbf{D} - \mathbf{CA}^{-1}\mathbf{B}$ is called a Schur complement of **A**. The first step is the same as multiplying on the left by a block matrix:

$$\left[\begin{array}{cc} \mathbf{I} & \mathbf{0} \\ -\mathbf{CA}^{-1} & \mathbf{I} \end{array}\right] \left[\begin{array}{cc} \mathbf{A} & \mathbf{B} \\ \mathbf{C} & \mathbf{D} \end{array}\right] = \left[\begin{array}{cc} \mathbf{A} & \mathbf{B} \\ \mathbf{0} & \mathbf{D} - \mathbf{CA}^{-1}\mathbf{B} \end{array}\right]$$

The second step is the same as multiplying on the right by a block matrix:

$$\left[\begin{array}{cc} \mathbf{A} & \mathbf{B} \\ \mathbf{0} & \mathbf{D} - \mathbf{CA}^{-1}\mathbf{B} \end{array}\right] \left[\begin{array}{cc} \mathbf{I} & -\mathbf{A}^{-1}\mathbf{B} \\ \mathbf{0} & \mathbf{I} \end{array}\right] = \left[\begin{array}{cc} \mathbf{A} & \mathbf{0} \\ \mathbf{0} & \mathbf{D} - \mathbf{CA}^{-1}\mathbf{B} \end{array}\right]$$

Putting these two block matrix equations together, we have

$$
\begin{bmatrix} I & 0 \\ -CA^{-1} & I \end{bmatrix} \begin{bmatrix} A & B \\ C & D \end{bmatrix} \begin{bmatrix} I & -A^{-1}B \\ 0 & I \end{bmatrix} = \begin{bmatrix} A & 0 \\ 0 & D - CA^{-1}B \end{bmatrix}
$$

Because the inverses of these lower and upper block matrices are obtained by changing the sign of the off-diagonal block matrices, we have

$$
\begin{bmatrix} A & B \\ C & D \end{bmatrix} = \begin{bmatrix} I & 0 \\ CA^{-1} & I \end{bmatrix} \begin{bmatrix} A & 0 \\ 0 & D - CA^{-1}B \end{bmatrix} \begin{bmatrix} I & A^{-1}B \\ 0 & I \end{bmatrix}
$$

and

$$
\begin{bmatrix} A & B \\ C & D \end{bmatrix}^{-1} = \begin{bmatrix} I & -A^{-1}B \\ 0 & I \end{bmatrix} \begin{bmatrix} A^{-1} & 0 \\ 0 & (D - CA^{-1}B)^{-1} \end{bmatrix} \begin{bmatrix} I & 0 \\ -CA^{-1} & I \end{bmatrix}
$$

$$
= \begin{bmatrix} A^{-1} + A^{-1}B(D - CA^{-1}B)^{-1}CA^{-1} & -A^{-1}B(D - CA^{-1}B)^{-1} \\ -(D - CA^{-1}B)^{-1}CA^{-1} & (D - CA^{-1}B)^{-1} \end{bmatrix}
$$

Here we must assume that $D - CA^{-1}B$, the Schur complement of A, is invertible.

A second approach is to eliminate the $(1, 2)$ block first and then the $(2, 1)$ block, obtaining

$$
\begin{bmatrix} A & B \\ C & D \end{bmatrix}^{-1} = \begin{bmatrix} I & 0 \\ -D^{-1}C & I \end{bmatrix} \begin{bmatrix} (A - BD^{-1}C)^{-1} & 0 \\ 0 & D^{-1} \end{bmatrix} \begin{bmatrix} I & -BD^{-1} \\ 0 & I \end{bmatrix}
$$

$$
= \begin{bmatrix} (A - BD^{-1}C)^{-1} & -(A - BD^{-1}C)^{-1}BD^{-1} \\ -D^{-1}C(A - BD^{-1}C)^{-1} & D^{-1} + D^{-1}C(A - BD^{-1}C)^{-1}BD^{-1} \end{bmatrix}
$$

Here we assume that the Schur complement of D is invertible; it is $A - BD^{-1}C$.

By equating terms in the final result from these two approaches, one obtains the **Woodbury matrix identity**:

$$
(A - BD^{-1}C)^{-1} = A^{-1} + A^{-1}B(D - CA^{-1}B)^{-1}CA^{-1}
$$

A special case of this is the **Woodbury formula**

$$
(A + UV)^{-1} = A^{-1} - A^{-1}U(I + VA^{-1}U)^{-1}VA^{-1}
$$

where $\mathbf{U}$ and $\mathbf{V}$ are matrices of the appropriate sizes. This is useful in some numerical computations when one has already computed $\mathbf{A}^{-1}$ and wants to compute $(\mathbf{A} + \mathbf{UV})^{-1}$. Another special case is the **Sherman-Morrison formula**

$$(\mathbf{A} + \mathbf{uv}^{T})^{-1} = \mathbf{A}^{-1} - (1 + \lambda)^{-1}(\mathbf{A}^{-1}\mathbf{u})(\mathbf{v}^{T}\mathbf{A}^{-1})^{T}$$

where $\lambda = \mathbf{v}^{T}\mathbf{A}^{-1}\mathbf{u}$ and $\mathbf{uv}^{T}$ is the outer product of column vectors $\mathbf{u}$ and $\mathbf{v}$.

Application: Linear Least-Squares Problem

We now show how to solve a least-squares problem by three different approaches. Consider

$$\begin{cases} 3x_1 + 2x_2 = 1 \\ x_1 + 4x_2 = 3 \\ 2x_1 + x_2 = 1 \end{cases}$$

In matrix notation, we have

$$\begin{bmatrix} 3 & 2 \\ 1 & 4 \\ 2 & 1 \end{bmatrix} \begin{bmatrix} x_1 \\ x_2 \end{bmatrix} = \begin{bmatrix} 1 \\ 3 \\ 1 \end{bmatrix}$$

First, we find a least-squares solution using calculus. The total squared error is

$$E = (3x_1 + 2x_2 - 1)^2 + (x_1 + 4x_2 - 3)^2 + (2x_1 + x_2 - 1)^2$$

This is a nonlinear function of x_1 and x_2. We want E to be a minimum. Using calculus, we set $\partial E/\partial x_1 = 0$ and $\partial E/\partial x_2 = 0$. Taking these partial derivatives, we find

$$2(3x_1 + 2x_2 - 1)3 + 2(x_1 + 4x_2 - 3)1 + 2(2x_1 + x_2 - 1)2 = 0$$

$$2(3x_1 + 2x_2 - 1)1 + 2(x_1 + 4x_2 - 3)4 + 2(2x_1 + x_2 - 1)1 = 0$$

The collecting of terms and simplifying lead to this linear system:

$$\begin{cases} 7x_1 + 6x_2 = 4 \\ 4x_1 + 7x_2 = 5 \end{cases}$$

Gaussian elimination is straightforward to find the solution $x_1 = -2/25$ and $x_2 = 19/25$. We find that the resulting error is $E = 6/25 = 0.24$.

Next, we solve via the **normal equation**. Writing the original system in matrix–vector form $\mathbf{Ax} = \mathbf{b}$, we arrive at the normal equation:

$$\mathbf{A}^T\mathbf{Ax} = \mathbf{A}^T\mathbf{b}$$

$$\begin{bmatrix} 3 & 1 & 2 \\ 2 & 4 & 1 \end{bmatrix} \begin{bmatrix} 3 & 2 \\ 1 & 4 \\ 2 & 1 \end{bmatrix} \begin{bmatrix} x_1 \\ x_2 \end{bmatrix} = \begin{bmatrix} 3 & 1 & 2 \\ 2 & 4 & 1 \end{bmatrix} \begin{bmatrix} 1 \\ 3 \\ 1 \end{bmatrix}$$

$$\begin{bmatrix} 14 & 12 \\ 12 & 21 \end{bmatrix} \begin{bmatrix} x_1 \\ x_2 \end{bmatrix} = \begin{bmatrix} 8 \\ 15 \end{bmatrix}$$

It is no accident that the normal equations are the same as those found from the first approach above! Again, the solution is

$$\begin{bmatrix} x_1 \\ x_2 \end{bmatrix} = \begin{bmatrix} -2/25 \\ 19/25 \end{bmatrix}$$

Finally, we find the least-squares solution using the QR-factorization. We let

$$\begin{bmatrix} \mathbf{a}_1 & \mathbf{a}_2 \end{bmatrix} \mathbf{x} = \begin{bmatrix} 3 & 2 \\ 1 & 4 \\ 2 & 1 \end{bmatrix} \begin{bmatrix} x_1 \\ x_2 \end{bmatrix} = \begin{bmatrix} 1 \\ 3 \\ 1 \end{bmatrix}$$

The first step is to orthogonalize the columns $\mathbf{a}_1$ and $\mathbf{a}_2$ of the coefficient matrix. Let $\mathbf{q}_1 = \mathbf{a}_1$ and

$$\mathbf{q}_2 = \mathbf{a}_2 - \frac{\mathbf{a}_2 \bullet \mathbf{q}_1}{\mathbf{q}_1 \bullet \mathbf{q}_1}\mathbf{q}_1 = \begin{bmatrix} 2 \\ 4 \\ 1 \end{bmatrix} - \frac{6}{7}\begin{bmatrix} 3 \\ 1 \\ 2 \end{bmatrix} = \frac{1}{7}\begin{bmatrix} -4 \\ 22 \\ -5 \end{bmatrix}$$

Now we find that these vectors $\mathbf{q}_1 = \begin{bmatrix} 3 \\ 1 \\ 2 \end{bmatrix}$ and $\mathbf{q}_2 = \begin{bmatrix} -4 \\ 22 \\ -5 \end{bmatrix}$ form an orthogonal set because: $\mathbf{q}_1 \bullet \mathbf{q}_2 = 0$. Normalize them to get

$$\mathbf{Q} = \begin{bmatrix} (1/\sqrt{14})\begin{bmatrix} 3 \\ 1 \\ 2 \end{bmatrix}, & \begin{bmatrix} 1/(5\sqrt{21}) \end{bmatrix}\begin{bmatrix} -4 \\ 22 \\ -5 \end{bmatrix} \end{bmatrix} = \begin{bmatrix} 3/\sqrt{14} & -4/(5\sqrt{21}) \\ 1/\sqrt{14} & 22/(5\sqrt{21}) \\ 2/\sqrt{14} & -5/(5\sqrt{21}) \end{bmatrix}$$

Recall that if $\mathbf{A} = \mathbf{QR}$ and $\mathbf{Q}^T\mathbf{Q} = \mathbf{I}$, then $\mathbf{Q}^T\mathbf{A} = \mathbf{Q}^T\mathbf{QR} = \mathbf{R}$. Consequently, the system $\mathbf{Ax} = \mathbf{b}$ becomes $\mathbf{Rx} = \mathbf{Q}^T\mathbf{b}$ and $\mathbf{x} = \mathbf{R}^{-1}\mathbf{Q}^T\mathbf{b}$, which is the solution of the least-squares problem. In this example, we have

$$\mathbf{R} = \mathbf{Q}^T\mathbf{A} = \begin{bmatrix} 3/\sqrt{14} & 1/\sqrt{14} & 2/\sqrt{14} \\ -4/(5\sqrt{21}) & 22/(5\sqrt{21}) & -5/(5\sqrt{21}) \end{bmatrix} \begin{bmatrix} 3 & 2 \\ 1 & 4 \\ 2 & 1 \end{bmatrix}$$

$$= \begin{bmatrix} \sqrt{14} & (\frac{6}{7})\sqrt{14} \\ 0 & (\frac{5}{7})\sqrt{21} \end{bmatrix}$$

$$\mathbf{Q}^T\mathbf{b} = \begin{bmatrix} 3/\sqrt{14} & 1/\sqrt{14} & 2/\sqrt{14} \\ -4/(5\sqrt{21}) & 22/(5\sqrt{21}) & -5/(5\sqrt{21}) \end{bmatrix} \begin{bmatrix} 1 \\ 3 \\ 1 \end{bmatrix} = \begin{bmatrix} (\frac{4}{7})\sqrt{14} \\ (19/25)\sqrt{21} \end{bmatrix}$$

Then we find

$$\mathbf{R}\widetilde{\mathbf{x}} = \mathbf{Q}^T\mathbf{b}$$

$$\begin{bmatrix} \sqrt{14} & (\frac{6}{7})\sqrt{14} \\ 0 & (\frac{5}{7})\sqrt{21} \end{bmatrix} \begin{bmatrix} \widetilde{x}_1 \\ \widetilde{x}_2 \end{bmatrix} = \begin{bmatrix} (\frac{4}{7})\sqrt{14} \\ (19/25)\sqrt{21} \end{bmatrix}$$

Solving, we obtain $\widetilde{\mathbf{x}} = [-\frac{2}{25}, \frac{19}{25}]^T = [-0.08, 0.76]$. Here $\widetilde{\mathbf{x}}$ is the solution to the least-squares problem $\mathbf{Ax} = \mathbf{b}$, and the error is $||\mathbf{A}\widetilde{\mathbf{x}} - \mathbf{b}||_2$. How close is $\mathbf{A}\widetilde{\mathbf{x}}$ to $\mathbf{b}$? The solution found above leads to

$$\mathbf{A}\widetilde{\mathbf{x}} - \mathbf{b} = \begin{bmatrix} 3 & 2 \\ 1 & 4 \\ 2 & 1 \end{bmatrix} \begin{bmatrix} -2/25 \\ 19/25 \end{bmatrix} - \begin{bmatrix} 1 \\ 3 \\ 1 \end{bmatrix} = \begin{bmatrix} 7/25 \\ -1/25 \\ -10/25 \end{bmatrix}$$

Its Euclidean norm is

$$||\mathbf{A}\widetilde{\mathbf{x}} - \mathbf{b}||_2 = \sqrt{(7/25)^2 + (1/25)^2 + (10/25)^2} = \tfrac{1}{5}\sqrt{6} \approx 0.49$$

Mathematical Software

When using mathematical software to obtain the LU-factorization of a matrix, one must be aware that the algorithm may swap rows to produce strong pivot elements. For example, consider the following matrix:

$$\mathbf{A} = \begin{bmatrix} 3 & 5 & 3 \\ 2 & 6 & -4 \\ 5 & 4 & -2 \end{bmatrix}$$

We enter the matrix **A** into the chosen computer software and then request the LU-factorization with the appropriate commands. In MATLAB, we use the command shown.

MATLAB
A = [3,5,3;2,6,-4;5,4,-2]
[L,U,P] = lu(A)

We obtain the LU-factorization of a permuted version of **A**:

$$\mathbf{P} = \begin{bmatrix} 0 & 0 & 1 \\ 0 & 1 & 0 \\ 1 & 0 & 0 \end{bmatrix}, \qquad \mathbf{PA} = \begin{bmatrix} 5 & 4 & -2 \\ 2 & 6 & -4 \\ 3 & 5 & 3 \end{bmatrix}$$

$$\mathbf{L} = \begin{bmatrix} 1.0000 & 0 & 0 \\ 0.4000 & 1.0000 & 0 \\ 0.6000 & 0.5909 & 1.0000 \end{bmatrix}, \qquad \mathbf{U} = \begin{bmatrix} 5.0000 & 4.0000 & -2.0000 \\ 0 & 4.4000 & -3.2000 \\ 0 & 0 & 6.0909 \end{bmatrix}$$

The pivot elements are taken in the order $(3, 2, 1)$. One can verify that

$$\mathbf{PA} = \mathbf{LU}$$

Here are the corresponding Maple and Mathematica commands:

Maple	Mathematica

with(LinearAlgebra):
 A := Matrix([[3,5,3],[2,6,-4],
     [5,4,-2]]);
 (P,L,U) := LUDecomposition(A);
	A = {{3,5,3},{2,6,-4},
	     {5,4,-2}}
	 {LU,P,CondNo} =
	     LUDecomposition[A]

In Maple, we discover that the order of the pivot elements is not changed from the natural order $(1, 2, 3)$. In Mathematica, we find that the pivot elements are selected in the order $(2, 1, 3)$ and the matrices **L** and **U** are stored in the array **LU** by omitting the diagonal 1's from the unit lower triangular matrix **L**.

EXAMPLE 11   We can use the mathematical software in MATLAB, Maple, and Mathematica to compute the Cholesky decomposition $\mathbf{A} = \mathbf{LL}^T$ for a typical example such as this matrix:

$$\mathbf{A} = \begin{bmatrix} 4 & -1 & -1 & 0 \\ -1 & 4 & 0 & -1 \\ -1 & 0 & 4 & -1 \\ 0 & -1 & -1 & 4 \end{bmatrix}$$

SOLUTION   This matrix is obviously real and symmetric. It is also positive definite. In MATLAB, we use the commands below:

```
MATLAB
A = [4,-1,-1,0;-1,4,0,-1;-1,0,4,-1;0,-1,-1,4]
L = chol(A)
L*L'
```

We find the Cholesky factorization to be $\mathbf{LL}^T$, where

$$\mathbf{L}^T = \begin{bmatrix} 2.0000 & -0.5000 & -0.5000 & 0 \\ 0 & 1.9365 & -0.1291 & -0.5164 \\ 0 & 0 & 1.9322 & -0.5521 \\ 0 & 0 & 0 & 1.8516 \end{bmatrix}$$

One can check this independently by computing $\mathbf{L}^T\mathbf{L}$. In Maple, the corresponding commands are

```
Maple
with(LinearAlgebra):
A := Matrix([[4,-1,-1,0],[-1,4,0,-1],[-1,0,4,-1],
    [0,-1,-1,4]]);
L := LUDecomposition(A,method='Cholesky');
L.Transpose(L)
```

We obtain the exact form of the matrix $\mathbf{L}$:

$$\mathbf{L} = \begin{bmatrix} 2 & 0 & 0 & 0 \\ -\frac{1}{2} & \left(\frac{1}{2}\right)\sqrt{15} & 0 & 0 \\ -\frac{1}{2} & (-1/30)\sqrt{15} & (2/15)\sqrt{210} & 0 \\ 0 & (-2/15)\sqrt{15} & (-4/105)\sqrt{210} & \left(\frac{2}{7}\right)\sqrt{42} \end{bmatrix}$$

and we can verify this by computing $\mathbf{LL}^T$. In Mathematica, we use these commands:

```
Mathematica
<<LinearAlgebra`Cholesky`
A = {{4,-1,-1,0},{-1,4,0,-1},{-1,0,4,-1},{0,-1,-1,4}}
L = CholeskyDecomposition[A];
MatrixForm[L.Transpose[L]]
```

Again, we obtain the exact form of the matrix $\mathbf{L}^T$ in a slightly different form:

$$\mathbf{L}^T = \begin{bmatrix} 2 & -\frac{1}{2} & -\frac{1}{2} & 0 \\ 0 & (\frac{1}{2})\sqrt{15} & -1/(2\sqrt{15}) & -2/\sqrt{15} \\ 0 & 0 & 2\sqrt{14/15} & -(\frac{4}{7})\sqrt{14/15} \\ 0 & 0 & 0 & 2\sqrt{\frac{6}{7}} \end{bmatrix}$$

One can easily show that all of these factorizations are the same. ∎

EXAMPLE 12 Using the mathematical software in MATLAB, Maple, and Mathematica, find the least-squares solution of this system by using the QR-factorization of the coefficient matrix:

$$\begin{cases} 4x + 7y - 3z = 1 \\ 2x - 4y + 6z = 5 \\ -3x + 2y + 5z = 7 \\ x - 2y + 4z = -3 \end{cases}$$

SOLUTION In MATLAB, we invoke these commands:

```
MATLAB
A = [4.,7., -3.;2.,-4.,6.;-3.,2.,5.;1.,-2.,4.]
b = [1.;5.;7.;-3.]
[Q, R] = qr(A)
x = R\Q'*b
e = A'*R
```

The results are $\mathbf{Q}, \mathbf{R}$, numerical values for the solution vector, the residual vector, and the error vector:

$$
\mathbf{Q} = \begin{bmatrix} -0.7303 & 0.6539 & -0.1964 & -0.0228 \\ -0.3651 & -0.5812 & -0.5180 & -0.5104 \\ 0.5477 & 0.3875 & -0.7366 & -0.0855 \\ -0.1826 & -0.2906 & -0.3880 & 0.8554 \end{bmatrix}
$$

$$
\mathbf{R} = \begin{bmatrix} -5.4772 & -2.1909 & 2.0083 \\ 0 & 8.2583 & -4.6741 \\ 0 & 0 & -7.7537 \\ 0 & 0 & 0 \end{bmatrix}
$$

$$
\mathbf{x} = \begin{bmatrix} -0.2752 \\ 0.6561 \\ 0.8742 \end{bmatrix}, \qquad \mathbf{r} = \mathbf{Ax} - \mathbf{b} = \begin{bmatrix} -0.1309 \\ -2.9294 \\ -0.4910 \\ 4.9096 \end{bmatrix}
$$

$$
\mathbf{A}^T \mathbf{r} = \begin{bmatrix} -0.0977 \\ -0.3730 \\ 0.2132 \end{bmatrix} \times 10^{-13}
$$

Here are the Maple and Mathematica commands:

```
Maple
with(LinearAlgebra):
A := Matrix([[4,7,-3],[2,-4,6],[-3,2,5],[1,-2,4]]);
b := Vector([1.,5.,7.,-3.]);
(Q,R) := QRDecomposition(A);
xt := LinearSolve(R,Transpose(Q).b);
x := evalf(LeastSquares(A,b));
r := A.x-b
e := Transpose(A).r
```

```
Mathematica
A = {{4.,7.,-3.},{2.,-4.,6.},{-3.,2.,5.},{1.,-2.,4.}}
b = {1.,5.,7.,-3.}
{Q, R} = QRDecomposition[A]
xt = LinearSolve[R,Q.b]
x = N[xt]
r = A.x-b
e = Transpose[A].r
```

It is informative to compare the results from these three software systems. In each system, we can obtain exact results, usually as quotients of integers.

These, in turn, can be expressed as floating-point numbers, with 15 decimal places:

$$\mathbf{x} = [-33853/123005, 80703/123005, 107536/123005]^T$$

$$\approx [-0.2752, 0.6561, 0.8742]^T \qquad ■$$

EXAMPLE 13 Using the mathematical software in MATLAB, Maple, and Mathematica, find the singular values of this matrix and verify the factorization.

$$\mathbf{A} = \begin{bmatrix} 2.0 & 2.0 & 2.0 & 2.0 \\ 1.7 & 0.1 & -1.7 & -0.1 \\ 0.6 & 1.8 & -0.6 & -1.8 \end{bmatrix}$$

SOLUTION In MATLAB, we use these commands:

```
MATLAB
A = [2.0,2.0,2.0,2.0;1.7,0.1,-1.7,-0.1;
     0.6,1.8,-0.6,-1.8]
[P,D,Q] = svd(A)
A = P*D*Q
```

and we find this factorization:

$$\mathbf{A} = \begin{bmatrix} 1.0 & 0.0 & 0.0 \\ 0.0 & 0.6 & -0.8 \\ 0.0 & 0.8 & 0.6 \end{bmatrix} \begin{bmatrix} 4 & 0 & 0 & 0 \\ 0 & 3 & 0 & 0 \\ 0 & 0 & 2 & 0 \end{bmatrix}$$

$$\begin{bmatrix} -0.5 & -0.5 & -0.5 & -0.5 \\ -0.5 & -0.5 & +0.5 & +0.5 \\ -0.5 & +0.5 & +0.5 & -0.5 \\ -0.5 & +0.5 & -0.5 & +0.5 \end{bmatrix} = \mathbf{PDQ}$$

Here are the Maple and Mathematica commands:

```
Maple
with(LinearAlgebra):
A := Matrix([[2,2,2,2],[1.7,0.1,-1.7,-0.1],
     [0.6,1.8,-0.6,-1.8]]);
P,Diag,Q := SingularValues(A,output=['U','S','Vt']);
P.DiagonalMatrix(Diag[1..3],3,4).Q;
```

```
Mathematica
A = {{2,2,2,2},{1.7,0.1,-1.7,-0.1},{0.6,1.8,-0.6,-1.8}}
{P,Diag,Q} = SingularValues[A]
MatrixForm[Transpose[P].DiagonalMatrix[Diag].Q]
```

We find similar results in using Maple and Mathematica.   ■

SUMMARY 8.2

- A typical permutation matrix:

$$\mathbf{P} = \begin{bmatrix} 0 & 0 & 1 & 0 & 0 \\ 0 & 0 & 0 & 0 & 1 \\ 1 & 0 & 0 & 0 & 0 \\ 0 & 1 & 0 & 0 & 0 \\ 0 & 0 & 0 & 1 & 0 \end{bmatrix}$$

It embodies a reordering of the rows of an identity matrix $\mathbf{I}$.

- $\mathbf{A}$ is positive definite if $\mathbf{x}^T\mathbf{Ax} > 0$ for all $\mathbf{x} \neq \mathbf{0}$, when $\mathbf{A}$ is real. In the complex case, use $\mathbf{x}^H\mathbf{Ax} > 0$.

- LU-factorization: $\mathbf{A} = \mathbf{LU}$, where $\mathbf{L}$ is unit lower triangular and $\mathbf{U}$ is upper triangular. Here $\mathbf{L} = \mathbf{E}_1^{-1}\mathbf{E}_2^{-1}\cdots\mathbf{E}_k^{-1}$, where $\mathbf{E}_i$ are elementary matrices (Gaussian elimination).

- One can solve $\mathbf{Ax} = \mathbf{b}$ by solving (in order) $\mathbf{Ly} = \mathbf{b}$ and $\mathbf{Ux} = \mathbf{y}$. Here $\mathbf{A} = \mathbf{LU}$.

- PLU-factorization: $\mathbf{PA} = \mathbf{LU}$, where $\mathbf{P}$ is a permutation matrix, $\mathbf{L}$ is a unit lower triangular matrix, and $\mathbf{U}$ is upper triangular.

- LL^T-factorization: $\mathbf{A} = \mathbf{LL}^T$, where $\mathbf{L}$ is lower triangular and $\mathbf{A}$ is symmetric (Cholesky factorization).

- LDL^T-factorization: $\mathbf{A} = \mathbf{LDL}^T$, where $\mathbf{L}$ is lower triangular, $\mathbf{D}$ is diagonal, and $\mathbf{A}$ is symmetric.

- Multiplication of partitioned matrices: $\mathcal{C} = (\mathbf{C}_{ij}) = \mathcal{AB}$, where $\mathbf{C}_{ij} = \sum_{s=1}^{n}\mathbf{A}_{is}\mathbf{B}_{sj}$ for $1 \leq i, j \leq n$.

- 2×2 block LU-factorization:

$$\begin{bmatrix} \mathbf{A} & \mathbf{B} \\ \mathbf{B}^T & \mathbf{0} \end{bmatrix} = \begin{bmatrix} \mathbf{I} & \mathbf{0} \\ \mathbf{B}^T\mathbf{A}^{-1} & \mathbf{I} \end{bmatrix}\begin{bmatrix} \mathbf{A} & \mathbf{B} \\ \mathbf{0} & -\mathbf{B}^T\mathbf{A}^{-1}\mathbf{B} \end{bmatrix}$$

- 2×2 block LDL^T-factorization:

$$\begin{bmatrix} \mathbf{A} & \mathbf{B} \\ \mathbf{B}^T & \mathbf{0} \end{bmatrix} = \begin{bmatrix} \mathbf{I} & \mathbf{0} \\ \mathbf{B}^T\mathbf{A}^{-1} & \mathbf{L}_2 \end{bmatrix}$$
$$\times \begin{bmatrix} \mathbf{A} & \mathbf{0} \\ \mathbf{0} & -\mathbf{D}_2 \end{bmatrix}\begin{bmatrix} \mathbf{I} & \mathbf{A}^{-1}\mathbf{B} \\ \mathbf{0} & \mathbf{L}_2^T \end{bmatrix}$$

where the factorization of the reduced system is $\mathbf{B}^T\mathbf{A}^{-1}\mathbf{B} = \mathbf{L}_2\mathbf{D}_2\mathbf{L}_2^T$.

- Theorems:

  - If $\mathbf{A}$ is real, symmetric, and positive definite, then $\mathbf{A} = \mathbf{LL}^T$, where $\mathbf{L}$ is a real lower triangular matrix having positive elements on its diagonal.

  - Let $\mathbf{A}$ be $m \times n$. Let $\mathbf{Q}$ be $m \times n$. Let $\mathbf{R}$ be $n \times n$. The factorization $\mathbf{A} = \mathbf{QR}$ is useful if the columns of $\mathbf{Q}$ form an orthonormal set and $\mathbf{R}$ is upper triangular and invertible. To get the least-squares solution of $\mathbf{Ax} = \mathbf{b}$, solve $\mathbf{Rx} = \mathbf{Q}^T\mathbf{b}$ for $\mathbf{x}$. (QR-factorization.)

  - Let the matrix $\mathbf{A}$ be $m \times n$ and have rank n. Let $\mathbf{Q}$ be $m \times n$ and have an orthonormal set of columns. Let $\mathbf{R}$ be an $n \times n$ upper triangular matrix. If $\mathbf{A} = \mathbf{QR}$, then this factorization is called the QR-factorization of $\mathbf{A}$.

  - Let $\mathbf{A}$ be $m \times n$, let $\mathbf{P}$ be $m \times n$, let $\mathbf{D}$ be $n \times n$, and $\mathbf{Q}$ be $n \times n$, where $\mathbf{P}$ and $\mathbf{Q}$ are unitary and $\mathbf{D}$ is diagonal. If $\mathbf{A} = \mathbf{PDQ}$, then this is the SVD-factorization of $\mathbf{A}$.

KEY CONCEPTS 8.2

Permutation matrices, forward and backward substitution, elementary matrices, positive definite matrices, factoring matrices, factorizations: $\mathbf{A} = \mathbf{LU}$ (LU-factorization, Gaussian elimination), $\mathbf{A} = \mathbf{LL}^T$ (Cholesky LL^T-factori- zation), $\mathbf{A} = \mathbf{LDL}^T$ (LDL^T-factorization), $\mathbf{A} = \mathbf{QR}$, (QR-factorization), and $\mathbf{A} = \mathbf{PDQ}$ (singular-value decomposition), block matrices, application: linear least-squares problem, normal equations

GENERAL EXERCISES 8.2

1. Find the LU-factorizations of these matrices:

a. $\mathbf{A} = \begin{bmatrix} 2 & -4 \\ 10 & -17 \end{bmatrix}$

b. $\mathbf{B} = \begin{bmatrix} -3 & 1 & 5 \\ -9 & -1 & 16 \\ -6 & -6 & 14 \end{bmatrix}$

2. (Continuation.) Using the factorization of $\mathbf{B}$ found previously, solve this system of equations

$$\begin{cases} -3x_1 + x_2 + 5x_3 = -31 \\ -9x_1 - x_2 + 16x_3 = -49 \\ -6x_1 - 6x_2 + 14x_3 = -62 \end{cases}$$

3. Find the LU-factors of these matrices:

a. $\mathbf{A} = \begin{bmatrix} 3 & 0 & 3 \\ 0 & -1 & 3 \\ 1 & 3 & 0 \end{bmatrix}$

b. $\mathbf{B} = \begin{bmatrix} 4 & -1 & -1 & 0 \\ -1 & 4 & 0 & -1 \\ -1 & 0 & 4 & -1 \\ 0 & -1 & -1 & 4 \end{bmatrix}$

4. Find the Cholesky factorization of

$$\begin{bmatrix} 1 & 3 & 5 \\ 3 & 13 & 23 \\ 5 & 23 & 77 \end{bmatrix}$$

5. Find the Cholesky factorization of

$$\begin{bmatrix} 1 & 4 & 5 \\ 4 & 20 & 32 \\ 5 & 32 & 70 \end{bmatrix}$$

6. Repeat Example 3 using $\mathbf{P} = \begin{bmatrix} 0 & 0 & 1 \\ 1 & 0 & 0 \\ 0 & 1 & 0 \end{bmatrix}$

7. Find the Cholesky factorization of

$$\begin{bmatrix} 1 & 3 & 4 \\ 3 & 13 & 10 \\ 4 & 10 & 18 \end{bmatrix}$$

8. Find the QR-factorization of

$$\mathbf{A} = \begin{bmatrix} 2 & 2 & -1 \\ 2 & 8 & 2 \\ 1 & 7 & -1 \end{bmatrix}$$

9. Solve $\mathbf{Ax} = \mathbf{b}$ by using the LU-factorization

where $\mathbf{A} = \begin{bmatrix} 2 & 1 & -4 \\ 4 & 3 & -5 \\ -4 & 1 & 19 \end{bmatrix}$ and

$$\mathbf{b} = \begin{bmatrix} -10 \\ -11 \\ 53 \end{bmatrix}$$

10. Find the QR-factorization of

$$A = \begin{bmatrix} -1 & -1 & 1 \\ 1 & 1 & -4 \\ -1 & -1 & 1 \\ 3 & 7 & -6 \end{bmatrix}$$

11. Find the Cholesky factorization of

$$\begin{bmatrix} 1 & 5 & 4 \\ 5 & 32 & 20 \\ 4 & 20 & 70 \end{bmatrix}$$

12. Find the LU-factorization of

$$A = \begin{bmatrix} 3 & 1 & -1 & 2 \\ 6 & 4 & 1 & 1 \\ -9 & 1 & 7 & -11 \\ 12 & 0 & -14 & 17 \end{bmatrix}$$

13. (Continuation.) Using the factorization LU, solve $\mathbf{Ax} = \mathbf{b}$, where $\mathbf{b} = [9, 29, 13, 0]^T$.

14. Find the LU-factorization of

$$A = \begin{bmatrix} 2 & -6 & 6 \\ -4 & 5 & -7 \\ 3 & 5 & -1 \\ -6 & 4 & -8 \\ 8 & -3 & 9 \end{bmatrix}$$

15. Using the LU-factorization, determine the inverse of $\mathbf{A} = \begin{bmatrix} 2 & -1 & 2 \\ -6 & 0 & -2 \\ 8 & -1 & 5 \end{bmatrix}$

16. Find the LU-factorization of each of these matrices, if possible:

a. $\begin{bmatrix} 0 & 5 \\ 3 & 7 \end{bmatrix}$ **b.** $\begin{bmatrix} 2 & 1 & 2 \\ 4 & 3 & 1 \\ 6 & -1 & 1 \end{bmatrix}$

c. $\begin{bmatrix} 1 & 2 \\ 3 & 4 \\ 5 & 6 \end{bmatrix}$ **d.** $\begin{bmatrix} 1 & 2 & 3 \\ 4 & 5 & 6 \end{bmatrix}$

17. This exercise involves the LU-factorization of a nonsquare matrix. If $\mathbf{A}$ is an $m \times n$ matrix, we seek an $m \times m$ matrix for the factor $\mathbf{L}$ and an $m \times n$ matrix for the factor $\mathbf{U}$. Find such a factorization for this matrix:

$$A = \begin{bmatrix} 2 & 1 & 3 \\ 4 & 1 & 8 \\ -4 & -4 & 1 \\ 6 & 4 & 16 \end{bmatrix}$$

18. Let $\mathbf{A} = \begin{bmatrix} 1 & 2 & 3 \\ 2 & 7 & -5 \\ 3 & -5 & 4 \end{bmatrix}$

Verify that $\mathbf{A}$ is not positive definite but $\mathbf{A} + 4\mathbf{I}$ is. Elaborate on this: Is there a smallest number c for which $\mathbf{A} + c\mathbf{I}$ is nonnegative definite? What is the general case, when $\mathbf{A}$ is $n \times n$? Formulate and prove some theorems.

19. Verify that the matrix $\begin{bmatrix} 2 & 2 & 1 \\ 1 & 1 & 1 \\ 3 & 2 & 1 \end{bmatrix}$

cannot be factored into the product of a unit lower triangular matrix and an upper triangular matrix. Note that it does not suffice to show that our algorithm fails.

20. Explain that if $\mathbf{A}$ has been factored as $\mathbf{A} = \mathbf{LU}$, where the usual assumptions are made, then the determinant of $\mathbf{A}$ is easily computed.

21. A square matrix $\mathbf{A}$ is **positive definite** if $\mathbf{x}^T\mathbf{Ax} > 0$ when the vector $\mathbf{x}$ is not zero. Establish that the Gram matrix arising from a linearly independent set of vectors is positive definite. Establish that the Gram matrix is symmetric.

22. Verify in general that if a matrix is positive definite, then its eigenvalues are positive.

23. Confirm that if a matrix is positive definite, then its diagonal elements are positive. (Use vectors of the form $[0, \ldots, 0, 1, 0, \ldots, 0]$.)

24. (Continuation.) Find all the 2×2 symmetric positive definite matrices.

25. Establish that if a matrix $\mathbf{A}$ is positive definite, then its principal minors are also positive definite.

26. Establish the converse of Theorem 1:
If $\mathbf{L}$ is lower triangular and has only positive elements on its diagonal, then $\mathbf{LL}^T$ is invertible, symmetric, and positive definite.

27. Let $\mathbf{A}$ be an invertible, $n \times n$ matrix. Factor it as $\mathbf{A} = \mathbf{QR}$ as described in this section. Explain why $\mathbf{RQ}$ is similar to $\mathbf{A}$.

28. (Micro-research Project.) Let $\mathbf{B}$ be an $m \times n$ matrix such that $\mathbf{B}^T\mathbf{B}$ is a diagonal matrix. Let $\mathbf{R}$ be an invertible $n \times n$ matrix. How can we use these matrices to solve the system $\mathbf{Ax} = \mathbf{b}$ in the least-squares sense, if $\mathbf{A} = \mathbf{BR}$? (You should discover an alternative to the use of a QR-factorization in solving least-squares problems. Certainly, the normalization of the columns in $\mathbf{Q}$ can be avoided.)

29. Find the Cholesky factorization of a general 2×2 symmetric positive definite matrix.

30. Find the LDL^T-factorization of a general 2×2 symmetric positive definite matrix where $\mathbf{L}$ is unit lower triangular and $\mathbf{D}$ is diagonal.

31. Explain why a lower triangular matrix $\mathbf{L}$ is positive definite if its diagonal elements are positive.

32. Establish that if the $n \times n$ real matrix $\mathbf{A}$ is positive definite, and if $\mathbf{D}$ is a diagonal matrix having positive elements on its diagonal, then $\mathbf{DA}$ is also positive definite.

33. In the subsection on 2×2 block matrices, verify the results for each of the following:

a. The inverse of the block matrix is

$$\begin{bmatrix} \mathbf{A} & \mathbf{B} \\ \mathbf{B}^T & \mathbf{0} \end{bmatrix}^{-1} = \begin{bmatrix} \mathbf{0} & \mathbf{B}^{-T} \\ \mathbf{B}^{-1} & -\mathbf{B}^{-1}\mathbf{AB}^{-T} \end{bmatrix}$$

when $\mathbf{A}$ and $\mathbf{B}$ are invertible.

b. The block LU-factorization:

$$\begin{bmatrix} \mathbf{A} & \mathbf{B} \\ \mathbf{B}^T & \mathbf{0} \end{bmatrix} =$$

$$\begin{bmatrix} \mathbf{I} & \mathbf{0} \\ \mathbf{B}^T\mathbf{A}^{-1} & \mathbf{I} \end{bmatrix} \begin{bmatrix} \mathbf{A} & \mathbf{B} \\ \mathbf{0} & -\mathbf{B}^T\mathbf{A}^{-1}\mathbf{B} \end{bmatrix}$$

c. The block LDL^T-factorization:

$$\begin{bmatrix} \mathbf{A} & \mathbf{B} \\ \mathbf{B}^T & \mathbf{0} \end{bmatrix} =$$

$$\begin{bmatrix} \mathbf{I} & \mathbf{0} \\ \mathbf{B}^T\mathbf{A}^{-1} & \mathbf{L}_2 \end{bmatrix} \begin{bmatrix} \mathbf{A} & \mathbf{0} \\ \mathbf{0} & -\mathbf{D}_2 \end{bmatrix} \begin{bmatrix} \mathbf{I} & \mathbf{A}^{-1}\mathbf{B} \\ \mathbf{0} & \mathbf{L}_2^T \end{bmatrix}$$

where the factorization of the reduced system is $\mathbf{B}^T\mathbf{A}^{-1}\mathbf{B} = \mathbf{L}_2\mathbf{D}_2\mathbf{L}_2^T$.

34. Consider the block 3×3 matrix

$$\mathcal{A} = \begin{bmatrix} \mathbf{A}_{11} & \mathbf{A}_{12} & \mathbf{B}_1^T \\ \mathbf{A}_{21} & \mathbf{A}_{22} & \mathbf{B}_2^T \\ \mathbf{B}_1 & \mathbf{B}_2 & \mathbf{0} \end{bmatrix} \text{ where } \mathbf{B}_1 \text{ is invertible.}$$

Verify that this matrix can be factored as $\mathcal{A} = \mathcal{L}\mathcal{G}\mathcal{L}^T$, where

$$\mathcal{L} = \begin{bmatrix} \mathbf{I} & \mathbf{0} & \mathbf{0} \\ (\mathbf{B}_1^{-1}\mathbf{B}_2)^T & \mathbf{I} & \mathbf{0} \\ \mathbf{0} & \mathbf{0} & \mathbf{I} \end{bmatrix}$$

$$\mathcal{G} = \begin{bmatrix} \mathbf{A}_{11} & \mathbf{C} & \mathbf{B}_1^T \\ \mathbf{C}^T & \mathbf{D} & \mathbf{0} \\ \mathbf{B}_1 & \mathbf{0} & \mathbf{0} \end{bmatrix}$$

Here $\mathbf{C} = -\mathbf{A}_{11}(\mathbf{B}_1^{-1}\mathbf{B}_2) + \mathbf{A}_{12}$,

$\mathbf{D} = \mathbf{A}_{22} - \mathbf{A}_{21}(\mathbf{B}_1^{-1}\mathbf{B}_2) - (\mathbf{B}_1^{-1}\mathbf{B}_2)^T\mathbf{C}$

35. (Continuation.) Verify that if this matrix is symmetric, then it can be factored as $\mathcal{A} = \mathcal{HGH}^T$, where

$$\mathcal{H} = \begin{bmatrix} \mathbf{B}_1^T & \mathbf{0} & \mathbf{0} \\ \mathbf{B}_2^T & \mathbf{I} & \mathbf{E} \\ \mathbf{0} & \mathbf{0} & \mathbf{I} \end{bmatrix}$$

$$\mathcal{G} = \begin{bmatrix} \mathbf{D}_1 & \mathbf{0} & \mathbf{I} \\ \mathbf{0} & \mathbf{D}_2 & \mathbf{0} \\ \mathbf{I} & \mathbf{0} & \mathbf{0} \end{bmatrix}$$

Here $\mathbf{D}_1 = \mathbf{B}_1^{-T}\mathbf{A}_{11}\mathbf{B}_1^{-1}$,
$\mathbf{D}_2 = \mathbf{A}_{22} - \mathbf{B}_2^T\mathbf{D}_1\mathbf{B}_2 - \mathbf{E}\mathbf{B}_2 - \mathbf{B}_2^T\mathbf{E}$
$\mathbf{E} = \mathbf{A}_{21}\mathbf{B}_1^{-1} - \mathbf{B}_2^T\mathbf{D}_1$

36. (Continuation.) Repeat the previous instructions, where

$$\mathcal{H} = \begin{bmatrix} \mathbf{B}_1^T & \mathbf{0} & \mathbf{L}_1 \\ \mathbf{B}_2^T & \mathbf{L}_2 & \mathbf{E} \\ \mathbf{0} & \mathbf{0} & \mathbf{I} \end{bmatrix}, \quad \mathcal{G} = \begin{bmatrix} \mathbf{D}_1 & \mathbf{0} & \mathbf{I} \\ \mathbf{0} & \mathbf{D}_2 & \mathbf{0} \\ \mathbf{I} & \mathbf{0} & \mathbf{0} \end{bmatrix}$$

Here we choose an $m \times m$ matrix $\mathbf{L}_1$ and $(n-m) \times (n-m)$ invertible matrix $\mathbf{L}_2$ such that
$\mathbf{D}_1 = \mathbf{B}_1^{-T}\mathbf{A}_{11}\mathbf{B}_1^{-1} - \mathbf{B}_1^{-T}\mathbf{L}_1 - \mathbf{L}_1^T\mathbf{B}_1^{-1}$
$\mathbf{D}_2 = \mathbf{L}_2^{-1}(\mathbf{A}_{22} - \mathbf{B}_2^T\mathbf{D}_1\mathbf{B}_2 - \mathbf{E}\mathbf{B}_2 - \mathbf{B}_2^T\mathbf{E}^T)\mathbf{L}_2^{-T}$
$\mathbf{E} = \mathbf{A}_{21}\mathbf{B}_1^{-1} - \mathbf{B}_2^T\mathbf{D}_1 - \mathbf{B}_2^T\mathbf{L}_1^T\mathbf{B}_1^{-1}$
$\mathbf{L}_2\mathbf{D}\mathbf{L}_2^T = \mathbf{A}_{22} - \mathbf{B}_2^T\mathbf{D}_1\mathbf{B}_2 - \mathbf{E}\mathbf{B}_2 - \mathbf{B}_2^T\mathbf{E}^T$
This is Schilders' factorization.

37. Find the determinants of the block matrices

$$\begin{bmatrix} \mathbf{A} & \mathbf{0} \\ \mathbf{0} & \mathbf{I} \end{bmatrix} \text{ and } \begin{bmatrix} \mathbf{A} & \mathbf{0} \\ \mathbf{0} & \mathbf{B} \end{bmatrix}$$

38. Let $\mathbf{A}$ be an $n \times n$ invertible matrix, and let $\mathbf{u}$ and $\mathbf{v}$ be two vectors in $\mathbb{R}^n$. Find the necessary and sufficient conditions on these vectors in order that the matrix $\mathcal{A} = \begin{bmatrix} \mathbf{A} & \mathbf{u} \\ \mathbf{v}^T & 0 \end{bmatrix}$ be invertible, and give a formula for the inverse when it exists.

39. Let $\mathcal{A}$ have the block-partitioned form

$$\begin{bmatrix} \mathbf{A} & \mathbf{B} \\ \mathbf{C} & \mathbf{I} \end{bmatrix}$$

a. Argue that if $\mathbf{A} - \mathbf{BC}$ is invertible, then so is $\mathcal{A}$.

b. Argue the stronger result that the dimension of the null space of $\mathcal{A}$ is no greater than the dimension of the null space of $\mathbf{A} - \mathbf{BC}$.

c. What can you say about the case when $\mathbf{I} - \mathbf{CA}^{-1}\mathbf{B}$ is invertible?

40. a. Let $\mathcal{A} = \begin{bmatrix} \mathbf{A}_{11} & \mathbf{A}_{12} \\ \mathbf{A}_{21} & \mathbf{A}_{22} \end{bmatrix}$, $\mathcal{B} = \begin{bmatrix} \mathbf{B}_{11} & \mathbf{B}_{12} \\ \mathbf{B}_{21} & \mathbf{B}_{22} \end{bmatrix}$

Verify that

$$\mathcal{AB} = \begin{bmatrix} \mathbf{A}_{11}\mathbf{B}_{11} + \mathbf{A}_{12}\mathbf{B}_{21} & \mathbf{A}_{11}\mathbf{B}_{12} + \mathbf{A}_{12}\mathbf{B}_{22} \\ \mathbf{A}_{21}\mathbf{B}_{11} + \mathbf{A}_{22}\mathbf{B}_{21} & \mathbf{A}_{21}\mathbf{B}_{12} + \mathbf{A}_{22}\mathbf{B}_{22} \end{bmatrix}$$

b. Let $\mathcal{A} = \left[\begin{array}{cc|cc} 1 & 2 & -3 & 2 \\ 4 & 1 & -1 & 3 \\ \hline 3 & 2 & 1 & -4 \end{array}\right]$

$$\mathcal{B} = \left[\begin{array}{c|c} -2 & 3 \\ 3 & 2 \\ \hline 4 & -1 \\ 1 & -4 \end{array}\right]$$

Compute $\mathcal{AB}$ using these submatrices.

c. Verify that

$$\mathcal{AB} = \begin{bmatrix} \mathbf{A}_{11} & \mathbf{A}_{11} \\ \mathbf{A}_{22} & \mathbf{A}_{22} \end{bmatrix} \odot \begin{bmatrix} \mathbf{B}_{11} & \mathbf{B}_{12} \\ \mathbf{B}_{21} & \mathbf{B}_{22} \end{bmatrix}$$
$$+ \begin{bmatrix} \mathbf{A}_{12} & \mathbf{A}_{12} \\ \mathbf{A}_{21} & \mathbf{A}_{21} \end{bmatrix} \odot \begin{bmatrix} \mathbf{B}_{21} & \mathbf{B}_{22} \\ \mathbf{B}_{11} & \mathbf{B}_{12} \end{bmatrix}$$

(Here $\odot$ is the *direct product* of multiplying corresponding submatrices.)

d. Using the approach given in Part **c**, compute $\mathcal{AB}$ with the numerical submatrices in Part **b**.

e. Show that

$$\mathcal{AB} = \begin{bmatrix} \mathbf{A}_{11} & \mathbf{A}_{12} \\ \mathbf{A}_{22} & \mathbf{A}_{21} \end{bmatrix} \odot \begin{bmatrix} \mathbf{B}_{11} & \mathbf{B}_{22} \\ \mathbf{B}_{21} & \mathbf{B}_{12} \end{bmatrix}$$
$$+ \begin{bmatrix} \mathbf{A}_{12} & \mathbf{A}_{11} \\ \mathbf{A}_{21} & \mathbf{A}_{22} \end{bmatrix} \odot \begin{bmatrix} \mathbf{B}_{21} & \mathbf{B}_{12} \\ \mathbf{B}_{11} & \mathbf{B}_{22} \end{bmatrix}$$

f. Using the approach given in Part **e**, compute $\mathcal{AB}$ with the numerical submatrices in Part **b**. [*Note:* Parts **c** and **e** are examples of parallel matrix–matrix multiplication algorithms that would be particularly useful on parallel computer systems. Here the $\odot$ operations could be done in parallel on a 2D grid of four processors.]

41. If **L** is a lower triangular matrix with positive elements on its diagonal, does it follow that $\mathbf{LL}^T$ is positive definite?

42. Explain why or find a counterexample for the assertion that the nonzero eigenvalues of a skew-symmetric matrix are pure imaginary.

43. Find the inverse of this block matrix
$$\begin{bmatrix} \mathbf{A} & \mathbf{b} \\ \mathbf{b}^T & c \end{bmatrix} \text{ for } m \times m \text{ matrix } \mathbf{A}, m\text{-vector } \mathbf{b},$$
and scalar c.

44. Let $\mathbf{P} = \begin{bmatrix} \mathbf{E} & \mathbf{F} \\ \mathbf{0} & \mathbf{G} \end{bmatrix}$. Find $\mathbf{P}^{-1}$ and write it as a factorization of two block matrices.

45. Consider $\begin{bmatrix} \mathbf{0} & \mathbf{I} \\ -\mathbf{I} & \mathbf{0} \end{bmatrix}$ and $\begin{bmatrix} \mathbf{A} & \mathbf{B} \\ \mathbf{C} & \mathbf{D} \end{bmatrix}$. In order that these matrices commute, what conditions must there be on $\mathbf{A}, \mathbf{B}, \mathbf{C}, \mathbf{D}$?

46. Let $\mathbf{S}_{\mathbf{A}} = (\mathbf{D} - \mathbf{CA}^{-1}\mathbf{B})$ and $\mathbf{S}_{\mathbf{D}} = (\mathbf{A} - \mathbf{BD}^{-1}\mathbf{C})$. Verify each of these results

a. $\begin{bmatrix} \mathbf{A} & \mathbf{B} \\ \mathbf{C} & \mathbf{D} \end{bmatrix}^{-1}$
$$= \begin{bmatrix} \mathbf{A}^{-1} + \mathbf{A}^{-1}\mathbf{BS}_{\mathbf{A}}^{-1}\mathbf{CA}^{-1} & -\mathbf{A}^{-1}\mathbf{BS}_{\mathbf{A}}^{-1} \\ -\mathbf{S}_{\mathbf{A}}^{-1}\mathbf{CA}^{-1} & \mathbf{S}_{\mathbf{A}}^{-1} \end{bmatrix}$$

b. $\begin{bmatrix} \mathbf{A} & \mathbf{B} \\ \mathbf{C} & \mathbf{D} \end{bmatrix}^{-1}$
$$= \begin{bmatrix} \mathbf{S}_{\mathbf{D}}^{-1} & -\mathbf{S}_{\mathbf{D}}^{-1}\mathbf{BD}^{-1} \\ -\mathbf{D}^{-1}\mathbf{CS}_{\mathbf{D}}^{-1} & \mathbf{D}^{-1} + \mathbf{D}^{-1}\mathbf{CS}_{\mathbf{D}}^{-1}\mathbf{BD}^{-1} \end{bmatrix}$$

47. (Continuation.) In the 2×2 scalar case, verify that both inverses reduce to
$$\begin{bmatrix} a & b \\ c & d \end{bmatrix}^{-1} = \Delta \begin{bmatrix} d & -b \\ -c & a \end{bmatrix} \text{ where } \Delta = ad - bc.$$

48. Carry out the details in the second procedure to obtain the inverse of a 2×2 block matrix in factored form.

49. Show how the Woodbury matrix identity is obtained.

50. Derive the Sherman-Morrison formula from the Woodbury-Morrison formula.

51. Verify that the inverse of **B** is the identity $\mathbf{B}^{-1} = \mathbf{A}^{-1} - \mathbf{B}^{-1}(\mathbf{B} - \mathbf{A})\mathbf{A}^{-1}$.

52. Verify that for an $n \times n$ upper (lower) block triangular matrix $\mathcal{A}$ with diagonal blocks $\mathbf{A}_{ii}$, we have $\text{Det}(\mathcal{A}) = \prod_{i=1}^{n} \text{Det}(\mathbf{A}_{ii})$ using
$$\begin{bmatrix} 2 & 3 & 4 & 7 & 8 \\ -1 & 5 & 3 & 2 & 1 \\ 0 & 0 & 2 & 1 & 5 \\ 0 & 0 & 3 & 1 & 4 \\ 0 & 0 & 5 & 2 & 6 \end{bmatrix}$$
In other words, verify that
$$\text{Det}\begin{bmatrix} \mathbf{A} & \mathbf{B} \\ \mathbf{0} & \mathbf{D} \end{bmatrix} = \text{Det}(\mathbf{A})\,\text{Det}(\mathbf{D}).$$
Find an example to show $\text{Det}\begin{bmatrix} \mathbf{A} & \mathbf{B} \\ \mathbf{C} & \mathbf{D} \end{bmatrix} \neq$
$\text{Det}(\mathbf{A})\,\text{Det}(\mathbf{D}) - \text{Det}(\mathbf{C})\,\text{Det}(\mathbf{B}).$

53. Find two forms of the solution of the 2×2 block linear system $\begin{bmatrix} \mathbf{A} & \mathbf{B} \\ \mathbf{C} & \mathbf{D} \end{bmatrix}\begin{bmatrix} \mathbf{x} \\ \mathbf{y} \end{bmatrix} = \begin{bmatrix} \mathbf{b} \\ \mathbf{c} \end{bmatrix}$

54. Let $\mathbf{A} = \begin{bmatrix} \mathbf{B} & \mathbf{C} \\ \mathbf{0} & \mathbf{I} \end{bmatrix}$ have the block form shown in which the submatrices are $n \times n$. Prove that if $(\mathbf{B} - \mathbf{I})$ is invertible, then

a. $\mathbf{A}^k = \begin{bmatrix} \mathbf{B}^k & (\mathbf{B}^k - \mathbf{I})(\mathbf{B} - \mathbf{I})^{-1}\mathbf{C} \\ \mathbf{0} & \mathbf{I} \end{bmatrix}$

b. Repeat for $\mathbf{A}^T$.

COMPUTER EXERCISES 8.2

1. Program and test the algorithm for the Cholesky factorization.

2. Let **A** be a square matrix whose eigenvalues are to be computed. Use the QR commands in MATLAB, Maple, or Mathematica to get $\mathbf{A} = \mathbf{QR}$. Define $\mathbf{B} = \mathbf{RQ}$, and then replace **A** by **B**. Repeat this process several times, getting new versions of **A**. For testing purposes, use $\mathbf{A} = \begin{bmatrix} 3 & 12 & 8 \\ -10 & 37 & 26 \\ 12 & -36 & -25 \end{bmatrix}$ whose eigenvalues are $3, 5, 7$. This algorithm is the QR-algorithm for computing eigenvalues.

3. If you have access to MATLAB, Maple, or Mathematica, verify the results of the examples in this section.

4. For a given square matrix, compute the factorizations $\mathbf{A} = \mathbf{PDP}^{-1}$ and $\mathbf{A} = \mathbf{PDQ}$. (The matrix **P** is not assumed to be the same in each factorization.) What can you conclude?

5. Use mathematical software to compute the QR-factorization of $\begin{bmatrix} 1 & 3 & 2 \\ 5 & -2 & 1 \\ 0 & 3 & 4 \\ 3 & 0 & 1 \end{bmatrix}$

6. Find the QR-factorization of a general 2×2 symmetric positive definite matrix.

8.3 ITERATIVE METHODS FOR LINEAR EQUATIONS

Perhaps the greatest paradox of them all is that there are paradoxes in mathematics.

—E. KASNER AND J. NEWTON

Although this may seem a paradox, all exact science is dominated by the concept of approximation.

—BERTRAND SHAW (1872–1970)

Introduction

In this chapter, iterative procedures for solving systems of linear equations are introduced. The term *iterative* refers to a process that is repeated over and over again, using as the input in one step the output from the preceding step. Examples of this strategy have already been seen in dynamical systems: from a starting vector $\mathbf{x}^{(0)}$, one can generate a sequence of vectors using the formula $\mathbf{x}^{(k)} = \mathbf{Ax}^{(k-1)}$, for $k = 1, 2, 3$, and so on. Typically, the number of steps needed in an iterative algorithm will be far less than the number of variables.

Iterative methods are frequently used to solve large sparse systems of linear equations where the number of equations and unknowns is enormous. (A large sparse matrix contains primarily zero values and relatively few nonzero entries.) Problems can be encountered where there are hundreds of millions of equations and hundreds of millions of unknowns. Such problems often originate in the numerical solution of partial differential equations. One advantage of iterative methods, in general, is that if the coefficient matrix has relatively few nonzero entries, the algorithms use only these nonzero entries with no fill-in of nonzero values for the zero values. ("*Fill-in*" means that an entry in the coefficient matrix that is zero is changed to a nonzero value.) Furthermore, iterative methods produce a sequence of approximate solutions that improve as the iteration continues. The practitioner can stop the computer program when sufficiently accurate solutions have been obtained. Of course, the familiar row-reduction algorithm does *not* have this wonderful property!

Richardson Iterative Method

Here is an attempt to design an iterative scheme for solving the equation

$$\mathbf{Ax} = \mathbf{b}$$

The procedure takes the form

$$\mathbf{x}^{(k)} = \mathbf{Gx}^{(k-1)} + \mathbf{c} \qquad (k = 1, 2, 3, \ldots)$$

Naturally, the choice of $\mathbf{G}$ and $\mathbf{c}$ must depend on $\mathbf{A}$ and $\mathbf{b}$. If this iterative scheme succeeds, then the sequence of iterates $\mathbf{x}^{(k)}$ will converge to a vector $\mathbf{x}^*$, say. By passing to the limit in the previous equation, one arrives at

$$\mathbf{x}^* = \mathbf{Gx}^* + \mathbf{c} \qquad \text{or} \qquad (\mathbf{I} - \mathbf{G})\mathbf{x}^* = \mathbf{c}$$

Because the objective is to solve $\mathbf{Ax} = \mathbf{b}$, we can try $\mathbf{I} - \mathbf{G} = \mathbf{A}$ and $\mathbf{c} = \mathbf{b}$. The resulting iteration formula is

$$\mathbf{x}^{(k)} = (\mathbf{I} - \mathbf{A})\mathbf{x}^{(k-1)} + \mathbf{b}$$

This is called the **Richardson iterative method**.

   To see that this method is not far-fetched, an example will be given where it works very well. (There are many examples where this process does *not* work at all!)

EXAMPLE 1 Solve the following system $\mathbf{Ax} = \mathbf{b}$. Use the Richardson iterative method just described.

$$\begin{bmatrix} 1.0 & -0.3 & -0.2 \\ -0.3 & 1.0 & -0.1 \\ -0.2 & -0.1 & 1.0 \end{bmatrix} \begin{bmatrix} x_1 \\ x_2 \\ x_3 \end{bmatrix} = \begin{bmatrix} 7 \\ 5 \\ 3 \end{bmatrix}$$

SOLUTION The Richardson iteration formula is $\mathbf{x}^{(k)} = (\mathbf{I} - \mathbf{A})\mathbf{x}^{(k-1)} + \mathbf{b}$. A computer program can be written to initialize the vector $\mathbf{x} = (0, 0, 0)$ and then repeat the process 25 times as in the following pseudocode, for example.

$$x_1 \leftarrow 0, \quad x_2 \leftarrow 0, \quad x_3 \leftarrow 0$$
$$\textbf{for } k = 1 \textbf{ to } 25$$
$$\begin{bmatrix} x_1 \\ x_2 \\ x_3 \end{bmatrix} \leftarrow \begin{bmatrix} 0 & 0.3 & 0.2 \\ 0.3 & 0 & 0.1 \\ 0.2 & 0.1 & 0 \end{bmatrix} \begin{bmatrix} x_1 \\ x_2 \\ x_3 \end{bmatrix} + \begin{bmatrix} 7 \\ 5 \\ 3 \end{bmatrix}$$
$$\textbf{end for}$$

The successive values of $\mathbf{x}$ can be printed or displayed as they are computed. When that is done, there is no need to save the intermediate values. A computer program can do all of this for us. The first few iterations produce $\mathbf{x}^{(0)} = (0, 0, 0)$, $\mathbf{x}^{(1)} = (7, 5, 3)$, and $\mathbf{x}^{(2)} = (9.1, 3.4, 4.9)$. A MATLAB program to carry out this process is given at the end of this chapter. Using this program, we find that the approximate solutions get better and better, so that by the 15th step we have, in rounded form, $\mathbf{x}^{(15)} = (10.8726, 8.8679, 6.0613) = \mathbf{x}^{(14)}$. At this stage there are already five significant figures in the answer. ∎

Jacobi Iterative Method

In Example 1, if the diagonal elements in the matrix $\mathbf{A}$ were not all equal to 1, then each equation could be divided by its nonzero diagonal entry a_{ii} before applying the Richardson iterative method. In other words, the linear system could be **scaled** to become

$$\mathbf{D}^{-1}\mathbf{A}\mathbf{x} = \mathbf{D}^{-1}\mathbf{b}$$

where $\mathbf{D} = \text{Diag}(\mathbf{A})$. Here $\text{Diag}(\mathbf{A})$ means the diagonal matrix whose diagonal equals the diagonal of $\mathbf{A}$. Then let

$$\mathbf{B} = \mathbf{I} - \mathbf{D}^{-1}\mathbf{A} \quad \text{and} \quad \mathbf{c} = \mathbf{D}^{-1}\mathbf{b}$$

The resulting procedure is called the **Jacobi iterative method**.

$$\mathbf{x}^{(k)} = \mathbf{B}\mathbf{x}^{(k-1)} + \mathbf{c} \qquad (k = 1, 2, \ldots)$$

We know that if the iterates $\mathbf{x}^{(k)}$ converge (say to $\mathbf{x}^*$), then $\mathbf{x}^*$ will be a solution of the system. What remains is to prove the convergence of the process, under suitable hypotheses on the matrix $\mathbf{A}$. This will be accomplished in Theorem 4 and its corollaries.

EXAMPLE 2 In Example 1, change all the diagonal entries from 1 to 2:

$$\begin{bmatrix} 2.0 & -0.3 & -0.2 \\ -0.3 & 2.0 & -0.1 \\ -0.2 & -0.1 & 2.0 \end{bmatrix} \begin{bmatrix} x_1 \\ x_2 \\ x_3 \end{bmatrix} = \begin{bmatrix} 7 \\ 5 \\ 3 \end{bmatrix}$$

Now apply the Jacobi iterative method. (The Richardson iterative method does not work well for this example!)

SOLUTION The Jacobi iteration formula is $\mathbf{x}^{(k)} = \mathbf{B}\mathbf{x}^{(k-1)} + \mathbf{D}^{-1}\mathbf{b}$, where $\mathbf{B} = \mathbf{I} - \mathbf{D}^{-1}\mathbf{A}$ and $\mathbf{D} = \text{Diag}(\mathbf{A})$. The calculation can be described as follows:

$$x_1 \leftarrow 0, \quad x_2 \leftarrow 0, \quad x_3 \leftarrow 0$$

for $k = 1$ **to** 25

$$\begin{bmatrix} x_1 \\ x_2 \\ x_3 \end{bmatrix} \leftarrow \begin{bmatrix} 0 & 0.15 & 0.1 \\ 0.15 & 0 & 0.05 \\ 0.1 & 0.05 & 0 \end{bmatrix} \begin{bmatrix} x_1 \\ x_2 \\ x_3 \end{bmatrix} + \begin{bmatrix} 3.5 \\ 2.5 \\ 1.5 \end{bmatrix}$$

end for

This can be written in terms of the individual variables:

$$x_1 \leftarrow 0, \quad x_2 \leftarrow 0, \quad x_3 \leftarrow 0$$
$$\textbf{for } k = 1 \textbf{ to } 25$$
$$y_1 \leftarrow 0.15x_2 + 0.1x_3 + 3.5$$
$$y_2 \leftarrow 0.15x_1 + 0.05x_3 + 2.5$$
$$y_3 \leftarrow 0.1x_1 + 0.05x_2 + 1.5$$
$$x_1 \leftarrow y_1, \quad x_2 \leftarrow y_2, \quad x_3 \leftarrow y_3$$
$$\textbf{end for}$$

In either case, the approximate solution $(4.1930, 3.2330, 2.0810)$ is obtained after nine iterations. A sample MATLAB program is given at the end of this section. ∎

Gauss–Seidel Method

A powerful and easily programmed iterative procedure is the **Gauss–Seidel**[7] **Method**. Let the system be, as usual,

$$\mathbf{Ax = b} \quad \text{or} \quad \sum_{j=1}^{n} a_{ij}x_j = b_i \quad (1 \leq i \leq n)$$

Assume that all the diagonal elements in the matrix **A** are nonzero. One can write the system in the form

$$a_{ii}x_i = -\sum_{\substack{j=1 \\ j \neq i}}^{n} a_{ij}x_j + b_i \quad (1 \leq i \leq n)$$

Once all the data have been entered, the Gauss–Seidel iteration uses this pseudocode:

> **for** $k = 1$ **to** k_max
> > **for** $i = 1$ **to** n
> >
> > $$x_i \leftarrow (b_i/a_{ii}) - \sum_{\substack{j=1 \\ j \neq i}}^{n} (a_{ij}/a_{ii})x_j$$
> >
> > **end for**
>
> **end for**

The unknowns are *updated* one-by-one, using always the most recent values for each variable. The replacement step in the pseudocode is called the *updating formula*. The code is very easy to write, but the matrix form of the procedure is more complicated, requiring **A** to be split into the sum of a diagonal matrix, a lower triangular matrix, and an upper triangular matrix. Specifically, **D** is the diagonal matrix containing the diagonal elements of **A**, **L** is the lower triangular part of $-\mathbf{D}^{-1}\mathbf{A}$, and **U** is the upper triangular part of $-\mathbf{D}^{-1}\mathbf{A}$. Hence, we have $\mathbf{D}^{-1}\mathbf{A} = \mathbf{I} - \mathbf{L} - \mathbf{U}$. Then the Gauss–Seidel method can be written as

$$\mathbf{x}^{(k)} = \mathbf{Lx}^{(k)} + \mathbf{Ux}^{(k-1)} + \mathbf{c}$$

where $\mathbf{c} = \mathbf{D}^{-1}\mathbf{b}$.

---

[7] Philip Ludwig von Seidel (1821–1896) submitted two completely different doctoral theses only 6 months apart—the first on astronomy and the second on mathematical analysis.

EXAMPLE 3 Repeat the task in Example 2 using the Gauss–Seidel iterative method.

SOLUTION Because the new values are to be used as soon as possible, we omit the **y**-variables. (We do not want to postpone updating the **x**-variables until the end of each loop, as is done in the Jacobi method.) The Gauss–Seidel method can be described in terms of the individual variables:

$$x_1 \leftarrow 0; \quad x_2 \leftarrow 0; \quad x_3 \leftarrow 0$$
for $k = 1$ **to** 25
$$x_1 \leftarrow 0.15x_2 + 0.1x_3 + 3.5$$
$$x_2 \leftarrow 0.15x_1 + 0.05x_3 + 2.5$$
$$x_3 \leftarrow 0.1x_1 + 0.05x_2 + 1.5$$
end for

After only three iterations, an approximate solution (4.1930, 3.2330, 2.0810) emerges, and it is correct to five significant figures. To obtain full precision in the MATLAB program, it took 13 iterations. In this example, the Gauss–Seidel method required approximately one-half as many iterations as the Jacobi method to arrive at the same degree of precision. This relationship between the Jacobi and Gauss–Seidel methods turns out to be true for many examples. ∎

Successive Overrelaxation (SOR) Method

In some cases, the Gauss–Seidel method can be accelerated by the introduction of a *"relaxation factor"* ω as in this formula:

$$\mathbf{x}^{(k)} = \omega\left\{\mathbf{L}\mathbf{x}^{(k)} + \mathbf{U}\mathbf{x}^{(k-1)} + \mathbf{c}\right\} + (1 - \omega)\mathbf{x}^{(k)}$$

Notice that when $\omega = 1$ we recover the Gauss–Seidel method. When $1 < \omega < 2$, this procedure is called the **successive overrelaxation (SOR) method**. The optimal ω is often near 2. In his dissertation, Young [1950] first presented the formula $(\lambda + \omega - 1)^2 = \lambda\omega^2\mu^2$, sometimes called **Young's equation**.[8] It connects the eigenvalues μ of the Gauss-Seidel iteration

---

[8] David M. Young, Jr., (1923–2008) established the mathematical foundation for the SOR method in his Ph.D. thesis at Harvard University. Ever since this groundbreaking research, iterative methods have been used on a wide range of scientific and engineering applications for solving large sparse systems of linear equations with many new iterative methods (called *preconditioners*) having been developed.

matrix with the eigenvalues λ of the SOR iteration matrix (assuming certain conditions on the matrix $\mathbf{A}$). This equation can be used to find the *best* relaxation factor $\omega_b = 2/(1 + \sqrt{1 - \rho^2})$, where ρ is the largest eigenvalue of the Jacobi method $\mathbf{B}$ in absolute value. The SOR method is one of the fundamental iterative methods in linear algebra and has been widely used in applications.

Conjugate Gradient Method

In theory, the conjugate gradient iterative method solves a system of n linear equations in at most n steps, if the matrix $\mathbf{A}$ is symmetric and positive definite. Moreover, the nth iterative vector $\mathbf{x}^{(n)}$ is the unique minimizer of the quadratic function $q(\mathbf{x}) = \frac{1}{2}\mathbf{x}^T\mathbf{A}\mathbf{x} - \mathbf{x}^T\mathbf{b}$. When the conjugate gradient method was introduced by Hestenes and Stiefel [1952], the initial interest in it waned once it was discovered that this finite-termination property was not obtained in practice. But two decades later, there was renewed interest in this method when it was viewed as an iterative process by Reid [1971] and others. In practice, the solution of a system of linear equations can often be found with satisfactory precision in a number of steps considerably less than the order of the system. For many very large and sparse linear systems, preconditioned conjugate gradient methods are now the iterative methods of choice! Here is a pseudocode for the **conjugate gradient algorithm**:

$$k \leftarrow 0; \ \mathbf{x} \leftarrow 0; \ \mathbf{r} \leftarrow \mathbf{b} - \mathbf{Ax}; \ \gamma \leftarrow \langle \mathbf{r}, \mathbf{r} \rangle$$
$$\textbf{while} \quad \sqrt{\gamma} > \varepsilon\sqrt{\langle \mathbf{b}, \mathbf{b} \rangle} \ \textbf{and} \ \left(k < k_{\max}\right)$$
$$\qquad k \leftarrow k + 1$$
$$\qquad \textbf{if} \ k = 1 \ \textbf{then}$$
$$\qquad\qquad \mathbf{p} \leftarrow \mathbf{r}$$
$$\qquad \textbf{else}$$
$$\qquad\qquad \beta \leftarrow \gamma \,/\, \gamma_{\text{old}}$$
$$\qquad\qquad \mathbf{p} \leftarrow \mathbf{r} + \beta\mathbf{p}$$
$$\qquad \textbf{end if}$$
$$\qquad \mathbf{w} \leftarrow \mathbf{Ap}$$
$$\qquad \alpha \leftarrow \gamma \,/\, \langle \mathbf{p}, \mathbf{w} \rangle$$
$$\qquad \mathbf{x} \leftarrow \mathbf{x} + \alpha\mathbf{p}$$
$$\qquad \mathbf{r} \leftarrow \mathbf{r} - \alpha\mathbf{w}$$
$$\qquad \gamma_{\text{old}} \leftarrow \gamma$$
$$\qquad \gamma \leftarrow \langle \mathbf{r}, \mathbf{r} \rangle$$
$$\textbf{end while}$$

Young was one of the pioneers in modern numerical analysis and scientific computing. His car license plate read *Dr. SOR*.

Here ε is a parameter used in the convergence criterion (such as $\varepsilon = 10^{-5}$) and k_{max} is the maximum number of iterations. Usually, the number of iterations needed is much less than the size of the linear system. We save the previous value of γ in the variable γ_{old}. If a good guess for the solution vector $\mathbf{x}$ is known, then it should be used as an initial vector instead of zero. The variable ε is the desired convergence tolerance. The algorithm produces not only a sequence of vectors $\mathbf{x}^{(i)}$ that converges to the solution, but also an orthogonal sequence of residual vectors $\mathbf{r}^{(i)} = \mathbf{b} - \mathbf{Ax}^{(i)}$ and an $\mathbf{A}$-orthogonal sequence of search direction vectors $\mathbf{p}^{(i)}$; namely, $\langle \mathbf{r}^{(i)}, \mathbf{r}^{(j)} \rangle = 0$ if $i \neq j$ and $\langle \mathbf{p}^{(i)}, \mathbf{Ap}^{(j)} \rangle = 0$ if $i \neq j$.

The main computational features of the conjugate gradient algorithm are complicated to derive, but the final conclusion is that in each step only *one* matrix–vector multiplication is required and only a few dot-products are computed. These are extremely desirable attributes in solving large and sparse linear systems. Also, unlike Gaussian elimination, there is no fill-in, so only the nonzero entries in $\mathbf{A}$ need to be stored in the computer memory. For some partial differential equation problems, the equations in the linear system can be represented by stencils that describe the nonzero structure within the coefficient matrix. Sometimes these "*stencils*" are used in a computer program rather than storing the nonzero entries in the coefficient matrix.

> **EXAMPLE 4** Repeat the task in Example 2 using the conjugate gradient method.

SOLUTION Programming the pseudocode using MATLAB, we obtain the iterates $\mathbf{x}^{(1)} = (4.3484, 3.1013, 1.8638)$, $\mathbf{x}^{(2)} = (4.1903, 3.2419, 2.0723)$, $\mathbf{x}^{(3)} = (4.1930, 3.2330, 2.0810)$. In only three iterations, we have the answer accurate to full machine precision, which illustrates the finite termination property. The matrix $\mathbf{A}$ is symmetric positive definite and the eigenvalues of $\mathbf{A}$ are 1.5887, 2.0911, 2.3202. This simple example may be a bit misleading because one cannot expect such rapid convergence in realistic applications. (The rate of convergence depends on various properties of the linear system.) In fact, the previous example is too small to illustrate the power of advanced iterative methods on very large and sparse systems. ∎

Diagonally Dominant Matrices

Here we focus attention on some theorems that will identify cases in which the iterative methods will work well. (Look ahead to Theorem 4 and its corollaries.) The following concept will play a role.

DEFINITION

A square matrix is **diagonally dominant** if each diagonal element in absolute value is greater than the sum of the absolute values of the off-diagonal elements in that row. If the matrix in question is **A** and has generic elements a_{ij}, the diagonal dominance is expressed by

$$\sum_{\substack{j=1 \\ j \neq i}}^{n} |a_{ij}| < |a_{ii}| \qquad (1 \leq i \leq n)$$

Such matrices are automatically invertible, as we now prove.

THEOREM 1

Every diagonally dominant matrix is invertible.

PROOF Let **A** be a diagonally dominant $n \times n$ matrix. We shall prove that $\mathbf{Ax} \neq \mathbf{0}$ if $\mathbf{x} \neq \mathbf{0}$. In other words, the null space of **A** contains only the zero vector. Thus, we let **x** be any nonzero vector, and set $\mathbf{y} = \mathbf{Ax}$. It is to be proved that $\mathbf{y} \neq \mathbf{0}$. Select i so that $|x_i|$ is the largest component of **x**, in absolute value: $|x_j| \leq |x_i| \neq 0$ for all j. Then we have

$$y_i = a_{ii}x_i + \sum_{\substack{j=1 \\ j \neq i}}^{n} a_{ij}x_j \qquad \text{or} \qquad a_{ii}x_i = y_i - \sum_{\substack{j=1 \\ j \neq i}}^{n} a_{ij}x_j$$

Applying absolute values and our knowledge of $|x_i|$, we have

$$|a_{ii}||x_i| \leq \left| y_i - \sum_{\substack{j=1 \\ j \neq i}}^{n} a_{ij}x_j \right| \leq |y_i| + \sum_{\substack{j=1 \\ j \neq i}}^{n} |a_{ij}||x_i| = |y_i| + |x_i| \sum_{\substack{j=1 \\ j \neq i}}^{n} |a_{ij}|$$

Finally, we obtain

$$|x_i| \left(|a_{ii}| - \sum_{\substack{j=1 \\ j \neq i}}^{n} |a_{ij}| \right) \leq |y_i|$$

Using the diagonal dominance property, we conclude that $y_i \neq 0$ and $\mathbf{y} \neq \mathbf{0}$.   ∎

Gerschgorin's Theorem

A useful consequence of the preceding theorem is Gerschgorin's Theorem[9] concerning the location of the eigenvalues in the complex plane for a given square matrix with real or complex elements. It says, roughly, that each eigenvalue may not be far from a diagonal element when the off-diagonal entries are small in norm!

> **THEOREM 2   Gerschgorin's Theorem**
>
> *If λ is an eigenvalue of an $n \times n$ matrix $\mathbf{A} = (a_{ij})$, then, for some index i, we have*
>
> $$|\lambda - a_{ii}| \leq \sum_{\substack{j=1 \\ j \neq i}}^{n} |a_{ij}|$$

PROOF The matrix $\mathbf{A} - \lambda \mathbf{I}$ is noninvertible because λ is an eigenvalue of $\mathbf{A}$. By Theorem 1, $\mathbf{A} - \lambda \mathbf{I}$ cannot be diagonally dominant. Thus, for some index i, we must have the inequality in the theorem.   ∎

The **Gerschgorin row discs** for an $n \times n$ matrix $\mathbf{A} = (a_{ij})$ are given by

$$\mathcal{G}_i(\mathbf{A}) = \left\{ z : z \in \mathbb{C} \quad \text{and} \quad |z - a_{ii}| \leq \sum_{\substack{j=1 \\ j \neq i}}^{n} |a_{ij}| \equiv r_i \right\} \qquad (1 \leq i \leq n)$$

For each row i, a Gerschgorin row disc is a closed disc in the complex plane with center at a_{ii} (the diagonal element), and its radius r_i is the sum of the absolute values of the off-diagonal entries in the ith row. We let $\mathcal{G}_i(\mathbf{A})[a_{ii}; r_i]$ denote the ith Gerschgorin row disc with center a_{ii} and radius r_i.

Gerschgorin's Theorem is useful in finding bounds on the spectrum, $\sigma(\mathbf{A})$, of eigenvalues for a given matrix $\mathbf{A}$. We want to **localize** the eigenvalues of $\mathbf{A}$. In other words, we want to quickly find regions in the complex plane containing the eigenvalues, and to determine approximate values for them. Remember that the eigenvalues of a matrix may turn out to be complex numbers, even if the matrix is real.

---

[9] Semyon Aranovich Gerschgorin (1901–1933) worked in the Leningrad Mechanical Engineering Institute. In 1931, he published a paper containing his now famous Circle Theorem. It was his only paper not in Russian, but in German. Unfortunately, he died at a rather young age before he could make many other major contributions. Corresponding to transliterations of the Yiddish, there are various spellings of the sir name of this mathematician from Belarus.

Even when a matrix is *not* diagonally dominant, Gerschgorin's Theorem can be useful in determining whether a matrix is invertible—when zero is *not* an eigenvalue. Also, if the region containing the eigenvalues lies entirely to the left or right of the imaginary axis, the theorem provides information about the sign of the real parts of all the eigenvalues. Gerschgorin's Theorem has many uses in applications such as in control theory (stability/instability) and scientific computing (stability of difference schemes). When solving a linear system of equations with a large condition number, Gerschgorin's Theorem can be useful in finding a preconditioning matrix.

There are several consequences of Gerschgorin's Theorem.

COROLLARY 1

Every eigenvalue of $\mathbf{A}$ lies within at least one of the Gerschgorin row discs of $\mathbf{A}$.

The eigenvalue of a matrix $\mathbf{A}$ and its transpose $\mathbf{A}^T$ are the same because they have the same characteristic equation $\text{Det}(\mathbf{A} - \lambda\mathbf{I}) = \text{Det}(\mathbf{A}^T - \lambda\mathbf{I})$. (Recall Theorem 3, from Section 4.2.) Consequently, we can apply Gerschgorin's Theorem to both $\mathbf{A}$ and $\mathbf{A}^T$. So we can obtain the Gerschgorin *column* discs as well as *row* discs. By doing so, we obtain pairs of discs with the same centers, but one of them may have a smaller radii!

For the matrix $\mathbf{A}$, the **Gerschgorin column discs** are the closed discs

$$\mathcal{G}_i(\mathbf{A}^T) = \left\{ z : z \in \mathbb{C} \quad \text{and} \quad |z - a_{ii}| \le \sum_{\substack{i=1 \\ i \ne j}}^{n} |a_{ij}| \equiv c_i \right\} \qquad (1 \le i \le n)$$

For each column i, a Gerschgorin column disc is a closed disc in the complex plane centered at a_{ii} with radii c_i, which is the sum of the absolute values of the off-diagonal elements in *column i*. We let $\mathcal{G}_i(\mathbf{A}^T)[a_{ii}; c_i]$ denote the ith Gerschgorin column disc with center a_{ii} and radius c_i.

COROLLARY 2

The eigenvalues of $\mathbf{A}$ lie within the Gerschgorin column discs of $\mathbf{A}$.

The special case of diagonal matrices is of interest.

COROLLARY 3

The Gerschgorin discs of $\mathbf{A}$ coincide with the eigenvalue spectrum if and only if $\mathbf{A}$ is a diagonal matrix.

Gerschgorin's Theorem asserts that all of the eigenvalues of an $n \times n$ matrix **A** are contained within the union of the n row discs in the complex plane:

$$\bigcup_{i=1}^{n} \mathcal{G}_i(\mathbf{A}) \tag{8.1}$$

In other words, the eigenvalues are trapped within the union of the n row discs. Also, all of the eigenvalues are contained within the union of the n column discs:

$$\bigcup_{i=1}^{n} \mathcal{G}_i(\mathbf{A}^T) \tag{8.2}$$

Since the eigenvalues of **A** are contained within both the union of the row discs (1) and the union of the column discs (2), they must be in the intersection of these two regions

$$\left(\bigcup_{i=1}^{n} \mathcal{G}_i(\mathbf{A}) \right) \bigcap \left(\bigcup_{i=1}^{n} \mathcal{G}_i(\mathbf{A}^T) \right) \tag{8.3}$$

Thus, we have obtained another region containing all of the eigenvalues of the matrix **A**.

By Theorem 3, if a Gerschgorin disc is disjoint from the other Gerschgorin discs, then it contains precisely one of the eigenvalues of **A**. In addition, if the union of k Gerschgorin discs does *not* touch any of the other $n - k$ discs, then there are exactly k eigenvalues (counting multiplicities) in the union of these k discs.

COROLLARY 4

*For the matrix **A**, if the union of k discs is disjoint from the union of the other $n - k$ discs, then the former union contains exactly k eigenvalues of **A** and the latter union contains exactly $n - k$ eigenvalues of **A**.*

EXAMPLE 5 Use Gerschgorin's Theorem to localize the eigenvalues of

$$A = \begin{bmatrix} 2 & 5 \\ -3 & 4 \end{bmatrix}$$

SOLUTION The first Gerschgorin row disc $\mathcal{G}_1(\mathbf{A})[2; 5]$ has center 2 and radius 5. The second row disc $\mathcal{G}_2(\mathbf{A})[4; 3]$ has center 4 and radius 3. The first Gerschgorin column disc $\mathcal{G}_1(\mathbf{A}^T)[2; 3]$ has center 2 and radius 3.

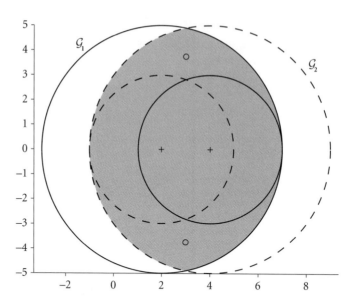

FIGURE 8.4 Example 5 Gerschgorin discs.

The second column disc $\mathcal{G}_2(\mathbf{A}^T)[4; 5]$ has center 4 and radius 5. Figure 8.4 shows the row discs with solid circles, and the column discs with dashed circles. Using (1), we take the union of the two row discs $\mathcal{G}_1(\mathbf{A}) \cup \mathcal{G}_2(\mathbf{A})$. Since the second row disc is contained within the first row disc, this union is the larger one $\mathcal{G}_1(\mathbf{A})$. Using (2), we take the union of these two column discs $\mathcal{G}_1(\mathbf{A}^T) \cup \mathcal{G}_2(\mathbf{A}^T)$. Since the first column disc is contained within the second column disc, the union of these two discs is the larger one $\mathcal{G}_2(\mathbf{A}^T)$. Applying (3), we find $\mathcal{G}_1(\mathbf{A}) \cap \mathcal{G}_2(\mathbf{A}^T)$, which is the shaded oval region as shown in Figure 8.4. We can check these results using the characteristic equation $\mathrm{Det}\begin{bmatrix} 2-\lambda & 5 \\ -3 & 4-\lambda \end{bmatrix} = \lambda^2 - 6\lambda + 23 = 0$. Hence, the eigenvalues are $\lambda = 3 \pm i\sqrt{14} \approx 3 \pm 3.7417i$. Clearly, the shaded oval-shaped region contains this pair of the complex conjugate eigenvalues. ∎

EXAMPLE 6 Use Gerschgorin's Theorem to localize the eigenvalues of

$$\begin{bmatrix} 2 & 1 & -1 \\ -2 & 4 & 1 \\ 1 & -3 & 11 \end{bmatrix}$$

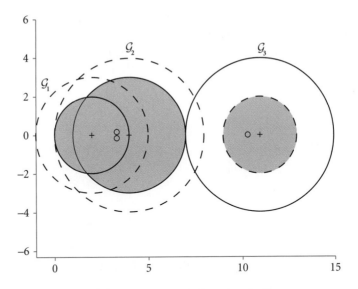

FIGURE 8.5 Example 6 Gerschgorin discs.

SOLUTION The Gerschgorin row discs are $\mathcal{G}_1(\mathbf{A})[2;2]$, $\mathcal{G}_2(\mathbf{A})[4;3]$, and $\mathcal{G}_3(\mathbf{A})[11;4]$. The Gerschgorin column discs are $\mathcal{G}_1(\mathbf{A}^T)[2;3]$, $\mathcal{G}_2(\mathbf{A}^T)[4;4]$, and $\mathcal{G}_3(\mathbf{A}^T)[11;2]$. Figure 8.5 shows the row discs with solid circles and the column discs with dashed circles. Using (3), we find that the complex conjugate pair of eigenvalues are in the left shaded region $(\mathcal{G}_1(\mathbf{A}) \cup \mathcal{G}_2(\mathbf{A})) \cup (\mathcal{G}_2(\mathbf{A}^T) \cap \mathcal{G}_3(\mathbf{A}))$, and a single real eigenvalue is in the disjoint shaded region $\mathcal{G}_3(\mathbf{A}^T)$, on the righthand side. A computer program confirms that the eigenvalues are $3.3314 \pm 0.1626i$ and 10.3371. ∎

EXAMPLE 7 Use Gerschgorin's Theorem to localize the eigenvalues of

$$\begin{bmatrix} 9 & 1 & 2 \\ 1-i & 14+9i & 2 \\ -1 & 5 & 10-5i \end{bmatrix}$$

SOLUTION The Gerschgorin row discs are $\mathcal{G}_1(\mathbf{A})[9;3]$, $\mathcal{G}_2(\mathbf{A})[14+9i;2+\sqrt{2}]$, and $\mathcal{G}_3(\mathbf{A})[10-5i;6]$. The Gerschgorin column discs are $\mathcal{G}_1(\mathbf{A}^T)[9;2+\sqrt{2}]$, $\mathcal{G}_2(\mathbf{A}^T)[14+9i;6]$, and $\mathcal{G}_3(\mathbf{A}^T)[10-5i;4]$. Figure 8.6 (p. 538) shows the row discs with solid circles and the column discs with dashed circles. By (3),

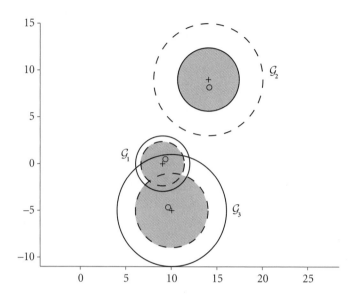

FIGURE 8.6   Example 7 Gerschgorin discs.

two of the eigenvalues are in the shaded region $\mathcal{G}_1(\mathbf{A}^T) \cup \mathcal{G}_3(\mathbf{A}^T)$, and one eigenvalue is in the disjoint shaded region $\mathcal{G}_2(\mathbf{A})$. The actual eigenvalues are $9.3133 + 0.4865i$, $9.5902 - 4.6468i$, and $14.0966 + 8.1604i$. ∎

EXAMPLE 8   Use Gerschgorin's Theorem to localize the eigenvalues of

$$\begin{bmatrix} 5 & -2 & 1 \\ 1 & 11 & -3 \\ -1 & 1 & 19 \end{bmatrix}$$

SOLUTION   The Gerschgorin row discs are $\mathcal{G}_1(\mathbf{A})[5;3]$, $\mathcal{G}_2(\mathbf{A})[11;4]$, and $\mathcal{G}_3(\mathbf{A})[19;2]$. The Gerschgorin column discs are $\mathcal{G}_1(\mathbf{A}^T)[5;2]$, $\mathcal{G}_2(\mathbf{A}^T)[11;3]$, and $\mathcal{G}_3(\mathbf{A}^T)[19;4]$. Figure 8.7 (p. 539) shows the row discs with solid circles and the column discs with dashed circles. Clearly, all of the eigenvalues are in the disjoint discs $\mathcal{G}_1(\mathbf{A}^T)$, $\mathcal{G}_2(\mathbf{A}^T)$, and $\mathcal{G}_3(\mathbf{A})$. So each disc contains a single eigenvalue. By examining the characteristic polynomial, we can determine that the eigenvalues of the matrix are real. Consequently, we know that these eigenvalues are in the intervals [3,7], [8,14], and [17,21]. The actual eigenvalues are 5.3488, 11.1653, and 18.4857. ∎

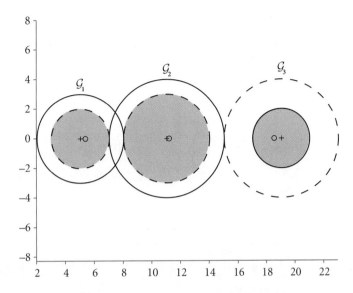

FIGURE 8.7 Example 8 Gerschgorin discs.

Interest in this topic began with the seminal paper by Gerschgorin in 1931, and the sharpening of his Circle Theorem via irreducibility by Taussky-Todd in 1949. It continues to fascinate researchers because of the beautiful results, and their ease of use in applications. Consequently, there is an ever-growing body of new developments. See, for example, Brualdi and Mellendorf [1994] and Varga [2004]. For additional details, see Faddeev and Faddeeva [1963, pp. 114–115], Golub and van Loan [1985], Horn and Johnson [1985, 343–364], Meyers [2000], and Wilkinson [1965].

Infinity Norm

Although there are many matrix norms, we will use the one that is easiest to compute.

DEFINITION

*The **infinity norm** of an $m \times n$ matrix $\mathbf{A}$ is defined to be the quantity*

$$\|\mathbf{A}\|_\infty = \max_{1 \le i \le m} \sum_{j=1}^{n} |a_{ij}|$$

A special case of this definition occurs when the matrix is $m \times 1$, in other words, a column vector, $\mathbf{x}$. In this case, the preceding formula still applies and is easy to calculate:

$$\|\mathbf{x}\|_\infty = \max_{1 \le i \le m} |x_i|$$

EXAMPLE 9 Compute the infinity norm of a vector and a matrix.

SOLUTION If $\mathbf{x} = (3, 7, -2)$, then $\|\mathbf{x}\|_\infty = 7$. If $\mathbf{A} = \begin{bmatrix} 2 & -7 \\ 3 & 4 \end{bmatrix}$, then $\|\mathbf{A}\|_\infty$ is the larger of the two sums $2 + 7 = 9$ and $3 + 4 = 7$. Hence, we obtain $\|\mathbf{A}\|_\infty = 9$. ∎

Lemma 1

If $\mathbf{A}$ is an $m \times n$ matrix and $\mathbf{x}$ is a column vector having n components, then

$$\|\mathbf{Ax}\|_\infty \le \|\mathbf{A}\|_\infty \cdot \|\mathbf{x}\|_\infty$$

PROOF We have

$$\|\mathbf{Ax}\|_\infty = \max_{1 \le i \le m} |(\mathbf{Ax})_i| = \max_{1 \le i \le m} \left| \sum_{j=1}^n a_{ij} x_j \right|$$

$$\le \max_{1 \le i \le m} \sum_{j=1}^n |a_{ij} x_j| \le \max_{1 \le i \le m} \sum_{j=1}^n |a_{ij}| \cdot \|x\|_\infty$$

$$= \|\mathbf{A}\|_\infty \cdot \|\mathbf{x}\|_\infty$$ ∎

Lemma 2

If $\mathbf{A}$ is a square matrix such that $\|\mathbf{A}\|_\infty < 1$, then $\mathbf{I} - \mathbf{A}$ is diagonally dominant.

PROOF The hypothesis $\|\mathbf{A}\|_\infty < 1$ leads to $\sum_{j=1}^n |a_{ij}| < 1$ for each i. It follows that

$$\sum_{\substack{j \ne i \\ j=1}}^n |a_{ij}| < 1 - |a_{ii}| \le |1 - a_{ii}| \qquad (1 \le i \le n)$$

This inequality displays the criterion for diagonal dominance in the matrix $\mathbf{I} - \mathbf{A}$. ∎

Convergence Properties

The numerical analysis literature is a source of many theorems establishing convergence properties of iterative methods. Here is one basic theorem from this far-ranging subject.

THEOREM 3

If $\mathbf{G}$ is an $n \times n$ matrix such that $\left\| \mathbf{G} \right\|_\infty < 1$, then the iteration formula

$$\mathbf{x}^{(k)} = \mathbf{G}\mathbf{x}^{(k-1)} + \mathbf{c}$$

produces a sequence of vectors $\mathbf{x}^{(k)}$ that converges to $(\mathbf{I} - \mathbf{G})^{-1}\mathbf{c}$ from any starting point, $\mathbf{x}^{(0)}$.

PROOF The matrix $\mathbf{I} - \mathbf{G}$ is invertible by Lemma 2 and Theorem 1. Define $\mathbf{x} = (\mathbf{I} - \mathbf{G})^{-1}\mathbf{c}$. Then $(\mathbf{I} - \mathbf{G})\mathbf{x} = \mathbf{c}$ and $\mathbf{x} = \mathbf{G}\mathbf{x} + \mathbf{c}$. Consequently, we have

$$\mathbf{x}^{(k)} - \mathbf{x} = (\mathbf{G}\mathbf{x}^{(k-1)} + \mathbf{c}) - (\mathbf{G}\mathbf{x} + \mathbf{c}) = \mathbf{G}(\mathbf{x}^{(k-1)} - \mathbf{x})$$

Hence by Lemma 1, we have

$$\left\| \mathbf{x}^{(k)} - \mathbf{x} \right\|_\infty \leq \left\| \mathbf{G} \right\|_\infty \cdot \left\| \mathbf{x}^{(k-1)} - \mathbf{x} \right\|_\infty$$

By repeating this argument k times, we arrive at

$$\left\| \mathbf{x}^{(k)} - \mathbf{x} \right\|_\infty \leq \left\| \mathbf{G} \right\|_\infty^k \cdot \left\| \mathbf{x}^{(0)} - \mathbf{x} \right\|_\infty$$

The righthand side of this inequality converges to 0 because $\left\| \mathbf{G} \right\|_\infty < 1$. ∎

COROLLARY 5

If $\left\| \mathbf{I} - \mathbf{A} \right\|_\infty < 1$, then the Richardson iteration, defined by

$$\mathbf{x}^{(k)} = (\mathbf{I} - \mathbf{A})\mathbf{x}^{(k-1)} + \mathbf{b}$$

produces a sequence that converges to the solution of the equation $\mathbf{A}\mathbf{x} = \mathbf{b}$.

PROOF Define $\mathbf{G} = \mathbf{I} - \mathbf{A}$, and use Theorem 3. We see that $\left\| \mathbf{G} \right\|_\infty < 1$. By Theorem 3, the iterates converge to $(\mathbf{I} - (\mathbf{I} - \mathbf{A}))^{-1}\mathbf{b} = \mathbf{A}^{-1}\mathbf{b}$. ∎

COROLLARY 6

If **A** is diagonally dominant, then the Jacobi iteration

$$\mathbf{x}^{(k)} = \mathbf{B}\mathbf{x}^{(k-1)} + \mathbf{c}$$

produces a sequence that converges to the solution of the equation $\mathbf{A}\mathbf{x} = \mathbf{b}$. The definitions are

$$\mathbf{B} = \mathbf{I} - \mathbf{D}^{-1}\mathbf{A}, \qquad \mathbf{c} = \mathbf{D}^{-1}\mathbf{b}, \qquad \mathbf{D} = \text{Diag}(\mathbf{A})$$

PROOF In the Jacobi method, we have $\mathbf{G} = \mathbf{B} = \mathbf{I} - \mathbf{D}^{-1}\mathbf{A}$ and $\mathbf{c} = \mathbf{D}^{-1}\mathbf{b}$, where $\mathbf{D}$ is the diagonal of $\mathbf{A}$. In order to use Theorem 3, we need to know that $\|\mathbf{G}\|_\infty < 1$. We have

$$\|\mathbf{G}\|_\infty = \|\mathbf{I} - \mathbf{D}^{-1}\mathbf{A}\|_\infty = \max_{1 \leq i \leq n} \sum_{j=1}^{n} |\delta_{ij} - a_{ij}/a_{ii}| = \max_{1 \leq i \leq n} \sum_{\substack{j=1 \\ j \neq i}}^{n} |a_{ij}/a_{ii}| < 1 \qquad \blacksquare$$

COROLLARY 7

If the coefficient matrix in a system of equations is diagonally dominant, then the Gauss–Seidel iteration converges for any starting vector.

The proof requires a more complicated spectral radius argument, and the fact that the Gauss–Seidel iterative method converges for symmetric positive definite matrices. This type of matrix arises in the discretization of elliptic partial differential equations. Informally, we say that the more diagonally dominant the matrix is, the more rapid will be the convergence of the Jacobi and Gauss–Seidel methods. However, there can be exceptions. See Golub and van Loan [1996].

Power Method for Computing Eigenvalues

The problem of computing the eigenvalues of a matrix is at least as hard as finding the roots of a given polynomial. This problem inevitably brings in approximate methods, as there are no purely algebraic formulas for the roots of general polynomials of degree five or higher. (Abel–Ruffini Impossibility Theorem, see Appendix B.) Using higher transcendental functions to compute roots of a polynomial is possible but not a competitive strategy!

To see how iterative methods can produce approximate eigenvalues of a matrix $\mathbf{A}$, we make some simplifying assumptions. Suppose that $\mathbf{A}$ is an $n \times n$ matrix whose eigenvalues satisfy these inequalities:

$$|\lambda_1| > |\lambda_2| \geq |\lambda_3| \geq \cdots \geq |\lambda_n|$$

Assume, further, that there is a *complete* set of eigenvectors:

$$\mathbf{A}\mathbf{u}^{(j)} = \lambda_j \mathbf{u}^{(j)} \qquad (1 \leq j \leq n)$$

We mean by this that these vectors form a basis for $\mathbb{R}^n$. For the iteration, select any initial vector $\mathbf{x}^{(0)}$ that has a component in the direction of $\mathbf{u}^{(1)}$. Thus, for suitable coefficients,

$$\mathbf{x}^{(0)} = c_1 \mathbf{u}^{(1)} + c_2 \mathbf{u}^{(2)} + \cdots + c_n \mathbf{u}^{(n)}$$

and we assume that $c_1 \neq 0$. Because these coefficients c_i can be absorbed into the eigenvectors $\mathbf{u}^{(i)}$, we can write simply

$$\mathbf{x}^{(0)} = \mathbf{u}^{(1)} + \mathbf{u}^{(2)} + \cdots + \mathbf{u}^{(n)}$$

To get the next vector in the sequence, we multiply both sides of this equation by the matrix $\mathbf{A}$, and make use of the fact that these vectors $\mathbf{u}^{(j)}$ are eigenvectors. The result is then

$$\mathbf{x}^{(1)} = \mathbf{A}\mathbf{x}^{(0)} = \lambda_1 \mathbf{u}^{(1)} + \lambda_2 \mathbf{u}^{(2)} + \cdots + \lambda_n \mathbf{u}^{(n)}$$

This process is repeated, and the general case is

$$\mathbf{x}^{(k)} = \mathbf{A}^k \mathbf{x}^{(0)} = \lambda_1^k \mathbf{u}^{(1)} + \lambda_2^k \mathbf{u}^{(2)} + \cdots + \lambda_n^k \mathbf{u}^{(n)}$$

By simple algebra, this becomes

$$\mathbf{x}^{(k)} = \lambda_1^k \left[\mathbf{u}^{(1)} + \left(\frac{\lambda_2^k}{\lambda_1^k} \right) \mathbf{u}^{(2)} + \cdots + \left(\frac{\lambda_n^k}{\lambda_1^k} \right) \mathbf{u}^{(n)} \right]$$

The terms $(\lambda_j/\lambda_1)^k$ converge to zero because we have assumed that $|\lambda_1| > |\lambda_j|$ for $j > 1$. Thus, we can write

$$\mathbf{x}^{(k)} = \lambda_1^k [\mathbf{u}^{(1)} + \mathbf{e}_k]$$

where the term $\mathbf{e}_k$ accounts for all the terms that converge to $\mathbf{0}$. Select a vector $\mathbf{v}$ and use dot-products so that we will be dealing with numbers instead of vectors:

$$\langle \mathbf{v}, \mathbf{x}^{(k)} \rangle = \lambda_1^k \langle \mathbf{v}, \mathbf{u}^{(1)} \rangle + \langle \mathbf{v}, \mathbf{e}_k \rangle$$

Forming ratios brings us to

$$\frac{\langle \mathbf{v}, \mathbf{x}^{(k+1)} \rangle}{\langle \mathbf{v}, \mathbf{x}^{(k)} \rangle} = \lambda_1 \left[\frac{\langle \mathbf{v}, \mathbf{u}^{(1)} \rangle + \langle \mathbf{v}, \mathbf{e}_{k+1} \rangle}{\langle \mathbf{v}, \mathbf{u}^{(1)} \rangle + \langle \mathbf{v}, \mathbf{e}_k \rangle} \right] \to \lambda_1 \qquad \text{as} \quad k \to \infty$$

Because λ_1 is the largest eigenvalue in absolute value, all the terms except the first inside the brackets converge to $\mathbf{0}$.

In carrying out such a calculation, it is advisable to normalize the vectors $\mathbf{x}^{(k)}$ at each step. Otherwise the sequence may converge to zero or infinity. Here is a segment of pseudocode for the **power method**:

$$\mathbf{y} \leftarrow \mathbf{Ax}$$
$$\mathbf{r} \leftarrow \langle \mathbf{v}, \mathbf{y} \rangle / \langle \mathbf{v}, \mathbf{x} \rangle$$
$$\mathbf{x} \leftarrow \mathbf{y} / \|\mathbf{y}\|_\infty$$

Here we could use the vector $\mathbf{v} = \mathbf{e}_1 = (1, 0)$ as in the following example.

EXAMPLE 10 Find one of the eigenvalues of the following matrix, using the power method.

$$\begin{bmatrix} 1 & 3 & 5 \\ 7 & -8 & 1 \\ 2 & 4 & -2 \end{bmatrix}$$

SOLUTION We repeatedly execute the steps in this pseudocode:

$$\mathbf{x} \leftarrow [1, 1, 1]^T$$
$$\textbf{for } k = 1 \textbf{ to } 50$$
$$\mathbf{y} \leftarrow \begin{bmatrix} 1 & 3 & 5 \\ 7 & -8 & 1 \\ 2 & 4 & -2 \end{bmatrix} \mathbf{x}$$
$$r \leftarrow y_1 / x_1$$
$$\alpha \leftarrow 1 / \|\mathbf{y}\|_\infty$$
$$\mathbf{x} \leftarrow \alpha \mathbf{y}$$
$$\textbf{end for}$$

In each iteration, new values for the vectors $\mathbf{x}$ and $\mathbf{y}$ are computed. At the same time, the ratio y_1/x_1 converges to the dominant eigenvalue of $\mathbf{A}$. It starts with $r = 9$ and then $r = 3.222$ and does not settle down to -8.5730 until after 47 iterations. A computer program is needed to accomplish this and is given at the end of this section. ∎

Application: Demographic Problems, Population Migration

Linear algebra is one of the principal tools in demography. For example, it can be used in predicting how the distribution of the population will change yearly within a certain region of a country. The broad strategy is to study the past in order to discover a mathematical model, and then to use the model in predicting the future or in controlling the future by appropriate public policy.

Let us say that the region in question has a major city at its *hub*, and this city is surrounded by a *suburban* area. Beyond these two areas there is an outer region with population loosely associated culturally and economically with the city and its suburbs. We call it the *rural* area. Because of shifts in population (*migration*), the numbers of people in each of the three regions change over time. By taking surveys or consulting census records, one knows the population in the three areas at some instant in time, and one knows the migration patterns. Here we are assuming a *closed system* in which no one enters or leaves this group of three regions!

To make a concrete numerical case, we assume that the initial populations of these three areas are 3 million in the *hub* (core area), 4 million in the *suburbs*, and 6 million in the *rural* region. These regions are disjoint and the total population of 13 million remains constant, but there are annual movements of people from one region to another.

Suppose that in any year, 20% of the people in the hub move to the suburbs and 10% of the people in the hub move to the rural region. The remaining 70% stay where they are in the hub.

Further, suppose that in any year, 12% of the people in the suburban area move to the hub, and 8% of the people in the suburban area move to the rural region. The remaining 80% stay in the suburban area.

Finally, suppose that 30% of the population in the rural region moves to the suburban area, and 20% of the population in the rural region moves to the hub. The remaining 50% of the people in the rural region stay where they are.

At a given instant, the population distribution can be represented by a vector $\mathbf{x}$ having three components: h (for *hub*), s (for *suburban* area), and r (for outer *rural* area). Thus, we let $\mathbf{x} = (h, s, r)$. From the preceding information, these details emerge about the change of h over a one-year period:

1. 70% of the people in the hub remain there.
2. 12% of the people in the suburban area move to the hub.
3. 20% of the population in the rural region moves to the hub.

Similar results for s and r can be deduced. From the data, for each component in our vector we have

$$\begin{cases} h_{\text{new}} = 0.70h + 0.12s + 0.20r \\ s_{\text{new}} = 0.20h + 0.80s + 0.30r \\ r_{\text{new}} = 0.10h + 0.08s + 0.50r \end{cases}$$

The *initial condition* of the model is described by the vector $\mathbf{x}^{(0)} = (3, 4, 6)$, where the units are 1 million people. The yearly migration can be described by a matrix

$$\mathbf{x}^{(1)} = \mathbf{A}\mathbf{x}^{(0)}$$

$$\begin{bmatrix} h_{\text{new}} \\ s_{\text{new}} \\ r_{\text{new}} \end{bmatrix} = \begin{bmatrix} 0.70 & 0.12 & 0.20 \\ 0.20 & 0.80 & 0.30 \\ 0.10 & 0.08 & 0.50 \end{bmatrix} \begin{bmatrix} h_{\text{old}} \\ s_{\text{old}} \\ r_{\text{old}} \end{bmatrix}$$

Notice that row one in the matrix $\mathbf{A}$ pertains to the hub. It retains 70% of its population, and gains 12% of the population from the suburbs, as well as 20% of the people from the outer rural region.

On the diagonal of the matrix the numbers are relatively strong, reflecting the fact that most people do not move during any one year. Also note that the columns sum to 1, because column one accounts for the migration of all the people in the hub, and the other columns can be interpreted similarly.

Put in the starting values of the vector $\mathbf{x}^{(0)} = (3, 4, 6)$ and compute

$$\mathbf{x}^{(1)} = \mathbf{A}\mathbf{x}^{(0)} = \begin{bmatrix} 0.70 & 0.12 & 0.20 \\ 0.20 & 0.80 & 0.30 \\ 0.10 & 0.08 & 0.50 \end{bmatrix} \begin{bmatrix} 3 \\ 4 \\ 6 \end{bmatrix} = \begin{bmatrix} 4.38 \\ 5.00 \\ 3.62 \end{bmatrix}$$

This shows the population of the three sectors after one year. In order to see what happens over a number of years, mathematical software can be put to use, computing $\mathbf{x}^{(0)}, \mathbf{x}^{(1)}, \mathbf{x}^{(2)}$, and so on. The vector $\mathbf{x}^{(0)}$ is the initial distribution of population, and the preceding computation gives

$$\mathbf{x}^{(1)} = \mathbf{A}\mathbf{x}^{(0)}$$

The dynamical system is described recursively by the equation $\mathbf{x}^{(k)} = \mathbf{A}\mathbf{x}^{(k-1)}$. In a few years, the populations of the three sectors reach equilibrium. For example, we obtain $\mathbf{x}^{(8)} = (4.093, 6.969, 1.939)$, and eight years later we have $\mathbf{x}^{(16)} = (4.083, 6.983, 1.934)$.

In such a dynamical system, there is also the capability of reversing time to look into the past. For this, one can use the formulas $\mathbf{y}^{(0)} = (3, 4, 6)$ and $\mathbf{y}^{(k+1)} = \mathbf{A}^{-1}\mathbf{y}^{(k)}$.

Population models such as this simple one have obvious weaknesses. For example, the total population of the three regions has been assumed to be constant: there are no deaths, no births, and no migration into or out of the region. However, more elaborate models can grow from simple ones, and modifications can be incorporated to improve the predictions.

The behavior of a dynamical system can be analyzed in great detail by using the eigenvalues, eigenvectors, and diagonalized form of the matrix that describes the system. This procedure was explained in Section 6.1, and here we show how it applies to the migration problem discussed previously. In particular, the long-term behavior of the dynamical system can be understood once we have the eigenvalues in hand. Recall that

$$\mathbf{x}^{(0)} = c_1\mathbf{p}_1 + c_2\mathbf{p}_2 + c_3\mathbf{p}_3$$

where the eigenvectors are the columns of $\mathbf{P} = [\mathbf{p}_1, \mathbf{p}_2, \mathbf{p}_3]$. Also, we have

$$\mathbf{x}^{(k)} = \mathbf{A}^k\mathbf{x}^{(0)} = c_1\lambda_1^k\mathbf{p}_1 + c_2\lambda_2^k\mathbf{p}_2 + c_3\lambda_3^k\mathbf{p}_3$$
$$\approx c_1\lambda_1^k\mathbf{p}_1 = c_1\mathbf{p}_1 \qquad \text{as} \quad k \to \infty$$

since $\lambda_1 = 1, |\lambda_2| < 1$, and $|\lambda_3| < 1$.

Let mathematical software do the work in the preceding example to obtain the eigenvalues and eigenvectors of the matrix $\mathbf{A}$. The eigenvectors are columns of $\mathbf{P}$, each having been normalized.

$$\mathbf{D} = \begin{bmatrix} 1.0000 & 0 & 0 \\ 0 & 0.5894 & 0 \\ 0 & 0 & 0.4106 \end{bmatrix}$$

$$\mathbf{P} = \begin{bmatrix} -0.4909 & -0.6646 & -0.3853 \\ -0.8396 & 0.7431 & -0.4308 \\ -0.2325 & -0.0784 & 0.8161 \end{bmatrix}$$

$$\mathbf{P}^{-1} = \begin{bmatrix} -0.6399 & -0.6398 & -0.6398 \\ -0.8775 & 0.5477 & -0.1252 \\ -0.2666 & -0.1296 & 1.0310 \end{bmatrix}$$

Now recall the equation

$$\mathbf{A} = \mathbf{PDP}^{-1}$$

which is the *diagonalization* of **A**. The eigenvalues of **A** are in the diagonal matrix **D**. The eigenvectors (normalized) are the columns of **P**. The purpose of this diagonalization is to make available the simple formula

$$\mathbf{A}^k = \mathbf{PD}^k\mathbf{P}^{-1}$$

Because two of the eigenvalues are less than 1 in absolute value,

$$\lim_{k\to\infty} \mathbf{D}^k = \begin{bmatrix} 1 & 0 & 0 \\ 0 & 0 & 0 \\ 0 & 0 & 0 \end{bmatrix} = \mathbf{Q}$$

Therefore, we obtain

$$\lim_{k\to\infty} \mathbf{x}^{(k)} = \mathbf{PQP}^{-1}\mathbf{x}^{(0)} = \begin{bmatrix} 4.0826 \\ 6.9835 \\ 1.9339 \end{bmatrix}$$

This calculation confirms that the vectors $\mathbf{x}^{(k)}$ are settling down to a constant vector.

Application: Leontief Open Model

An economic model that is more elaborate than the one presented in Section 6.1 takes into account the existence of sectors in our economy that consume but do not produce output to be sold. Consumer activity can be interpreted as one such sector. The matrix problem that arises from this model is

$$\mathbf{x} = \mathbf{Ax} + \mathbf{b}$$

where **A** and **b** are known numerically. Again the question is whether there is a solution, and, of course, the answer depends upon what we wish to assume about **A** and **b**. In this model, **x** is a vector whose components are the *production levels* of the various industries. It should be a nonnegative vector; that is, $x_i \geq 0$ for all i. Likewise, the matrix **A**, now having to do with consumption of the various commodities, should be nonnegative. We write $\mathbf{x} \geq \mathbf{0}$ and $\mathbf{A} \geq \mathbf{0}$ to express these requirements, and refer to *nonnegative vectors* and *nonnegative matrices*.

> **THEOREM 4**
>
> Let **A** be a nonnegative $n \times n$ matrix and let **b** be a nonnegative vector in $\mathbb{R}^n$. If all column sums of **A** are less than 1, then the system
>
> $$\mathbf{x} = \mathbf{Ax} + \mathbf{b}$$
>
> has a unique solution, **x**, and **x** is nonnegative.

PROOF Put $\mathbf{E} = (\mathbf{I} - \mathbf{A})^T$. We will show that **E** is diagonally dominant. Then **E** will be invertible by Theorem 1. From that, it will follow that $\mathbf{E}^T$ is invertible and $\mathbf{I} - \mathbf{A}$ is invertible. Then the system in question has the (unique) solution $\mathbf{x} = (\mathbf{I} - \mathbf{A})^{-1}\mathbf{b}$. The nonnegativity is not proven here, but can be found in more advanced books. To prove the diagonal dominance of **E**, we write

$$\sum_{\substack{j=1 \\ j \neq i}}^{n} |E_{ij}| = \sum_{\substack{j \neq i \\ j=1}}^{n} |-a_{ji}| = \sum_{\substack{j \neq i \\ j=1}}^{n} a_{ji}$$

$$= \sum_{j=1}^{n} a_{ji} - a_{ii} < 1 - a_{ii} = E_{ii}$$

Consequently, **E** is diagonally dominant. ∎

The economic model described here is known as the **Leontief Open Model** of an economy having n **sectors**. The vector **x** is the **production vector**. It gives the amounts of each product produced in one year. The **open sector** has no production, since it only consumes. The vector **b** is the final **demand vector**. It is the *consumer demand* and lists the values of goods and services consumed by the open sector. The model leads to the vector **x** that satisfies the equation, and this in turn reveals the correct level of production in each sector.

The correct **x**-vector shows the level of production for each industry that is needed to satisfy consumer demand plus the needs of the various industries. These are related by the matrix–vector equation

$$\mathbf{x} = \mathbf{Ax} + \mathbf{b}$$

The vector **Ax** gives the amounts of each product consumed in the manufacturing phase. It represents the resources needed in production, which economists call the **intermediate demand**. The product **Ax** accounts for the consumption required just to reach the level of output **x**. In the matrix **A**, each entry a_{ij} is the output of industry i needed by industry j. Each column in the matrix $\mathbf{A}$ gives the inputs needed to produce one unit. This column is the **unit consumption vector**. The matrix $\mathbf{A}$ has special properties: its elements are nonnegative ($a_{ij} \geq 0$), and each column sum is less than 1 ($\sum_{i=1}^{n} a_{ij} < 1$ for all j). The column vector of $\mathbf{Ax}$ is

$$x_1 \begin{bmatrix} a_{11} \\ a_{21} \\ \vdots \\ a_{n1} \end{bmatrix}$$

Here $x_1 a_{11}$ is the amount of the first product required in manufacturing process number one. In the same way, $x_1 a_{21}$ is the amount of product number two required in manufacturing process number two, and so on. In general, we have

$$(\mathbf{Ax})_i = \begin{bmatrix} a_{i1} & a_{i2} & \cdots & a_{in} \end{bmatrix} \begin{bmatrix} x_1 \\ x_2 \\ \vdots \\ x_n \end{bmatrix} = \sum_{j=1}^{n} a_{ij} x_j$$

which is the total intermediate demand for the ith product. We can write the total production as the sum of the intermediate demand and the consumer demand (or final demand), thus arriving at

$$\mathbf{x} = \mathbf{Ax} + \mathbf{b}$$

This is the same as

$$(\mathbf{I} - \mathbf{A})\mathbf{x} = \mathbf{b}$$

Recall that the inequality $\mathbf{A} \geq \mathbf{0}$ means that all entries in the matrix $\mathbf{A}$ are nonnegative. Theorem 5 asserts that if $\mathbf{A} \geq \mathbf{0}$ and $\mathbf{b} \geq \mathbf{0}$ and $\sum_{i=1}^{n} a_{ij} < 1$ for each j, then $(\mathbf{I} - \mathbf{A})^{-1}$ exists and $\mathbf{x} = (\mathbf{I} - \mathbf{A})^{-1}\mathbf{b}$ and $\mathbf{x} \geq \mathbf{0}$.

Notice that

$$(\mathbf{I} - \mathbf{A})(\mathbf{I} + \mathbf{A}) = \mathbf{I} - \mathbf{A}^2$$

$$(\mathbf{I} - \mathbf{A})(\mathbf{I} + \mathbf{A} + \mathbf{A}^2) = \mathbf{I} - \mathbf{A}^3$$

In general, we have

$$(\mathbf{I} - \mathbf{A})(\mathbf{I} + \mathbf{A} + \mathbf{A}^2 + \cdots + \mathbf{A}^m) = \mathbf{I} - \mathbf{A}^{m+1}$$

If $\mathbf{A}^{m+1}$ converges to $\mathbf{0}$ as m goes to infinity, then we obtain the **Neumann Series**:

$$(\mathbf{I} - \mathbf{A})^{-1} = \mathbf{I} + \mathbf{A} + \mathbf{A}^2 + \cdots$$

Here we are glossing over some technical matters that belong to the study of functional analysis.

EXAMPLE 11 Suppose we have a table of values

Sell\Buy	Agr.	Mfg.	Ser.
Agr.	0.3	0.1	0.3
Mfg.	0.2	0.3	0.4
Ser.	0.4	0.5	0.2

Suppose the final demand vector is $[12, 18, 15]^T$. Find the required production levels.

SOLUTION This table corresponds to the consumption matrix $\mathbf{A}$. In this example, to produce a unit of manufactured goods requires 0.1 units of agriculture (Agr.) goods, 0.3 units of manufacture (Mfg.) goods, and 0.5 units of services (Ser.) goods.

We have

$$\mathbf{x} = \mathbf{A}\mathbf{x} + \mathbf{b} = \begin{bmatrix} 0.3 & 0.1 & 0.3 \\ 0.2 & 0.3 & 0.4 \\ 0.4 & 0.5 & 0.2 \end{bmatrix} \begin{bmatrix} x_1 \\ x_2 \\ x_3 \end{bmatrix} + \begin{bmatrix} 12 \\ 18 \\ 15 \end{bmatrix}$$

Notice that the elements of the matrix $\mathbf{A}$ are nonnegative and the column sums are less than 1. The system $(\mathbf{I} - \mathbf{A})\mathbf{x} = \mathbf{b}$ must be solved for $\mathbf{x}$. The augmented matrix is

$$[\mathbf{I} - \mathbf{A} \mid \mathbf{b}] = \begin{bmatrix} 0.7 & -0.1 & -0.3 & \mid & 12 \\ -0.2 & 0.7 & -0.4 & \mid & 18 \\ -0.4 & -0.5 & 0.8 & \mid & 15 \end{bmatrix}$$

$$\sim \begin{bmatrix} 7 & -1 & -3 & \mid & 120 \\ -2 & 7 & -4 & \mid & 180 \\ -4 & -5 & 8 & \mid & 150 \end{bmatrix}$$

$$\sim \begin{bmatrix} 1 & 0 & 0 & \mid & 115 \\ 0 & 1 & 0 & \mid & 154 \\ 0 & 0 & 1 & \mid & 176 \end{bmatrix}$$

The production levels turn out to be 115 units of agriculture goods, 159 units of manufacturing goods, and 176 units of service goods. ∎

Mathematical Software

Here is a short MATLAB program pertaining to Example 1. It will carry out 50 steps of the Richardson iteration once the values of $\mathbf{G}$, $\mathbf{b}$, and $\mathbf{x}^{(0)}$ have been given to the program as input.

```
MATLAB
G = [0.0,0.3,0.2;0.3,0.0,0.1;0.2,0.1,0.0];
b = [7.0;5.0;3.0];
x = [0.0,0.0,0.0];
z = [0,x']
for k=1:50
   x = G*x + b;
   z = [k,x']
end
```

In the MATLAB code, the $\mathbf{z}$-array is used only for facilitating the display of the results. By using the command `format long`, one obtains a display of all available digits in the calculations. This means about 15 digits of precision. By examining all 15 digits in the computed vectors, one finds that the solution vector has been computed to full MATLAB precision by the 42nd step. In other words, further iteration will not change the result (except for rounding errors), if one examines only the first 15 digits. A check on the work consists in computing $\mathbf{Ax}$. This leads to the vector $\mathbf{b}$, accurate to 15 decimal places.

Sample Maple and Mathematica codes for Example 1 are presented here:

```
Maple
with(LinearAlgebra):
G := Matrix([0.0,0.3,0.2],[0.3,0.0,0.1],[0.2,0.1,0.0]);
b := Vector(7.0,5.0,3.0);
x := Vector(0.0,0.0,0.0)
for k from 1 to 50 do
  x := G.x + b
od
```

```
Mathematica
G = {{0.0,0.3,0.2},{0.3,0.0,0.1},{0.2,0.1,0.0}};
b = {7.0,5.0,3.0}
x = {0.0,0.0,0.0}}
For[k=1, k<<=50, k++
   x = G.x + b
]
```

In the remainder of this chapter, only MATLAB codes are given because it is straightforward to write similar codes in Maple and Mathematica.

For Example 2, the MATLAB code could be written as follows:

```
MATLAB
B = [0.0,0.15,0.1;0.15,0.0,0.05;0.1,0.05,0.0]
c = [3.5;2.5;1.5]
x = [0.0;0.0;0.0]
z = [0,x']
for k=1:25
  x = B*x + c;
  z = [k,x']
end
```

This code could be written as follows in terms of the individual variables:

```
MATLAB
x1 = 0.0; x2 = 0.0; x3 = 0.0;
z = [0,x1, x2, x3]
for k=1:25
  y1 = 0.15*x2 + 0.1*x3 + 3.5;
  y2 = 0.15*x1 + 0.05*x3 + 2.5;
  y3 = 0.1*x1 - 0.05*x2 + 1.5;
  x1 = y1; x2 = y2; x3 = y3;
  z = [k,x1,x2,x3]
end
```

To obtain full machine precision, it took 24 iterations.

In Example 3, the MATLAB code for the Gauss–Seidel method is especially simple. It can be written as follows:

```
MATLAB
x1 = 0.0; x2 = 0.0; x3 = 0.0;
z = [0,x1,x2,x3]
for k=1:25
  x1 = 0.15*x2 + 0.1*x3 + 3.5;
  x2 = 0.15*x1 + 0.05*x3 + 2.5;
  x3 = 0.1*x1 + 0.05*x2 + 1.5;
  z = [k,x1,x2,x3]
end
```

In Example 10, a MATLAB program can be written to carry out this calculation:

```
MATLAB
A = [1 3 5; 7 -8 1; 2 4 -2]
x = [1; 1; 1]
for k=1:50
  y = A*x;
  r = y(1)/x(1);
  x = y/norm(y);
  z = [k,x']
end
```

This program generates 50 new values for the vectors $\mathbf{x}$ and $\mathbf{y}$. At the same time, the ratio $y(1)/x(1)$ is being recorded. This is the quantity that should converge to the dominant eigenvalue of $\mathbf{A}$. The output from the program starts with $r = 9$ and $r = 3.222$. By the 47th step, however, the value of r has settled down to -8.5730.

Of course, MATLAB has commands for obtaining the eigenvalues immediately. For this exercise, we input the matrix $\mathbf{A}$ and then use the command `eig(A)`. The result of doing this is the set of three values: 5.5054, -8.5730, -5.9324. These eigenvalues are available in 15-decimal precision. For example, the value we were seeking is $c = -8.57299818380420$.

Is there still another check that can be applied to guarantee the precision of the eigenvalue? Yes, we can compute the determinant of $\mathbf{A} - c\mathbf{I}$, where c is the 15-digit value quoted previously. The MATLAB command to do this is `det(A - c * eye(3))`, and the result is -9.474×10^{-14}.

SUMMARY 8.3

- Richardson Iterative Method:
  $\mathbf{x}^{(k)} = (\mathbf{I} - \mathbf{A})\mathbf{x}^{(k-1)} + \mathbf{b}$

- Scaled System: $\mathbf{D}^{-1}\mathbf{A}\mathbf{x} = \mathbf{D}^{-1}\mathbf{b}$, where $\mathbf{D}^{-1}\mathbf{A} = \mathbf{I} - \mathbf{B} = \mathbf{I} - \mathbf{L} - \mathbf{U}$. Here $\mathbf{D} = \text{Diag}(\mathbf{A})$, $\mathbf{L} =$ lower triangular part of $\mathbf{D}^{-1}\mathbf{A}$, $\mathbf{U} =$ upper triangular part of $\mathbf{D}^{-1}\mathbf{A}$, and $\mathbf{c} = \mathbf{D}^{-1}\mathbf{b}$. These formulas define $\mathbf{B}$.

- Jacobi Iterative Method: $\mathbf{x}^{(k)} = \mathbf{B}\mathbf{x}^{(k-1)} + \mathbf{c}$, where $\mathbf{c}$ and $\mathbf{B}$ are as noted previously.

- Gauss–Seidel Method: $(\mathbf{I} - \mathbf{L})\mathbf{x}^{(k)} = \mathbf{U}\mathbf{x}^{(k-1)} + \mathbf{c}$

- Successive Overrelaxation (SOR) Method:
  $(\mathbf{I} - \omega\mathbf{L})\mathbf{x}^{(k)} = \{\mathbf{U}\mathbf{x}^{(k-1)} + \mathbf{c}\} + (1 - \omega)\mathbf{x}^{(k)}$
  The relaxation factor ω is between 0 and 2.

- Conjugate Gradient Method: If $\mathbf{A}$ is an $n \times n$ matrix, then the nth vector in the algorithm, $\mathbf{x}^{(n)}$, is the unique minimizer of the quadratic function $q(\mathbf{x}) = \frac{1}{2}\mathbf{x}^T\mathbf{A}\mathbf{x} - \mathbf{x}^T\mathbf{b}$ as well as the solution of $\mathbf{A}\mathbf{x} = \mathbf{b}$, when $\mathbf{A}$ is symmetric positive definite.

- A **diagonally dominant** $n \times n$ matrix satisfies $|a_{ii}| > \sum_{\substack{j=1 \\ j \neq i}}^{n} |a_{ij}|$ for all i.

- Infinity norm of an $m \times n$ matrix: $\|\mathbf{A}\|_\infty = \max_{1 \le i \le m} \sum_{j=1}^n |a_{ij}|$ where $\mathbf{A}$ is $m \times n$.
- The infinity norm of a vector is a special case, where $m = n$, $\|\mathbf{x}\|_\infty = \max_{1 \le i \le n} |x_i|$. Here $\mathbf{x}$ is in $\mathbb{R}^n$.

  - Every diagonally dominant matrix has an inverse.
  - Gerschgorin Theorem: If λ is an eigenvalue of $\mathbf{A}$, then $|\lambda - a_{ii}| \le \sum_{\substack{j=1 \\ j \ne i}}^n |a_{ij}|$ for some index i.
  - Nonoverlapping Gerschgorin discs contain exactly one eigenvalue each.
  - $\|\mathbf{A}\mathbf{x}\|_\infty \le \|\mathbf{A}\|_\infty \cdot \|\mathbf{x}\|_\infty$
  - If $\|\mathbf{G}\|_\infty < 1$, then the iteration $\mathbf{x}^{(k)} = \mathbf{G}\mathbf{x}^{(k-1)} + \mathbf{c}$ produces a sequence $\mathbf{x}^{(k)}$ that converges to $(\mathbf{I} - \mathbf{G})^{-1}\mathbf{c}$ from any starting point.
  - The Richardson iteration converges to the solution of $\mathbf{A}\mathbf{x} = \mathbf{b}$, when $\|\mathbf{I} - \mathbf{A}\|_\infty < 1$.

- The Jacobi iteration converges to the solution of $\mathbf{A}\mathbf{x} = \mathbf{b}$, when $\mathbf{A}$ is diagonally dominant.
- The Gauss–Seidel iteration converges to the solution of $\mathbf{A}\mathbf{x} = \mathbf{b}$, when $\mathbf{A}$ is diagonally dominant.
- If $\mathbf{A} \ge 0$, $\mathbf{b} \ge 0$, and $\sum_{i=1}^n a_{ij} < 1$ for each j, then $(\mathbf{I} - \mathbf{A})^{-1}$ exists and the vector $\mathbf{x} = (\mathbf{I} - \mathbf{A})^{-1}\mathbf{b}$ solves the equation $\mathbf{x} = \mathbf{A}\mathbf{x} + \mathbf{b}$ and $\mathbf{x} \ge 0$.
- Power Method: One can select a vector $\mathbf{x}^{(0)}$ and carry out the process $\mathbf{x}^{(k+1)} = \mathbf{A}\mathbf{x}^{(k)} / \|\mathbf{A}\mathbf{x}^{(k)}\|_\infty$ under favorable circumstances. Here, k runs through the sequence $1, 2, 3, \ldots$. The vectors $\mathbf{x}^{(k)}$ will converge to an eigenvector of $\mathbf{A}$.
- Leontief Open Model: $\mathbf{x} = \mathbf{A}\mathbf{x} + \mathbf{b}$
- Neumann Series: If $\|\mathbf{A}\|_\infty < 1$, then $(\mathbf{I} - \mathbf{A})^{-1} = \mathbf{I} + \mathbf{A} + \mathbf{A}^2 + \ldots$. There is a limiting process involved here, as indicated by the three dots, but the meaning and justification are outside the scope of this book.

KEY CONCEPTS 8.3

Iterative algorithms: Richardson method, Jacobi method, Gauss–Seidel method, successive overrelaxation (SOR) method, conjugate gradient methods, diagonally dominant matrices, Gerschgorin's Theorem, Gerschgorin discs, infinity matrix norm, nonnegative systems, computing eigenvalues using an iterative method, applications (demographic problems, population migrations, and economic models)

GENERAL EXERCISES 8.3

1. Show that the eigenvalues of an $n \times n$ matrix $\mathbf{A}$ lie in the union of n discs in the complex plane described as follows:
$$\mathcal{G}_i = \left\{ \lambda \,:\, |\lambda - a_{ii}| \le \sum_{\substack{j=1 \\ j \ne i}}^n |a_{ji}| \right\}$$
Notice that the sum in the formula uses the elements in the ith column of $\mathbf{A}$, not the ith row.

2. Explain why it is true or find a counterexample: For $n \times n$ matrices, we have $\|\mathbf{A}\mathbf{B}\|_\infty \le \|\mathbf{A}\|_\infty \cdot \|\mathbf{B}\|_\infty$.

3. Find a small interval on the real line that contains the eigenvalues of this matrix:

$$\begin{bmatrix} 1 & 5 & 3 & -4 \\ 5 & 2 & 1 & 7 \\ 3 & 1 & 3 & 1 \\ -4 & 7 & 1 & 4 \end{bmatrix}$$

Do not do any work other than using Gerschgorin's Theorem. Take advantage of the symmetry of the matrix **A**.

4. Use Gerschgorin's Theorem to deduce that a diagonally dominant matrix cannot have 0 as an eigenvalue, and thus conclude that every diagonally dominant matrix is invertible. (Do not use Theorem 1.)

5. Confirm that the function $\mathbf{A} \mapsto \left\|\mathbf{A}\right\|_\infty$ acting on the linear space of all $n \times n$ matrices has these properties:

a. $\left\|\mathbf{A}\right\|_\infty > 0$ if $\mathbf{A} \neq \mathbf{0}$

b. $\left\|\alpha\mathbf{A}\right\|_\infty = |\alpha|\left\|\mathbf{A}\right\|_\infty$

c. $\left\|\mathbf{A} + \mathbf{B}\right\|_\infty \leq \left\|\mathbf{A}\right\|_\infty + \left\|\mathbf{B}\right\|_\infty$

6. Show that for two $n \times 1$ matrices, $\mathbf{x}$ and $\mathbf{y}$, we have $|\mathbf{x}^T\mathbf{y}| \leq n\left\|\mathbf{x}\right\|_\infty \cdot \left\|\mathbf{y}\right\|_\infty$.

7. Establish that all eigenvalues of a matrix **A** lie in the disc of radius $||\mathbf{A}||_\infty$ centered at 0 in the complex plane.

8. Affirm that if $\left\|\mathbf{A}\right\|_\infty < c$, then $c\mathbf{I} \pm \mathbf{A}$ is invertible.

9. Solve the Leontief Open Model problem using these data

a. $\mathbf{A} = \begin{bmatrix} 0.1 & 0.6 & 0.6 \\ 0.3 & 0.2 & 0.0 \\ 0.3 & 0.1 & 0.1 \end{bmatrix}$, $\mathbf{b} = \begin{bmatrix} 0 \\ 18 \\ 0 \end{bmatrix}$

b. $\mathbf{A} = \begin{bmatrix} 0.5 & 0.4 & 0.2 \\ 0.2 & 0.3 & 0.1 \\ 0.1 & 0.1 & 0.3 \end{bmatrix}$, $\mathbf{b} = \begin{bmatrix} 50 \\ 30 \\ 20 \end{bmatrix}$

10. Establish that for any matrix **A** there is a vector **x** such that $\left\|\mathbf{x}\right\|_\infty = 1$ and $\left\|\mathbf{Ax}\right\|_\infty = \left\|\mathbf{A}\right\|_\infty$.

11. In the proof of Lemma 2, we used the inequality $1 - |a| \leq |1 - a|$. Establish this. Is it true for complex numbers? Is it true if we generalize the inequality to $b - |a| \leq |b - a|$?

12. What is the largest value that $||\mathbf{Ax}||_\infty$ can have if **A** is given and fixed, while **x** is constrained only by $||\mathbf{x}||_\infty \leq 1$?

13. Consider the matrix $\mathbf{A} = \begin{bmatrix} 2 & 1 & 0 \\ 1 & -3 & 1 \\ 1 & 0 & 5 \end{bmatrix}$

Apply Gerschgorin's Theorem and show the three discs in the complex plane whose union contains the eigenvalues. Note that these discs are disjoint from one another, and each disc contains an eigenvalue. This illustrates the refinement of Gerschgorin's Theorem described in Theorem 3.

14. Let **A** be a real, $n \times n$ matrix. Show that if $\alpha + i\beta$ is an eigenvalue of **A** (α and β being real), then we have

$$|\beta| \leq \max_{1 \leq i \leq n} \sum_{\substack{j=1 \\ j \neq i}}^{n} |a_{ij}|.$$

15. Establish that $(\mathbf{I} - \mathbf{B}^k)(\mathbf{I} - \mathbf{B})^{-1} = \mathbf{I} + \mathbf{B} + \mathbf{B}^2 + \cdots + \mathbf{B}^{k-1}$ for $k \geq 1$.

16. Let $\mathcal{A}$ have the block form $\begin{bmatrix} \mathbf{B} & \mathbf{C} \\ \mathbf{0} & \mathbf{I} \end{bmatrix}$ in which the submatrices are $n \times n$. Establish that if $(\mathbf{B} - \mathbf{I})$ is invertible, then

a. $\mathcal{A}^k = \begin{bmatrix} \mathbf{B}^k & (\mathbf{B}^k - \mathbf{I})(\mathbf{B} - \mathbf{I})^{-1}\mathbf{C} \\ \mathbf{0} & \mathbf{I} \end{bmatrix}$

b. Repeat for $\mathcal{A}^T$.

17. Let $\mathbf{A}$ be an $n \times n$ matrix. Define $p_i = \sum_{\substack{j=1 \\ j \neq i}}^{n} |a_{ij}|$, $q_i = \sum_{\substack{i=1 \\ i \neq j}}^{n} |a_{ij}|$, and $r_i = \min\{p_i, q_i\}$. Show that each eigenvalue of $\mathbf{A}$ is in one of the discs
$$\mathcal{C} = \{z : z \in \mathbb{C}, |z - a_{ii}| \leq r_i\}$$

18. Let $\mathbf{A} = \begin{bmatrix} 1 & 3 \\ 2 & 2 \end{bmatrix}$.

Use the power method and $\mathbf{x}^{(0)} = (1, 0)$ to find an eigenvalue of $\mathbf{A}$.

19. Let $\mathbf{A} = \begin{bmatrix} -3 & 1 & s \\ 2 & -4 & 1 \\ s & 1 & -2 \end{bmatrix}$.

Find conditions on s so that $\mathbf{A}$ is diagonally dominant.

20. Let $\mathbf{B} = \begin{bmatrix} 0 & 1 & 0 \\ 0 & -1 & 2 \\ 1 & -1 & 1 \end{bmatrix}$.

From the Gerschgorin discs, what can you conclude about the eigenvalues of $\mathbf{B}$? Sketch these discs and compare to the true eigenvalues.

21. Let $\mathbf{C} = \begin{bmatrix} 5 & 4 & 1 \\ 2 & 6 & 5 \\ 2 & 1 & 7 \end{bmatrix}$.

This matrix is *not* diagonally dominant, but it is invertible! What does this say with regard to Theorem 1?

22. If either a matrix or its transpose is diagonally dominant, then both of them are invertible. Explain.

23. A nonzero diagonal matrix is diagonally dominant and invertible. Explain.

24. The Gerschgorin discs coincide with the eigenvalue spectrum if and only if the matrix is diagonal. Explain.

25. Let $\mathbf{A} = \begin{bmatrix} 3 & t & 2 \\ 1 & 3 & 1 \\ t & 1 & -5 \end{bmatrix}$.

Find conditions on t so that $\mathbf{A}$ is diagonally dominant.

26. Let $\mathbf{H} = \begin{bmatrix} -5 & 2 & 2 \\ 1 & -2 & 0 \\ 3 & 5 & 8 \end{bmatrix}$.

$\mathbf{H}$ is *not* diagonally dominant, but it is invertible. Explain.

27. Let $\mathbf{A} = \begin{bmatrix} 5 & 1 & 1 \\ 1 & 7 & 0 \\ 1 & 1 & 9 \end{bmatrix}$.

What can you say about the eigenvalues of $\mathbf{A}$?

28. Let $\mathbf{G} = \begin{bmatrix} 5 & 2 \\ 4 & 7 \end{bmatrix}$.

When the power method is applied to the matrix $\mathbf{G}$, the result is a sequence of vectors that settle down to a vector of the form $[r, 1]^T$ where $|r| < 1$. Find approximate eigenvalue–eigenvector pairs of $\mathbf{G}$.

29. Let $\mathbf{A} = \begin{bmatrix} 0.3 & 0.1 & 0.0 \\ 0.0 & 0.2 & 0.1 \\ 0.0 & 0.0 & 0.3 \end{bmatrix}$

$$\mathbf{b} = \begin{bmatrix} 5 \\ 15 \\ 7 \end{bmatrix}$$

In the Leontief Open Method, we have $(\mathbf{I} - \mathbf{A})\mathbf{x} = \mathbf{b}$. Find a numerical value for $\mathbf{x}$.

30. (Continuation.) Repeat the previous exercise with $\mathbf{B} = \begin{bmatrix} -1 & -1 & 0 \\ 0 & -1 & -1 \\ 0 & 0 & 0 \end{bmatrix}$

$$\mathbf{c} = \begin{bmatrix} 3 \\ 12 \\ 2 \end{bmatrix}$$

31. If $A = (a_{ij})$ is a real $n \times n$ diagonally dominant matrix with all the diagonal elements a_{ii} being positive/negative, then the real part of its eigenvalues are positive/negative. Explain why.

32. Sometimes the definition of a diagonally dominant matrix is called **strict diagonal dominant** because the strict inequality $>$ is used, and is called **weak diagonal dominant** when the inequality $\geq$ is used. Examine these matrices for being strict or weak diagonally dominant. Are they invertible?

a. $A = \begin{bmatrix} 5 & -2 & 1 \\ 2 & 4 & -1 \\ 1 & -2 & -6 \end{bmatrix}$

b. $B = \begin{bmatrix} -2 & 1 & 0 \\ -1 & 2 & 1 \\ -2 & 2 & 3 \end{bmatrix}$

c. $C = \begin{bmatrix} 4 & 2 & -1 \\ -2 & -5 & 1 \\ 1 & -2 & -3 \end{bmatrix}$

33. Let $\begin{bmatrix} 4 & 1 \\ 1 & 3 \end{bmatrix} \begin{bmatrix} x_1 \\ x_2 \end{bmatrix} = \begin{bmatrix} 1 \\ 2 \end{bmatrix}$

Perform two steps of the conjugate gradient method starting with $x^{(0)} = 0$. Compare the results to the true solution.

34. Let $\begin{bmatrix} 2 & -1 & 0 \\ -1 & 2 & -1 \\ 0 & -1 & 2 \end{bmatrix} \begin{bmatrix} x_1 \\ x_2 \\ x_3 \end{bmatrix} = \begin{bmatrix} 1 \\ 0 \\ 1 \end{bmatrix}$

Perform two steps of the conjugate gradient method starting with $x^{(0)} = (0, 0, 0)$. Compare the results to the true solution.

35. Establish that for a square matrix A, zero is not an eigenvalue if and only if A is invertible.

36. Use Gerschgorin's Theorem to provide an alternative proof for Theorem 1.

37. Find $||x||_\infty$ when $x = (1, -3, 0, -2)$.

38. Determine $||A||_\infty$, when

$$A = \begin{bmatrix} -3 & 3 & 4 & 4 \\ 5 & 1 & 2 & -3 \\ -1 & 4 & -3 & -4 \\ -3 & -2 & 4 & -2 \end{bmatrix}$$

39. Using the previous two exercises, verify that $||Ax||_\infty \leq ||A||_\infty \cdot ||x||_\infty$.

40. The set of all eigenvalues of a square matrix A is the eigenvalue **spectrum** and is denoted $\sigma(A)$. Explain why a cheap, but rather crude, upper bound on all eigenvalues $\lambda \in \sigma(A)$ is:
$$|\lambda| \leq ||A||_\infty$$
Apply to A in General Exercise 38.

41. (Continuation.) Explain why the **spectral radius** of matrix A is $\rho(A) = \max_{\lambda \in \sigma(A)} |\lambda|$. Then
$$\rho(A) \leq ||A||_\infty$$
Apply to A in General Exercise 38.

42. Let $A = \begin{bmatrix} 1 & -2 & 0 \\ 1 & 2 & 0 \\ 0 & 0 & -1 \end{bmatrix}$

Find bounds on the eigenvalues of A in the following ways.

a. First, find a crude estimate using General Exercise 40.

b. Second, an estimate using the union of the Gerschgorin row discs.

c. Then, an estimate using the union of the Gerschgorin column discs.

d. Finally, use the intersection of the union of the row discs and the union of the column discs.

e. Illustrate with sketches of the Gerschgorin discs.

f. Compare to the true eigenvalues.

43. (Continuation.) Repeat for this matrix:

$$\mathbf{B} = \begin{bmatrix} 5 & 1 & 1 \\ 0 & 6 & 1 \\ 1 & 0 & -5 \end{bmatrix}$$

44. Explain why the Successive Overrelaxation method can be considered a generalization of the Gauss-Seidel method.

45. Let $\mathbf{A} = \begin{bmatrix} a & b \\ c & d \end{bmatrix}$

Show that $\mathbf{I} - \mathbf{A}$ is diagonally dominant if $||\mathbf{A}||_\infty < 1$.

46. Does Lemma 2 hold for the identity matrix?

47. Combine Lemma 2 and Theorem 1 to produce another result. Explain.

48. Explain how to obtain the optimum relaxation parameter ω_b from Young's equation.

49. Under what conditions on the matrix $\mathbf{A}$ does it follow that $\mathbf{A}$ is diagonally dominant if and only if $\mathbf{A}^T$ is diagonally dominant?

50. Using Gerschgorin discs, find bounds on the eigenvalues of these matrices.

a. $\mathbf{A} = \begin{bmatrix} 4 & -\frac{1}{2} & 0 \\ \frac{3}{5} & 5 & -\frac{3}{5} \\ 0 & \frac{1}{2} & 3 \end{bmatrix}$

b. $\mathbf{B} = \begin{bmatrix} 2 & 1 & 0 \\ -\frac{1}{2} & 6 & -\frac{1}{2} \\ 2 & 0 & 8 \end{bmatrix}$

COMPUTER EXERCISES 8.3

1. Write code to solve Example 2 using the following iterative methods:
   a. Jacobi
   b. Gauss–Seidel
   c. SOR with ω_b
   d. Conjugate gradient

2. Let $\mathbf{A} = \begin{bmatrix} 0.3 & 0.1 & 0.3 \\ 0.2 & 0.3 & 0.4 \\ 0.4 & 0.5 & 0.2 \end{bmatrix}$

Use mathematical software to compute $(\mathbf{I} - \mathbf{A})^{-1}$ and verify the solution in Example 11. Compute $\mathbf{A}^2$, $\mathbf{A}^3$, $\mathbf{A}^{10}$, and $\mathbf{A}^{20}$. For increasing values of m, does $\mathbf{A}^m \to \mathbf{0}$?

3. (Continuation.) Show that

$$(\mathbf{I} - \mathbf{A})^{-1} = \frac{5}{131} \begin{bmatrix} 56 & 23 & 21 \\ 16 & 44 & 6 \\ 21 & 6 & 47 \end{bmatrix}$$

so that the exact solution $(\mathbf{I} - \mathbf{A})^{-1}\mathbf{b}$ with

$$\mathbf{b} = \begin{bmatrix} 0 \\ 18 \\ 0 \end{bmatrix}$$ agrees with the approximate

solution given in the examples.

4. Program the Jacobi iterative method to solve the problem $\mathbf{Ax} = \mathbf{b}$, where

$$\mathbf{A} = \begin{bmatrix} 3 & 1 & 1 \\ 4 & 2 & 0 \\ 7 & -1 & 5 \end{bmatrix} \quad \text{and} \quad \mathbf{b} = \begin{bmatrix} 1 \\ 3 \\ 4 \end{bmatrix}$$

Draw a conclusion from the output of your program. Then run the same program using as the initial vector $[2, -2.5, -2.5]$ and with at least 200 steps. Explain what is happening.

5. (Continuation.) Change the matrix to

$$A = \begin{bmatrix} 3 & 1 & -1 \\ 0 & 2 & 1 \\ 1 & 3 & 5 \end{bmatrix}$$

and use the Jacobi method again. (Probably 40 or more steps will be needed to get a reasonable approximate solution.)

6. Let

$$\begin{bmatrix} 1 & 0.1 & -0.2 \\ -0.3 & 1 & 0.1 \\ -0.2 & 0.3 & 1 \end{bmatrix} \begin{bmatrix} x_1 \\ x_2 \\ x_3 \end{bmatrix} = \begin{bmatrix} 7 \\ 5 \\ 3 \end{bmatrix}$$

Write mathematical software to solve it using the following iterative methods:

a. Jacobi
b. Gauss–Seidel
c. SOR with ω_b
d. Conjugate gradient

7. Consider $\begin{cases} 2x_1 - x_3 = 2 \\ -x_1 + 3x_2 = -3 \\ x_2 + 4x_3 = 4 \end{cases}$

Starting with $\mathbf{x}^{(0)} = (0, 0, 0)$, calculate $\mathbf{x}^{(4)}$ by the Jacobi method and $\mathbf{x}^{(4)}$ by the Gauss–Seidel method.

8. Consider the augmented matrix

$$\begin{bmatrix} 2 & 1 & | & 7 \\ 1 & 3 & | & 11 \end{bmatrix}$$

Starting with $\mathbf{x}^{(0)} = (2, 3)$, calculate $\mathbf{x}^{(4)}$ by the Jacobi method and $\mathbf{x}^{(2)}$ by the Conjugate gradient method.

9. Verify calculations and assertions in Examples 5, 6, 7, and 8. Give details and justifications.

10. Consider the model elliptic partial differential equation known as the **Poisson equation** $\partial^2 u(x, y)/\partial x^2 + \partial^2 u(x, y)/\partial y^2 = f(x, y)$ on the unit square $0 < x, y < 1$ with boundary conditions $u(x, y) = g(x, y)$. Let $f(x, y) = 9(x^2 + y^2)$ and $g(x, y) = x - y$. With a uniform mesh subdivision of the square region, the finite difference approximations at mesh point (x_i, y_j) may be written $u_{i,j} - u_{i+1,j} - u_{i-1,j} - u_{i,j+1} - u_{i,j-1} = -h^2 f(x_i, y_j)$. With spacing $h = \frac{1}{4}$, we obtain nine unknowns and a 9×9 system of linear equations. In Figure 8.8, we indicate the ordering for the unknown vector $\mathbf{u}$ by first numbering the mesh points of the problem solution region in two different ways. Let the kth component $\mathbf{u}_k$ of the vector $\mathbf{u}$ be the unknown corresponding to the mesh point marked k. Consider the system $\mathbf{Au} = \mathbf{b}$, where

$$\begin{bmatrix} 4 & -1 & 0 & -1 & 0 & 0 & 0 & 0 & 0 \\ -1 & 4 & -1 & 0 & -1 & 0 & 0 & 0 & 0 \\ 0 & -1 & 4 & 0 & 0 & -1 & 0 & 0 & 0 \\ -1 & 0 & 0 & 4 & -1 & 0 & -1 & 0 & 0 \\ 0 & -1 & 0 & -1 & 4 & -1 & 0 & -1 & 0 \\ 0 & 0 & -1 & 0 & -1 & 4 & 0 & 0 & -1 \\ 0 & 0 & 0 & -1 & 0 & 0 & 4 & -1 & 0 \\ 0 & 0 & 0 & 0 & -1 & 0 & -1 & 4 & -1 \\ 0 & 0 & 0 & 0 & 0 & -1 & 0 & -1 & 4 \end{bmatrix}$$

$$\times \begin{bmatrix} u_1 \\ u_2 \\ u_3 \\ u_4 \\ u_5 \\ u_6 \\ u_7 \\ u_8 \\ u_9 \end{bmatrix} = \begin{bmatrix} g_{11} + g_{15} - h^2 f_1 \\ g_{12} - h^2 f_2 \\ g_{13} + g_{16} - h^2 f_3 \\ g_{17} - h^2 f_4 \\ -h^2 f_5 \\ g_{18} - h^2 f_6 \\ g_{19} + g_{22} - h^2 f_7 \\ g_{23} - h^2 f_8 \\ g_{20} + g_{24} - h^2 f_9 \end{bmatrix}$$

This is the **natural ordering** with the unknown points $u_1, u_2, u_3, u_4, u_5, u_6, u_7, u_8, u_9$ in the lexicographical order (left-to-right and bottom-to-top). Write mathematical software to solve it using the following iterative methods:

a. Jacobi
b. Gauss–Seidel
c. Conjugate gradient
d. SOR with ω_b

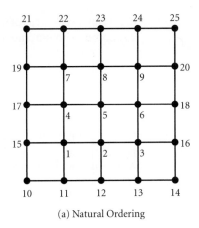

(a) Natural Ordering

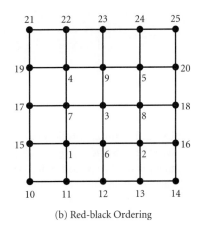

(b) Red-black Ordering

FIGURE 8.8

11. (Continuation.) Reorder the equations and unknowns in the system $\mathbf{Au} = \mathbf{b}$ to have this form,

$$\begin{bmatrix} 4 & 0 & 0 & 0 & 0 & -1 & -1 & 0 & 0 \\ 0 & 4 & 0 & 0 & 0 & -1 & 0 & -1 & 0 \\ 0 & 0 & 4 & 0 & 0 & -1 & -1 & -1 & -1 \\ 0 & 0 & 0 & 4 & 0 & 0 & -1 & 0 & -1 \\ 0 & 0 & 0 & 0 & 4 & 0 & 0 & -1 & -1 \\ -1 & -1 & -1 & 0 & 0 & 4 & 0 & 0 & 0 \\ -1 & 0 & -1 & -1 & 0 & 0 & 4 & 0 & 0 \\ 0 & -1 & -1 & 0 & -1 & 0 & 0 & 4 & 0 \\ 0 & 0 & -1 & -1 & -1 & 0 & 0 & 0 & 4 \end{bmatrix}$$

$$\times \begin{bmatrix} u_1 \\ u_2 \\ u_3 \\ u_4 \\ u_5 \\ u_6 \\ u_7 \\ u_8 \\ u_9 \end{bmatrix} = \begin{bmatrix} g_{11} + g_{15} - h^2 f_1 \\ g_{13} + g_{16} - h^2 f_2 \\ -h^2 f_3 \\ g_{19} + g_{22} - h^2 f_4 \\ g_{20} + g_{24} - h^2 f_5 \\ g_{12} - h^2 f_6 \\ g_{17} - h^2 f_7 \\ g_{18} - h^2 f_8 \\ g_{23} - h^2 f_9 \end{bmatrix}$$

This is the **red-black ordering** with the red unknown points u_1, u_2, u_3, u_4, u_5 and the black unknowns u_6, u_7, u_8, u_9.

12. Let $\begin{cases} 9x_1 + x_2 - x_3 = 18 \\ x_1 + 15x_2 + x_3 = -12 \\ -x_1 + x_2 + 20x_3 = -17 \end{cases}$

Using the initial condition $\mathbf{x}^{(0)} = (1, 1, 1)$, write the pseudocode for these iterative methods. Write a computer program that carries out a number of iterations. Compare the results to the true solution.

a. Jacobi

b. Gauss–Seidel

13. Write the pseudocode for approximating the eigenvalue of largest magnitude of these matrices by using the power method with the starting vectors given. Write a computer program that carries out a number of iterations. Compare the results to the true eigenvalues.

a. $\mathbf{A} = \begin{bmatrix} -1.2 & 1.1 \\ 3.6 & -0.8 \end{bmatrix}$

$\mathbf{x}^{(0)} = \begin{bmatrix} 1 \\ 1 \end{bmatrix}$

b. $\mathbf{B} = \begin{bmatrix} 1 & 1 & -1 \\ 1 & 2 & 1 \\ 2 & -1 & 1 \end{bmatrix}$

$$\mathbf{x}^{(0)} = \begin{bmatrix} 1 \\ 1 \\ 1 \end{bmatrix}$$

14. Let $\begin{cases} 4x_1 + 2x_2 - x_3 = 5 \\ 2x_1 + 5x_2 + 2x_3 = 7 \\ x_1 + x_2 + 3x_3 = 9 \end{cases}$

Using the initial condition $\mathbf{x}^{(0)} = (0, 0, 0)$, write the pseudocode for these iterative methods. Write a computer program that carries out a number of iterations. Compare the results to the true solution.
a. Jacobi
b. Gauss–Seidel

15. Let $\mathbf{A} = \begin{bmatrix} 2 & -1 & 0 \\ -1 & 2 & -1 \\ 0 & -1 & 2 \end{bmatrix}$

$$\mathbf{b} = \begin{bmatrix} 1 \\ 0 \\ 1 \end{bmatrix}$$

Write pseudocode for the SOR iterative methods with $\omega = 1.7$ and starting with $\mathbf{x}^{(0)} = (0, 0, 0)$. Carry out a number of iterations and compare the approximate solution to the true solution.

16. Consider $\begin{bmatrix} 16 & 3 \\ 7 & -11 \end{bmatrix} \begin{bmatrix} x_1 \\ x_2 \end{bmatrix} = \begin{bmatrix} 11 \\ 13 \end{bmatrix}$

Write the pseudocode and carry out a number of iterations of the following iterative methods starting with $\mathbf{x}^{(0)} = (0.8, -0.6)$. Compare the approximate solution to the true solution.
a. Jacobi method
b. Gauss–Seidel method

17. Carry out several steps of the power method and compare to the true results.

a. $\mathbf{A} = \begin{bmatrix} 1 & 1 \\ 2 & 0 \end{bmatrix}$

$$\mathbf{x}^{(0)} = \begin{bmatrix} 1 \\ 0 \end{bmatrix}$$

b. $\mathbf{B} = \begin{bmatrix} 0 & 5 & -6 \\ -4 & 12 & -12 \\ 2 & -2 & 10 \end{bmatrix}$

$$\mathbf{x}^{(0)} = \begin{bmatrix} 1 \\ 1 \\ 1 \end{bmatrix}$$

18. Write and test a computer program that carries out the following scheme proposed by Brualdi and Mellendorf [1994] to make Gerschogrin's Theorem stronger.
For any $n \times n$ matrix $\mathbf{A}$, let $r_i = \sum_{\substack{i=1 \\ i \neq j}}^{n} |a_{ij}|$ and let k be an integer with $1 \leq k \leq n$ and let $s_j^{(k-1)}$ be the sum of the magnitude of the $r - 1$ largest off-diagonal elements in column j. Then each eigenvalue of $\mathbf{A}$ is either in one of the discs

$$\{z : |z - a_{jj}| \leq s_j^{(k-1)}\}$$

or in one of the regions

$$\{z : \sum_{i \in P} |z - a_{ii}| \leq \sum_{i \in P} r_i\}$$

where P is any subset of $\{1, 2, \ldots, n\}$ such that $|P| = k$.
For a given real or complex matrix $\mathbf{A}$, plot the resulting discs to identify a region in the complex plane that contains all of the eigenvalues of the matrix.

Deductive Reasoning and Proofs

> " *Education is the lighting of the fire, not the filling of a bucket.*
> —WILLIAM BUTLER YEATS (1865–1939, Nobel Prize in Literature, 1923)
>
> *A mind is a fire to be kindled, not a vessel to be filled.* "
> —PLUTARCH (AD 45–125, Priest of the Delphic Oracle)

A.1 Introduction

One of your objectives in studying this book should be to improve your skill in deductive reasoning. This activity ought to be enjoyable, as it involves real thinking of an analytic nature. Finding a convincing justification for a suspected truth after a long search is very good for your ego! It is creative activity akin to solving a puzzle. But it is better than that, because in some cases you may have established a fact for the first time in history, and thus have actually added to civilization's store of knowledge.

Strings of logical arguments are essential in mathematics. They can put a stamp of validity on a conjecture that was only guessed at the start. Or they can lay to rest as incorrect some other conjecture. As you probably know, there are many conjectures in mathematics that have impressive evidence for their validity, yet have no complete justification. For example: Is it true that there are infinitely many prime numbers, n, such that $n + 2$ is also prime? These pairs are called **twin primes**. Examples are $(3, 5)$, $(5, 7)$, $(11, 13)$, $(17, 19)$, $(29, 31)$, and so on. Nobody knows whether this string of examples comes to a halt or continues forever.

There are many opportunities to practice the art of deductive reasoning in the subject of linear algebra. However, at the beginning, we can discuss some aspects of deduction in other more familiar contexts, such as in calculus, geometry, or algebra.

A.2 Deductive Reasoning and Direct Verification

One method of establishing the validity of some result is the use of **direct verification**. Usually this means that the proving process has a natural starting place with something known, and proceeds from there on a straight logical path to the desired conclusion. Often in such a procedure, there is no inventiveness required—only a calculation. Let us give an example.

THEOREM 1

If x and y are real numbers, and if $a = y^2 - x^2$, $b = 2xy$, and $c = x^2 + y^2$, then $a^2 + b^2 = c^2$.

PROOF With a, b, c as given we have

$$
\begin{aligned}
a^2 + b^2 &= (y^2 - x^2)^2 + (2xy)^2 \\
&= y^4 - 2y^2x^2 + x^4 + 4x^2y^2 \\
&= y^4 + 2y^2x^2 + x^4 \\
&= (y^2 + x^2)^2 \\
&= c^2
\end{aligned}
$$
■

In the previous demonstration, a small black box has been printed to signify the end of the proof. This notation is used throughout this book to indicate the end of proofs or examples.

In Theorem 1, if x and y are integers satisfying the inequality $0 < x < y$, then a and b are the integer sides of a right triangle and c is the hypotenuse. For example, $(3, 4, 5)$ is such a triple, obtained by taking $x = 1$ and $y = 2$. If a, b, and c are positive integers such that $a^2 + b^2 = c^2$, then we call (a, b, c) a **Pythagorean[1] triple**.

---

[1] The Greek mathematician and savant Pythagoras flourished around 500 BC. He was born on the island of Samos, but emigrated to Croton in southern Italy around 531 BC. He is credited with discovery of the ratios among pleasing musical intervals. In Croton, he drew about him disciples in a new religion open to both men and women, which was unusual for the time. His group seems to have grasped the notion that the earth is spherical. He or his group discovered irrational numbers and the so-called Pythagorean Theorem about right triangles. In ancient times these matters were of practical importance in building, because they made it easy to construct right angles.

Why don't you try an example (unless you are already skilled in this art)? Here is one to test yourself on.

> ## THEOREM 2
>
> *Every integer that can be expressed in the form $14n - 3$ can be expressed in the form $7k + 4$, where n and k are integers.*

A.3 Implications

In mathematics, we are often trying to establish implications. This means proving that one assertion implies another. These implications have the form "If P then Q," where P and Q are assertions. A typical example of an implication (one that happens to be true) is

> "*If x is a real number, then x^2 is greater than or equal to 0.*"

The symbol $\Rightarrow$ is often used to mean **implies**. Thus, the preceding example can be abbreviated by writing

$$x \in \mathbb{R} \Rightarrow x^2 \geq 0$$

An assertion $P \Rightarrow Q$ can be read as (P implies Q) or (if P is true, then Q is true). There are several other variations in the phraseology:

> "*P is sufficient for Q,*" or "*Q is necessary for P,*" or "*Q if P,*"
>
> "*P only if Q,*" or "*if Q is false, then so is P,*" or
>
> "*if P is true, then so is Q.*"

Consider a statement of the form $P \Rightarrow Q$. The **contrapositive** of this statement is $\neg Q \Rightarrow \neg P$. The symbol $\neg$ means *negation*. The contrapositive is logically equivalent to the original statement. In other words, if one is true then so is the other.

The **converse** of $P \Rightarrow Q$ is $Q \Rightarrow P$. The converse is *not* equivalent to the original statement, $P \Rightarrow Q$.

The symbol $\Leftrightarrow$ means **if and only if**. In this case, there are two implications, going in opposite directions. As an example, let P stand for "$x^2 - 3x + 2 = 0$," and let Q stand for "*either $x = 1$ or $x = 2$.*" In this case, we have $P \Leftrightarrow Q$. The terminology is that P and Q are logically equivalent to each other. We also say, "*P is true if and only if Q is true.*" Another similar expression is "*P is a necessary and sufficient condition for Q.*" Because the relationship is **reciprocal**, we can also say "*Q is necessary and sufficient for P, or Q is true if and only if P is true.*"

Despite the great difference between the two implications $P \Rightarrow Q$ and $Q \Rightarrow P$, we often begin a search for a proof of $P \Rightarrow Q$ by asking, *What follows from assuming Q?* Thus, we might start with Q and draw conclusions from it. If we can arrive at P in a logical manner, then we can try to reverse all the implications, to get a proof of $P \Rightarrow Q$. Here is an illustration of this modus operandi.

EXAMPLE 1 Find the solutions of the equation $\sqrt{2x-5}+4-x=0$.

SOLUTION This sort of problem arises in many guises, and most mathematicians would start the search by stating the equation in the form $x-4 = \sqrt{2x-5}$. In other words, we begin by assuming the validity of the equation, and we will see what might follow from that premise. By elementary algebra, these consequences follow:

$$x^2 - 8x + 16 = 2x - 5, \qquad x^2 - 10x + 21 = 0, \qquad (x-7)(x-3) = 0$$

Thus, we obtain $x = 7$ or $x = 3$. So far we have proved that if x satisfies the given equation, then x is either 7 or 3. It is not legitimate to conclude that 7 and 3 are solutions, because one of the implications in the chain is not reversible. It is the step where we have put $x = 3$ and want to get back to the equation $x-4 = \sqrt{2x-5}$ with that value. Testing 3 as a possible solution reveals that 3 is *not* a solution of the original equation. ∎

Do not make the mistake of thinking that $P \Rightarrow Q$ is the same as $Q \Rightarrow P$. Each of these implications is the **converse** of the other. For example, the converse of

"*If n is divisible by 2, then n is divisible by 6*"

is the statement

"*If n is divisible by 6, then n is divisible by 2*"

The first statement is false, whereas the second (the converse of the first) is true. For another example, consider this proposition:

"*If a and b are the legs of a right triangle and the hypotenuse is c, then $a^2 + b^2 = c^2$.*"

The converse of this assertion is

"*If $a^2 + b^2 = c^2$, then the triangle having sides $|a|, |b|, |c|$ is a right triangle.*"

Are both of these implications true?

A.4 Method of Contradiction

Suppose that we want to prove a theorem of the form $P \Rightarrow Q$. We can assume that P is true and Q is false and see whether that leads to an obvious contradiction. If it does, then the hypothesis P and not Q is not tenable. In other words, if P is true then so is Q.

EXAMPLE 2 Use the method of contradiction to prove that for real numbers x and y we always have $|x+y| \le |x|+|y|$.

SOLUTION Suppose that for some pair of numbers x and y we have

$$|x+y| > |x|+|y|$$

We seek some consequence of that assumption that is immediately seen to be false. Let $\sigma = 1$ if $x+y \ge 0$ and $\sigma = -1$ if $x+y < 0$. Then $|x| \ge \sigma x$ and $|y| \ge \sigma y$. Hence, we have

$$\sigma x + \sigma y = \sigma(x+y) = |x+y| > |x|+|y| \ge \sigma x + \sigma y$$

This is plainly false, as no number can be greater than itself. This contradiction establishes the desired inequality. ∎

A.5 Mathematical Induction

An important technique in establishing theorems is called **mathematical induction**. This is often used when a proposition depends on a positive integer n. Call it $P(n)$. We want to establish $P(n)$ for all values of n. That is, $n = 1, n = 2, n = 3, n = 4$, and so on.

THEOREM 3   Mathematical Induction
For each natural number n, let $P(n)$ be a mathematical statement. If $P(1)$ is true and if the implication $P(n) \Rightarrow P(n+1)$ is true for $n = 1, 2, 3, \ldots$, then $P(n)$ is true for all natural numbers n.

PROOF We use the method of contradiction. Thus, we assume the hypotheses are true while the conclusion is false. We then expect to arrive at a contradiction. Suppose that $P(n)$ is sometimes false. Let m be the first integer for which $P(m)$ is false. Thus, $P(1), P(2), \ldots, P(m-1)$ are true but $P(m)$ is false. Here, we must invoke another principle of logic: It asserts that every nonempty set of positive integers must contain a least element. Note that $m > 1$ because $P(1)$ is true. Note also that $P(m-1)$ is true. But $P(m-1)$ implies $P(m)$. So by our second hypothesis $P(m)$ must be true, contradicting our choice of m. ∎

EXAMPLE 3 Use mathematical induction to establish that for all natural numbers, n,
$$1 + 4 + 9 + \cdots + n^2 = \tfrac{1}{6}n(n+1)(2n+1)$$

SOLUTION Use $P(n)$ to stand for this equation. First, we verify that $P(1)$ is true. This is easy, as $P(1)$ simply states that $1 = \tfrac{1}{6}(2)(3)$. Now we establish the implication $P(n) \Rightarrow P(n+1)$. In this part of our work, $P(n)$ becomes our hypothesis, and n is fixed. Hence, we can write

$$1 + 4 + \cdots + n^2 + (n+1)^2 = \tfrac{1}{6}n(n+1)(2n+1) + (n+1)^2$$
$$= \tfrac{1}{6}(n+1)\left[n(2n+1) + 6(n+1)\right]$$
$$= \tfrac{1}{6}(n+1)\left[2n^2 + 7n + 6\right]$$
$$= \tfrac{1}{6}(n+1)(n+2)(2n+3)$$

This last expression is exactly the formula we want for $P(n+1)$. ∎

Induction can be defined in two different forms: *weak* induction and *strong* induction.

Definition of Weak Induction

If $P(1)$ *is true* (*base case*) and $P(n) \Rightarrow P(n+1)$ for $n = 1, 2, \ldots$, (*inductive step*), then $P(n)$ is true for all positive integers n.

Definition of Strong Induction

If $P(1)$ is *true* (*base case*) and whenever $P(1), P(2), \ldots, P(n)$ are all true $\Rightarrow P(n+1)$ is true (*inductive step*), then $P(n)$ is true for all positive integers n.

Obviously, the difference is in the inductive step. In ordinary or weak induction, we proceed from one step to the next step as on rungs of a ladder. In strong induction, one must know that all the rungs below the current step are true before concluding that the current case is true. From a practical point of view, they are logically the same. Nevertheless, strong induction may make the inductive step easier to prove.

Induction can be generalized by taking $P(k)$ as the first step (assumed to be true). Then we establish that when $P(n)$ is true $P(n+1)$ is also true (inductive step). Here $n = k, k+1, \ldots$. The conclusion is that $P(n)$ is true for all positive integers n greater than or equal to k.

A.6 Truth Tables

Sometimes, in complicated arguments, one needs **truth tables**. The simplest example of a truth table is the one for logical **not**:

P	$\neg P$
T	F
F	T

(Here the notation $\neg P$ is used for *not-P*.) This table is interpreted as follows: If P is true then not-P is false, and if P is false then not-P is true.

The truth table for **implication** is especially important:

P	Q	$P \Rightarrow Q$
T	T	T
T	F	F
F	T	T
F	F	T

Notice that the only circumstance in which the implication $(P \Rightarrow Q)$ is false is that P is true while Q is false. For example, the following statement is false: "*If* $5^2 = 25$, *then Caesar wrote The Merchant of Venice.*"

Here is the truth table for P **and** Q (denoted by $P \wedge Q$).

P	Q	$P \wedge Q$
T	T	T
T	F	F
F	T	F
F	F	F

A.7 Subsets and de Morgan Laws

A common occurrence in proofs is the need to establish that two sets, say S and T, are equal. This task can be broken into two parts: First, we show that every member of S is a member of T, and second we show that every member of T is a member of S. A moment's thought should convince you that with these two facts established, the sets must be identical. Both assertions are needed, however. Using the symbol $\subseteq$, we say that we must prove $S \subseteq T$ and $T \subseteq S$. (The symbol $\subseteq$ is rendered in English as "*is a subset of*" or "*is contained in.*") We are using this principle of logic:

$$[S = T] \Leftrightarrow [S \subseteq T \text{ and } T \subseteq S]$$

EXAMPLE 4 Show that these two sets are equal:

$$A = \{n \ : \ n = 13k + 2 \text{ for some integer } k\}$$
$$B = \{n \ : \ n = 13k - 11 \text{ for some integer } k\}$$

SOLUTION We have

$$n \in A \implies n = 13k + 2 = 13(k+1) - 11 \implies n \in B$$
$$n \in B \implies n = 13k - 11 = 13(k-1) + 2 \implies n \in A \qquad \blacksquare$$

Recall the definitions of **union, intersection,** and **difference** of sets:

$$S \cup T = \{x \ : \ x \in S \text{ or } x \in T\} \qquad \textbf{(union)}$$
$$S \cap T = \{x \ : \ x \in S \text{ and } x \in T\} \qquad \textbf{(intersection)}$$
$$S \smallsetminus T = \{x \ : \ x \in S \text{ and } x \notin T\} \qquad \textbf{(set difference)}$$

THEOREM 4   de Morgan Laws
For two sets A and B in the universal set U, we have

$$U \smallsetminus (A \cup B) = (U \smallsetminus A) \cap (U \smallsetminus B)$$
$$U \smallsetminus (A \cap B) = (U \smallsetminus A) \cup (U \smallsetminus B)$$

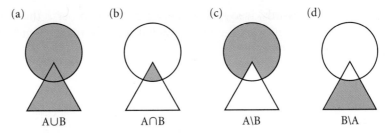

FIGURE A.1 de Morgan laws and Venn diagrams. The diagrams illustrate $(A \smallsetminus B) \cup (B \smallsetminus A) \cup (A \cap B) = A \cup B$.

PROOF We prove the first and leave the second to the reader. We have this string of equivalences:

$$x \in [U \smallsetminus (A \cup B)] \Leftrightarrow [x \in U \text{ and } x \notin (A \cup B)]$$
$$\Leftrightarrow [x \in U \text{ and } x \notin A \text{ and } x \notin B]$$
$$\Leftrightarrow [x \in (U \smallsetminus A) \text{ and } x \in (U \smallsetminus B)]$$
$$\Leftrightarrow [x \in [(U \smallsetminus A) \cap (U \smallsetminus B)]] \qquad \blacksquare$$

In Figure A.1 there are illustrations of the de Morgan laws.

A.8 Quantifiers

Many mathematical assertions have quantifiers in them. There are two types of quantifier: "*For all x, …*" and "*For some x, ….*" The first is called a **universal quantifier** and the second is called an **existential quantifier**. Typical examples are these two (true) statements:

1. For all integers n, $n^3 - n$ is divisible by 3.
2. There exists a real number x such that $x^2 + 3x - 7 = 0$.

Often a shorthand is used for these two quantifiers. The symbol $\forall$ means **for all**, and $\exists$ means **there exists**. These are particularly useful for homework and informal proofs. However, neither of these should be used in formal mathematical exposition unless absolutely necessary for clarity. When there are several quantifiers in one statement, it sometimes helps to use these symbols. Consider an example where we are working with real numbers and integers (n, m).

$$(\forall \varepsilon > 0)(\exists m \in \mathbb{N})(\forall n > m) \left\{ \left| \frac{1+n}{1+3n} - \frac{1}{3} \right| < \varepsilon \right\}$$

In words, it says: For every positive number ε there exists a positive integer m with the property that for any integer n greater than m, we have $\left|\dfrac{1+n}{1+3n}-\dfrac{1}{3}\right|<\varepsilon$. This is the mathematically exact way of stating that

$$\lim_{n\to\infty}\frac{1+n}{1+3n}=\frac{1}{3}$$

A.9 Denial of a Quantified Assertion

To create the denial of a statement involving quantifiers, follow these patterns, where the symbol $\neg$ means **not**, and $P(x)$ is some statement involving x:

$$\left[\neg(\forall x)(P(x))\right]\Leftrightarrow\left[(\exists x)(\neg P(x))\right]$$
$$\left[\neg(\exists x)(P(x))\right]\Leftrightarrow\left[(\forall x)(\neg P(x))\right]$$

These symbolic statements have simple verbalizations. The first is:

"*Asserting that $P(x)$ is not always true is logically equivalent to asserting that for some x, $P(x)$ is false.*"

The second is:

"*Asserting that no x satisfies $P(x)$ is equivalent to asserting that for all x, $P(x)$ is false.*"

For example, if we wish to deny the statement about limits above, we write

$$(\exists\varepsilon>0)(\forall m)(\exists n>m)\left\{\left|\frac{1+n}{1+3n}-\frac{1}{3}\right|\geq\varepsilon\right\}$$

A.10 Some More Questionable "*Proofs*"

In the remainder of Appendix A, we shall examine some proofs and pseudo-proofs. We shall use the term pseudoproof to designate a line of reasoning that falls short of being a genuine proof because of some lapse in logic.

One very common mistake is to attempt a proof that $P\Rightarrow Q$ by an argument that in fact establishes $Q\Rightarrow P$. In such a case, if the logical steps can be reversed, a proof of $P\Rightarrow Q$ emerges. But this last step of reversing the argument is essential. Here are two illustrations.

EXAMPLE 5 We want to prove that $5=7$.

SOLUTION It is natural to start with $5 = 7$. Then add -6 to both sides, getting $-1 = 1$. Upon squaring both terms in the last equation, we get $1 = 1$. What does this short argument actually prove? It proves that if $5 = 7$, then $1 = 1$. That is a true statement, but not of any value. (Refer to the truth table for implication to see that a false statement implies any other statement, true or false.) What we really should have proved is the argument in reverse: If $1 = 1$, then $5 = 7$. (That's much harder, isn't it?) ∎

EXAMPLE 6 Here is another such *pseudoproof.* We want to prove that π is really 0.

SOLUTION Start with the equation $\pi = 0$. Take the sine function of both sides of the equation, getting $\sin \pi = \sin 0 = 0$. ∎

EXAMPLE 7 We want to prove that if x and y are positive integers, then so is $(x^5 + y^5)/(x + y)$. Because of that division, this assertion is certainly suspicious.

SOLUTION We note that

$$(x + y)(x^4 - x^3 y + x^2 y^2 - xy^3 + y^4)$$
$$= x^5 - x^4 y + x^3 y^2 - x^2 y^3 + xy^4 + x^4 y - x^3 y^2 + x^2 y^3 - xy^4 + y^5$$
$$= x^5 + y^5$$

It follows that

$$\frac{x^5 + y^5}{x + y} = x^4 - x^3 y + x^2 y^2 - xy^3 + y^4$$

Can you find a flaw in this argument? ∎

SUMMARY APPENDIX A

- **Pythagorean triple** (a, b, c): $a^2 + b^2 = c^2$, where a, b, and c are positive integers.
- If $a = y^2 - x^2$, $b = 2xy$, and $c = x^2 + y^2$, where $x, y \in \mathbb{R}$, then $a^2 + b^2 = c^2$.

- **Mathematical Induction:** For $n \in \mathbb{N}^+$, let $P(n)$ be a mathematical statement. If $P(1)$ is true and if the implication $P(n) \Rightarrow P(n+1)$ is true for all $n \in \mathbb{N}^+$, then $P(n)$ is true for all $n \in \mathbb{N}^+$.

- **Union:** $S \cup T = \{x : x \in S \text{ or } x \in T\}$
  Intersection: $S \cap T = \{x : x \in S \text{ and } x \in T\}$
  Difference: $S \smallsetminus T = \{x : x \in S \text{ and } x \notin T\}$

- **de Morgan Laws:** For sets A and B in the universal set U, we have
  $$U \smallsetminus (A \cup B) = (U \smallsetminus A) \cap (U \smallsetminus B)$$
  $$U \smallsetminus (A \cap B) = (U \smallsetminus A) \cup (U \smallsetminus B)$$

- **Function:** If f maps X into Y, then X is the domain of f, and Y is the codomain of f. If y is the image of x by the mapping f, then $y = f(x)$. The range of f is the set $\{f(x) : x \in X\}$.

- The composition of two mappings f and g is defined by the equation $(f \circ g)(x) = f(g(x))$.

KEY CONCEPTS APPENDIX A

Twin primes, deductive reasoning, direct verification, Pythagorean triples, pseudoproofs, implication, contrapositive, converse, inverse, method of contradiction, logical equivalence, weak and strong induction, hypothesis, conclusion, algebra of sets, truth tables, de Morgan's Laws, union, difference, and intersection of sets, functions, mappings, domain, codomain, range, composition of mappings, universal and existential quantifiers, denial of a quantified assertion, more pseudoproofs

GENERAL EXERCISES APPENDIX A

1. Establish this factorization:
   $$x^n - y^n = (x-y)(x^{n-1} + x^{n-2}y + x^{n-3}y^2 + \cdots + x^2 y^{n-3} + xy^{n-2} + y^{n-1})$$
   Begin by writing in detail the first three cases: $n = 2, 3$, and 4, and verifying that they are correct. Then you should see how to verify the general case. Mathematical induction is not recommended; the problem requires only elementary algebra.

2. (Continuation.) Establish that for every positive integer n, $7^n - 4^n$ is an integer multiple of 3. (*Hint:* Use General Exercise 1.)

3. Compute the first few partial sums in the expression $1^3 + 2^3 + 3^3 + 4^3 + \cdots$ and observe that the results are perfect squares. Establish that this is true for all partial sums. Try to find a formula for the partial sums in question.

4. Justify the claim that for positive real numbers x and y we have the **arithmetic–geometric mean property:** $\sqrt{xy} \le \frac{1}{2}(x+y)$

5. Although an easier method is available, use induction to establish the formula $1 + 2 + 3 + \cdots + n = \frac{1}{2}n(n+1)$, for $n \ge 1$.

6. **Pythagorean triples** (a, b, c) are three integers that satisfy $a^2 + b^2 = c^2$, which was known to Euclid and used by Diophantus. A way to generate these triples is to let n and m be integers $n > m$ and define $a = n^2 - m^2$, $b = 2nm$, and $c = m^2 + n^2$. Can you establish that the resulting three numbers a, b, and c always form a Pythagorean triple?

7. (Continuation.) Find all the Pythagorean triples (a, b, c) in which $b = a + 1$. Here, a theorem should be formulated and justified with a suitable argument.

8. a. Establish the assertion that if a prime number greater than 3 is divided by 6, then the remainder is either 1 or 5.

b. If you succeed with this, go on to establish that if p is a prime number greater than 3, then either $p+1$ or $p-1$ is a multiple of 6.

c. If you succeed with this, justify the assertion that if p is a prime greater than 3 such that $p+2$ is also prime, then $p=6n-1$ for some integer n.

d. Finally, take the giant step of establishing that there are infinitely many integers n having the property that $6n-1$ and $6n+1$ are prime.

9. Establish that for arbitrary sets A_i contained in a set X, the **de Morgan Laws** are valid:

$$U \smallsetminus \bigcup_{i=1}^{n} A_i = \bigcap_{i=1}^{n} (U \smallsetminus A_i)$$

$$U \smallsetminus \bigcap_{i=1}^{n} A_i = \bigcup_{i=1}^{n} (U \smallsetminus A_i)$$

Note that $\bigcup_i A_i$ is the set of all x that belong to at least one of the sets A_i. The expression $\bigcap_i A_i$ is similar. It is the set of all entities that belong to each A_i.

10. Find the relationship between $f(\bigcap_i A_i)$ and $\bigcap_i f(A_i)$. Establish your answer with suitable deductions and give examples to convince one that no stronger relationship is valid in general.

11. Establish that for arbitrary sets, if $A \subseteq B$ and $B \subseteq C$, then $A \subseteq C$.

12. Establish the famous theorem from elementary geometry that the altitudes of a triangle meet at a point. (A line segment is an **altitude** of a triangle if it starts at a vertex and ends at the side opposite that vertex, meeting that side in a right angle. The side

in question may have to be extended to the point of intersection mentioned.)

13. Assume that x is not 0 or a negative integer. Find a formula for

$$\frac{1}{x(x+1)} + \frac{1}{(x+1)(x+2)} + \frac{1}{(x+2)(x+3)}$$
$$+ \cdots + \frac{1}{(x+n-1)(x+n)}$$

Let $n \to \infty$ in your formula and find the limit of the preceding expression. A **partial fraction decomposition** familiar from calculus can be used.

14. Discover and justify this formula:
$(1)(2)+(2)(3)+(3)(4)+\cdots+(n)(n+1)$.
Perhaps there is a formula of the type $a+bn+cn^2+dn^3+en^4+\cdots$.

15. Here is the technical definition of continuity of a function f at every point: $(\forall x)(\forall \varepsilon > 0)(\exists \delta > 0)(\forall y)\big[|x-y| < \delta \Rightarrow |f(x)-f(y)| < \varepsilon\big]$. What is the denial of this assertion in symbols?

16. A median of a triangle is a line segment joining a vertex to the midpoint of the opposite side. Verify that the three medians of any triangle meet at a point.

17. What is the negation (denial) of the assertion that for every nonzero real number x, x^4 is positive?

18. Establish that for every positive integer, $n^3 - n$ is divisible by 3.

19. Show that if (a,b,c) is a **Pythagorean triangle**[‡] and if the three integers a, b, and c have no common factor, then a and b have opposite polarity; that is, one is odd and the other is even.

[‡]Olga Taussky-Todd (1906–1995) gave her International Linear Algebra Society's Noether Lecture, which honors women who have made fundamental and sustained contributions to the mathematical

(*continued*)

20. Discover and justify a formula for the sum:
$1+3+5+7+\cdots+(2n-1)$

21. Affirm that the two roots of a quadratic equation, $ax^2+bx+c=0$, add up to $-b/a$. Is there a similar result for the roots of a cubic equation, $ax^3+bx^2+cx+d=0$? What generalization can you establish?

22. Establish that for any two sets A and B, we have $A=B$ if and only if $A \subseteq B$ and $B \subseteq A$.

23. Show that for three sets, C, D, and E, if $C \subseteq D$ and $D \subseteq E$, then $C \subseteq E$.

24. Explain why if $B = A \cap B$, then $B \subseteq A$, and conversely.

25. Argue that if $B = A \cup B$, then $A \subseteq B$, and conversely.

26. Show that real numbers obey this rule: $|x+y| \le |x|+|y|$. This is the **triangle inequality**. (Do not copy the proof used in illustrating the Method of Contradiction.)

27. Explain why or give a counterexample to the assertion that $|x| \le |y|$ if and only if $-y \le x \le y$.

28. The **triangular numbers** are the integers generated by the partial sums of the series $1+2+3+4+5+6+\cdots$. The first few are therefore 1, 3, 6, 10, ... Sometimes a triangular number is a perfect square. For example, 1 and 36 are triangular and perfect squares. Are there any others? Can you find a way of getting all of them? (See Silverman [2001].)

29. The result of General Exercise 4 has a generalization: If the average of n positive numbers is raised to the nth power, the result is *not* less than the product of those numbers. Establish this. (See Hall and Knight [1948].)

30. We write a sequence of real numbers in the following way: $\mathbf{x} = [x_0, x_1, x_2, \ldots]$. If all the differences $x_1 - x_0,\ x_2 - x_1,\ x_3 - x_2 \ldots$ are equal, the sequence is called an **arithmetic progression**. Show that, for any arithmetic progression, there is a formula $x_n = a + bn$, where a and b are two constants, and n is a positive integer.

31. (Continuation.) Given an arithmetic progression, as described in the preceding exercise, find a formula for the sum of the first n terms.

32. (Continuation.) Let $x_0, x_1, x_2, \ldots$ be an arithmetic progression. If the sum of the first seven terms is 49 and the sum of the first 17 terms is 289, what is the formula for the sequence?

33. (Continuation.) If the sum of the first n terms in an arithmetic progression is $n(5n-3)$, what is the formula for the progression?

34. A sequence $\mathbf{x} = [x_0, x_1, x_2, \ldots]$ is called a **geometric progression** if the ratios x_1/x_0, $x_2/x_1 \ldots$ are all the same. Affirm that such a sequence must obey a rule of the form $x_n = ar^n$, for $n = 0, 1, 2, \ldots$.

35. (Continuation.) Find and prove a rule for the sum of n terms in a geometric progression. Use it to find the sums of

---

sciences, on "*The many aspects of Pythagorean triangles.*" See Taussky-Todd [1982], which was coauthored with her husband John Todd. She wrote "*... both in the work of others and in my own work I look for beauty, and not only for achievement.*"

the geometric progression whose first two terms are $1/\sqrt{2}$ and -2.

36. Confirm that:
$$1^3 + 2^3 + \cdots + n^3 = \tfrac{1}{4}n^2(n+1)^2$$

37. Use induction to establish the **Binomial Theorem:** $(x+y)^n = \displaystyle\sum_{k=0}^{n} \binom{n}{k} x^{n-k} y^k$, for $n \geq 1$ and $x \in \mathbb{R}$. The binomial coefficients are defined by $\dbinom{n}{k} = \dfrac{n!}{k!(n-k)!}$.

38. Verify that for every integer n greater than 2, the number $n^4 + 2n^3 - n^2 - 2n$ is divisible by 24.

39. Consider this assertion: $(\forall x \in \mathbb{R})(x^2 \geq 0)$. Is its denial $(\forall x \in \mathbb{R})(x^2 < 0)$?

40. Critique this "*solution*" of $\sqrt{x} + x = 12$:
$\sqrt{x} = 12 - x \Rightarrow x = 144 - 24x + x^2 \Rightarrow x^2 - 25x + 144 = 0 \Rightarrow (x-9)(x-16) = 0 \Rightarrow x = 9$ or 16.

41. If $x_i \geq 1$ for $i = 1, 2, \ldots, n$, how large can $\sum_{i=1}^{n} x_i^{-2}$ be? Answer the same question for $\sum_{i=1}^{n} i x_i^{-2}$.

42. Establish the formula $\sum_{k=1}^{n}(kx-1) = \tfrac{1}{2}nx(n+1) - n$.

43. Critique this "*proof*": For $n = 0, 1, 2, \ldots$, define f_n by the equation $f_n(x) = x^n$. We shall prove by induction that these functions are all constant. The assertion is obviously correct for $n = 0$ because $f_0(x) = 1$. If the assertion is correct for $n = 0, 1, 2, \ldots, k$, then it is true for $n = k+1$ by the following reasoning. First, $f_{k+1} = f_1 f_k$. Second, $f_1' = 0$, since f_1 is constant (by the induction hypothesis). Third, $f_k' = 0$ because the induction hypothesis tells us that f_k is con-

stant. Then $f_{k+1}' = (f_1 f_k)' = f_1' f_k + f_1 f_k' = 0$. (See Barbeau [2000].)

44. Under what circumstances is this implication *not* true? $\big[(P \text{ or } Q) \text{ and } (P \text{ or } R)\big] \Rightarrow (Q \text{ or } R)$.

45. Is this a theorem about propositions? $\big\{(P \text{ or } Q) \text{ and } (P \text{ or } R) \text{ and } \big[\text{not } (Q \text{ or } R)\big]\big\} \Rightarrow P$

46. Examine the accompanying triangular figure. It shows how some right triangles can be dissected and arranged in two different ways and produce two answers for the total area. Explain the fallacy.

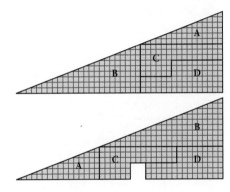

47. Find the truth table for $[P \text{ or } (Q \text{ and } R)]$.

48. The **harmonic mean** of n numbers $x_1, x_2, \cdots, x_n$ is $n(x_1^{-1} + x_2^{-1}, \cdots + x_n^{-1})^{-1}$. Show that the harmonic mean of n positive numbers is not greater than their **arithmetic mean** (**average**).

49. Some integers can be represented as a sum of an even number of consecutive integers. For example, $3 = 1 + 2$, $10 = 1 + 2 + 3 + 4$, $14 = 2 + 3 + 4 + 5$. Find some examples of integers that are *not* representable in this way. Can you prove a theorem about those integers that are not representable in the way described? (See Halmos [1991].)

50. If $P \Rightarrow Q$, does it follow that P is true whenever Q is true? Provide the truth table for $(P \Rightarrow Q) \Rightarrow (Q \Rightarrow P)$.

51. If P implies Q, does it follow that not-Q implies not-P? Give the truth table for the latter.

52. Critique this "*proof*" that 1 is the largest positive integer. Let n be the largest positive integer. We want to prove that $n = 1$. Since n^2 is a positive integer, and n is the largest positive integer, $n^2 \leq n$. Divide this inequality by n to get $n \leq 1$. This says that the largest positive integer is less than or equal to 1. (See the book by Barbeau [2000].)

53. (Continuation.) Critique this "*proof*" by induction, that every positive integer n satisfies the equation $(n-2)^2 = n^2$. First, we note that the formula is correct for $n = 1$. Assuming the formula to be true for n, we prove it for $n+1$. That is the assertion $((n+1)-2)^2 = (n+1)^2$. Equivalently, $(n-1)^2 = (n+1)^2$. Further equivalent equations then are $n^2 - 2n + 1 = n^2 + 2n + 1$, and $-4n^2 = 0$. Next, we have $-4n = 0$. Multiply this equation by $1 - 1/n$ to get $-4n + 4 = 0$. Add n^2 to get $n^2 - 4n + 4 = n^2$, or $(n-2)^2 = n^2$. Since this is the induction hypothesis, it is true by assumption. Hence the case $n+1$ is established. (Here again we refer to Barbeau [2000].)

54. Write down a number of true assertions using these symbols correctly:
$\{\cdots : \cdots\}, \Leftarrow, \Leftrightarrow, \subseteq, \cup, \cap, \smallsetminus, \exists, \forall, \in, \mathbb{Z}, \mathbb{N}$.

55. Refer to General Exercise 30. If all the successive differences in a sequence $[x_n]$ obey the formula $x_{n+1} - x_n = cn$, what is the closed formula for x_n?

56. Establish this formula, known as **summation by parts**:
$$\sum_{k=1}^{n} a_k b_k = S_n b_n + \sum_{k=1}^{n-1} S_k (b_k - b_{k+1})$$
where $S_i = a_1 + a_2 + \cdots + a_i$, $S_o = 0$, and $1 \leq n$. This formula is the discrete analog of integration-by-parts in calculus.

57. For real numbers, we have $(a+b) - a = b$. The corresponding assertions for sets would be $(A \cup B) \smallsetminus A = B$. Is it true?

58. An **arithmetic progression** is a sequence $a, a+b, a+2b, \ldots, a+nb$. Find and prove a formula for the sum of these terms.

59. Critique this "*proof*" that a symmetric transitive relation is an equivalence relation: Let $x \equiv y$. Then $y \equiv x$ by symmetry and $x \equiv x$ by transitivity. (See Birkhoff and MacLane [1941].)

60. Define an equivalence relation as follows: $[a, b, c, d] \equiv [a', b', c', d']$ means that $|c| + |d| > 0$, $|c'| + |d'| > 0$, $[c', d']$ is a multiple of $[c, d]$, and $[a' - a, b' - b]$ is a multiple of $[c, d]$. Establish that this is an equivalence relation and explain its connection to lines in $\mathbb{R}^2$.

61. A real number is said to be **rational** if it can be expressed as the quotient of two integers. In the contrary case, the number is said to be **irrational**. Substantiate the correct statements and give suitable counterexamples for the incorrect ones:
  a. The sum of any two rational numbers is rational.
  b. The sum of any two irrational numbers is irrational.
  c. The product of any two rational numbers is rational.

d. The product of any two irrational numbers is irrational.

e. The sum of a rational number and an irrational number is irrational.

f. The product of a rational number and an irrational number is irrational.

62. Establish the items that are true and find counterexamples to those that are false:
  a. $x > 7 \Rightarrow x > 6$
  b. $x^2 = 9 \Rightarrow x = 3$
  c. $\sin x = 0 \Rightarrow x = 0$
  d. $e^x = 0 \Rightarrow x > 0$
  e. $x > e \Rightarrow x > \pi$

63. Define a function $f : \mathbb{N} \to \mathbb{N}$ by these two equations: $f(1) = 1$ and $f(n+1) = f(n) + n + 1$ for $n = 1, 2, 3, \ldots$. (This is an example of a **recurrence relation**.) Find a *closed* formula for $f(n)$ and prove its correctness.

64. (Continuation, but a more challenging problem.) Define a function $f : \mathbb{N} \to \mathbb{N}$ by these two equations: $f(1) = 3$ and $f(n+1) = n + 1 + f(n)$ for $n = 1, 2, 3, \ldots$. Find a *closed* formula for f. Establish, by induction, that your formula is correct.

65. Show that, for every positive integer n, the number $n(n+1)(n+2)$ is divisible by 6. Is there any better conclusion of that type?

66. If an integer n is divisible by two different integers, k and m, does it follow that n is divisible by km? Give examples and formulate theorems about this question. You may assume the theorem that every positive integer n is the product $p_1 p_2 \ldots p_k$, where each p_k is a prime number. (The prime numbers in the product can be repeated.) For example, $18 = 2 \cdot 3 \cdot 3$.

67. Explain this puzzle: We let $x = 1 + \frac{1}{3} + \frac{1}{5} + \frac{1}{7} + \cdots$ and $y = \frac{1}{2} + \frac{1}{4} + \frac{1}{6} + \cdots$. Then we have $2y = 1 + \frac{1}{2} + \frac{1}{3} + \frac{1}{4} + \cdots = x + y$ so that $y = x$. But each term in the series $x - y$ is positive, and therefore $x > y$. (This example is from the book by Maxwell [1959].)

68. Critique a student's "*solution*" of the equation $(x+3)(2-x) = 4$ that goes like this: Since the equation is in factored form, either $x + 3 = 4$ or $2 - x = 4$. Thus, we get the solutions $x = 1$ or $x = -2$. To verify the work we first substitute $x = 1$ in the original equation, getting $(1+3)(2-1) = 4$. Then we substitute $x = -2$ in the equation, getting $(-2+3)(2-(-2)) = 4$. The method used is correct because we checked the answers. (This example is due to Edwin A. Maxwell [1959].)

69. Show that $3^{200} > 2^{300}$. Avoid using a calculator or the use of logarithms, if possible.

70. Affirm that if $x > 0$ then $x + (1/x) > 2$.

71. Show that, for any three numbers, x, y, z, we have $x^2 + y^2 + z^2 \geq xy + yz + zx$. Can you generalize this to any number of variables? Under what conditions do we get the strict inequality ($>$) instead of the weak version?

72. Critique this sequence of logical deductions: From the well-known equation $\cos^2 x + \sin^2 x = 1$, we obtain $\cos x = (1 - \sin^2 x)^{\frac{1}{2}}$. Hence, we have $1 + \cos x = 1 + (1 - \sin^2 x)^{\frac{1}{2}}$. Let $x = \pi$, so that $\cos x = -1$ and $\sin x = 0$. The result is $0 = 2$. (This example is from Maxwell [1959].)

73. In proving the principle of mathematical induction, we invoked the more basic theorem that every nonempty set of positive integers contains a smallest member. What other sets of real numbers have this property? Theorems and examples are needed.

74. Solve the equation $\sqrt{x} + 2x = 3$, and check your work in an independent manner. Explain any unexpected results.

75. Invent a proposition $P(n)$ that involves the positive integers and that is true for $n = 1, 2, 3, \ldots, 50$, but fails for $n = 51$.

76. Establish that the square of every odd integer is odd. Is the same true for the cubes of odd integers? What can you say about other powers, $n = 4, 5, \ldots$?

77. Substantiate this case of the **Cauchy–Schwarz inequality**. For arbitrary real numbers a, b, x, y, we have
$$(ax + by)^2 \leq (a^2 + b^2)(x^2 + y^2).$$

78. Show that if r is a real number such that $rx + x^2 \geq 0$ for all real values of x, then $r = 0$.

79. The **triangular numbers** are the integers generated by the recursive formulas $x_0 = 0$ and $x_{n+1} = x_n + n$. Find a formula for x_n. (This should be easy.)

80. Use induction to prove that $n^2 - n$ is even for all $n \in \mathbb{N}$. What can you discover about negative values of n?

Complex Arithmetic

> *The shortest path between two truths in the real domain passes through the complex domain.*
>
> —JACQUES HADAMARD

> *The different branches of arithmetic—ambition, distraction, uglification, and derision.*
>
> —LEWIS CARROLL in *Alice in Wonderland*

B.1 Complex Numbers and Arithmetic

Let us outline the rules for complex arithmetic. A **complex number** z is of the form $z = x + iy$, where x and y are real numbers and i is defined by the requirement $i^2 = -1$. We call x the **real part** of this complex number, and y the **imaginary part**. An **imaginary number** is iy, where y is a real number. Addition, subtraction, and multiplication are defined by the equations

$$(x + iy) \pm (a + ib) = (x + a) \pm i(y + b)$$
$$(x + iy)(a + ib) = (xa - yb) + i(xb + ya)$$

(In both equations, a, b, x, and y are real. Sometimes we forget to emphasize this.) A **reciprocal** is calculated like this:

$$\frac{1}{x + iy} = \frac{x - iy}{(x + iy)(x - iy)} = \left(\frac{x}{x^2 + y^2} \right) - i \left(\frac{y}{x^2 + y^2} \right)$$

The absolute value or **modulus** of the complex number $z = x + iy$ is defined to be $|z| = \sqrt{x^2 + y^2}$. The **conjugate** of the same complex number z is defined to be $\bar{z} = x - iy$. Consequently, $z = \bar{z}$ if and only if z is real. Also, one then verifies quickly that $z\bar{z} = |z|^2$. Thus, the **reciprocal** of z as calculated previously is more easily obtained by writing

$$\frac{1}{z} = \frac{\bar{z}}{z\bar{z}} = \frac{\bar{z}}{|z|^2}$$

It follows that a **quotient** of complex numbers w and z can be computed as

$$\frac{w}{z} = \frac{w\bar{z}}{|z|^2}$$

Other properties are $|wz| = |w|\,|z|$ and $\overline{wz} = \bar{w}\,\bar{z}$. We can write a complex number z in **polar coordinates** as

$$z = |z|(\cos\theta + i\sin\theta)$$

with an angle θ and a magnitude $|z|$. If $w = |w|(\cos\varphi + i\sin\varphi)$, then $zw = |z|\,|w|[\cos(\theta + \varphi) + i\sin(\theta + \varphi)]$. As a special case, when $z = w$, we obtain

$$z^2 = |z|^2[\cos(2\theta) + i\sin(2\theta)]$$

In general, we have

$$z^k = |z|^k[\cos(k\theta) + i\sin(k\theta)]$$

which is proved by induction. The equation

$$(\cos\theta + i\sin\theta)^n = \cos(n\theta) + i\sin(n\theta)$$

is known as **de Moivres' Theorem.**[1] The famous equation

$$e^{i\theta} = \cos\theta + i\sin\theta$$

---

[1] Abraham de Moivres lived from 1667 to 1754. He discovered the cited theorem around 1707.

is called **Euler's Equation**. It was discovered by the great mathematician Leonhard Euler in 1748.[2]

B.2 Fundamental Theorem of Algebra

Now that complex numbers are available, we can state an important historical theorem:

THEOREM 1

Every polynomial of degree one or more has a root in the complex field.

Note that this theorem does not say that every real polynomial has a real root. The simplest examples show that this is not so. Thus, the polynomial $x^2 + 1$ has no real root, even though its coefficients are real.

The preceding theorem is called the **Fundamental Theorem of Algebra**. It was first conjectured by Girard in 1629. The first proof was by Gauss in 1799. That indicates an astonishing delay in establishing such a crucial result. The history of this theorem is related in detail in the book by Kline [1968].

Once this theorem has been established, it is a simple matter to prove that a polynomial of degree n has a factorization of the following type:

$$p(z) = a_0 + a_1 z + a_2 z^2 + a_3 z^3 + \cdots + a_n z^n = (z - r_1)(z - r_2) \cdots (z - r_n)$$

In this equation, the roots r_j of the polynomial may be complex; they are not necessarily different from each other. If one factor is repeated k times in the factorization, then there will be a term of the form $(z - r_j)^k$. We say that r_j is a root having multiplicity k.

B.3 Abel–Ruffini Theorem

The following result is considered one of the jewels of mathematics. It was established by Ruffini in 1813 and by Abel [1826]. For more details, see Wells [1986, p. 59] and Pesic [2003].

---

[2] Leonhard Euler made contributions to virtually all branches of mathematics. He wrote 866 articles and books on calculus, planetary motion, the calculus of variations, and mechanics, to name a few topics in which he was the acknowledged master. Euler was a complete genius: His memory was phenomenal and he could recite many poems, including the Aeneid. He could carry out complicated calculations in his head. He spent many years in Russia at the Petersburg Academy, and served there at the request of Catherine the Great.

> **THEOREM 2   Abel–Ruffini Impossibility Theorem**
>
> *The roots of some polynomial equations of degree five or higher are incapable of algebraic solutions by a finite number of additions, multiplications, divisions, and root extractions operating on the coefficients.*

The second, third, and fourth degree polynomial equations can always be solved by radicals. However, $x^5 - x + 1 = 0$ cannot be solved by radicals (although $x^5 - x^4 - x + 1 = 0$ can).

Answers/Hints for General Exercises

> *It is not the answer that enlightens, but the question.*
>
> —EUGENE IONESCO DECOUVERTES (1909–1994)

> *Sometimes the questions are complicated and the answers are simple.*
>
> —DR. SEUSS (1904–1991)

> *Judge others by their questions rather than their answers.*
>
> —FRANCOIS-MARIE AROUET VOLTAIRE (1694–1778)

General Exercises 1.1

1. $\mathbf{x} = (x_1, x_2, x_3) = (2, 5, 7)$.

3. $\mathbf{x} = (x_1, x_2) = (-3/10, -\frac{4}{5}) = (-0.3, -0.8)$.

5. $a_{21} = 2, a_{12} = 4$.

7. No solutions other than $(1, 3, 2)$.

9. $y = 3.1$ and $x = -2.7$.

11. The system is inconsistent. No solutions.

13. These lines meet at $(3, -1)$.

15. $\begin{bmatrix} 1 & 0 & 0 & | & -\frac{2}{7} & 11/14 & -1/14 \\ 0 & 1 & 0 & | & -\frac{1}{7} & 5/14 & -3/14 \\ 0 & 0 & 1 & | & \frac{3}{7} & -13/14 & 5/14 \end{bmatrix}$

17. Original system has the unique solution $(1, 2)$. Second system is inconsistent and has no solution. Final system has infinitely many solutions.

19. For arbitrary righthand-side vector (a, b), the solution vector is $x_1 = (4a+b)/11$ and $x_2 = (-3a+2b)/11$. When $a = 11$ and $b = -11$, solution is $(3, -5)$.

21. $n^2/2$ positive terms if n is even and $(n^2 + 1)/2$ if n is odd.

23. From $[f(x)]^2 = [g(x)]^2$ we conclude only that either $f(x) = g(x)$ or $f(x) = -g(x)$.

25. $x = 23/(5a+3c)$ and $y = (4a+7c)/(5a+3c)$.

27. $\begin{bmatrix} 11 & 0 & 0 & 34 \\ 0 & 11 & 0 & -93 \\ 0 & 0 & 11 & -17 \end{bmatrix}$

29. $x = 3$ and $y = e^4$.

31. Solution $(500, 300, 100)$.

33. Simplify to $15x - 12y = 13$.

35. $\mathbf{x} = (1, 0, -3)$.

37. $v = -27/58$ and $u = 97/58$; so $x = u - v = 62/29$ and $y = u + v = 35/29$.

39. $\begin{bmatrix} 1 & 5 & 0 \\ 0 & 0 & 1 \\ 0 & 0 & 0 \end{bmatrix} \mapsto \begin{bmatrix} 1 & 0 \\ 0 & 1 \\ 0 & 0 \end{bmatrix}$

41. Solution is $(x, y) = (7, 5)$.

43. $x = 7$ and $y = 5$.

45. $x_1 = 3$, $x_2 = -2$, and $x_3 = 5$.

47. $\begin{bmatrix} 5 & 4 & 3 & 2 & 1 \end{bmatrix}$ is not in reduced row echelon form.

49. Yes.

51. residual vector: $\widehat{x} = A\widehat{x} - b$, error vector: $\widehat{x} = \widehat{x} - x$, where $x = (1, -1)$.

General Exercises 1.2

1. $x = (-2, 5)$; inconsistent.

3. Yes.

5. Both matrices are upper triangular.

7. Reduced row echelon form: $\begin{bmatrix} 1 & 3 & 0 & 5 & 6 \\ 0 & 0 & 1 & -2 & 2 \\ 0 & 0 & 0 & 0 & 0 \end{bmatrix}$.

9. $\begin{bmatrix} x_1 \\ x_2 \\ x_3 \\ x_4 \\ x_5 \\ x_6 \end{bmatrix} = \begin{bmatrix} 3 \\ 0 \\ 6 \\ 0 \\ 2 \\ 0 \end{bmatrix} + x_2 \begin{bmatrix} -3 \\ 1 \\ 0 \\ 0 \\ 0 \\ 0 \end{bmatrix} + x_4 \begin{bmatrix} -5 \\ 0 \\ 2 \\ 1 \\ 0 \\ 0 \end{bmatrix}$
$+ x_6 \begin{bmatrix} 1 \\ 0 \\ -4 \\ 0 \\ -3 \\ 1 \end{bmatrix}$

11. No solution!

13. $\mathbf{r}_p \leftarrow \mathbf{r}_p - \alpha\mathbf{r}_q$; $\mathbf{r}_q \leftarrow \alpha^{-1}(-\mathbf{r}_p + \mathbf{r}_q)$;
$\mathbf{r}_q \leftarrow \mathbf{r}_p$, $\mathbf{r}_q \leftarrow \beta^{-1}\mathbf{r}_q$.

15. Different size matrices.

17. $br = as$ and $bt = cs$. (Assume all six coefficients are nonzero.)

19. $\begin{bmatrix} x_1 \\ x_2 \\ x_3 \end{bmatrix} = \begin{bmatrix} 0 \\ 7 \\ 2 \end{bmatrix} + x_1 \begin{bmatrix} 1 \\ -1 \\ 1 \end{bmatrix}$

21. $x = e^3$ and $y = \sqrt{2}$.

23. $\begin{bmatrix} x_1 \\ x_2 \\ x_3 \\ x_4 \\ x_5 \end{bmatrix} = \begin{bmatrix} 0 \\ \frac{3}{2} \\ \frac{5}{3} \\ 0 \\ 0 \end{bmatrix} + x_1 \begin{bmatrix} 1 \\ 0 \\ 0 \\ 0 \\ 0 \end{bmatrix} + x_4 \begin{bmatrix} 0 \\ \frac{7}{2} \\ -\frac{5}{3} \\ 1 \\ 0 \end{bmatrix}$
$+ x_5 \begin{bmatrix} 0 \\ 2 \\ -\frac{2}{3} \\ 0 \\ 1 \end{bmatrix}$

25. We can interchange summation signs.

27. Begin by letting $\mathbf{E}$ be the reduced row echelon form of $\mathbf{A}$.

29. $\mathbf{x}^T\mathbf{y}$ is a scalar and $\mathbf{x}\mathbf{y}^T$ is an $n \times n$ matrix.

31. $x_2 = 4 - 2x_3 - 3x_5$, $x_4 = 5 - 7x_5$. The independent variables are x_3 and x_5.

33. $\mathbf{A}$ is upper triangular.

35. No.

37. When m is even, the answer is $n(m/2)$.

39. Minimize $[35 + 2t]^2 + [90 + 27t]^2 + [17t]^2$.

41. Let $\mathbf{x} = \mathbf{e}_j$, which is a standard unit vector in $\mathbb{R}^n$.

43. Four missing steps.

45. a. $\begin{bmatrix} 4 \\ -1 \\ 8 \end{bmatrix}$.   **b.** 8.

47. a. $x_1 = -1 - x_5, x_2 = 1 + 3x_5, x_3 = $ free,
$x_4 = -4 - 5x_5, x_5 = $ free.
b. $x_1 = $ free, $x_2 = -\frac{3}{4}x_3 - \frac{34}{28}x_5 - \frac{69}{84}$,
$x_3 = $ free, $x_4 = \frac{5}{21} - \frac{2}{7}x_5, x_5 = $ free, $x_6 = \frac{2}{3}$.

49. a. $x_1 = -2, x_2 = 3, x_3 = -1, x_4 = $ free.
b. $x_1 = -0.9333 - 0.8x_2, x_2 = $ free,
$x_3 = 1.444 - 1.333x_4, x_4 = $ free, $x_5 = \frac{1}{3}$.

51. a. Let $r = -20$.   **b.** Let $s = 6$.

General Exercises 1.3

1. There is no solution since the system is inconsistent.

3. $[57, -17, 0]^T + x_1[-54, 15, 1]^T$.

5. $r[-5, -7, 3/2, 1, 0]^T + s[69, -225, 75, 0, 1]^T$.

7. $\mathbf{X} = \frac{1}{7} \begin{bmatrix} -303 & -296 & 627 & 951 & -296 \\ -123 & -137 & 295 & 404 & 88 \\ 107 & 107 & -214 & -321 & 107 \end{bmatrix}$,

$\mathbf{X} = \frac{1}{7} \begin{bmatrix} -22 & -1 & 31 \\ -9 & -2 & 13 \\ 8 & 1 & -10 \end{bmatrix}$.

9. $\mathbf{x} = 0$.

11. a. $(1, -1)$ **b.** $(2, 1)$ **c.** $(2, 1, -1)$.

13. Suppose
$\alpha(\mathbf{v}_1 + \mathbf{v}_2) + \beta(\mathbf{v}_2 + \mathbf{v}_3) + \gamma(\mathbf{v}_3 + \mathbf{v}_1) = 0$.

15. Solutions are $(x_1, x_2) = (2, 4)$ and $(y_1, y_2) = (5, -3)$.

17. Yes.

19. Review the discussion in other exercises.

21. Let $\mathbf{Au} = \mathbf{b}$ and $\mathbf{Av} = \mathbf{b}$ with $\mathbf{u} \neq \mathbf{v}$.

23. Same reduced row echelon form.

25. $\mathbf{Ax}$ is the linear combination of the columns of $\mathbf{A}$.

27. Use General Exercise 1.3.26.

29. System has more variables than equations.

31. Rank$(\mathbf{A}) = n$.

33. The pivot positions are $(1, 2)$ and $(2, 3)$.

35. Maximum number of pivot positions is m. Least number of non-pivot columns is $n - m$. Least number of variables is $n - m$.

37. No.

39. No.

41. $\mathbf{A}$ is inconsistent.

43. Reduced row echelon form has two zero rows.

45. Consider the case $n = 3$.

47. Yes! Suppose on the contrary they are linearly dependent for a fixed nonzero constant λ.

49. Yes. For a fix nonzero constant λ, there exists $a\mathbf{v}_1 + b(\mathbf{v}_2 + \lambda\mathbf{v}_1) = 0$ with $|a| + |b| > 0$.

51. $x = \frac{7}{4}w, y = \frac{1}{2}w$, and $z = \frac{1}{2}w$. A convenient solution is $x = 7, y = 2, z = 2, w = 4$.

53. $h = -9$.

55. Do not mix regular and technical English.

57. Rank of $\mathbf{A}$ is 2.
$\begin{bmatrix} x_1 \\ x_2 \\ x_3 \end{bmatrix} = x_3 \begin{bmatrix} 10 \\ -4 \\ 1 \end{bmatrix}$

59. Rank is 2. $\begin{bmatrix} x_1 \\ x_2 \\ x_3 \end{bmatrix} = \begin{bmatrix} 2 \\ 0 \\ 4 \end{bmatrix} + x_2 \begin{bmatrix} -3 \\ 1 \\ 0 \end{bmatrix}.$

For the homogeneous system, $\begin{bmatrix} x_1 \\ x_2 \\ x_3 \end{bmatrix}$

$= c \begin{bmatrix} -3 \\ 1 \\ 0 \end{bmatrix}.$

61. Let $AX_1 = B$ and $AX_2 = B$.

63. At least, one free variable.

65. Rank$(A) = 2$. $\begin{bmatrix} x_1 \\ x_2 \\ x_3 \end{bmatrix} = \begin{bmatrix} 2 \\ 0 \\ 4 \end{bmatrix} + x_2 \begin{bmatrix} -3 \\ 1 \\ 0 \end{bmatrix},$

where x_2 is a free variable.

67. Let the augmented matrix be $\begin{bmatrix} 1 & 1 & | & 1 \\ 1 & 1 & | & 0 \end{bmatrix}$.

69. Yes.

71. Several of them are obvious.

73. Linearly dependent.

75. a. Linearly independent.
b. Linearly dependent.
c. Linearly dependent.

77. Every row echelon form of the matrix will have a zero row.

79. Do if and only-if parts separately.

81. A maximum of one additional pivot position is added.

General Exercises 2.1

1. $\mathbf{w} - \mathbf{u}$ must be a multiple of $\mathbf{v} - \mathbf{u}$. Concrete case: Yes!

3. Missing coefficient $c = -5$.

5. $-1(1, 3, 2) + 2(4, 1, 1) = (7, -1, 0)$.

7. $x_1 = -9, x_2 = 1, x_3 = -8$.

9. a. $(\frac{2}{3}, \frac{5}{3})$.   **b.** $(0, \frac{3}{4}, 0)$.

11. The set of polynomials spans $\mathbb{P}_3$.

13. $\mathbf{p} - \mathbf{v} = \mathbf{w}, \mathbf{q} = -\frac{1}{3}\mathbf{w}$.

15. Writing $\mathbf{u} = (u_1, u_1)$ and $\mathbf{v} = (v_1, v_1)$, we find slope: v_2/v_1; vertical intercept: $u_2 - (u_1/v_1)v_2$; slope-intercept form: $y = (v_2/v_1)x + [u_2 - (u_1/v_1)v_2]$; intercepts form: $x/[u_1 - (v_1/v_2)u_2] + y[u_2 - (v_2/v_1)u_1] = 1$; point-slope form: $(y - u_2) = (v_2/v_1)[x - u_1]$.

17. *Hint:* Use General Exercise 2.1.16.

19. No!

21. No. All true except for two properties.

23. Look at the contrapositive.

25. a. Yes.   **b.** Yes.   **c.** No.
d. No.   **e.** Yes.

27. Center of mass $\mathbf{c} \to \mathbf{x}_k$, when the free weight $w_k \to \infty$.

29. Similar to $\mathbb{R}^{m+n}$ except the $m+n$ components are arranged in a rectangle rather than in a column.

31. Center of mass is $(a/2, a/(2\sqrt{3}))$.

33. -14.

35. If $\beta \neq 41$, this system is not solvable!

37. $x_1 = x_4 + 2, x_2 = x_4 + 4, x_3 = x_4 - 1$.

39. Component-wise operations apply to all elements of doubly infinite-length vectors.

41. $(-\frac{2}{7}, 0)$.

43. $I_1 = 2, I_1 = 7$.

45. If $9a - 4b - 7c \neq 0$, this system is not solvable!

47. No.

49. $x = 2, y = 2$.

51. Component-wise operations apply to all elements of infinite-length vectors.

General Exercises 2.2

1. $(-5, 6) + t(12, -3)$. No. Another form is $(7, 3) + s(-12, 3)$.

3. We need $h = 0$ for the system to be consistent. Yes, the lines intersect at $(\frac{9}{2}, 2, 2)$ when $t = \frac{3}{2}$ and $s = -\frac{1}{2}$.

5. $t = 3 - 4s$.

7. $\begin{bmatrix} x_1 \\ x_2 \\ x_3 \end{bmatrix} = \begin{bmatrix} 0 \\ \frac{7}{3} \\ 0 \end{bmatrix} + t \begin{bmatrix} 0 \\ 5 \\ 3 \end{bmatrix}$

9. No, they are not even parallel.

11. Yes, $s = 0$ and $t = 3$.

13. $\begin{bmatrix} x_1 \\ x_2 \\ x_3 \\ x_4 \end{bmatrix} = \begin{bmatrix} -5a - 3b \\ 2a + b \\ 0 \\ 0 \end{bmatrix} + s \begin{bmatrix} 13 \\ -5 \\ 1 \\ 0 \end{bmatrix} + t \begin{bmatrix} 16 \\ -6 \\ 0 \\ 1 \end{bmatrix}$

15. $\begin{bmatrix} x_1 \\ x_2 \\ x_3 \end{bmatrix} = \begin{bmatrix} 2 \\ 0 \\ 0 \end{bmatrix} + s \begin{bmatrix} 3 \\ 2 \\ 0 \end{bmatrix} + t \begin{bmatrix} -7 \\ 0 \\ 2 \end{bmatrix}$

17. $11x + 2y = 47$.

19. Yes.

21. Yes to both.

23. $(a_1 - b_1)x_1 + (a_2 - b_2)x_2 = \frac{1}{2}(a_1^2 - b_1^2) + \frac{1}{2}(a_2^2 - b_2^2)$.

25. True.

27. a. Yes. **b.** No.

29. a. Yes. **b.** Yes.

31. a. Yes. **b.** No.

33. No, for $(2, -3, 0)$. Yes, for $(0, 10, 9)$.

35. $x + 2y + 2z = 5$.

37. $x_1 + 6x_3 + 8x_4 = -\frac{1}{2}, x_2 - \frac{9}{2}x_3 - 6x_4 = 0$.

39. We obtain $10y + 7z = 1$, which is a line!

41. Yes, if they are both parallel and intersect.

43. a. $a = 36, b = 7$. **b.** $c = 6, d = 16$.

45. Yes, $t = -2, t = -3$.

47. Yes, $4x_1 - 3x_2 + 5x_3 = 0$.

49. $x - y - 2z = 3$.

51. Normal equations:

$$\begin{bmatrix} n & \sum_{i=1}^{n} i & \sum_{i=1}^{n} i^2 \\ \sum_{i=1}^{n} i & \sum_{i=1}^{n} i^2 & \sum_{i=1}^{n} i^3 \\ \sum_{i=1}^{n} i^2 & \sum_{i=1}^{n} i^3 & \sum_{i=1}^{n} i^4 \end{bmatrix} \times \begin{bmatrix} a \\ b \\ c \end{bmatrix}$$
$$= \begin{bmatrix} \sum_{i=1}^{n} d_i \\ \sum_{i=1}^{n} i d_i \\ \sum_{i=1}^{n} i^2 d_i \end{bmatrix}$$

General Exercises 2.3

1. No linear mapping possible!

3. Check corner points of square.

5. Yes.

7. Domain is $\mathbb{R}^4$,
co-domain is $\mathbb{R}^3$,
kernel is $s(0,0,0,1)$,
not 1-1 (injective) by Theorem 2.3.5,
onto (surjective) by Theorem 2.3.4,
range is span of the original three columns
in $\mathbf{A}^T$.

9. No.

11. No.

13. Yes.

15. a. Yes.   **b.** No.

17. Cannot be done!

19. a. one-to-one.   **b.** not one-to-one.
  c. one-to-one.   **d.** one-to-one.

21. a. Reflection through the x_1-axis.
  b. Reflection through the x_2-axis.
  c. Reflection through the line $x_1 = x_2$.
  d. Reflection through the line $x_1 = -x_2$.
  e. Reflection through the origin.

23. a. Shear horizontal left.
  b. Shear vertical down.
  c. Shear horizontal right.
  d. Shear vertical up.

25. a. Contraction by factor c.
  b. Dilation by factor c.

27. T is not linear.

29. Yes. Find some $\mathbf{y}$ so that $\mathbf{A}\mathbf{y} \neq \mathbf{0}$ but $\mathbf{A}^2\mathbf{y} = \mathbf{0}$.

31. Find a 2×2 matrix $\mathbf{A}$ so that $\mathbf{A}\mathbf{x} \neq \mathbf{0}$ and
$\mathbf{A}^2\mathbf{x} = \mathbf{0}$.

33. Case 1. $\{\mathbf{r}, \mathbf{u}, \mathbf{w}\}$ is linearly dependent.
  Case 2. $\{\mathbf{r}, \mathbf{u}, \mathbf{w}\}$ is linearly dependent.
  Case 2a. $\mathbf{w}$ is a linear combination of $\mathbf{r}$
  and $\mathbf{u}$.
  Case 2b. $\mathbf{u}$ and $\mathbf{w}$ are multiples of $\mathbf{r}$.

35. Yes, T is linear with $\mathbf{A} = \begin{bmatrix} 0 & 1 \\ 1 & 0 \end{bmatrix}$.

37. Let $\mathbf{A} = \begin{bmatrix} 1 & -1 \\ 0 & 0 \end{bmatrix}$.

39. Let $\mathbf{A} = \begin{bmatrix} 0 & 0 & 0 \\ 0 & 1 & 0 \\ 0 & 0 & 0 \end{bmatrix}$.

41. Yes, in general, one will find many pairs.

43. Let $\mathbf{A} = \begin{bmatrix} 1 & 0 & -3 \\ 1 & 2 & 0 \\ 1 & 1 & 1 \\ -1 & 0 & 1 \end{bmatrix}$.

The transformation T is linear and one-to-
one but not surjective (onto).

45. Area of the parallelogram is the sum of two
right triangles.

47. No.

49. $\mathbf{A} = \text{UpperBiDiagonals}(\frac{1}{2}, \frac{1}{2})$,
  $\mathbf{B} = \text{TriDiagonal}(\frac{1}{4}, \frac{1}{2}, \frac{1}{4})$.

51. Yes, easy!

53. Contrapositive.

55. Consider $f(\mathbf{x}) = x^2$.

57. Easy.

59. No.

61. All, $\mathbf{0}$, all, $\mathbf{0}$. Horizontal line through $(0,0)$,
horizontal line through $(0,0)$, $x_1 = x_2$. $\mathbf{0}$ unless $\alpha = c(2\pi)$.

63. Second part does not require the one-to-
one property.

65. Let $\sqrt{x} = a - x$.

67. For $x \mapsto x^2$, $(-\infty, \infty) \mapsto [0, \infty)$.

69. Chain rule is $F'(x) = f'(g(x))g'(x)$.

General Exercises 2.4

1. Linearly independent.

3. Linearly independent

5. a. Linearly independent.
b. Linearly dependent.
c. Linearly dependent.

7. Yes, linearly dependent.

9. $\text{Span}\{\mathbf{u}, \mathbf{v}\} \cap \text{Span}\{\mathbf{w}\} = \text{Span}\{\mathbf{u}, \mathbf{v}\}$.
$\text{Span}\{\mathbf{u}, \mathbf{v}\} + \text{Span}\{\mathbf{w}\} = \text{Span}\{\mathbf{u}, \mathbf{v}\}$.

11. No. Axioms **7, 8,** and **10** fail!

13. a. Kernel is the span of any constant.
b. Kernel is the span of any constant.

15. Yes.

17. Consider all possible vectors of length four with two zero entries and all other entries 1.

19. Use Axioms **4, 7, 2,** and **5.**

21. No, discontinuous functions do not form a vector space.

23. Give two proofs.

25. Let c_m be the last nonzero coefficient.

27. a–b. Use definitions of linear dependence and a set spanning a vector space.
c–d. Give counterexamples to show that the assertions do not hold.

29. Use Theorem 2.4.10.

31. Let $V = (0, \infty), u \oplus v = uv, \alpha \otimes v = v^\alpha$, and verify axioms for a vector space.

33. Let $V^{\mathbb{R}}$ consist of all functions $f : \mathbb{R} \to V$ and check axioms.

35. Show that all ten axioms are satisfied.

37. A zero element $\mathbf{z}$ should have the property $\mathbf{u} + \mathbf{z} = \mathbf{u}$ for all $\mathbf{u}$.

39. Disprove.

41. $\text{Span}(\text{Span}(X)) = \text{Span}(X)$.

43. Let $\mathbf{x} \in \text{Span}(S)$.

45. Easy.

47. No.

49. Straightforward.

51. a. No.   **b.** Yes.

General Exercises 3.1

1. $\mathbf{A}_{2 \times 3}, \mathbf{B}_{3 \times 3}$.

3. No. Find a 2×2 matrix example to disprove it.

5. $\begin{bmatrix} a & b \\ c & d \end{bmatrix}$ with $a = -c + d$ and $b = \frac{2}{3}c$ with any values of c and d.

7. Only one is not an elementary matrix.

9. $\mathbf{A}^T = \begin{bmatrix} 1 & 4 \\ 2 & 5 \\ 3 & 6 \end{bmatrix}$, $\mathbf{A}\mathbf{x} = \begin{bmatrix} 2 \\ 5 \end{bmatrix}$ but $\mathbf{A}^T\mathbf{x}$ is undefined.

11. $\begin{bmatrix} c+d & 3c/2 \\ c & d \end{bmatrix}$

13. Carry out row reduction on $[\mathbf{A} \,|\, \mathbf{b}]$.

15. Extreme cases:
$\mathbf{x} \approx (17.70, 5.91, 0.0, 0.88)$
$\mathbf{x} \approx (0.0, 1.10, 8.27, 9.35)$.

17. $\begin{bmatrix} 0 & 0 & -1 \\ 4 & -2 & 1 \\ 2 & -1 & 0 \end{bmatrix}$

19. $\mathbf{A} = \begin{bmatrix} a & b \\ b/2 & a+b \end{bmatrix}$

21. $\mathbf{A}_{ij} = 0$ and $\mathbf{B}_{ij} = 0$ if $j < i$.

23. $\mathbf{A}$ is symmetric if and only if $\mathbf{A}^T = \mathbf{A}$.

25. $\mathbf{A} = \mathbf{D} + \mathbf{L} + \mathbf{U}$ where $\mathbf{D}$ is diagonal, $\mathbf{L}$ is strictly lower triangular, $\mathbf{U}$ is strictly upper triangular.

27. Look at a generic element $\mathbf{A}_{ij}$.

29. If you cannot prove it, try to find a 2×2 example to disprove it. Alternatively, it could be true with additional conditions.

31. $\mathbf{A}$ commutes with itself.

33. Let $\mathbf{A}^2 = (a+d)\mathbf{A} + (bc - ad)\mathbf{I}$.

35. Yes.

37. Suppose $\mathbf{A}^T = -\mathbf{A}$ and $\mathbf{B}^T = -\mathbf{B}$.

39. Let $\mathbf{A} = \begin{bmatrix} a & b \\ c & d \end{bmatrix}$.

41. Rank is 1.

43. Scalars commute.

45. Two are $\mathbf{A}$ and two are not possible.

47. Scalars commute but not matrices.

49. $\oplus$ commutative but not associative.

51. $\mathbf{y} = \mathbf{A}\mathbf{x}$.

53. $\mathbf{C} = \mathbf{E}_k \cdots \mathbf{E}_2 \mathbf{E}_1$.

55. If $\mathbf{s} = [1, 1, \ldots, 1]^T$ then $\mathbf{s}\mathbf{A}^k = c^k \mathbf{s}$.

57. $\mathbf{A} = \mathbf{A}^T, \mathbf{B} = \mathbf{B}^T$. Yes. No.

59. $(\mathbf{A} + \mathbf{A}^T)^T = \mathbf{A}^T + \mathbf{A}$.

61. Consider dimensions of $\mathbf{A}\mathbf{A}^T$ and $\mathbf{A}^T\mathbf{A}$. No.

63. Yes.

65. $x_1 = \frac{1}{3}, x_2 = \frac{1}{9}, x_3 = 1$.

67. Let $\mathbf{A} = \begin{bmatrix} a & b \\ b & c \end{bmatrix}$ and $\mathbf{B} = \begin{bmatrix} d & e \\ e & f \end{bmatrix}$.
$a - c = d - f$ if $b = e = 1$.

69. Think of Ω as the matrix
$\mathbf{E} = \mathbf{E}_m \mathbf{E}_{m-1} \cdots \mathbf{E}_2 \mathbf{E}_1$.

71. It maps $\mathbb{R}^{r \times s} \mapsto \mathbb{R}^{s \times r}$.

General Exercises 3.2

1. a. $\begin{bmatrix} -1 & 4 & 2 \\ 1 & -7 & -4 \\ -1 & 2 & 1 \end{bmatrix}$ **b.** $\begin{bmatrix} 2 & -11 & 5 \\ 1 & -3 & 2 \\ 1 & -4 & 2 \end{bmatrix}$

c. $\begin{bmatrix} -2 & 6 & 5 \\ 1 & -9 & 2 \\ -1 & 4 & 3 \end{bmatrix}$ **d.** $\begin{bmatrix} \frac{7}{8} & \frac{1}{2} & \frac{5}{8} \\ -1 & -1 & 1 \\ \frac{3}{8} & \frac{1}{2} & \frac{1}{8} \end{bmatrix}$

3. General left inverse is

a. $\begin{bmatrix} -\frac{5}{3} + s & \frac{4}{3} - 2s & s \\ \frac{2}{3} + t & -\frac{1}{3} - 2t & t \end{bmatrix}$

b. $\begin{bmatrix} -\frac{5}{3} + (\frac{8}{3})r & \frac{4}{3} - (10/3)r & r \\ \frac{2}{3} + (\frac{8}{3})s & -\frac{1}{3} - (10/3)s & s \end{bmatrix}$

5. If $ab = bc$, then $\begin{bmatrix} a & b \\ c & d \end{bmatrix}$ is not invertible.

7. a. Finding right inverse of $\begin{bmatrix} 1 & 2 & 1 \\ 3 & 5 & 1 \end{bmatrix}$ or left

inverse of $\begin{bmatrix} 1 & 3 \\ 2 & 5 \\ 1 & 1 \end{bmatrix}$.

b. Solving three linear systems with the matrix coefficient $\begin{bmatrix} 1 & 2 \\ 3 & 5 \end{bmatrix}$ and three different right-hand sides.

c. Solving four linear systems with the matrix coefficient $\begin{bmatrix} 1 \\ 3 \end{bmatrix}$ and four different right-hand sides. These are all inconsistent systems!

9. a. $\begin{bmatrix} 17 & 26 & 31 \\ 11 & 17 & 20 \\ 7 & 11 & 13 \end{bmatrix}$

b. $\begin{bmatrix} 8 & 3 & 1 \\ 10 & 4 & 1 \\ 3.5 & 1.5 & 0.5 \end{bmatrix}$ **c.** $\begin{bmatrix} 5 & -4 & 2 \\ -2 & 2 & -1 \\ -8 & 7 & -3 \end{bmatrix}$

11. $\begin{bmatrix} 6 & 5 & -2 \\ -9 & -7 & 3 \\ 4 & 3 & 1 \end{bmatrix}$

13. No!

15. General right inverse is

$$\begin{bmatrix} -\frac{5}{3}+s & \frac{2}{3}+t \\ \frac{4}{3}-2s & -\frac{1}{3}-2t \\ s & t \end{bmatrix}$$

17. $\begin{bmatrix} 5 & -2 & -8 \\ -4 & 2 & 7 \\ 2 & -1 & -3 \end{bmatrix}$

19. $3 - (13/3)t + 5t^2 + (\frac{1}{3})t^3$.

21. $\begin{bmatrix} e-1 & 1-2e & e \\ 2+f & -1-2f & f \end{bmatrix}$, where e and f are free parameters.

23. Use Theorem 3.2.1 and Theorem 1.2.4.

25. $\begin{bmatrix} -\frac{5}{3} & \frac{2}{3} \\ \frac{4}{3} & -\frac{1}{3} \\ 0 & 0 \end{bmatrix} + s \begin{bmatrix} \frac{1}{3} & 0 \\ -\frac{2}{3} & 0 \\ 1 & 0 \end{bmatrix}$

$+ t \begin{bmatrix} 0 & \frac{1}{3} \\ 0 & -\frac{2}{3} \\ 0 & 1 \end{bmatrix}$

27. This problem is meant to illustrate the dangers of solving an equation $f(x) = 0$ by starting with that equation and drawing consequences. The logic has to go the other way.

29. It is *not* true! Let $\mathbf{A} = \begin{bmatrix} 4 & -9 \\ -2 & 6 \end{bmatrix}$ and $\mathbf{B} = \begin{bmatrix} 2 & 3 \\ 1 & 2 \end{bmatrix}$.

31. Matrix $\mathbf{A}$ invertible implies $\mathbf{A}^{-1}$ exists.

33. $\mathbf{A}^{-1} = \mathbf{A}^3$.

35. Use direct verification.

37. Use Theorem 1.2.4.

39. Given $\mathbf{AA}^{-1} = \mathbf{A}^{-1}\mathbf{A} = \mathbf{I}$, $\lambda\lambda^{-1} = 1$, and $\lambda \neq 0$.

41. Use $\sum_{k=1}^{n} \mathbf{A}_{ik}\mathbf{B}_{kj} = \delta_{ij}$.

43. Yes, it is correct.

45. Begin with 2×2 matrices.

47. $\mathbf{AB} = \mathbf{I} = \mathbf{BA}$.

49. Suppose not.

51. $\mathbf{A}_{m \times j}\mathbf{B}_{j \times m} = \mathbf{I}_{m \times m}$, $\mathbf{C}_{n \times k}\mathbf{A}_{k \times n} = \mathbf{I}_{n \times n}$.

53. No. No.

55. The assertion is never correct if $n > 1$.

57. Simple counterexample

$$\mathbf{A} = \begin{bmatrix} 2 & 0 \\ 0 & 1 \end{bmatrix}, \quad \mathbf{B} = \begin{bmatrix} 1 & 1 \\ 2 & 1 \end{bmatrix}$$

59. $\mathbf{A} = \mathbf{B}^{-1}$.

61. It is possible if $\mathbf{I} - \mathbf{A}^{-1}$ is invertible.

63. Yes.

65. Let $\mathbf{B} = \mathbf{E}_1 \mathbf{E}_2 \cdots \mathbf{E}_m$.

67. $1 + 2\cos t - 3\sin t$.

69. Use change of variables.

71. No.

73. Use Theorem 3.2.13.

75. $f(t) = 3/(t+1) - 4/(t-2) + 5/(t+2)$.
No because $f(-1) = \infty, f(-2) = \infty$,
$f(2) = -\infty$.

77. $a = 2, b = 3$, and $c = -1$.

79. Use Theorem 3.2.4.

81. If $\mathbf{A}$ is invertible, $\mathbf{A}^{-1} = \mathbf{E}_m \mathbf{E}_{m-1} \cdots \mathbf{E}_1$.

83. Let $\mathbf{B} = (\mathbf{A}^k)^{-1}$.

85. Use contrapositive.

87. $x_1 = 1 + x_3$, $x_2 = -2 - 2x_3$. Example 3.2.5
uses a different free variable.

89. $[\mathbf{A} \,|\, \mathbf{I}] \sim [\mathbf{I} \,|\, \mathbf{A}^{-1}]$.

General Exercises 4.1

1. a. 12.   **b.** 12.   **c.** 30.   **d.** 0.

3. 57/2.

5. Linearly dependent, $\mathrm{Det}(\mathbf{A}) = 0$.

7. 24.

9. $x = -2, 3$.

11. 50.

13. -116.

15. $t = -1/36$.

17. 31.

19. -21.

21. Show that $f(\alpha \mathbf{x} + \beta \mathbf{y}) = \alpha f(\mathbf{x}) + \beta f(\mathbf{y})$ and
also for g.

23. Yes, twice.

25. Use Theorem 4.1.7.

27. $|u_1 v_2 - v_1 u_2|$.

29. 5.

31. Yes. 11/2.

33. First, move $\mathbf{u}$ parallel to $\mathbf{v}$ onto the first
coordinate axis. Then move $\mathbf{v}$ parallel to the
first coordinate axis to the second coordinate axis.

35. Determinant undefined.

37. The triangle determined by these three
points is degenerate.

39. Let $\mathbf{A} = \begin{bmatrix} a & b \\ c & d \end{bmatrix}$ and consider various
cases.

41. $-bc, ad - bc$.

43. $\beta \neq \pm \sqrt{3}$.

45. a. $x = 1$.   **b.** $t = 4, t = 1$.

47. $y = 0$, x any value.

49. a. $B^2 - 4AC = 4 > 0$, $A + C = 0$: it is a rectangular hyperbola.
b. $a - b + c \neq 0$.

General Exercises 4.2

1. $29x_1 - 19x_2 + 23x_3$.

3. Involves single digit numbers only.

5. $A^{-1} = -1/20 \begin{bmatrix} -3 & 1 & 7 \\ -10 & 10 & 10 \\ 1 & -7 & -9 \end{bmatrix}$

7. $x_1 = 86/35$, $x_2 = -103/35$, and $x_3 = -\frac{2}{7}$.

9. -28.

11. 38.

13. $\mathrm{Det}(A) = 21$, $\mathrm{Det}(B) = -4$.

15. $A^{-1} = \begin{bmatrix} 1 & -1 & 1 \\ 2 & 0 & -1 \\ -4 & 1 & 1 \end{bmatrix}$,

$B^{-1} = \begin{bmatrix} 0 & -1 & 1 \\ -3 & 1 & 1 \\ 2 & 0 & -1 \end{bmatrix}$

17. $C = \begin{bmatrix} -17 & 13 & 4 \\ -9 & -9 & 24 \\ 32 & 1 & -13 \end{bmatrix}$

19. $C = \begin{bmatrix} -1 & -1 & 1 \\ -2 & 0 & 2 \\ 5 & 1 & -7 \end{bmatrix}$, $A^{-1} = -\frac{1}{2}C^T$

21. 24.

23. -72.

25. -80.

27. Use Theorem 4.2.4.

29. Use Theorems 4.2.2 and 4.2.3.

31. Consider A^T.

33. Use cofactor expansion using a row.

35. Use Theorem 4.2.6.

37. Use mathematical induction and Theorem 4.2.2.

39. Look at replacement elementary operation.

41. Yes, if cofactor is nonzero.

43. Look at $n = 2$ case.

45. $\mathrm{Det}(AB) = 34$, $\mathrm{Det}(BA^T) = 34$, $\mathrm{Det}(2A) = 17 \cdot 2^n$, $\mathrm{Det}(A^{-1}) = 1/17$, $\mathrm{Det}(B^2) = 4$.

47. Let $d = 1$. Not colinear.

49. $\mathrm{Det}(u, v, w) = 0$. No.

51. a. $a = (y_0t_1 - y_1t_0)/(t_1 - t_0)$ and $b = (y_1 - y_0)/(t_1 - t_0)$.
b. $c = y_0$ and $d = (y_1 - y_0)/(t_1 - t_0)$.

53. -36.

General Exercises 5.1

1. Yes.

3. Use Theorem 5.1.1.

5. Yes.

7. Consider vector components.

9. $T[U] = \mathbb{R}^2$.

11. $\begin{bmatrix} 1 & 0 \\ 1 & 0 \end{bmatrix}, \begin{bmatrix} 1 & 0 \\ 0 & 0 \end{bmatrix}$

13. U is x-axis and W is y-axis.

15. No.

17. $Q \subset \mathbb{P}_n \subset \mathbb{P}.$

19. Use Theorem 5.1.7.

21. Find a counterexample, where **A** and **B** are noninvertible but **A** + **B** is invertible.

23. Use Theorem 5.1.1 and linearity of transformation.

25. Show $a\mathbf{w} + b\mathbf{z} = c\mathbf{u} + d\mathbf{v}$.

27. Think about derivatives of polynomials.

29. No.

31. Think of **A** as column vectors and **B** as row vectors.

33. No.

35. Find 2×2 matrices such as in General Exercise 5.1.11.

37. Yes.

39. a. $\begin{bmatrix} 5 \\ -1 \end{bmatrix}, \begin{bmatrix} 0 \\ 1 \end{bmatrix}$

   b. Yes.

   c. $\begin{bmatrix} 1 & 0 & 0 & 2 \end{bmatrix}, \begin{bmatrix} 1 & 0 & 1 & 6 \end{bmatrix}$

   d. No.

   e. $\begin{bmatrix} 0 \\ 1 \\ 0 \\ 0 \end{bmatrix}, \begin{bmatrix} -2 \\ 0 \\ -6 \\ 1 \end{bmatrix}$

   f. Yes.

41. Consider $at + bt^2 + ct^5 + dt^7$.

43. $\mathbf{x} \in f[\text{Span}(W)] \Leftrightarrow \mathbf{x} \in \text{Span}(f[w])$.

45. Null space of **A** is the span of $[-2, 1, 0]^T$ and the null space of $\mathbf{A}^T$ is the span of

$[1, 3, -3]^T$. Null space of **B** is the span of $[-2, 1]^T$ while the null space of $\mathbf{B}^T$ is the span of $\{[-\frac{2}{3}, 1, 0]^T, [-1, 0, 0]^T\}$.

47. $U + W$: 2×2 matrices whose lower right entry is zero.
$U \cap W$: 2×2 matrices whose second row and second column are zeros.

49. Basis of the row space of **A** is
$\{(1, 2, 1, 3, 1, 2), (0, 1, 3, 0, 2, 1), (0, 0, 0, 1, 1, 2)\}$.
Basis of the column space of **A** is
$\{[1, 2, 3, 1, 2]^T, [2, 5, 7, 5, 6]^T, [3, 6, 11, 8, 11]^T\}$.

General Exercises 5.2

1. No.

3. $p = (190/3)P_0 - (588/5)P_1 + (200/3)P_2 - (\frac{2}{5})P_3$.

5. The coordinate vector is $(2, -5, 3)$.

7. $\text{Null}(\mathbf{A}) = \text{Span}\{[-3, 1, 0, 0]^T, [4, 0, 1, 0]^T\}$.

9. Show that $L(p) = p + p'$ is injective and surjective.

11. Yes.

13. No inconsistent case.

15. $\{t, t^2, 1 + t^2\}$ is a basis.

17. Yes.

19. $\begin{bmatrix} 22 \\ 19 \end{bmatrix}$

21. $\sim \mathbf{I}_4$.

23. Determinants: **a.** 1.   **b.** -1.   **c.** 74.

25. Basis for the column space
$\{[-1, 1, 0]^T, [4, 2, 3]^T\}$ with dimension 2.

27. $\{[1, -1, 0, -2, 4], [0, 0, 1, 3, -1]\}$.

29. Range basis $\{[1, 2, 3, 1]^T, [3, 3, 5, 2]^T\}$,
Kernel basis $\{[1, 1, 1]^T\}$,
$\text{Dim}(\text{Ker}(\mathbf{A})) = 1$,
$\text{Dim}(\text{Range}(\mathbf{A})) = 2$,
$\text{Dim}(\text{Domain}(\mathbf{A})) = 3$.

31. $\text{Dim}(\text{Span}(S))$ equals the number of elements in any basis.

33. See Theorem 5.2.6.

35. $n^2 - 1$.

37. Yes. n^2 if $\mathbf{B}$ is noninvertible.

39. Show $S \circ T$ is linear, injective, and surjective.

41. 3.

43. Use the basis $\{\mathbf{u}_1, \mathbf{u}_2, \ldots, \mathbf{u}_m\}$.

45. Look at the reduced row echelon form.

47. See General Exercise 5.2.16.

49. a_i are unique.

51. Use definition of basis of a vector space.

53. An isomorphism preserves all algebraic operations.

55. Suppose not.

57. Do cases: $n > m$ and $n < m$.

59. $\text{Dim}(\mathbb{M}^{n \times m}) = \text{Dim}(\mathbb{M}^{p \times q})$.

61. $\sum_{i=1}^{n} c_i p_i = 0$ implies $c_i = 0$ for all i.

63. Obvious.

65. Two-dimensional subspace.

67. Yes.

69. $\text{Dim}(\text{Domain}(\mathbf{A})) = 7$,
$\text{Dim}(\text{Range}(\mathbf{A})) = 5$,
$\text{Dim}(\text{Ker}(\mathbf{A})) = 2$,
$\text{Dim}(\text{Row}(\mathbf{A})) = 5$,
$\text{Dim}(\text{Col}(\mathbf{A})) = 5$.

71. n linearly independent columns.

73. $\text{Dim}(\text{Null}(L)) = 2$,
$\text{Dim}(\text{Domain}(L)) = 5$,
$\text{Dim}(\text{Range}(L)) = 3$.

75. $\text{Dim}(\text{Domain}(\mathbf{A})) = n = \text{Dim}(\text{Range}(\mathbf{A}))$.

77. $\text{Dim}(\text{Domain}(T)) = 8$,
$\text{Dim}(\text{Range}(T)) = 8$,
$\text{Dim}(\text{Null}(T)) = 0$.

79. $\text{Dim}(\text{Col}(\mathbf{A})) = \text{Dim}(\text{Row}(\mathbf{A}))$.

81. Let $\mathbf{A}$ be 8×7.

83. Consider cases: $m \leq n$ and $m > n$.

85. No.

87. $\mathbf{x}^T \mathbf{A}^T \mathbf{y} = \mathbf{y}^T \mathbf{A} \mathbf{x}$.

89. Let $\mathbf{A}$ be $[k + (n - k)] \times [k + (m - k)]$.

General Exercises 5.3

1. Transition matrix is $\mathbf{P} = \begin{bmatrix} 1 & 3 \\ -1 & -2 \end{bmatrix}$.

The vector $\mathbf{x} = 5\mathbf{u}_1 + 3\mathbf{u}_2 = 14\mathbf{v}_1 - 11\mathbf{v}_2$.

3. $\mathbf{A}^{-1} = \mathbf{A}^T = \begin{bmatrix} \cos\theta & \sin\theta \\ -\sin\theta & \cos\theta \end{bmatrix} = \mathbf{P}$.

5. $\begin{bmatrix} 1 & 0 & -1 & 0 & 1 & 0 \\ 0 & 1 & 0 & -3 & 0 & -5 \\ 0 & 0 & 2 & 0 & -8 & 0 \\ 0 & 0 & 0 & 4 & 0 & -20 \\ 0 & 0 & 0 & 0 & 8 & 0 \\ 0 & 0 & 0 & 0 & 0 & 16 \end{bmatrix}$

7. $\mathbf{P} = \begin{bmatrix} 60/79 & 63/9 & -6/79 \\ 26/79 & -28/79 & 29/79 \\ 101/79 & 31/79 & -18/79 \end{bmatrix}$

9. $\mathbf{P} = \begin{bmatrix} 1 & -2 \\ 2 & 4 \end{bmatrix}$

11. $\mathbf{P} = \begin{bmatrix} 2 & 0 & -3 \\ -1 & 3 & 0 \\ 1 & 1 & 2 \end{bmatrix}$ and $[\mathbf{x}]_{\mathcal{D}} = (-4, -7, 3)$.

13. $T = [T(\mathbf{u}_1), T(\mathbf{u}_2), T(\mathbf{u}_3)] = \begin{bmatrix} 0 & 0 & 0 \\ 0 & 0 & -1 \\ 0 & 1 & 0 \end{bmatrix}$,

$T(\mathbf{x}) = \begin{bmatrix} 0 \\ -7 \\ -2 \end{bmatrix}$

15. $\begin{bmatrix} 0 & 1 & 1 \\ 2 & 0 & -1 \end{bmatrix}$

17. $\begin{bmatrix} 2 & 1 \\ -3 & 4 \end{bmatrix}$

19. $\mathbf{P} = \begin{bmatrix} \alpha d & b \\ 0 & d \end{bmatrix}$

21. $\mathbf{u}_i = 0\mathbf{u}_1 + \cdots + 0\mathbf{u}_{i-1} + 1\mathbf{u}_i$
$\qquad\qquad\qquad + 0\mathbf{u}_{i+1} + \cdots + 0\mathbf{u}_n$.

23. $\mathbf{A}$ and $\mathbf{A}^{-1}$.

25. $\mathbf{A} \simeq \mathbf{B}$ is $\mathbf{A} = \mathbf{SBS}^{-1}$.

27. **Case 1. B** is invertible
Case 2. A is invertible.

29. $\mathbf{P} = \begin{bmatrix} (a - d + b)/c & 1 \\ & 1 & 1 \end{bmatrix}$

31. No.

33. See General Exercise 5.3.32.

35. $\begin{bmatrix} a_{11} - b_{11} & a_{12} & -b_{21} & 0 \\ a_{12} & a_{22} - b_{11} & 0 & -b_{12} \\ -b_{21} & 0 & a_{11} - b_{22} & a_{21} \\ 0 & -b_{21} & a_{12} & a_{22} - b_{22} \end{bmatrix}$

is noninvertible.

37. Claim: $[x_1, x_2, x_1 + x_2, x_1 + 2x_2, 2x_1 + 3x_2,$
$\dots,] = x_1[1, 1, 2, 3, 5, \dots] + (x_2 - x_1)$
$[0, 1, 1, 2, 3, \dots]$.

39. **a.** No.   **b.** Yes.

41. **a.** $\mathbf{Q} = \begin{bmatrix} 1 & 1 & 1 & 1 \\ 0 & 1 & 1 & 1 \\ 0 & 0 & 1 & 1 \\ 0 & 0 & 0 & 1 \end{bmatrix}$

b. $\mathbf{Q} = \begin{bmatrix} 1 & 3 & 2 \\ -1 & 0 & 1 \\ 1 & 1 & -2 \end{bmatrix}$

43. **a.** No.   **b.** Yes.

45. **a.** $\mathbf{P} = \mathbf{Q}^{-1} = \begin{bmatrix} \frac{1}{3} & \frac{1}{3} \\ -\frac{1}{3} & \frac{2}{3} \end{bmatrix}$

b. $\mathbf{P} = \mathbf{Q}^{-1} = \begin{bmatrix} -8 & -11 \\ 3 & 4 \end{bmatrix}$

c. $\mathbf{P} = \mathbf{Q}^{-1} = \begin{bmatrix} \frac{3}{2} & \frac{2}{3} & \frac{1}{3} \\ -\frac{5}{3} & \frac{11}{3} & \frac{1}{3} \\ -\frac{1}{3} & -\frac{1}{3} & \frac{1}{3} \end{bmatrix}$

47. $\mathbf{Q} = \mathbf{P}^{-1} = \begin{bmatrix} -2 & -2 & 3 \\ 2 & 1 & -2 \\ -1 & 0 & 1 \end{bmatrix}$

49. **a.** $\mathbf{P} = \begin{bmatrix} -2 & -3 \\ 1 & 1 \end{bmatrix}$, $\mathbf{P}^{-1} = \begin{bmatrix} 1 & 3 \\ -1 & -2 \end{bmatrix}$

b. $\mathbf{P} = \begin{bmatrix} 1 & 2 \\ 3 & 5 \end{bmatrix}$, $\mathbf{P}^{-1} = \begin{bmatrix} -5 & 2 \\ 3 & -1 \end{bmatrix}$

c. $\mathbf{P} = \begin{bmatrix} -1 & 1 & 4 \\ 8 & 4 & -1 \\ 1 & -1 & 0 \end{bmatrix}$,

$\mathbf{P}^{-1} = \frac{1}{48} \begin{bmatrix} 1 & 4 & 17 \\ 1 & 4 & 31 \\ 12 & 0 & 12 \end{bmatrix}$

51. a. Yes.   **b.** No.   **c.** No.   **d.** Yes.

53. a. $[(a,b)]_\mathcal{T} = [-8a+3b, 3a-b]^T$.

b. $\mathbf{Q} = \begin{bmatrix} -14 & -36 \\ 5 & 13 \end{bmatrix}$

c. $\mathbf{P} = \begin{bmatrix} -9 & -18 \\ \frac{3}{8} & 7 \end{bmatrix}$

d. $\mathbf{Q} = \begin{bmatrix} -14 & -36 \\ 5 & 13 \end{bmatrix}$

e. $\begin{bmatrix} -14 & -36 \\ 5 & 13 \end{bmatrix} \begin{bmatrix} -9 & -18 \\ \frac{3}{8} & 7 \end{bmatrix} =$

$\begin{bmatrix} 1 & 0 \\ 0 & 1 \end{bmatrix}$

55. a. Not similar!

b. Yes, similar: $\mathbf{P} = \begin{bmatrix} i & 1 \\ 1 & i \end{bmatrix}$,

$\mathbf{P}^{-1} = -\frac{1}{2} \begin{bmatrix} i & -1 \\ -1 & i \end{bmatrix}$

General Exercises 6.1

1. Eigenvalues of $\mathbf{A}$: $1.3, -7.4$.
Eigenvalues of $\mathbf{B}$: $5, 7$.

3. Eigenvalues/Eigenvectors:
$0, (-b, a, 0); 0, (-c, 0, a); a+b+c, (1, 1, 1)$.

5. $\mathbf{P}^{-1}\mathbf{AP} = \begin{bmatrix} \frac{1}{2} & i/2 \\ \frac{1}{2} & -i/2 \end{bmatrix} \begin{bmatrix} 0 & -1 \\ 1 & 0 \end{bmatrix} \begin{bmatrix} 1 & 1 \\ -i & i \end{bmatrix}$

$= \begin{bmatrix} i & 0 \\ 0 & -i \end{bmatrix} = \mathbf{D}$.

7. Consider a or b complex.

9. $\{[1,1]^T, [3,2]^T\}$.

11. Column sums are 3.

13. $\lambda = \frac{1}{2}(-1 \pm i\sqrt{3})$.

15. $\mathbf{P} = \begin{bmatrix} 1 & 1 \\ 1 & 2 \end{bmatrix}, \mathbf{D} = \begin{bmatrix} 5 & 0 \\ 0 & 7 \end{bmatrix}$.

17. Eigenvalues/Eigenvector:
$-2, (1, -4); 3, (1, 1)$.

19. $(0, 1, 0, 0)$.

21. a. $a \pm bi, (1, \pm i)$.
c. A clockwise rotation through an angle θ followed by a scaling by r.

23. a. $(a-d)^2 + 4bc \geq 0$.

25. a. $a \neq d$ and $(a-d)^2 + 4bc = 0$.
c. $\mathbf{A} = \begin{bmatrix} 1 & 3 \\ 0 & 1 \end{bmatrix}$

27. For $\mathbf{A} = \begin{bmatrix} a & b \\ c & d \end{bmatrix}$, we find $a(d-1) = bc + d - 1$.

29. $\mathbf{Ax} = 0\mathbf{x}$ has no nonzero solutions.

31. Show that $\mathbf{A}^{-1}\mathbf{x} = \lambda^{-1}\mathbf{x}$.

33. Show they have the same characteristic polynomial.

35. If $\mathbf{x}^{(r)} = [a, b]^T$, then the sum of components $\mathbf{x}^{(r+1)}$ is $a + b$.

37. Yes, the eigenvalues d_i can be assigned freely.

39. $\mathbf{A} = \begin{bmatrix} 0 & 3 \\ -3 & 0 \end{bmatrix}$

41. a. $3, (-1, 1); 5, (1, 1)$.
b. $3 \pm \sqrt{2}, (1/(-1 \pm \sqrt{2}), 1)$.
c. $\frac{5}{2} \pm \frac{1}{2}i\sqrt{3}, (1/(-\frac{1}{2} \pm \frac{1}{2}i\sqrt{3}), 1)$.

d. $3, 3, (-1, 1)$.
e. $-2, (-\frac{1}{2}, 1); 3, (2, 1)$.
f. $\pm i, (1, \pm i)$.
g. $1, 1, (1, 0)$.

43. Yes, each vector in V is an eigenvector of $\mathbf{A}$.

45. $\mathbf{A} = \begin{bmatrix} 1 & 2 \\ -1 & 1 \end{bmatrix}$

47. $\mathrm{Det}(A - \lambda I) = \prod_{i=1}^{n}(a_{ii} - \lambda)$.

49. $\lambda = \pm\sqrt{a^2 + b^2 - c^2}$.

51. $\lambda = 0$.

53. $\prod_{i=1}^{n}(a_{ii} - \lambda)$ is the product of the diagonal entries of $\mathbf{A} - \lambda\mathbf{I}$.

55. Let $\mathbf{V} = [\mathbf{v}_1, \mathbf{v}_2, \dots, \mathbf{v}_n]$ and $\mathbf{D} = \mathrm{Diag}(\lambda_1, \lambda_2, \dots, \lambda_n)$.

57. $\frac{1}{3}; 1, 1$.

59. $\mathbf{AB} = \mathrm{Diag}(a_i b_i)$, $\mathbf{AB}^{-1} = \mathrm{Diag}[1/(a_i b_i)]$ assuming $a_1 a_2 \cdots a_n \neq 0$ and $b_1 b_2 \cdots b_n \neq 0$.

61. Divergent.

63. Yes.

65. See Sandefur [1993].

67. $t = \lambda$.

69. $\mathbf{u}$ is an eigenvector for $\mathbf{A}^T$ corresponding to the eigenvalue 1.

71. $\mathbf{A} = \begin{bmatrix} 1 & 1 \\ 1-i & 1+i \end{bmatrix} \begin{bmatrix} 2-i & 0 \\ 0 & 2-i \end{bmatrix}$
$\times \begin{bmatrix} (1-i)/2 & i/2 \\ (1+i)/2 & -i/2 \end{bmatrix}$

73. $\mathbf{x}^T\mathbf{x} > 0$.

75. $\mathbf{A}^2 = \begin{bmatrix} 8 & 1 \\ 4 & 5 \end{bmatrix}$ and $\mathbf{A}^3 = \begin{bmatrix} 20 & 7 \\ 28 & -1 \end{bmatrix}$.

77. Consider the characteristic polynomial.

General Exercises 7.1

1. $(a, b) \cdot (b, -a) = 0$.

3. $-(13/19)\mathbf{u} + \mathbf{v} - (8/19)\mathbf{w} = \mathbf{x}$.

5. Use $||\mathbf{x} - \mathbf{y}||^2 = \langle \mathbf{x} - \mathbf{y}, \mathbf{x} - \mathbf{y}\rangle$.

7. $\mathbf{0}$.

9. Linear.

11. a. Yes.   **b.** No.   **c.** Yes.

13. $(3.8997, 0.0564, -1.3375)$.

15. Yes, converse true.

17. $\sqrt{59}, \sqrt{41}$.

19. Proof by contrapositive.

21. Let $w_1 = 1$, $w_2 = -1$, and $n = 2$.

23. Both straightforward.

25. $\sqrt{74}$.

27. Show $\mathbf{x} \in [\mathrm{Col}(\mathbf{A})]^{\perp} \Longleftrightarrow \mathbf{x} \in \mathrm{Null}(\mathbf{A}^T)$.

29. Use formula for the angle between two vectors.

31. Nothing special about $\mathbf{B}$.

33. All straightforward.

35. Let $\langle \mathbf{x}, \mathbf{y}\rangle = ia$ for a real.

37. Straightforward use of definitions.

39. Show $\mathbf{x} \in Q^T$ implies $\mathbf{x} \in S^T$.

41. $\{\mathbf{u}_1, \mathbf{u}_2, \ldots, \mathbf{u}_n\}$ is either linearly independent or orthonormal.

43. Let $\mathbf{u} = \mu\mathbf{x}$ and $\mathbf{v} = 2\mathbf{y}$ with $\mu = 4\lambda$.

45. Case 1. $\mathbf{x} - \mathbf{b}$ is not a multiple of $\mathbf{c}$.
Case 2. $\mathbf{x} - \mathbf{b}$ is a multiple of $\mathbf{c}$.

47. Use $\langle f, g \rangle = \int_{-1}^{1} f(t)g(t)\, dt$.

49. Let $\mathbf{x} = (x_1, x_2)$ and $\mathbf{y} = (y_1, y_2)$.

51. Use the Cauchy–Schwarz Inequality.

53. Straightforward.

55. Use projection of $\mathbf{x}$ into $\mathbf{v}$.

57. Yes.

59. Suppose that $\mathbf{w} \in V$ and $(\mathbf{x} - \mathbf{w}) \perp V$.

61. $\theta = 90°$.

63. Straightforward.

65. No.

67. Use $\langle \mathbf{x}, \mathbf{y} \rangle \leq 0$.

69. Suppose n is odd and $\mathbf{x} \cdot \mathbf{y} = 0$.

71. For an acute angle, $0 \leq \mathbf{x}^T\mathbf{y} \leq \pi/2$.

73. $\mathbf{P} = \text{proj}_\mathbf{u}\, \mathbf{x}$, $\mathbf{Q} = \text{proj}_\mathbf{v}\, \mathbf{x}$.

75. Quadratic function $f(t) = at^2 + bt + c$ in the complex plane $\mathbb{C}$ corresponds to a parabola.

77. The converse is false!

79. Try interchanging $\mathbf{x}$ and $\mathbf{y}$.

81. Use triangle inequality.

83. Assume $(\mathbf{x} \cdot \mathbf{y}/\mathbf{x} \cdot \mathbf{x})\, \mathbf{x} = \mathbf{x}$.

85. Try a geometric argument.

87. The point $(3, 9)$.

89. $50000/\sqrt{3154}$.

General Exercises 7.2

1. $[8/21, 13/70]^T$.

3. $[653/209, -337/209]^T$.

5. $\mathbf{s} = (8/21, 13/70)$.

7. $\mathbf{u}_1 = (1, 2, 1)$, $\mathbf{u}_2 = \frac{2}{3}(1, -1, 1)$,
$\mathbf{u}_3 = \frac{3}{2}(1, 0, -1)$.

9. $\mathbf{u}_1 = [1, 1, -1, 1]/2$, $\mathbf{u}_2 = [1, -1, 1, 1]/2$,
$\mathbf{u}_3 = [0, -1, 1, 0]/\sqrt{2}$.

11. $1102/505 \approx 2.18$.

13. $\mathbf{p} = (-1.1412, -0.4116, -1.5411)$.

15. Any even function is orthogonal to any odd function.

17. No.

19. $\frac{1}{2}(3, 1)$.

21. Inductive step $\langle \mathbf{u}_{k+1}, \mathbf{u}_i \rangle = 0$.

23. By Cramer's rule,
$$\alpha = \left[m\sum_{i=1}^{m} y_i t_i - \left(\sum_{i=1}^{m} t_i \right)\left(\sum_{i=1}^{m} y_i \right) \right] \Big/ D,$$
$$\beta = \left[\left(\sum_{i=1}^{m} t_i^2 \right)\left(\sum_{i=1}^{m} y_i \right) - \left(\sum_{i=1}^{m} t_i \right)\left(\sum_{i=1}^{m} y_i t_i \right) \right] \Big/ D,$$
$$D = m\sum_{i=1}^{m} t_i^2 - \left(\sum_{i=1}^{m} t_i \right)^2.$$

25. $(3, 2, 1) = (92, 59, 25)/30 + (-2, 1, 5)/30$.

27. Use Theorem 7.2.9.

29. $\mathbf{x} = \mathbf{0}$.

31. Let $\mathbf{y} = \alpha\mathbf{x}$.

33. Let $\mathbf{x} \mapsto \text{Proj}_U\mathbf{x}$.

35. Yes.

37. $(\mathbf{A}^T)^T = \mathbf{A}$.

39. Use General Exercise 7.2.38.

41. Suppose $\mathbf{A}\mathbf{A}^T$ is noninvertible.

43. Multiply by $\mathbf{x}^T$.

45. $\{\mathbf{w}_{r+1}, \mathbf{w}_{r+2}, \cdots, \mathbf{w}_n\}$.

47. **a.** $\mathbf{v}$ must be a unit vector and $\mathbf{x}$ must be a multiple of $\mathbf{v}$.
b. Fixed points are all $\mathbf{x}$ that are multiples of $\mathbf{v}$.

49. $\mathbf{B} = \begin{bmatrix} 3/\sqrt{14} & 2/\sqrt{13} & 3/\sqrt{182} \\ 1/\sqrt{14} & 0 & -13/\sqrt{182} \\ 2/\sqrt{14} & -3/\sqrt{13} & 2/\sqrt{182} \end{bmatrix}$,

where $\mathbf{B}\mathbf{B}^T = \mathbf{I} = \mathbf{B}^T\mathbf{B}$.

51. Take inner products/dot products.

53. Use linearity.

55. **a.** Use one norm.
b. Use Euclidean norm.

General Exercises 8.1

1. $\mathbf{P} = \begin{bmatrix} 1 & 1 \\ 1 & -1 \end{bmatrix}$, $\mathbf{D} = \begin{bmatrix} 5 & 0 \\ 0 & -1 \end{bmatrix}$, $\mathbf{P}^{-1} = \frac{1}{2}\mathbf{P}$

3. $\frac{1}{2}(3 - 5i) \pm \frac{1}{2}\sqrt{-4 + 10i}$.

5. Yes.

7. $\mathbf{A}^H = \mathbf{A}$.

9. $a_{ii} = \bar{a}_{ii}$.

11. $\begin{bmatrix} 2 & 6-c \\ c & 5 \end{bmatrix}$

13. $\lambda = 0, (1, 1); \lambda = 2c, (1, -1)$.

15. $\mathbf{A} = \begin{bmatrix} \alpha & \delta \\ -\delta & \beta \end{bmatrix}$, where δ is arbitrary.

17. $\frac{1}{2}(\mathbf{A} + \mathbf{A}^H)$ is Hermitian.

19. Let $\mathbf{x} = \mathbf{e}_k = (0, \ldots, 0, 1, 0, \ldots, 0)$ with 1 in the kth entry.

21. No.

23. $\mathbf{D}^H\mathbf{D} = \mathbf{D}\mathbf{D}^H$ for $\mathbf{D} = \text{Diag}(d_i)$.

25. $(\mathbf{A}^H)^H = \mathbf{A}$.

27. Yes.

29. Let $f(\mathbf{x}) = \mathbf{A}\mathbf{x}$.

31. Show for $m = 2$.

33. Consider $f_i(\mathbf{x}) = \langle \mathbf{x}, \mathbf{v}_i \rangle \mathbf{v}_i$.

35. Use Theorem 8.1.4.

37. Let $\mathbf{U}\mathbf{A}\mathbf{U}^T = \mathbf{D}$.

39. Use Cayley–Hamilton Theorem: $p(\mathbf{A}) = \mathbf{0}$.

41. See Example 10.

43. Consider dimensions of $\mathbb{M}^{n \times n}$.

45. See Example 10.

47. Circular logic.

49. If $\mathbf{A} = \begin{bmatrix} a & b \\ c & d \end{bmatrix}$, $\mathbf{P} = \begin{bmatrix} (a-d+b)/c & 1 \\ 1 & 1 \end{bmatrix}$.

51. Yes.

53. Suppose $\mathbf{AA}^T$ is not invertible.

55. No.

57. See Cayley–Hamilton Theorem.

59. No.

61. $x_1^2 + 5x_2^2 + 8x_3^2 + 6x_1x_2 + 10x_1x_3 + 14x_2x_3.$

63. $x_1 = \frac{1}{\sqrt{2}}(-y_1 + y_2), x_2 = \frac{1}{\sqrt{2}}(y_1 + y_2).$

65. True.

67. $x_1 = \left[-2\left[2(5-\sqrt{5})\right]^{-\frac{1}{2}}\right]y_1$

$$+ \left[-2\left[2(5+\sqrt{5})\right]^{-\frac{1}{2}}\right]y_2,$$

$$x_2 = \left[(1-\sqrt{5})\left[2(5-\sqrt{5})\right]^{-\frac{1}{2}}\right]y_1$$

$$+ \left[(1+\sqrt{5})\left[2(5+\sqrt{5})\right]^{-\frac{1}{2}}\right]y_2$$

General Exercises 8.2

1. a. $\mathbf{A} = \begin{bmatrix} 2 & -4 \\ 10 & 17 \end{bmatrix} = \begin{bmatrix} 1 & 0 \\ 5 & 1 \end{bmatrix}\begin{bmatrix} 2 & -4 \\ 0 & 3 \end{bmatrix} = \mathbf{LU}$

b. $\mathbf{B} = \begin{bmatrix} -3 & 1 & 5 \\ -9 & -1 & 16 \\ -6 & -6 & 14 \end{bmatrix} = \begin{bmatrix} 1 & 0 & 0 \\ 3 & 1 & 0 \\ 2 & 2 & 1 \end{bmatrix}$

$$\times \begin{bmatrix} -3 & 1 & 5 \\ 0 & -4 & 1 \\ 0 & 0 & 2 \end{bmatrix} = \mathbf{LU}$$

3. $\mathbf{A} = \begin{bmatrix} 3 & 0 & 3 \\ 0 & -1 & 3 \\ 0 & 0 & 8 \end{bmatrix} = \begin{bmatrix} 1 & 0 & 0 \\ 0 & 1 & 0 \\ \frac{1}{3} & -3 & 0 \end{bmatrix}$

$$\times \begin{bmatrix} 3 & 0 & 3 \\ 0 & 1 & 0 \\ \frac{1}{3} & -3 & 0 \end{bmatrix} = \mathbf{LU}$$

$$\mathbf{B} = \begin{bmatrix} 4 & -1 & -1 & 0 \\ -1 & 4 & 0 & -1 \\ -1 & 0 & 4 & -1 \\ 0 & -1 & -1 & 4 \end{bmatrix}$$

$$= \begin{bmatrix} 1 & 0 & 0 & 0 \\ -\frac{1}{4} & -1 & 0 & 0 \\ -\frac{1}{4} & -1/15 & 1 & 0 \\ 0 & -4/15 & -\frac{2}{7} & 1 \end{bmatrix}$$

$$\times \begin{bmatrix} 4 & -1 & -1 & 0 \\ 0 & 15/4 & -\frac{1}{4} & -1 \\ 0 & 0 & 56/15 & -16/15 \\ 0 & 0 & 0 & 24/7 \end{bmatrix}$$

$$= \mathbf{LU}$$

5. $\mathbf{L} = \begin{bmatrix} 1 & 0 & 0 \\ 4 & 2 & 0 \\ 5 & 6 & 3 \end{bmatrix}$

7. $\mathbf{L} = \begin{bmatrix} 1 & 0 & 0 \\ 3 & 2 & 0 \\ 4 & -1 & 1 \end{bmatrix}$

9. $\mathbf{L} = \begin{bmatrix} 1 & 0 & 0 \\ 2 & 1 & 0 \\ -2 & 3 & 1 \end{bmatrix}, \mathbf{U} = \begin{bmatrix} 2 & 1 & -4 \\ 0 & 1 & 3 \\ 0 & 0 & 2 \end{bmatrix},$

$$\mathbf{x} = \begin{bmatrix} 1 \\ 0 \\ 3 \end{bmatrix}$$

11. $\mathbf{L} = \begin{bmatrix} 1 & 0 & 0 \\ 4 & 2 & 0 \\ 5 & 6 & 3 \end{bmatrix}$

13. $\mathbf{x} = \begin{bmatrix} 3 \\ 1 \\ 5 \\ 2 \end{bmatrix}$

15. $\mathbf{A}^{-1} = \begin{bmatrix} \frac{1}{3} & -\frac{1}{2} & -\frac{1}{3} \\ -\frac{7}{3} & 1 & \frac{4}{3} \\ -1 & 1 & 1 \end{bmatrix}$

17.
$$\begin{bmatrix} 1 & 0 & 0 & 0 \\ 2 & 1 & 0 & 0 \\ -2 & 2 & 1 & 0 \\ 3 & -1 & 3 & 1 \end{bmatrix} \begin{bmatrix} 2 & 1 & 3 \\ 0 & -1 & 2 \\ 0 & 0 & 3 \\ 0 & 0 & 0 \end{bmatrix}$$

19.
$$\begin{bmatrix} 2 & 2 & 1 \\ 1 & 1 & 1 \\ 3 & 2 & 1 \end{bmatrix} = \begin{bmatrix} 1 & 0 & 0 \\ \ell_{21} & 1 & 0 \\ \ell_{31} & \ell_{32} & 1 \end{bmatrix}$$
$$\begin{bmatrix} u_{11} & u_{12} & u_{13} \\ 0 & u_{22} & u_{23} \\ 0 & 0 & u_{32} \end{bmatrix}$$

21. Gram matrix $\mathbf{G}_{ij} = \langle \mathbf{u}_i, \mathbf{u}_j \rangle$.

23. Consider $\mathbf{e}_i^T \mathbf{A} \mathbf{e}_i$.

25. If $\mathbf{x} = [x_1, x_2, \dots, x_k, 0, 0, \dots, 0]^T$,
let $\mathbf{y} = [x_1, x_2, \dots, x_k]^T$.

27. $\mathbf{Q}^T \mathbf{A} \mathbf{Q}$.

29. For $\mathbf{A} = \begin{bmatrix} a & b \\ b & d \end{bmatrix}$, $\mathbf{L} = \begin{bmatrix} \sqrt{a} & 0 \\ b/\sqrt{a} & \sqrt{c - b^2/a} \end{bmatrix}$.

31. $\ell_{ii} > 0$.

33. Multiply block matrix factors.

35. Show $\mathcal{H}\mathcal{G}\mathcal{H}^T = \mathcal{A}$.

37. $\text{Det}(\mathbf{A})$ and $\text{Det}(\mathbf{A})\,\text{Det}(\mathbf{B})$.

39. Let $\mathbf{D} \leftarrow \mathbf{I}$ in the 2×2 block matrix inverses.

41. Yes.

43. Let $\mathbf{S}_A = c - \mathbf{b}^T \mathbf{A}^{-1}\mathbf{b}$ and $\mathbf{S}_c = \mathbf{A} - c^{-1}\mathbf{b}^T\mathbf{b}$.

45. $\mathbf{B} = -\mathbf{C}$ and $\mathbf{A} = \mathbf{D}$.

47. Use $\mathbf{S}_a = \Delta a^{-1}$ and $\mathbf{S}_d = \Delta d^{-1}$, where $\Delta = ad - bc$.

49. Equate the $(1, 1)$ block entries of each matrix form of the 2×2 block matrix inverse.

51. $\mathbf{B}[\mathbf{A}^{-1} - \mathbf{B}^{-1}(\mathbf{B} - \mathbf{A})\mathbf{A}^{-1}] = \mathbf{I}$.

53. $\mathbf{y} = \mathbf{S}_A^{-1}(\mathbf{c} - \mathbf{C}\mathbf{A}^{-1}\mathbf{b})$, $\mathbf{x} = \mathbf{A}^{-1}(\mathbf{b} - \mathbf{B}\mathbf{y})$.
$\mathbf{x} = \mathbf{S}_D^{-1}(\mathbf{b} - \mathbf{B}\mathbf{D}^{-1}\mathbf{c})$, $\mathbf{y} = \mathbf{D}^{-1}(\mathbf{c} - \mathbf{C}\mathbf{x})$.
$\mathbf{S}_A = \mathbf{D} - \mathbf{C}\mathbf{A}^{-1}\mathbf{B}$, $\mathbf{S}_D = \mathbf{A} - \mathbf{B}\mathbf{D}^{-1}\mathbf{C}$.

General Exercises 8.3

1. $\mathbf{A}$ and $\mathbf{A}^T$ have the same eigenvalues.

3. $(-11, 16)$.

5. Use definition of $\|\mathbf{A}\|_\infty$.

7. Use $|z| = |z - a_{ii} + a_{ii}|$.

9. a. $[33.333, 35, 15]^T$.
b. $[225.9259, 118.5785, 77.7778]^T$.

11. Let $1 = a + (1 - a)$ and $b = a + (b - a)$.

13. $|z - 2| \leq 1$, $|z + 3| \leq 2$, $|z - 5| \leq 1$.

15. Use Neumann series.

17. Use Gerschgorin's Theorem.

19. $-2 < s < 2$.

21. Theorem 8.3.1 does not say that if a matrix is *not* diagonally dominant, then it is *not* invertible!

23. By Theorem 8.3.1, $\mathbf{D}$ is invertible.

25. $-4 < t < 4$.

27. Gerschgorin's row discs: $\mathcal{G}_1(\mathbf{A})[5; 2]$, $\mathcal{G}_2(\mathbf{A})[7; 1]$, and $\mathcal{G}_3(\mathbf{A})[9; 1]$.
Gerschgorin's column discs: $\mathcal{G}_1(\mathbf{A}^T)[5; 1]$, $\mathcal{G}_2(\mathbf{A}^T)[7; 2]$, and $\mathcal{G}_3(\mathbf{A}^T)[9; 1]$.

29. $\mathbf{x} = \begin{bmatrix} 10 \\ 20 \\ 10 \end{bmatrix}$

31. Union of Gerschgorin discs does not contain the origin.

33. $\mathbf{x} = \begin{bmatrix} 7/11 \\ 1/11 \end{bmatrix}$

35. $\text{Det}(\mathbf{A} - 0\mathbf{I}) \neq 0$.

37. $||\mathbf{x}||_\infty = 3$.

39. $20 = ||\mathbf{Ax}||_\infty \leq ||\mathbf{A}||_\infty \cdot ||\mathbf{x}||_\infty = 14 \cdot 3 = 42$.

41. $|\lambda| \leq ||\mathbf{A}||_\infty$, for all eigenvalues λ of $\mathbf{A}$.

43. a. Crude estimate: $|\lambda| \leq ||\mathbf{A}||_\infty = 7$, for all $\lambda \in \sigma(\mathbf{A})$.
    b. Eigenvalues are in/on $\mathcal{G}_1(\mathbf{A})[5; 2] \cup \mathcal{G}_2(\mathbf{A})[6; 1] = \mathcal{G}_1(\mathbf{A})[5; 2]$.
    c. Eigenvalues are in/on $\mathcal{G}_1(\mathbf{A}^T)[5; 1] \cup \mathcal{G}_2(\mathbf{A}^T)[6; 1]$.
    d. Best estimate: eigenvalues are in/on $\mathcal{G}_1(\mathbf{A})[5; 2] \cap (\mathcal{G}_2(\mathbf{A}^T)[5; 1] \cup \mathcal{G}_2(\mathbf{A}^T)[6; 1]) = \mathcal{G}_1(\mathbf{A})[5; 2]$.
    e. See figure below:

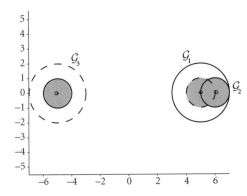

    f. True eigenvalues: $\lambda_1 = 5.0000$ and $\lambda_{2,3} = \frac{1}{2}(1 \pm 5\sqrt{5}) \approx 6.0902, -5.0902$.

45. $\mathbf{A}$ is diagonally dominant if $|b| < |1 - a|$ and $|c| < |1 - d|$.

47. $\mathbf{B}$ is diagonally dominant and invertible.

49. When $\mathbf{A}$ is a symmetric matrix.

General Exercises Appendix A

1. Expand righthand side.

3. Use induction.

5. Let $P(n): \sum_{i=1}^{n} i = n(n+1)/2$.

7. Pythagorean triples of the form $(n, n+1, m)$ less than 1000 are: $(3, 4, 5)(14, 15, 29)$, $(119, 120, 169), (696, 697, 985)$.

9. Straightforward.

11. Let $\mathbf{x} \in A$.

13. Use the identity
$$1/[a(a+1)] = 1/a - 1/(a+1).$$

15. $(\exists x)\ (\exists \varepsilon > 0)\ (\forall \delta > 0)\ (\exists y)$ $[|x - y| < \delta \quad \text{and} \quad |f(x) - f(y)| \geq \varepsilon]$.

17. The only solution is $x = 0$.

19. Some Pythagorean triples are:

	$m = 1$	2	3	4
$n = 2$	$(3, 4, 5)$			
3	$(8, 6, 10)$	$(5, 12, 13)$		
4	$(15, 8, 17)$	$(12, 16, 20)$	$(7, 24, 25)$	
5	$(24, 10, 26)$	$(21, 20, 29)$	$(16, 30, 34)$	$(9, 40, 41)$

21. Prove $r_1 + r_2 = -b/a, r_1 + r_2 + r_3 = -b/a$, and $r_1 + r_2 + \cdots + r_n = -b/a$.

23. Let $x \in C$.

25. Draw Venn diagram.

27. Disprove.

29. First, do the case $n = 2$. See Hall and Knight [1948, p. 211].

31. Recall $x_n = nd + x_0$.

33. $[2, 12, 22, 32, 42, \ldots]$

35. $S = a(1-r^n)/(1-r)$. $S = \left[1-(-2\sqrt{2})^n\right]/$
$(\sqrt{2}+4)$

37. Pascal's Identity:
$$\binom{n+1}{k} = \binom{n}{k} + \binom{n}{k-1}$$

39. No.

41. $n, \frac{1}{2}n(n+1)$.

43. Be sure that the implication from f_0 to f_1 is valid.

45. Yes.

47.

P	Q	R	$Q \wedge R$	$P \vee (Q \wedge R)$
T	T	T	T	T
T	T	F	F	T
T	F	T	F	T
T	F	F	F	T
F	T	T	T	T
F	T	F	F	F
F	F	T	F	T
F	F	F	F	F

49. $1+2+3=6$.

51. Yes.

53. If you start with a false statement you can prove anything!

55. General formula: $x_{n+1} = x_1 + \frac{1}{2}n(n+1)c$.

57. No.

59. This assumes that there are at least two elements.

61. All are correct except for two of them.

63. $f(n) = 2 + \frac{1}{2}n(n+1)$.

65. The example $5 \cdot 6 \cdot 7$ is instructive.

67. Two divergent series.

69. $9 > 8$.

71. $0 \le (x-y)^2 + (y-x)^2 + (x-z)^2$.

73. Closed interval.

75. Consider roots of a polynomial.

77. Let LHS $= a^2x^2 + 2axby + b^2y^2$ and RHS $= a^2x^2 + a^2y^2 + b^2x^2 + b^2y^2$.

79. $x_{n+1} = \frac{1}{2}n(n+1)$.

References

Abel, N.H. [1826]. "Beweis der Unmöglichkeit, algebraische Gleichungen von höheren Graden als dem vierten allgemein aufuzulösen," *J. reine angew. Math.* 1, 65. (Reprinted by Abel, N.H. [1881]. (L. Sylow and S. Lie, Eds.) Christiania, Norway, Olso. Reprinted by Johnson Reprint Corp., New York, 66–87, 1988.

Althoen, S.C., R. McLaughlin [1987]. "Gauss-Jordan reduction: A brief history." *American Mathematical Monthly.* Vol. 94, 130–142.

Andrilli, S., D. Hecker [2003]. *Elementary Linear Algebra*, 3rd Ed. New York: Elsevier/Academic Press.

Axler, S. [1995]. "Down with determinants." *American Mathematical Monthly.* Vol. 102, 139–154.

Barbeau, E.J. [2000]. *Mathematical Fallacies, Flaws, and Flimflam.* Washington, D.C.: Mathematical Association of America.

Bell, E.T. [1937]. *Men of Mathematics.* New York: Simon and Schuster.

Bernstein, D.S. [2005]. *Matrix Mathematics: Theory, Facts, and Formulas with Application to Linear Systems Theory.* Princeton, New Jersey: Princeton University Press.

Berry, M.W., M. Browne [2005]. *Understanding Search Engines: Mathematical Modeling and Text Retrieval*, 2nd Ed. Society of Industrial and Applied Mathematics, Pennsylvania: Philadelphia.

Birkhoff, G., S. MacLane [1999]. *Survey of Modern Algebra.* New York: Chelsea. (reprint of Macmillan, 1941.)

Boyer, C. [1968]. *A History of Mathematics.* Princeton, New Jersey: Princeton University Press.

Brin, S., L. Page [1998]. "The anatomy of a large-scale hyper-textual web search engine." *Computer Networks and ISDN Systems.* Vol. 33, 107–117.

Brualdi, R.A., and Mellendorf, S. [1994]. "Regions in the Complex Plane Containing the Eigenvalue of a Matrix," *American Mathematical Monthly*, Vol. 101, 975–985.

Cheney, E.W. [1982]. *Introduction to Approximation Theory*, 2nd Ed. Rhode Island: Providence. American Mathematical Society.

Cheney, W., D. Kincaid [2008]. *Numerical Mathematics and Computing*, 6th Ed. Belmont, California: Thomson Brooks/Cole.

Crilly, T. [2005]. *Arthur Cayley: Mathematician Laureate of the Victorian Age.* Baltimore, Maryland: Johns Hopkins University Press.

Daintith, D., R.D. Nelson [1989]. *Dictionary of Mathematics.* New York: Penguin.

Davis, P.J. [1965]. *The Mathematics of Matrices: A First Book of Matrix Theory and Linear Algebra.* New York: Blaisdell.

Dollar, H.S., A. J. Wathen. [2006] "Incomplete-factorization constraint pre-conditioners for saddle-point matrices." *Society of Industrical and Applied Mathematics Journal Numerical Analysis.* Vol. 27, 1555–1572.

Dörrie, H. [1940]. *Determinanten.* Oldenburg, Munich, Germany.

Faddeev, D.K., V.N. Faddeeva [1963]. *Computational Methods of Linear Algebra.* San Francisco: W.H. Freeman. (translated from Russian by C D. Benster, 1959.)

Gerschgorin, S. [1931]. "Uber die Abgrenzung der Eigenwerte einer Matrix" Izv. Akad. Nauk, USSR Otd. Fiz-Mat. Nuak 8, 749–754.

Golub, G.H, and W. Kahan [1965]. "Calculating the singular-values and pseudo-inverse of a matrix," *SIAM J. Numer. Analy.,* **2**, Ser. B, 205–224.

Golub, G., C. van Loan [1996]. *Matrix Computations,* 3rd Ed. Baltimore, Maryland: Johns Hopkins University Press.

Hall, H.S., S.R. Knight [1948]. *Higher Algebra.* New York: Macmillan.

Halmos, P.R. [1991]. *Problems for Mathematicians, Young and Old.* Dolciani Mathematical Expositions No. 12, Washington, D.C.: Mathematical Association of America.

Harter, H.L. [1974–76]. "The method of least squares." *International Statistical Review,* Vol. 42–44. (Six separate chapters of 225 pages.)

Hestenes, M.R., E. Stiefel [1952]. "Methods of conjugate gradients for solving linear systems." *J. Res. Natl. Bur. Stand..* Vol. 49, 409–436.

Horn, R.A., C.R. Johnson [1985]. *Matrix Analysis.* New York: Cambridge University Press.

Ikramov, H.D. [1983]. *Linear Algebra: Problem Book.* (translated from Russian by O. Efimov, 1975), MIR, Moscow.

Kincaid, D., W. Cheney [2002]. *Numerical Analysis: The Mathematics of Scientific Computing,* 3rd Ed. Providence, Rhode Island: American Mathematics Society.

Klein, J. [1968]. *Greek Mathematical Thought and the Origin of Algebra.* (translated by E. Brann), MIT, Massachusetts: Cambridge.

Kleinberg, J. [1998]. "Authoritative sources in a hyper-lined environment." *Journal Association of Computing Machines.* Vol. 46.

Kline, M. [1972]. *Mathematical Thought from Ancient to Modern Times.* New York: Oxford University Press.

Langville, A.N., C.D. Meyer [2005]. *Understanding Web Search Engine Ranking: Google's PageRank, Teoma's HITS, and Other Ranking Algorithms.* Princeton, New Jersey: Princeton University Press.

Langville, A.N., C.D. Meyer [2006]. *Google's PageRank and Beyond: The Science of Search Engine Rankings.* Princeton, New Jersey: Princeton University Press.

Lawson, C.L., R.J. Hanson, D.R. Kincaid, F.T. Krogh [1979]. "Basic linear algebra subprograms for Fortran usage," *ACM Transactions on Mathematical Software.* Vol. 5, 308–323.

Lay, D.C. [2003]. *Linear Algebra and its Applications,* 3nd Ed. (Updated), New York: Addison-Wesley.

Leontief, W. [1941]. *The Structure of the American Economy 1919–1929: An Empirical Application of Equilibrium Analysis.* Cambridge, Massachusetts: Harvard University Press.

Leontief, W. [1985]. *Input-Output Economics,* 2nd Ed. New York: Oxford University Press.

Lipschutz, S. [1989] *Schaum's Solved Problems Series, 3000 Solved Problems in Linear Algebra* McGraw-Hill, New York.

Littlefield, D.E. [1960]. *The Skeleton Key of Mathematics: A Simple Account of Complex Algebraic Theory.* Mineola, New York: Dover.

Lynch, S. [2004]. *Dynamical Systems with Applications*. Boston: Birkhäuser.

Madden, M.A. (ed.) [1981]. *Maybe He's Dead*. New York: Random House.

Maxwell, E.A. [1959]. *Fallacies in Mathematics*. New York: Cambridge University Press.

Meyers, C.D. [2000]. *Matrix Analysis and Applied Linear Algebra*, Society of Industrial and Applied Mathematics, Pennsylvania: Philadelphia.

Midonic, H.O. (ed.) [1965]. *The Treasury of Mathematics*. New York: Philosophical Library.

Miller, R.E., P. Blair [1985]. *Input Output Analysis: Foundations and Extension*. Englewood, New Jersey: Prentice-Hall.

Mirsky, L. [1990]. *An Introduction to Linear Algebra*. New York: Mineola, Dover. (reprint of 1955.)

Muir, T. [1960]. *The Theory of Determinants in Historical Order*. New York: Mineola, Dover. (reprint of Macmillan, 1906).

Muir, T. [2005]. *A Treatise on the Theory of Determinants*. New York: Mineola, Dover. (reprint of 1933.)

Nakos, G., D. Joyner [1998]. *Linear Algebra with Applications*. Belmont, California: Thomson Brooks/Cole.

Nelson, D. (ed.) [2003]. *The Penguin Dictionary of Mathematics*, 3rd Ed. New York: Penguin Putman.

Noble, B.N., J.W. Daniel [1988]. *Applied Linear Algebra*, 3rd Ed. Englewood Cliff, New Jersey: Prentice-Hall.

Parshall, K.H. [2006]. *James Joseph Sylvester: Jewish Mathematician in a Victorian World*. Baltimore, Maryland: John Hopkins University Press.

Pesic, P. [2003]. *An Essay on the Sources and Meanings of Mathematical Unsolvability*. Cambridge, Masschusetts: MIT Press.

Poole, D. [2006]. *Linear Algebra, A Modern Introduction*, 2nd Ed. Belmont, California: Thomson Brooks/Cole.

Reid, J. [1972]. "The Use of Conjugate Gradients for Systems of Linear Equations possessing 'Property A'," *Society Industrial Applied Mathematics J. Numerical Analysis*. Vol. 9, 325–332, 1972. (Harwell Report TP 445, March 1971)

Rose, N. (ed.) [1988]. *Mathematical Maxims and Minims*. Raleigh, North Carolina: Rome Press.

Sadun, L. [2001]. *Applied Linear Algebra: the Decoupling Principle*. New Jersey: Prentice-Hall.

Sandefur, J.T. [1993]. *Discrete Dynamical Systems, Theory and Applications with Applications*. New York: Oxford University Press.

Sauer, T. [2006] *Numerical Analysis*. New York: Pearson/Addison-Wesley.

Sierpinski, W. [2003]. *Pythagorean Triangles*. Mineola, New York: Dover.

Silverman, J.H. [2006]. *A Friendly Introduction to Number Theory*. 3rd Ed. Englewood, New Jersey: Prentice-Hall.

Stoll, R.R. [1952]. *Linear Algebra and Matrix Theory*. New York: McGraw-Hill.

Stoll, R.R., E.T. Wong [1968]. *Linear Algebra*. New York: Academic Press.

Strang, G. [1988]. *Linear Algebra and Its Applications*, 3rd Ed. Saunder, Pennsylvania: Philadelphia.

Strang, G. [2003]. *Introduction to Linear Algebra*, 3rd Ed. Wellesley, Massachusetts: Wellesley-Cambridge Press.

Taussky-Todd, O. [1949]. "A Recurring Theorem on Determinants," *American Mathematical Monthly*, Vol. 56, 672–676.

Taussky-Todd, O. [1982]. "The many aspects of Pythagorean triangles," *Linear Algebra and Applications*, Vol. 43, 285–295.

Varga, R.S. [2004]. *Geršgorin and His Circles*. Heidelberg, Germany: Springer-Verlag.

Wells, D. [1986]. *The Penguin Dictionary of Curious and Interesting Numbers.* Middlesex, England: Penguin Books.

Wilkinson, J. [1965]. *The Algebraic Eigenvalue Problem.* London: Clarendon/Oxford Press.

Young, D.M. [1950]. *Iterative Methods for Solving Partial Differential Equations of Elliptic Type.* Ph.D thesis, Harvard University, Massachusetts: Cambridge.

Young, D.M. [1971]. *Iterative Solutions of Large Linear Systems.* New York: Academic Press.

Yuster, T. [1984]. "The reduced row echelon form of a matrix is unique: A simple proof." *Mathematical Magazine.* Vol. 57, 93–94.

Index